T0255728

CAMBRIDGE LIBRARY COLLECTION

Books of enduring scholarly value

Botany and Horticulture

Until the nineteenth century, the investigation of natural phenomena, plants and animals was considered either the preserve of elite scholars or a pastime for the leisured upper classes. As increasing academic rigour and systematisation was brought to the study of 'natural history', its subdisciplines were adopted into university curricula, and learned societies (such as the Royal Horticultural Society, founded in 1804) were established to support research in these areas. A related development was strong enthusiasm for exotic garden plants, which resulted in plant collecting expeditions to every corner of the globe, some-times with tragic consequences. This series includes accounts of some of those expeditions, detailed reference works on the flora of different regions, and practical advice for amateur and professional gardeners.

Flora Capensis

This seminal publication began life as a collaborative effort between the Irish botanist William Henry Harvey (1811–66) and his German counterpart Otto Wilhelm Sonder (1812–81). Relying on many contributors of specimens and descriptions from colonial South Africa – and building on the foundations laid by Carl Peter Thunberg, whose *Flora Capensis* (1823) is also reissued in this series – they published the first three volumes between 1860 and 1865. These were reprinted unchanged in 1894, and from 1896 the project was supervised by William Thiselton-Dyer (1843–1928), director of the Royal Botanic Gardens at Kew. A final supplement appeared in 1933. Reissued now in ten parts, this significant reference work catalogues more than 11,500 species of plant found in South Africa. Opening with a preface which clarifies the project's original scope, Volume 1 covers Ranunculaceae to Connaraceae.

Cambridge University Press has long been a pioneer in the reissuing of out-of-print titles from its own backlist, producing digital reprints of books that are still sought after by scholars and students but could not be reprinted economically using traditional technology. The Cambridge Library Collection extends this activity to a wider range of books which are still of importance to researchers and professionals, either for the source material they contain, or as landmarks in the history of their academic discipline.

Drawing from the world-renowned collections in the Cambridge University Library and other partner libraries, and guided by the advice of experts in each subject area, Cambridge University Press is using state-of-the-art scanning machines in its own Printing House to capture the content of each book selected for inclusion. The files are processed to give a consistently clear, crisp image, and the books finished to the high quality standard for which the Press is recognised around the world. The latest print-on-demand technology ensures that the books will remain available indefinitely, and that orders for single or multiple copies can quickly be supplied.

The Cambridge Library Collection brings back to life books of enduring scholarly value (including out-of-copyright works originally issued by other publishers) across a wide range of disciplines in the humanities and social sciences and in science and technology.

Flora Capensis

Being a Systematic Description of the Plants of the Cape Colony, Caffraria & Port Natal

VOLUME 1:
RANUNCULACEAE TO CONNARACEAE

WILLIAM H. HARVEY *ET AL.*

CAMBRIDGE
UNIVERSITY PRESS

University Printing House, Cambridge, CB2 8BS, United Kingdom

Cambridge University Press is part of the University of Cambridge.
It furthers the University's mission by disseminating knowledge in the pursuit of education, learning and research at the highest international levels of excellence.

www.cambridge.org
Information on this title: www.cambridge.org/9781108068062

This edition first published 1894
This digitally printed version 2014

ISBN 978-1-108-06806-2 Paperback

FLORA CAPENSIS.

FLORA CAPENSIS:

BEING A

Systematic Description of the Plants

OF THE

CAPE COLONY, CAFFRARIA, & PORT NATAL.

BY

WILLIAM H. HARVEY, M.D., F.R.S.

PROFESSOR OF BOTANY IN THE UNIVERSITY OF DUBLIN, ETC., ETC., ETC.

AND

OTTO WILHELM SONDER, PH. D.

OF HAMBURGH.

MEMBER OF THE IMPERIAL LEOP. CAROLINE ACADEMY NATURÆ CURIOSORUM, ETC., ETC., ETC.

VOLUME I.

RANUNCULACEÆ TO CONNARACEÆ.

L. REEVE & CO., LTD.

LLOYDS BANK BUILDINGS, BANK ST., ASHFORD, KENT.

1894.

TO HIS EXCELLENCY

SIR GEORGE GREY, K. C. B.

D. C. L. OXON. &c. &c. &c.

Governor & Commander-in-chief of the Colony of the Cape of Good Hope,

THE FLORA CAPENSIS,

WHICH OWES ITS EXISTENCE MAINLY TO HIS FOSTERING PATRONAGE,

Is gratefully Dedicated

BY HIS EXCELLENCY'S

FAITHFUL AND OBEDIENT HUMBLE SERVANTS,

W. H. HARVEY,
W. SONDER.

PREFACE.

In undertaking the Flora Capensis, the authors propose to furnish to the colonists in the British South African provinces a clear and concise descriptive catalogue of the vegetable productions of their adopted country. As the colonies have no very definite limits to the northward, neither have the authors been anxious to fix a boundary line to this Flora. Generally speaking, the Cape Flora is limited on the North by the Gariep or Orange River, and on the East by the Tugela—boundaries more convenient than natural, for the Orange River at its western extremity rather flows through, than bounds the peculiar Desert Flora of Namaqualand; and the Tugela merely limits the British Colony of Natal, while the characteristic vegetation of Kafferland, of which Natal is a section, extends northward at least to Delagoa Bay, gradually assuming the features of Tropical African vegetation. Whilst therefore our Flora will be found tolerably complete for the old-established colonial districts, both of the Western and Eastern provinces, it presents little more than an outline sketch of the Northern and North Eastern Regions, and of the Natal Colony; and still more imperfectly pourtrays the vegetation of Great Namaqualand, Betchuana-land, the Orange River Free State, and the Transvaal Republic, all lying beyond the Gariep.

The authors have diligently availed themselves of every accessible collection of plants from the last named regions; but so few botanical travellers have yet explored them, save in some scattered spots, that their vegetation is as yet all but unknown. From what we know of the plants of Transvaal, especially of its mountains and high plateaus, that country promises to the botanist the richest harvest yet ungathered in South Africa; and the long mountain range that divides Kaffraria from the Western regions, while it limits the distribution of the greater portion of the subtropical types that mingle in the Cape Flora, probably still retains in its unexplored wilds multitudes of interesting plants. This we infer from the fact that almost every small package of specimens received from the Natal, or the Transvaal district, contains not only new species but new genera ; and some of the latter are of so marked and isolated a character, as to lead us to infer the existence in the same region of unknown types that may better connect them with Genera or Orders already known.

It is not intended at present to enter into details on the geographical distribution of South African plants, or the relations between the South African Flora and that of other countries having a nearly similar climate. Whilst the work is in progress, new collections of plants continue to be received, rendering it impossible to prepare a satisfactory geographical introduction, until the descriptive portion has been finished. At the conclusion of the work the authors propose to give a general introduction, which will embrace the geographical relations of the Flora, and include a summary of the labours of botanical explorers in South Africa, and of the various treatises that have been written on South African plants.

It is scarcely possible to say definitely to what number of volumes the descriptive matter may extend. That it will require at least five volumes of the size of this first volume seems probable ; and these can scarcely be completed in less than ten years. Within that time the authors anticipate, from their numerous friends and correspondents in the Colonies, so much additional information on South African botany, that probably a considerable supplement may be needed to the earlier volumes. If their undertaking meet with approval after its completion equal to the encouragement with which they have already been favoured by the Colonial Government, and by private individuals in the colony, they will have received the best and most grateful reward of their labours.

The authors desire here most prominently to record their deep sense of the confidence reposed in them, and the essential aid extended to their work, by the Parliament of South Africa, which, on the proposal of the Governor General, has, by a liberal grant of money (at the rate of £150 per volume), relieved them from much of the cost of publication; and at the same time has left the whole impression at their disposal. Few works of the kind have a large sale, and without such patronage the present Flor could not be carried out. Nor has the Governor's thoughtful kindness been limited to obtaining the parliamentary grant. His Excellency, by a government notice (No. 387.—1857) has invited contributions of dried specimens of plants from persons residing within the Colonies, or the neighbouring Free States, and has undertaken to forward the same, if sent to the Colonial Office, Capetown, to the authors, free of expence. Already this notice has elicited the active co-operation of several obliging collectors of specimens, and furnished much valuable material for this work. The authors now earnestly solicit from their *still unknown friends*, collections, large or small, of specimens, dried according to the plain directions which will be found at page xxiv of the "*Intr. to Botany*" that follows this preface; which collections shall be duly and thankfully acknowledged in future volumes. And, as they may serve to render the Flora more *complete* than it could otherwise be, the senders will be doubly rewarded; they will have the satisfaction of having forwarded the cause of science, and will reap the advantage promised to those that cast bread on the waters,—they will themselves gain information by imparting it to others.

The authors have now thankfully to acknowledge their obligations:—

To MRS. F. W. BARBER of Queenstown district, and her brother HENRY BOWKER, ESQ., for several very interesting parcels of the rarer plants of the Winterberg, and of the country extending thence eastward beyond the Kei.

To GENERAL BOLTON, R. E., for collections made in the neighbourhood of Grahamstown.

To HENRY HUTTON, Esq. of Grahamstown, for considerable collections made in Albany.

To JOHN SANDERSON, Esq. of D'Urban, for very interesting and valuable collections from the Natal Colony and from Transvaal, containing many new genera and species.

To DR. SUTHERLAND, Surveyor-General, Natal, for small but

carefully selected collections made in various parts of his district, during hasty professional visits; in one of which expeditions he discovered the GREYIA SUTHERLANDI,* one of the most remarkable of South-east African shrubs.

To R. HALLACK, Esq., Port Elizabeth, for interesting information on plants of the Natal Colony, accompanied by specimens obtained on hasty journeys when the traveller could not burden his horse with paper or vasculums. These plants having been collected and preserved under difficulties, deserve the more regard; and especially as most of them were new to us. Mr. Hallack is well acquainted with plants, and we look forward to his future visits to Natal with considerable expectation.

To MRS. HOLLAND, of Port Elizabeth, for well-executed outline drawings, accompanied by dried specimens, of Orchideæ and other plants of Uitenhage. Mrs. Holland has most kindly undertaken to furnish Dr. Harvey with sketches from the life of plants desirable to figure in "*Thesaurus Capensis.*"

To DR. ROSER, of Gnadendal, for an interesting series of well dried specimens from that rich botanical region.

To DAVID ARNOT, Esq. of Colesberg, for a box of living succulent plants (forwarded to Kew Gardens).

To CHARLES WRIGHT, Esq. of the United States Japan Expedition, for upwards of five hundred species of plants collected whilst the vessels were detained in Simon's Bay. Strange to say, this collection contains some species not received from other collectors.

To ANDREW WYLEY, Esq., late Geological-Surveyor, for several parcels of specimens collected chiefly along the Orange River, from Colesberg to the mouth, and in Great Namaqualand. These parcels, received while the last sheet of this volume was passing through the press, contain several new plants; and among other things of interest is a specimen (leaf and flowers) of the "*Elephant's Trunk,*" that most singular of Namaquan plants, and which proves to be a species of *Adenium* (*A. Namaquanum* Wyl.), a genus found also in Senegal and Arabia.

* Now introduced to Britain from a few seeds sent by Dr. Sutherland in a letter to Sir William Hooker, and successfully raised by Mr. Moore, of the Botanic Gardens, Glasnevin, Dublin.

Lastly, but not least, the authors offer their best acknowledgments to DR. PAPPE, Colonial-Botanist, for several very valuable collections already received, and for his expressed intention of continuing to assist them, as the work proceeds, with materials for each volume. Dr. Pappe's long residence at the Cape, and long familiarity with its botany, have given him peculiar qualifications for filling the post of Colonial Botanist with honor to himself and advantage to the public. The Cape government is fortunate in being able to secure, on the spot, the services of a gentleman so fully competent to render assistance; and the authors of this Flora deem themselves equally fortunate in participating in the fruits of Dr. Pappe's old and new explorations. It is a real pleasure to receive a packet, large or small, of Dr. Pappe's *personal* collections; for not only are the specimens themselves *well selected and complete*, but they are most carefully dried,—flattened without being squeezed, and never tangled or interwoven. Besides his own extensive collections, Dr. Pappe is in possession of the whole of the dried plants left by the lamented CHARLES ZEYHER; from the duplicates of which he has furnished the authors with many rarities.

Turning a moment from South African friends, the authors have now to express gratitude to those who in Europe have favoured their undertaking. And first and specially they are most deeply indebted to SIR WILLIAM HOOKER, not only for throwing open to them, in the freest manner, the unrivalled Kew Herbarium, permitting them to study the specimens at Dublin and at Hamburgh, and to compare and authenticate with their own collections;—but also for the great interest he has from the commencement shown in this undertaking. To him primarily it is due that the work was set on foot. He it was who suggested it, and assisted in devising the plan on which it should be moulded; he also introduced the authors to many of the valued South African correspondents to whom they have recorded their obligations; and, lastly, his strong recommendation of the undertaking to Sir George Grey was mainly instrumental in obtaining the grant from the Colonial Parliament.

Warm thanks are also due to the following distinguished

botanists and curators of museums, for allowing access to their Herbaria and to museums under their charge, viz. :—

To DR. ELIAS FRIES, Professor at Upsala, for access to the original specimens of Thunberg's "*Flora Capensis.*"

To DR. N. J. ANDERSSON, Professor at Stockholm, for access to the plants of *Bergius, Thunberg, Sparmann, &c.*

To DR. F. KLOTZSCH, Conservator of the Berlin Herbarium, for access to Willdenow's, Link's, Kunth's, Chamisso and Schlechtendahl's, and Mundt's plants.

To DR. E. FENZL, Professor at Vienna, for access to Jacquin's original specimens, and other important collections in the Imperial Herbarium.

To DR. C. F. MEISNER, Professor at Basle, for plants from the collection of Krauss, &c.

To DR. FERDINAND KRAUSS, Professor at Stuttgart, for a collection of *Diosmeæ*, named by Steudel.

To DR. GRISEBACH, Professor at Goettingen, for the *Diosmeæ* of the Herbarium of Bartling.

To H. WENDLAND, Director of the Royal Garden at Herrenhausen, for authenticated *Diosmeæ, Rhamneæ, Geraniaceæ, and Byttneriaceæ.*

To DR. JOSEPH ROEPER, Professor at Rostoch, for *Oxalideæ, Diosmeæ, and Celastrineæ*, named by Lamarck.

To GEORGE BENTHAM, Esq., V.P.L.S., for the free use of his rich Herbarium (now deposited at Kew).

To DR. J. D. HOOKER, F.R.S. Assistant-Director of the Royal Gardens, Kew, for facilitating our access to the collections, and aiding us from time to time by personal references to scattered plants.

To N. B. WARD, Esq. F.R.S., for the donation of specimens of the rarer heaths, collected by Masson and other early explorers.

To DR. ALEXANDER PRIOR, of Halse House and Regent's Park, for access to his Herbarium, and the donation of several interesting specimens of Cape plants.

To Dr. W. J. BURCHELL, of Fulham, the celebrated traveller, for the verification of several of the *Polygalæ* and *Muraltiæ*, first described from his specimens.

To this copious list two distinguished names should have been gratefully added, had not both been removed by death during the publication of this volume. The late Professors, LEHMANN, of Hamburgh, and E. MEYER of Konigsberg, who possessed rich collections of South African plants, have largely contributed to the completeness of this Flora by placing in the authors' hands, for examination and description, many rarities not otherwise accessible; and Professor Lehmann, especially, as Dr. Sonder's early friend and first instructor in botany, felt a lively interest in the labours of his former pupil.

A few words may be useful to the student or amateur in guiding him to the use of this FLORA. In the present stage of the publication we must *suppose* that he has some previous knowledge, sufficient to tell him whether the plant he wishes to name be one described in this first volume or not. This volume contains only the *Thalamifloræ* (or polypetalous Exogens with *hypogynous* stamens) and six Orders of *Calycifloræ* (a great group, which will extend over vols. 2 and 3). Unless therefore his plant be *exogenous, polypetalous,* and with *hypogynous,* or somewhat *perigynous,* but not *epigynous* stamens, he will not find it in this first volume. Supposing it be *polypetalous,* with *hypogynous* stamens, and that he does not know to what Order it should be referred, (if he have no manual of Systematic Botany at hand) let him turn to the table headed "Sequence of natural orders, &c." page xxxiv. There, partly by reading the characters of the Orders, and comparing with his specimen; and partly by passing over such Orders as he knows it cannot be referred to, he may arrive—after a little practice—at a knowledge of what Order his plant belongs to. This is at first a difficult process, and the unassisted student will make many blunders; but if he have patience to struggle with this difficulty, and memory to retain any step once fairly gained, the rest of the book will be comparatively easily mastered. For, once the student is *quite sure* that he knows the proper natural Order of his specimen by turning to that Order in the body of the work, and consulting the "*Table of the South African Genera*" which follows the description of the Order, he will find the genus indicated in the fewest words, and contrasted with other allied genera. Thus, suppose his specimen be a *Dianthus* (or Wild Pink), and that he has found out that it belongs to the *Caryophylleæ*; turning to page 120, he sees by the "*Table*" that there are six sub-orders, under one or other of which his plant will be placed. To ascertain the proper sub-order, he must carefully look at the characters of each; and as the characters given are *absolute*, except where the contrary is stated, failure in one character will exclude a genus from any sub-order. Of sub-order (1) *Sileneæ*, the first character is, "*calyx tubular*;" in all the other sub-orders, except (4) *Mollugineæ*, we have "*calyx 5-parted or 4-parted*;" consequently if his plant have a *tubular*, or a *cleft* (not *parted*) calyx, it must belong either to *Sileneæ* or to *Mollugineæ*. To determine to *which* of these it belongs, he must compare it with the other characters given. If it have no petals, or if it have numerous (*more than* 5) linear petals, or alternate, tufted or whorled leaves, or stipules, it cannot belong to *Sileneæ*. These are *obvious* characters that strike the eye; but if still doubtful, let him cross-cut the ovary, and see whether it be "*unilocular*," or 3–5-*celled*:" this character is absolute, and will determine his plant (supposed a *Dianthus*) to belong to *Sileneæ*. Under *Sileneæ* he has three genera contrasted; and a glance at the base of the calyx will tell him that there are *bracts*, and consequently that his plant must be a *Dianthus*. Having thus determined the genus, let him turn to page 122, and read the full generic character there given, and the remarks in small print under it; he will thus be satisfied that he is right in his use of the table, or will discover his error, if he have made one. Still supposing he has hold of a *Dianthus*, he now further wishes to ascertain which *species* it is. The nine species described are grouped under three sub-sections, thus:—

* *Stem simple, one flowered.* (Sp. 1–2).
** *Stem paniculately branched. Petals entire or toothed* (Sp. 3–6).
*** *Stems branched. Petals deeply digitate or pinnatifid* (Sp. 7–9).

These characters are so strongly contrasted, that no difficulty will be found in referring our Dianthus to one or other. If the stem be one-flowered, the plant must either be *D. cæspitosus* or *D. scaber*. To decide between these, observe the words which are in the specific character printed in *italics*; these indicate the points to be specially noticed in the description, being those by which the two species chiefly differ from each other. Thus the "stems *glabrous*," and bracts "2–3 *times shorter than the (1½–2 inch long) tube*," contrast with "stems *scabrid*" and bracts "*twice or thrice as short as the* (*inch long*) *tube*." The points to be noted, therefore, are whether the stem be quite smooth or rough, and whether the calyx be only an inch long, or approaching 2 inches. The further remarks in small print under each species, and the *habitats* or stations where the plant has been gathered, will afford additional helps to the student, and it is hoped will enable him to decide on the name.

An explanation of the principal contractions used may be useful. Thus, after the localities of each species, appears some such formula as (Herb. Thunb., Hook., T.C.D., Vind., Sond., Lehm., r. Ber.) This would mean that the author had personally consulted and compared specimens of the species in the *Herbaria* of Thunberg; Hooker; Trinity College, Dublin; Vienna; Sonder; Lehmann; and Berlin: thus enabling the future student, by reference to any of these Herbaria, to ascertain the plant intended to be described in the text.

Authors' names are variously contracted. It does not seem necessary to explain all, but some of those most constantly used are, *Eck. and Zey*, or *E. & Z.*=*Ecklon and Zeyher*, the well known collectors and distributors of Cape plants. *E. Mey. and E. M.*=the late *Prof. Ernest Meyer*, author of a commentary on part of *Drege's* collections, and the authority for the names of most of Drege's distributed plants. *DC.*=*De Candolle; Linn.* or *L.*=*Linnæus;* *Hook* or *Hk.*=*Hooker;* *Thunb.* or *Th.*= *Thunberg; Burm.*=*Burmann; Burch.*=*Burchell; Bartl. and Wendl.* or *B. & W.*=*Bartling and Wendland; Endl. Gen.*= *Endlicher's Genera Plantarum; Bot. Mag.*=*Botanical Magazine; Jacq.*=*Jacquin; L'Her.*=*L'Heritier; Harv.* or *H.*=*Dr. Harvey; Sond.* or *Sd.*=*Dr. Sonder.*

The work throughout has been written in the English language, in order that it may be useful as a book of reference to the widest circle of the Colonial public, to many of whom it would be comparatively useless if composed partly or wholly in Latin.

The authors have endeavoured to avoid unnecessary technical words or phrases, where a common English expression conveys as definite a meaning. In the majority of cases, however, they have used the ordinary botanical terms, as being the briefest and clearest for the purpose; but as most or all of these terms will be found explained in the "*Outline Introduction to Botany,*" and are referred to in the "*Index of terms*" which follows, the student may soon become familiar with those in most constant use; and having learned to contrast "*glabrous*" with "*smooth,*" (the former being used for smoothness as distinguished from *hairiness,* the latter smoothness as contrasted with inequality of surface, or roughness) he will see the necessity of having definite terms with accurately discriminated significations.

We should have added *English names* to the genera and principal species, had we not found it impossible to do so in a practically useful manner; and this for the following reasons: 1st. *English names* of plants are of no certain application, and often differ in different districts of England; and, for aught the authors know, may have acquired new meanings in the Colony. 2nd. The number of Cape genera having established English names is extremely few; and often the same English name is applicable to several genera. Thus, the colonial name "*Milkbush*" signifies not merely a *Gomphocarpus,* but any Asclepiadeous plant, and would probably also be given to a shrubby *Euphorbia* or any milky-juiced plant. "*Zuurebesjies*" is given to several distinct and widely separated shrubs or trees, which happen to agree in having acid, edible fruits. "*Blumbosch*" has a still wider range of meaning; and so of other colonial names. 3rd.—English names for genera being comparatively few; if we had adopted the practice of always giving *English names,* we could only accomplish the feat by *inventing* colloquial names or soubriquets, i. e. introducing new *barbarous* words, a practice we do not think desirable to follow. If colonial names do exist for a large number of genera, colonial botanists must communicate them to the authors before the latter can be expected to know them, or their application. We can determine the *botanical* name, if it have one, of any unnamed plant sent to us; but no amount of sagacity or learning could discover for us the proper Colonial or local name. Wherever our correspondents have furnished us with these local names, we have given them; where they have not done so, we have only abstained from *inventing* "nicknames." We make these observations in reference to a friendly criticism which has reached us from the Cape, and we trust this explanation will satisfy our critic that it is less easy than he imagines to comply with his requirements.

A word or two respecting the part each author has taken in the work may fitly close this preface. As one lives in Dublin, and the other in Hamburgh, joint action in every case is impossible. Feeling this difficulty, the authors determined to divide the work between them; each taking those Orders for his share with which he was most familiar, and working them out independently without reference to his brother author. The Orders severally elaborated by each are indicated at the head of the page by the (Harv.) or (Sond.) following the heading; and the author's name is printed in full immediately above the description of the Orders. To ensure uniformity of plan, the Ordinal characters and subtending remarks are all written by Dr. Harvey, who has also had the oversight of the printing, and the general editorship; and who is therefore responsible for certain errors of typography which have escaped notice, and which, so far as observed, will be found corrected in the list of addenda and corrigenda.

Trinity College, Dublin,
May 10, 1860.

ADDENDA AND CORRIGENDA.

Page.

22, for **A? nudicaulis** read **A? nudiuscula,** *E. Mey.!* *A. nudiuscula,* E. Mey. is the same as *Barbaræa præcox?* (Herb. Paris) from Abyssinia, and of Schimper's Abyss. Pl. No. 113 and 674, fide *J. D. Hook. in litt.*

24, **S. capense,** var. α = Drege 7538, ex pte. var. γ, = Drege 7539, ex pte.

26, line 11, for "bent" read "flexuous."

28, under **L. myriocarpa,** add:

Stem erect, 2 feet or more high. Radical leaves wanting in our specimens; the middle ones 2–3 inches long, 1–2 lines wide; the upper smaller. Panicle 1 foot or longer: racemes many flowered, 1½–3 inches long. Pedicels 2 lines long. Flowers white, very small. Silicule ½ line long.

29, line 20, add, *Drege,* 7543.

30, line 11, add, *Drege,* 7542, b.

32, line 7, add, *Drege,* 7547.

55, line 34, in place of 1683 read 1834.

59, line 33, read "*Ovary* stipitate, unilocular: ovules numerous."

61, after **Boscia caffra** insert:

2. B. angustifolia (Harv.); leaves lanceolate-linear, much attenuate at base, subpetiolate, mucronulate, coriaceous, veiny, glabrous, *stipulate;* peduncles axillary and terminal, much shorter than the leaves, densely racemose, several flowered; sepals villoso-ciliate; flowers polyandrous.

HAB. About Jackalsberg and Missionary Drift, Namaqualand, *Andrew Wyley, Esq.* (Herb. T.C.D., Hook., Sond.)

A virgate, slender, glabrous shrub, with pale yellowish-green twigs and leaves. Leaves 2–2½ inches long, 1½–2 lines wide, rigid, flat or inrolled, scattered, subdistant, spreading. Peduncles ½–1½ inches long, 6–10 flowered; bracts at the base of the pedicels deciduous, linear, each with two minute, persistent stipellæ. Stipules toothlike, minute. Pedicels 2–4 lines long. Calyx tube conical, densely glandular at the throat. Sepals oblong, blunt, with woolly margins. Ovary on a long stalk, oval; fruit not seen. "The wood is hard and close grained." *A. W.*

62, line 10, add, *Drege,* 7535, *Zey.* 1910.

65, line 6, add, *Drege,* 2940.

67, line 1, add, *Gueinzius,* 96 and 100.

68, after **P. Mundtii,** add *Zey.* 3785; after **P. Eckloni,** *Zey.* 3783; after **P. Zeyheri,** *Zey.* 3784.

69, after **T. trinervis,** add *Zey.* 3989, *Drege,* 2352, *E. & Z. Urtic.* 2; after **T. alnifolia,** add *Drege,* 4613, *E. & Z. Urtic.* 3.

89, after **P. affinis,** insert:

16.* P. Carmichaelii (Harv.); suffruticose, diffuse, patently pubescent, with long, trailing branches; leaves minutely petioled, *oblongo-lanceolate, very acute, midribbed, flat,* pubescent; racemes opposite the leaves, short, few-flowered, spreading or reflexed; *peduncles filiform;* bracts persistent, lanceolate, acute; alæ elliptic-ovate, acute, *glabrous;* ant. sepals ovate, acute, *glabrous;* keel crested; lat. petals oblong, 2–lobed; capsule obcordate, as wide as the purple alæ.

HAB. Cape, *Carmichael!* (Herb. T.C.D., Hook., Sond.)

Root woody. Stems many from the crown, 2–3 feet long, procumbent, with many slender branches. The whole plant is patently pubescent. Leaves 3 lines long, 1 line wide, quite flat. Peduncles uncial, with 3–4 flowers near the extremity. Alæ 1½ line long.

Page.
89, under **P. Lehmanniana**, add :
var. β, **pteropus** (*Harv.*) ; densely pubescent, and more robust ; *peduncles rather broadly winged upwards,* two-edged toward the slender base. (Hab.—Cape, *Capt. Carmichael,* in Herb. T.C.D., Hook., Sond.)
102, after **M. asparagifolia**, insert :

16.* M. acerosa (Harv.) ; cæspitose, many stemmed ; stems short, virgate, subsimple, tomentulose ; leaves fascicled, very narrow-subulate, sub-erect, glabrous, pungent ; flowers minute, sessile ; sepals lanceolate, acuminate ; petals linear, subacute, shorter than the amply lobed keel ; capsule . . . . ?

HAB. Swellendam, *Dr. Pappe, No.* 76 (Herb. T.C.D., Hook., Sond.)

Root thick and woody. Stems very numerous, simple, 3–5 inches high, densely leafy. Leaves pale yellow-green, very slender, but much broader than those of *M. asparagifolia.* Flowers purple, a line long. Intermediate between *M. asparagifolia* and *M. acicularis,* but seemingly distinct from both : much more glabrous than the latter.

114, to **F. capitata**, var. α, add *Zey.* 1955 ; var. δ, *Zey.* 1954.
156, erase **G. miltus** (Fenzl.) altogether. The plant so named by Moquin in DC. Prod. proves to be merely a discoloured specimen of *G. pentadecandra, E. Mey.*
175, line 11, add syn. *H. cuneifolius,* Garke.
230, erase **S. Zeyheri** (Planch.) altogether. The shrub so named proves to be an abnormal condition of **Elæodendron croceum** (page 468.)
319, line 22, in place of " 5–10 angled," read " 5–6 angled."
321, to **O. linearis**, var. γ, add syn. *O. aretioides,* Turcz. Bull. Mosc. 1858. No. 2, p. 436.
324, to **O. laburnifolia**, *Jacq.* add syn. *O. arthrophylla,* Turcz. l. c. 433.
325, change **O. albida**, Sond. to **O. leucotricha**, Turcz. l. c.
325, line 36, " *Petals* yellow, with blackish margins."
327, to **O. ramigera**, var. α, Sond. add syn. *O. gymnoclada,* Turcz. l. c. ; to var. β. add *O. minutifolia,* Turcz. l. c.
332, to **O. pulchella**, Jacq. add syn. *O. foveolata,* Turcz. ex pte.
333, to **O. calligera**, Sond. add syn. *O. foveolata,* Turcz. ex pte.
334, to **O. punctata**, Linn. add syn. *O. favosa,* Turcz. l. c. ; and add to the habitats. *Zey.* 236, 238, 336, 2104, 2118.
336, line 1, after " keeled," add " leaflets."
336, change **O. glaucovirens**, Sond. to **O. stenoptera**, Turcz. l. c.
336, to **O. minima**, Sond. add syn. *O. nidulans,* Turcz. l. c., non E. & Z.
337, change **O. Uitenhagensis**, Sond. to **O. psilopoda**, Turcz. l. c.
337, to **O. imbricata**, E. Z. add syn. *O. multifolia,* Turcz. l. c.
337, to **O. imbricata**, E. Z. var. β, Sond. add syn. *O. elegantula,* Turcz. l. c.
339, to **O. natans**, L. add syn. *O. rugulosa,* Turcz. l. c.
339, line 38, " corolla white or whitish with a yellowish tube."
341, to **O. heterophylla**, DC. add syn. *O. stenodactyla,* Turcz. l. c.
344, **O. densifolia**, Sond. is the same as **O. densifolia**, Turcz. l. c.
347, line 11, for " Flowers 4½ inch long," read " Flowers about 1 inch long."
347, line 17, for Thunb. Diss. N. 22, read No. 25.
348, line 22, for 113 read 13.
348, line 36, for 539 read 659.
353, line 25, for petiole read peduncle.
353, line 41, for Wright read Wight.
355, after Augea, insert :

III.* FAGONIA, Tourn.

Calyx 5-parted, deciduous. *Petals* 5, clawed, longer than the calyx. *Stamens* 10, hypogynous, equal ; filaments filiform, naked at base, erect ; anthers cordate. *Ovary* acutely 5-angled, 5-celled. *Style* 5-angled, continuous with the ovary ; stigma acute. *Capsule* pyramidal, 5-sided, of 5-cocci, falling away at maturity from a persistent axis. —*Endl. Gen. No.* 6034.

Herbaceous plants, natives of the dry Mediterranean regions, and of the African and Asiatic deserts. Leaves 3-foliolate, opposite; stipules pungent, spreading. Peduncles 1-flowered, axillary. Flowers purple or violet, rarely yellow. Name in honour of M. Fagon, a great patron of botany under Louis XIV.

1. **F. cretica** (Linn.) var. **glandulosa** (Harv.); glandularly pubescent; leaflets rhombic-ovate or ovato-lanceolate, the lateral unequal-sided, spinoso-mucronate; stipules subulate, pungent, patent; capsule pubescent or glandular, broader than long.

HAB. Namaqualand, *Mr. A. Wyley.* (Herb. T.C.D., Hook., Sond.)

Stems woody at base, procumbent, much branched. Petioles 4-7 lines long; the leaflets as long or shorter, 3-4 lines wide, the medial twice the width of the lateral. Flowers lilac-purple, with bright orange stamens. Except in pubescence this does not differ from the plant of the North African desert. We have equally pubescent specimens from India.

Page.
357, line 7, "seeds blackish."
358, line 39, erase syn. *Z. capense*, Lam.
359, to **Z. flexuosa**, var. β, add *Drege*, 7168.
360, to **Z. morgsana**, L. add syn. *Z. capense*, Lam.! Herb.
361, line 26, for Drege 1265, read 7165; line 36, add "capsule half-inch long, 5-6 lines wide."
369, after *Melianthus Dregeana*, insert:

5. **M. pectinata** (Harv.); leaflets *narrow-linear, very entire, with revolute margins*, glabrous and furrowed above, albo-tomentose beneath; stipules subulate; racemes erect, flowers approaching in whorls; bracts ovate-acuminate; upper calyx segment cuspidate; capsule?

HAB. Namaqualand, *Andr. Wyley, Esq.* (Herb. T.C.D.)

A rigid shrub. Leaves densely set on short twigs, pectinato-pinnate, the common petiole with narrow, glabrous, revolute wings between the leaf-pairs. Leaflets 8-10 pairs, 1-1½ inch long, 1 line wide, patent, exactly linear, pale green, rigid. Stipules broad at base, subulate-attenuate. Peduncles purple, puberulous, 3-4 inches long, with 3-4 imperfect whorls of flowers. Bracts equalling the pedicels, 2-3 lines wide. Flowers dull red. Posterior sepal 5 lines long, concave, broad at base, ending in 2 short, blunt lateral lobes, and one medial, subulate cusp: lateral (inner) sepals broadly subulate; two lower sepals 9-10 lines long, oblong, acute, striate. Petals 4, on long claws; lamina undulate, lanceolate, reflexed, the two lower toothed at the base of the lamina. Shorter stamens connate at base; longer not half as long as the lower sepals. A very distinct species, easily known by its narrow and quite entire leaflets. Only one specimen seen.

373, to **Euch. dubia** Sond. add syn. *Acmadenia cassiopoides*, Turcz. l. c.
377, line 35, add *Drege* 2250.
378, line 47, add *Zey.* 3775 (Gnidia).
383, to **A. assimile**, Sond. add syn. *Euchætis lævigata*, Turcz.
415, for **A. craspedata**, read **A. craspedota.**
439, to **M. villosa**, add syn. *Mac. Sieberi*, Turcz. l. c.
449, to **A. pulchra**, add *Zey.* 302.
450, to **C. cymosa**, add Zey. 536.
468, line 4, for *Chloroxylon* read *Crocoxylon.*

TABLE OF THE CLASSES AND SUB-CLASSES.

CLASS I. **DICOTYLEDONES** v. **EXOGENÆ.**

Stem having a central pith, surrounded by one or more concentric rings of woody and vascular tissue, and coated by a separable bark. *Leaves* usually articulated with the stem, and traversed by branching and anastomosing veins. *Embryo* with two (or more) opposite cotyledons; the young stem rising between the cotyledons.

Sub-Class 1. **THALAMIFLORÆ.**—*Calyx and corolla* (generally) present. *Petals* separate, inserted, as are also the *stamens*, on the receptacle (i.e. *hypogynous*). *Ovary* free.

Sub-Class 2. **CALYCIFLORÆ.**—*Calyx and corolla* (generally) present. *Calyx* gamosepalous. *Petals* separate, or united into a monopetalous *corolla*, either perigynous or epigynous. *Stamens* inserted on the calyx *(perigynous)*, or on the tube of a perigynous, or epigynous corolla. *Ovary* free, or more or less adnate to the calyx-tube.

Sub-Class 3. **COROLLIFLORÆ.**—*Calyx and corolla* both present. *Petals* united in a monopetalous, perigynous corolla. *Stamens* inserted on the corolla. *Ovary* free or nearly so.

Sub-Class 4. **MONOCHLAMYDEÆ.**—*Perianth* single (a calyx, or calyx and corolla soldered together, coloured or green) or more or less imperfect, or altogether absent.

CLASS II. **MONOCOTYLEDONES** or **ENDOGENÆ.**

Stem not distinguishable into pith, wood, and bark, but consisting of bundles of woody and vascular tissue separately imbedded in cellular tissue, and encased in a firmly adherent outer rind. *Leaves* usually sheathing at base, and traversed by sub-parallel, unbranched veins, running from the base to the apex, and connected by straight, cross veinlets. *Embryo* with one cotyledon; the young stem starting from a cavity in the side.

CLASS III. **ACOTYLEDONES** or **CRYPTOGAMÆ.**

Plants destitute of true flowers, or seed-producing organs:—propagated by *spores*, i. e. reproductive cells, not containing any embryo. *Spores* variously evolved, and fertilized in various ways. [This class comprises Ferns and Filicoid-plants, Characeæ, Mosses, Hepaticæ, Fungi, Lichens, and Algæ or Seaweeds].

SEQUENCE OF ORDERS CONTAINED IN VOL. I. WITH BRIEF CHARACTERS.

Sub-Class I. Thalamifloræ. Orders I.–XLI. [Several orders include apetalous genera. In *Bixaceæ*, *Caryophylleæ*, and *Phytolacceæ*, the stamens are sometimes perigynous; and in *Bixaceæ* the ovary is sometimes partially adnate.]

I. RANUNCULACEÆ (page 1). *Flowers* bisexual. *Stamens* indefinite; anthers adnate; filaments subulate. *Carpels* numerous, separate. (*Slender climbers or herbs. Clematis, Anemone, Buttercup.*)

II. ANONACEÆ (page 7). *Fl.* bisexual. *Sepals* 3, valvate. *Petals* 6. *Stamens* indefinite; anthers adnate; filaments thickened upwards. *Carpels* numerous, separate. (*Trees or shrubs, with simple, entire, exstipulate leaves. Custard-apple.*)

III. MENISPERMACEÆ (page 9). *Fl.* minute, unisexual, green. *Stamens* definite (few) monadelphous. *Carpels* 1–3, separate. (*Slender, climbing suffrutices, with alternate, simple, netted-veined leaves. Fl. in axillary cymes, racemes, or umbels. Davidjes.*)

IV. NYMPHÆACEÆ (page 13). *Fl.* bisexual, large and showy. *Petals* numerous, in several rows. *Stamens* indefinite. *Carpels* numerous, sunk in a fleshy *torus*, and thus combined into a plurilocular ovary. (*Water-lilies. Leaves on long stalks, peltate or cordate.*)

V. PAPAVERACEÆ (page 14). *Sepals* 2–3, deciduous. *Petals* 4–6, equal, spreading, crumped in the bud. *Stamens* indefinite. *Ovary* 1*–celled*, with 2 or several parietal placentæ. (*Poppy.*)

VI. FUMARIACEÆ (page 15). *Sepals* 2, minute, scale-like. *Petals* 4, connivent in pairs: one or both of the two outer, spurred or saccate at base. *Stamens* 6, diadelphous. (*Glabrous, climbing herbs, with much divided leaves, and small purple, white, or yellow flowers. Fumitory. Bladder-weed.*)

VII. CRUCIFERÆ (page 19). *Sepals* 4, deciduous. *Petals* 4, clawed, equal, cruciate. *Stamens* 6; 4 long and 2 short. *Ovary* bilocular, with parietal placentæ. *Fruit* a pod or pouch. (*Wall-flower, Cabbage, Mustard, Water-cress, &c.*)

VIII. CAPPARIDEÆ (page 54). *Sepals* 4. *Petals* 4–8, or more, clawed, often unequal. *Stamens* 4, 6 or many. *Ovary* unilocular, with 2 parietal placentæ. *Fruit* a dry capsule or a fleshy berry. (*Trees, shrubs, or herbs, with alternate, simple or compound leaves. Pubescence often glandular and fœtid. Caper-bush.*)

IX. RESEDACEÆ (page 63). *Flowers* small, green or whitish. *Sepals* several, persistent. *Stamens* 3 or several. *Ovary* unilocular, *open at the summit*, with 3–6 parietal placentæ. *Fr.* a gaping capsule. (*Small herbs or suffrutices, with alternate, entire or cut leaves, and racemose or spiked flowers. Mignionette, Woad.*)

X. BIXACEÆ (page 65). *Fl.* small, often unisexual, regular. *Calyx* often gamosepalous, persistent. *Petals* sometimes wanting. *Stamens* definite or indefinite. *Ovary* unilocular, with parietal placentæ. (*Trees or shrubs with alternate, simple leaves.*

XI. VIOLARIEÆ (page 72). *Sepals* 5, persistent. *Petals* 5, unequal, the lower one spurred at base. *Stamens* 5, filaments broad and flat; anthers adnate, conniving round the stigma. *Ovary* unilocular, with 3 parietal placentæ. *Capsule* 3–valved, bearing the seeds on the middle of the valve. (*Violet and Pansy.*)

XII. DROSERACEÆ (page 75). *Sepals* 5, persistent. *Petals* 5, equal. *Stamens* 5, on slender filaments. *Ovary* unilocular, with 3–5 parietal placentæ. *Capsule* splitting. (*Herbs or suffrutices, covered with viscidly glandular hairs. Sundew. Catch-fly.*)

XIII. POLYGALEÆ (page 79). *Flowers* irregular. *Sepals* 5, unequal, the two lateral often coloured like petals. *Petals* 3; the two upper small, the lower (*keel* or *carina*) large, enclosing the stamens and ovary, and often crested in front. *Stamens* 8, monadelphous. *Ovary* 2-celled; ovules solitary, pendulous. (*Small shrubs, suffrutices, or herbs. Leaves simple, entire, exstipulate. Flowers racemose or spiked, mostly purple or pink. Milkwort.*)

XIV. FRANKENIACEÆ (page 113). *Fl.* regular. *Calyx* tubular, 4–5-toothed, ribbed, hardening after flowering. *Petals* 4–5, with long claws, deciduous, *Stamens* mostly 6. *Ovary* 1-celled, with 3–4 parietal placentæ and numerous ovules. *Capsule* enclosed in the calyx, many-seeded. (*Small, perennial or half shrubby plants, with densely crowded, minute, heath-like leaves. Flowers purple or white, fugacious. Sea-heath.*)

XV. ELATINACEÆ (page 115). *Flowers* minute, regular. *Sepals* 5, separate. *Petals* 5; *stamens* 10. *Ovary* 5-celled; styles 5; ovules numerous, on axile placentæ. *Capsule* 5-valved. (*Minute, herbaceous or half-shrubby plants. Leaves opposite, entire, with membranous stipules. Bergia.*)

XVI. HYPERICINEÆ (page 117). *Fl.* regular, yellow. *Sepals* 5, persistent, imbricate. *Petals* 5, unequal sided, spirally twisted in the bud, and often black-dotted at margin. *Stamens* numerous, united in 3–5 parcels (*polyadelphous*). *Ovary* imperfectly 3–5 celled, with numerous ovules. (*Shrubs or herbs, with opposite, very entire, pellucid-dotted, exstipulate leaves. Fl. in cymes. St. John's Wort.*)

XVII. TAMARICINEÆ (page 119). *Flowers* minute, regular, 4–5-merous, spiked or racemose. *Stamens* 4–10, united at base into a ring. *Ovary* 1-celled, with parietal placentæ, and numerous ovules; styles 3. (*Shrubs, with minute, scale-like, crowded or imbricate leaves. Tamarisk.*)

XVIII. CARYOPHYLLEÆ (page 120). *Fl.* regular, cymose, often without petals. *Calyx* 4–5 cleft or parted, persistent. *Petals* clawed. *Stamens* 4–10, sometimes perigynous. *Ovary* unilocular, or more or less 3–5 celled, with 3–5 styles; ovules one or many, on axile placentæ. *Fruit* a capsule or an achene. *Seeds* reniform, with a marginal embryo and floury albumen. *Herbs or small suffrutices, with opposite or scattered, entire leaves. Stipules none or membranous. Pink or Carnation. Chickweed.*)

XIX. PHYTOLACCEÆ (page 151). *Fl.* as in *Caryophylleæ*. *Ovary* of two or several, separate or connate, 1-ovuled carpels, with separate styles. *Fruit* dry or fleshy, breaking up into one-seeded cocci. *Seed* as in *Caryophylleæ*. (*Herbs or shrubs, with alternate, entire, mostly exstipulate leaves, and small, green or white spiked, racemose, or cymose flowers. Limeum. Virginian Poke.*)

XX. MALVACEÆ (page 157). *Fl.* regular, mostly large and showy. *Calyx* 5-fid, with *valvate* æstivation, mostly involucrate at base. *Petals* strongly twisted in æstivation, diliquescent, attached at base to the staminal tube. *Stamens* indefinite, united in a tube, enclosing the ovary and styles: anthers 1-celled. *Ovary* of 5 or many carpels whorled round a central column. (*Trees, shrubs or herbs, with alternate, many-nerved leaves, and stipules. Mallow. Hibiscus.*)

XXI. STERCULIACEÆ (page 178.) (A tree, with digitate leaves. Only one Cape species, *Sterculia Alexandri*. *Flowers* unisexual, without petals. *Calyx* campanulate, coloured, 5–7-cleft, valvate. *Stamens* united in a column; anthers 2-celled. *Carpels* 5, partly connate, many ovuled. The *Baobab* and *silk-cotton tree* also belong to this Order.)

XXII. BYTTNERIACEÆ (page 179). *Fl.* regular. *Calyx* 4–5 cleft, valvate. *Petals* 4–5 twisted. *Stamens* definite, 5, 10, 15, 20, &c. some often sterile, united at base into a ring or tube; anthers 2-celled, introrse. *Ovary* 5–10 celled; ovules in pairs, axile; style single, with 5–10 stigmas. (*Shrubs or suffrutices, rarely herbs, with stellate pubescence, and simple stipulate leaves. Hermannia, &c.*)

XXIII. TILIACEÆ (page 223). Like *Byttneriaceæ;* but stamens not connate at base, and usually numerous: anthers 2-celled, introrse. (*Trees, shrubs, and herbs, with stellate pubescence and stipulate, simple leaves. Sparmannia.*)

XXIV. HIPPOCRATEACEÆ (page 229). *Flowers* regular, small, greenish, 5 parted. *Stamens* 3, inside a fleshy, perigynous disc. *Ovary 3-celled,* with a short, trifid style. (*Shrubs or trees resembling Celastrus; differing by the three hypogynous stamens*).

XXV. MALPIGHIACEÆ (page 231). *Flowers* perfect, regular. *Calyx* 5-parted, imbricate, with external fleshy glands. *Petals* 5, clawed, spreading. *Stamens* 10 (5 sometimes abortive), connate at base. *Ovary* 3-2 celled; ovules solitary. (*Shrubs, erect or climbing, with opposite, simple, mostly entire leaves, and corymbose or racemose red or yellow flowers. Pubescence silky. Natal Country.*)

XXVI. ERYTHROXYLEÆ (page 233). *Fl.* as in *Malpighiaceæ;* but, *calyx* not glanded; *petals* with a bifid scale in front; *stigmas* capitate. (*Two Natal shrubs*).

XXVII. OLACINEÆ (page 234). *Fl.* regular. *Calyx* small, truncate or toothed *Petals* 4-6, with valvate æstivation. *Stamens* as many or twice as many as the petals, free (or connate). *Ovary* unilocular; ovules 2-4, pendulous from a free central placenta, or the side of the ovary. *Albumen* copious. (*Trees or shrubs, with simple entire leaves. 2 S. African species*).

XXVIII. SAPINDACEÆ (page 236). *Flowers* often unisexual. *Calyx* 4-5 parted, imbricate. *Petals* 4-5, or none, imbricate. *Disc* fleshy, outside the often excentrical or unsymmetrical stamens. *Ovary* 2-4 celled, with axile placentæ; ovules 1-3, rarely more. *Fruit* fleshy or capsular. (*Trees and shrubs, rarely herbs, with pinnate or trifoliolate, rarely with simple leaves. Stipules none. Flowers small, in racemes or panicles*).

XXIX. MELIACEÆ (page 244). *Fl.* regular. *Calyx* small, toothed or cleft. *Petals* longer than the calyx, sessile, with valvate or sub-imbricate æstivation. *Stamens* 4-10 united into a tube. *Ovary* plurilocular, on a fleshy disc; ovules 1-2, axile; style simple. *Fruit* a berry, drupe or capsule. (*Trees or shrubs. Leaves pinnate or simple, alternate, exstipulate. Flowers small, in racemes or panicles.*)

XXX. AMPELIDEÆ (page 248). *Fl.* regular, minute. *Calyx* cup-shaped. *Stamens* opposite the concave, valvate petals. *Ovary* free, 2 celled; ovules in pairs. *Fruit* a berry. (*Shrubs, rarely erect, mostly climbing by tendrils. Branches swollen at the joints, brittle. Leaves simple or compound. Vine.*)

XXXI. GERANIACEÆ (page 254). *Fl.* perfect, showy. *Sepals* 5, unequal, strongly imbricate. *Petals* 5 (or fewer), clawed, with convolute æstivation, caducous. *Stamens* definite, mono- or poly-adelphous; filaments subulate. *Ovary* of 5 one-ovuled carpels cohering round an awl-shaped or beak-like torus, to which the styles adhere. (*Geranium. Pelargonium. Monsonia*).

XXXII. LINEÆ (page 308). *Fl.* symmetrical. *Sepals* 4-5, persistent, imbricate *Petals* caducous, convolute. *Stamens* 8-10, slightly connate at base. *Ovary* 4-5 celled, with 4-5 styles; ovules 2 in each cell, axile. *Capsule* dry, 8-10 celled. (*Herbs or small undershrubs. Leaves simple, entire, veinless, exstipulate. Flowers yellow or blue, rarely red. Flax*).

XXXIII. BALSAMINEÆ (page 311). *Fl.* very irregular. *Calyx* coloured, spurred at base. *Petals* connate; one (*labellum*) much larger. *Stamens* 5. *Ovary* 5 celled, ovules indefinite. *Fruit* a capsule, splitting with elasticity. *Succulent herbs, with fragile stems, and generally alternate, exstipulate leaves. Flowers axillary, or in terminal racemes. Balsam.*)

XXXIV. OXALIDEÆ (page 312). *Fl.* regular. *Sepals* 5, imbricate. *Petals* convolute, deciduous. *Stamens* 10, five shorter than the others, slightly connate at base. *Ovary* 5-lobed, 5-celled, few or many ovuled; styles 5, filiform, with capitate stigmas. *Fruit* capsular. (*Stemless or caulescent, often tuberous rooted. Leaves compound, exstipulate. Oxalis* or "*Wood-sorrel.*")

XXXV. ZYGOPHYLLEÆ (page 351). *Fl.* regular. *Calyx* 4–5 parted. *Petals* 4–5, clawed, with twisted æstivation, rarely none. *Stamens* 8–10. *Ovary* on an annular disc, 4–5 angled and celled; ovules axile, 2 or more. *Styles* connate, rarely free. *Fruit* capsular or fleshy. *(Herbs, suffrutices or shrubs. Leaves opposite, compound, rarely simple, stipulate. Flowers solitary, axillary or terminal, yellow or white.*

XXXVI. MELIANTHEÆ (page 366). *Fl.* more or less irregular. *Calyx* 5 parted, imbricated. *Petals* clawed. *Stamens* 4–5, within an annular or horse-shoe-shaped disc. *Ovary* 4–5 celled; cells 2 ovuled. *Fr.* capsular. *(Shrubs or undershrubs. Leaves alternate, impari-pinnate, the petiole often winged, stipulate. Fl. racemose or axillary. Melianthus.)*

XXXVII. RUTACEÆ (page 369). *Fl.* regular. *Calyx* 4–5 parted, imbricate. *Petals* 4–5 (rarely none), twisted or valvate. *Disc* cup-shaped, free or attached to the calyx. *Stamens* outside the disc, as many or twice as many as the petals; 5 opposite the petals sterile *(staminodia)*. *Ovary* of 3–5 carpels, rarely syncarpous. *Style* single. *Fruit* of 1–5, one seeded cocci or capsules; the dry walls of the capsule splitting into an inner and outer shell. *(Buku-bushes. Leaves with pellucid, glandular dots, exstipulate).*

XXXVIII. PITTOSPOREÆ (page 443). *Fl.* regular. *Calyx* 5 parted, imbricate. *Petals* 5, imbricate. *Stamens* 5, free. *Ovary* syncarpous, 2–5 celled, ovules numerous; style single. *Fruit* capsular or fleshy. Seeds lying in pulp. *(One S. African shrub, with entire leaves and yellow-green flowers.)*

XXXIX. AURANTIACEÆ (page 444). *Fl.* regular. *Calyx* very short, toothed or entire. *Petals* 4–5, broad at base, imbricate. *Stamens* as many or twice as many as the petals. *Ovary* on a short gynophore, 3–5, or many celled; ovules 1–2; style single. *Fruit* pulpy, indehiscent. *(Trees or shrubs, with pellucid-dotted leaves, &c. Orange and Lemon. One S. African species.)*

XL. XANTHOXYLEÆ (page 445). *Fl.* unisexual. *Calyx* 4–5 parted. *Petals* twisted in æstivation. *Stamens* 4–5 or 8–10, free. *Carpels* 1–3–5, on a gynophore, separate or more or less connate; styles united or distinct, or very short. *Fruit* fleshy or follicular. *(Shrubs or trees with pellucid-dotted leaves, often prickly. Flowers small. Wild-Cardamon.)*

XLI. OCHNACEÆ (page 448). *Fl.* perfect, regular, showy. *Calyx* 4–6 parted, persistent, imbricate. *Petals* 4–6. *Stamens* numerous; the anthers hard and dry, erect, opening by pores or slits. *Ovary* of 4–6 separate carpels. *Fruit* drupaceous. *(Glabrous shrubs, with glossy, simple leaves, and yellow or orange flowers.)*

Sub-Class II. Calycifloræ. Orders XLII.–XLVII.

XLII. CHAILLETIACEÆ (page 450). *Fl.* small, regular. *Calyx* 5 parted, coloured within. *Petals* 5, bifid. *Stamens* 5, inserted with the petals, in the base of the calyx, and opposite its lobes. *Ovary* 2–3 celled; ovules in pairs, pendulous; styles distinct. *Fruit* capsular or drupaceous. *(Trees or shrubs. One S. African species, a very dwarf shrub.)*

XLIII. CELASTRINEÆ (p. 451). *Fl.* small, regular. *Calyx* 4–5 lobed, persistent, imbricate. *Petals* inserted under the edge of a fleshy, perigynous disc. *Stamens* 4–5, alternate with the petals and inserted with them. *Ovary* 2–5 celled; cells 1 or several ovuled. *Fruit* capsular or drupaceous. *(Trees or shrubs. Leaves simple, coriaceous, mostly glabrous. Flowers small, greenish or whitish in axillary cymes or panicles).*

XLIV. AQUIFOLIACEÆ (page 472). *Flowers* regular. *Calyx* small, 4–6 cleft, imbricate. *Corolla* mostly monopetalous, rotate, 4–5 lobed. *Stamens* 4–5, on the corolla, and alternate with its lobes. *Ovary* sessile, fleshy, 1–2–6–8 or several celled; ovules solitary, pendulous; stigma subsessile. *Fruit* fleshy. *(Holly).*

XLV. RHAMNEÆ (page 475). *Fl.* regular, small. *Calyx* tubular, 4–5 toothed, with valvate æstivation. *Petals* in the throat of the calyx, small. *Stamens* opposite the petals. *Ovary* 2–4 celled; ovules solitary. *Fr.* fleshy or capsular. *(Trees or shrubs, with simple, mostly alternate leaves. Flowers in cymes, panicles or heads. Phylica).*

XLVI. TEREBINTACEÆ (page 502). *Fl.* regular, small, often unisexual. *Calyx* 3–5 parted, imbricate. *Petals* 3–5, round a fleshy disc, spreading, imbricate. *Stamens* alternating with the petals, or twice as many. *Ovary* of one carpel (or of one perfect and 2–4 abortive carpels); ovules solitary; styles or stigmas as many as the carpels. *Fruit* drupaceous. *(Trees or shrubs, with compound, rarely simple leaves, resinous juices and minute flowers. Rhus.)*

XLVII. CONNARACEÆ (page 527). *Fl.* perfect and regular. *Calyx* 3 cleft. *Petals* 5, in the bottom of the calyx, imbricate. *Stamens* 10, inserted with the petals. *Ovary* of 5 separate carpels (several often obortive); ovules in pairs; styles terminal. *Capsules* follicular, 5 or fewer. *Radicle* remote from the hilum. *(Subtropical shrubs and trees. 1 sp. at Natal).*

[The Orders of CALYCIFLORÆ will be continued in Vol. II.]

FLORA CAPENSIS.

ORDER I. RANUNCULACEÆ. D.C.

(By W. H. HARVEY).

(Ranunculi, Juss. Gen. p. 231. Ranunculaceæ, D.C. Prod. 1. p. 2. Endl. Gen. Pl. No. clxxviii. Lindl. Veg. King. cliv.)

Sepals 3–20, mostly 5, separate, deciduous, rarely persistent, green or petaloid and coloured. *Petals* 5–15 (often wanting,) separate, hypogynous, in one or several rows. *Stamens* indefinitely numerous, hypogynous, free ; anthers consolidated with the filament, erect, two-celled, opening laterally by a longitudinal slit. *Ovaries*, many or few, often very numerous, separate (rarely cohering by the ventral sutures) ; *ovules* one or several, sutural. *Fruit* either dry achenia ; succulent drupes or berries ; or many-seeded follicles. *Seeds* without arillus, anatropous, with copious albumen, and a minute, sub-basal embryo.

Herbs or twining shrubs, with colourless, acrid juice. Leaves, except in *Clematis*, alternate ; petioles concave, expanded into an imperfect sheath at the base, and clasping the stem ; lamina usually much cut, or multipartite. Flowers gaily coloured, solitary or panicled, pedunculate.

With the exception of the first Tribe *(Clematideæ)*, the plants of this extensive and widely-distributed Order are herbaceous. They abound throughout the northern Temperate Zone, particularly in its colder and moister climates. They are compatively rare in S. America and Australia, and still less frequent in Africa. Of the 40 genera and 1200 species known to botanists, but 5 genera, comprising 18 species, fall within the limits of our Flora : and only one genus *(Knowltonia)* is peculiar to South Africa. Acridity and causticity in the juices are prevailing characteristics of *Ranunculaceæ*. Some are violent poisons ; but in many the poisonous principle is volatile, disappearing in the process of drying. Of the S. African species, but two have a place in the pharmaceutical list ; namely, *Knowltonia vesicatoria*, a common, rustic blistering application ; and *Ranunculus pinnatus*, whose juice is prescribed for cancerous ulcers.

TABLE OF THE SOUTH AFRICAN GENERA.

A. twining shrubs, with *opposite* leaves.

I. **Clematis.**—*Sepals* valvate. *Carpels* with feathery tails.

B. Herbaceous plants with *alternate* or *radical* leaves.

* *Sepals* coloured. *Petals* none.

II. **Thalictrum.**—*Sepals* 4–5, shorter than the stamens. *Carpels* without tails, dry.

III. **Anemone.**—*Sepals* numerous, longer than the stamens. *Carpels* dry, tailed.

** *Sepals* green. *Petals* present.

IV. **Knowltonia.**—*Sepals* 5. *Petals* numerous. *Carpels* fleshy and juicy.

V. **Ranunculus.**—*Sepals* 3–5. *Petals* 5–10, each with a scale or pit near the base. *Carpels* dry.

I. CLEMATIS, Linn.

Sepals 4–8, coloured, valvate in the bud. *Petals* none, or very small. *Carpels* numerous, dry, one-seeded, with hairy or feathery tails. *DC. Prod.* 1. *p.* 2.

Climbing, rarely erect, slender, vinelike *shrubs*, with opposite, mostly decompound leaves. Flowers white, purple or blue, solitary or in cymes or panicles. Natives of the warmer parts of the N. & S. Temperate Zones. One species is wild in England, and many are cultivated in gardens. The colonial name for the Cape species is "*Klimop.*" None of the S. African species have petals, and all have 4 sepals and feathery seed-tails. They are confined to the districts east of Swellendam. The generic name *Clematis* is derived from *κλημα, the shoot of a vine.*

1. **C. Thunbergii** (Steud.;) pubescent, leaves sub-bipinnately parted; pinnæ distant, leaflets petiolate, ovate, lanceolate or trifid or 3 toothed, the segments mucronulate; panicles shorter than the leaf; *flower buds ovate, acute;* sepals spreading, *lanceolate-acuminate;* filaments hairy at base; anthers glabrous, *linear. C. triloba, Thunb. Cap. fide Eck. & Zey. Enum. No.* 2. *Harv. Thes. t.* 8.

HAB. Woods of Adow, Uitenhage, E. & Z. 1. (v. s. in Herb. T.C.D., Hook., Sond.)

A rambling climber, more glabrous than *C. brachiata*, with less compound leaves, and readily known from other Cape species by its lanceolate sepals, and pointed and slightly twisted flower buds.

2. **C. brachiata** (Thunb. Cap. p. 441.;) pubescent; leaves bipinnately or tripinnately parted; pinnæ distant, trifoliolate, leaflets petiolate, ovate-acuminate, toothed, the teeth mucronulate; panicles elongate, spreading; *flower-buds very obtuse;* sepals spreading, *elliptical, obtuse;* filaments *flat*, hairy at base, anthers *linear. DC. Prod.* 1. *p.* 6. *Ker. Bot. Mag. t.* 96. *E. & Z. No.* 1. *Drege, No.* 7594, 7595, 7596, 7597. *C. Kerrii, Steud. C. Massoniana, DC. Prod.* 1. *p.* 3.

HAB. Frequent in the country from Swellendam to Port Natal, *E. & Z.! Burke! Drege! Gueinzius!* &c. (Herb. T.C.D., Hook, Sond.)

Climbing over trees and bushes. The pubescence varies greatly in different specimens, and the old leaves are frequently nearly smooth. Flowers in long terminal or axillary, naked panicles, whose branches spread at right angles with the stem. Known from *C. Thunbergii* by its obtuse buds and sepals; and from *C. Oweniæ*, besides other marks, by the filaments and anthers. The carpels are glabrescent, orbicular, compressed and margined, with long, feathery tails.

3. **C. Oweniæ** (Harv. Thes. t. 9;) *densely* pubescent; leaves bipinnately parted, pinnæ 3–4 pairs, with an odd one, distant, trifoliolate, leaflets ovate-acuminate, coarsely toothed, the lateral ones small and subsessile, the middle one stalked, the teeth mucronulate; panicles axillary, shorter than the leaf; *flower-buds obtuse;* sepals spreading, *elliptic-lanceolate;* filaments *filiform*, hairy at base, anthers *oval.*

HAB. Port Natal, *Miss Owen!* (Herb. T.C.D.)

Slender, densely pubescent in all parts. Leaves horizontally patent at their insertion, the petiole reflexed. Flowers small, in much branched panicles. Anthers scarcely longer than broad.

4. **C. Stanleyi** (Hook. Ic. t. 589); densely *albo-tomentose;* leaves tripinnately parted, pinnæ 3–4 pairs, *multipartite*, the laciniæ *narrow-linear*, simple or cloven; flowering branches paniculate, the panicle leafy; peduncles single-flowered, longer than the leaves; flower-buds obtuse; sepals spreading, *broadly obovate*, obtuse or emarginate.

HAB. Macallisberg, *Burke!* Port Natal, *Miss Owen* (Herb. T.C.D., Hook.)

A tall, stout, much branched, suberect shrub, thickly clothed in all parts with pale, silky, spreading hairs. Leaves densely set, and cut into slender shreds. Panicles terminal, and corymbose. The peduncles are often bracteate in the middle. The flowers are 1½ inches in diameter and seemingly purple (or blue ?). A noble species, named in compliment to the late Lord Derby by whose collector it was first sent to Europe. It was, however, first found by Miss Owen, in the Zooloo Country, in 1840.

II. THALICTRUM, Tourn.

Sepals 4–5, coloured, imbricate in the bud, caducous. *Petals*, none. *Carpels* 4–15, dry, one seeded, tipped with a short beak. *DC. Prod.* 1, *p.* 11.

Herbaceous perennials, with annual, often hollow, erect, branching stems ; alter nate, decompound leaves, and panicled flowers of small size ; the stamens conspicuous. The species are numerous, dispersed throughout Europe, the temperate parts of Asia, North America, and on high mountains within the Tropics. Our only S. African species is also a native of a considerable portion of the Northern Temp. Zone. The generic name is derived from θαλλω, *to flourish*.

1. **T. minus** (Linn.) ; glabrous or pubescent ; leaves 3–4-pinnate ; segments roundish or wedge shaped, variously cleft ; panicle diffuse, much branched ; anthers mucronate ; carpels 6–8, sessile, oval, strongly ribbed, tipped with the thickened style. *DC. Prod.* 1. *p.* 13. *Eng. Bot. t.* 11. *T. caffrum, E. & Z. ! No.* 3. *T. gracile, E. Mey ! in Herb. Drege !*

HAB. Kaffirland, *E. & Z. ! Drege !* Orange River, *Burke !* (Herb. T.C.D., Hook., Sond.)

Stems 2-3 feet high, erect, leafy ; branches ending in nearly naked spreading panicles of small, pale, purplish flowers. Stamens conspicuous, on hairlike filaments, with large, linear, yellow anthers. Leaves very compound ; leaflets glaucous on the under side. I cannot distinguish the Cape specimens from the glabrous and glaucous form of the common *T. minus* of Europe, Asia, and N. Africa.

III. ANEMONE, Hall.

Involucre, 2–3 leaved, remote from the flower. *Sepals*, 5–20, coloured, imbricate, deciduous. *Petals* none. *Carpels* very numerous, densely capitate, dry, one seeded, with (or without) feathery tails. *DC. Prod.* 1. *p.* 16.

Herbaceous perennials, stemless or short-stemmed. Leaves usually much cut. Peduncles terminal, naked, whorled beyond the middle with 2-3 subsessile, *involucral* leaves, one or many flowered. Flowers large and showy ; in the Cape species rosy-white. Chiefly alpine or subalpine plants, abounding in Europe, Asia, and North America ; a few straggling into woods, or scattered over the sunny plains of the Mediterranean region. Still fewer are found in the Southern Hemisphere ; a few occur on the Andes, and in Southern Chile ; and one in Tasmania. The two Cape species belong to Decandolle's section, "*Pulsatilloides*," characterised by very hairy carpels, with short, glabrous tails, numerous sepals and small involucres. The generic name is derived from ανεμος, the wind ; because the flowers are easily blown to pieces.

1. **A. capensis** (Linn.) ; stem short, densely leafy ; leaves *bi-ternately cut*, glabrescent, the segments stalked, wedge shaped, 3-lobed, or 3-parted, deeply cleft ; petioles villous ; flower stalk 1–2 flowered, villous below, woolly above the involucre ; sepals numerous, *silky*. *DC. Prod.* 1. *p.* 18. *Bot. Mag. t.* 716. *Pulsatilla Africana, E. & Z. ! No.* 5. *Atragene capensis, Thunb. Fl. Cap. p.* 440. *Andr. Rep. t.* 9.

Var. *β.* **tenuifolia** ; leaves triternately cut, segments multifid, very narrow ; flowers smaller. *A. tenuifolia, DC. l. c. Atragene tenuifolia, Linn. f. E. & Z. ! No. 6.*

HAB. Table Mountain, Thunberg, W.H.H. &c. (common.) β. Swellendam *E. & Z., Drege !* (Herb. T.C.D., Hook., Sond.)

Stems 6-8 inches high, densely covered with leaves, and produced into a long, one or two-flowered, scapelike peduncle, 2 feet high or more. Leaves on long petioles, rigid, twice or thrice parted and variously cut. β. is generally held for a distinct species ; but I find, in specimens from the same locality, and even on the same root, that the leaflets vary much in the degree of incision.

2. A. caffra (Eck. & Zey. Enum. No. 4) ; stemless ; leaves rigid, glabrescent, *5–7-lobed,* the lobes biserrate ; petioles villous ; flower stalk one flowered, villous below, densely woolly above the involucre ; sepals numerous, lanceolate, *nearly smooth. Harv. Thes. t. 7. A. alchemillæfolia, E. Mey ! in Herb. Drege.*

HAB. Eastern districts and Caffraria, on grassy hills, *E. & Z. ! Col. Bolton !* and others. (Herb. T.C.D., Hook., Sond.)

Leaves several from the woody rootstock. Petioles 3-4 inches long ; lamina subpalmatifid. Flowers 1½ inches diameter, rosy white ; sepals 12-18, sparsely silky or glabrescent.

IV. **KNOWLTONIA,** Salisb.

Involucre none. *Sepals* 5, green, imbricate, deciduous. *Petals* numerous (5-15) flat, with naked claws. *Carpels* very numerous, capitate, one-seeded ; when ripe becoming fleshy and juicy. *Style* deciduous. *Seed* suspended. *DC. Prod.* 1. *p.* 23.

Perennial, stemless, *herbs,* with radical, *rigid,* ternately decompound leaves ; branching, cymose or umbellate scapes ; and greenish or yellowish flowers. This genus is exclusively South African. The species are extremely acrid and the commonest *(K. vesicatoria* and *K. rigida)* are popular colonial remedies for rheumatism, &c. The bruised leaves, applied to the skin, raise an effective blister. Popular name "*Brand-Blaren,*" *(see Pappe's Fl. Cap. Med. Prod. p.* 1.*)* The generic name is in memory of Mr. Th. Knowlton, a meritorious cultivator, and formerly curator of Sherard's famous garden at Eltham.

1. K. vesicatoria (Bot. Mag. t. 775) ; subglabrous ; leaves biternate, thick, segments ovate or cordate, *serrulate or nearly entire,* the lateral ones obliquely truncate at base, subsessile ; peduncles spreading, umbellate, the umbel compound, many rayed ; pedicels villous ; petals spathulate, obtuse ; ovaries glabrous, *as long as the subulate style. DC. Prod* 1. *p.* 23. *Bot. Reg. t.* 936. *E. & Z. ! En. No.* 8.

HAB. Common near Capetown, and throughout the Colony ? also in Kaffirland, *E. & Z. !* &c. (Herb. T.C.D., Hook., Sond.)

Leaves a foot or more broad ; the leaflets 3 inches long and 2 broad, often nearly entire. Flowers green. Berries blackish purple. The umbel, is said, by authors, to be unbranched and few flowered ; but it is more frequently doubly and sometimes trebly compound, as is well represented in the figure in Bot. Reg. t. 936.

2. K. rigida (Salisb. Prod. 372) ; sparsely hairy ; leaves biternate, thick, subglabrous, segments ovate or cordate *sharply serrate,* the lateral ones obliquely truncate, or cordate at base, subsessile ; peduncles patent, umbellate, the umbel many rayed ; pedicels pubescent or villous ; petals subspathulate, obtuse ; ovaries glabrous, *shorter than the subulate style. DC. Prod.* 1. *p.* 23. *Anamenia coriacea, Vent. Malm! t.* 22.

Var. *β*. **ternata**; leaves ternate; segments cordate, serrate.

Var. *γ*. **simplicifolia**; leaves simple, cordate, serrate, *Drege* 7604.

HAB. In rocky mountain ground, Round Table Mt., &c.; and at the Paarl, *Drege*, W.H.H. &c., Caledon, *E. & Z.* ! Uitenhage and Albany, *Zeyher*, &c. (Herb. T.C.D., Hook., Sond.)

Leaves 6-9 inches broad, on long, somewhat hairy petioles; the leaflets 1-2 inches long and nearly glabrous. Umbel varying much in composition and pubescence.—A smaller and less common plant than *K. vesicatoria*, with more sharply serrate leaves, smaller fruits and longer styles. Vars. *β*. and *γ*. are remarkable forms, seemingly depauperated.

3. **K. gracilis** (DC. Syst. I. p. 219); sparsely hispid; leaves biternate, segments subglabrous, ovate, acute, sharply serrate, the lateral ones subsessile, obliquely truncate at base; peduncles *tall, erect*, branched near the summit; petals broadly linear; ovaries *densely setose*, half as long as the filiform styles. *DC. Prod.* I. *p.* 23. *Deless. Ic.* I. *t.* 19. Eck. & *Zey* ! *En. No.* 9. *Drege, No.* 7600, 7601 !

HAB. Near Constantia, *E. & Z.* ! at the Paarl and French Hoek, *W.H.H.* Dutoits Kloof and Drackenstein, *Drege* ! (Herb. T.C.D., Hook. Sond.)

Petioles and scapes thinly sprinkled with soft, spreading hairs; leaves nearly glabrous, except along the veins. Scapes 2 feet high or more. Ovaries densely clothed with rigid, white bristles; by which character this species is readily known from the preceding.

4. **K. hirsuta** (DC. Syst. I. p. 220); *thickly hairy* with spreading hairs; leaves biternate, segments ovate, cuneate at base, sharply serrate, the lateral ones unequal-sided; peduncles umbellato-cymose, their branches long and spreading, simple or divided; petals *very narrow, acute*; ovaries *nearly glabrous*, as long as the styles. *DC. Prod.* I. *p.* 23. *E. & Z.* ! *No.* 10. *Burm. Afr. t.* 51 ! *Adonis capensis, Thunb. Cap. p.* 441. *K. daucifolia, E. & Z.* ! *No.* 11 *(non DC.)*

HAB. Near Capetown, in dry, scrubby places, *E. & Z.* ! W.H.H. &c. (Herb. T.C.D., Hook., Sond.)

Every part of the plant densely hairy. Petioles 4-5 inches long, spreading, as well as the short flower stalks. The ovaries have a few bristles, which soon fall off. The flowers are a dingy, greenish yellow.

5. **K. daucifolia** (*DC. Syst.* I. *p.* 220); sparsely pubescent; leaves biternate, rigid, thick, glabrescent, segments *bipinnatifid and much incised*, the laciniæ *linear*, with a thickened margin; flower stalks erect, umbellate, the umbel compound; petals obovate, obtuse. *DC. Prod.* I. *p.* 23.

HAB. S. Africa, Herb. *Hook* ! near George, *Dr. Alexander* ! (Herb. Hook., Alex.)

Root thick and woody. Radical leaves numerous, on laxly villous petioles, 3-4 inches long, the segments petiolate and very much divided in a pinnate or bipinnate manner, their divisions very narrow. Peduncle much taller than the leaves. Petals fewer and broader than in the other species, from which this is readily known by its leaves being as finely divided as those of the narrow-leaved variety of *Anemone capensis*. It appears to be of rare occurrence, and was not gathered by Ecklon and Zeyher, whose specimens, distributed under this name, belong to *K. hirsuta*.

V. RANUNCULUS, Hall.

Sepals 3–5, green, imbricate in bud, deciduous. *Petals* 5–10, flat, with a minute, fleshy scale, or a pit near the base, inside. *Carpels* very numerous, capitate, dry, one-seeded, mucronate, or horned. *Seed* erect. *DC. Prod.* 1. *p.* 26.

Annual or perennial *herbs*, stemless or caulescent, with alternate, multifid or entire leaves, and yellow or white, rarely red or purple flowers. Inflorescence various. An immense genus, found in all parts of the world, but most abundant in the Northern Hemisphere, where they cover the fields, in the low lands, or ascend to the snow line of the loftiest mountains, or migrate to the shores of the Polar Sea. Several are water plants, either erect or floating, and most others grow in moist spots. Few occur within the tropics, and then, chiefly on high mountains. The Cape species are mere weeds. One of them *(R. pinnatus)* called "Kankerblaren" is a colonial remedy for cancerous sores. A naturalized species *(R. sceleratus)* sometimes found about towns in wet ditches in spring, has still more active qualities, its juice rapidly blistering the skin, and forming a sore difficult to heal. The generic name *Ranunculus*, is formed from Rana, a *frog;* because many of the species are amphibious.

1. R. aquatilis (Linn.); *submerged and floating;* the submersed leaves *multipartite*, with *narrow-linear*, repeatedly forked segments; the floating leaves *(when present)* reniform, trifid or tripartite, their segments cuneate and crenate; flowers, *white;* fruits transversely wrinkled, minutely pubescent; receptacle hispid. *DC. Prod.* 1. *p.* 26. *E. & Z.! No.* 12. *Drege*, 7605! *E. Bot. t.* 101. *& R. fluitans, t.* 2870! *R. rigidus, Godr. Flora*, 24. *p.* 174.

HAB. In rivers, lakes, ponds, and ditches. Swellendam & Graaf Reynet, *E. & Z.!* Kraairivier, *Drege!* (Herb. T.C.D., Hook., Sond.)

When growing in ponds or in gently flowing rivers, this plant has both floating and submersed leaves; in rapid streams only the latter. Several pseudo-species have been made in Europe from varieties of this common plant, which, under one form or other, is found throughout Europe, Temperate Asia, N. America, Abyssinia and Algeria, and in Australia.

2. R. pinnatus (Poir. Dict. 6. p. 126); *tall, branching, villous;* radical leaves *pinnately* or bi-pinnately cut, hairy, with broadly wedge-shaped, trifid or laciniate segments; stem-leaves three-parted, the upper ones simple; flowers panicled; sepals reflexed; fruits obovate, compressed; margined, *minutely tuberculed* on the disc, with a short beak. *DC. Prod.* 1. *p.* 42. *R. pubescens, Thunb. Cap. p.* 443. *Eck. & Zey.! No.* 14.

HAB. In moist grassy places. Near Capetown, *W. H. H.* Uitenhage, *Zeyher! Drege!* Albany, *T. Williamson*. (Herb. T.C.D., Hook., Sond.)

Root fibrous. Radical leaves numerous, on long petioles, either 3 parted with the middle lobe petiolate, or pinnate, 2 pairs and an odd one, or sub-bipinnate, always hairy. Stems diffuse, cymoso-paniculate. Carpels sometimes nearly smooth.

3. R. plebeius (R. Br.); *tall, branching, villous;* radical leaves on long petioles *ternately* cut, with sessile or petiolate 3–5–fid, toothed and laciniate segments; stem-leaves similar, the upper ones sessile, trifid or simple; flowers panicled, on furrowed peduncles; sepals reflexed; petals 5, obovate; fruits obovate, compressed, margined, *quite smooth*, with a short hooked beak; receptacle hairy. *DC. Prod.* 1. *p.* 39. *Hook. Fl. Nov. Zeal.* 1. *p.* 9.

HAB. Southern slopes of the Schneeberg, *Burke!* (Herb. Hook., Sond.)

Nearly (we fear too nearly) related to the preceding, from which it is chiefly known by its perfectly smooth fruits. It is a native also of Australia and New Zealand.

4. **R. capensis** (Thunb. Cap. p. 442); *small, subsimple, villous;* radical leaves few, 3-lobed or 3-parted, with wedgeshaped trifid segments; stem-leaves one or two, petiolate, 3-parted and cut; flower-stalks opposite the leaves, slender, one flowered; sepals *spreading;* fruits orbicular, subcompressed, margined, *quite smooth,* with a short style. *Eck. & Zey. ! En. No.* 13.

HAB. In moist places, among grass. Aug. Sep. Among the *Euphorbia Caput Medusæ,* at Greenpoint, Capetown, *Ecklon! W.H.H!* Also in Clanwilliam, *E. & Z. !* (Herb. T.C.D., Sond.)

A small annual. Radical leaves 3 or 6. Stems scapelike, 3–6 inches high, scarcely longer than the root-leaves; mostly bearing 1–2 leaves and a solitary small flower.

5. **R. sceleratus** (Linn.); *glabrous, erect,* panicled; radical leaves on long petioles, roundish, tripartite, with wedgeshaped, incised, obtuse lobes; stem-leaves tripartite, sessile; sepals reflexed; fruits subrotund, smooth, in an oval-oblong head; receptacle hairy. *DC. Prod.* 1. *p.* 34. *E. Bot. t.* 681.

HAB. Wet ditches, near Capetown *(introduced)* in Sep.–Oct.

Enumerated by Ecklon and Zeyher, as having been found "on the banks of the Zwartkops River, Uitenhage;" but the specimen so ticketed in Ecklon's Herbarium (now Herb. Sonder.) belongs to *R. pinnatus.* I have seen this plant formerly in ditches near Capetown, but neglected to preserve specimens. It is a very common European and Asiatic weed.

6. **R. Meyeri** (Harv.); nearly *glabrous, creeping by runners;* radical leaves on long petioles, fleshy, cordate or *reniform, crenate,* glabrous; petioles villous; peduncles *scapelike,* one flowered; sepals reflexed; petals *several, linear-lanceolate, acute;* fruits smooth, ovate, subcarinate, shortly beaked. *Ficaria radicans, E. Mey. ! in Herb. Drege.*

HAB. Kat-berg, in grassy places 3–4000 ft. November. *Drege!* (Herb. T.C.D., Hook.)

Crown emitting prostrate runners which root at the joints. Petioles 2–3 inches long, more or less hairy; leaves few. Scapes about as long as the leaves, quite naked. Carpels few, convex. This has much of the habit of a *Casalea.*

ORDER II. ANONACEÆ. Rich.

(By W. SONDER.)

(Anonæ, Juss. Gen. 283. Anonaceæ, Rich.—DC. Prod. 1. p. 83.—Endl. Gen. Pl. No. clxxiv. Lindl. Veg. King. clii.)

Flowers perfect, or rarely unisexual. *Sepals* 3, hypogynous, mostly valvate in the bud, separate or united at base. *Petals* six, in two rows (rarely only three, or altogether wanting,) valvate or imbricate, separate or united at base. *Stamens* very numerous, in many rows; filaments very short; anthers consolidated with the thickened and produced apex of the filament, erect, two celled, opening longitudinally. *Ovaries* numerous (rarely definite or solitary,) sessile, separate or cohering, one celled; ovules one or several, sutural, anatropous; styles scarcely any; stigmas separate or united. *Fruit* various: *carpels* sessile or stalked, separate, or confluent into a multicellular, dry or pulpy fruit, mostly indehiscent. *Seeds* with copious, ruminated albumen, and a minute basal embryo

Trees, or shrubs; the latter erect, or climbing, or trailing; leaves alternate, simple, entire, without stipules; flowers terminal or axillary, solitary, or few together, green or brown

An Order consisting almost exclusively of Tropical trees and shrubs, remarkable for a strong aromatic taste and smell; and many of them yielding edible fruits. It is easily recognised by the invariable ternary composition of the floral envelopes; the densely packed and very short stamens, the numerous ovaries, and the ruminated albumen: and by the no less constant characters of the foliage. An admirable analysis is given by Hooker and Thomson in Flora Indica, p. 86, where these authors estimate the number of species at about 600; of which about 250 may be Eastern, 250 American, and 100 from Tropical and sub-Tropical Africa. The most Northern species is *Asimina pygmæa*, found on the South coast of Lake Erie; and the most southern appears to be the curious *Eupomatia laurina*, found in N. S. Wales (Lat. 35° S.) Only two species have yet been detected in extra tropical S. Africa.

Among the cultivated *fruits* of the Order, the *Custard apples* of the East and West Indies and the *cherimoyer* of Peru are the most remarkable; the latter is one of the most delicious of subtropical fruits. It would probably succeed well at the Cape in sheltered places with a warm exposure, as it ripens freely in Mr. MacLeay's garden, at Sydney, N.S. Wales.

TABLE OF THE SOUTH AFRICAN GENERA.

I. **Uvaria.**—*Petals* imbricate. *Ovaries* multiovulate. *Fruits* pulpy or dry, mostly many seeded.

II. **Guatteria.**—*Petals* valvate. *Ovaries* uniovulate. *Fruits* dry, one seeded.

I. UVARIA. L. Endl.

Sepals 3, valvate in æstivation, often connate at base. *Petals* 6, in two rows, equal or unequal, imbricated in æstivation, often cohering at base. *Stamens* indefinite, in many rows, compressed, oblong or linear-oblong, the anther-cells linear, the connective produced into an oblong, expanded or truncate and short process. *Torus* scarcely elevated, truncate, pubescent, often densely tomentose between the ovaries. *Ovaries* indefinite, straight, linear-oblong, angular, furrowed on the inner side, crowned with a continuous, truncate style, with involute margins, viscid on the stigma; ovules indefinite, in two rows. *Carpels* many seeded, sometimes, by abortion, few, or one seeded. *Hook. & Thoms. Fl. Ind. p.* 95. *Endl. Gen. No.* 4717.

Climbing or trailing shrubs, with stellate pubescence. Flowers mostly opposite the leaves, rarely axillary. A large genus, confined to the tropical and subtropical regions of the Eastern Hemisphere. The name is an alteration of *Uva*, a cluster of grapes.

1. **U. caffra** (E. Mey. in Herb. Drege); young twigs, petioles and peduncles covered with stellate tomentum, otherwise glabrous; branches and branchlets spreading; leaves on short petioles ovate-oblong or oblong, acute at each end, subcoriaceous, with netted veins; peduncles opposite the leaves, solitary, pedicels few, recurved; berries oval or roundish, 2–3 seeded; seeds plano-convex.

HAB. In woods, near Port Natal. April. *Drege.* (Herb, Sond.)

Branches and twigs round. Leaves like those of a laurel, 2½–3 inches long, inch wide; petiole 1–2 lines long. Peduncles 6–8 lines long, thickish. Pedicels 2 lines long. Torus thickish. Berries fleshy, as large as cherries. Seeds horizontal, shining, 3½ lines long.

II. GUATTERIA. R. & P. Endl.

Sepals 3, rounded or ovate, small. *Petals* 6, in two rows valvate in æstivation, flat, ovate, oblong or linear. *Stamens* indefinite, broadly wedgeshaped, with a truncate, capitate connective; anther-cells dorsal, remote. *Ovaries* numerous, oblong; ovule solitary, basal, erect. *Style* oblong, furrowed at base, on the inside. *Torus* little raised, plano-convex, sometimes hollow in the centre. *Carpels* dry, with a thin and often brittle pericarp. *Seed* erect. *Hook. & Thoms. Fl. Ind. p.* 138. *Endl. Gen. No.* 4721.

Trees, often of large size, or shrubs—(sometimes climbing ?) various in habit; leaves obliquely nerved. Flowers axillary or opposite the leaves. A large tropical and subtropical genus, with species in both hemispheres. Name in honour of *J. B. Guatteri*, an early Italian botanist, once Professor at Parma.

1. G. caffra. (Sond.); twigs, petioles and nerves of the leaves at first appressedly hispidulous; leaves on short petioles, oblong, obtuse, acute or acuminate, obtuse at base, paler on the under side; peduncles opposite the leaves, solitary, very short, bifid, the longer branch bracteate; pedicels 2–4, recurved; carpels oval-oblong, apiculate, glabrous, somewhat longer than the stalk, one-seeded. *Unona caffra, E. Mey. in Pl. Drege.*

HAB. Woods at Omsamwubo and Port Natal; Feb. Apl. *Drege, Gueinzius* (Herb. Sond., T.C.D.)

Branches glabrous, tuberculated with white warts: twigs flexuous, round. Petioles erectopatent, 2 lines long. Leaves 3–4 inch long, 1–1¼ inch wide, the lower ones on the branches smaller, very entire, membranaceous, netted veined, glabrous above, the nerves prominent below, and at first covered with close-pressed reddish hairs. Peduncles opposite the leaves or supra-axillary, a line long, divided into two spreading arms ½ inch long. Pedicels 2 lines long, thickish. Flowers unknown. Torus dilated, 2 lines wide. Carpels 2–4. glabrous, greenish, minutely granulated, 3 lines long, minutely apiculate, with a thin pericarp. Seed erect.—Not unlike *G. suberosa*, Dun. Probably, the supposed *Popowia* from Natal, alluded to by Hook. & Thoms. (*Fl. Ind. p.* 105) is the same as our plant.

ORDER III. MENISPERMACEÆ, DC.

(By W. H. HARVEY.)

(Menispermeæ, Jus. Gen. 284. Menispermaceæ, D.C. Prod. 1. p. 95. Endl. Gen. Pl. No. clxxii. Lindl. Veg. King. civ.)

Flowers minute, unisexual, usually dioecious. *Sepals* 2–12, in one or two or several rows, separate, or rarely united at base, almost always imbricate. *Petals* mostly shorter than the sepals, equalling them in number, or fewer, imbricate, sometimes none, sometimes united into a cup-shaped corolla. *Stamens* (in the male flowers) as many, or twice, or thrice as many as the sepals; filaments free or united into a column; anthers opening outwards, either vertical or horizontal, adnate with the filament. *Ovaries* (in the female flowers) mostly three, rarely solitary, sometimes six or more, separate or connate; ovules solitary, mostly amphitropous; styles terminal subulate. *Fruits* drupaceous, oblique or crescent-shaped; putamen indurated, horseshoe-shaped, the seed filling up the cavity. *Albumen* variable in quantity, sometimes none; embryo curved, cotyledons either divergent or close.

Climbing or twining, slender, shrubby plants, with alternate, simple, often palmate-nerved, reticulately veined, entire leaves, without stipules. *Flowers* very small, green or white, in axillary cymes or racemes, rarely solitary, and almost always unisexual.

These plants are chiefly natives of the tropical parts of Asia and America; very few comparatively are African. One is found in Canada, and two more in the United States of N. America; one in Eastern Siberia, and a few in China and Japan. Several species occur in Australia. Dr. Hooker (*Fl. Ind. p.* 174) computes the whole number at about 200. These are distributed by Mr. Miers (*An. Nat. Hist. 2nd ser. Vol. VII. p.* 33.) into forty genera, of which three, belonging to the tribe CISSAMPELIDEÆ are found in S. Africa. I am indebted to Mr. Miers for MS. descriptions of *Homocnemia* and *Antizoma*, which I have freely used in drawing up the characters and descriptions.

TABLE OF THE SOUTH AFRICAN GENERA.

I. **Homocnemia.**—*Fem. flower* with four petals and four sepals.
II. **Cissampelos.**—*Fem. fl.* with one petal, opposite the single sepal.
III. **Antizoma.**—"*Fem. fl.* with two petals." *Miers.*

I. HOMOCNEMIA. Miers.

"*Flowers* dioecious. *Male fl.* unknown. *Female*: *Sepals* 4, obovate, hairy on the outside, opposite in pairs, imbricate in æstivation. *Petals* 4, much shorter than the sepals, roundish, fleshy, hypogynous. *Stamens* none. *Ovary* solitary, ovate, on a short stipe, compressed, with a longitudinal furrow on one side, unilocular, uniovulate; the ovule attached to a ventral placenta. *Style* very short, obtusely emarginate." *Miers' MS. in litt.*

A vine-like *twiner*, with peltate leaves; and flowers in axillary, compound umbels. The generic name is compounded of *ομος, like* and *κνημια, the spoke of a wheel;* in allusion to the umbellate inflorescence.

1. H. Meyeriana (Miers MSS.) and in An. Nat. Hist. Ser. 2. vol. VII. p. 40. *Cissampelos umbellata, E. Mey.! in Herb. Drege.*

HAB. On the Omsamwubo, Natal 1,000–2,000ft. Feb. *Drege!* (Herb. Hooker.)
Stems voluble, distantly branched, striate, densely clothed with short, rusty pubescence, becoming subglabrous. Leaves on long petioles broadly peltate, ovate-orbicular, subacute, mucronulate, pubescent, many nerved, and reticulated on the lower surface. Peduncles of the fem. fl. axillary, shorter than the petioles, umbellate; umbel few-rayed, twice compounded, the pedicels tomentose. Flowers minute; sepals broadly obovate or rhomboid, subacute, keeled at back, thrice as long as the reniform, very obtuse petals. Young drupes orbicular, glabrous and quite even.—Readily known from the S. African plants of the Order by its peltate leaves and umbellate flowers.

II. CISSAMPELOS. Linn.

Flowers dioecious. *Male*: *Sepals* 4, separate. *Corolla* cupshaped, nearly entire, shorter than the sepals (composed of 4 confluent petals.) *Stamens* united into a central column, peltate at the summit and bearing 4–12 anther-lobes, which open horizonally and outwards. *Female; Sepal* one, anterior. *Petal* one (or two-confluent) in front of the sepal, half clasping the ovary. *Ovary* single, one-ovuled, style trifid. *Drupe* kidney-shaped; nut compressed and wrinkled at edges. *DC. Prod.* I. *p.* 100.

Suberect or twining, slender, shrubby plants. Leaves simple petiolate, very entire, minutely reticulate, ovate or reniform or cordate, often peltate. Male flowers in axillary *cymes*; females (on separate roots) racemose, densely tufted in the axils of leafy bracts. Natives of the tropics of both hemispheres; a few straggling into the warmer parts of the temperate zone. Name from κισσος, *the ivy*, and αμπελος, *the vine*; aptly expressing the aspect of these plants.

1. C. Pareira (Linn. Sp. 1473;) voluble, *pubescent* or densely tomentose; leaves reniform or *cordate*, *mucronulate*, pubescent; male cymes pedunculate, much branched; racemes of female flowers elongated, with large, leafy, cordate bracts; drupes hispid. *DC. Prod.* I. *p.* 100. *Hook. and Thoms! Fl. Ind.* I. *p.* 198. *C. apiculata, Hochst! Walp.* 5. *p.* 17.

HAB. Port Natal, *Krauss* (232)! *Gueinzius!* (Herb. Hook., T.C.D., Sond.)

Very variable in the amount of its pubescence, &c. The Natal specimens examined have cordate-reniform leaves, pubescent on the upper and densely velvetty on the lower surface. The number of anther lobes varies in the same cyme from 4–8. The corolla is 4 toothed. The sepals hairy, obovate, with inflexed points.

2. C. torulosa (E. Mey! in Pl. Drege;) voluble, *subglabrous*; leaves *broadly* reniform, *pointless*, glabrous or sparsely pubescent, pale underneath and conspicuously *reticulated*; male cymes pedunculate, few flowered, sepals nearly *glabrous*; racemes of female flowers elongate, with large, leafy, reniform bracts; drupes pedicellate, glabrous, tuberculated. *Menispermum capense,* *Thunb. Cap. p.* 402. *E. & Z! No.* 18.

HAB. Eastern Districts. Caffirland to Port Natal, *E. & Z! Drege! Gueinzius!* (Herb. T.C.D., Hook., Sond.)

A slender, distantly branched, vinelike climber, either quite glabrous or thinly sprinkled, with close pressed hairs, especially on the lower surface of the leaves. Petioles 2–3 inches long. Inflorescence of both kinds supra-axillary, with a gland below the base of the peduncle.

3. C. Capensis (Thunb. Prod. p. 110); shrubby, *densely branched*, partly voluble; leaves petioled, *ovate* or *roundish*, obtuse or subacute, glabrous or pubescent; male cymes much shorter than the leaves; umbels of female flowers axillary, sessile; drupes glabrous. *Thunb. Cap. p.* 501. *E. & Z.! No.* 16. *Drege.* 7591, 7592, 7593. *DC. Prod.* I. *p.* 102. *C. fruticosa, Thunb. l. c. p.* 500. *E. & Z.! No.* 17. *C. humilis, Poir, DC. l.c.*

Var. *β.* **pulverulenta**; leaves pubescent on both sides.

HAB. Frequent throughout the colony, in stony and bushy places (Herb. T.C.D.)

A small, erect or spreading, much and closely branched shrub; the upper branches trailing or twining round other shrubs. Young twigs downy. Leaves $\frac{1}{2}$-$\frac{3}{4}$ inch long, on slender, downy petioles. Flowers very minute, densely woolly, crowded in the axils of the leaves.—Colonial name, *Davidjès*: the roots are emetic and purgative; the leaves poisonous to cattle *(Pappe)*.

III. ANTIZOMA, Miers.

"*Flowers* diœcious. *Males*: *Sepals* 4, obovate-wedge-form, thrice as long as the petal. *Petal* single, cup-shaped, depressed, crenulate at the margin, somewhat fleshy. Staminal column as in *Cissampelos:* anther-lobes 4-10. *Female: Sepals* 2, opposite, ovate, very concave, fleshy, slightly imbricate in æstivation. *Petals* 2, opposite the sepals, minute, scale-like, orbicular, fleshy, hypogynous. *Ovary* single, obovate, sub-compressed, conical above. *Style* none. *Stigma* obsolete, or obtusely 2-lobed. The fruit unknown."—*Mier's MSS. in litt.*

Small, South African shrubs, erect or somewhat twining; the branches mostly virgate. Leaves alternate, sometimes very small, linear or oblongo-lanceolate, very entire, opaque, leathery, on short petioles; the petiole with a short spine at base, on the outside. Inflorescence axillary; flowers very minute. The generic name is composed of *αντι, opposite,* and *ζωμα, a vesture;* "from the position of the sepals in regard to each other, and from the petals being again opposed to them."—*Miers.*

1. A. calcarifera (Miers); "erect, branching; twigs virgate, striate, the younger downy; leaves elliptic-oblong, rounded at both extremities, emarginate or mucronulate, with revolute margins, leathery; nerves few, parallel, patent-oblique, uniting toward the margin, immersed; petiole nearly obsolete, pubescent, the infra-petiolar spine very short, acute, reflexed; male flowers few, fascicled on a very short axillary peduncle."—*Miers. Cissampelos calcarifera, Burch. in DC. Prod.* 1. *p.* 102.

HAB. S. Africa, *Burchell.* (Cat. 1795 & 2529.) (Herb. Burchell.)
The leaves are 5-10 lines long, very minutely glandularly-rugose and sparingly cinereo-puberulent on each side, 3 lines broad, rounded at the extremities; the petiole scarcely ¼ line long.—*Miers.*

2. A. Burchelliana. (Miers); "somewhat climbing, twigs slender, pubescent; leaves lanceolate, the upper ones narrower, obtuse at each end, mucronulate, the subsinuate margin scarcely revolute; nerves few, very slender, arching obliquely from a subprominent midrib; petiole rather short, pubescent; the spine obtuse, strong (inch long) subreflexed, much longer than the petiole." *Miers.*

HAB. S. Africa, *Burchell.* (Cat. 1795. bis.). (Herb. Burch.)
"In habit this plant is very different from the preceding species, that being a low erect shrub, while this has slender, scandent branches: in the former species the nervures are immersed and scarcely perceptible, short and very patent; here they are more parallel with the leaf and slightly prominent. The leaves are comparatively large, 3 inches long, and from ½ to ⅝ inch broad, brownish-green, glandularly-rugulose and thinly pubescent at each side: the petiole is ⅝-inch long; the very obtuse spine is about an inch in length; the distance of the internodes about an inch." *Miers.*

3. A. Harveyana (Miers); "stem erect, slightly scandent; branches virgate, striate, glabrous; leaves lanceolate, the upper ones gradually narrower, obtuse at the apex, mucronate, thickish, the young puberulent, the older glabrous on both sides and glandularly rugulose, the underside glaucescent, the margin revolute; nerves obsolete; petiole very short, armed at base with a short, acute, reflexed spine; male inflorescence axillary, peduncle solitary, twice as long as the petiole, flowers capitate-crowded." *Miers. MSS.*

HAB. Interior of S. Africa. Crocodile River, *Burke.* (Herb. Hook., T.C.D.).
"This species is sufficiently distinct from the two preceding. It appears to have a simple and erect stem, seemingly but little scandent. Its leaves are 10-18 lines

long, 2-3 lines broad, the petiole being scarcely a line in length; the flowers are male, with four smooth sepals, a small cupuli-form petal, with a crenulated border, and a ten-lobed, peltate anther." *Miers.*

4. **A. angustifolia** (Miers); "climbing, the stem striate, glabrous, leaves broadly linear, rounded at point, often emarginate, mucronate, leathery, quite glabrous at both sides; petiole short, armed with a short, obtuse, reflexed spine; peduncles in pairs, axillary, very short, glabrous, one flowered." *Miers.*

HAB. S. Africa, Burchell (Cat. No. 1717). (Herb. Burch.)

"The leaves are 16 lines long and 2 lines broad, on a glabrous petiole, a line in length." *Miers.*

5. **A. Miersiana** (Harv.); suberect, glabrous and glaucous; the stem striate; leaves lineari-cuneate, tapering to an acute base, rounded at point, either emarginate or mucronulate, leathery, with recurved margins, smooth above, cinereo-rugulose below, with immersed veins; petiole armed with a blunt, conical, hardened tubercle; peduncles (of male flowers) longer than the petiole, cymose, several-flowered. *Cissampelos angustifolia, Drege Pl.*

HAB. Between Zwartdoorn river and Groenrivier, under 1000 f., *Drege,* (Herb. T.C.D., Hook.)

A much branched, small shrub, with rigid and thick leaves, scarcely an inch long and two lines broad at the point, tapering much toward the base; the margins thickened; the under-surface slightly concave. Male peduncles twice as long as the petiole; the cyme branched. Sepals ovate, warted externally; anther-lobes usually four. The spine or spur in this species is degenerated into a mere, hardened tubercle.

ORDER IV. NYMPHÆACEÆ, Salisb.

(By W. H. HARVEY).

(Salisb. Ann. Bot. 2. p. 69. DC. Prod. 1. p. 113. Endl. Gen. Pl. No. clxxxv. Lindl. Veg. Kingd. cxlviii.)

Flowers bisexual, of large size. *Sepals* 4–5, separate or united at base, free, or adhering to the fleshy receptacle *(torus)* which surrounds the ovary. *Petals* numerous, in several rows, the inner ones narrower and shorter, gradually assuming the appearance of stamens. *Stamens* inserted within the petals, indefinite, in several rows; filaments flat and petaloid; anthers adhering to the face of the filament, two celled, opening longitudinally inwards. *Carpels* numerous, immersed in the fleshy receptacle, and thus united into a plurilocular ovary, crowned with radiating sessile, linear stigmas, alternating with the dissepiments. *Ovules* very numerous, anatropal, affixed to both surfaces of the dissepiments. *Fruit* baccate, many celled, indehiscent. *Seeds* with much flowery albumen, and a minute embryo, lodged within a proper sac, near the base of the seed.

Water plants with prostrate, rooting and rootlike submerged stem, and floating cordate or peltate leaves, on long footstalks.

These plants are commonly known as *Water-lilies,* and species occur throughout the temperate and tropical zones of both hemispheres. Of the five genera known to botanists three are tropical and two belong to the temperate zone. The giant Waterlily of South America (*Victoria Regia*) has peltate leaves, six feet in diameter, so admirably buoyed up by a system of ribs and veins of peculiar structure that each leaf is capable of supporting on the surface of the water the weight of a full-grown man. The seeds of all the Order are edible, containing much fecula. The root-stocks are astringent, and have been used for tanning leather. The flowers of all are remarkably handsome, and of many are sweetly scented.

I. NYMPHÆA, Linn.

Sepals 4, inserted at the base of a fleshy, bottle-shaped receptacle in which the carpels are immersed. *Petals* and *Stamens* numerous, in several rows, covering the sides of the receptacle. *Ovary* many celled; stigmata sessile, radiating. *Berry* leathery, irregularly bursting; seeds indefinite, lodged in pulp, albuminous. *DC. Prod.* 1. *p.* 114.

Water plants, with submerged, prostrate *rhizomes*, throwing up leaves and flowers to the surface. *Leaves* on long, terete petioles, cordate or peltate. *Flowers* on simple peduncles, large and showy, white, red or blue, never yellow; floating, or standing out of the water. Natives of the temperate zones, rare within the tropics. Name from Νυμφαια, the flower of the *nymphs*, because found in clear waters.

1. **N. stellata** (Willd. Sp. Pl. 2. p. 1153); leaves orbicular or elliptical, deeply cordate at base, sinuately dentate or entire, veiny below; sepals lanceolate, nerved; petals lanceolate, acute; anthers with subulate points; stigma 12–20 rayed, the rays prolonged into short horns. *Fl. Ind.* 1. 431. *Bot. Mag. t.* 2058. *N. scutifolia, DC. Prod.* 1. *p.* 114. *N. Capensis, Thunb. Cap. p.* 431. *E. & Z.! No.* 19. *N. cærulea, Andr. Rep. t.* 197. *Bot. Mag. t.* 552.

HAB. In rivers and lakes. Zeekoe Valley, Cape; Zwartkops R. Uitenhage, &c. *E. & Z.! W.H.H.* Schonstone, *Burke!* (Herb. T.C.D., Hook, Sond.)

Leaves leathery, 9-12 inches long and nearly as wide, the basal lobes often overlapping, sometimes elegantly scolloped, sometimes nearly or quite entire, usually purple on the lower surface. Flowers standing out of the water, blue or rarely white, sweetly scented. Petals variable in breadth and sharpness.—I willingly follow the authors of Flora Indica in uniting the above synonyms.

ORDER V. PAPAVERACEÆ, Juss.

(By W. H. HARVEY).

(Papaveraceæ, Juss. Gen. 236. DC. Prod. 1. p. 117. Endl. Gen. Pl. No. clxxx. Lindl. Veg. Kingd. No. clvi.)

Sepals 2, rarely 3, separate, deciduous. *Petals* 4–6, hypogynous, rarely wanting. *Stamens* mostly indefinite, rarely definite, adhering in parcels to the base of the petals; filaments slender; anthers two celled, basifixed, erect, opening lengthwise. *Ovary* free, one celled, composed of 2 or many carpels; placentæ parietal, often projecting far into the cavity, and in *Romneya* nearly meeting in the centre; ovules numerous, anatropal or amphitropal; style single or none; stigmas as many as the carpels, radiating. *Fruit* a dry capsule (rarely berried) variously dehiscent, or indehiscent. *Seeds* numerous; albumen copious, between fleshy and oily; embryo minute, basal.

Herbaceous or very rarely shrubby plants, with a coloured, narcotic juice. Leaves alternate, simple or multifid, without stipules. Flowers mostly solitary, on long peduncles, sometimes panicled; petals white, red, or yellow, never blue.

Natives chiefly of the Northern Hemisphere, common throughout the temperate zone, especially in the Eastern Continent. Several genera are peculiar to North America. Very few are found south of the equator. *Argemone Mexicana,* originally from the new world, has become naturalized throughout the tropics and sub-tropics of both hemispheres. Opium is the well known product of the Poppy: and similarly narcotic properties prevail throughout the Order. Some are violently acrid poisons. Only one species occurs in S. Africa.

I. PAPAVER, Linn.

Sepals 2–3, convex, deciduous. *Petals* 4–6, crumpled in the bud. *Stamens* indefinite. *Ovary* obovate, crowned with 4–20, radiating, linear, sessile stigmata. *Capsule* oblong, dry, opening by small pores under the stigmata; placentæ projecting into the cavity and dividing it into several incomplete chambers. *Seeds* very numerous. *DC. Prod.* 1. *p.* 117.

Annual or perennial herbs, with milky juice, often prickly, roughly hispid. Leaves pinnatifid, variously cut; peduncles axillary, one-flowered. Flowers red, yellow, white or purple, or parti-coloured. *Poppies* abound in Europe and Temperate Asia; one *(P. nudicaulis)* is found within the Arctic Circle; one in South Africa, and another, very similar to it, in Australia. Named, it is said, because *Opium* (we hope not always) "is administered to children with *pap (papa* in Celtic) to induce sleep."

1. **P. aculeatum** (Thunb. Fl. Cap. p. 431); capsules glabrous, oblong-obovate; sepals hispid; stem erect, branched, densely covered with spreading, rigid, unequal bristles; leaves setoso-hispid, sinuately pinnatifid, with spine-tipped laciniæ. *E. & Z.! No.* 20. *P. Gariepinum, Burch. in DC. Prod.* 1. *p.* 119.

HAB. Sandy ground near rivers, in the Northern and Eastern Districts. Orange River, *Burchell, E. & Z.! Burke!* Uitenhage, *E. & Z.!* On the Cowie, Albany, *T. Williamson!* Outeniquas, George, *Thunb.* (Herb. T.C.D.)

Root annual. *Radical* leaves numerous, rosulate, tapering at base into a broad, flat petiole, sinuate or deeply pinnatifid, setose, and armed on the nerves with strong, erect, yellow bristles. Stemleaves sessile, clasping, very hispid. Stem 1-4 feet high, rough with bristles and rigid hairs. Flowers on a long naked peduncle. *Petals* a "scarlet-orange" *(Burch.).*

ORDER VI. FUMARIACEÆ. DC.

(By W. H. HARVEY).

(Fumariaceæ, DC., Syst. 2. p. 105. Prodr. 1. p. 125. Endl. Gen. Pl. under Papaveraceæ, p. 858. Lindl. Veg. Kingd. No. clviii.)

Sepals 2, deciduous, squamæform. *Petals* 4, cruciate, one or both of the outer pair saccate at base; the inner pair callous and hooded at the apex, where they cohere and enclose the anthers and stigma. *Stamens* diadelphous, 3 in each parcel, opposite the outer petals. *Ovary* free, one celled, one or many ovuled; ovules parietal, amphitropal; style filiform. *Fruit* various, dry; either a one-seeded nut or a many seeded bivalve or indehiscent pod. *Seeds* often crested, with copious, fleshy albumen, and a minute, excentric embryo.

Herbaceous plants, with fibrous or tuberous roots; weak, straggling, brittle, succulent stems, and a watery juice. Leaves alternate, multipartite, without stipules, the petioles often changed into branching tendrils. *Flowers* racemose (in the S. African species) of small size, white, pink or purple, or yellow.

A small group, often considered as a suborder of *Papaveraceæ*, which they closely approach in technical characters, but from which they differ very much in aspect, and in sensible properties. They are completely destitute of the narcotic juices so characteristic of *Papaveraceæ* ; and their flowers (except in *Hypecoum*) are highly irregular. Each parcel of stamens consists of one whole and two half stamens ; the normal number is therefore four, not six ; but the latter is the apparent number, except in *Hypecoum*. They are chiefly natives of the Northern Temperate zone, especially of the Eastern Hemisphere. About 120 species, grouped under 12 (or 15) genera are known to Botanists. Of the four South African genera, two (*Cysticapnos and Discocapnos*) are peculiar to the Cape. *Corydalis* and *Fumaria* are nearly cosmopolitan. None are of much importance to mankind but several are cultivated in Europe as ornamental plants.

TABLE OF THE SOUTH AFRICAN GENERA.

* *Fruit a many-seeded, dehiscent capsule.*

I. **Cysticapnos.**—*Capsule* bladdery, sub-globose.
II. **Corydalis.**—*Capsule* lanceolate, compressed.

** *Fruit a one-seeded, indehiscent utricle.*

III. **Discocapnos.**—*Utricle* flattened, orbicular.
IV. **Fumaria.**—*Utricle* subglobose.

I. CYSTICAPNOS, Boerh.

Petals 4, the posterior one spurred at base. *Capsule* 2-valved, bladdery ; epicarp inflated, spongy within ; endocarp delicately membranous, supported by slender filaments in the centre of the cavity, and bearing, at the margins, many seeded placentæ. *Seeds* compressed, beaked, shining. *DC. Prod.* 1. *p.* 126.

A succulent climbing herb, with decompound cirrhiferous leaves, peculiar to the Cape. The seed bag is curiously suspended within an inflated membranous capsule by means of cords. Name from κυστις a *bladder* and καπνος, *smoke*, or (in botanical language) the herb *Fumitory*.

1. C. africana, (Gaertn. Fruct. 2. p. 161. t. 115) ; *DC. Syst.* 2. *p.* 112. *Prod.* 1. *p.* 126. *E. & Z.* ! *No.* 21. *Fumaria vesicaria, Thunb. Cap. p.* 554.

HAB. In bushy places. Common near Capetown. Stellenbosch and Zwellendam, *E. & Z.* ! (Herb. T.C.D., Hook., Sond.)

Annual. Stem voluble, very long, climbing through bushes, irregularly branched. Leaves pinnately decompound, ending in a branched tendril ; pairs of pinnæ about three, the lowest pair close to the stem, stalked, once or twice ternate, with broadly cuneate and deeply lobed segments. Peduncles opposite the leaves, gradually lengthening. Flowers rosy white. Sepals 2, deltoid, acuminate, cordate at base, toothed or entire. Capsule pendulous, an inch in diameter, globose, pointed, splitting into two concave valves, which are spongy on the inside, the inflated portion partly filled with a fibrous net work.

II. CORYDALIS, DC.

Petals 4, the posterior one spurred at base. *Capsule* podshaped, compressed, one celled, bivalve ; valves separating from a persistent, placentiferous *replum* (or frame.) *Seeds* numerous, lenticular, beaked. *DC. Prod.* 1. *p.* 126. *Phacocapnos, Bernh. Linn.* 12. *p.* 664. *Endl. Gen.* 4838.

Fibrous rooted annuals, or bulbous perennials, with climbing or erect succulent stems, decompound, cirrhiferous leaves, and racemose white, pink or yellow flowers. The species are common in Europe and Temperate Asia ; more rare in N.

America. The South African species have been separated from the rest by Bernhardi, under the name *Phacocapnos,* because their seeds are destitute of arillus or strophiolus. We think this separation unnecessary. From the base of the *synema* (or compound filament) a horn-like body extends backwards into the cavity of the spur of the posterior petal; in the other genera this is represented by a single or double gland, absent in Fumaria. Name; κορυδαλις, the Greek word for *Fumitory.*

1. C. Cracca (Schl. Linn. I. p. 567); climbing; leaves bipinnate, cirrhose; pinnæ tripartite, with cuneate or obovate, twice or thrice cut, obtuse, mucronulate, glaucous segments; petals all of *equal length, connivent and cucullate at the apex,* the *posterior one saccate at base;* pods lanceolate, pendulous. *E. & Z. ! No.* 22. *Cor. lævigata, E. Mey. ! also Drege,* 7586 ! 7587. *Phac. Cracca and Ph. Dregeana, Bernh. in Linn.* 12. *p.* 664.

HAB. Among shrubs and in shady places. At the Waterfall, Devil's Mountain, near Capetown, *W.H.H.* Caledon, Uitenhage and Albany, *E. & Z. ! Mrs. Barber.* (Herb. T.C.D., Hook., Sond.)

Annual. Stems weak and straggling, climbing among shrubs by means of branching tendrils. Leaves irregularly 2–3 pinnate. Flowers flesh coloured, with dark tips, small: the raceme at first short and nearly sessile, in fruit lengthened and longer than the leaf. Spur of the posterior petal very short and round. Seed without aril.

2. C. pruinosa (E. Mey. ! in Herb. Drege); climbing; leaves bi-tripinnate, cirrhose; pinnæ multifid, with cuneate or linear variously lobed, obtuse, glaucous segments; flowers *bilabiate, the outer petals longest,* the posterior one *ovate, expanded, with a reflexed oblong spur,* the anterior *obovate, pitted in the middle;* pods broadly lanceolate, acuminate, pendulous. *Phac. pruinosus, Bernh. l.c.*

HAB. On the Witberg, 4–5000 ft.; and near Enon, Uitenhage, 500 ft. *Drege !* Orange River, *Burke !* Novr. Janr. (Herb. T.C.D., Hook., Sond.)

In aspect much resembling *C. Cracca,* but with more finely decompound leaves, and differing essentially in the form of the petals. The flowers are larger than in *C. Cracca,* and apparently white. Posterior petal with a broadly ovate, reflexed limb, forming the upper lip of the flower, and an oblong very wide and blunt spur of ⅔ its length, bent upwards; the anterior petal, forming lower lip of flower, with a narrow linear claw, and ovate limb, in the middle of which is a deep pit on the inner side and a corresponding prominence on the outer. Lateral petals spoon-shaped, adhering at the apex and mucronulate.

3. C. Burmanni (Eck. and Zey. ! En. No. 23); climbing; leaves bi-tripinnate, cirrhose; pinnæ ternately parted, with broadly cuneate and deeply trifid, mucronulate segments, somewhat glaucous; racemes few-flowered; flowers *bilabiate, the outer petals very much expanded, obovate-orbicular,* the posterior one with a *minute gibbosity* at base; pods broadly lanceolate, pedunculate, *erect. Drege,* 7588. *also Cysticapnos grandiflora, E. Mey. ! in Herb. Drege.*

HAB. Saldanha Bay, and near Brackfontein, Clanwilliam, *E. & Z. !* Uit-Komst, 2–3000 ft. *Drege !* (Herb. Sond.)

This has the foliage and petals of *Cysticapnos africana,* but the fruit is that of a *Corydalis.* The flowers are much larger than in *Cysticapnos,* but otherwise similar. The segments of the leaves are shorter and broader than in *C. pruinosa.* The ripe pods are slightly warted, from the pressure of the seeds, and the latter are shining black.

III. DISCOCAPNOS, Ch. & Schl.

Petals 4, the posterior one spurred at base. *Fruit* (a *utricle*) orbicular, flattened, membranous, with a marginal wing and central nerve, indehiscent, one-seeded, tipped with the base of the style. *Seed* lenticular, beaked, shining. *Ch. & Sch. Linn. vol.* 1. *p.* 569. *Endl. Gen. No.* 4840.

A fibrous rooted annual, climbing by the branched tendrils of its decompound leaves. Flowers in racemes, flesh-coloured, with dark tips. Name from *δισκος*, a *disc* (whence *dish*) and *καπνος*.

1. D. Mundtii (Cham. & Schl. in Linn. 1. p. 569): *E. & Z. No.* 24.

VAR. *α*. **Mundtii**; racemes few-flowered; ovary elliptical, acute; fruits orbicular, scaberulous, with a broad, membranous, transversely costate wing. *Harv. Thes. t.* 10.

VAR. *β*. **Dregei**; racemes many flowered; ovary ovate-acuminate; fruits sub-elliptical, somewhat acute at each end, scabrous, with a narrow, even wing.

HAB. In shrubby places. Hills round Capetown, *Mundt. and Maire.* Near the Waterfall, Devil's Mt., *E. & Z.!* Camps Bay, W.H.H. *β.* at the Bosch River, George, *Drege!* (Herb. T.C.D., Hook., Sond.)

Root annual. Stems weak and straggling, succulent. Leaves on long petioles, bi-tri-pinnately decompound; the pinnæ alternate, tripartite or pinnate, with broadly cuneate, or obovate, incised, obtusely cut leaflets. Racemes opposite the leaves, at first very short, lengthening as the fruit ripens. Bracteæ scarcely as long as the pedicels. Flowers small. Petals all conniving at the point; the posterior with a short hooded point and an oblong, blunt, suberect spur of equal length. Appendix to the synema adhering to the spur. *Ovary* elliptical; style filiform, curved. Var. *β*. has smaller and more oblong fruits, with a narrower wing. It may possibly be specifically distinct. The general aspect is that of *Corydalis Cracca*; but the spur of the flower is longer and the fruits very different.

IV. FUMARIA, L.

Petals 4, the posterior one saccate at base. *Fruit* subglobose, at first fleshy, then dry, indehiscent, one-seeded. *DC. Prod.* 1. *p.* 129.

Annuals, of European Origin, common in cultivated ground and now naturalized throughout the temperate zones. The species have been needlessly multiplied, by the hair-splitting of novelty-seeking botanists, and I quite agree with Bentham in referring the majority of the so-called species to the old *F. officinalis*, which varies in size, in the colour of the flower, in the form of the fruit, the broader or narrower leaf segments, and the relative proportions of the sepals and petals. The generic name is derived from *fumus, smoke*; but why? Is it because most of the book-species are "all smoke"?

1. F. officinalis (Linn.) *DC. Prod.* 1. *p.* 130. *E. & Z. En. No.* 25.

VAR. **capensis**; stems diffuse, straggling; leaves on long petioles, bipinnate, the pinnæ petiolate, tripartite, with cuneate, sharply incised, mucronulate segments; racemes lax, few-flowered, petals thrice as long as the toothed sepals; fruit stalks patent; fruits globose, smooth and even. *F. capreolata β. Burchellii. DC. Prod.* 1. *p.* 130. *F. Lichtensteinii, Schl. Linn.* 1. *p.* 568. *F. Eckloniana, Sond. in litt.*; *F. muralis, Sond. in Koch. Syn. Fl. Germ.* 11. *p.* 1017?

HAB. Common in cultivated ground throughout the Colony, and assuredly introduced from Europe. (Herb. T.C.D.)

A common weed in gardens: the "*fumitory.*" My colleague Dr. Sonder regards the Cape plant as being probably the same as the *F. muralis* of Flora Germanica, by many Continental botanists regarded as a good species.

ORDER VII. CRUCIFERÆ, Juss.

(By W. SONDER.)

(Cruciferæ, Juss. Gen. 237. DC. Prod. I. p. 131. Endl. Gen. Pl. No. clxxxi. Brassicaceæ Lindl. Veg. King. No. cxxiii.)

Sepals 4, deciduous. *Petals* 4, cruciate, clawed. *Stamens* 6, of which 2, opposite the lateral sepals, are shorter than the other 4, which are placed, in pairs, opposite the anterior and posterior sepals. *Ovary* bilocular (with a spurious septum); stigmas two, subsessile, opposite the placentæ. *Fruit* a two-celled, two-valved pod *(silique* or *silicule)*; seeds pendulous, without albumen. *Embryo* having its cotyledons bent back upon the radicle.

Herbaceous, or rarely fruticose plants, with alternate, exstipulate leaves and racemose inflorescence. Flowers white, yellow, purple, brown, or sky-blue, usually without bracts. Juice colourless, generally pungent.

A large, important, widely dispersed, and most natural Order, easily known by its cruciate flowers, and *tetradynamous* stamens. It is equivalent to the 15th Class, Tetradynamia, in the Linnæan system. Nearly 2,000 species are known, comprised under about 200 genera; but the generic characters of many require revision. The student will find, at first, some difficulty in mastering the genera, which are distinguished by minute, and not always satisfactory characters. In many instances it is necessary (and is always useful) to be provided with ripe or nearly ripe fruit, in order to ascertain the name of a cruciferous plant; and special care must be taken to ascertain the relation of the radicle to the cotyledons, namely, whether the *edges* of the cotyledons face the radicle *(accumbent)*; or whether the *broad-side* be turned to the radicle *(incumbent)*; or whether the cotyledons be *plaited*, *spiral*, or *doubly folded*. In the S. African Flora the Order is very imperfectly represented: yet there are 6 genera, of which *Heliophila* is the most numerous in species, peculiar to the Cape. Many culinary plants, such as Cabbage, Turnips, Radishes, Cress, Mustard, Horseradish, &c., are familiar; and many more may be used, if necessary. None are poisonous; many are anti-scorbutic, and stimulant. Sulphur and potash are largely secreted in most.

ARTIFICIAL TABLE OF THE SOUTH AFRICAN GENERA.

A. SILIQUOSÆ.—Fruit-pod several times as long as broad; *(a Siliqua.)*

1. Siliqua with flat valves.

* Calyx equal at base.

† Seeds with flat, accumbent cotyledons (o =).

IV. **Turritis.**—Seeds in two rows.

V. **Arabis**—Seeds in one row.

VI. **Cardamine.**—Seeds in one row; pod opening with elasticity; seeds not margined.

† Seeds with linear, elongate, twice folded cotyledons (o || || ||).

XIX. **Heliophila.**—Pod long or short, linear, or moniliform, or lanceolate.

** Calyx two-spurred at base.

XIV. **Chamira.**—Pod lanceolate.

2. Siliqua compressed, two-edged, with keeled valves.

III. **Barbarea.**

3. Siliqua with convex, round-backed valves.

* Siliqua dehiscent at maturity.

† Seeds with flat, accumbent cotyledons (o =).

I. **Matthiola.**—Sepals erect, saccate at base.
II. **Nasturtium.**—Sepals spreading, equal at base.

†† Seeds with flat, incumbent cotyledons (o ‖).

VII. **Sisymbrium.**

††† Seeds with broad cotyledons folded over the radicle (o >>).

XII. **Brassica.**—Valves of pod one-nerved, and veiny.
XIII. **Sinapis.**—Valves of pod 3–5 nerved.

** Siliqua indehiscent.

XVIII. **Carponema.**—(Seeds as in *Heliophila).*

B. SILICULOSÆ.—Fruit-pod short, few seeded, not thrice as long as broad *(a silicula.)*

1. Silicula dehiscent at maturity.

* Silicula with flattish valves and a broad septum.

VII. **Alyssum.**—Silicle orbicular or oval. Cotyledons accumbent (o =).
XIX. **Heliophila.**—Cotyledons elongate, twice folded on the radicle (o‖‖‖).

** Silicula with keeled or boatshaped valves and a narrow septum.

X. **Lepidium.**—Seeds solitary in each cell.
XI. **Capsella.**—Seeds numerous in each cell.

2. Silicula indehiscent at maturity : seeds solitary.

* Silicula didymous or deeply 2-lobed.

IX. **Senebiera.**—Petals minute, scarcely equalling the sepals.
XV. **Brachycarpæa.**—Petals much longer than the sepals.

** Silicula orbicular or somewhat ovate.

XVI. **Cycloptychis.**—Silicula ovate, beaked, with convex wrinkled valves.
XVII. **Palmstruckia.**—Silicula orbicular, with flat valves.

Tribe I. PLEURORHIZEÆ.

Seeds with flat cotyledons, whose edges are directed to the radicle or *accumbent* (o =).

Sub-tribe I. ARRABIDEÆ.

Pod *(siliqua)* long or shortish, linear, cylindrical or compressed, many seeded.

I. MATTHIOLA. R. Br.

Sepals erect, the two lateral ones saccate at base. *Siliqua* sub-terete, elongate, with round-backed valves. *Stigma* thickened, bidentate. *Seeds* compressed, mostly margined, in a single row. *DC. Prod.* I. *p.* 132.

Herbs or suffrutescent plants, mostly natives of the South of Europe and North Africa, hoary or rough, with short, stellate, thickly set pubescence. Leaves entire or sinuate-toothed ; racemes terminal. Flowers white or purple. The Common Garden Stock (or July-flower) is the type of the genus. The name is in honour of *P. A. Matthiolus,* an Italian physician and botanist of the 16th century.

1. **M. torulosa** (DC. Syst. II. p. 169) ; stem erect, slightly branched, rough with stellate hairs ; leaves linear, subentire, tomentose ; pods

subtorulose, downy, and rough with stalked glands. *DC. Prod.* I. *p.* 133. *Cheiranthus torulosus, Thunb. Prod. p.* 108. *Fl. Cap. p.* 493.

VAR. *β.* **tricornis**; leaves twice as broad, subspathulate, sinuate, the cauline ones pinnatifid or entire; pods tricuspidate, the points obtuse, longer than the stigma. *M. stelligera, Sond. in Linn.* 23. *p.* 1.

HAB. Sandy and grassy places. At the Gauritz and Gariep, *Burchell*; Zeekoe-rivier, and in the Nieuweveld, *Drege*. Caledon-river, *Burke and Zeyher* (17.) Nov.—May. (Herb. Thunb., Sond., T.C.D.)

A foot high, canescent. Radical leaves crowded, subsinuate, tapering at base, 1½–2 inches long, 2 lines wide. Racemes 3–5 inches long. Pedicels very short and thick. Calyx 3 lines long. Petals purplish, the claw as long as the calyx, the lamina oblong-oval. Pod widely spreading, 2–2½ inches long; stigmata thickened at back.

II. NASTURTIUM R. Br.

Sepals spreading, equal at base. *Siliqua* nearly cylindrical, sometimes short, the valves rounded at back. Stigma somewhat bilobed. *Seeds* not margined, irregularly in two rows. *DC. Prod.* I. *p.* 137.

Water or marsh plants, dispersed over the world, glabrous or rough, with simple, rigid hairs. Leaves pinnatipartite or pinnatifid. Flowers small, white or yellow. The common *Water-cress* is a well known example of this genus. The name is traced to *Nasus-tortus*, a *wrinkled nose*; and is supposed to allude to the pungent qualities of these plants.

* *Flowers white.*

1. **N. officinale** (R. Br. Hort. Kew. 2. vol. 4. p. 100); leaves pinnatipartite, the upper in 3–7 pairs; the lower ternate; leaflets repand, the side ones elliptical, the terminal ovate-subcordate; pods linear, curved, about as long as the pedicels. *DC. l. c. p.* 137. *Sisymbrium Nasturtium, Lin. Sp.* 916. *E. Bot. t.* 885. *Schkuhr Handb. t.* 187.

HAB. In streams and ditches, near Capetown, and at Krakakamma, Uitenhage. (Herb. Sond.)

"Water-cress." Root perennial, creeping; stem ascending. Petals white, stamens and pistil purple.

** *Flowers yellow.*

2. **N. fluviatile** (E. Mey.! in Herb. Drege); leaves pinnatifid, the upper smaller; laciniæ oblong, toothed or sub-pinnatifid, the terminal one larger, obtuse; pods oblong, longer than the pedicel; style elongate.

VAR. *β.* **caledonicum**; stem elongate, lower leaves elongato-lanceolate, pinnatifido-dentate, upper linear entire. *N. Caledonicum, Sond. in Linn. Vol.* 23. *p.* 2.

VAR. *γ.* **brevistylum**; style very short, stigma thickened. *N. elongatum, E. Mey.*

HAB. In waterpools. Zwartkops and Boschman's Rivers, *E. & Z.*! Grahamstown, Fish River, Zeekoe River and Stormberg, *Drege*! *β*, Caledon River, *Burke & Zeyher*. *γ*, Fish River, at Trompetersdrift, *Drege*! *Albany, Miss Bowker*. (Herb. Sond., T.C.D., Hook.)

Stem erect, 1–2 feet high, somewhat branched. Lower leaves on long petioles 2–4 inches long; upper generally smaller. Raceme 6 inches long or more; fruiting pedicels thickened. Petals twice as long as the calyx, obovate. Pods turgid, 4–12 lines long; style 1–2 lines. β is remarkable for its taller, and more slender stem, and chiefly for its elongated, less deeply cut leaves. In var. γ. the stem is straight, all the leaves pinnatifid, and the pods an inch long.

III. BARBAREA, R. Br.

Sepals erect, equal at base. *Siliqua* 4-sided, 2-edged, valves keeled at back, awnless at the apex. *Stigma* capitate. *Seeds* not margined, in a single row. *DC. Prod.* 1 *p.* 140.

Biennial or perennial herbs, dispersed throughout the temperate zones, smooth or rough with scattered, simple hairs. Leaves lyrato-pinnatifid or rarely undivided. Flowers small, yellow, in terminal racemes. Name in honor of *St. Barbara*, to whom these plants were formerly dedicated.

1. **B. præcox** (R. Br. Hort. Kew. ed. 2. 4. p. 109); lower leaves lyrate, the terminal lobe ovate; upper leaves pinnati-parted, the lobes linear-oblong, quite entire. *DC. Prod.* 1. *p.* 140. *Sm. E. Bot. t.* 1129.

HAB. A weed, in cultivated ground, introduced from Europe. (Herb. Sond.)
Plant 1–1½ feet high, erect, biennial.

IV. TURRITIS. L.

Sepals spreading, equal at base. *Siliqua* linear, compressed, with flat valves. *Seeds* very numerous, in a double row. *DC. Prod.* 1 *p.* 141.

Biennial herbs, natives of the temperate zones, glabrous or rough with forked hairs. Leaves alternate, amplexicaul, entire. Flowers small, white, in long, terminal racemes. Name, from *turris*, a *tower;* the typical species is called in England "*tower-mustard.*"

1. **T. Dregeana** (Sond. in Linn. 23. p. 2); glabrous, somewhat glaucous; stem straight; branches elongate, appressed; leaves sessile, lanceolate, cordato-sagittate at base, very entire; pods straight, much longer than the pedicel; seeds winged. *Drege*, 7537.

HAB. Witbergen, *Drege!* Jan. (Herb. Sond.)
Stem 4 feet high or more, terete, with many branches. Radical leaves not seen; cauline 2½–3 inches long, 6 lines wide at base. Fruiting racemes 1–2 feet long. Flowers whitish; petals 1 line long. Pod 3 inches long, 1 line wide; stigma sessile, orbicular; valves one-nerved. Seeds oval, bordered with a very narrow membrane.

V. ARABIS L.

Sepals erect, equal or two lateral ones saccate at base. *Siliqua* as in *Turritis*, but the seeds in a single row. *DC. Prod.* 1. *p.* 142.

A large genus, very abundant in the northern temperate zones, rare in the southern. Annual or perennial plants, rarely suffruticose, usually rough with forked or simple hairs. Leaves entire or lyrate, often amplexicaul. Flowers in terminal racemes, white or rosy. Name, *αραβις*; applied by Dioscorides to *Lepidium Draba.*

1. **A? nudicaulis** (E. Mey. in Hb. Drege); glabrous, leafy below, naked above; cauline leaves sessile, radical crowded, petiolate, obovate-oblong, irregularly toothed, pinnatifid at base, the teeth obtuse, mucronulate; raceme nude, pedicels erecto-patent, as long as the pod.

HAB. Zondag River, Graafreynet; Sneuwebergen and Uitflugt at Limoenfontein, 2–6000 ft. *Drege.* Sep.—Dec. (Herb. Sond.)

A foot high, with the habit of a *Nasturtium.* Root leaves 2 inches long, 6–8 lines wide, with a flat petiole, ½ inch long; the upper teeth ½–1 line long, lower 1–2 lines; the uppermost leaf below the middle of the stem, an inch long. Raceme 3–4 inches. Flowers small, white. Pod (unripe) half an inch long, straight or slightly curved: stigma sessile, thickened.

VI. CARDAMINE. L.

Sepals erect or patulous, equal at base. *Siliqua* linear; the valves flat, nerveless, often opening with elasticity. Seeds ovate, not margined, on slender stalks, in a single row. *DC. Prod.* 1. *p.* 149.

Annual, biennial, or perennial herbs, dispersed over the globe, glabrous or rarely sprinkled with hairs. Leaves petiolate, the upper ones sometimes sessile, either undivided, lobed or pinnatipartite. Flowers white or purple, in terminal, leafless racemes. Name, from καρδια, the *heart,* and δαμαω, to *fortify*; from supposed strengthening properties.

1. C. africana (Lin. Sp. 914); leaves glabrous, or sparsely hispid on the upper surface, ternate, the leaflets petiolulate, ovate, acuminate, toothed; pods linear. *DC. Prod.* 1. *p.* 151. *Thunb. Fl. Cap. p.* 497. *E. & Z.* ! 29. *C. Auteniquana, Burch. in DC. l. c.*

HAB. In woods. Grootvadersbosch, *Thunberg!* Zeyher! Waterfall, on the Devil's Mount, Capetown, *Ecklon, Drege, &c.* June—Sep. (Herb. Thunb., Sond., T.C.D.)

Root perpendicular, long. Stem decumbent or erect, ½–1 foot high. Leaves on long petioles; leaflets an inch long, or sometimes 2 inches, ovate, acute or acuminate, the side ones oblique at base, the terminal mostly a little larger, equal and cordate or obtuse at base. Racemes 6–12-flowered. Fl. small, white. Pods smooth, 12–15 lines long, a line wide; pedicels erectopatent, 3–6 lines long. It occurs either with glabrous leaves, or with leaves hairy at both sides and ciliolate, or at least sparsely pilose on the upper surface; the smooth form is *C. africana, DC.*; the hairy one is *C. auteniquana, Burch.*

Sub-tribe II. ALYSSINEÆ.

Pod *(silicula)* roundish or oval, with a broad septum, and flat or convex valves, dehiscing. Seeds compressed, often margined. Cotyledons accumbent.

VII. ALYSSUM. L.

Sepals erect, equal at base. *Petals* entire. *Stamens* toothed or entire. *Silicula* orbicular or elliptical; the valves flat or convex in the centre. Seeds 1–4 in each cell, compressed, sometimes with a membranous, winglike margin. *DC. Prod.* 1. *p.* 160.

Herbs, natives of the warmer temperate zones, chiefly of the northern hemisphere, annual, biennial, or perennial, covered with short, whitish stellate pubescence. Leaves entire. Racemes terminal; flowers white or yellow. Name from *a, privitive* and λυσσα, *rage*; from the ancient reputation of a plant called *alyssum.*

Sub-gen. **1. Adyseton** (DC.) *Flowers* yellow. *Stamens* toothed.

1. A. glomeratum (Burch. in DC. Prod. 1. p. 163); annual, clothed with soft, greyish, stellate pubescence; stems ascending; leaves lanceolate, obtuse, villous; raceme very dense; calyces persistent; silicules orbicular, glabrous.

HAB. Roggeveld, near Riet River, *Burchell.*
Stem slender, 3 inches long. Lower leaves petiolate, oval. Raceme surrounded with leaves: pedicels 2–3 lines long. Calyx persistent, nearly as long as the petals. Silicule $1\frac{1}{2}$ lines long, tipped with a style $\frac{1}{2}$ line long. Seeds 2 in each loculus.

Sub-gen. 2. **Lobularia** (DC.) *Flowers* white. *Stamens* toothless.

2. **A. maritimum** (Lam. Dict. 1. p. 98); stems suffruticose at base, procumbent; leaves linear-lanceolate, acute, canescent; silicules oval, glabrous, tipped with a short style. *DC., l. c. p.* 164. *E. & Z.! No.* 30. *E. Bot. t.* 1729. *Clypeola, Lin. Koniga, R. Br.*

HAB. Sandy places near Capetown, not uncommon. April. (Herb. Sond.)
Stem 3–12 inches long. Leaves more or less hoary. Flowers small, sweetly scented. Silicule $1\frac{1}{2}$ lines long. Seeds one in each loculus.

Tribe II. NOTORHIZEÆ.

Seeds with flat cotyledons, whose backs are directed to the radicle, or *incumbent* (o ||).

Sub-tribe I. SISYMBRIEÆ.

Pod (*siliqua*) elongate, usually bilocular, with flattish or convex valves. Seeds numerous, in one or two rows.

VIII. SISYMBRIUM, All.

Sepals patulous, equal at base. *Siliqua* subterete, sessile; valves convex, usually three-nerved. *Seeds* ovate or oblong: cotyledons incumbent or oblique. *Stamens* toothless. *DC. Prod.* 1. *p.* 190.

Annuals or perennials, chiefly *weeds*, inhabiting the temperate zones of both hemispheres, glabrous or hairy. Leaves either simple, pinnatisect or decompound. Flowers white or yellow, mostly in terminal, leafless racemes. Name σισυμβρῖον; given by the ancients to several plants, one of which is supposed to be a cress.

* *Flowers white.*

1. **S. capense** (Thunb. Fl. Cap. p. 497); perennial, glabrous, or hairy; stem erect, branched above; lower leaves largest, runcinate or pinnatifid, lobes toothed; upper leaves narrower, pinnatifid or lanceolate-toothed; racemes glabrous; pods turned upwards, glabrous.

VAR. α. **latifolium**; stem, glabrous, hairy below; leaves runcinato-pinnatifid, the upper toothed or pinnatifid; pods very long. *S. strigosum, Thunb. Herb. fol. e. S. Burchellii, E. & Z.! No.* 32 (*ex parte*) *non DC. S. argutum E. Mey.!*

VAR. β. **montanum**; pubescent with short hairs; leaves mostly lanceolate-toothed, or the lower ones pinnatifid; racemes subglabrous; pods shorter. *S. montanum, E. Mey.! in Herb. Drege. S. Burchellii, E. & Z.! ex parte.*

VAR. γ. **angustifolium**; stem smooth or hairy; almost all the leaves pinnatifid, the lobes oblong or linear, entire or toothed; raceme glabrous; pods very long. *S. capense, Thunb.! Herb.*

HAB. Among shrubs. α., Zwartkops, Kat and Konab Rivers; Kenko River, Gauritz River and Langekloof, *E. & Z.! & Zey.* 1893. Albany, *Drege., Mrs.*

Barber! Natal, *Gueinzius.* β, Caledon River, *Burke & Zey.* Vische River, *Drege* γ. Swellendam, *Thunb.!* Caledon River, *B. & Z.* Nieuweveld, *Drege.* Oct.—Dec. (Herb, Thunb., Sond., T.C.D., Hook.)

Stem, 2 feet or more in height, striate, with long spreading branches. Lower leaves 4–7 in. long, (in α) 1½–2 inch wide, with ovate, toothed lobes ½ inch long or less; in γ 1–2 inch wide, deeply pinnatifid, the rachis 1–1½ lines wide, lobes 6–10 at each side, horizontally patent, ½–1 inch long, with short teeth or quite entire. Upper leaves 1 inch long. Raceme 6–12 inches; flower stalks 3–4, fruit stalks 4–6 lines long. Petals longer than calyx, 2–2½ lines long. Pods 2–3 inches, in β 1–1½ inches long, ½ line wide, with 3-nerved valves: style ½–1 line long. Seeds in a single row.

2. S. Thalianum (Gay, An. Sc. Nat. VII.); annual, hairy with simple or 2–3 forked hairs; stem erect, branched above; leaves toothed or entire, the radical ones rosulate, narrowed into the petiole, ovate-oblong, cauline narrowed at base, uppermost nearly linear; pods glabrous. *Arabis Thaliana, Linn. Sp.* 929. *E. Bot. t.* 901. *Sckr. Handb. t.* 195. *Conringia Th., Reich. Arabis Zeyheriana, Turcz. Animadv.* 1854. *p.* 22 *(No.* 1157.)

HAB. Cape Flats, *Ecklon!* Caledon, *Zeyher* 1899. Orange River and Caledon River, *E. & Z.!* Paarlberg and Schneewberg, *Drege.* Sep. Oct. (Herb. Sond., T.C.D.)

A small, weak-growing plant 6 inches high or more, hispid below, glabrous above. Radical leaves ½–2 inches long, 3–8 lines wide; the cauline few and sessile. Racemes lax. Younger pedicels hispidulous. Petals oblong, obtuse. Pods 6–8 lines long. It varies with a stem hispid to the summit. The Cape and European specimens are quite similar.

** *Flowers yellow: leaves pinnately lobed, the lobes entire or toothed.*

3. S. lyratum (Burm. Fl. Cap. 17); stem erect, terete, sprinkled at base with simple hairs; lower leaves hairy, *lyrato-runcinate,* the lobes toothed, *upper leaves oblong, glabrous, toothed; pedicels filiform;* pods erectopatent, nearly smooth; seeds in a single row. *DC. Syst.* 2. *p.* 471. *Deless. Ic. Sel.* 2. *t.* 64.

HAB. Sands by the sea shore. Cape L'Agulhas, Swell., *E. & Z.* July. (Hb. Sond.)

1–1½ feet high, lower leaves 3 inches long, 8–9 lines wide, petiolate, roughly hairy on both sides, pinnati-partite; the lobes ovate, toothed, the basal ones distinct, the upper confluent, the terminal oblong, toothed obtuse. Racemes elongating. Pods 2 inch long, ½ line wide, hairy; the valves flattish, 3 nerved; stigma obtuse. *Sisymb.* 7540, of Drege's Herb. is a little different, but does not seem to be specifically distinct.

4. S. Burchellii (DC. l. c. p. 472); stem *hispid with spreading hairs,* somewhat branched; leaves pinnati-partite, *hairy,* the lobes oblong, angulato-repand; pedicels short, thick; pods suberect, roughly hairy, the seeds in two rows. *DC. Prod.* 1. *p.* 193.

HAB. Sackriver, *Burchell.* Between Shiloh and Windvogelsberg; at Zeekoe river; at Beaufort, and Rhinosterkopf, *Drege!* Oct. Nov. (Herb. Sond., T.C.D.)

Stems suffruticose, 6–12 inches long, radical and lower leaves 2–2½ inch, petiolate, with 7–8 lobes at each side; the upper ½–1 inch long, sessile. Racemes elongating. Fruitstalks 1–2 lines long, densely pubescent. Petals not much longer than the calyx. Pods terete, 10–12 lines long, ½ line wide, valves 3-nerved; style short. Seeds very rarely uniseriate.

5. S. Gariepinum (Burch. Cat. Geogr.); stem covered with long, simple, and *short branching hairs, slightly branched; the leaves pinnati-partite, lobes oblong, toothed,* subacute, *rough with branching hairs;* pods patulous, rough; pedicels short and thick; seeds in one row. *DC. Prod.* I. *p.* 193.

VAR. *α*, **apricum**; *Burch. Cat. No.* 2080.

β. **nemorosum**; Burch. Cat. 2558; pubescence of the stem a little less close, and pods more patent.

HAB. In open places. *β*, in woods, at Kosifontein, *Burchell.* Calcareous hills, at Springbokkeel; Rhinosterkop and Bitterfontein, *Burke and Zeyher!* March. (Herb. Sond., Hook., T.C.D.)

Nearly related to the preceding, but differing in the glaucous colour; fasciculately branched pubescence mixed with longer hairs, and more patent pods. Stem mostly bent; branches rigid. Stem leaves sessile, an inch long, with 4–6 subacute lobes at each side, each 2–3 toothed. Pedicels 1½–2 lines long. Flowers as in the preceding. Pods rigid, rough with stellate bristles, 12–15 lines long. Seeds rarely in two rows.

6. S. exasperatum (Sond. Linn. 23. p. 3.); stem and branches *hairy with simple hairs*; leaves pinnati-parted, lobes oblong, hairy, *sharply toothed;* pods *suberect*, roundish, *glabrous*, 3–4 *times longer* than the pedicel; seeds in one row.

HAB. Sandy places. Orange river, *Zeyher* (18). Feb. (Herb. Sond.)

Stem 1–1½ feet high, terete, with erecto-patent branches. Leaves 1½–2 inch. long, upper uncial, 3 lines wide; the lower petiolate, with 5–7 sharp lobes on each side, each with 2–3 angular teeth. Fruiting raceme 3–4 inches long. Fl. small. Pedicels 2–3 lines long. Pods uncial, ½ line wide; style very short; stigma 2-lobed.

7. S. Turczaninowii (Sond.); *densely hairy*, with simple hairs; stem erect, slightly branched; leaves pinnati-partite, with ovate *obtuse and toothed* lobes; pods turned up, terete, *densely covered with short bristles*, thrice as long as the pedicel; seeds in two rows. *Tricholobos capensis, Turcz. Anim. No.* 1101.

HAB. Caledon River, *Zeyher!* Oct. (Herb. Sond., T.C.D.)

A foot high. Lower leaves petiolate, crowded, 2–4 inches long, with 7–9 lobes at each side, the lower lobes sub-distant, the upper close, the middle 3–5 lines long, toothed; stem leaves gradually smaller. Fruiting racemes, 3–4 inches long; pedicels very hairy. Sepals a line long. Claw of petals as long as the calyx, lamina oval. Pods 6–8 lines long, ¾ line wide; valves 3 nerved; style very short; stigma thick, bilobed. Seeds minute, with incumbent cotyledons, not "accumbent," as Turcz. describes them. Allied to *S. Burchellii and S. Gariepinum*, from both which it differs in the longer leaves, and especially in the shorter, densely setose pods.

*** *Flowers yellow; leaves bi-tripinnati-parted, lobes toothed or pinnatifid.*

8. S. tripinnatum (DC. Syst. 2. p. 475); stem herbaceous, erect, branching, velvetty, with stellate pubescence; leaves velvetty, tripinnatisect, the lobules oblong-linear, subdentate; pods suberect, slender, glabrous. *DC. Prod.* I. *p.* 194. *Sinapis? tripinnata, Burch. Cat.* 1640.

HAB. Dry places near the Gauritz R., *Burchell*, Orange River, *Zeyher*, Sept. (Herb. Sond.)

Near *S. Sophia.* Stem flexuous, 1–2 feet high, smooth above. Leaves subcanescent, the lower 3 inches long, petiolate, 2–3 pinnate, the primary segments petiolate, 5–6 at each side, alternate or opposite; the secondary 3–4 at each side, subsessile; lobes oblong or linear oblong, deeply toothed. Upper leaves less compound. Racemes 3–5 inches long: pedicels downy, 4 lines long. Petals as long as the calyx. Pods 8–10 lines long, ¼ line wide; valves convex, 3 nerved, torulose; seeds in a single row.

Sub-tribe II. LEPIDINEÆ.

Pod *(silicula)*, with a very narrow septum; the valves either keeled or very convex. Seeds one or several in each cell, ovate, immarginate.

IX. SENEBIERA, DC.

Sepals patent, equal at base. *Petals* very short. *Stamens* 2–4–6. *Silicula* didymous, subcompressed, indehiscent; valves subglobose, rugose or crested, cells one seeded. *Cotyledons** long and linear, obliquely curved. *DC. Prod.* 1. *p.* 202.

Widely dispersed, littoral, annual or biennial herbs, often prostrate. Leaves entire or pinnatisect. Racemes opposing the leaves, short; the flowers minute, white. Whole plant strongly pungent. Name in honour of *M. Senebier*, a distinguished Genevese physiologist.

* *Pods deeply emarginate at the apex.*

1. S. linoides (DC. Syst. 2. p. 522, excl. syn.); *glabrous;* stem *erect*, branched; leaves *linear, entire, acute;* pods subcompressed, didymous, netted with raised lines. *Coronopus linoides, E. Mey! in Herb. Drege b.*

HAB. Banks of the Gariep, at Verleptpram, *Drege.* Sep. (Herb. Sond., T.C.D.)
More than a foot high, somewhat glaucous, with rodlike branches. Leaves lanceolate-linear, narrowed at base, nerveless, the lower 1½ inches long, 1–1½ lines wide; upper shorter and narrower. Racemes 3–4 inch long. Stamens 2. Silicle depressed, as in *S. didyma*, but smaller; valves turgid, delicately reticulate. Very similar in aspect to *Lepidium linoides*, Thunb.; but with very different fruit.

2. S. Heleniana (DC. l. c. p. 523); *pubescent;* stem *prostrate*, branching; *lower leaves pinnati-partite*, with cut lobes, upper linear or divided in few lobes; pods subcompressed, didymous, netted with raised lines. *C. linoides, α, E. Mey.!*

HAB. Dry hills at Ebenezer, *Drege!* Nov. (Herb. Sond., T.C.D.)
A small annual. Stems many from the same crown, 2–5 inches long, sprinkled with thickish but very short hairs or glabrescent. Leaves glabrous; the radical uncial, with a linear rachis; cauline 6 lines long, some entire wider at the tip, others lobed, the lobes linear. Fruiting racemes 1–2 inch long. Fl. and fr., as in preceding.

3. S. didyma *(Pers. Ench.* 2. *p.* 185); *hairy*; stem prostrate, branching; leaves *all pinnati-partite*, their lobes *oblong* or *somewhat cut;* pods compressed didymous, netted with raised lines. *S. pinnatifida, DC. l. c. p.* 523. *E. & Z.! No.* 36. *Lepidium didymum, Linn. Mant.* 92. *Coron. didyma, E. Bot. t.* 248. *C. incisa, Hornm.*

HAB. Waysides and on rubbish, about Capetown, Oct. (Herb. Sond., T.C.D.)
Stems somewhat hairy, pilose or glabrescent, 6–15 inches long. Lobes of the leaves ovate or oblong, entire, toothed or deeply cut. Racemes 1–2 inches long. Petals very small or none. Stamens 2, 4, or 6. Pod as in the preceding, but larger and more coarsely netted.

** *Pods entire (not emarginate) at the apex.*

4. S. Coronopus (Poir. Dict. 7, 76.); leaves pinnati-lobed, the lobes

---

* The structure of the seed in this genus is something like that of one of the *Spirolobeæ;* but the genus is retained among *Lepidineæ*, because in habit and pungency it resembles the cresses; while it is very unlike *Brachycarpæa.*

entire, toothed, or pinnatifid; pods subacute, cordate at base, compressed, the *valves ridged and crested at back. Cochlearia Coronopus. L. Sp.* 904. *Schk. hb. t.* 181. *Coronopus Ruellii, All. E. Bot. t.* 1660.

HAB. Waysides and on rubbish heaps, &c. Common near Capetown. (Herb. T.C.D.)

Whole plant prostrate, pressed to the ground. Leaves variously lobed. Racemes short and many flowered. Flowers very minute. Easily known by its strongly ridged and crested fruit.

X. LEPIDIUM L.

Sepals equal at base. *Silicula* ovate or subcordate; the valves keeled (or rarely ventricose); loculi one-seeded. Seeds three-cornered or compressed. *DC. Prod.* I. *p.* 203.

Herbs or undershrubs, dispersed over the globe, very various in habit and foliage. Racemes terminal, elongating. Flowers small, white; often deficient in the number of petals or stamens. A large genus, chiefly inhabiting the S. of Europe and borders of Asia: with outlying species in most countries. The garden cress (*Lepidium sativum*) is well known. The name is from λεπις, a *scale*; from the form of the pod.

* *All the leaves entire, or the upper ones toothed.*

1. L. flexuosum (Thunb. Prod. 107.); glabrous, glaucescent; stems *decumbent*, flexuous, slightly branched; leaves oblong or *oblongo-linear, subobtuse*, narrowed at base; the radical l. petiolate, cauline semiamplexicaul; pods ovate, *tipped with an evident style. Thunb. Fl. cap.* 490, *Syst.* 2. *p.* 552.

HAB. Sandy fields, near the sea shore, Verlooren Valley, *Thunb.* Oct. (Herb. Holm.)

Stems numerous, 6–12 inches long. Leaves thickish, several-nerved at base, the radical on long stalks, the petiole keeled, dilated at base, 2 inches long; lamina uncial, 2–3 lines wide; cauline ½ inch narrower. Racemes an inch or more long, pedicels 2 lines. Petals white, as long as the calyx. Silicule 1 line long; style ⅓ lin. and stigma bilobed. Known by its longer style from the other species.

2. L. myriocarpum (Sond. Linn. 23. p. 4); glabrous, pale green; *stem erect*, branches spreading; leaves *linear acuminate*, narrowed at base, the uppermost very narrow; racemes slender, at length paniculate; pods *elliptic-ovate*, with a *very short* style.

HAB. Shady places, on the Banks of Caledon R. *Burke and Zeyher.* In Glenfilling, *Drege*, 7541. Dec.—Jan. (Herb. Hook., Sond.)

3. L. linoides (Thunb. Prod. 107.); glabrous; stem erect, branches virgate; leaves *lanceolate*, narrowed or linear; pods *elliptical, obtuse, emarginate. Thunb. Fl. cap. p.* 490. *E. & Z.!* 38. (*excl. syn. Cand.*)

VAR. β. **subdentatum**; leaves sharply toothed near the point, the upper entire. *S. subdentatum, Burch, ap. DC. l. c. p.* 545. *L. linoides* β. *E. & Z.* 38.

VAR. γ. **pumilum**; 2–3 inches high; leaves toothed near the point, minutely downy on the lower surface, as is also the stem.

HAB. In shrubby places. Tulbagh Valley, *E. & Z.* β. near rivulet in the Roggeveld. Karroo, *Burchell*, Grootepost, *E. & Z.* γ. Winterveld, 3–4000 ft. *Drege* (Nov. —April) (Herb. Sond.)

Stem generally 1–2 feet high. Radical leaves narrowed into a petiole; lower cauline 1½ 2 inches long, 1–1½ lines wide, acuminate; in *β* 3 lines wide, and with 2–4 teeth near the point; upper shorter. Fruiting racemes often much elongated. Pedicels spreading, 1½–2 lines long. Flowers minute. Pods 1½ lines long; style very short, not projecting beyond the tips of the cells. Known from *L. myriocarpum*, by its thrice larger and subemarginate silicules.

** *All the leaves, or at least the lower ones, pinnatifid, the upper toothed or entire.*

4. L. Capense (Thunb. Prod. 107.); *very thinly pubescent;* stem terete, decumbent at base, then erect and branching; *radical leaves oblong serrate;* lower cauline pinnatifid, middle ones *serrated*, uppermost entire; pods elliptic-ovate, subemarginate, with a very short style. *Thunb. Fl. cap.* 491. *DC. l. c. p.* 552. *L. Eckloni, Schrad. E. & Z. No.* 41. *L. flexuosum, E. & Z.* 42. *Un. It. No.* 459.

VAR. *β.* **sylvaticum**; stem taller and more erect; cauline leaves wider, serrate, narrowed at base. *L. sylvaticum, E. & Z.* 37. *L. subdentatum, Meisn. Pl. Krauss.*

HAB. Hills and waste places, near Capetown, common. *Thunb. Ecklon*, &c. Cape L'Agulhas *Mundt.*; Port Elizabeth, *E. & Z.* Zondag river, *Drege*; Albany, *T. Williamson.* *β.* hills near Adow, *Zeyher*, Zitzekamma, *Krauss.* Omsamwubo, *Drege.* Jun.—Dec. (Herb. Thunb., Meisn. Sond., T.C.D.)

Stems 6–12 inches long, branches spreading, root-leaves petioled, 2 inches long; lower cauline 2–3 inches, deeply pinnatifid, the lobes ovate or oblong, acute, toothed; the middle one lanceolate, narrowed at base, 1½–2 lines wide, toothed or serrate; the uppermost entire or somewhat toothed. Fruiting racemes 3–4 inches long; pedicels 1½ lines long, spreading. Fl. minute. Young pod ovate, apiculate; the ripe elliptical-ovate, minutely notched; style as long as the points of valves, or a little longer. Var. *β.* is taller, with broader, mostly obtuse and serrate leaves.

5. L. Africanum (DC. l. c. p. 552); glabrous or sparsely puberulous; stem *erect*, terete, somewhat angled above, branching; *radical leaves lyrato-pinnatisect*, lobes cut, the terminal one very large, *middle ones pinnatisect* or deeply cut, upper entire; pods elliptic-ovate, scarcely emarginate, with a very short style. *Deless. Ic.* 2. *t.* 73. *L. subdentatum, E. & Z. !* 39, *non. Burch. L. capense, E. & Z.* 43, *non Thunb. L. divaricatum, E. & Z.* 44, *non Ait. Sisymb. serratum, Thunb. Fl. Cap. p.* 496 *& Herb.*

HAB. In fields and shrubby places. Lion's Mt., *Sieber*, Zwartkops R. and at the Knysna, George; and Winterfeld, Beauf., *E. & Z.* Nov. (Herb. Thunb., Sond.)

Very near the last, and chiefly distinguished by the taller, generally glabrous stem, more divided radical, and more deeply cut cauline leaves. Stem 1½ feet. Radical leaves 2–3 inch long, the terminal lobe ovate, serrate, the lower lobes 2–4 pairs, ovate-toothed or inciso-multifid: cauline leaves 1½ inches long, some pinnatifid, some inciso-serrate. Uppermost leaves 6–8 lines long, narrow. Fruiting raceme 3–6 inches long: ripe pouch 1½ lines long.

6. L. desertorum (E. & Z. No. 40); annual, small, glabrous or minutely puberulent; stems *ascending, erect*, slightly branched; radical leaves and lowermost stem-leaves petiolate, pinnatifid; upper linear, entire or toothed; pods *cordate-ovate*, tipped with a stigma; the septum open (fenestrate). *L. fenestratum, E. Mey. ! Herb. Drege.*

HAB. In high Karroo districts, near Gauritz river, Swell., *E. & Z.* ! Springbokkeel, March, *Zeyher* (34.) Zilverfontein, Oct.; and Zwarteberg, June, *Drege.* (Herb. Sond., T.C.D.

A small plant, 4–5 inches high. Root leaves 1½–2 inch long, pinnatifid, with 3–4 subdentate lobes at each side, the terminal larger, 3–6 lines long. Raceme ½–1 inch: pedicels 1 line long. Pouches scarcely ¾ line long, as long as broad. Differs from the others by the shape of its pouches.

7. L. pinnatum (Thunb. Prod. 107); tall, glabrous, stem erect, divaricately branched; leaves *all pinnatifid*, the lobes *divaricate, acute*, subdentate, the uppermost entire or toothed; pods *oval, subemarginate, Thunb. Fl. Cap. p.* 491.; *L. divaricatum, Ait. Kew.* 1. *vol.* 2. *p.* 441. *Thlaspi divaricatum, Poir.*

HAB. Fields, near Capetown, *Thunberg, W.H.H.* Heerelogementsberg, *Zeyher.* Paarlberg, 1–2000 ft., *Drege (L. capense, E. Mey.!)* June. (Herb. Thunb., Sond., T.C.D.

1–2 feet high. Leaves mostly fasciculate, 1½–2 inch long, petiole uncial, rachis narrow, lobes 3 at each side, patent, entire or few toothed, 1–2 lines wide; the stem leaves similar, uppermost subentire. Racemes elongating; pedicels 1–1½ lines long. Flowers very minute. Pouches 1 line long, subemarginate, the style as long as the notch.

8. L. trifurcum (Sond. Linn. 23. p. 4.); suffruticose, erect, quite glabrous; leaves *linear-subulate, channelled, three forked*, the lobes erecto-patent, acute; pods oblong-ovate, slightly emarginate, with a very short style, equalling the notch. *Zey.* 23.

HAB. Rocky places at Modder River, Bekuana Land, Feb., *Burke and Zeyher.* Winterveld and Nieuweveld, Nov., *Drege.* (Herb. Hook., Sond., T.C.D.)

Stem 1–1½ feet, with virgate branches, ramulose at the end. Stem leaves mostly crowded, 3-forked, the lateral lobes 1–1½ lines, terminal 2–3 lines long; rarely with 5 lobes, pinnately disposed. Upper leaves trifid or entire. Fruiting racemes 2–3 inches long; pedicels 1–1½ lines. Pouch 1 line long. Fl. minute.

9. L. hirtellum (Sond.); suffruticose; *stem and leaves rough with very short hairs; lower leaves oblong, serrate*, narrowed into the petiole, middle and upper ones pinnatipartite, the lobes and rachis linear, very entire; pods elliptical, slightly emarginate, with a very short style, equalling the notch. *L. pinnatum, E. & Z.!* 45, non Thunb.

HAB. Near pools, "Valleyen," in Quaggasflats, Uitenhage. Jul. *E. Z.* (Herb. Sond.)

A small, slightly branched plant. Lower leaves uncial, 2 lines wide, serrated; all the rest pinnatifid, uncial, the rachis scarcely ½ line wide, with 3–4 lobes at each side, 2 lines long. Racemes 2 inches long, the rachis and pedicels hairy, at length glabrescent. Pouch glabrous, 1 line long.

10. L. bipinnatum (Thunb. Prod. 107); stems erect, *decumbent at base*, simple, velvetty; radical leaves *bipinnati-partite, with filiform lobes;* cauline linear, trifid at the summit, the uppermost entire; pods (young) evidently apiculate. *Thunb. Fl. Cap.* 491. *DC. l. c. p.* 553.

HAB. Onderste Roggeveld, Dec., *Thunberg*, (Herb. Thunb.)

Stems 3 or more, terete, compressed at base, 4–8 inches long. Radical leaves crowded, 2–2½ inches long; the petiole 1–1½ inch, canaliculate, dilated at base, the lamina uncial, multipartite, the lobes petiolulate, secund, lobules subulate, 1 line long. Cauline leaves remote, 4–2 lines. Flowering racemes corymbose, short, pedicels 1–1½ lines long. Sepals obtuse, membrane-edged. Petals minute. Young pod ovate, tipped with a short style and obtuse stigma.

"Lepidium capense," E. Mey.! in Herb. Drege is perhaps a new species, allied to *L. bipinnatum;* but the specimens we have seen are incomplete.

XI. **CAPSELLA**, Vent.

Sepals flattish, equal at base. *Silicula* triangular-wedge shaped, the valves boat-shaped, wingless; cells many seeded. *DC. Prod.* 1. *p.* 177.

Small annuals, one of them of European origin, now dispersed over the globe. Root perpendicular. Radical leaves rosulate, toothed or lobed; cauline sagittate, oblong. Racemes terminal. Flowers small, white. Name from *capsula, a capsule* or little box.

1. C. Bursa-Pastoris (Moench, Meth, 271); leaves runcinato-pinnatifid, laciniæ ovate-triangular, acute and toothed; stem leaves sagittate at base, the uppermost undivided. *E. & Z. No.* 31.

HAB. In cultivated ground, throughout the colony. Sept.—Oct.

A common weed, "Shepherd's Purse;" introduced from Europe. Hairy or glabrescent. Pedicels much longer than the purses. It varies with entire sinuate or pinnatifid leaves; the flowers often want petals.

Tribe III. ORTHOPLOCEÆ.

Seeds with broad and short incumbent, plicate cotyledons, clasping round the radicle (o >>).

Sub-tribe I. BRASSICEÆ.

Siliqua dehiscing longitudinally; the septum linear. Seeds globose; cotyledons conduplicate.

XII. **BRASSICA**, Koch.

Siliqua linear or oblong, with convex, one nerved and netted-veined valves. *DC. Prod.* 1. *p.* 213.

Biennials, rarely annual or perennial, natives of the temperate zones of both hemispheres. Radical leaves petioled, lyrate or pinnatifid; cauline sessile or amplexicaul, subentire. Racemes long, leafless. Flowers bright yellow. *Brassica oleracea (Cabbage, Cauliflower, &c.,) B. Rapa (Turnip)* and *B. Napus (Rape)* are universally known. The name is derived from the Celtic *bresic* (or *praiseach*); the wild cabbage or navew of English fields.

1. B. strigosa (DC. Syst. 2. 603); stem erect, branching, angular, *rough with reflexed rigid hairs,* glabrescent above; leaves *hispid, the lower ones lyrato-pinnatifid,* lobes ovate, toothed, the terminal one larger, oblong, upper lobes lanceolate, toothed or pinnatifid at base; pods erecto-patent, glabrous, four times as long as their beak. *Sisymb. strigosum, Thunb. Fl. Cap.* 496, & *Herb. fol.* β. & γ.

VAR. β. **glabratum**; lower leaves sinuate-dentate, lanceolate-elongate, and the stem subglabrous. *Br. erosa, Turz. Anim. p.* 41. *Sisymb. erosum, E. Mey.*

HAB. Doornriver, *Thunberg,* Orange R., *Zeyher* (8). Nieuweveld, 3-4000 ft., *Drege* β. Moojee river and Karreebosch, *B. & Z.* 22. Zitzekamma and near Uitenhage *E. & Z.* Buffelvalley, *Drege.* March—April. (Herb. Thunb., Sond., Hook, T.C.D.)

Stem flexuous, a foot or more in height. Leaves scabrid, the lower petiolate, 4-6 inches long, 2-3 inches wide; uppermost uncial or smaller. Raceme 6-12 inches long; pedicels 4-6 lines long, patent. Fl. yellow: petals longer than calyx. Pods uncial, tetragonous-compressed, with a conical beak.

2. B leptopetala (Sond.); stem terete, striate below, *glabrous or* *sparsely pilose*, young branches hispid; leaves *glabrous, runcinato-pinnatifid*, hispid on the petiole, the upper ones gradually smaller; pods patent, glabrous, 4–6 times as long as the beak. *Sinapis leptopetala, DC. l. c. p.* 610. *Deless Ic. Sel.* 2. *t* 87. *E. & Z. No.* 49. *Sinapis retrorsa E. & Z.* 50, non Burch. *Sisymb. strigosum, fol.* α & δ *Thunb. Herb.!*

HAB. Oliphants-riversbad, *Thunberg*. Hills between the Coega and Sondags River, Uitenhage, and Adow, *E. & Z.* Zwartkops, *Zeyher*, &c., June—Sep. (Herb. Thunb., Sond., Hook! T.C.D.)

1–1½ feet high, more slender than the preceding, to which it is closely allied. Lower leaves petiolate, 2–3 inches long, 6–10 lines wide, lobes ovate, obtuse, dentate, the upper lobes longer and coalescent. Racemes glabrous; pedicels 4–6 lines long, patent in fruit. Petals longer than the calyx; either narrower or broader than the sepals. Pods uncial, 4 angled-compressed, with a conical beak.

3. B. nigra (Koch, Syn. Fl. Germ. Ed. 2. p. 59); all the leaves petioled, the lower *lyrate-toothed*, the terminal lobe very large, lobed; upper leaves lanceolate entire; *pods closely pressed to the rachis. Sinapis nigra, Linn. E. Bot. t.* 969. *E. & Z. No.* 47.

Var. β. **lævigata**; leaves and stem smooth. *S. lævigata, Burm. Prod. p.* 18. *ex DC.*

HAB. In cultivated ground. Varschevalley, near Salt-river, Cap. *E. & Z.* Common *wild mustard*. Introduced from Europe.

XIII. SINAPIS, Koch.

Siliqua linear or oblong, with convex, 3–5 nerved valves. *DC. Prod.* 1. *p.* 217.

Biennials, scattered over the globe; smooth or hairy. Leaves lyrate or toothed and cut. Racemes terminal, leafless; flowers yellow. Closely allied to the preceding genus, with which it is united by some botanists. Name, *σιναπι, mustard.*

1. S. retrorsa (Burch. in DC. Syst. 2. p. 609); stem sub-angular, erect, branching, retrorsely pubescent; leaves lyrato-pinnatifid, the lowest lobes stipulæform, appressedly pubescent; upper leaves sessile, erose; pods long, spreading, glabrous, slightly rough; style short. *S. Burmanni, E. & Z.!* 48. *Sisymb. lyratum, Meisn. in Pl. Krauss.*

HAB. Banks of Sondag's river, *Burchell*. Konab and Key river, *E. & Z.* Caledon River and Buffelvalley, Orange R., *Zeyher*. (7) Fort Beaufort, *Drege*. Port Natal, *Krauss* (412) Gueinzius, 518. July—Sep. (Herb. Meisn., Sond., T.C.D.)

1–2 feet high, leafy. Lower leaves 5 inches long, the lobes decreasing toward the base, the upper lobes confluent. Racemes 6 inches long; pedicels 4-8 lines, scabrid. Fl. yellow, petals oblong, exceeding the calyx. Pod 3–4 inches long, valves subcompressed, with 3–5 raised nerves. Style ½–1 line long.

Sub-tribe II. CHAMIREÆ.

Siliqua with flattish, nerveless valves. Seeds compressed; cotyledons conduplicate, twice inflexed (o >> >>).

XIV. CHAMIRA, Thunb.

Calyx two-spurred at base. *Siliqua* substipitate, oblong, compressed, with a subulate beak; valves flattish. Seeds compressed, immarginate, *DC. Prod.* 1. *p.* 131.

A glabrous, South African herb, with petiolate, cordate leaves, and leafless racemes of white flowers. Readily known by having the two lateral sepals produced at base into a spur. Name, from χαμαι on the ground ; a low growing plant.

1. C. cornuta (Thunb. Nov. Gen. 2. 48 ; Fl. Cap. p. 496). *Sond. in Hamburg Nat. Abhandl.* 1. *p.* 269. *t.* 29. *DC. Prod.* 1. *p.* 141.

HAB. In fissures of rocks, Witteklipp, Swartland, *Thunberg.* Sand dunes at Saldanha Bay, *Drege.* Sep. (Herb. E. Mey., Sond.)

A weak growing, glabrous annual, 2 feet high, branched above. Lower leaves opposite, sessile, reniform-cordate, 2 inches long, 3-4 inches wide ; middle leaves mostly alternate, deeply cordate, acute, 2 inches long and wide ; uppermost smaller, cordate-acuminate. Raceme few flowered ; the flower bearing pedicels 1-2 lines, the fruiting 4 lines long. Flowers small, white. Pods 7–10 lines long, 2–3 lines wide, with nerveless striate valves. Seeds 2–4 in each loculus, compressed, ovate, 2 lines long, 1½ line wide, with a mucilaginous epidermis. Cotyledons wider than their length, when opened out 4 lines broad and 2 lines long, twice bent inwards laterally ; the radicle opposite the plication, filiform. A rare, curious, and little known plant.

Tribe IV. SPIRILOBEÆ.

Seeds with long, linear, incumbent cotyledons, spirally rolled on the radicle. (o ‖ ‖).

XV. BRACHYCARPÆA, DC.

Sepals equal at base. *Silicula* indehiscent, didymous, with a very narrow dissepiment ; the valves very ventricose ; loculi one seeded. *DC. Prod.* 1. *p.* 236.

Glabrous, suffrutescent perennials, peculiar to S. Africa. Leaves oblong or linear, entire. Racemes long and leafless. Flowers large, yellow or purple. Name from βραχυς, *short*, and καρπος, *fruit.*

1. B. varians (DC. Syst. 2 p. 698); leaves lanceolate, glabrous ; petals oval-oblong. *Deless Ic.* 2 *t.* 100. *Cleome juncea, Thunb. Fl. Cap. p.* 497. *Coronopus anomalus, Spr. Syst. Veg.* 2. *p.* 853. *Br. varians and B. polygaloides, E. & Z.* 52, 51.

VAR. α **flava** ; flowers yellow. *Heliophila flava, Lin. f. sup.* 297.

VAR. β **purpurascens** ; flowers purplish. *Polygala bracteolata, Burm.*

HAB. Sandy fields. Swartland, *Thunberg.* Hills at Brackfontein, Clanwilliam, and near Klapmuts and Groenekloof, *E. & Z.* Klippberg, *Zeyher.* Wupperthal, *Wurmb,* Paarlberg, 1–2000, *Drege.* Sep.-Oct. (Herb. Thunb., Sond., T.C.D.).

A virgate, glabrous undershrub, resembling a *Genista,* 3 feet high or more, with long, striolate branches. Leaves sessile, the lower uncial, 2-3 lines wide ; upper smaller and recurved, all obtuse or mucronulate, narrowed toward the base. Racemes long and naked ; pedicels 3–6 lines long. Petals 6–8 times longer than the calyx, clawed. Anthers scarcely longer than the calyx. Fruit somewhat netted with ridges, 4 lines wide. Style thickish.

2. B. laxa (Sond.) ; leaves linear ; petals lanceolate.

VAR. α **laxa**; stem ascending, branches incurved. *Cleome laxa, Thunb. Fl. Cap.* 498. *Br. linifolia, E. & Z. No.* 53.

VAR. β **stricta** ; stem and branches straight, erect. *B. emarginata, E. Mey. in Herb. Drege. Sond. Rev. Hel. t.* 29 *(fr.).*

HAB. Var. α, Sandy fields, near Verloren Valley and Langekloof, *Thunb.* Klein river, Caledon, *E. & Z.* and *Zeyher* 1912. β Grom riv. and Waterval, Honigvallei, Kamiesberg and Leliefontein, *Drege* 7584. (Herb. Thunb., Lehm., Sond., T.C.D.)

More slender than the last; the stem 1-1½ feet high, branches filiform, ramulous. Leaves ½ inch long, a line wide; flowers purplish or yellow, half the size of those of *B. varians*. Petals 4 times as long as the Calyx. Fruit half the size of that of *B. varians*.

Tribe V. Diplecolobeæ.

Seeds with long, linear, incumbent cotyledons twice folded transversely on the radicle. (o ‖ ‖ ‖).

Sub-tribe I. Cycloptychideæ.

Silicula nucamentaceous, indehiscent, orbicular-ovate, rostrate; septum orbicular; valves somewhat convex, with elevated ridges, radiating from the more prominent, keeled centre: seeds solitary in each cell.

XVI. CYCLOPTYCHIS, E. Mey.

Character the same as that of the sub-tribe.

Suffruticose or herbaceous plants, peculiar to S. Africa, with virgate branches; simple, entire, narrow, sessile, scattered leaves, and long, naked, terminal racemes of purple flowers. They resemble *Brachycarpœa* in aspect, but differ both in the form of the seed vessel and structure of the seed. The name is compounded of κυκλος *a circle* and πτυξ a *plait*, or *ridge*, from the radiating ridges on the fruit.

1. C. virgata (E. Mey. in Herb. Drege); suffruticose, erect, quite glabrous, glaucous; branches virgate, terete; leaves coriaceous, lanceolate, mucronulate; racemes elongate; pods glabrous. *Cleome virgata*, *Thunb. Cap. p.* 498.

Hab. Interior districts, *Thunberg*. Piquetberg, at Groenvalley, Oct.; Pikenierkloof, Jan.; Gift Berg, 1000-2500 f., Nov., *Drege*. Tulbagh, *E. & Z.!* Sep. (Herb. Sond., T.C.D.).

2 feet or more high, very similar, when in flower, to *Brachycarpœa*. Branches simple, filiform. Leaves remote, alternate, sessile, very entire, one nerved below; the lower ones oblong-lanceolate, uncial, 2–3 lines wide, upper gradually smaller. Racemes long, many flowered; pedicels erecto-patent, 2–4 lines long. Sepals oblong, 3 nerved, 2 lines long. Petals oblong, clawed, 7–8 lines long, purple. Stamens 6, as long as the calyx; anthers linear, notched at base. Ripe pouch brownish, the orbicular valves 3 lines in diameter, the smooth beak equalling them in length. Seeds 1 line long.

2. C. polygaloides (Sond.); herbaceous, very thinly pubescent; stem weak, branched; branches filiform; leaves oblong or oblong-lanceolate, acute, narrowed at base; raceme short; pods (young) pubescent.

Hab. Tulbagh, Nov. *Zeyher* (Herb. Sond.).

Decumbent, with the aspect of *Polygala vulgaris*, a span long, slender. Leaves alternate, closely set, 4–6 lines long, 1–1½ line wide. Raceme 4–8 flowered; pedicels erecto-patent pubescent, 6–8 lines long. Sepals broadly oblong, puberulous on the outside, 2 lines long. Petals more than twice as long as the calyx, obovate, purplish. St. 6. Ovary 3-lines long, ovate, rostrato-acuminate. Ripe fruit unknown.

Sub-tribe II. Palmstrukieæ.

Silicula indehiscent, orbicular, one celled, one seeded; valves flat.

XVII. PALMSTRUCKIA, Sond.

Silicula sessile, orbicular, compresso-plane, indehiscent, unilocular, one seeded. *Seed* orbicular, compressed, with a membranous, marginal wing. *Cotyledons* linear, incumbent, twice folded.

A glabrous, erect herb. Leaves linear-filiform. Flowers racemose, cernuous; pedicels filiform, ebracteate. Name in honour of *J. W. Palmstruck*, editor of "*Svensk Botanik.*"*

1. P. Capensis. (Sond.) *Peltaria Capensis, Thunb. Fl. Cap. p.* 490. (*non. Lin. f.*).

HAB. Onderste Roggeveld, Nov. (deflorata) *Thunberg.* (Herb. Thunb.) Stem herbaceous, somewhat decumbent at base, then erect, terete, a foot high, alternately branched. Leaves remote, according to Thunb. "filiform." Fruiting racemes elongate; pedicels 6–9 lines long. Ripe pods 7–8 lines long and wide, rounded at top or somewhat retuse, tipped with a very short style; the valves flattish, nerveless, veined. Septum none. Seed 3 lines in diameter, emarginate. Not found by any collector since Thunberg. The flowers are unknown.

Sub-tribe III. HELIOPHILEÆ.

Siliqua dehiscent or rarely indehiscent, elongate, or rarely oblong or oval, with a linear or oval septum; valves flat or somewhat convex. Seeds several.

XVIII. CARPONEMA, Sond.

Siliqua sessile, *indehiscent*, linear, terete, tapering to each end, tipped with a conical style, somewhat constricted between the seeds, with a very thin septum, two-celled; one cell smaller and empty, the other seed-bearing: valves hardened, nerveless. *Seeds* in a single row, oblong, terete, immarginate, separated by transverse partitions. *Sond. in Hamb. Naturwus. Abhandl.* I. *p.* 178.

An annual herb, 12–18 inches high, glabrous or sparsely pilose; stem terete, branched; leaves linear, the lower long. Racemes elongate; pedicels filiform, erect in flower. Flowers blue or purplish. Pods pendulous, an inch long. Name, *καρπος fruit* and *νημα a thread*, from the slender pods.

1. C. filiforme (Sond. l. c. p. 179. t. xvii). *Heliophila filiformis, Lin. f. Supp.* 296. *DC. Syst.* 2. 679. *Eck. & Zey. No.* 54.

HAB. Cape Flats, in sandy places: Riedvalley, Doornhooghde, &c., Aug. *Eck. & Zey. Drege* 2813 b. (Herb. Sond.).

XIX. HELIOPHILA, Burm. L.

Siliqua sessile or pedicellate, *dehiscent*, compressed or subterete, the margins either straight, or sinuated, constricted between the seeds, and moniliform; with a membranous septum, two-celled, bivalve. Seeds in a single row, compressed, often bordered with a wing. *DC. Prod.* I p. 231.

Annuals or suffruticose perennials, natives exclusively of South Africa. Leaves various. Racemes elongate, leafless; pedicels filiform. Flowers yellow, white, rosy

---

* *Palmstruckia*, M. C. Retz. Obs. Bot. 1810.=*Chænostoma*, Benth., a name now universally adopted, and which it seems undesirable to alter. The genus now proposed will equally preserve *Palmstruck's* name to a Cape plant.

or sky-blue. Name, ἥλιος the *Sun*, and φιλεω to *love;* the species inhabit sunny places. The following table is intended to assist the student:—

ANALYSIS OF THE SPECIES.

SEC. 1. LEPTORMUS. Pods linear, moniliform; the beading *oval. Herbs.*

* Style thickened, nodose, acute ... ... (1) **dissecta.**

** Style nearly filiform:

† Pods erect:

(a) *quite glabrous.*

Leaves entire or tripartite ... ... (5) **longifolia.**

Leaves pinnati-partite ... ... (2) **sonchifolia.**

(b) *hairy in the lower part.*

Stem hollow; pods erecto-patent ... ... (8) **fistulosa.**

Stem solid; pods closely appressed ... ... (3) **caledonica.**

(c) *hairy all over.* ... ... (7) **pubescens.**

†† Pods reflexed or pendulous:

Style short, obtuse ... ... (6) **affinis.**

Style long, filiform ... ... (4) **Eckloniana.**

SEC. 2. ORMISCUS. Pods linear, moniliform; beadings *orbicular. Herbs.*

* All the stamens toothless, or the two shorter ones toothed.

Leaves oblong, or somewhat lanceolate ... ... (9) **amplexicaulis.**

Leaves linear, undivided ... ... (10) **pusilla.**

Leaves linear, *pinnati-partite:*

(a) *glabrous species.*

† Glaucescent (blueish-green) ... ... (15) **monticola.**

†† Green:

the shorter stamens toothed ... ... (16) **trifida.**

all the stamens toothless:

style as long as the pedicel ... ... (12) **concatenata.**

style shorter than the pedicel:

(1) Petals obovate:

leaf-lobes flat; pods sub-erect ... (11) **rivalis.**

leaf-lobes furrowed; pods pendulous (14) **pendula.**

(2) Petals oblong ... ... (13) **variabilis.**

(b) *Pubescent species* ... ... (17) **coronopifolia.**

** All the stamens toothed ... ... (18) **dentifera.**

SEC. 3. SELENOCARPÆA. Pods oval or sub-orbicular. *Herbs.*

Pods compressed; oblong, 4–8 seeded ... ... (19) **diffusa.**

Pods compressed, ovate-orbicular, 2-3 seeded ... (20) **Peltaria.**

Pods somewhat inflated ... ... (21) **flacca.**

SEC. 4. ORTHOSELIS. Pod linear, with straight margins. *(In H. elata, H. cornuta and H. refracta somewhat torulose. Herbs* or *Shrubs.*

§ Subsection 1: Herbaceous species.

* Leaves pinnati-partite:

(A) Pods *oblongo*-linear.

Style thickened, cylindrical ... ... (23) **macrostylis.**

Style short, filiform ... ... (22) **latisiliqua.**

(B) Pods exactly linear:

(a) *Glabrous species.*

† The shorter stamens toothed:

Pods patent or pendulous ... ... (24) **Meyeri.**

Pods erect; leaves subsessile; stem solid ... (30) **viminalis.**
Pods erect; leaves on long petioles; stem hollow ... ... (31) **tenuifolia.**

†† Stamens toothless: (pods pendulous, or refracted).

Pods compressed, 1 nerved; seeds orbicular, margined ... ... (32) **seselifolia.**
Pods compressed, 3 nerved; seeds suborbicular, immarginate ... ... (25) **pectinata.**
Pods *torulose;* seeds ovate, immarginate ... ... (33) **refracta.**

(b) *Hairy or velvetty species.*

† Stamens toothless:

Plant velvetty, pubescent ... ... (26) **crithmifolia.**
Plant hairy below, glabrous above ... ... (27) **chamæmelifolia.**

†† The shorter stamens toothed.

Plant somewhat hairy; pods compressed ... (28) **foeniculacea.**
Plant velvetty pubescent; pods subterete ... (29) **gracilis.**

** Leaves trifurcate (the lobes narrow linear) ... ... (34) **trifurca.**

*** Leaves entire (*in H. pilosa* sometimes lobed).

Plant blueish-glaucous; leaves linear lanceolate; pods upright ... ... (36) **stricta.**
Plant green; leaves lanceolate, narrowed at base; pods spreading ... ... (37) **linearis.**
" " leaves linear ... ... (35) **divaricata.**
" " lower leaves crowded, spathulate ... ... (38) **graminea.**
" " *hairy* (or glabrous); leaves oblong ... ... (39) **pilosa.**

§§ Subsection 2. Shrubby species.

* Sepals horned at the apex ... ... (44) **cornuta.**

** Sepals simple (not horned):

(A) Leaves lobed ... ... (42) **abrotanifolia.**

(B) Leaves undivided:

(a) Leaves amplexicaul: oblongo-lanceolate ... ... (40) **brassicæfolia.**
" " ovate, acute ... ... (41) **reticulata.**

(b) " sessile:

† Racemes few flowered:

leaves linear-subulate ... ... (54) **scoparia.**
leaves linear-spathulate ... ... (56) **trachycarpa.**
leaves ovate (small), muricate ... ... (57) **Dregeana.**

†† Racemes elongate, many flowered:

leaves ovate ... ... (53) **virgata**
leaves oblong ... ... (43) **glauca.**
leaves lanceolate... ... ... (55) **callosa.**
leaves linear or filiform: with submoniliform pods (45) **elata.**
leaves linear or filiform, with straight-edged pods:

Style as long as the pedicel or longer.

Pods linear, pendulous ... ... (52) **stylosa.**
" lanceolate, pendulous ... ... (49) **rigidiuscula.**
" linear, suberect ... ... (50) **fascicularis.**

Style half as long as the pedicel, or shorter.

Leaves lin-subulate, pods spreading, narrowed at base ... ... (46) **suavissima.**
leaves lin-subulate, pods spreading, not narrowed at base ... ... (48) **subulata.**
leaves linear-filiform; pods pendulous (47) **succulenta.**
leaves linear-acute; pods erect ... (51) **linearifolia.**

SEC. 5. PACHYSTYLUM. Pubescent suffrutices. Pods linear, tipped with a short and thickened style.

Leaves spathulate ... ... (58) **incana.**
Leaves linear ... ... (59) **arenaria.**

SEC. 6. LANCEOLARIA. Glabrous shrubs, with lanceolate pods.

Leaves linear-spathulate, mucronate; raceme short ... ... (60) **florulenta.**
Leaves linear-acute; raceme elongate (61) **macrosperma.**

SEC. 1. LEPTORMUS (DC.). *Siliqua* sessile, subcompressed, very slender, submoniliform, that is, contracted between the seeds; the beadings ovate-oblong. Annuals. (Sp. 1–8).

1. **H. dissecta** (Thunb. Prod. p. 108); herbaceous, glabrous; pods slender, submoniliform, tipped with a thick, nodose, acute style; radical leaves *very narrow*, entire, trifid or tripartite. *Thunb. Cap. p.* 495. *DC. Syst.* 2. 680. *Sond. l. c. t.* 18. *Leptormus dissectus, E. & Z.* 56. *L. trifidus, E. & Z.* 64 *(ex parte)*. *Pachystylum glabrum, E. & Z. No.* 100.

VAR. β. **albiflora**; stem much branched, flowers white. *Sond. l. c.*

VAR. γ. **simplex**; very small; all the leaves entire. *H. tenella, Banks. DC. l. c. Lept. tenellus, E & Z.* 58.

HAB. Sandy and stony places. Zwartland, *Thunberg*. Capeflats, and mts. round Capetown; Hott. Holland; Pottberg, Caledon, *E. & Z.* Nieuwekloof, *Drege!* 7352. Klipfontein, *Zeyher*, 47 and 48. Oct.–Nov. (Herb. Thunb., Vahl, Sond.).

Stem erect, simple or branched, 6 inches to 2 feet high, with subvirgate branches. Leaves crowded in the lower part of the stem, 2–6 inches long. Rachis of the pinnati-partite leaves ½–1 line broad, lateral lobes 5–7, erecto-patent, 3–5 lines long, rarely uncial. Upper leaves smaller, undivided. Flowering racemes short, few flowered; fruiting elongated. Flowers either *blue, with a yellowish centre, lilac, yellowish or white.* Sepals 2 lines long. Petals obovate, twice as long as the calyx. Shorter stamens toothed at base. Pods erect or somewhat spreading, 1–1½ inches long, ½–¾ line broad, terete or subcompressed, obsoletely 1–3 nerved. Style thickened, cuspidate, ½–1 line long. Seeds oval, not margined.

2. **H. sonchifolia** (DC. Syst. 2. 681); herbaceous, glabrous; pods erecto-patent, slender, subcompressed, torulose, tipped with a *conico-filiform style;* radical leaves pinnati-partite, lobes 2–3 pair, *linear;* lateral stamens toothed. *Sond. l. c. p.* 204. *Lept. trifidus, E. & Z.* 64, *ex parte.*

HAB. Stony places. On the mts. round Capetown and Hott. Holl., *Masson, Bergius, E. & Z.* Oct. (Herb. Reg. Ber., Sond.)

Very similar to the preceding; but differs in the broader leaf-lobes and compressed, subacuminate pods, with a filiform style. Stem a foot high, somewhat flexuous, with virgate branches. Radical leaves crowded, 3–4 inches long, lobes opposite, subacute, 4–6 lines long, ½–¾ line wide, with interspaces of 2–3 lines. Cauline trifid, or the upper ones undivided. Flowers *blue*, similar to those of *H. dissecta.* Pods 1–1½ inch long, ¾ line wide. Seeds oval, compressed, with a very narrow marginal wing.

3 **H. Caledonica** (Sond. l. c. p. 205); herbaceous, *hairy at base;* pods *appressed* submoniliform, tipped with a conico-filiform style; leaves pinnately or bipinnately parted, lobes *linear-setaceous;* lateral stamens toothed. *Lept. Caledonicus, E. & Z. No.* 60.

HAB. Zwarteberg, near Caledonbaths, *E. & Z.* Rietvalley, *Bergius.* Aug. (Herb. R. Berl., Sond.)

Stem 2 feet high, terete, branched beyond the middle. Lower leaves often crowded, glabrous or velvetty with very short hairs, 3–4 inches long, with a long petiole; the lobes ½–1 inch long. Upper leaves trifid, glabrous; rameal undivided. Racemes elongate. Petals obovate, *bright blue*. Pods compressed, 2 inches long, ¾ line wide. Style 1 line long. Seeds compressed, immarginate.

4. H. Eckloniana (Sond. 4. c. p. 206. t. 20); herbaceous, glabrous; pods submoniliform, *reflexed*, tipped with a *long, filiform style*; leaves linear, entire or partite; stamens *toothless*. *Lept. acuminatus, E. & Z.* 57.

HAB. Stony and rocky places. Mountains near Klapmutz 3d alt, Stell., *E. & Z.* Between Eikenboom and Riebeckskasteel, below 1000f. *Drege* 7553! Paarlberg? *Drege* 7752 (fl.) Oct. (Herb. Sond., T.C.D.)

A foot high or more. Lower leaves 2 inches and longer, undivided or 3–5 fid at the apex, lobes linear, unequal. Upper leaves entire, or rarely 2–3 lobed. Flowers *white, small*. Pods linear, uncial; with 1 nerved valves. Style 2 lines long. Seeds not margined.

5. H. longifolia (DC. Syst. 2. 681); herbaceous, glabrous; pods *very slender*, compressed, submoniliform, *patent*, tipped with a *punctiform* stigma; leaves linear, entire or tripartite; petals broadly obovate; *lat. stamens toothed. Sond. l. c. p.* 207. *t.* 19. *H. filiformis, Lam. Dict.* 3. *p.* 91. *Illustr. t.* 563. *f.* 3. *H. liniflora, E. & Z.* 75.

HAB On hills. Brackfontein, alt. 2, Clanw., *E. & Z.* near Hellrivier, Drege 3171. Jul. Aug. (Herb. Vahl, Lamk. (*Roper,*) Sond.)

Known from *H. dissecta* by its more slender habit, broader irregularly divided leaves, larger flowers and stigma. Stem 1–2 feet high, branched about the middle; branches virgate. Leaves 2–4 inches long, 1 line wide, the lobes unequal, middle one generally longer; upper leaves undivided. Racemes slender, the fruiting ones 3–4 inches long. Flowers *large, blue,* with a yellowish centre. Pods 1½–2 inches long, with compressed three-nerved valves. Seeds (immature) compressed, not margined.

6. H. affinis (Sond. l. c. p. 208); herbaceous, glabrous; pods very slender, *pendulous*, tipped with a *short* obtuse style; leaves linear, entire; stamens *toothless*. *H. longifolia E. Mey. Herb. Drege, non DC.*

HAB. Mountains. Between Uitkomst and Geelbeks—Kraal; Haasenkraals River, 2–3000f., and Camiesberg, Kl. Namaqualand, *Drege!* Aug. (Herb. Hook. Sond. T.C.D.)

A span long, branched from the base. Lower leaves 2–3 inches long, ½ line wide. Racemes distantly flowered. Pedicels erect, deflexed after flowering. Flowers *small, yellowish*. Pods compressed, 1–1½ inches long, ½ line wide.

7. H. pubescens (Burch. Trav. 1. p. 259); herbaceous, *hairy*; pods patent, compressed, submoniliform, tipped with a short style; leaves pinnati-partite, lobes in 4–5 pairs, linear; stamens toothless. *Sond. l. c. p.* 208. *H. hirsuta, E. Mey. Hb. Drege.*

HAB. Rocky places. Rhinoster river, *Burchell*. South side of Zuurplas Mt. Sneeuwberg, 4–5000ft *Drege*, Aug. (Herb. Burch., Sond., T.C.D.)

A span long, branched from the base. Leaves pectino-pinnate, clothed with white hairs; rachis a line wide in the lower leaves, 2 inches long; in the upper half that length; lobes 3–5 at each side, acute, 4–6 lines long, a line wide, generally opposite. Fl. very small, pale rosy or white. Racemes at length elongate. Fruiting pedicels 1 line long. Sepals 1 line long; petals a little longer. Pods 6 lines long, ¾ line wide: with three-nerved valves. Seeds suborbicular, not margined.

8. H. fistulosa (Sond. l. c. p. 209): herbaceous, hairy at base; pods linear, *elongate*, erecto-patent, submoniliform, compressed, tipped with a filiform style; leaves *narrow-linear*, the lower ones mostly trifid at the points; stem *hollow;* lateral stamens toothed. *Lept. tripartitus E. & Z.* 61. *Hel. sphærostigma, Kunze, Ind. Sem. Hort. Lips.* 1846.

HAB. Sandy places. Heerelogement, Clanw. *E. & Z.* Haasenkraals River, *Drege* 7573*a*. Oct. (Herb. Sond.)

2 feet high, glabrous from the middle to the top, striate, below generally somewhat inflated; branches slender. Lower leaves hairy or glabrous, on long petioles. trifid, rarely undivided; lobes shorter than the 2-inch-long petioles. Racemes elongate. Fl. *blue, rather large.* Pods 3 inches long, 1 line wide. Style a line long, Seeds suborbiculate,

Sect. 2. ORMISCUS (DC.) Pods sessile, much compressed, moniliform, that is, the margins between the seeds sinuately-contracted; the beadings generally one-seeded, orbicular. Seed orbicular, much compressed. Annuals. (Sp. 9–18.)

9. H. amplexicaulis (Linn. f. Suppl. 296); herbaceous, glabrous or somewhat hairy; pods moniliform, spreading; lowest leaves opposite, obtuse, upper alternate *cordato-amplexicaul, oblong or lanceolate, acute*, very entire. *Thunb. Fl. Cap.* 494. *Jacq. fragm.* 49. *t.* 64. *t.* 2. *DC. Syst.* 2. *p.* 682. *Sond. l. c. p.* 210. *Trentepohlia integrifolia Roth. Cat. Bot.* 2. 76. *Orm. amplex. E. & Z. No.* 65.

VAR. β. **grandiflora**; flowers twice as large, *white.*

VAR. γ. **spathulata**; stem weak, 3–4 inches long, leaves sessile spathulate; fl. smaller. *H. spathulata, E. Mey. Herb. Drege.*

HAB. On hills, in sandy soil. Saldanha Bay; Zwartland and elsewhere, *Thunberg.* Brackfontein, Clan, *E. & Z.* Oliphant's R.; between Zwartdoorn R. and Groen river; in Langvallei, and at Leliefontein at the foot of Mt. Ezelskop, 1–5000f., *Drege* 7550. β. Wupperthal, *Drege* 7551. γ. Camiesberg, &c. 3–4000ft. *Drege.* Jul.–Nov. (Herb. Sond. T.C.D.)

1–2 feet high, glaucous. Stem terete, fistular. Lower branches opposite; upper alternate. Leaves 5-nerved on the under side, the lower opposite, subcordate at base, sessile, uncial, 2–3 lines wide; upper 1–2 inches long, 3–4 lines wide. Racemes elongating: pedicels filiform, 8–12 lines, or 4–6 lines long. Flowers *purple, yellowish* or *white.* Pods uncial, tapering to base and apex; style filiform.

10. H. pusilla (Lin. f. Suppl. 295); herbaceous, glabrous; pods moniliform, patent; leaves linear-setaceous, entire. *Thunb. Cap.* 493. *DC. Syst.* 2.684. *Sond. l. c.* 210. *Un. It. No.* 390 *and* 404. *H. tenuisiliqua DC. l. c.* 680. *Deless. Ic.* 2. *t.* 96. *Orm. pusillus and tenuisiliqua, E. & Z.!* 66, 67. *Drege* 7554, 7555.

HAB. Sandy and stony ground; mts. and plains near Capetown, and Stellenbosch, *Thunb., E. & Z. Drege, &c.* Aug.–Sept. (Herb. Thunb., Vahl., Hook., Sond., T.C.D.)

2–12 inches high, stem simple or branched. Leaves scattered, 6–12 lines long, ½–1 line wide. Fruiting racemes 1–3 inches long. Flowers *small, white.* Pods ½–1 inch long, tipped with a short style; the beadings either remote and orbicular, or near together and ovate-orbicular. Seeds with a very narrow margin.

11. H. rivalis (Burch. Cat. No. 5496.); herbaceous, glabrous, green; pods moniliform, *suberect*, tipped with a *moderate* style *thrice as short as the pedicel;* leaves pinnati-partite, lobes rather remote, 3–5 pair, linear, acute; petals obovate; stamens toothless. *DC. Syst.* 2. 682. *Sond. l. c. p.* 213.

HAB. In wet places. Melkout-kraal, Knysna, *Burchell*. Paarlberg, 1-2000f. *Drege* 7577. Nov.–Dec. (Herb. Burch., Sond.)

Stem 1–2 feet high, striate, with spreading branches. Leaves subdistant, their rachis 2–3 inches long, a line wide, with 3–5 opposite or alternate lobes at each side, 3–4 lines between each, 6–12 lines long, acute. Upper leaves smaller, trifid or entire. Racemes 6 inches long. Fl. white, purplish or lilac. Petals twice as long as the calyx. Pods 1½ inches long, a line wide; style filiform 1–2 lines long. Seeds immarginate.

12. H. concatenata (Sond. l. c. p. 214.); herbaceous, glabrous; pods moniliform, *pendulous*, tipped with a *long-filiform* style equalling the pedicel; leaves pinnati-partite, lobes in 5–7 pair, linear, subacute; petals obovate, clawed; stamens toothless. *Drege*, 7576, *a.*

HAB. Rocky and stony places. Paarlberg, *Drege*. Aug.–Oct. (Herb. Sond., T.C.D.)

Stems several, 1–2 feet high, flexuous, with slender branches. Leaves 2–3 inches long, their rachis a line wide, lobes 5–8 lines long, opposite or alternate. Upper leaves trifid or entire. Fl. lilac. Petals twice as long as the calyx. Pods 1½ inches long, with 10–14 ovate-orbicular, often rather remote beadings. Seeds immarginate.

13. H. variabilis (Burch. Cat. 1249.); herbaceous, glabrous, green; pods moniliform, *suberect*, tipped with a *short* style; leaves pinnati-partite, lobes 3–5 pair, linear, acute, the terminal elongated; petals *oblong*; stamens toothless. *DC. Syst.* 2.683. *Sond. l. c.* 214.

VAR. β. **tenuifolia**, leaves subfiliform; pedicels elongate. *H. dissecta*, a, *Herb. Drege.*

HAB. Desert places. Roggeveld, near Jack Riv. *Burchell*. β. near Goedemans-kraal, Rustbank and Kookfontein in Kaus, 3–4000ft. *Drege*. Sept.–Oct. (Herb. Burchell. β. in Herb Sond., T.C.D.)

A foot high, resembling *H. chamaemelifolia*. Rachis of the leaves 2–3 inches long, ½ line wide; the lobes uncial, the terminal one longest. Raceme long; pedicels 5–6 lines long, those of the fruit 7–8 lines. Fl. white or rosy. Pods (young) 1 inch. β. has leaves twice as narrow; fruit pedicels 1 inch long; and twice larger, white flowers.

14. H. pendula (Willd. Sp. III. 529.); herbaceous, glabrous, *green*; pods moniliform, *pendulous*, tipped with a short style; leaves pinnati-partite, lobes 3–5 pair, linear, acute, *channelled* above; petals *obovate*; stamens toothless. *DC. Syst.* 2. *p.* 683. (*excl. syn. Lam.*) *Sond. l. c.* 215. *H. pennata*, *Vent. Malm. t.* 113 (*excl. syn.*) *R. Br. Hort. Kew* 2. *vol.* 4. *p.* 102 (*non. Lin. f.*) *Ormiscus pinnatus*, *E. Z.* 68. *Drege* 7579.

HAB. Grassy fields. Zwartkops and Krakakamma, Uitenhage, *E. & Z.* Swellendam, *Mundt*. Between Brede river and Hassagaiskloof, *Zeyh.* 1898 (Herb. Willd., Sond., T.C.D.)

Stems mostly several, the middle ones erect, 1–2 feet high, lateral spreading, with erecto-patent branches. Leaves somewhat fleshy, the lobes mostly opposite. 6–12 lines long, ½ line wide or shorter. Fruiting racemes 6 inches long: flower stalks erect, fruit-st. recurved. Sepals 1 line long; petals 2 lines, pale yellow Pods uncial, with about 12, rarely confluent beadings.

15. H. monticola (Sond. l. c. p. 216.); herbaceous, glabrous, *glaucous*; pods moniliform *erecto-patent*; lower leaves pinnati-partite, lobes 3–5 pair, linear-subulate; uppermost trifid; *lateral stamens toothed. H. pendula, E. Mey. Herb. Drege.*

HAB. Hills at Mierenkasteel, below 1000f. *Drege*. Aug, (Herb. Sond., T.C.D.)

A slender annual, 4–8 inches high. Stem flexuous, branched from the base.

Leaves 1–1½ inch, lobes opposite, 6–10 lines long; upper similar or rarely trifid, never undivided. Flower stalks 5–6 lines, in fruit 9–12 lines long. Fl. small, *violet.* Pods uncial, a line wide, with a short style. Habit of the preceding; but more nearly related to the following species.

16. H. trifida (Thunb. prod. 108.); herbaceous, glabrous, *green;* pods moniliform, *spreading or pendulous;* lower leaves *trifid* or rarely pinnately 5-parted, with *subfiliform lobes;* upper *entire;* lateral stamens toothed. *Thunb. Fl. Cap. p.* 495. *DC. Syst.* 2. *p.* 683. *Sond. l. c.* 217. *H. pinnata, L. f. Sup.* 297.

HAB. Sandy places. Cape Flats, *Thunb., Preiss, W.H.H., Ecklon.* Assagaiskloof, *Zeyher.* 1902, *ex parte.* Sep.–Nov. (Herb. Thunb., DC., Sond., T.C.D.)

Very like *H. pusilla.* Stem ½–1 foot high, simple or much branched, the branches ascending. Leaves uncial, rarely biuncial, the lobes equalling the petioles or shorter; the middle one longest. Racemes 6–12 flowered. Fl. small purplish, like those of *Arabis verna.* Flower stalks erect, 3–4 lines, in fruit spreading 5–6 lines long. Pods 1–1½ inch long, or shorter; the beadings often confluent, tipped with a short style.

17. H. coronopifolia (Lin. Sp. 927. excl. syn. Pluk.); herbaceous, *velvetty-pubescent,* or in the upper part, glabrous; pods moniliform, patulous or pendulous, tipped with a short style; leaves pinnati-partite, lobes *close together* 4–7 pair, linear, subacute; stamens *toothless. DC. Syst.* 2. 687. *E. & Z.* 79. *Sond. l. c.* 218 *Thunb. in Herb.* (*non Prod. nec Fl. Cap.*)

HAB. Moist places. East side of Devil's Mountain, *E. & Z.* Drakensteenberg, 1–2000f. *Drege,* 7576. *and H. pusilla, c.* Sept. (Herb. Thunb., Sond.)

A small, slightly branched annual; minutely pubescent or somewhat hairy at base. Leaves glabrous or downy, 2-inches long, pectinate, lobes 1–2 lines apart, 3–5 lines long, ½–1 line wide; upper similar, smaller. Racemes 8–16 flowered. Flowers middle size, purplish. Flower stalks 2–3 lines, in fruit 5–6 lines long. Pods uncial, a line wide, style 1 line long.

18. H. dentifera (Sond. l. c. 219); herbaceous, glabrous; pods (young) suberect, submoniliform, tipped with a filiform style; leaves pinnati-partite, the lobes 6–8 pair, *linear, obtuse;* all the stamens toothed, the tooth ciliate.

HAB. Devil's Mt., 2 alt. *E. & Z.* Nov. (Herb. Sond.)

1½ foot high, quite glabrous, resembling the preceding species, but more robust, the leaflobes twice as wide; the flowers larger and white, and the stamens toothed. Lower leaves crowded, 2-3 inches long, petioled, their rachis and lobes 1–1½ lines wide; lobes conspicuously 3-nerved and veiny: the cauline smaller, with narrower lobes. Ripe fruit not known.

Sec. 3. SELENOCARPÆA DC. Pods *oval, or suborbicular, few-seeded,* tipped with a style. Seeds much compressed, orbicular. Glabrous annuals. (Sp. 19–21.)

19. H. diffusa (DC. Syst. 2. p. 685); herbaceous, glabrous; pods oblong, compressed, 4-8 *seeded, much longer than the style;* leaves pinnati-partite, with linear filiform lobes. *Sond. l. c. p.* 220. *t.* 21. *Lunaria diffusa, Thunb. Cap.* 491. *Farsetia dif. Desv. Journ. Bot.* 3. 173. *Selen. diffusa, E. & Z. No.* 71. *Hel. lepidioides, Link. En.* 2. 174.

HAB. Sandy and stony places. Foot of Devil's Mt. *E. & Z.* Roodesand, between Nieuwekloof & Slangenteuvel. *Drege* Aug.–Oct. (Herb. Thunb. R. Berl. Sond. T.C.D.)

Stems decumbent, a foot or more long, weak, with filiform, erect branches. Lower leaves 1–2 inches long, on each side with 2–3 linear patent lobes, 4–6 lines long. Upper subsessile, similar or trifid. Racemes lax. Flowers small, white. Fruit stalks 2–3 lines long. Pods 3–5 lines long, 1½ lines wide, their style ½–1 line long; valves obtuse at each end, one nerved, veiny. Seeds 4–8, rarely 3 in each pod, with a very narrow margin.

20. H. Peltaria (DC. Syst. 2. 685); herbaceous, glabrous; pods *sessile, oval-orbicular*, compressed, 2-3 *seeded*, about *twice* as long as the filiform style; leaves pinnati-partite, with linear lobes. *Sond. l. c. p.* 222. *t.* 22. *f.* 2. *Peltaria capensis, L. f. suppl.* 296 *(non Thunb.) Lunaria pinnata. Thunb. Fl. Cap.* 491. *Aurinia capensis, Desv. l. c. p.* 162. *and Farsetia pinnata, p.* 173. *Selenocarpæa pinnata, E. & Z. No.* 70.

HAB. Stony and rocky places. Among the *Silver Trees* on Devil's Mt., *E. & Z. Drege.*, Hills round Capetown, *W.H.H.* Sept.–Oct. (Herb. Thunb., Sond., T.C.D.)

Smaller and more slender than the preceding, with much shorter pods. Stems 6 inches long. Leaves, racemes and flowers as in *H. diffusa.* Pods 2 lines long, 1½ line wide; style 1 line. Seeds 2, rarely 3.

21. H. flacca (Sond. l. c. p. 223. t. 22. f. 1.); herbaceous, glabrous; *pods on short pedicels*, ovate, *somewhat swollen, two-seeded*, tipped with a style of *equal* length to the pod; leaves pinnati-parted, lobes linear-setaceous. *Selen. Peltaria, E. & Z.* 69. *(excl. Syn.) Zeyher* 1894.

HAB. Among shrubs, 2 alt. Caledon Baths, Zwarteberg, *E. & Z.* Aug. (Herb. Sond.)

Easily known from the preceding, by its more slender, mostly erect stem, setaceous leaves and inflated pedicellate pods. Lower leaves petiolate, their rachis uncial, lobes 3–5 at each side, opposite or alternate. Flowers small, white. Pods 1½ lines long, a line wide, valves ventricose; cells one seeded.

SEC. 4. ORTHOSELIS, DC. Pods sessile or subsessile, compressed, linear, with perfectly ***straight*** or ***sub-sinuate***, (rarely somewhat ***moniliform***) margins; tipped with a style. (Sp. 22–57.)

Sub-section 1. *Stems herbaceous, annual.* (Sp. 22–39.)

22. H. latisiliqua (E. Mey. Herb. Drege); herbaceous, glabrous, or hairy at base; pods subpendulous, oblongo-linear, one-nerved, netted-veined, tipped with a short, filiform style; leaves somewhat fleshy, pinnati-partite, lobes 3–4 pair, filiform, ***close together;*** stamens toothless. *Sond. l. c. p.* 224. *Lunaria elongata, Thunb. Cap.* 492. *Carpopodium Thunbergii, E. & Z. No.* 103.

HAB. Stony and sandy places. Between Verlooren Valley and Lange Kloof, *Thunberg.* Tulbagh Kloof, and Witsenberg Mt., *E. & Z.* Between Hexriverberg and Bokkeveld, 3-4000f. *Drege* Sep.–Oct. (Herb. Thunb., E. Mey., Sond.)

A foot high. Stem short, branches flexuous, ascending, elongate, simple, or slightly branched, with few leaves. Leaves pectinate; their obtuse rachis uncial, ½ line wide, the lobes 2–3 lines long, ½–1 line apart. Cauline leaves smaller. Racemes few-flowered: fl. stalks 4–6 lines, in fruit 8–11 lines long. Petals *purplish,* 2 lines long. Pods sessile or shortly stalked, 1–1½ inches long, 3 lines wide, often purplish, with a broad, green medial line. Seeds broadly winged.

23. H. macrostylis (E. Mey. in Herb. Drege); herbaceous glabrous; pods pendulous, oblong-linear, one-nerved, somewhat veiny, tipped with *a thick, cylindrical* style; leaves pinnati-partite, lobes 3–4 pair, linear, elongate, ***remote; lateral stamens toothed.*** *Sond. l. c. p.* 225.

HAB. Between Kooperberg and Silverfontein. *Drege.* Sep. (Herb. E. Mey., Sond.)

Differs from the preceding chiefly by its taller stem ($1\frac{1}{2}$-2 feet high); larger and more numerous leaves, but particularly by the *thickened* style. Root leaves none; cauline somewhat glaucous, $3\frac{1}{2}$–4 inches long, their rachis $\frac{3}{4}$ lines wide, acute, the lobes 1-$1\frac{1}{2}$ inches long, $\frac{1}{2}$ line wide. Pedicels as in preceding. Petals *white,* $2\frac{1}{2}$ lines long, pods $1\frac{1}{2}$ inches long, 4 lines wide; style $1\frac{1}{2}$–2 lines long, $\frac{1}{2}$–$\frac{3}{4}$ line wide.

24. H. Meyeri (Sond. l. c. p. 226); herbaceous, glabrous; pods *linear,* one-nerved, spreading or pendulous, *not tapering at base,* tipped with a *filiform* style; leaves pinnati-partite, the lobes 4–5 pair, pectinate, linear, subdistant; lateral stamens *toothed. H. pectinata, E. Mey., non Burch.*

HAB. Stony places in the shade. Near the river at Gnadendhal, 2-3000f. *Drege.* Oct. (Herb. E. Mey., Sond.)

Stems several, erect or ascending; branches erecto-patent. Root leaves crowded, their rachis 3–4 inches long, acute, a line wide; lobes sub-remote, 4–6 lines long, a line wide. Upper stem leaves trifid or entire. Flowers mediocre, *yellowish,* or *white.* Pods uncial, a line wide, valves with a raised medial line and two lateral, obsolete nerves. Style 1-$1\frac{1}{2}$ lines long.

25. H. pectinata (Burch. Trav. 1. p. 260); herbaceous, glabrous; pods *narrow-linear, three nerved, attenuate at base,* pendulous; leaves pinnati-partite, the lobes 3–5 pair, linear, *near together;* stamens *toothless. DC. Syst. 2. p.* 688. *Sond. l. c. p.* 227. *H. inconspicua, E. Mey. Herb. Drege.*

HAB. In the Roggeveld, near Riet river, *Burchell.* Mts. between Hexriviersberg and Bokkeveld, in stony places, 3–4000f. *Drege.* Aug.–Sep. (Herb. Burch., E. Mey., Sond., T.C.D.)

A delicate little plant, 3–6 inches high, much resembling *H. Peltaria* in habit, but with very different fruit. Stem erect, flexuous, with spreading branches. Root leaves linear-spathulate and trifid; the rest pinnati-partite, the lobes acute, opposite, $1\frac{1}{2}$–2 lines apart; 2–3 or 5 lines long, $\frac{1}{2}$–1 line wide. Upper leaves trifid or entire. Racemes distantly flowered; fl. stalks 2 lines long. Flowers *very small, white.* Pods 6–8 lines long, $\frac{1}{3}$–$\frac{1}{2}$ line wide, with a very short style. Seeds immarginate.

26. H. crithmifolia (Willd. En. 2. 682); herbaceous, *velvetty pubescent;* pods linear, one-nerved, pendulous, *opaque,* tipped with a very short style; leaves pinnati-partite, somewhat fleshy, the lobes 2–4 pair, subremote, *semiterete, furrowed above;* stamens toothless. *DC. Syst.* 2. 689. *Del. Ic.* 2. *t.* 97. *Sond. l. c. p.* 228. *H. seselifolia, E. & Z.* 81. *ex parte. Sisymb. crithmifolium, Roth.*

VAR. β. **parviflora,** Burch: fl. smaller, white.

VAR. γ. **lævis,** Sond.; stem taller, velvetty or pilose below, glabrous in the upper part; leaf-lobes 1–$1\frac{1}{2}$ inch long, glabrous. *H. lævis, E. Mey. Herb. Drege.*

HAB. In sandy fields and hillsides. Roggeveld Karroo, near Tuch river, *Burchell.* Oliphant's river, *E. & Z.* Hassagaiskloof, and R. Zonderende, *Zeyher* 1901 *ex parte.* Zilverfontein (3165) and Mieren Kasteel, *Drege.* Aug.–Sep. (Herb. Willd., E. Mey., Sond.; β. in Herb. T.C.D.)

6-12 inches high or more. Leaves 2–3 inches long, their rachis linear, channelled, acute. Racemes elongating. Pedicels glabrous, 5–6 lines long. Fl. small, *purple,* or in var β. *white;* in var. γ. *yellowish.* Pods 15–20 lines long, $1\frac{1}{2}$–2 lines wide, valves marked with a raised, medial line. Seeds narrow margined.

27. H. chamæmelifolia (Burch. Trav. 1. p. 222, 225); herbaceous, *hairy below, glabrous above;* pods linear, erect or spreading, one-nerved, *shining,* tipped with a very short style; leaves pinnati-partite, fleshy, the lobes 2–4 pair, subremote, *linear*; stamens toothless. *DC. Syst.* 2. *p.* 689. *Sond. l. c. p.* 229. *H. seselifolia, E. & Z. pro parte. H. crithmifolia Drege, ex parte.*

HAB. Roggeveld Karroo, at Ongeluch and Tuchrivers, *Burchell.* Fields near, Oliphants R., *E. & Z.* Hills at Mierenkasteel, *Drege.* Jul.–Aug. (Herb. Burch. Sond. T.C.D.)

Differs from the preceding by the stem, hairy in the lower part; the longer and shining pods and the smaller seeds: perhaps a mere variety? Flowers lilac or whitish. Pods 1½–2 inches long, 1½ lines wide.

28. H. fœniculacea (R. Br. Hort. Kew. Ed. 2. vol. 4. p. 100); herbaceous, somewhat hairy; pods linear, one-nerved, spreading, tipped with a very short style; *leaves pinnately or bipinnately-parted, the lobes filiform, elongate; lateral stamens toothed. DC. Syst.* 2. *p.* 689. *Sond. l. c. p.* 230. *H. seselifolia, E. & Z. No.* 81. *ex parte.*

HAB. In fields, near Brackfontein, Clanw., *E. & Z.* Near Simonstown, *Mr. C. Wright* 75. Aug. (Herb. Sond. T.C.D.)

Stems 1–1½ foot high, erect, branched, mostly glabrous at top. Leaves hairy or the upper glabrous, their very slender lobes 8–12 lines long; cauline leaves often trifid. Racemes 6–8 flowered. Flower-stalks erect, 4–6 lines; in fruit 8–12 lines long, glabrous or minutely pubescent. Fl. *purplish,* small. Pods 1½ inches long, a line wide; margins parallel.

29. H. gracilis (Sond. l. c. p. 230); herbaceous, velvetty; pods *subterete, pendulous,* tipped with *a long, subulate style equalling the pedicel*; leaves *sessile*, pinnati-partite, lobes 4–6 pair, capillary; lateral stamens toothed at base. *Carponema aggregata, E. & Z. No.* 55.

HAB. Sandy places. Vogelvalley and Swartland; and at Berg River, *E. & Z.* Sep. (Herb. Sond.)

Very slender, 1–1½ feet high, branched from the middle, branches virgate. Leaves uncial, the lower often trifid, the rest pinnatisect; lobes opposite or alternate, 3–5 lines long, the lower ones closer together, and shorter. Racemes 10–20 flowered. Pedicels 2–3 lines long. Flowers purplish small; calyx downy. Pods 1½ inches long, ½ line wide, scarcely compressed; style 2–3 lines long. Seed minute, oval.

30. H. viminalis (E. Mey. in Hb. Drege.); herbaceous, glabrous and glaucous; pods *erect*, narrow-linear, *three-nerved* tipped with *a short conical style*: leaves *sessile or minutely petiolate,* pinnati-partite; lobes 3–6 pair, suberect, opposite or alternate, filiform, thickish; lateral stamens toothed, seeds ovate immarginate; *stem solid. Sond. l. c. p.* 231.

HAB. Rocky, mountain situations. Nieuwekloof, 1-2000f., on a sandy, moist soil. *Drege.* Oct. (Herb. E. Mey., Sond.)

Stems several, 1½ feet high, virgate; the lateral ascending, leafy, branched. Lower leaves 1 inch long or more, the lobes 3–6 lines long; rameal similar, but smaller. Racemes 6–12 inches long; pedicels 3–4 lines long. Fl. middle size, pallid. Pods 1½–2 inches long, ½ line wide, compressed; valves subtorulose, the style ½–1 line long.

31. H. tenuifolia (Sond. l. c. p. 232); herbaceous, glabrous; pods *erect,* narrow-linear, subtrinerved, tipped with *a filiform cylindrical style;* leaves on *long petioles,* pinnati-partite, the lobes 2–4 pairs, distant, subfiliform, erect, alternate; lateral stamens toothed, seeds *ovate, immarginate*; stem *hollow. H. fœniculacea, E. & Z. No.* 80. (*non R. Br. Lept. pendulus, E. & Z. No.* 62. *excl. syn.*

HAB. Sandy places. Cape Flats; Berg R.; and between Pot R. and Langhoogde, Caledon, *E. & Z.* Rietvalley, *Mundt and Maire.* Aug–Sep. (Herb. R. Berl., Sond.)

Stems straightly erect, terete, 1–1½ ft., slender, leafy, with erect branches. Leaves 2–2½ inches long; the lobes remote, 4–8 lines long: the rameal leaves mostly linear and undivided. Racemes lax: pedicels 3–4 lines long. Flowers small, purple. Pods 2 inches long, ½ line wide; valves subtorulose, style 1–1½ lines long.

32. H. seselifolia (Burch. Trav. I. p. 258.); herbaceous, glabrous, and glaucous; pods *sub-pendulous*, linear, *one-nerved*, tipped with a short style; leaves petiolate, pinnati-partite, the lobes 3–5 pair, patent, linear-subulate; lateral stamens toothed at base; seeds *orbicular, margined. DC. Syst.* 2. 684. *Sond. l. c.* 233. *H. coronopifolia, var. β. Lam. Dic.* 3. 90. *Illust. t.* 562. *f.* 2. *H. dissecta β. E. Mey.*

HAB. Roggeveld, near Jackalsfontein, *Burchell.* Between Koussiè and Zilverfontein, and between Z. fontein, Kooperberg and Kaus, 2–3000f. *Drege* 7580. Aug.–Sep. (Herb. Burch., Sond.)

6 inches high, slender, branched above. Lower leaves crowded, uncial, glaucous, the petiole 3–6 lines long; the lobes subfiliform, 3–5 lines long. Upper leaves often undivided. Racemes 8–12 fl. Pedicels 4–5 lines long; the fruiting ones rather longer. Flowers white or yellowish, mediocre. Pods horizontal or pendulous, uncial, ½ line wide; valves compressed; style ½–1 line long, the stigma discoid.

33. H. refracta (Sond. l. c. p. 234); herbaceous, glabrous, glaucous; pods elongate, linear, *very narrow, deflexed*, tipped with a short, obtuse style; leaves petiolate, pinnati-partite, lobes 2–3 pairs, subfiliform; lateral stamens toothed at base; the *narrow sepals horned at the apex; seeds ovate, immarginate. Lept. rivalis, E. & Z. No.* 63. *(excl. syn.)*

HAB. Sandy places. Cape Flats, near Doornhoogde, *E. & Z.* Aug. (Herb. Sond.)

1–2 feet high. Stems flexuous, leafy at the flexures; the branches slender. Leaves 2 inches long; petiole uncial; lobes suberect, 6–10 lines long; the upper undivided. Racemes 10–16 fl.; flower stalks suberect, 4–7 lines, in fruit refracted, 7–8 lines long. Fl. middle size, blue, with yellow centre. Pods 2–2½ inches long; stigma discoid. Seeds 40–50 in each pod. Remarkable for its very long, very slender and often submonilifirm pods, uniting the sections *Leptormus and Orthoselis.*

34. H. trifurca (Burch. Cat. No. 1487); herbaceous, glabrous; pods linear, deflexed; leaves *very narrow, three forked*, lobes linear, entire. *DC. Syst.* 2. 688. *excl. syn. Sond. l. c. p.* 235.

VAR. *β.* leaves trifid and pinnati-sect.

VAR. *γ.* **parviflora**, flowers twice as small, whitish. *H. pectinata E. & Z.* 78.

HAB. Near Sack River, *Burchell.* Winterhoeksberg, 1–2000f. *Drege* 7575. *β.* at Bitterfontein, Boscheman's Land, *Zeyher.* 42. *γ.* Table Mt., *E. & Z.* (Herb. Burch., Sond.)

Somewhat glaucous. Stems 1–2 feet high, striate, with virgate branches. Lower leaves trifid, rarely bifid or pinnatipartite; the lobes equalling the petiole, uncial, the middle lobe mostly longest. Upper leaves divided or undivided. Racemes lax; fl. stalks 5–6 lines long. Flowers purplish or lilac. Stamens toothless. Pods 1–1½ lines wide; valves one-nerved, with a very short style.

35. H. divaricata (Herb. Banks.); herbaceous, glabrous; pods (not known); leaves *linear, very entire;* branches spreading. *DC. Syst.* 2. 687. *Sond. l. c. p.* 236.

HAB. S. Africa, *Masson.*

Glabrous, resembling *Lepidium graminifolium.* Stem herbaceous, (?) terete, dividing into several, spreading, filiform branches. Leaves scattered, acute, 8–9 lines long, nearly a line wide. Pedicels filiform, 2–3 lines long. Flowers small, *yellowish?* Calyx spreading; sepals brown tipped.—Near *H. incisa*, but differs; in the undivided leaves and flowers half the size. *(DC. l. c.)*

36. H. stricta (Sond. l. c. p. 236); herbaceous, glabrous, cæsious; pods linear-elongate, *erect, appressed, 3–5 nerved;* leaves *from a broad base linear-lanceolate* or linear, *quite entire;* the stem tall, and branches virgate. *H. divaricata, E. & Z. No. 76. non. D.C. Drege,* 7549, 7571.

HAB. Sandy places. Mountains near Bergvalley, Worcester, *E. & Z.* Ribeckskasteel, on Mts. under 1000f. *Drege.* Cape Flats, at Kardow, *Zeyher* 48.; and 46 and 47 *ex parte.* Nov.–Jan. (Herb. Sond., T.C.D.)

Stems 2–4f. high, sparingly leafy, simple below. Leaves sessile on a broad base, obsoletely 3–5 nerved; the lower 2–4 inches long, 1–4 lines wide; the rameal smaller. Racemes often 1–2 feet long: flower stalks 3–4 lines long, often hairy; fruit stalks 5–6 lines long, glabrous. Fl. rather large, blue. Pods straight, 3–4 inches long, a line wide; the valves compressed; style 1–2 lines long.

37. H. linearis (DC. Syst. 2. p. 697); herbaceous, glabrous, or hairy at base; pods *spreading, linear, one-nerved* or obsoletely three-nerved; leaves *lanceolate, narrowed at both ends,* the upper ones linear. *Sond. l. c. p.* 238. *Cheiranthus linearis, Thunb. Cap. p.* 493. *H. falcata, E. & Z. No.* 77. *H. glauca β. E. & Z. No.* 91.

HAB. Cape. *Thunberg, Bergius.* Sandy hills at Zwartkops River, in Adow, at Sondag river, Uit., *E. & Z., Zeyher,* 1905. *Sieb. No.* 244. (Herb. Thunb., Sond.)

1–2 feet high, with virgate branches. Lower and medial leaves 1–1½ inches long, 1–2 lines wide; upper about an inch long. Racemes 3–5 inches long. Fl. stalks 4–5 lines long. Petals blue or pale purple, twice as long as the calyx. Shorter filaments toothed. Pods 1–1½ inches long, 1 line wide, compressed; style 1 line long. Seeds oval. Possibly *H. linearifolia,* Burch. (See No. 51) should be united with this species.

38. H. graminea (DC. Syst. 2. 697); herbaceous, glabrous; pods (young) linear, compressed, erecto-patent; lower leaves aggregate, *linear oblong, narrowed at base into long petioles;* cauline scattered, short, linear. *Sond. l. c. p.* 239. *Cheiranthus gramineus, Thunb. Cap.* 493.

HAB. Onderste Roggeveld, *Thunberg.* Oct.–Nov. (Herb. Thunberg.)

The specimen in Thunberg's Herbarium is a foot high, unbranched. The lower leaves are crowded, 3–4 inches long, 2–2½ lines wide near the apex; the cauline are 6–10 lines long. Flowers of middle size. Pedicels 6–8 lines long. The young pod is ½ inch long. This and *H. divaricata,* Banks, are perhaps merely glabrous varieties of *H. pilosa.*

39. H. pilosa (Lam. Dic. III. 90.); stems herbaceous, *rough with spreading hairs;* pods linear, erect or spreading; leaves *hairy,* either oblong or linear, entire, or sometimes lobed near the apex, cuneate at base. *DC. Syst.* 2. 686. *Sond. l. c. p.* 239. *H. rostrata, Presl. Bot. Bem. p.* 10.

VAR. *α,* **integrifolia**; leaves oblong or linear, entire. *Cheiranthus Africanus, L. Amoen. 6. p.* 90. *Hel. integrifolia, Lin. Sp.* 926. *excl. Pluk. Jac. Ic. Rar.* 1. *t.* 506. *Lamk. Illust. t.* 563. *f.* 1. *E. & Z. No.* 72. *H. integrifolia, and H. incana, Thunb. Cap.* 494. *H. stricta, Bot. Mag. t.* 2526.

VAR. *β.* **digitata**; lower leaves ovate, upper oblong, trifid at the point or palmately 5-lobed. *H. digitata, Lin. f. Thunb. Cap.* 495. *DC. Syst.* 2. *p.* 686.

VAR. γ. **incisa**; leaves linear-cuneate, trifid at the point, rarely 5-fid, the lobes linear or acuminate. *H. arabioides Bot. Mag. t.* 496. *E. & Z.* 73. *H. pilosa β. incisa. DC. l. c.*

VAR. δ. **glabrata**; stem suberect, glabrous above; leaves glabrous, oblong, cuneate at base, 3–7 lobed, the lobes narrow, acute. *H. incisa, Hb. Banks. DC. l. c. p.* 687. *E. & Z. No.* 74. *Var. α & γ.*

VAR. ε. **debilis**; stem weak, somewhat hispid below, divided into many long, filiform, glabrous, spreading branches; leaves oblong-linear or linear, entire, rarely trifid at the point. *H. incisa, β. E. & Z.* 74.

HAB. Sandy places. Common throughout the Cape District, *Thunberg, E. & Z. Krauss, W.H.H.* &c. Paarl, and other places in Stellenbosch, *Drege.* 7548, 9502. Vars δ, ε, at the Zwarteberg, Caledonbaths, *E. & Z.* Aug.–Feb. (Herb. Thunb., R. Berol., E. Mey., Lehm., Meisn., Sond., T.C.D.

A very variable species. Stem 6 inches to 1–2 feet high, erect or diffuse, simple or branched from the base. Lower leaves often opposite; the rest alternate, ovate, oblong, spathulate or lanceolate, ½–3 inches long, 2–6 lines wide. Racemes lax, at length elongated. Flower stalks 3–4, fruit stalks 6–8 lines long. Flowers rather large, *sky blue*, with a yellow centre. Pods at length smooth, 1–1½ inches long, the valves somewhat convex or compressed, three-nerved; style thick, short, or longish, subcylindrical. Var. δ. is more slender, glabrous toward the apex; the leaves 6–12 lines long, 2–3 lines wide, the flowers *lilac or yellow*. Var. ε. (perhaps the same as *H. divaricata*, DC.) has leaves 6–8 lines long, a line wide, and glabrous; smaller, *lilac or blue* flowers and a conical style.

Sub-section 2. *Stems suffruticose or frutescent: perennial.* (*Sp.* 40–57.)

40. H. brassicæfolia (E. & Z. No. 89); suffruticose, glabrous, pods (young) linear, *subreflexed;* leaves *amplexicaul, oblong-lanceolate,* at the tip *cucullate, contracted,* mucronate. *Sond. l. c. p.* 242.

HAB. Mountains, alt. iv. near Siloh, Tambukiland, *E. & Z.* Dec. (Herb. Sond.)

Stems erect, simple, 1½–2 feet high, terete, rather glaucous, leafy. Leaves 2 inches long, 6 lines wide, somewhat fleshy, with a raised midrib. Raceme elongate; fl. stalks 4–6 lines long, fr. st. remote, an inch long. Sepals obtuse, 2½ lines long, 1 line wide. Petals white, twice as long as calyx. Pods (unripe ones only known) an inch long, 1 line wide, with one-nerved valves, and a short style.

41. H. reticulata (E. & Z. No. 90); suffruticose, glabrous; pods *erect,* elongate, linear, three-nerved, tipped with a short style; leaves *eared and amplexicaul, ovate, acute, flat,* netted with veins. *Sond. l. c. p.* 243.

HAB. Sand Downs, near Cape L'Agulhas, Zwellendam, *E. & Z.* Nov. (Herb. Sond.)

Nearly allied to the preceding, but differing in its branching stem and leaves twice as broad, ovate-acute, not fleshy. Stem more than a foot high. Leaves with two small ear lobes clasping the stem, 1½–2 inches long, nearly an inch wide. Flowers not seen. Pods 4 inches long, 1¼ line wide, with 3 nerved valves, the middle nerve more prominent.

42. H. abrotanifolia (Herb. Banks.); suffruticulose, glabrous; pods linear, spreading; *leaves 3–5 lobed,* the lobes subulate, short; stem scapelike, nearly leafless. *DC. Prod.* 2. 690. *E. & Z. No.* 87. *Sond., l. c. p.* 244. *t.* 2?. *f.* 1. *H. chamæmelifolia & H. crithmifolia, E. & Z.* 86 & 88.

VAR. α., **tripartita**; leaves mostly trifid. *H. tripartita, Thunb. Cap.* 495.

VAR. β. **heterophylla**; leaves mostly pinnati-partite. *H. heterophylla, Thunb. l. c.*

VAR. γ. **tenuiloba**; leaf-lobes very slender.

HAB. Swartland, *Thunberg*. Mts. near Capetown. Langhoogde, Caledon; Vanstaadensberg, Uitenhage. Grahamstown and Fish R., *E. & Z.* Nieuwe Hantam, 4-5000f. *Drege*, 7564. γ. Hassagaiskloof, *Zeyher, No.* 1801. Feb.-Oct. (Herb. Thunb. Sond., T.C.D.)

Several stems from the same crown, a foot or more high, simple or branching. Leaves crowded below, erect, rigid, brittle when dry, filiform, some few undivided, the majority lobed; lobes 2-4 lines long, acute; petiole 1-2 inches long. Fruit pedicels 6 lines long. Petals lilac or flesh-coloured, $3\frac{1}{2}$ lines long. Pods 15-20 lines long, $1\frac{1}{2}$ lines wide, compressed, one-nerved, with a short style.

43. H. glauca (Burch. Cat. Geogr. 4782); suffruticose, glabrous, *glaucous; pods linear-oblong*, erect; leaves *oblong, rather thick*, the lower obtuse, upper acute, *DC. Syst.* 2. *p.* 698. *Sond. l. c. p.* 245. *Carpopodium cleomoides, EZ.* 102 (excl. syn.) *H. sarcophylla, Meisn. Lond. Journ. Bot.* 1. *p.* 463.

VAR. α **candida**; flowers white. Burch. l. c.

VAR. β. **purpurascens**; flowers purplish. Burch. l. c. 4969.

HAB. Near Loeri river, *Burchell*. Loeri and Kamtou river, Uit, *E. & Z.* Langekloof. George, *Krauss; Burchell*. Mar.-Dec. (Herb. Meisn., Sond., T.C.D.).

A shrub, 2 feet high. Stem as thick as pigeon's quill, greyish, with virgate branches. Leaves scattered or crowded at the base of the branches, the lower mostly broader, 8-10 lines long, 2-3 lines wide; upper narrow. Racemes long. Pedicels very erect, 2-3 lines long. Petals 2-3 times as long as the calyx. Stamens toothless. Pods 9-12 lines long, 2 lines wide, the valves one-nerved; style 1-2 lines long.

44. H. cornuta (Sond. l. c. p. 246. t. 28); fruticulose, glabrous; pods submoniliform, *pendulous*, tipped with a *very slender, incurved style;* leaves linear-filiform, acute; sepals horned at the apex. *Lept. longifolius, E. & Z. No.* 59. *excl. Syn. H. scoparia,* c, *E. Mey. Herb. Drege (non Burch.)*

HAB. Sandy places, on mountains at Brackfontein, Clanw., *E. & Z.* Wupperthal, *v. Wurmb, Drege*. Heerelogement, *Zeyher* 44. Jun.-Jul. (Herb. Sond.)

Stem 1 foot or more high, with spreading branches. Leaves 2-3 inch long, upper shorter. Pedicels 3-4 lines long; lax. Pods 2-$2\frac{1}{2}$ inches long, a line wide, the valves tapering at each end, one-nerved.

45 H. elata (Sond. l. c. p. 247); shrubby, glabrous; pods *erecto-patent*, narrow-linear, elongate, submoniliform, *tipped with a filiform straight style;* leaves linear-subulate.

HAB. Sandy hills. Ebenezer, below 500 f.; between Zwartdoorn river and Groen river. *Drege* 7566. Kardow, alt. 3, and Brandenberg, *Zeyher* 46 *and* 47 *partim.* Aug.-Jan. (Herb. Sond. T.C.D.).

Stem somewhat woody; branches 2 feet long, with spreading branchlets. Lower leaves $1\frac{1}{2}$-2 inches long; upper gradually smaller. Racemes lax; fr. st. 4 lines long. Flowers blue, of middle-size. Pods $2\frac{1}{2}$ inches long, scarcely $\frac{1}{2}$ line wide, valves one-nerved. Seeds minute, oval, margined.

46. H. suavissima (Burch. Cat. 2742); suffruticose, glabrous; pods *sublinear, tapering at base*, spreading, tipped *with a short style;* leaves linear-subulate, subacute. *DC. Syst.* 2. *p.* 291. *Sond. l. c. p.* 248. *t.* 25. *Zeyher No.* 1900.

VAR. β. **incana**; stem and leaves minutely downy, afterwards glabrous.

HAB. Grassy fields and on hills. Plettenberg's Bay, *Burchell.* Zwartkops R. Uit; and at Graaf Reynet, *E. & Z.* Somerset, *Mrs. Barber.* Gamke R, *Burke & Z.* Klaarstroom, Zwarteberg, *Drege.* β. Springbokkeel, *Zeyher.* Aug. Oct. and Feb. (Herb. Hook., Sond., T.C.D.).

Stem 1–2 feet high, branched at base, branches virgate, leafy in the lower part. Leaves 1–2 inches long, ½ line wide, often with axillary fascicles. Racemes 6 inches long, naked. Fruit stalks spreading. Flowers very sweet, rather large, violet or purple. Sepals 3 lines long. Style 1–2 lines long. Pod 1–1½ inches long, 1½ lines wide, with compressed valves, subsinuate, rarely straight-edged, somewhat three-nerved. Seeds 12–16 in each pod.

47. H. succulenta (Herb. Banks); suffruticose, glabrous; pods *linear, pendulous,* tipped with a short style; leaves fleshy, filiform-linear, *furrowed, obtuse. Sond. l. c. p.* 249. *Cheir. carnosus, Thunb. Cap. p.* 493. *H. platysiliqua, R. Br. Hort. Kew. Ed.* 2. *V.* 4. *p.* 99. *DC. Syst.* 2. *p.* 692.

HAB. On the shore, Verloren Valley, *Thunberg.* In forests, Krakakamma, Uit., *E. & Z.* Oct.—Jan. (Herb. Thunb., Sond.).

Stem a foot or more high, with long, nearly leafless branches. Leaves scattered or crowded, 1–1½ inches long, a line wide, semiterete, channelled, sometimes twice as wide near the point, and flattish. Racemes 6 inches long: fruit stalks uncial. Flowers purple, nearly as in preceding. Pod on a short stipes, 1½ inches long, 2 lines wide, obtuse at base; the valves flattish, with a prominent central nerve. Seeds 6–8 in each pod. Differs from *H. suavissima* in its fleshy leaves, and *stipitate* pods, not tapering at base.

48. H. subulata (Burch. Cat. 6214); fruticulose, *minutely downy;* pods linear, spreading, not tapering at base, tipped with a short style, leaves *linear-subulate, very acute. DC. Syst.* 2. *p.* 691. *Sond. l. c. p.* 250 *t.* 26. *H. subulata & H. pubescens, E. & Z.* 83 & 84. *Zeyher* 1910.

VAR. β. **glabrata**; leaves glabrous or nearly so. *Sond. l. c. t.* 26. *f. H. maritima, E. & Z.* 85.

HAB. Sandy or stony places. Hartenbosch, Mosselbay, *Burchell.* Between Breede and Duivenhoeks Rivers, Swell; and near Gauritz river and Langekloof, George; also Oliphantshoek, Uit.; and Zwarteberg, Caledon; and near Simonstown, *E. & Z.* Palmiet river; Howhoek; and near Bethelsdorp, *Drege.* 7560, 7565, 7570. Tulbagh, *Lichtenstein.* Port Natal, *Krauss.* β. Port Elizabeth, &c.; and in Swellendam. May–Dec. (Herb. Burch., r. Berol., Sond., T.C.D.).

Stem slender, 1–2 feet high, leaves ½–1 inch long, flattish or subterete, downy, or in β. glabrous. Flowering branches long and naked, ending in a many flowered raceme. Pedicels 4–6 lines long. Flowers purple. Pods spreading, in β. pendulous, 1½ inches long, 1 line wide, obtuse at each end; the valves compressed, three-nerved, the lateral nerves obsolete.

49. H. rigidiuscula (Sond. l. c. p. 251. t. 27); suffruticose, glabrous, *simple;* pods *pendulous, lanceolate, narrowed at base, tipped with a long, beak-like style;* lower leaves linear, upper filiform, acute. *H. virgata, E. & Z.,* 97 *(non Burch.) H. subulata and H. suavissima, E. Mey. Hb. Drege; No.* 5215, 3630.

HAB. Hills, near the Witte and Zwart Key Rivers, Tambukiland; also Wind-Vogelsberg, Caff., *E. & Z.* Kat and Klipplaats riv.; and Omtendo, *Drege,* May–Nov. (Herb. E. Mey, Sond., T.C.D.).

Glaucous. Stem 2 feet high, straight, strongly striate. Leaves thickish, 2–3 inches long, the lower ½–1 line wide, the rest very narrow. Raceme simple. Fl. rather large, violet. Fruit stalks 1 inch long. Pods 2 inches long, 2–2½ lines wide, beaked with a style 6–7 lines long; the valves flat, three-nerved, the lateral nerves obsolete. Seeds 4–8 in each pod.

50. H. fascicularis (Herb. Banks); fruticulose, glabrous; pods linear, sub-erect, *scarcely longer than the pedicel;* leaves filiform. *DC. Syst.* 2. *p.* 691. *Sond. l. c. p.* 252.

HAB. Cape of Good Hope, *Masson.*

Stems terete, erect, branched. Leaves erect, alternate, inch long, ½ line wide, subacute, with tufts in the axils. Racemes elongating. Pedicels filiform, uncial, erect both in fl. and young fruit, scarcely somewhat spreading. Pods compressed, linear, a line wide, about 1 inch long, with a pointed style.—Perhaps a mere variety of *H. linearifolia.*

51. H. linearifolia (Burch. Cat. No. 347 and 793); suffruticose, glabrous, or sparsely hairy; pods *erect, linear, three-nerved,* tipped with a subulate style; leaves linear, acute, entire; *lateral stamens toothed. DC. Syst.* 2. *p.* 692. *E. & Z. No.* 92. *Sond. l. c. p.* 252. *Cheir. elongatus, Thunb. Cap.* 493. *H ? elongata, DC. l. c. p.* 697. *E. & Z.* 93. *H. fascicularis, E. & Z.* 94. *Un. It.* 386.

VAR. β. **pilosiuscula**; stems sparsely hairy. *H. linearifolia, β hirsuta Burch. DC. H. platysiliqua, E. & Z.* 95.

VAR. γ. **lanceolata**; lower stem leaves linear-lanceolate. *H. linearifolia. E. Mey. Herb. Drege. No.* 122.

VAR. δ. **filifolia**; all the leaves linear-filiform, or the lower linear. *H. filifolia, Thunb. Cap.* 494. *Seiber, No.* 244.

HAB. Cape flats, near the shore at west side of Table Mt.; Langekloof, George; and Zwartkops R., Uit., *E. & Z.* β. at the Knysna, *Burchell.* Tulbagh, *E & Z.* γ, on Table Mt., *Drege.* δ. Swartland, *Thunberg.* Cape District, *Sieber, Bergius, W.H.H.* (Herb. Thunb., DC., Meisn., Sond., T.C.D.).

Stems erect or decumbent, 1–3 feet high, with slender branches. Leaves inch long, longer or shorter, a line wide; in γ. twice as wide; in δ much narrower, the younger glaucous. Racemes 10–16 flowered, the lower flowers remote. Pedicels 4–6 lines long. Petals rather longer than the calyx, blue, with yellow claws. Pods 1½–2 inches long, 1–1½ lines wide; in γ. not more than inch long. Style 1–1½ lines long.—A remarkable species, known from its immediate allies by its sharply three-nerved pods; but perhaps a form of *H. linearis DC. (See No. 37).*

52. H. stylosa (Burch. Cat. Geogr. 3291); shrubby, glabrous; pods *pendulous, linear, nearly nerveless,* tipped with a *filiform style longer than the pedicel;* leaves *linear or sublanceolate,* entire (or the lower pinnatifid); stamens toothless. *DC. Syst.* 2. *p.* 692; *Sond. l. c. p.* 254. *t.* 24. *H. platysiliqua and H. virgata, Meisn. Pl. Krauss. Lond. Journ.* 1. *p.* 462. *Zey.* 1907, 1911.

VAR. β. **lobata**; lower leaves pinnatifid, the lobes lanceolate, acute.

HAB. Kommedakka, *Burchell.* Stony places on mountains, Zuurebergs kette, Grahamstown and Assagaisbosch, Albany; also at Eland's Riv., Van Staadensberg, &c., Uit.; Winterberg, Caffr. *E. & Z.*, Katberg and Klipplaats river 3–4000f. and between Omtendo and Omsamculo, *Drege.* 3629, 5216, 7563. Winter Hoek and Sitzekamma, *Krauss,* 1253, 1244. β. near Grahamstown, *T. Williamson.* (Herb. Burch., Meisn., Sond., T.C.D.).

Distinguished from the preceding by its toothless stamens, and pendulous, nearly nerveless pods. 1–2 feet high. Branches long, nearly naked, dividing near the top. Leaves remote, linear or linear-lanceolate, subcoriaceous, 2–3 inches long, 1–2 lines wide; in β. half an inch long, with 2–4 spreading lobes at each side. Racemes elongate, 12–16 flowered. Flowers yellow or yellowish, tinted with red. Pods 2–3 inches long, 1–1½ lines wide, minutely stipitate, the valves obsoletely one-nerved; style, 4-6 lines long, rarely shorter.

53. H. virgata (Burch. Cat. Geogr.); suffruticose, glabrous; pods spreading or sub-deflexed, linear, *one-nerved*, tipped with a filiform style longer than the pedicel; leaves *ovate, entire or toothed;* stamens toothless. *DC. Syst.* 2 *p.* 693. *Sond. l. c. p.* 256.

VAR. α, **integrifolia**; leaves entire, flowers yellowish or white. *Burch. Cat.* 4605. *DC. l. c. H. glauca, E. & Z.* 91. *excl. var.* β.

VAR. β. **dentata**; leaves coarsely and sharply toothed; flowers white. *Burch. l. c.* 3933.

HAB. Krakakamma, *Burchell.* Vanstaadensberg and near Port Elizabeth, Uit., *E. & Z.* β. Riet Fontein, near the Kowie, Alb., *Burchell.* Oct. (Herb. Burch., Sond.,)

A slender suffrutex, 1-2 feet high, simply or slightly branched. Leaves thickish-leathery, ovate, acute, narrowed into a short petiole, 6-9 lines long, 3-4 lines wide, the upper narrower and smaller. Racemes elongate. Pedicels 3-5 lines long. Flowers mediocre. Pods 2-2½ inches long, a line wide. Style 4-6 lines long. Var. β., differing in having the leaves with 2-3 coarse and sharp teeth at each side, and a more branching and more woody stem, may be a distinct species.

54. H. scoparia (Burch, Cat. Geogr. 7887 and 8557.); shrubby, glabrous; pods erect, *linear, tapering into a short style;* leaves linear-subulate, rigid; racemes *axillary and terminal, few flowered. DC. Syst.* 2. *p.* 492. *Deless. Ic.* 2. *t.* 98. *E. & Z.* 99. *Sond .l. c. p.* 257. *Cheir. strictus, Lin. f. Thunb. Cap. p.* 492 *(excl. syn). Eck. Un. It. No.* 171.

HAB. Mountains round Capetown; Bavianskloof; Muysenberg; Brackfontein; Hott., Holl., and Houw Hoek, *Burchell, Thunberg, &c. E. & Z.* Breede river and Rondebosch, *Zeyher.* Mierenkasteel, Knakerberg 1000—1500f. and Kaus, 2-3000f. *Drege.* Ap.-Nov. (Herb. Thunb., Sond., r. Berol., T.C.D.).

A rigid, erect shrub, 1-2 feet high, branched from the base; the branches angular. Leaves very erect, thick and hard, subtrigonal, convex below, concave above, the points often incurved, 6-12 lines long, ½-1 line wide. Racemes 2-4 flowered, short. Pedicels shorter than the leaves; the fruiting ones 4-5 lines long. Sepals, 2 obtuse, 2 acute. Petals purplish or white, 4 lines long. Pods straight 2-2½ inches long, 1½ lines wide, one-nerved; style 1-2 lines long.

55. H. callosa (DC. Syst. 2. p. 696); pods linear, compressed, *stipitate;* leaves coriaceous, *lanceolate, acute, hard-pointed,* three-nerved below; stems shrubby, angular, glabrous. *Meisn. in Hook. L. Journ.* 1. *p.* 464. *Sond. l. c.* 258. *Cleome capensis, Lin. Sp.* 940. *Cheir. callosus, Lin. f. Sup.* 296. *Thunb. Cap. Fl.* 492. *H. cleomoides, DC. l. c.* 695. *Del. Ic.* 2. *t.* 99. *Drege* 7557.

HAB. In stony places, on the Mountains near Capetown, *Thunberg,* and most subsequent collectors. At the Waterfall, Devil's Mt., *W.H.H.* Ap.-Sep. (Herb. Thunb., r. Berol., Meisn., Sond., T.C.D.)

Shrubs 2 feet high or more. Stem as thick as a goose quill, angular, roundish above, branching. Branches subfastigiate, 6-12 inches long. Leaves erect, glabrous, rugulose above (when dry), three-nerved below; the lower or cauline 1½-2 inches long, 3-4 lines wide; upper linear-lanceolate, narrower and shorter. Fruiting racemes little elongate; pedicels 4-8 lines long. Outer sepals gibbous below the apex. Petals twice as long as the calyx, purplish. Pods 2-4 inches long, 2-3 lines wide; valves one-nerved. Stipe of the pod *(thecaphore)* sometimes 5-6 lines, sometimes 1½-2 lines long, cylindrical. Seeds 3 lines long or more, margined.

56. H. brachycarpa (Meisn. in Hook. Lond. Journ. 1. p. 465); pods (not known); ovary briefly stipitate, lanceolate-oblong; leaves *green, subpetiolate, linear-subspathulate, subacute, flat, one-nerved;* stem shrubby, terete, glabrous; raceme corymbose, few-flowered; petals with short claws. *Sond. l. c. p.* 260.

HAB. Clayey soil, at base of Winterhoek, Uitenhage, *Krauss*, 1254. April. (Herb. Meisn.)

Stem erect, the branches with raised striæ. Leaves 1 inch long, 1–1½ lines wide, one-nerved, the nerve more evident below, and branching. Pedicels spreading, 4–6 lines long. Petals twice as long as the calyx, yellowish white, oblong-spathulate. Ovary glabrous.

57. H. Dregeana (Sond. l. c. p. 260. t. 23. f. 2.); pods pedicellate, narrow linear, subfalcate, tipped with a short style; leaves *small, ovate, acute, thickish, muricate*; racemes terminal, few-flowered; stem shrubby, glabrous.

HAB. Wupperthal, *Von Wurmb, Drege* 7556. (Herb. Sond.)

Stem more than a foot high, with dark coloured bark; branches short, erect, scabrous, at length smooth. Leaves 2–2½ lines long, 1 line wide, rather concave above, below convex, one-nerved, mucronulate, green or reddish. Flowers 6–8, subcorymbose, apparently whitish. Lower pedicels 3 lines long. Petals twice as long as the calyx, 1½ lines wide. Pods 5 lines long, ½ line wide; *thecaphore* ½–1 line long.

SEC. 5. PACHYSTYLUM (DC.) *Siliqua* sessile, linear, subterete, *velvetty* tipped with a *thick, conical or cylindrical* glabrous style. Pubescent, shrubs, with entire leaves. (Sp. 58–59.)

58. H. incana (Ait. Hort. Kew. Ed. 1. vol. 2. p. 397); pods linear subterete, velvetty, tipped with a thick, conical, glabrous style; leaves *spathulate. DC. Syst. 2. p. 694. Sond. l. c. p. 261. H. frutescens, Lam. Dict. 3. p. 91.*

HAB. Cape of Good Hope, *Aiton*. August.

Stem 2 feet high; branches leafy, terete, villoso-pubescent. Leaves spreading, nerveless, thickish, canescent with a soft down, obtuse, the upper inch long, lower 2 inches. Racemes elongate. Pedicels pubescent, 4 lines long, erect. Calyx villous, obtuse; 2 sepals flat, 2 concave. Petals blue-purple, longer than the calyx, obovate. Lateral stamens toothed at base. Pods erect, villoso-tomentose, an inch or more long, scarcely a line wide, subincurved; the conical style 1½ lines long, ending in 2 minute, acute stigmata.

59. H. arenaria (Sond. l. c. p. 262); pods linear, subterete, torulose, velvetty, tipped with a thick, cylindrical style; leaves *linear*.

HAB. Sandy hills, under 500f. at Ebenezer, *Drege* 7568. (Herb. Sond.)

Stems 2 or more feet high; branches terete, velvetty, the lower ones spreading. Leaves thickish, velvetty, 1–1½ inches long, ½–1 line wide; the upper smaller. Racemes lax. Pedicels 3–5 lines long. Calyx velvetty, half as long as the blue petals. Pods velvetty, 2 inches long; ½ line wide, straight or curved, one-nerved; the thick style 1 line long. Seeds oval. Cotyledons sub-obliquely twisted, nearly as in *Carponema filiforme*.

SEC. 6. LANCEOLARIA, (DC.) *Siliqua* compressed, *lanceolate*, tapering into a short style, sessile or stipitellate. Seeds large. Cotyledons incumbent, linear, twice bent, the terminal part subspirally rolled round the other. Glabrous shrubs, with entire leaves. (Sp. 60–61.)

60. H. florulenta (Sond. l. c. p. 263); suffruticose, glabrous; pods lanceolate, attenuate at base, *raised on a short stipes*; leaves *linear-spathulate, mucronulate*; racemes *short, densely flowered*; petals spathulate, obtuse, clawed. *Carpopodium carnosum, E. & Z. No. 101. excl. syn.*

VAR. β. **obliqua**; pods oblique. *H. obliqua, E. Mey, Herb. Drege.*

HAB. Fields, among shrubs, near the Zwartkops River, and near Bethelsdorp Uitenhage, *E. & Z.* β. Witpoortberg, 2–3000f. *Drege.* Aug–Sep. (Herb. E. Mey., Sond., T.C.D.)

A shrub, 1 or more feet high, erect, with terete or angular branches and short branchlets. Leaves crowded on the upper branches, an inch long, 2 lines wide, subcoriaceous, flat or with revolute margins, one nerved. Raceme 1–2 inches long. Pedicels 3–4 lines long, the uppermost shorter. Flowers small, yellowish white. Pods uncial, tapering at base and apex, 3 lines wide in the middle, attenuated into a style 1–2 lines long. Seeds orbicular, compressed, 2 lines in diameter. In habit and leaves very similar to *H. brachycarpa*, Meisn., but differing in inflorescence, flowers, and in the shorter anthers, and especially in the ovary.

61. H. macrosperma (Burch. Cat. 3425); suffruticose, glabrous; pods *sessile*, lanceolate, tapering into a short style; leaves *linear, acute*; racemes *elongate lax.*; petals oblong. *DC. Syst.* 2. *p.* 695. *Sond. l. c. p.* 264.

HAB. In rocky places at Zwartwaterpost, *Burchell.* Uitenhage, *E. & Z.* Sep–and following months. (Herb. Sond.)

Stem somewhat angled; the branches terete, straight, virgate. Leaves often crowded at base, thickish or leathery, flat or subconcave, little narrowed at base, inch long, 1–1½ lines wide. Flower stalks 3–4 lines long; in fruit twice that length. Petals rosy, twice as long as the calyx. Pods sessile, or on a very short stipe, 1–1½ inches long, 2–3 lines wide, gradually narrowed upwards from a broader base; style 1–1½ lines long. Seeds 1½–2 lines long.

H. lyrata, Thunb. Fl. cap. p. 496, is not a species of this genus, but of *Sinapis* or *Brassica.*

H. mollugínea, DC. and **H. liniflora,** DC. (Syst. 2. p. 696), are not recognizable from the very bad figures so named.

W. S.

ORDER VIII. CAPPARIDEÆ. Juss.

(By W. SONDER.)

(Capparideæ, Juss. Gen. 242. DC. Prod. 1. p. 237. Endl. Gen. No. clxxxii. Lindl. Veg. Kingd. No. cxxv.

Sepals 4, separate or connate, mostly deciduous. *Petals* 4 or 8, or none, clawed, often unequal. *Stamens* 4, 6, or indefinite; generally some high power of 4. *Torus* often elongate, columnar, carrying the stamens and pistil. *Ovary* mostly stalked, one celled, with parietal placentæ and numerous ovules. *Fruit* either a podlike capsule, or a succulent or leathery, indehiscent berry. *Seeds* reniform, without albumen. *Embryo* curved or involute.

Shrubs or herbaceous plants, rarely trees, with alternate, exstipulate, simple or compound leaves and racemose inflorescence. Pubescence frequently glandular and fœtid. Flowers often large and handsome, white, yellow or pink, regular or irregular.

A considerable Order, chiefly tropical and subtropical. The shrubby and arborescent species, which have succulent, berry-like fruit, form the central group; the herbaceous species, with podlike fruit, and frequently definite stamens, nearly approach *Cruciferæ*, from which they differ in the unilocular ovary and want of *tetradynamous* stamens and mostly by their irregular petals. Few are of more than local use, in the countries they inhabit; with the exception of the common Caper-bush (*Capparis spinosa,*) a native of the Mediterranean region, whose unopened buds are a well known stimulant and pickle. This shrub might, no doubt, be freely cultivated at the Cape; it would grow in dry, rocky ground.

TABLE OF THE SOUTH AFRICAN GENERA.

Tribe I. Cleomeæ.—*Fruit* dry and capsular, dehiscent. (Herbs or undershrubs, often with composite leaves and glandular pubescence.)

I. **Gynandropsis.**—*Torus* elongate. *Stamens* 6, nearly equal, all fertile.
II. **Cleome.**—*Torus* hemispherical. *Stamens* 6, equal and fertile.
III. **Polanisia.**—*Torus* small. *Stamens* 8–32, fertile.
IV. **Dianthera.**—*Torus* small. *Stamens* 4; 2 long, fertile, 2 shorter and sterile.
V. **Tetratelia.**—*Torus* small. *Stamens* 8; 4 longer, fertile, 4 shorter, sterile.

Tribe II. Cappareæ. *Fruit* somewhat fleshy, indehiscent. (Shrubs or trees.)

* *Stamens* 4–5 *or* 8.

VI. **Schepperia.**—*Nectary* hood-shaped, hollow. *Stamens* 8. *Petals* none.
VII. **Cadaba.**—*Nectary* tongue-shaped. *Stamens* 4 to 5.

** *Stamens* 12–20 *or very numerous,* (rarely, in *Capparis* 8).

VIII. **Niebuhria.**—*Torus* cylindrical, short. *Petals* none or small. *Stamens* indefinite. *Leaves* trifoliolate.
IX. **Boscia.**—*Torus* short. *Petals* none. *Stamens* 12–20. *Leaves* simple.
X. **Capparis.**—*Torus* small. *Petals* 4. *Stamens* 8, or indefinite. *Leaves* simple.

Tribe I. Cleomeæ.

Fruit capsular, dehiscent.

I. GYNANDROPSIS. DC.

Sepals 4, short, spreading. *Petals* 4, clawed. *Stamens* 6, on the summit of a long stalklike *torus;* filaments subequal; anthers 2-celled, fertile. *Ovary* stipitate, with many ovules: stigma sub-sessile. *Capsule* podlike, unilocular, bivalve, many seeded. *Seeds* rugose. *DC. Prod.* I. *p.* 237. *Endl. Gen. No.* 4984.

Subtropical, annual or perennial herbs, natives of the Eastern and Western Hemispheres. Leaves compound, palmately 3–7-foliolate, the leaflets entire or toothed. Racemes terminal. Name from γυνη, *female,* ανηρ, *male,* and οψις, *resemblance;* in allusion to the position of the stamens with respect to the ovary.

1. G. pentaphylla (DC. Prod. I. p. 238.); glabrescent; middle leaves 5-foliolate, lowest and floral leaves 3-foliolate; leaflets obovate, ntir or subserrulate; pods linear. *Cleome pentaphylla, Lin. Sp.* 938. *Bot. Mag.* 1681. *C. Eckloniana, Schrad. Ind. Sem Gott.* 1683. *C. heterotricha, Burch. Trav.* I *p.* 537. *DC. l. c.*

Hab. On hills. Asbestos Mts., *Burchell.* Macallisberg, *Zeyher.* Interior regions, *Wahlberg.* (Herb. Holm. Sond.)

Annual, a foot or more high, branching, covered with long and short, and mostly glandular, clammy hairs. Lower and middle leaves on long petioles, upper subsessile, and much smaller. Leaflets ½–1 inch long, 4 lines wide. Pedicels ½ inch long, lengthening in fruit. Petals white or pale rosy. Pods 2 inches long or more, on a stipe ½ inch long, glandularly rough, tipped with a short style and broad stigma. Seeds reniform, rough with little pustules.

II. CLEOME, L.

Sepals 4, persistent or deciduous. *Petals* 4, sessile or clawed. *Stamens* 6, on a short, subglobose torus; filaments equal; anthers 2-

celled. *Ovary* sessile or stipitate, with many ovules; style short or none. *Capsule* podlike, unilocular, bivalve, many-seeded. Seeds usually rough. *DC. Prod.* 1 *p.* 238. *Endl. Gen. No.* 4985.

Herbs, mostly annual, common throughout the warmer zones of both hemispheres. Leaves simple or palmately 3–7 leaved. Flowers in terminal racemes. Name from κλειω, to *shut*; of uncertain application; adopted by Linnæus from Theodosius.

* *Leaves simple.*

1. **C. monophylla** (Linn. Sp. 940.); herbaceous, pubescent; leaves *simple*, petiolate, obtuse at base; pod roughish; thecaphore short.

VAR. *β*. **cordata**; leaves cordate or subcordate at base, *Sond. in Linn.* 23. *p.* 5. *Cl. cordata, Burch. in DC. Prod.* 1. *p.* 239. *C. subcordata, Stend.*

HAB. Macallisberg, *Burke & Zeyher*. Port Natal, *Gueinzius*. (Herb. Sond. T.C.D.)
Erect, branched, a foot or more in height, the whole plant glandularly pubescent and rough. Lower leaves broader and longer, oblong-lanceolate, 1–1½ inches long, the petiole ½ inch; upper subsessile, narrower and shorter. Racemes terminal; petals rosy, 4 lines long. Stamens 6, longer than the petals. Pods 1½–2 inches long; thecaphore 1 line long; style short.

** *Leaves 5–7-foliolate.*

2. **C. rubella** (Burch. Trav. 1. *p.* 543.); herbaceous, glandularly pubescent; leaves 5–7–foliolate; *leaflets lanceolate-linear, glabrous, glaucous;* pod roughish, on a very short thecaphore. *DC. Prod.* 1. *p.* 241.

HAB. Asbestos Mountains, *Burchell*. Port Natal, *Miss Owen*. (Herb. T.C.D.)
Annual, about 6 inches high, branching. Petioles terete, furrowed on the upper side, as long as the leaflets or longer, glandular. Larger leaflets 6–10 lines long, 1 line wide, obtuse; upper leaves smaller and on shorter petioles. Pedicels 4 lines long: flowers small, rosy. St. 6, fertile. Pods spreading, uncial, 1 line wide, roughish, at length smooth.

3. **C. rupestris** (Sond. Linn. 23. p. 6.); herbaceous, the stem, *leaves*, calyces and pods glandularly *pubescent;* leaves small, 5–7–foliolate; *leaflets linear, channelled above*, as long as the petiole; pods linear, pendulous, on a short thecaphore.

HAB. Rocky hills at Vaalrivier, *Zeyher* (Herb. Sond.)
Annual, about a foot high, much branched, clothed with yellow glandular hairs. Leaves, including the petiole, about 5 lines long, or the lower ones longer; leaflets subacute. Raceme secund; pedicels 4 lines long; fl. small, violate, petals obovate-oblong, somewhat clawed. Stamens 6. Pod uncial, 1½ lines wide; the style and thecaphore each 1 line long. Seeds smooth. Allied to the preceding, but differs in its smaller and glandular leaves, secund raceme and broader pods.

III. POLANISIA Rafin.

Sepals 4, lanceolate. *Petals* 4 sessile or clawed, often unequal. *Stamens* 8–32, on a short subglobose torus; filaments often unequal; anthers 2-celled, fertile. *Ovary* sessile or stipitate, with many ovules; style long or short. *Capsule* podlike, unilocular, bivalve, many-seeded. *Seeds* rugose. *DC. Prod.* 1. *p.* 242. *Endl. Gen. No.* 4988.

Annuals, of the warmer zones of both hemispheres, more numerous in America than on the Eastern Continents. Foliage glaucous, often glandular and disagreeably scented; leaves compound, palmately 3–9 leaved. Flowers in terminal racemes.

Name from πολυ, *many*, and ανισος, *unequal*; the stamens are numerous and very often unequal.

1. P. oxyphylla (DC. Prod. I. p. 242.); glandular and pubescent; leaves on long petioles, 3–7-foliolate; leaflets *oblong lanceolate, acute at both ends, glaucous;* stamens 8–12; pod *shortly stipitate*, striate, downy, pendulous. *Cleome oxyphylla, Burch. Trav. 2. p. 226.*

HAB. Klaarwater, *Burchell*. (Unknown to us.)

A foot and a half high, slightly branched, rough with sessile glands. Flowers yellow.

2. P. lutea (Sond.); herbaceous, the stem, leaves, calyces and pods glandularly rough; leaves 3–5-foliolate; leaflets *ovate-oblong*, flat, as long as the petiole or shorter; stamens 12; pod *on a long stipes*, linear, spreading. *Cleome lutea, E. Mey. Herb. Drege. Dianthera lutea, Klotsch, in Pet. Mozamb. Bot. p.* 160, *note.*

HAB. Orange River, near Verleptram, *Drege.* (Herb. Vind. Sond.)

1–1½ feet high, pale green, leafy, slightly branched, the whole plant rough with sessile glands. Lower leaves 5–, upper 3-foliolate; petioles terete, furrowed above, the lower ones uncial. Leaflets of intermediate leaves 8–10 lines long, 3–4 lines broad, acute, at each side glandular, especially at the margin; upper leaves gradually smaller. Calyx 3 lines long; petals more than twice as long, obovate, narrowed at base, yellow. Stamens in the specimens seen always 12, all fertile and of nearly equal length. Pods 2–3 inches long, a line wide, beaked with a style 1–1½ lines long; valves parallelly veined, the veins anastomosing; thecaphore 5–6 lines long. This seems to be very nearly related to the preceding.

IV. DIANTHERA, Klotsch.

Sepals 4, deciduous, subequal. *Petals* 4, the hinder-ones smaller, oblong, clawed; the front ones larger, obovate. *Stamens* 4–10, on a small torus, unequal; 2–8 short and sterile, clubshaped and often appendiculate at apex; 2 anterior alone fertile, very long, declinate: anthers oblong, 2-celled. *Capsule* podlike, linear, stipitate; style evident; the valves parallelly many veined, the veins here and there anastomosing. *Seeds* curved, reniform, minutely pitted, puberulent. *Kl. Peters. Mozamb. Bot. p.* 160.

African, branching, glaucous, glabrous or sparingly glandular herbs. Leaves 3–7 foliolate, leaflets linear. *Flowers* racemose. Name, δις, *two*, ανθηρ an *anther*, because 2 anthers alone are fertile.

1. D. Petersiana (Kl. l. c. tab. 27); glabrous, the branches and peduncles *sprinkled with minute, scattered glands;* lower leaves on long petioles, 5-*foliolate*, upper on short petioles, 3-foliolate; leaflets very narrow-linear, obtuse, with revolute margins, glaucous; petals, unequal, obovate; fertile stamens 2, sterile 6–8, thrice as short, clubshaped, with a *globose* appendicule; pods linear, subpendulous, glabrous.

HAB. Gamkerivier, near Bitterwater, *Burke & Zeyher*. Port Natal, *Mr. Hewitson*, Mosambique, *Dr. Peters*. (Herb. Sond., T.C.D.)

Two feet high or more, the stem and branches striate. Middle leaflets 6–10 lines long, ½ line wide, the lateral somewhat shorter. Petioles of the lower leaves as long as the leaflets. Peduncles in the axils of the upper leaves, 6–8 lines long, lengthening in fruit. Sepals oblong, acute, 2 lines long. Petals yellow, violet on the under side, the larger ones 8 lines long and 3–4 lines wide; smaller more than twice as narrow. Fertile stamens nearly an inch long. Pods 2–3 inches long, 1½–2 lines wide; thecaphore 4–6 lines long.

2. D. Burchelliana (Kl. l. c. p. 161. in note); glaucous, the *stem rough with minute-hooked prickles;* lower leaves on longish petioles, *7-foliolate*, upper on shorter petioles 3–5-foliolate; leaflets very narrow-linear, obtuse; petals unequal, obovate; fertile stamens 2, sterile 6, shorter, with a *conical-oblong* appendicule; pods linear, pendulous, glabrous. *Cleome diandra, Burch. Trav.* 1. *p.* 548. *Polanisia dianthera, DC. l. c. p.* 242.

HAB. Gattikamma, Feb., *Burchell.* Stony hills at Zwartbulletze and at Gamkeriver, 2500–3000f. *Drege.* (Herb. Sond.)

Annual, 2 feet high, slender, armed in the lower part with rigid, recurved prickles; branches erecto-patent, striate. Middle leaflets an inch long, ½ line wide. Flowers rosy (?), size of the preceding. Pedicels 8 lines long. Pods 2–2½ inches long, 1½ lines wide; style beaklike, 2 lines long; thecaphore 4 lines long. Seeds as in *D. Petersiana.*

3. D. semitetrandra (Kl. l. c. p. 162. in note); glabrous, the leaves trifoliolate; leaflets as long as the petiole, linear-filiform; upper leaves undivided; petals oblongo-spathulate or oblong; fertile stamens 2, sterile 2, twice as short; pods linear, curved, pendulous; thecaphore very short. *Cleome diandra, Burch ? E. Mey. Pl. Drege. Cl. semitetrandra, Sond. Linn.* 23. *p.* 5.

HAB. Stony places at Gamkeriver and Wilgeriver, March. *Burke & Zeyher, Drege.* (Herb. Sond., T.C.D.)

A small annual, 6–8 inches high, branching, resembling *Heliophila pinnata.* Leaves ½ inch long. Raceme lax; pedicels 4–6 lines long, horizontally patent in fruit. Flowers small, purplish. Calyx 1 line, petals 2 lines long. Fertile stamens as long as the petals; sterile more slender, obtuse. Pods uncial, 1¼ lines wide. Seeds downy.

V. TETRATELIA, Sond. (n. gen.)

Calyx-tube short, swollen; limb 4-parted, deciduous, subequal. *Petals* 4, on long claws, subinequal; 2 obovate-oblong, 2 oblong. *Stamens* 8, unequal, united at base into a tube; 4 sterile shorter, clavate, apiculate; 4 fertile elongate; anthers oblong, 2 celled. *Torus* small. *Pod* linear, stipitate, with a filiform style; valves parallelly 3-nerved, veinless, glabrous. *Seeds* curved, reniform, compressed, glabrous, transversely crested.

A South African, glabrous, branching annual, with alternate, petiolate, 3–5-foliolate leaves, the leaflets narrow linear. Racemes terminal, the bracts minute, setaceous. Name, *τετρα, four,* and *τελειος, perfect,* referring to the 4 perfect stamens.

1. T. maculata (Sond.) *Polanisia maculata, Sond. Linn.* 23. *p.* 6.

HAB. Stony and grassy hills. Moojiriver, Dec. *Zeyher.* Grahamstown, *Dr. Atherstone.* (Herb. Hook., Holm., Sond.)

A small plant, with the aspect of *Heliophila trifurca.* Stem 1 foot high, terete, striate. Stem-leaves on longish petioles; leaflets 1–1½ inches long, a line wide; branch-leaves about ½ inch long, subfiliform, all as long as the petiole, the younger ones rough with minute bristles, older glabrous. Racemes 2 inches long, corymbose: pedicels 3–4 lines long. Bracts scarcely conspicuous. Sepals lanceolate, 1 line long: petals 5 lines long, apiculate, violet, the longer ones uniformly coloured, the shorter with a yellow medial spot. Fertile stamens as long as the petals; sterile somewhat shorter, more slender, bearing a clavate, two-coloured abortive anther. Pod 2 inches long, a line wide; style 2–3 lines long: thecaphore 3–4 lines. Seeds small, yellowish.

Tribe II. Cappareæ.

Fruit fleshy, indehiscent.

VI. SCHEPPERIA, Neck.

Calyx coloured, 4-leaved, the two outer sepals keeled; the front sepal largest. *Petals* none. *Stamens* 8, on the summit of a long, filiform, curved *torus*, which has a hood-shaped *nectary* at its base, on the upper side. *Ovary* stipitate, ovate or oblong, unilocular; ovules numerous, on four parietal placentæ; stigma sessile. *Berry* cylindrical, glandular. *DC. Prod.* 1. *p.* 245. *Endl. Gen. No.* 4991.

A South African, leafless, twiggy shrub, with widely spreading, spiny-pointed branches. Flowers yellowish or purplish, in subcorymbose racemes: pedicels unibracteate at base. Named, it would seem, in honor of some obscure botanist, whose memory has otherwise passed away.

1. S. juncea (DC. l. c.); *Linnæa* 1., *p.* 255. *tab.* 3 *Cleome juncea, Sparm. Lin. Syst. p.* 605. *(non. Thunb. Fl. Cap.) Cl. aphylla, Thunb. Fl. Cap. p.* 497. *Macromerum junceum, Burch. Trav.* 1. *p.* 388. *Schep. aphylla and S. juncea, E. & Z. En. No.* 106, 107.

Hab. Karroo, beyond Hartequa's Kloof, *Thunberg*. Great Fish River, *Bergius*. Garriep, *Burchell*, Swellendam, Clanwilliam, Graaf Reynet and Uitenhage, *E. & Z.* In the same districts, and also in the far interior, *Drege, Krauss*. (Herb. Sond., Hook, T.C.D.)

A much branched, twiggy, pale-coloured shrub, 2 feet high. Branches alternate, rarely opposite or whorled and crowded, erecto-patent or widely spreading, terete, alternate, subspinous. Leaves none, except, on the young branches, acute leaf-scales, 1 line long. Racemes corymbose, lateral, short. Flowers yellow or purplish: peduncles 6–8 lines long, clammy. Sepals ovate. Stamens mostly 8. Fruit a sausage-like berry, 1½–2 inches long, as thick as a goose quill, terete, viscid, densely glandular, unilocular, many seeded. Seeds compressed, subcordate.

VII. CADABA, Forsk.

Sepals 4, unequal, two outer valvate, covering the two inner in the bud. *Petals* (4 or) none. *Stamens* 4–6, on the summit of a cylindrical, stipelike *torus*, which has a tongue-shaped, tubular *nectary* at its base. *Ovary* stipitate, unilocular, numerous: stigma sessile. *Berry* cylindrical, subtorulose. *DC. Prod.* 1. *p.* 244. *Endl. Gen.* 4993.

Unarmed shrubs, found in tropical and subtropical Asia and Africa, glabrous or glandular. Leaves alternate, simple, or trifoliolate. Flowers axillary solitary. Name, *Kadhab*, the Arabic name of *C. rotundifolia*.

1. C. Natalensis (Sond. Linn. 23. p. 8); unarmed, apetalous; leaves petiolate, oblong or obovate-oblong, obtuse or emarginate, mucronulate, coriaceous, glabrous; flowers axillary, on long peduncles, two outer sepals concave, 2 inner flat, suborbicular, mucronate; stamens 6, nectary lageniform, with a curved neck, split at the side; the mouth toothed; berry cylindrical, elongate.

Hab. Port Natal, *Gueinzius*, No. 87. (Herb. Sond.)

Branches long and smooth. Leaves 12–15 lines long, 4–5 lines wide at the tip; petioles 3 lines long. Peduncles subsolitary in the axils, somewhat racemose toward the ends of the branches, ¾–1 inch long. Sepals equal, the outer reddish; inner petaloid, white. Torus 1 inch long; nectary 4 lines long. Fruit inch long; its stipe 4 lines.

VIII. NIEBUHRIA. DC.

Calyx funnel-shaped, with a cylindrical, persistent tube and 4-parted, campanulate deciduous limb; valvate in æstivation. *Petals* none or very small. *Stamens* very numerous, on the summit of a cylindrical torus. *Ovary* stipitate, unilocular, with numerous ovules; stigma sessile. *Berry* ovoid or oblong; seeds few, reniform, lying in pulp. *DC. Prod.* 1. 243. *Endl. Gen. No.* 4995.

Shrubs or small trees, natives of S. Africa and tropical Asia. Leaves alternate, 3–5-foliolate, with minute, setaceous stipules. Flowers axillary, or terminal, corymbose, or racemose. Name in honor of *Carsten Niebuhr*, author of Travels in Arabia.

1. N. triphylla (Wendl. in Bartl. & Wendl. Beytr., 2. p. 29); leaves ternate, leaflets oblong, ovate-oblong or obovate, not conspicuously veined; raceme terminal, flowers *without petals. Capparis triphylla, Thunb. Fl. Cap. p.* 430. *Cratæva caffra & C. avicularis, Burch.—Nieb. caffra, avicularis, & oleoides, DC. Prod.* 1. *p.* 243, 244.

HAB. Banks of Kamtou and Zeekoe rivers, *Thunberg*. Caffraria, *Burchell*. In the forest of Adow, and by the Zwartkops R., Uitenhage, *E. & Z.*, Port Natal, *Gueinzius*. (Herb. Sond., Lehm., T.C.D.)

A glabrous shrub, 2 or more feet high, with yellowish branches; branchlets leafy. Leaves coriaceous: petiole-furrowed, 4 lines to 1½ inches long; leaflets articulated to the petiole, obtuse or retuse, with or without mucro, one nerved, obsoletely veined, dull green above, paler below, the middle one 1–2 inches long, lateral ones shorter. Flowers corymboso-racemose, axillary; peduncles 8–10 lines long. Calyx ½ inch long, subturbinate; segments oblong, acuminate, concave. Stamens flexuous, as long as calyx. Ovary on a stipe twice as long as the stamens; stigma discoid. Berry pyriform, ¾ inch long, shorter than its stipe.

2. N. nervosa (Hochst. in Flora, 1844); leaves ternate, leaflets oblong, ovate-oblong or obovate, evidently veiny on the under side; raceme terminal, corymbose; *petals four*.

HAB. Port Natal, in woods and on the sands, *Dr. Krauss*. (Herb. T.C.D.)

Known from *N. triphylla*, which it closely resembles in aspect, by its strongly nerved leaves, 4 petalled flowers and more obtuse and shorter calyx lobes. Sepals 4 lines, petals 2 lines long; stamens and ovary as in the preceding. It agrees with *Streblocarpus*, Arn. in its corolla, but differs in the shorter ovary and trifoliolate leaves.

3. N. rosmarinoides (Sond. Linn. 23. p. 7); glabrous; leaves 3–5-foliolate, leaflets much longer than the petiole, *linear*, mucronate, with revolute margins, pale underneath; corymbs terminal; petals 4.

HAB. Port Natal, *Gueinzius*. 467. (Herb. Sond.)

Branches terete, ramuli numerous, short. Leaves most often 3-foliolate, the petiole 4 lines long, petiolules very short; middle leaflet 1½–2 inches long, a line wide, lateral shorter. Lower peduncles axillary, 8–10 lines long. Calyx 3 lines long, sepals concave. Petals elliptical, clawed, ⅓ as long as calyx. Stamens about 16, twice as long as the calyx, on a short torus. Ovary cylindrical; stigma depressed.

IX. BOSCIA, Lam.

Sepals 4, concave, deciduous, valvate in æstivation. *Petals* none. *Stamens* 12–20, on a scarcely elevated torus; filaments monadelphous at base only. *Ovary* stipitate, ovoid, unilocular, with 4–5 ovules on a single, parietal placenta; style short. *Berry* globose, mostly one seeded, seed lying in pulp. *Lam. Illustr. t.* 395 *(non Thunb.) DC. Prod.* 1. *p.* 244. *Endl. Gen. No.* 4996.

African shrubs, unarmed, glabrescent. Leaves alternate, simple, coriaceous, very entire. Stipules setaceous minute. Flowers corymbose, small. Name in honour of *Louis Bosc*, formerly Professor of Agriculture in Paris, and author of several works.

1. B. caffra (Sond. Linn. 23. p. 8.); leaves on very short petioles, ovato-lanceolate, narrowed at base and apex, mucronulate, with undulate margins, veiny on both sides, glabrous; peduncles filiform, axillary, corymboso-racemose toward the ends of the branches; flowers polyandrous. *Capparis undulata, E. & Z. No.* 112. *Niebuhria acutifolia, E. Mey. N. pedunculosa, Hochst.*

HAB. In woods. Elands river, near Philipstown, and at Natal. June–Oct. *E. & Z.! Drege, Krauss*, &c. (Herb. Sond., Lehm., T.C.D.)

Branches and branchlets ashcoloured, glabrous. Leaves simple, the upper ones narrower; petiole 2 lines long. Peduncles uncial, shorter than the leaves or equalling them. Calyx 3 lines long, sepals obovato-spathulate, acute, concrete at base. Stamens 12 or more, more than twice as long as the calyx, on a torus 1 line long. Thecaphore longer than the stamens. Ovary ovate, with a short style and discoid stigma. Fruit pendulous, globose, larger than a pea, reddish, minutely dotted, unilocular, 1–2 seeded. Seeds crustaceous, shining. Embryo subconvolute, with an obtuse, terete radicle; cotyledons fleshly, broad, convolute. It varies with oblong-ovate or lanceolate, obtuse or emarginate, mucronulate, acute, or acuminate leaves, 1–3 inches long, ½–1 inch wide.

X. CAPPARIS. L.

Sepals 4, imbricated in æstivation. *Petals* 4, imbricated. *Stamens* numerous (rarely definite and few) on a small, hemispherical torus. *Ovary* stipitate, unilocular, with numerous ovules on two opposite, parietal placentæ. *Berry* globose or elongate, coriaceous, many seeded; seeds lying in pulp. *DC. Prod.* 1. *p.* 245. *Endl. Gen. No.* 5000.

Trees or shrubs, often climbing, and very frequently armed with spines: distributed through the warmer zones of both hemispheres. Leaves alternate, simple, entire or subentire, with spiny or setaceous stipules. Flowers solitary or racemose, panicled, or corymbose, generally white or cream coloured. Name, the Arabic *kabir*, (the name for *C. spinosa)*, in Greek, *καππαρις*, Lat. *capparis*; French *caprier*; English *caper*. The young flowerbuds of *C. spinosa* form the well known pickle "*capers*."—All the S. African species belong to De Candolle's section *Eucapparis*.

* *Pedicels one-flowered, supra-axillary, in vertical series.* (Sp. 1.)

1. C. Volkameriæ (DC. Prod. 1. p. 247); stipules spiny, hook-pointed; leaves ovate, with a hard mucro, with reddish pubescence along the nerves of both surfaces or of the lower; pedicels 2–3-seriate. *Volkameria capensis, Burm. Prod. Cap.* 17. *ex DC. E. & Z. En. No.* 108.

HAB. Dense woods on the Eastern Districts. Krakakamma and Adow, Feb. *E. & Z.* (Herb. Sond., Lehm., T.C.D.)

Branches flexuous, the younger covered with dense reddish tomentum, at length glabrous. Leaves on short petioles, 1½ inch long, inch wide, tipped with a short, obtuse, callous point. Flowers (not known to us): according to De Candolle the petals are ciliate and the stamens about 30. It varies with leaves subglabrous and green.

*** Pedicels in a corymb or raceme ; flowers polyandrous.* (Sp. 2–4.)

2. C. citrifolia (Lam. Dict. 1. p. 606); twigs downy or glabrous; stipules spiny, hooked; leaves *oval* or oblong, obtuse, mucronulate; thinly downy; pedicels terminal, umbellate; buds *glabrous*, petals ovate-oblong. *C. capensis, Thunb. Cap. p.* 430. *Bartl. & Wendl. Beytr.* 2 *p.* 31. *E. & Z. En.* 109.

VAR. β. **sylvatica**; branches and leaves glabrous; leaves oval or obovate, obtuse, emarginate.

HAB. In woods, Eastern Districts. Camtoos river, *Thunberg*, Uitenhage, Nov.–Apl. *E. & Z. !, Drege, Krauss, &c.* (Herb. Thunb., Lam., Sond., T.C.D.)

A shrub, 4 or more feet high, with rigid, spiny branches, and leafy twigs. Leaves alternate, on short petioles, very entire, with recurved margins, obtuse or retuse, with or without mucron, paler below, 1–1½ inches long, 6–10 lines wide. Flowers 4–10, subumbellate; peduncles filiform, pubescent, inch long. Calyx 2½ lines long, concave, obtuse. Petals 4, subciliate, villous at base and within, 3 lines long. Stamens numerous, twice as long as the petals. Ovary ovate, acute, glabrous; thecaphore as long as the pedicel. Berry spherical, apiculate, glabrous, as large as a small cherry, one seeded. *Cap. Drege* 7534, scarcely differs from the normal state of *C. citrifolia.*

3. C. corymbifera (E. Mey. in Herb. Drege); branches thinly tomentose; stipules spinous, subrecurved; leaves *oval-oblong, obtuse*, glabrous or subtomentose below; pedicels terminal, corymbose; buds *tomentose;* petals as long as the calyx. *C. hypericoides, Hochst. Fl.* 1844.

HAB. In primitive woods, Natal. June. *Drege, Krauss, Gueinzius.* (Herb. Sond., T.C.D.)

More robust than the preceding, with larger leaves (2–3 inches long, 14–16 lines wide), thicker peduncles, buds twice as large and tomentose; stamens 1½ inches long, and a globose (not *acute)* ovary. Thecaphore more slender and longer than the pedicel.

4. C. Gueinzii (Sond.); twigs downy; stipules spiny, incurved; leaves petiolate, *oblongo-lanceolate*, obtuse at each end, minutely emarginate, glabrous, the midrib downy; racemes *axillary*, as long as the leaves or little shorter; flowers polyandrous.

HAB. Port Natal, *Gueinzius.* (Herb. Sond.)

Branches slender, flexuous, terete. Leaves 2 inches long, ½ inch wide, one nerved, with inconspicuous veins; petiole 2–3 lines long. Racemes spreading, axillary; uppermost leafless, paniculate; rachis downy, upper pedicels corymbose, 3–4 lines long. Flowers yellow: sepals ovate, obtuse, 2 lines long; petals as long, stamens, about 30. Resembes *C. Zeyheri*, Turcz, from which it differs in the non-attenuate leaves, larger flowers and more numerous stamens.

** Pedicels axillary, mostly many flowered; stamens eight.* (Sp. 5–9.)

5. C. cluytiæfolia (Burch. Cat. 3881); unarmed; leaves oblong-cuneate, obtuse, mucronate, glabrous; pedicels axillary, *solitary, one flowered*, half as long as the leaf. *DC. l.c. p.* 248.

HAB. South Africa, *Burchell.* (Unknown to us).

6. C. oleoides (Burch. Cat. 4200); unarmed; leaves coriaceous, oblong, or linear-oblong, narrowed at base, retuse, mucronate, glabrous; *racemes* axillary, rather shorter than the leaf; thecaphore *shorter than* *the pedicel. DC. l. c. C. coriacea, Burch. Cat.* 2898. *DC. l. c. Bartl. & Wendl.* 2. *p.* 33. *E. & Z.* 110, 111.

HAB. Among shrubs, from Uitenhage to Caffraria, Oct. *Burchell, E. & Z., Drege.* (Herb. Sond., Lehm., T.C.D.)

A glabrous shrub; branches pale. Leaves erect, very entire, one-nerved, veinless, 1–2 inches long, 4–6 lines wide; petiole 2 lines long, furrowed. Flowers 6–10, racemoso-subcorymbose, ½–⅔ shorter than the leaf; pedicels 1 flowered, without bracts. Buds minute. Sepals ovate; reflexed; petals shorter, glabrous; stamens as long as sepals. Berry globose, as large as a small cherry; thecaphore 1 line long.

7. **C. albitrunca** (Burch. Trav. 1. p. 343); unarmed, with spreading branches; leaves oblong, obtuse or emarginate, attenuate at base, leathery, glabrous, glaucous below; racemes axillary, few flowered, shorter than the leaves; thecaphore *as long as the pedicel. DC. l. c. p* 248. *Pappe, Sylv. Cap. p.* 3.

HAB. At Gattikamma, at Sondag river, and in other parts of the Eastern districts. Oct.–Nov. *Burchell, Pappe.* (Not known to us.)

A tree 10–12 feet high, robust, with a white trunk. Flowers minute. Racemes sometimes springing from naked branches. Seemingly but little different from the preceding.

8. **C. punctata** (Burch. Trav. 1. p. 492); unarmed; leaves oblong, sub-attenuate at base, submucronate, glabrous, *with netted veins;* racemes axillary, much shorter than the leaves. *DC. l. c. p.* 248.

HAB. Klaarwater, Dec. *Burchell.* (Unknown to us.)

A shrub, 4-6 feet high, with spreading branches. Leaves narrow-lanceolate, very obtuse. Racemules very short, solitary, or in pairs, axillary. Fruit globose, smooth, netted and punctate. Scarcely of this genus: can it be our *Boscia caffra* ?

9. **C. Zeyheri** (Turcz. Animadv. p. 54); stipules *spiny, hooked;* twigs *downy;* leaves ovate or oblongo-lanceolate, tapering *to each end,* obtusely acuminate, tipped with a hard point, undulate at the margin, glabrous on the upper, covered with deciduous down on the lower surface; racemes axillary, shorter or longer than the leaf.

HAB. In the woods of Krakakamma, Feb., *Zeyher* (1915). Port Natal, *T. Williamson, Sanderson, Drege* 8505. (Herb. Hook., Sond,, T.C.D.)

Stem climbing; branches terete, green, smooth; twigs slender, flexuous, the young ones often rufescent. Leaves spreading, the lower 3 inches long, an inch wide; upper 2–2½ inches long, 5–6 lines wide or wider, venulose; the petioles 3–6 lines long, channelled above. Racemes axillary, often long and leafless, simple or paniculately branched. Buds globose, downy. Sepals 1 line long; petals glabrous, of equal length. Ovary ovate, acute. Berry globose, as big as a large pea; thecaphore as long or a little longer than the pedicel.

ORDER IX. RESEDACEÆ, DC.

(By W. H. HARVEY.)

(Resedaceæ, DC. Theor. 1. p. 214. Endl. Gen. No. clxxxiii. Lindl. Veg. Kingd. No. cxxiv.)

Sepals several (4–7), persistent. *Petals* irregular (2–7), entire or lacerated. *Disc* expanded, fleshy, unilateral. *Stamens* 3–40, inserted within the margin of the disc, free; anthers erect, two celled, opening longitudinally. *Ovary,* sessile or nearly so, one-celled, open at the summit, with 3–6 parietal placentæ, and numerous ovules. *Stigmas* sessile. *Fruit* a gaping, dry or succulent capsule, or apocarpous. *Seeds* reniform, without albumen. *Embryo* curved : radicle next the hilum.

Herbaceous or suffruticose small plants, mostly glabrous ; with alternate, exstipulate, entire or pinnatifid leaves, and racemose or spiked inflorescence. *Flowers* minute ; the petals frequently shorter than the sepals, white or greenish.

This Order consists of a few weeds inhabiting the temperate zones of both hemispheres ; about 50 species are known. Only two are of any celebrity ; the *migniomette (Reseda odorata)*, a native of the shores of the Mediterranean ; and the Weld *(Reseda luteola)* formerly much cultivated for its yellow dye in England, where it is a common weed. The few Cape plants of the Order are found on the Karroos.

I. OLIGOMERIS. Cambess.

Calyx 4–5–parted, the segments sometimes unequal. *Petals* two, alternate with the posterior sepals, flat, simple (not lobed), without appendage, separate, or confluent at base. Hypogynous disc none. *Stamens* 3–8, hypogynous ; filaments subulate, flat, united at base into a cup, persistent ; anthers deciduous. *Ovary* unilocular, 4 angled, with 4 conical points ; placentæ 4, parietal ; ovules numerous. *Capsule* membranous, inflated, open at the summit, 4-horned. *Endl. Gen. No.* 5012. *Resedella, Webb. & Bert. Ellimia, Nutt. Holopetalum, Turcz.*

Small, glabrous, slightly fleshy, annual or perennial plants, sometimes suffruticose at base. Leaves very narrow, undivided, not obviously veined. Flowers minute, white, in terminal spikes, bracteate. The few known species are widely scattered, being found in N. & S. Africa, the Canary Islands, Tropical Asia and California. From *Reseda* it is known by wanting the large fleshy disc, and by having undivided, uncrested petals. Turczanninow separates all the Cape species except *O. Dregeana*, to form his genus, *Holopetalum*, distinguished by more numerous and equally distributed stamens. But the number of stamens, in *Reseda* itself, and indeed in all the known genera of the Order, is notoriously variable, and this character seems insufficient to break up so natural and small a group. The name is compounded of ολιγος and μερις, in allusion to the depauperated flowers, as compared with those of *Reseda*.

Sub-genus 1. Resedella. *Stamens* 3–4, unilateral.

1. O. Dregeana (Presl. Bot. Bem. p. 8) ; diffuse or decumbent, flexuous ; leaves linear, elongate, subacute, with fascicles of smaller leaves in their axils ; bracts longer than the flower, subulate ; sepals subulate, acute, as long as the lanceolate petals ; stamens three or four, unilateral. *Reseda dipetala*, *E. Mey. ! in Herb. Drege.*

Hab. Zwartkey, by the River, 4000ft. Dec. *Drege !* (Herb. T.C.D., Hook. Sond.)

Branches densely clothed with leaves, which have usually innovations in their axils. Leaves ¾ inch long, slightly tapering at base and somewhat glaucous. Spikes rather lax, elongating. Lower bracts leaflike, much longer than the flowers, upper gradually shorter. Sepals veiny, with membranous edges. Stamens shorter than the petals.

Sub-genus 2. Holopetalum. *Stamens* 6–8, equally distributed.

2. O. capensis (Thunb. Cap. p. 402) ; erect and virgate, or diffusely branched and flexuous ; leaves *linear*, obtuse, or subacute ; spikes elongate, lax ; bracts much shorter than the flowers ; sepals 5, elliptic-oblong, or ovate, albo-marginate ; petals *linear-oblong*, very blunt, twice as long as the sepals ; stamens 6–8, equally distributed. *Holopetalum* *pumilum, Turcz. Bull. Acad. Mosc. XVI.* I. *p.* 51. *Reseda microphylla Presl. R.* 7533, *Drege. Reseda capensis, Thunb. l. c.*

VAR. *α*, **pumila**; diffuse, flexuous, with short, blunt leaves. *Drege* 7533*a*.

VAR. *β*. **virgata**; erect; flowering branches long and virgate; leaves acute or subacute. *R. dipetala, E. & Z.! No.* 113.

HAB. *γ*, Winterfeld, *Drege! β*. In the Karroo, among the Zwarteberg Mounts. and the Langekloof, George; and in the Winterfield, Beaufort, *E. & Z.!* Rhinosterkopf, *Burke!* Langekloof, *Thunberg.* (Herb. Hook., T.C.D., Sond.)

A small, woody suffrutex, variable in size and ramification; sometimes very short, scraggy and densely branched, spreading over the surface of the arid soil; sometimes (in moister places ?) tall, erect and virgate. Such variations are common among Karroo plants. There is no floral character to distinguish the above varieties. The flowers are often polygamous. Thunberg says that his plant is "3 feet high or more;" Drege's specimens are 3–12 *inches* only.

3. O. spathulata (E. Mey. MSS.); glabrous and glaucous, erect; leaves broadly *spathulate*, densely set; spikes dense; bracts shorter than the calyx, subulate, spreading; sepals elliptic-oblong, blunt, shorter than the *ovate* petals; stamens 6–8. *Holopetalum spathulatum, Turcz. Anim. p.* 60. *Muell. in Mohl. & Schl. Bot. Zeit.* 1856. *p.* 39.

HAB. On the hills, near the mouth of the Orange River, *Drege*. October (Herb. T.C.D.)

A small, somewhat glaucous, half woody plant. Branches erect, striate. Leaves thickly inserted, tapering at base, with a broad apex, ½–¾ inch long, thick and fleshy. Branches terminated with long, dense spikes of greenish flowers. Capsule globose, inflated, open, with four diverging points.

4. O. Burchellii (Muell. Bot. Zeit. 1856. p. 39); "dwarf, glabrous; branches straight, slender, elongate; leaves linear; petals obovate-oblong; capsules obovoid-oblong." *J. Muell. l. c. (Holopetalum.)*

HAB. South Africa, Burch, Cat. Geogr. 1850 and 2549. in Herb. DC.

Unknown to me. It is said to differ, by its longer capsules, from the other species.

ORDER X. BIXACEÆ, Endl.

(By W. H. HARVEY.)

(Bixaceæ, Endl. Gen. No. cxcv. Flacourtianeæ, A. Rich.—DC. Prod. I. p. 255. Bixineæ, Kunth.—DC. Prod. I. p. 259. Homalineæ, R. Br., Endl. No. cxcvi. DC. Prod. 2. p. 53. Flacourtiaceæ, Lindl. Veg. Kingd. No. cx. and Homaliaceæ, No. cclxxxiv.)

Flowers frequently, by abortion, unisexual. *Calyx* free, or more or less adnate with the lower part of the ovary; sepals 4–7–12–14, imbricate or valvate, persistent. *Stamens* hypogynous or perigynous, definite or indefinite, free; anthers introse, two-celled. *Ovary* sessile or half-inferior, unilocular (or *imperfectly* pluri-locular), with two or more parietal placentæ, and two or many anatropous ovules. *Fruit* either a pulpy-dehiscent capsule, or a fleshy berry. *Seeds* albuminous, with an axile embryo; leafy cotyledons; and radicle next the hilum.

Trees or shrubs, with alternate, simple, entire or toothed leaves, frequently pellucid-dotted, either without stipules, or with caducous ones. Flowers usually of small size, variously disposed, often greenish.

Found throughout the tropics, and in subtropical regions of both hemispheres, no where very plentiful. Perhaps 150 species may be known, arranged under 52 genera. The fruits of several are edible. The most generally known is *Bixa Orellana*, whose angular seeds are covered with an orange-red pulpy coat, which furnishes the substance called *Arnotto*, used in the preparation of chocolate, and for dying cheeses, &c. The affinities of this Order are generally considered to be towards Passifloraceæ, to which they are allied through *Smeathmannia*, and to Samydaceæ, from which they chiefly differ in the direction of the embryo in the seed. I venture to add *Homalineæ*, usually considered a distinct Order, but only to be known by its more or less perigynous stamens and adnate ovary, characters which are here of as little ordinal value as in Saxifragaceæ. Through *Dovyalis* and *Aberia* there is an unexpected passage almost into Euphorbiaceæ.

TABLE OF THE SOUTH AFRICAN GENERA.

Tribe I. PROCKIEÆ. *Flowers* bisexual, rarely polygamous. *Style* columnar or short. *Fruit* fleshy, indehiscent. *Ovary* superior.

I. **Oncoba.**—*Sepals* 5, deciduous. *Petals* 5, spreading, deciduous.

II. **Rawsonia.**—*Sepals* 4–5, persistent. *Petals* deciduous. A *petaloid-scale* opposite each petal. *Stigma* subsessile 4–5-fid.

III. **Phoberos.**—*Calyx* persistent, deeply 10–12-fid, in two rows; the inner segments smaller.

Tribe II. KIGGELARIEÆ. *Flowers* unisexual. *Ovary* superior. *Styles* as many as the carpels. *Fruit* dehiscent or indehiscent.

* Anthers opening by longitudinal slits.

IV. **Trimeria.**—*Calyx* 6–10 parted, in two rows.

V. **Dovyalis.**—*Calyx* 5–7 parted in a single row. *Seeds* smooth.

VI. **Aberia.**—*Calyx* 5–7 parted, in a single row. *Seeds* woolly.

** Anthers opening by terminal pores.

VII. **Kiggelaria.**—*Calyx* 5 parted. *Petals* 5.

Tribe III. HOMALINEÆ. *Flowers* bisexual. *Styles* as many as the carpels. *Ovary* half-inferior.

VIII. **Blackwellia.**—Perianth with a conical tube, and a 10–30–fid limb; the segments in two rows.

I. ONCOBA, Forsk.

Flowers bisexual. *Sepals* 5, deciduous, concave, strongly imbricated in the bud. *Petals* 5, hypogynous, clawed, obovate, spreading, deciduous. *Stamens* very numerous, inserted, in many rows, on a fleshy, hypogynous ring; filaments filiform; anthers linear, two-celled, basifixed, erect, opening at the sides. *Ovary* free, one-celled; placentæ parietal, 5–10, bearing numerous ovules. *Style* cylindrical, stigma dilated, notched. *Berry* leathery, pulpy within; seeds numerous. *Endl. Gen. No.* 5067. *Lam. Dict. t.* 471.

Shrubs or small *trees*, natives of tropical and subtropical Africa, spinous or unarmed, with alternate exstipulate leaves, and terminal, solitary white flowers of large size (for this Order). The name is an alteration of the Arab *Onkob*, by which the species of North Africa is known.

1. O. Kraussiana (Planch. in Herb. Hook.); unarmed; leaves elliptic-oblong, obtuse or subacute, downy, at length glabrous, very entire; peduncles terminal or opposite the leaves; anthers pointless. *Xylotheca Kraussiana, Hochst. in Pl. Krauss. No.* 352.

HAB. Port Natal, *Dr. Krauss! Mr. Plant! Mr. Sanderson* (Herb. T.C.D., Hook.)

A much-branched shrub (without thorns?) Young branches pubescent, older with a rough ash-coloured bark. Leaves 2 inches long, midribbed and penninerved, with netted veins, rather pale on the under side. Peduncles 2–3 inches long. Flowers solitary, erect, more than an inch broad, white. *Calyx* pubescent; sepals roundish and very concave. Petals twice as long, spreading, with narrow claws, cuneate at base, broadly obovate, with scattered, woolly hairs. Ovary hairy; stigma 5–6-rayed.

II. RAWSONIA, Harv. & Sond.

Flowers perfect, or abortively unisexual. *Calyx* 4–5-parted, the sepals very unequal, concave, imbricate, persistent. *Petals* 4–5, deciduous, unequal, concave (like the sepals), imbricate in æstivation. *Petaloid-scales* opposite the petals and longer than them, hypogynous, each with a 2-lobed fleshy gland at base. *Stamens* very numerous, in several rows, the inner hypogynous, the outer attached to the base of the petaloid scales. *Anthers* sagittate, erect. *Ovary* on a convex torus, unilocular, with 4–5 parietal multi-ovulate placentæ. *Stigma* subsessile, 4–5-parted. *Fruit* a berry?

A South African shrub, with glabrous and glossy, alternate, exstipulate, serrated leaves, and axillary sub-capitate spikes of (yellow?) flowers. The generic name is bestowed in honour of RAWSON W. RAWSON, Esq., C.B., Secretary to Government, Cape of Good Hope; a gentleman strongly attached to Natural History, and joint author of a "*Synopsis Filicum Africæ Australis*"; and to whom the authors of the *Flora Capensis* wish to express their sense of obligation for countenance and assistance afforded to their undertaking.

1 R. lucida (Harv. & Sond.)

HAB. Colony of Port Natal, *Mr. Sanderson*. (Herb. T.C.D., Hook., Sond.)

A shrub or small tree, nearly or quite glabrous. Leaves alternate, oblongo-lanceolate, acute or acuminate, cuneate at base, 3–4½ inches long, 1½–2 inches wide, rigid, glabrous and glossy, penninerved and reticulately veined, sharply serrulate, the serratures 1½ lines apart, directed towards the apex of the leaf, callous. Stipules none. Petioles 2–3 lines long, channelled above. Spikes axillary, scarcely twice as long as the petiole, on peduncles of 2–3 lines, few-flowered. Flowers subsessile, seemingly brownish or greenish yellow. *Sepals, petals,* and *petaloid-scales* all, among themselves, unequal: the longest sepals shorter than the petals; the petals generally shorter than their scales. *Stamens* numerous, 40–60, longer than the floral envelopes. Perfect and imperfect (male) flowers occur together in the same spike.

III. PHOBEROS, Lour.

Flowers bisexual. *Calyx* persistent, with a short, conical tube, and a 10–12-parted limb; the segments in two rows, the inner ones smaller. *Disc* fleshy, filling the calyx tube; its margin, opposite the bases of the outer calyx segments cut into numerous, glandular lobules. *Stamens* very many, in several rows, within the margin of the disc, slightly perigynous; filaments capillary; anthers two-celled, acuminate or horned, splitting. *Ovary* free, sessile, one-celled, formed of 2–3 carpels, with inflexed edges; style single, columnar; stigma subcapitate, bifid; placentæ parietal, riblike; ovules few. *Berry* fleshy, 3–4 seeded. *Wight and Arn. Prod. Vol.* 1. *p.* 29. *Endl. Gen. No.* 5068, *Eriudaphus, Nees. Harv. Gen. S. A. Pl. p.* 296. *Adenogyrus, Klotsch.*

Arborescent *shrubs* or small *trees,* frequently spiny. *Leaves* rigid, glabrous, alternate, entire or toothed; the teeth callous. Flowers small, in axillary racemes. Found in Tropical Asia as well as South Africa. In the Cape species the inner calyx lobes are much smaller in proportion to the outer than they are in the Asiatic; the margin of the disc is conspicuously glandular, and its surface woolly; the calyx tube is more conical and the stamens more perigynous. Name from φοβερος, *formidable*; from the stout spines with which many of the species are armed.

1. **P. Mundtii** (Arn. in Hook. Journ. 3. p. 150.); *unarmed;* leaves ovato-lanceolate, *acuminate,* shining, *sharply calloso-dentate;* racemes and calyces *glabrous,* sepals *acute. Eriudaphus Mundtii, Nees, in E. & Z! En. No.* 1755. *Drege* 3576. *Adenogyrus Krebsii, Kl.*

Hab. In forests. Swellendam, *Mundt!* Uitenhage, *E. & Z!* (Herb. T.C.D.)

A tree 25–35 feet high, with rough ash-coloured bark, glabrous in every part. Leaves petiolate, cuneate at base, somewhat rhomboid, sharply and equally serrate, rigid and glossy. Racemes 6–8 flowered; pedicels spreading, subdistant. Inner calyx segments minute, lanceolate, sometimes wanting. Anthers roundish, the thickened connective produced into a short horn. Berries fleshy, tipped with the persistent style.

2. **P. Ecklonii** (Arn. l. c.); *unarmed;* leaves rhomboid, cuneate at base, obtuse, *entire* or repando-dentate; racemes and calyces *glabrous,* sepals oblong, *acute,* ciliate. *Eriudaphus Ecklonii, Nees in Eck. & Zey! En. No.* 1754. *Adenogyrus Braunii, Kl. Walp. An.* 4. *p.* 227.

Hab. In mountainous woods. Kat River, *Ecklon.* Van Staaden's mts., Uitenhage, *Zeyher!* (Herb. T.C.D.)

A tree 20–35 feet high, with rough ashen bark. Leaves 2–3 inches long, tapering much at the base, somewhat angular at the sides, entire or very obscurely repand beyond the middle. Racemes few flowered. Crest of the anthers bidentate.

3. **P. Zeyheri** (Arn. l. c.); generally *armed* with spreading *spines;* leaves roundish or ovate or obovate, very obtuse, entire or somewhat crenated; *racemes and calyces minutely velvetty,* sepals very *obtuse. Eriudaphus Zeyheri, Nees, in E. & Z! En. No.* 1756.

Hab. In forests. Uitenhage, *C. Zeyher!* Albany, *Mrs. Barber, &c.* (Herb. T.C.D)

A tree 15–20 feet high, generally bristling with axillary, divergent spines, 2–3 inches in length. Leaves about an inch long, very variable in outline, but always blunt. Racemes 6–8-flowered, minutely but equally pubescent in all parts.

IV. TRIMERIA, Harv.

Flowers dioecious. *Male: perianth* 6–10 parted, the segments concave, imbricated in two rows, the inner ones largest. *Disc* bearing marginal glands opposite to each of the outer segments of the perianth. *Stamens* 9–12, perigynous, inserted in parcels of 3 or 4, alternating with the glands of the disc. *Female:* perianth as in the male, but smaller and without glands. *Ovary* free, sessile, unilocular, formed of three valvate carpels; styles 3, short, persistent; placentæ parietal, each bearing near its base a single, ascending, anatropous ovule. *Capsule* dry, three valved, 1–3 seeded; placentæ cordlike, in the middle of each valve. *Harv. Gen. S. A. Pl. Suppl. p.* 417. *Monospora, Hochst.*

Shrubs or trees, natives of South Africa, with alternate, many-nerved, exstipulate, dentate leaves, and minute, axillary spiked or paniculate flowers. Name from τρις, *three* and μερις, *a part;* because, in the first found species, all the parts of the flowers are in threes or multiples of three.

1. T. trinervis (Harv. Gen. p. 417.) leaves *ovate*, acute or obtuse, serrate, nearly glabrous, 3–5-nerved at base; *male* and female spikes *unbranched*, perianth trimerous; stamens 9; *filaments smooth*.

HAB. Forests of the Van Staaden's mts., *Zeyher*. (Herb. T.C.D., Hook.)

A much branched tree 15–20 feet high, with the habit of a *Rhamnus*. Young twigs pubescent. Leaves an inch long, variable in shape; the serratures callous. Both male and female flowers in minute, simple spikes. Bracts 3, at the base of each flower, minute, scale-like; 3 inner segments of perianth linear and petaloid, smooth; stamens in parcels of 3, opposite the inner segments. Capsules $\frac{1}{8}$ inch long, 3 cornered.

2. T. alnifolia (Planch. in Herb. Hook.); leaves *orbicular* or *obovate*, very obtuse or immarginate, serrate, glabrescent, many-nerved at base; *male* spikes *much branched*, female simple; perianth 4–5-merous; stamens 10–12, filaments *hairy. Antidesma ? alnifolia, Hook. Ic. t.* 481. *Monospora grandifolia, Hochst. Pl. Kraus.* (160).

HAB. Eastern districts. Knysna, *Mr. Bowie*. Caffirland, *Rev. J. Brownlee*. Port Natal, *Krauss, Gueinzius*. (Herb. T.C.D., Hook., Sond.)

A large arborescent shrub, 10–12 feet high. Young twigs pubescent. Leaves 2 inches long and rather more in breadth, sub-rotund, coarsely but irregularly serrate, sprinkled with small hairs or glabrescent. Male spikes much branched and twice as long as petiole. Perianth 8–10 parted, in two rows. Capsule 3-cornered, longer than its breadth, turbinate; seed solitary, the testa elegantly pitted.

V. DOVYALIS. E. Mey.

Flowers dioecious. *Male: calyx* deeply 5-cleft, its segments slightly imbricate in the bud. *Petals* none. *Receptacle* covered with fleshy glands. *Stamens* 12–20; filaments filiform; anthers didymous. *Female: Calyx* 5–7-parted, segments bordered with stalked glands. *Perigynous* disc fleshy, adherent to the base of the calyx, its margin deeply lobed. *Ovary* free, one-celled, composed of two (rarely three) carpels, with inflexed edges; placentæ prominent, marginal, each bearing a single ascending ovule. *Styles* 2, rarely 3, divergent, channelled on the upper side, stigma simple. *Fruit* fleshy and pulpy, 1–2 seeded; seeds glabrous. *W. Arn. in Hook. Journ. Bot. Vol.* 3. *p.* 251. *Sond. in Linn. Vol.* 23. *p.* 12.

Rigid, coarse-growing, spiny shrubs, peculiar to South Africa. Leaves alternate exstipulate, glabrous, entire or denticulate. Flowers small and green, axillary; the males in branching clusters, shorter than the leaves; the females solitary. The generic name has not been explained.

1. Dovyalis rhamnoides (Burch. and Harv.); branches *whitish*; leaves *thin* (not coriaceous), ovate, sub-rotund, entire or denticulate, three-nerved, reticulate; peduncle of the female flowers *as long as* the calyx or longer; sepals linear, *much enlarged in fruit. Flacourtia rhamnoides, Burch ! Cat.* 4012. *DC. Prod.* 1. *p.* 256. *E. & Z. No.* 115. *Dovyalis zizyphoides, E. Mey.—Sond. l. c.*

HAB. In the districts of Uitenhage and Albany, *Burchell, Drege, Zeyher, &c.* District of George, *Dr. Alexander Prior*. (Herb. T.C.D., Hook., Sond.)

A rigid, spinous shrub, with spreading branches, mostly armed with patent spines 1–2 inches long. Leaves soft, pale green, shining. Flowers minute, green; the males in branched, axillary clusters, the females solitary, on simple pedicels. Berries "delicious, making a very fine preserve," *(Mrs. Barber)*. I have compared with an original specimen of Burchell's No. 4012; and finding it to agree with our plant, have preserved the earlier specific name.

2. D. rotundifolia (Thunb. and Harv.); branches *ash-coloured;* leaves rigid, *leathery,* roundish or obovate, obtuse or emarginate, entire, 3-nerved at base, reticulate; peduncle of the female flowers *shorter* than the calyx, sepals ovate-oblong, persistent, *not enlarged* in fruit. *D. celastroides, Sond. in Linn.* 23. *p.* 12. *Celastrus rotundifolius, Thunb. Prockia rotundifolia, E. & Z! No.* 119.

HAB. About the Zwartkops' River, Uitenhage, *Zeyher!* Near the sea shore, *Mrs. Barber.* (Herb. T.C.D, Hook., Sond.)

A coarse shrub armed with thorns 2–3 inches long. It is readily known from *D. rhamnoides* by its dark-green, leathery leaves and ash-coloured bark. There are also important differences in the female perianth, which remains unchanged under the ripe fruit. In the present species the lobes of the fleshy disc alternate with the calyx segments, in *D. rhamnoides* they oppose them. The fruit is as good as in *D. rhamnoides.* Both are called "*Zuurebesjies.*"

VI. ABERIA, Hochst.

Flowers dioecious. *Male: calyx* 4–5-parted, its segments nearly valvate in the bud. *Petals* none. *Stamens* indefinitely numerous, on a fleshy receptacle; filaments very short, anthers erect, basifixed, 2-celled, opening outwards. *Receptacle* covered with fleshy glands. A rudimentary ovary. *Female: calyx* 5–7 parted, persistent. *Petals* none. *Ovary* free, sessile, on a lobed, fleshy disc, 2-celled (rarely 3 or 1-celled); the dissepiment sometimes incomplete: ovules solitary, on the inflexed margins of the carpels. *Styles* 2–3, divergent. *Fruit* fleshy, indehiscent, 2-celled, 2-seeded; seeds covered with dense woolly hairs. *Hochst. Bot. Zeit.* 27. 2. *besond. beil. p.* 2.

Shrubs or small trees, unarmed or spiny, natives of Abyssinia and Caffraria. Leaves alternate, simple, entire or denticulate, exstipulate. Male fl. minute, on short simple peduncles; female inflorescence similar, the calyx enlarging as the fruit ripens. The genus is closely allied to *Dovyalis,* from which it differs in the two-celled fruit and woolly seeds. The generic name is taken from Mount Aber, in Abyssinia, where the first discovered species was found.

1. A. Zeyheri (Sond. Linn. xxiii. p. 10.); arborescent, *thorny;* branches clothed with *yellow* hairs; leaves obovate, obtuse, narrowed at the base, 3-nerved, crenate, the younger ones pubescent, the older glabrous; male flowers in clusters of 3–4, their calyx 5-cleft; female solitary, calyx 5–7-parted, sepals ovato-lanceolate; annular disc lobed, villous; fruit fleshy, ovate, tomentose, crowned with the persistent styles. *Sond. l. c.*

HAB. Crocodile River, *Burke & Zeyher!* (Herb. T.C.D., Hook., Sond.)

A middle-sized tree, with greyish, warted branches, armed with sharp axillary spines 1–2 inches long. Leaves, 1–1½ inches long, 8–10 lines broad, remotely crenulate, with minute ciliary processes at each crenature. Stamens very numerous, with very short filaments. Fruit yellowish, tomentose, oblong-ovate, crowned with the styles; seeds densely clothed with long, white hairs.

2. A. tristis (Sond. l. c.) shrubby, *unarmed;* branches *ash-coloured,* somewhat warted, *glabrous,* branchlets *pubescent;* leaves coriaceous, obovate, obtuse or emarginate, the margin sub-revolute, very entire or few toothed, glabrous, 3-nerved, glossy above, pale underneath; female flowers solitary, on axillary peduncles 2–3-times as long as the leaf-stalks; calyx tomentose, sepals 5–6, oblong, acute, spreading; disc 10-lobed, the lobes rounded, villous; ovary ovate, tomentose, with 2 very short styles. *Sond. l. c.* *Royena, n. sp.* (15) *E.&Z.!*

HAB. Philipstown, Kat River, 2–3000ft. *Eck.* & *Zey.!* (Herb. Sond.)

Branches round, ashy, with whitish warts. Leaves 1 inch long, 8–10 lines wide; petioles 2 lines long. Peduncles 3 lines, pubescent. Male flowers unknown.

VII. KIGGELARIA. Linn.

Flowers dioecious. *Calyx* 5-parted, deciduous; *sepals* valvate in the bud. *Petals* 5, imbricate, coriaceous, each with a fleshy gland at its base, inside. *Male: stamens* 10, crowded in the centre of the flower; filaments short, anthers hard and dry, two-celled, opening by terminal pores. *Female: ovary* sessile, one-celled, with 2–5 parietal placentæ; ovules numerous; styles 2–5, short. *Capsule* globose, pubescent, leathery, many seeded, bursting imperfectly into 2–5 valves. *DC. Prod.* 1. *p.* 257.

South African *shrubs* or small *trees*, without spines. Leaves scattered, petiolate, simple, exstipulate. Pubescence stellate, minute. Male flowers in axillary cymes, with long pedicels; females solitary, pedunculate. Named in honour of Francis Kiggelaer, an old Dutch botanist, author of a garden-catalogue, published in 1690.

1. K. africana (Lin. Sp. 1466); leaves ovato-lanceolate, *serrulate*, acute, membranaceous, thinly tomentose on the lower surface, reticulate; styles 5. *DC. Prod.* 1. *p.* 257. *Lam. Ill. t.* 821. *Thunb. Cap. p.* 395. *E. & Z.! No.* 116.

HAB. In hedges and waste places. Common about Capetown. Tulbagh, *Eck. & Zey.!* (Herb. T.C.D., Hook., Sond.)

A much branched, erect shrub, 10–15 feet high, becoming almost a tree. Young twigs thinly tomentose; older with a rough, striate bark. Leaves 2–3 inches long, distantly toothed. Venation-pinnate, with netted intermediate veinlets obvious on both sides.

2. K. Dregeana (Turcz. Animad. p. 63); leaves either lanceolate and acute at both ends or elliptic-oblong, and obtuse, *entire*, membranaceous, green and glabrescent above, pale and minutely canescent below; male cymes lax, petals longer than the sepals, their glands ovate, free above.

VAR. α. **acuta**; leaves acute at both ends, *Drege No.* 6722!

VAR. β. **obtusa**; leaves mostly obtuse, *K. integrifolia, E. & Z.* 111.

HAB. Zuureberge, 2000ft. *Drege!* Sitzekamma, Oliphant's Hoek, and Kaffraria, *E. & Z.!* (Herb. T.C.D., Hook., Sond.)

Leaves of smaller size and thinner substance than in *K. africana;* their margin *quite entire.* The young leaves have a few scattered stellate hairs on the upper surface; the under side is always whitish with minute stellate down. The form is very variable, even on the same bush. Turczaninow says "floribus octandris," but I find *ten* stamens both in Drege's and Ecklon's specimens. The styles have fallen on our specimens, but there are 5 scars on the vertex of the fruit.

Kiggelaria integrifolia, Jacq. is a nonentity, as appears by reference to *Jac. Ic. Rar. Vol.* 3. *p.* 19, where this author states that on a re-examination of the flowers of his plant he finds that the corolla is *monopetalous;* and then refers his supposed *Kiggelaria* to *Royena polyandra,* L. f. *(Euclea elliptica, DC.)!* It is strange that this should have escaped the notice of recent writers who continue to quote Jacquin as an authority.

3. K. ferruginea (E. & Z.! Enum. No. 118); leaves oblong or ovato-lanceolate, coriaceous, entire or denticulate, *covered with rust-coloured pubescence* on both sides; styles (according to E. & Z.) two. *K. africana, E. Mey.! in Herb. Drege (non Linn).*

HAB. Dry places on the Kamiesberg, *E. & Z.!* Paarlberg; Dutoits Kloof, Leliefontein; and near Beaufort, *Drege!* (Herb. T.C.D., Hook., Sond.)

Every part covered with rusty, stellate pubescence. Leaves very variable in shape, stamens sometimes 11–12. I have seen no female flowers. Glands of the petals dark, adhering by their backs to the face of the petal. Eck. & Zey. attribute but *two* styles to this species, and I cannot contradict them, though I think I detect the scars of *five* on the old capsules.

VIII. BLACKWELLIA, Comm.

Flowers bisexual. *Calyx* persistent, with a conical tube and multipartite (10–30-cleft) limb; segments in two rows, the inner ones largest. A *gland* opposite the base of each of the outer segments. *Stamens* perigynous, alternating with the glands, singly, or in parcels of 2 or 3; filaments filiform; anthers didymous, opening longitudinally. *Ovary* half-inferior, one-celled, with 3–5 parietal placentæ; styles 3–5, subulate, divergent; ovules few, pendulous. *Capsule* one or few seeded. *DC. Prod. 2. p. 54.*

Shrubs or small trees, natives of Mauritius, Madagascar and Tropical Asia. Leaves alternate, petiolate, exstipulate, toothed or entire, glabrous or pubescent, penninerved. Flowers in axillary or terminal spikes, racemes or panicles, small. The name is given in honour of Mrs. Eliz. Blackwell, author of "A curious Herbal, containing 500 cuts of the most useful plants, which are now used in the practice of Physic, London, 1737;" a work of much merit, which has been translated into German and Latin. An account of the authoress may be found in *Pulteney's Sketches, Vol. 2. p. 251.*

1. B. rufescens (E. Mey.!–Harv.); leaves elliptic-oblong, entire or denticulate; panicles axillary, longer than the leaves; perianth 16–18-parted; stamens 8–9. *Pythagorea rufescens, E. Mey.*

HAB. Port Natal, *Drege! Gueinzius!* (Herb. T.C.D., Hook., Harv.)

A much branched, nearly glabrous shrub. Leaves 1–1½ inches long, obtuse or subacute, prominently ribbed, penni-nerved and reticulate below; on short petioles. Panicles axillary and terminal, divaricately branched; peduncles minutely pubescent; pedicels as long as the flowers. Perianth tomentose, its limb about 16-parted; laciniæ linear-oblong, obtuse, ciliate. Stamens half as many as the segments of the perianth, alternating with as many fleshy glands; anthers globose, didymous. Ovary very hairy, unilocular, half sunk in the calyx tube. Flowers seemingly reddish.

ORDER XI. VIOLARIEÆ, DC.

(By W. SONDER.)

(Violarieæ, DC. Prod. 1. p. 287. Endl. Gen. No. cxc. Violaceæ, Lindl. Veg. Kingd. No. cxvi.)

Flowers mostly irregular. *Sepals* 5, persistent, imbricate, often produced at base. *Petals* 5, mostly unequal; one spurred, marcescent. *Stamens* 5, alternate with the petals; filaments short and broad, connate at base, hypogynous; anthers introrse, adnate, the connective prolonged into a crest beyond the loculi, (two often spurred at base).

Ovary unilocular, free, with 3 parietal placentæ and numerous ovules; style simple, with a hollow or lobed stigma. *Capsule* usually splitting into three valves, each carrying a medial placenta; rarely indehiscent. *Seeds* with fleshy albumen, and a straight, axile embryo; radicle next the hilum.

Herbs, undershrubs or shrubs, with alternate, simple, entire or cut leaves; usually with large, leafy stipules. *Flowers* axillary, solitary or variously arranged.

The common garden Violet is the well known type of this Order, which also includes a number of tropical and subtropical shrubs *(Alsodineæ)* with small, regular flowers, little resembling the type in aspect, though agreeing in technical characters. Upwards of 300 species, arranged in 15 genera, are known. Species of *Viola*, the largest genus, are found in all parts of the world, but are most numerous in Europe Extra-tropical Asia and North America; a few are Australian, and several occur, along the higher Andes of Peru, descending to the sea level in Southern Chili. The roots of most species contain emetic properties: several of the S. American species are often used as a substitute for ipecacuanha. The petals of the sweet scented Violet are a children's medicine, in frequent use. Several other species are in local repute in the countries where they occur.

TABLE OF THE SOUTH AFRICAN GENERA.

I. **Viola.**—*Sepals* eared at base.
II. **Ionidium.**—*Sepals* not eared at base.

I. VIOLA, L.

Sepals 5, nearly equal, produced at base into earlike lobes. *Petals* 5, unequal, the under one *(labellum)* spurred or saccate at base. *DC. Prod.* 1. 291. *Endl. Gen.* 5040.

Herbs or suffrutices very generally dispersed throughout the Northern temperate zones, rare within the tropics and in the Southern temperate zone. Stem short or long; Leaves alternate, petiolate, stipulate. Peduncles axillary, one-two flowered, bibracteolate, curved, but seldom jointed. Flowers blue, white or yellow, or particoloured, sometimes sweetly scented. The garden Violet *(V. odorata)*, and the Pansy *(V. tricolor)* are familiar examples. The name is of Celtic origin: *fail* meaning a *smell*, and *fail-chuach*, a *violet*:—also said to be derived from the Greek *ιον*, a *violet*.

1. V. decumbens (Linn. f. Suppl. p. 397); suffruticose, stems procumbent, much branched; *leaves linear, very narrow*, entire, crowded; stipules subulate-linear, adnate; *spur tubular, straight, nearly as long as the sepal. Thunb. Fl. Cap. p.* 186. *V. decumbens, α, tenuis, Bartl. Linn.* 7. 540. *E. & Z.* 120. *V. decumbens, β. longifolia, E. Mey.!*

Hab. Sandy places. Hott. Holl. Berg, near Palmiet River, and Klynrivier-berg *E. & Z., Zeyher* 1923, *Drege!* (Herb. Sond., Lehm., T.C.D.)

Stems many from the same root, filiform, brownish, minutely downy, 3–6 inches long. Leaves glabrous, 1–1½ inch long or shorter, ½–¾ line wide, much longer than the internodes. Stipules toothed at base, 1–1½ line long. Flowers terminal, pedunculate; peduncles one or several, flexuous, glabrous, bibracteate near the top. Corolla blue, yellow within. Sepals acuminate. Petals about 5 lines long. Stigma hooked.

2. V. scrotiformis (DC. Prod. 1. p. 299); suffruticose; stems branched; *leaves sub-falcato-lanceolate*, acute, narrowed at base, entire, the lowest ones remote; *stipules lanceolate-subulate; spur saccate, twice as short as the sepal. V. decumbens β. stipulacea, Bartl. l. c. E. & Z. No.* 121. *V. decumbens, α E. Mey.!*

HAB. Among stones, on the Zwarteberg, Caledon, *E. & Z.!* Zey. 1922. Gnadendahl, *Drege! Krauss!* (Herb. Sond., T.C.D.)

Very similar to the preceding. Stem 6–12 inches long, brown or reddish. Leaves, especially the lower-ones, subremote, as long as or rather longer than the internodes, 6–8 lines long, 1 line wide or wider. Stipules 3–4 lines long, toothed at base. Peduncles and flowers as in the preceding, but the spur shorter and more inflated.

3. V. arvensis (Murray); *annual;* stems diffuse, angular; leaves ovate-oblong, crenate-toothed; stipules *pinnatifid. E. Bot. Suppl. t.* 2712.

HAB. A weed in cultivated ground, throughout the Colony and in Caffirland, *E. & Z.!* (Herb. Sond., T.C.D.)

Introduced from Europe.

II. IONIDIUM, Vent. (ex parte) DC.

Sepals 5, unequal, not produced at base. *Petals* 5, very unequal, the under one *(labellum)* much larger than the rest, clawed, the claw dilated and concave, or shortly spurred or saccate at base. *DC. Prod.* 1. *p.* 307. *Endl. Gen.* 5041.

Herbs and undershrubs, chiefly found within the tropics, especially of the American continent, rare in the warmer temperate zone. Leaves alternate or opposite, serrate or entire, stipulate; the stipules lateral, entire or laciniate and multifid. Flowers axillary, or in terminal racemes, usually nodding, the peduncle often jointed below the curved portion. Name, from *ιον*, a *violet* and *ειδος*, *like.*

* *Flowers without spurs.*

1. I. capense (R. & Sch. 5. p. 393); suffruticulose; stems erect, very thinly downy, as are also the leaves; leaves with very short petioles, obovate, the margin recurved, subserrato-dentate; stipules subulate; peduncles axillary, one-flowered; sepals ovate, acute, pubescent; labellum subcordate, roundish. *DC. Prod.* 1. *p.* 308. *E. & Z.!* 123. *Viola capensis, Thunb. Fl. Cap. p.* 186.

HAB. In woods. Galgebosch, *Thunberg!* Hill sides near Port Elizabeth, Krakakamma, &c., *Uitenhage, E. & Z.!, Zeyher* 1919. Near Port Natal, *Drege!* (Herb. Sond., T.C.D.)

Stems ascending, 2–3 inches long. Leaves 8–12 lines long, 3 lines wide, subobtuse, narrowed into the petiole, paler on the under side. Peduncles solitary, filiform, bibracteolate. Sepals nearly 2 lines long. Corolla whitish; upper and lateral petals rather longer than the calyx; labellum clawed, its lamina 4 lines long. Capsule minutely downy, thrice as long as the calyx.

** *Flowers with (very short) spurs.*

2. I. caffrum (Sond. Linn. 23. p. 23); suffruticulose; stems erect, as well as the leaves, *pubescent and scabrid;* leaves on very short petioles, *ovate or oblong-ovate,* acute, the margin recurved, subserrate; stipules subulate; peduncles axillary, one flowered; sepals lanceolate, glabrous at the point; labellum very large, *somewhat 4-angled,* with a very short spur.

HAB. Port Natal, *Gueinzius, Sanderson.* (Herb. Sond., T.C.D., Hook.)

Stems a span-long, somewhat branched, with spreading pubescence. Leaves 8–12 lines long, 5–6 lines wide, longer than the internodes, penninerved. Stipules minute. Peduncles as long as the leaves, or longer, at length nodding. Sepals 2–3 lines long. Labellum pale rosy, 6 lines long and wide; its spur ½ line long.

3. I. thymifolium (Presl. Bot. Bem. p. 11); suffruticulose; stems erect, *glabrous*; leaves subsessile, *oblong-lanceolate or lanceolate*, acute, obsoletely serrate, glabrous, the margin ciliate, narrowed at base; stipules subulate; peduncles axillary, one flowered; sepals lanceolate, glabrous; labellum very large, *transversely-oblong, mucronate*, with a very short spur.

HAB. In grassy places. Omsamwubo, Omsamcaba, Omtendo, towards Port Natal, *Drege*. Gathered also by *Sieber*; station not given. (Herb. Sond., T.C.D.)

6–12 inches high, angulately branching. Leaves 8–12 lines long, 4 lines wide, coriaceous, green, glabrous on both sides, the lower on very short petioles, upper sessile. Peduncles longer than the leaves, compressed, downy, nodding, bibracteolate. Sepals 2 lines long. Labellum thrice as long as the lateral petals, clawed, the limb 4 lines long, 5–6 lines wide, cordate at base, subretuse at apex. Calcar ½ line long.—Known from the last by its want of pubescence, by the smaller, transversely oblong labellum and the narrower leaves.

ORDER XII. DROSERACEÆ, DC.

(By W. SONDER.)

(Droseraceæ, DC. Prod. I. p. 317. Endl. Gen. No. clxxix. Lindl. Veg. Kingd. No. clvii).

Flowers regular. *Sepals* 5, distinct, or connate, imbricate, persistent. *Petals* 5, hypogynous or adnate to the sepals, imbricate, equal. *Stamens* as many as the petals and alternate with them, or 2–4-times as many; filaments free, filiform; anthers extrorse, erect, and fixed, or versatile. *Ovary* free, unilocular, with 3–5 parietal, or a single basal, placenta; rarely 2–3-locular, with axile placentæ, styles 3–5, distinct or partly, or wholly confluent, often forked (sometimes multifid); stigmata capitate. *Ovules* numerous, anatropous. *Capsule* girt with the persistent filaments, dry, splitting into valves; seeds containing much albumen, and a minute basal embryo.

Herbs, or suffrutices; often stemless, sometimes twining plants, more or less covered with glandular hairs, exuding a clammy fluid. Leaves alternate, simple, entire, or rarely cloven, with involute vernation sometimes with stipules. Flowers solitary, or in secund, circinate racemes, gradually unrolling during anthesis, white, pink, or purple, soon withering. Petals very thin and fragile. Several, in Australia particularly, have tuberous roots.

Species of *Drosera* (Sundew,) are scattered over most parts of the world, usually frequenting exposed subalpine bogs and marshy places; but in Australia many are found in the driest ground, where they lie dormant for the greatest portions of the year, reviving with the first rains. Several contain a reddish brown dye, and others a brilliant purple in their roots. None, as yet, have been made available. Though retained in the present work, in the vicinity of Violarieæ, where this group is placed by De Candolle, we quite agree with Dr. Hooker, that these plants are more naturally allied to Saxifrageæ, with which Order many Botanists now unite them.

TABLE OF THE SOUTH AFRICAN GENERA.

I. **Drosera.**—*Styles* 3–5, bifid or bipartite. *Ovary* unilocular.
II. **Roridula.**—*Style*, one, simple, stigma capitate. *Ovary* trilocular.

I. DROSERA L.

Calyx 5-parted, equal. *Petals* 5, obovate. *Stamens* 5; anthers adnate, slitting. *Ovary* unilocular, with 3–5, many ovuled, parietal placentæ.

Styles 3–5, bifid or bipartite, the branches undivided or multifid. *Capsule* membranous, 3–5-valved, manyseeded. *DC. Prod.* 1. *p.* 317. *Endl. Gen.* 5033.

Stemless and scapigerous, or caulescent herbs, mostly perennial, often with tuberous roots, found in all parts of the world; very numerous in Australia. Leaves alternate, scattered or rosulate, clothed on the upper surface and margin with gland-tipped hairs, circinate when young, either exstipulate, or furnished with an axillary membranous, simple, or multifid scale, composed of two confluent stipules. Flowers in scorpioid unilateral cymes, or secundly racemose, rarely solitary; rosy, purple, or white. Petals very delicate, and soon withering. Name, from δροσος *dew*; the glands of the leaves exuding dew-like drops of fluid, which glitter in sunshine: whence, the English name "Sundew."

SECT. I. **Rossolis.** *Styles* 3, bifurcate near the base; the branches club-shaped, *undivided* or bilobulate. *Placentæ* 3, many ovuled. *Rossolis and Crypterisma, Pl. Ann. Sc. Nat.* 1848, *p.* 92. (Sp. 1–6).

**Stemless; leaves radical, rosulate.* (Sp. 1–3).

1. D. trinervia (Spreng Anleit, 1. p. 298); small; all the leaves radical, sessile; *spathulato-cuneate*, 3–5-nerved; stipules *scarcely any*, except two filaments attached to the base of the petiole; scapes few flowered, with the pedicels and calyces glanduloso-pubescent. *DC. Prod.* 1. *p.* 318. *Planch. l. c. p.* 191. *D. cuneifolia.* α & γ, *Thunb. Dissert. p.* 5. *Bartl. Linn.* 7. *p.* 620. *E. & Z.* 124. *Un. Itin.* 254, 33.

HAB. Table mountain. *Thunb.*, *E & Z.! W. H. H. Pappe, &c.* Hott. Holl, & Tulbagh Mts. and on the Piquetberg, *Zeyher* 53. Paarlberg & Drackensteen, *Drege*, 7258. (Herb, Sond. Lehm. T.C.D.)

About two inches high; rarely 3–4 inches. Leaves rosulate, 4–8 lines long, 1½ 3 lines wide, glabrous on the lower surface. Scape slender, 2–4 flowered: rarely 6–8 flowered; pedicels 1–2 lines long. Calyx lobes ovate. Petals white, twice as long as the calyx. Anthers ovate. Stigmata dilated, cuneate, obsoletely palmatifid.

2. D. Burkeana (Planch. l. c. p. 192); dwarf; all the leaves radical, small, *with a subrotund lamina* shorter than the petiole; stipules connate into a single intra-axillary piece, conspicuous, each bifid; scapes slender, racemose, the pedicels and calyces glandularly puberulous.

HAB. Aapjes-rivier, Macallisberg, *Zeyher & Burke!* Natal, *Sanderson.* (Herb., Hook., Sond., T.C.D.)

Leaves ½ inch long, the lamina shorter than the petiole, 2 lines wide, naked on the lower side; the marginal ciliæ longish. Scape 3–4 inches long, 2–6 flowered; pedicels 1 line long, the lower often elongate. Calyx obtuse, in flower 1 line, in fruit 1½ lines long. Petals white or rosy, twice as long as the calyx.

3. D. cuneifolia (Thunb. Fl. Cap. p. 278); small; all the leaves radical, sessile, *cuneiform-obovate*, 3–5-nerved, *stipules connate into a single* 10–*cleft intra-axillary piece*, conspicuous, scapes slender, racemose 5–10 flowered; the pedicels and calyx glanduloso-pubescent. *D. cuneifolia* β. *Thunb. Planch. l. c. p.* 195. *E & Z. No.* 125. *Hb. Un. It.* 253.

HAB. Sandy, moist places. Table Mt.; sides of Winterhoeksberg, near Tulbagh, and in Zwartland, *Thunb.*, *E & Z*, *Drege, &c.* (Herb. Sond. Lehm., T.C.D).

From 3–6 inches high, to a foot or more. Raceme in the largest specimens 12–16 flowered. Flowers purple, larger than in *D. trinervia*, from which it also differs in the broader leaves, the evident stipules, and generally the longer raceme.

** *Caulescent; leaves scattered along the stem.* (Sp. 4–6.)

4. D. Capensis (Linn. Sp. 403); sub-caulescent; leaves sub-radical, *oblong-linear, obtuse,* tapering into a long, flat, glabrous, or somewhat hairy petiole, glabrous on the lower surface; stipules *connate into a single ovato-lanceolate rather large intra-axillary piece*; scape ascending, angulate, roughly hairy, longer than the leaves; corolla twice as long as the calyx. *Burm. Afr. t.* 75 *f.* 1. *Berg. Cap. p.* 81. *Thunb. Diss. p.* 6. *Fl. Cap. p.* 620. *DC. l. c. p.* 318. *E. & Z.* 126. *Planch. l. c. p.* 196. *Hb. Un. It.* 252.

HAB. Wet places, in subalpine situations, near Capetown; Dutoit's-kloof: Paarlberg, and Tulbagh, &c. *Thunb. E. & Z. &c.* Herb. Sond., T.C.D).

Stem sometimes scarcely any, sometimes 1–2 inches long. Leaves with their petiole, 4–6 inches long, the lamina 1½-3 lines wide, about as long as the petiole, which is a line wide. Stipules pale, 3-4 lines long, entire or sublacerate at the point. Scape from an ascending base erect, 6-12 inches high. Raceme 6–20 flowered. Flowers purple; pedicels as long as the calyx, or longer. Capsule with 3 obcordate valves. It varies with the scape and peduncles more or less slender, hairy or glabrous, and flowers larger or smaller.

5. **D. hilaris** (Cham. & Schl. Linn. 1. p. 548); caulescent, tall; leaves *spathulato-lanceolate* obtuse, narrowed into *a wide, tomentose petiole,* glabrous on the lower surface; stipules connate into a *deeply cut setaceo-laciniate intra-axillary piece;* scape erect, or subascending, hairy, glandulose above, much longer than the leaves; corolla thrice as long as the calyx. *E. & Z.* 127. *Planch. l. c. p.* 201.

HAB. East side of Devil's & Table Mts., and near Constantia. *Bergius, Mundt & Maire, E. & Z. W.H.H,* &c. (Herb. R. Ber., Sond., T.C.D).

Stem 1–3 inches long, clothed, as well as the leaves, with a reddish tomentum. Lower leaves at length reflexed, spathulate, 1–2 inches long; intermediate spreading or recurved, somewhat longer; upper rosulate, about 3 inches long, 3–5 lines wide, all with long marginal cilia. Stipules 1½–2lines long, of the colour and substance of those of *D. ramentacea,* adnate to the inner face of the petiole; in the middle region of the stem, on account of the petioles being appressed to the stem, they are often inconspicuous, but easily seen in the upper portion. Scapes 6–12 inches high, hairy, and without glands at base, upwards towards the inflorescence glandular, together with the pedicels, bracts and calyx. Raceme 4–8 flowered, pedicels 3 lines long, bracts minute. Calyx 3 lines long; sepals broadly ovate, subacute. Corolla dark purple, 12–15 lines across; petals obovate. Styles 3, bifid at the very base; the branches sometimes again bifid, stigmata clavate, emarginate. Near *D. Capensis* and *D. ramentacea,* with which it agrees in the caulescent habit; but from both which it is readily known (at least when dry) by the denser, more reddish pubescence, the broader leaves, and the wider and generally shorter tomentose petiole.

6. **D. ramentacea** (Burch. Cat. Geogr. No. 7692); stem *elongate, covered with old (withered) deflexed leaves;* leaves crowded round the apex of stem; the younger erect, with a *narrow-obovate* lamina, somewhat hairy on the lower side, half as long as the semi-terete petiole; stipules connate into a deeply cut setaceo-laciniate, intra-axillary piece; scapes from an ascending base erect, much longer than the leaves, glabrous below, above, with the calyces and pedicels, glandularly hairy; corolla twice as long as the calyx. *Planch. l. c. p.* 197.

VAR. α, **Burchelliana**; stem tall; petiole glabrous on the upper, strigose on the lower side and ciliated, equalling or exceeding the linear-oblong or obovate lamina. *D. ramentacea, DC. l. c. p.* 318.

VAR. β. **glabripes**; stem tall; petiole glabrous, about twice as long as the lamina. *Harv. Thes. t.* 26.

VAT γ. **curvipes**; stem short; petiole flat or with the margins revolute, and thus sulcate below, strigoso-pilose, twice as long as the lamina, flowers smaller. *D. curvipes, Pl. l. c. p.* 196.

HAB. Interior districts, *Masson, Burchell.* Summit of Table Mountain: eastern side, *Ecklon, Preiss*; β same situation, *Pappe, W.H.H.*; γ, at Macallisberg, *Burke & Zeyher.* (Herb. Sond. Hook., T.C.D).

Stems a foot or more high, very leafy; in γ, $1\frac{1}{2}$ inches long. Leaves $1\frac{1}{2}$ 2 inches long; the lamina $\frac{1}{2}$–1 inch long, 2 lines wide, the upper surface covered with long, bright red, gland-tipped hairs. Stipules scarious, somewhat horny, fulvous, shining, 4 lines long, very deeply six-cleft, the laciniæ subulate, attenuate, unequal; rarely undivided or shortly cut. Scape 4–8 inches long, sub-compressed and furrowed. Raceme 4–12 flowered; pedicels bracteate, the lower ones longest. Calyx obtuse, 2–3 lines long. Petals dark purple. Anthers oblong. Capsule longer than the calyx, with obcordate valves.

SECT. 2. **Ptycnostigma,** *Pl. l. c. p.* 92. *Styles* 3, bifurcate or bipartite, the branches *flabellato-multifid*, the subdivisions gradually, and not greatly dilated from base to apex. (Sp. 7–8.)

7. D. pauciflora (Banks: DC. Prod. 1. p. 317) *stemless;* leaves sessile, spathulato-cuneiform, 3-nerved; stipules none; scape ciliate, with glandular hairs, about one-flowered. *Pl. l. c. p.* 202. *D. grandiflora, Bartl. Linn.* 7. *p.* 620.

VAR. β. **minor**; all parts more slender, flowers smaller; pale or white.

VAR. γ. **acaulis**; scape shorter than the leaves, flowers smaller, whitish. *D. acaulis, Thunb. Prod. p.* 57. *Fl. Cap. p.* 278.

HAB. Wet places below the baths on the Zwarteberg, Caledon, *E. & Z. Zey.* 1921. Paarlberg, *Drege! W.H.H.* β, in the same places, *Zeyher,* 1920. γ, Koude Bokkeveld, beyond Eland's kloof, *Thunberg.* (Heb. Thunb. Sond., T.C.D).

Leaves 6–8 lines long, 1–3 lines wide, with bright red glandular hairs as long as the breadth of the leaf. Scape erect, leafless, or rarely with one leaf, 3–6 inches high; in β, 1–2 inches; in γ, 1–4 lines long, never altogether wanting, glandular, the glands more copious towards the summit, and about the calyx. Flower one, sometimes two, and more rarely 3–4. Calyx obtuse, 2–3 lines long. Petals obovate-cuneate, sub-retuse, rosy, with a dark purple spot at the base; in the varieties, paler or white, in α, 8–12 lines, in β & γ. 4 lines long. Anthers ovate. Styles bifid, half as long as the petals; the arms capillaceo-multifid. This has the habit of *D. trinervia* & *D. cuneifolia,* but it is easily known by its larger flowers and multifid styles.

8. D. cistiflora (Lin. Amoen. 6. p. 85); *stem erect, simple, leafy*; leaves lanceolate or linear-lanceolate, the radical ones rosulate and subspathulate; flowers terminal, solitary or few, pedicellate. *Burm. Afr.,* 75. *f.* 2. *Thunb. Fl. Cap. p.* 279. *Un. It.* 251. *E & Z.* 129. *Pl. l. c.* 202.

VAR. α. **alba**; flowers white, petals spotted at base, *Thunb. l. c.*

VAR. β, **violacea**; flowers rosy, purple, or red. *Thunb. l. c. D. violacea, Willd.*

HAB. Moist Sandy places near Capetown; on the Cape Flats, & Hott. Holl. common. Also Berg-river, Saldanha bay, Brackfontein, Clanw., Klyn-river, Caledon; under Vanstaadenberg Mts. Uitenhage. (Herb. Thunb., Sond. T.C.D. &c).

Root fasciculate, succulent. Stem 3 inches to a foot high, covered with short, glandular, clammy, mostly reddish hairs. Radical leaves spathulate, obtuse or oblong subacute, narrowed at base, 6–8 lines long; cauline alternate, sessile, spreading, 1–$1\frac{1}{2}$ inches long, 1–2 lines wide. Flowers solitary, or 2–6 in a raceme. Calyx lobes acute. Petals obovate, subretuse, in some specimens 5–6 lines, in others

1 inch long, striate, mostly with a deep-colored spot at base. Styles $\frac{1}{2}$ as long as petals, deeply bipartite; the arms flabellately multifid. The smaller, few-leaved, less glandularly hairy specimens constitute *D. speciosa*. *Presl. Bot. Bem. p.* 14. *Pl. l. c. p.* 202; and the taller many leaved, 3–6 flowered, the *D. helianthemum*, *Pl. l. c. p.* 203.

II. RORIDULA, Linn.

Calyx 5-parted, equal. *Petals* 5, oval or oblong. *Stamens* 5; anthers adnate, opening by terminal pores. *Ovary* trilocular; ovules solitary or in pairs, pendulous from the summit; *style* simple, stigma capitate. *Capsule* 3 celled, 3–valved; seeds solitary. *DC. Prod.* 1. *p.* 320. *Endl. Gen.* 5038.

Suffruticose or shrubby, glandularly hairy, and viscid plants, natives of South Africa. The name is a diminution of *ros, roris*, dew: because of the dew-like drops that exude from the hairs of the leaves. *R. dentata* is hung up in country houses, (according to *Thunberg*) for the purpose of catching flies.

1. R. dentata (Lin. gen. p. 567); leaves linear-lanceolate, subulate, acuminate, *pinnatifido-dentate*, the teeth filiform, glandularly ciliate; flowers racemose; pedicels *longer* than the bract; sepals lanceolate. acuminate, glandular at the margin, *as long as the obtuse petals. Lam, Ill.* 1. 141. *DC. Prod.* 1. 320. *Pl. l. c. p.* 307. *Drosera Roridula Thunb.*

HAB. Mountain tops. Rodesand, Bokkeveld, and elsewhere, near streams, *Thunb.* Stellenbosch, on high mountains, between Nieuwekloof and Ylandskloof; Clanwilliam, on the Blauberg and near Honig Valley, *Drege!* (Herb. Sond. T.C.D).

The whole plant viscid. Stem shrubby, 3–6 feet high, branches and ramuli brownish, glabrous. Leaves crowded at the end of the branchlets, one nerved, ciliate with long and short hairs, and pinnatifid with patent ciliæform, 2–3 lineal subulate teeth, black when dry, 2–2$\frac{1}{2}$ inches long, 1$\frac{1}{2}$–2 lines wide at base. Racemes ending the branchlets, villous, 4–6 flowered; pedicels 6–12 lines long, bibracteolate in the middle, and subtended by a leafy bractea of their own length or shorter. Sepals from an ovato-lanceolate base subulate, acuminate. Petals oval, pale rosy or white. Anthers oblong. Capsule valves ovate.

2. R. Gorgonias (Pl. l. c. p. 307); leaves linear-lanceolate, subulate, acuminate, *entire*, densely glandularly ciliate, racemes (in flower) spiciform. *Pedicels more than twice as short as the bract;* sepals lanceolate, setaceo-acuminate, villoso-ciliate at the margin, *longer than the acute petals. R. dentata. E. & Z.* 130.

HAB. In high moist places, on the mountains, near Tulbagh; and in similar situations near Riv. Zonderende, Swell., *E & Z.* (Herb. Sond).

It differs from the preceding in the more slender habit, leaves not pinnatifido-dentate, and especially in the spicato-racemose inflorescence, and sepals without glands and with a white-woolly margin. The denuded branches bear at the summit crowded, taper-pointed leaves, 2$\frac{1}{2}$ inches long, 1$\frac{1}{2}$ lines wide at base. Flowering pedicels very short, bracteolate at base; the fruiting ones longer ($\frac{1}{2}$ inch long). Capsule valves oblong.

ORDER XIII. POLYGALEÆ, Juss.

(By W. H. HARVEY.)

(Polygaleæ, Juss. An. Mus. 14. p. 386. DC. Prod. 1. p. 321. Endl. Gen. No. ccxxxiii. Polygalaceæ, Lindl. Veg. Kingd. No. cxxxiii.)

Flowers irregular *Sepals* 5, (rarely 4–3), distinct, unequal, strongly imbricated, three exterior, two lateral (the *alæ* or wings) interior and often petaloid. *Petals* 3; one in front (keel or *carina)* larger, concave, enclosing the stamens and ovary and very often crested; two much smaller, toothlike, lateral and adnate to the staminal tube; rarely 5. *Stamens* 8, hypogynous, their filaments united into a split tube (very rarely free); anthers erect, fixed, one celled, opening at the summit. *Ovary* free, compressed, bilocular, the valves fore and aft; ovules solitary, pendulous, anatropous. *Style* single, thickened upwards. *Fruit* a dry capsule or a berry; seeds solitary, with fleshy albumen, and an axile embryo; the radicle next the hilum.

Herbs, undershrubs, shrubs or even trees, with scattered (rarely opposite), or fascicled, simple, entire, exstipulate leaves. Flowers solitary, racemose or spiked, commonly purple or pink, rarely yellow, blue or white; the pedicels tribracteate at base.

A considerable Order, comprising between 500 and 600 species, dispersed throughout the tropics and the warmer parts of the temperate zone, with a few outliers in the colder zones. The flowers are remarkable for great irregularity. The calyx, in *Polygala*, is partly coloured, and its two lateral sepals often form the most conspicuous part of the flower; the petals are generally confluent into a single, boat-shaped piece, to which the staminal tube is more or less adnate; the whole blossom is almost papilionaceous. The general properties of the Order are bitterness and acridity; some are valuable tonics, and others emetics and cathartics. Few are inert. The *Kramerias* (Rhatany-roots) are strongly astringent, and their coloured juices used to adulterate Port wine. *Polygala serpentaria* has a Colonial reputation, as a remedy for the bite of snakes.

TABLE OF THE SOUTH AFRICAN GENERA.

* *Sepals very unequal, the two lateral winglike.*

I. **Polygala.**—*Capsule* membranous, oblong or obcordate.
II. **Mundtia.**—*Fruit* a juicy *drupe.*

* *Sepals nearly equal, of similar form.*

III. **Muraltia.**—*Capsule* membranous, mostly 4-horned or 4-tubecled.

I. POLYGALA. Tourn.

Sepals 5, the two lateral *(alæ)* much larger than the rest, winglike, and coloured. *Petals* 3–5, united at base and attached to the staminal tube; the lower one keelshaped, usually with a multifid crest below the apex; lateral petals small, simple or bifid; posterior frequently wanting. *Stamens* 8, united into a slit tube, and hidden within the anterior petal. *Style* bent upwards; stigma oblique. *Capsule* membranous, compressed, elliptical, obovate, or obcordate, often notched; seeds generally pubescent. *DC. Prod.* 1. *p.* 321.

Shrubs, undershrubs or *herbs,* with alternate, rarely opposite, simple, entire leaves, and racemose, spiked or capitate, terminal or lateral inflorescence. Pedicels tribracteate at base. An immense genus, common in the Northern temperate zone, and in the tropics of Asia and America, as well as in S. Africa. Name, πολυ, *much* and γαλα, *milk:* but why?

The Cape species are numerous and difficult to determine. They appear to me to have been needlessly multiplied by authors describing from solitary specimens. I have endeavoured, as far as my materials allow, to reduce the spurious species to their proper heads. In some cases I may not have gone far enough; in others it may be thought that I have gone too far. If those who think so will give me the means of forming a juster estimate of the value of reputed species, by sending me

well-dried specimens, I shall be glad to reconsider any of my present decisions. The following artificial table refers to the species I have actually examined :—

ARTIFICIAL ANALYSIS OF THE SPECIES.

A. *Shrubs or undershrubs (rarely subherbaceous).*

(a) *Alæ* coloured, broader than the capsule.
  * *Leaves* opposite ... ... (1) **oppositifolia.**
  ** *Leaves* alternate ; *lateral petals* bifid or bilobed.
    † *Racemes* terminal, or sublateral, erect or erecto-patent.
      Leaves ovate or lanceolate, oblong or linear *(flowers showy.)*
        Shrubs.
          Leaves flat, ovate or lanceolate ... ... (2) **myrtifolia.**
          Leaves with revolute margins, linear, truncate at base ... ... (4) **teretifolia.**
          Leaves with revolute margins, linear, tapering at base ... ... (3) **pinifolia.**
        An *undershrub* (leaves and pubescence variable) ... (6) **bracteolata.**
      Leaves subulate, very narrow *(flowers rather small)* ... (12) **ericæfolia.**
    †† *Racemes* lateral, horizontally patent or deflexed.
      Leaves linear, the margin reflexed or revolute.
        A *shrub*, with showy flowers ... ... (5) **peduncularis.**
        An *undershrub*, with small flowers ... ... (16) **affinis.**
      Leaves subulate, or linear-lanceolate, flat.
        Leaves linear-lanceolate ; *carina* with a minute, undivided crest... ... ... (19) **Lehmanniana.**
        Leaves subulate ; *carina* with a multifid crest ... (17) **refracta.**
        Leaves few and very minute ; *stems* filiform ... ... (18) **macra.**
  *** *Leaves* alternate ; *lat. petals* oblong, obovate or obcordate *(not bifid).*
    † *Racemes* terminal, or sublateral, erect.
      (1) *Plants* glabrous or nearly so.
        Anterior sepals connate into a bilobed sepal ... ... (15) **tenuifolia.**
        Ant. sepals distinct ; *leaves* ciliato-denticulate ... (13) **ciliatifolia.**
        Ant. sepals distinct ; *leaves* not ciliate or denticulate.
          Stems sharply triquetrous ; *alæ* ovate, acute ... (7) **triquetra.**
          Stems terete, shrubby ; *alæ* obtuse *(flowers showy)* (8) **virgata.**
          Stems terete, suffruticose ; *alæ* obtuse *(flowers smaller).*
            Capsules obcordate.
              Branches erect, virgate ; *seeds* albo-tomentose (9) **hottentotta.**
              Branches erect, virgate ; *seeds* half-glabrous (10) **seminuda.**
              Branches divaricate, much divided, rigid ... (11) **leptophylla.**
            Capsules oblong, bidentate ... ... (14) **Garcini.**
      (2) *Plants* tomentose, villous, or pubescent.
        Pedicels shorter than the flower ; *alæ* pubescent ... (22) **pubiflora.**
        Pedicels shorter than the flower ; *alæ* glabrous, obovate ... ... (24) **hispida.**
        Pedicels shorter than the flower ; *alæ* glabrous, ovate (25) **Ohlendorfiana.**
        Pedicels as long as the flower, or longer ... ... (23) **gracilipes.**
    †† *Racemes* lateral, patent, or reflexed.
      Leaves subulate, glabrous ... ... (20) **Ludwigiana.**
      Leaves minute, ovate-oblong or oblong, mucronulate, pubescent ... ... (21) **brevifolia.**

(b) *Alæ* greenish or green, narrower than the capsule.
  Racemes erect, terminal or lateral, pluri-flowered ... ... (26) **Bowkeræ.**
  Racemes lateral, patent or reflexed ; *leaves* obtuse or mucronulate, elliptic, oblong, or linear ... ...
    Alæ elliptical, mucronate, albomarginate ; *petals* clawed, cuneate, truncate ... ... (27) **Serpentaria.**
    Alæ obliquely oblong, curved, acuminate, green ; petals oblong, obtuse ... ... (28) **asbestina.**

Racemes lateral, patent or reflexed; leaves acute at both ends, lanceolate.
Decumbent; branches angular; *leaves* linear lanceolate; *alæ* oval-oblong, acute; capsule obcordate ... (29) **illepida.**
Erect; branches filiform; *leaves* lanceolate, acuminate, pungent; *alæ* ovato-lanceolate, acute; capsule oval-oblong ... ... (30) **amatymbica.**

B. *Herbaceous plants, with minute flowers in terminal spiked-racemes.*

Leaves linear, obtuse; *bracts* oval; *carina* longer than the petals; *capsule* obcordate ... ... (31) **pallida.**
Leaves linear-subulate, acute; *bracts* subulate; *carina* shorter than the petals; *capsule* orbicular ... ... (32) **capillaris.**

Group 1. OPPOSITIFOLIÆ. Shrubs, with opposite leaves. (Sp. 1).

1. P. oppositifolia (Linn.); glabrous or pubescent, shrubby; leaves *opposite, subsessile,* spreading or reflexed, cordate or ovate or acuminate, mucronate; racemes subterminal, few-flowered, the pedicels much longer than the ovate, keeled, one nerved bracts; alæ broadly elliptic-ovate, oblique, mucronulate; anterior sepals ovate, obtuse; lateral petals bilobed, the posterior lobe earshaped, reflexed, the anterior deltoid or cuspidate; capsule obcordate. *DC. Prod.* 1. *p.* 322.

VAR. *α.* **nummularia**; glabrous; leaves sub-orbicular, cordate at base, mucronulate. *P. nummularia, Burch. DC. l. c. E. & Z. !* 131.

VAR. *β.* **cordata**; glabrous or downy; leaves broadly cordate, acute, or acuminate. *P. cordifolia, Thunb. E. & Z. !* 133. *P. attenuata, Lodd. E. & Z. !* 134. *P. Zeyheri, Spreng. E. & Z. !* 135.

VAR. *γ.* **cuspidata**; glabrous; leaves cordate, cuspidato-acuminate; branches more or less 4-angled. *P. tetragona, Burch. DC. l. c. E. & Z. !* 132.

VAR. *δ.* **trigonoides**; glabrous; leaves large, broadly ovate or sub-triangular, acute. *P. oppos. var. trigonoides, E. Mey. !*

VAR. *ε.* **latifolia**; pubescent; leaves broadly ovate, cuspidate-acuminate. *P. latifolia, Ker. Bot. Reg.* 645. *E. & Z. !* 136.

VAR. *ζ.* **borboniæfolia**; pubescent; leaves ovate, acute. *P. borboniæfolia, Burch. DC. l. c. E. & Z. !* 138.

VAR. *η.* **rhombifolia**; glabrous or pubescent; leaves rhombic ovate, trapeziform or ovato-lanceolate. *P. oppositifolia, E. & Z !* 137. *Drege ! & var. trapezoides, E. M. !* *P. rhombifolia, E. & Z. !* 139. *P. macrantha, Turcz. Anim. p.* 75.

VAR. *θ.* **lanceolata**; glabrous; leaves lanceolate ! *P. glauca, E. Mey. ! (pro parte).*

HAB. Subalpine woods and scrubs, from Swellendam eastwards to Port Natal and Delagoa Bay. Aug.–Dec. (Herb. T.C.D., Hook., Sond., &c.)

A tall, slender, virgate shrub, with conspicuous, brilliant purple flowers; frequently cultivated. The leaves are extremely variable in size and shape, and the flowers in size and number; but all the reputed species into which the old *P. oppositifolia* has been split may be reduced under *two* principal forms or races, namely, those with *cordate,* and those with *ovate* or *rhomboid* leaves. A careful study of a very large suite of specimens in several Herbaria, has convinced me that all may be further reduced to a single *species.*

Group 2. MYRTIFOLIÆ. Much branched, leafy shrubs, with alternate leaves. Racemes few-flowered, short, lateral or subterminal. Flowers large, brilliant purple, amply crested. (Sp. 2–5).

2. P. myrtifolia (Linn.); glabrous or pubescent, shrubby; leaves alternate, petiolate, *flat*, erect or spreading, dense, either elliptic-oblong, obovate, lanceolate or sublinear, obtuse or acute; racemes few flowered, *sessile*, *subterminal;* pedicels much longer than the bracts; alæ ovate-cordate, oblique, mucronulate; anterior sepals ovate (sometimes sharply keeled); lateral petals 2-lobed, the posterior lobe ear-shaped and reflexed, anterior acute, deltoid or cuspidate; capsule obcordate. *DC. Prod.* I. *p.* 322. *E. & Z. No.* 140–144.

VAR. *α.* **amoena**; leaves elliptic-oblong or obovate, obtuse or subacute. *P. myrtifolia and P. amoena, Thunb. P. grandiflora, Lodd. Cab. t.* 1227.

VAR. *β.* **Natalensis**; villoso-pubescent; leaves scattered, ovato-lanceolate, acute at each end and mucronate. *Sond.! Linn.* 23. *p.* 14.

VAR. *γ.* **Cluytioides**; glabrous or nearly so; leaves densely set, lanceolate or ovate-lanceolate, acute, mucronate. *P. Cluytioides, Burch. DC. l. c. P. glauca E.M.! (pro parte) E. & Z.!* 143.

VAR. *δ.* **ligularis**; pubescent; leaves linear, subacute, tapering to the base. *P. ligularis, Bot. Reg. t.* 637. *DC. l. c. Eck. & Zey.! No.* 144.

HAB. Common, in mountain gullies from Capetown to Port Natal. (Herb. T.C.D., &c.)

A densely branched, well covered shrub, 3–8 feet high, with large, showy, purple flowers at or near the ends of the branches. Leaves very variable in shape, mostly tapering to the base; sometimes very obtuse and even round-topped, sometimes acute or even acuminate. Among the above varieties, var. *δ. ligularis* is most like a species; but it runs insensibly into *γ.*, which passes by sundry gradations into *α.* The characters of var. *β.* appear to arise from its having grown in a very shady and moist situation. The floral characters are nearly as variable as those of the leaves; as above indicated.

3. P. pinifolia (Lam. Ill. t. 598, f. 2.); shrubby, the branches *nearly glabrous;* leaves alternate, crowded, erect, glabrous, *linear, the margins revolute,* the under surface with a central furrow, mucronate, *tapering at base;* racemes few-flowered, lateral and terminal, *subsessile;* pedicels shorter than the flower; bracts persistent, blunt; alæ cordate-ovate, oblique, obtuse; lateral petals two-lobed, the posterior lobe blunt, ear-shaped, reflexed, the anterior acute and cuspidate. *DC. Prod.* I. *p.* 322. *Drege! Herb. (not E. & Z.)*

HAB. Kendo, Zwarteberg, 3–4000ft., *Drege!* (Herb. T.C.D., Hook., Sond.)

A much branched, erect shrub, 2–3 feet high. Leaves longer and narrower than in *P. teretifolia,* and tapering (not truncate) at base. The whole plant is nearly glabrous. E. & Z.! confounded this plant with *P. teretifolia* in their joint collection; but the true plant occurs in Zeyher's recent collection.

4. P. teretifolia (Thunb.! prod. p. 120.); shrubby, the branches *albo-tomentose;* leaves alternate, very patent, glabrous or downy, *linear-terete, with revolute margins* and a central furrow, mucronate, *obtuse or truncate at base;* racemes lateral and terminal, few-flowered, *subsessile;* pedicels shorter than the flower; bracts persistent, deltoid;

alæ cordate-ovate, oblique, obtuse; lateral petals 2-lobed, both lobes acute and cuspidate. *Thunb.! Cap. p.* 554. *DC. Prod.* 1. *p.* 323. *E. & Z. No.* 153. *and* 154! *P. Eckloniana, Presl!*

HAB. Kamiesberg, *Drege! E. & Z.!* Gauritz river, *E. & Z.!* (Herb. Thunb., T.C.D., Hook., Sond.)

Nearly allied to *P. pinifolia*, but with squarrose leaves, white-woolly branches, &c. Much branched, 2–3 feet high, with the aspect of a *Phylica*. Flowers a palish purple, often turning green when dry.

5. P. peduncularis (Burch. Cat. 5163.); scabrous or glabrous, shrubby, with *angular branches;* leaves alternate, occasionally sub-opposite or ternate, approximate, *spreading*, linear, *with revolute margins*, mucronate, obtuse at base; racemes *lateral* and terminal, few-flowered, *conspicuously pedunculate;* peduncle angular, spreading, pedicels as long as the flower; bracts persistent, *subulate, acute;* alæ cordate-ovate, oblique, obtuse or sub-acute; lateral petals deeply bilobed, both lobes obtuse and nearly equal. *DC. Prod.* 1. *p.* 323 *(not of E. & Z.)*

VAR. *α.* **scabra**; stems hispidulous; leaves scabrous above. *Drege* 7188! *P. rosmarinifolia, E. & Z.!* 155. *(pro parte) P. hispidula, Presl.*

VAR. *β.* **glabra**; stem and leaves glabrous, the latter sometimes narrow-oblong. *P. intermedia, Drege! (not of E. & Z.) P. rosmarinifolia E. & Z. (p. parte.)*

HAB. Outeniquaberg, 3000ft., *Drege.!* Zeekoevalley, *E. & Z.!* *β.* between Pickenier's Kloof and Markuskraal, 1–1500ft. January. (Herb. T.C.D., Hook., Sond.)

An erect, sparingly branched shrub, 1–2 feet high, with virgate branches. Racemes mostly lateral, 2–6-flowered, on widely spreading peduncles 1–2 inches long; the flowers resembling those of *P. myrtifolia* and equally large. Our var. *α.*, appears to agree exactly with *P. peduncularis*, DC.; var. *β.* may be that author's *P. intermedia.* Its leaves are longer and sometimes much broader than in var. *α.*, and it looks sometimes suspiciously like *P. myrtifolia*, but differs from the narrowest leaved varieties of that species by the revolute leaf-margins and long peduncles. The lower leaves are often deflexed.

Group 3. VIRGATÆ. Virgate shrubs or under-shrubs, sparingly branched, with scattered, mostly linear or acicular leaves. Racemes *terminal*, many-flowered (rarely few-flowered), lengthening during anthesis. Flowers large or mediocre, crested; the alæ much *broader* than the capsule. (Sp. 6–15).

6. P. bracteolata (Linn.); glabrous or pubescent, suffruticose, with virgate branches; leaves alternate, lanceolate or linear, obtuse or acute, or cuspidate, flat, with a thickened or slightly reflexed margin; racemes terminal, many-flowered; pedicels *longer* than the flower; bracts *linear or lanceolate, persistent, the lowest longest;* alæ *broadly ovate, acute or acuminate;* anterior sepals ovate; lateral petals cuspidate, minutely eared at base. *DC. Prod.* 1. *p.* 322. *Bot. Mag.* 345. *Drege, No.* 7195, 7196! *Zeyher* 1940! *Burm. t.* 73. *f.* 2, *and* 5. *Thunb.! Cap. p.* 555.

VAR. *α.* **racemosa**; glabrous, or nearly so; leaves linear-lanceolate; racemes elongate. *P. bracteolata, E. & Z.!* 159. *P. Burmanni, E. & Z.!* 160 *(not of D.C.)* and *P. intermedia, E. & Z.* 158. *P. subulata, E.Mey!*

VAR. *β.* **umbellata**; pubescent or villous; leaves lanceolate or ovato-lanceolate, the margin conspicuously reflexed; racemes usually short and corymbose. *P. umbellata, Thunb. Cap.* 555. *P. pubiflora, E.&Z.!* 156. *(not of Burch.)* *P. calycina, Presl. !*

HAB. Moist, sandy flats and mountain sides; common throughout the colony (Herb. Thunb., T.C.D., Hook., Sond., &c.)

12–18 inches high, erect or spreading, mostly many branched from near the base, but often quite simple. Flowers bright purple, showy and amply crested. Intermediate states occur between the two varieties.

7. P. triquetra (Presl., Bot. Bem. p. 15.); glabrous, suffruticose, many-stemmed; stems sub-simple, virgate, *sharply triangular* or winged; leaves sessile, erect, rigid, callous and scaberulous at the margin, sharply mucronate, midribbed, the lower ones lanceolate, the upper linear, subulate; racemes terminal (or lateral) erect, densely many-flowered; pedicels as long as the alæ; bracts *deciduous*, subulate, acute; alæ clawed, *ovate, acute*, anterior sepals ovate; lateral petals longer than the amply-crested keel, erect, oblong-cucullate; capsule winged, elliptic-oblong, entire; seeds albo-setose. *Drege* 7193.

HAB. South Africa, *Drege!* (Herb. T.C.D., Hook., Sond.)

A remarkably rigid under-shrub, 12–18 inches high, scarcely branched. Flowers much smaller than in *P. bracteolata*, and less coloured: alæ greenish white in fruit. Drege *(Linn. Vol.* 19.) refers it to *P. Beiliana*, E. & Z.! but that, according to authentic specimens, is a *Muraltia!*

8. P. virgata (Thunb. Cap. p. 555.); *shrubby*, glabrous or puberulous, with virgate, *terete* branches; leaves scattered, distant or close, either cuneiform and mucronate, or linear, or lanceolate, obtuse or acute, mostly *tapering to the base*, the margin slightly reflexed; racemes terminal, elongating, many flowered; pedicels *shorter* than the flower; bracts subulate, *deciduous*, alæ *broadly elliptical or suborbicular, very obtuse*, ant. sepals elliptical, concave; keel amply crested, sub-acute, lat. petals clawed, flabelliform, sub-truncate, flattish or involute; capsule obcordate, winged. *P. simplex, P. speciosa, and P. genistoides, DC. Prod.* 1. *p.* 322–323. *P. cernua, Thunb. Cap.* 555 *(not E. & Z.)*

VAR. α. **decora**; branches and young leaves minutely downy; leaves lanceolate, and linear-lanceolate; racemes very long, and bracteæ much acuminate, and tardily deciduous. *P. decora, Sond.! Linn.* 23. *p.* 14, *Dietr. Fl. Univ. t.* 79! *P. bracteata, Drege!* *P. lasiopoda, Presl.!*

VAR. β. **intermedia**; glabrous; lower leaves very long, lanceolate, upper shorter and linear, mucronate. *Drege* 7211. *P. Burmanni, DC.? (fide Presl.); also P. longifolia, Presl.!*

VAR. γ. **speciosa**; glabrous or nearly so; lower leaves obovate, or cuneate, upper more linear, all obtuse or sub-truncate, mucronulate; racemes long and lax, bracteæ soon deciduous. *P. speciosa, Bot. Mag. t.* 1780. *DC. Prod.* 1. *p.* 323. *E. & Z.* 146. *P. virgata, E. & Z.!* 145. *Drege,* 7187! *P. simplex, DC. l. c.*

VAR. δ.? **Sprengeliana**; minutely downy or glabrescent; leaves linear, tapering at base, obtuse, mucronulate; racemes shorter, and flowers smaller than in var. γ *P. Sprengeliana, E. & Z.!* 147.

VAR. ε. **genistoides**; glabrous; leaves few, distant, very narrow-linear, acute or obtuse, sometimes subulate; racemes many-flowered, flowers larger or smaller. *P. cernua, Thunb.! (fide Herb. Holm.!)* *P.*

spartioides, P. simplex and P. ephedroides, E. & Z. 148, 149, 150. *P. macra, Drege (non. DC.). P. genistoides,* Poir. *DC. Prod.* 1. *p.* 323. *E. & Z. !* 151.

HAB. Subalpine places, among shrubs; Eastern Districts to Port Natal. (v. v. and Herb. T.C.D., Hook., Sond., &c.)

A shrub, or small tree, 2–5–15 feet high, with rod-like branches, terminating in racemes of handsome purple or flesh-coloured flowers. It varies extremely in different localities (as above indicated); but after a careful examination of multitudes of specimens in several Herbaria, I cannot find any constant characters to separate the book-species enumerated above. The 3 first varieties run insensibly into each other; and the 5th (ε) is obviously a starved state, from dry ground. Var. δ, which is smaller in all parts, with much shorter racemes, looks more like a distinct species; but I cannot find an absolute character.

9. **P. hottentotta** (Presl. Bem. p. 15.); rigid, *suffruticose* (or annual ?) erect, glabrous, with slender, virgate branches; leaves narrow-linear, or subulate, mucronate, scattered, sub-erect, flattish, tapering at base; racemes terminal, elongate, many flowered, pedunculate, the flowers often secund; bracts *deciduous, lanceolate, acute;* pedicels clavate, *much shorter* than the flower; alæ *obovate-elliptical, very blunt,* ant. sepals oval, very obtuse; keel acute, amply crested, lateral petals broadly obovate; capsule obcordate, winged; seeds *albo-tomentose. Drege No.* 7194 *! P. pedunculata, E. & Z.; No.* 152, *(non DC.) P. cernua, E. & Z. !* 174, *(non Thunb.). P. uncinata, Hochst. Zeyher No.* 1936.

HAB. Karroo places, Graaf Reynet and Uitenhage, *E. & Z. !* Riet River, and Macallisberg, *Burke !* Natal, *Krauss ! Sanderson !* (Herb. T.C.D., Hook., Sond.,)

A slender, erect, suffruticose or perhaps *annual* or biennial plant, nearly agreeing with *P. virgata* in floral characters, but much smaller and less woody.—To my eyes this plant resembles the much-disputed figure in Burman (tab. 74. fig. 4.) on which DC. founds his *P. Burmanni;* but, without reference to De Candolle's Herbarium, it would be unsafe to quote the latter name.

10. **P. seminuda** (Harv.); *herbaceous,* erect, minutely downy or glabrous, with flexuous branches; leaves few and distant, patent, linear, tapering at base, sub-petiolate, flattish; racemes terminal (and lateral), erect, elongating, many flowered; pedicels very short, scarcely exceeding the *elliptic, obtuse, sub-persistent* bracts; alæ *elliptical, oblique at base, obtuse, midribbed,* ant. sepals obovate; keel *very blunt,* crested, lat. petals broadly spathulate; capsule obcordate; seeds *half-glabrous. Zeyher. No.* 58 !

HAB. Springbok Keel and Bitterfontein, *Zeyher !* Pappe (No. 30.) (Herb. T.C.D. Hook., Sond.).

Root apparently annual ? Stem 8–12 inches high, irregularly branched, terete. Flowers small, 1–2 lines long; the keel flesh-coloured, the alæ pale, with a green rib. The seeds are naked for half their length, and densely albo-setose on the other (upper) half. It comes nearest to *P. leptophylla,* but differs in habit, and in the seeds, &c.

11. **P. leptophylla** (Burch. ! in DC. Prod. 1. p. 323.); *suffruticose, divaricately* much branched, glabrous or downy, with terete, straight, and rigid branches; leaves few and distant, patent, linear, obtuse or mucronulate, tapering at base, petiolate; racemes mostly terminal, erect, *rigid,* elongating, many-flowered; pedicels *very short* and flowers pendulous; bracts *elliptical, obtuse,* tardily deciduous; alæ elliptic-oblong, oblique at base, obtuse, midribbed, ant. sepals obovate; keel

crested, lat. petals broadly spathulate; capsule oblong-obcordate; seeds *densely albo-setose.* *P. rigens, E. Mey!* *P. recta, Presl.!*

HAB. Between Natvoet and the Gareep and Camdebo. *Drege!* Namaqualand, *Zeyher!* (Herb. Burch! T.C.D., Hook., Sond.)

A small, rigid, much branched undershrub, 1–2 feet high; the peduncles, after the flowers fall, indurating, but not spinous. Leaves $\frac{1}{2}$–$\frac{3}{4}$ inch long; flowers 2–3 lines long, pale, the alæ whitish, with green veins. I describe chiefly from Drege's and Zeyher's specimens, which I have compared with an authentic fragment, received from Dr. Burchell.

12. P. ericæfolia (DC. Prod. I. p. 323.); suffruticose, glabrous or nearly so, with virgate branches; leaves *crowded, erect, rigid, linear-subulate* or lanceolate, *acutely mucronate,* flattish, ribbed or keeled below, with the margin microscopically denticulate or rough, the younger leaves ciliate; racemes terminal (or sublateral) *short, corymbose, few flowered;* pedicels shorter or longer than the flower; bracts *persistent, ovate, keeled;* alæ *ovate-elliptical, obtuse, mucronulate,* ant. sepals ovate, obtuse, ciliolate; keel crested, lat. petals either short and subtruncate, or unequally bilobed, the after-lobe minute, ovate, the anterior sword-shaped; capsule oblong-obcordate. *E. & Z.! No.* 164. *Drege* 7200!

VAR. α. **ericoides**; leaves flattish, conspicuously denticulate; lateral petals commonly bilobed. *P. ericæfolia E. & Z.*

VAR. β. **Eckloniana**; leaves narrower and more subulate, somewhat channelled, round-backed or keeled; petals commonly bilobed. *P. Eckloniana, Lehm. E. & Z.!* 162. *P. pungens, E. & Z.!* 165. *P. acerosa, E. Mey.* *P. No.* 5614, *Burchell!*

VAR. γ. **Mundtiana**; leaves broader, and more lanceolate, midribbed; petals short and truncate. *P. Mundtiana, E. & Z.!* 167.

VAR. δ. **microlopha**; more branching, with fewer and shorter, smooth-edged leaves; smaller flowers, less divided crests to the keel, and unequally bilobed petals. *P. microlopha, E. & Z.!* 168. (an DC.? l. c.) *P. pungens, E. Mey!*

HAB. Grassy places, and among small shrubs. From Capetown to Kaffirland. Swellendam and Uitenhage, E. & Z.! β. Cape Flats, *Ecklon, W.H.H.* &c. γ. Swellendam, *Mundt! Sparmann!* (in Herb. Holm.) (Herb. T.C.D., Hook., Sond.)

A slender, virgulate, many-stemmed undershrub, 6–12–18 inches high. Leaves generally very erect, variable in abundance, length, breadth, and roughness of the edges, &c. The petals vary, as above described, in specimens from the same locality and which are otherwise identical. Extreme forms look somewhat different, but the larger the suit of specimens examined, the more difficult will it be found to separate the above varieties. I have not seen any authentic specimen of Burchell's *P. microlopha.*

13. P. ciliatifolia (Turcz. Anim. No. 2671.); suffruticose, glabrous, or nearly so; leaves densely set, erecto-patent, rigid, linear-subulate, acute, mucronate, flat, midribbed, *conspicuously ciliate-denticulate;* racemes terminal, erect, *subumbellate,* few flowered; pedicels filiform, longer than the flowers; bracts *subulate, acute;* alæ *ovate, acute,* ant. sepals ovate, acuminate, not ciliolate; keel moderately crested, lat. petals small, linear-oblong, truncate, capsule . . . ? *P. macra, E & Z.* 166, *(non DC.) P. ciliolata, Harv.*

HAB. Mountain slopes near the Pot, Steenbocks, and Klyn-rivers, Caledon, *E. & Z.* (Herb. T.C.D., Hook., Sond.)

Closely related to *P. ericæfolia*, but differing as follows: The bracts are much longer, and more acute, the alæ and ant.-sepals and lat.-petals are very different, and the leaves are much more decidedly denticulate. The inflorescence is more umbellate than in any other Cape species.

14. P. Garcini (DC. Prod. 1. p. 323.); glabrous, suffruticose, with terete, virgate branches; leaves *linear-subulate*, scattered, *channelled above*, *keeled below*, erect, mucronate; racemes terminal, many-flowered; pedicels much shorter than the flower; bracts deciduous, ovate, keeled; alæ oval, obtuse, ant. sepals ovate-orbicular, very obtuse; keel shorter than the ample crest, lat. petals as long as the keel, spathulate, truncate; capsule *oblong*, *sharply bidentate*, margined, but not winged. *E. & Z.* 161. *Burm. t.* 73. *f.* 3. *Drege* 7203, 7210. *P. bracteolata*, γ, *Linn. Thunb. in Herb. Holm.*

HAB. Mts. and Hills round Capetown; also Hott. Holland, and Saldanha Bay. *E.* & *Z. Pappe*, &c. common. (Herb. T.C.D., Hook. Sond. &c.)

1-1½ feet high, divided near the root; branches sub-simple. Flowers bright purple. Alæ longer than the crest and much enlarged in fruit, becoming obovate. The capsule is usually notched, very rarely quite entire.

15. P. tenuifolia (Link, Hort. Berol. 2. p. 220); suffruticose, glabrous, or nearly so, with *angular*, virgate branches; leaves *linear* or *obovato-linear* or *oblong*, approximate, flattish, with slightly *revolute* margins, obtuse, mucronulate, midribbed; racemes terminal, many-flowered, elongating; pedicels clavate, shorter than the flower; bracts deciduous, subulate; alæ oval, obtuse, *anterior sepals connate into a single, bidendate calyx segment*; keel oblong, with a small crest, lat.-petals as long as the keel, obovate or obcordate; capsule oblong, bidentate, margined. *E. & Z.* 163. *P. tenuis, Dietr. (non DC.)*

VAR. α. **latifolia**; leaves varying from linear to elliptic-oblong or oval.

VAR β. **linearis**; leaves narrow linear, erect, and straight, *P. linearis, E. Mey.*

VAR. γ. **uncinata**; leaves obovato-linear, or linear, recurved at the point. *P. uncinata, E. Mey. P. rigens, E. & Z.* 173. *(not of Drege, or Burchell.)*

HAB. Var. α, Port Natal, *Sanderson*. β & γ. Eastern Districts, Caffraria and Natal; *Various collectors*. (Herb. T.C.D., Hook., Sond.)

6–12 inches high, more or less branching. Leaves very variable in number and shape. Flowers with the perfume of violets; the keel dark purple, the wings pale, with dark green, branching veins, This species is readily known by the confluent anterior sepals, whose edges cohere for more than ⅔ of the whole length.

Group 4. REFRACTÆ. Slender suffrutices, with scattered leaves, glabrous or thinly downy. Racemes *lateral*, *pedunculate*, few flowered, *very patent or reflexed*. Keel either crested or nearly nude. (Sp. 16–21).

16. P. affinis (DC. Prod. 1. p. 322); half herbaceous or suffruticose, erect or diffuse, *thinly tomentose*, with filiform stems; leaves linear or lanceolate, obtuse or mucronulate, pubescent, midribbed, with *reflexed or revolute margins;* racemes opposite the leaves, short and few-flowered, spreading or reflexed; the flowers subdistant, on deflexed short pedicels; bracts persistent, ovate, acute; alæ elliptic-ovate, subacute, mucronulate,

hispidulous, ant. sepals ovate, ***hairy***; keel crested, lat. petals obovate, bilobed or bidentate; capsule obcordate, ***wider*** than the alæ. *E. & Z.* 172. *Drege, (pro parte) also P. hispida, E. Mey. (non Burch).*

Hab. South Africa, *Masson*, in Herb. Banks, Paarlberg, *Dr. Alexander*. Mountains near Caledon, and the Winterhoeksberg, Tulbagh, *E. & Z.* River Zonderende, *Zeyher*. Hex-riviers Kloof, *Drege*. Formerly cultivated at Kew. (Herb. Banks., Hook., Sond).

Root slightly branched. Stems 6–12 inches high, slender, branching. Leaves ½–¾ inch long, sometimes with a reflexed point, ½ to 1 line broad. Flowers pale, 4–8 in a lax, deflexed raceme, the wings with dark veins. This has the foliage and pubescence of some starved states of *P. hispida* and the hairy sepals of *P. pubiflora*, but a very different inflorescence.

17. P. refracta (D.C. Prod. I. p. 323); suffruticose, erect, glabrous, branching, the branches angular; leaves scattered, subulate, keeled, erecto-patent. acute, mucronate, flattish above, racemes lateral, horizontally patent or reflexed, few-flowered, pedicels clavate, as long as the flower, deflexed, bracts deciduous, lanceolate, acute; alæ broadly ovate, mucronulate; keel with a short *multifid* crest, lat. petals sharply bifid; capsule obcordate, narrower than the alæ.

Var. β. **Steudeliana**; leaves very remote, and shorter. *E. & Z. No.* 180.

Hab. On the Cape Flats, *W.H.H. &c. Ecklon*, Grahamstown, *T. Williamson*, (Herb. T.C.D., Hook. Sond).

A slender, straggling, slightly branched undershrub, 12–16 inches high, erect or spreading. Flowers purple, turning greenish when dry.

18. P. macra (DC. Prod. I. p. 323); suffruticose, glabrous, many-stemmed, with sub-simple, slender, acute-angular branches; leaves *few and distant, minute*, subulate, acute, erect, the uppermost reduced to mere scales; racemes lateral, patent or reflexed, few-flowered, pedicels about as long as the flower, bracts deciduous, subulate; alæ ovate, acute, broader than the capsule; keel with a *multifid crest;* lateral petals sharply bifid; capsule obcordate. *P. Pappeana, E. & Z.* 176. *P. restiacea, E. Mey, in Herb. Drege. also Drege, No.* 7205.

Hab. Among Restios, on the Mts., near the Waterfall, Tulbagh, *E. & Z.* Dutoits Kloof, 2–3000 ft. *Drege.* (Herb. Hook., Benth., Sond.)

Stems very slender, diffuse or decumbent, curved, often nearly leafless. Leaves very minute, and close pressed. Flowers rather small, and pale purple, with shorter or longer pedicels. This so closely agrees with De Candolle's diagnosis, and answers so aptly to the trivial name *macra* (meagre) that I have little doubt of the name. *E. & Z.* incorrectly call the inflorescence an *umbel*, and the keel *beardless;* their own specimens have racemes and an amply crested keel.

19. P. Lehmanniana (E. & Z.! No. 177); suffruticose, glabrous or pubescent, slightly branched, with angular branches; leaves *closely approximate, linear-lanceolate, acute, mucronate, flat, midribbed;* racemes lateral, horizontally patent or deflexed, few flowered; pedicels shorter than the flowers, deflexed; bracts subpersistent, subulate, acute; alæ ovate-elliptical, *obtuse*, ant. sepals ovate-oblong, acute, oblique; keel *with a flabelliform, undivided crest;* lat. petals *deeply bi-lobed*, the posterior lobe broad, oblong, the anterior narrow linear.

Hab. Shrubby and grassy places round the Lion's Head, Capetown; and on Mountain sides, Tulbagh, *E. & Z.! Dr. Pappe!* (Herb. T.C.D., Sond., Benth).

Stem 6–12 inches high, very slender, usually minutely downy. Leaves ½–¾ inch long, ¾ line broad, twice or thrice as long as the internodes, erecto-patent. The *undivided* crest of the keel is remarkable, and peculiar to this species and the following.

20. P. Ludwigiana (E. & Z. ! No. 175.) ; suffruticose, glabrous, diffuse, with slender, angular branches ; leaves *distantly scattered, subulate, channeled,* acutely mucronate ; racemes lateral, horizontally patent or deflexed, pedunculate, short and few-flowered ; pedicels very much shorter than the flower ; bracts persistent, subulate, downy ; alæ broadly ovate, *mucronulate,* ant. sepals roundish-ovate, acute, oblique ; keel *with a very minute, undivided crest,* lat. petals *broadly spathulate, subentire ;* capsule oblong-obcordate, scarcely indented. *Drege No.* 7201 !

HAB. Among grass, on the mountains at Brackfontein, Clanwilliam, *E. & Z. !* Nieuwekloof, *Drege !* (Herb. Hook., Sond. & Benth.)

A very slender, half herbaceous, slightly branched plant, with a few, very narrow, distant, erect leaves, and subumbellate, deflexed racemes. The crest of the keel is extremely minute, quite entire, and situated just below the apex, and may easily escape notice. The flowers are bright purple. I have chiefly described from Drege's specimens. A single fragment from E. & Z. in Dr. Sonder's Herbarium appears to be deformed, having the peduncles imbricated for ⅔ their length with closely set, empty bracts.

21. P. brevifolia (Harv.) ; suffruticose, minutely downy, *divaricately much branched,* with angular, rigid branches ; leaves *minute, scattered, downy,* ovate-oblong or oblong, sharply mucronate, *flat ;* racemes lateral, horizontally patent, 1–2-flowered ; pedicels deflexed, clavate, much shorter than the flower ; bracts subulate, persistent ; alæ broadly ovate, subacute, oblique, ant. sepals ovate, acute ; keel (without crest ?), lat. petals . . ? ; capsules narrow-obcordate. *Drege No.* 7202.

HAB. On the Blaauwberg, 3000–4000ft., *Drege !* (Herb. Hook., Sond., Benth.)

Seemingly prostrate. Stems 6–12 inches long, with straight branches. Leaves 1–2 lines long, few and far between. Peduncles supra-axillary, ½–¾ inches long, bracteate near the apex, and in our specimens bearing a solitary flower. I have not seen the lat. petals, and only imperfectly the keel.

Group 5. HISPIDÆ. Small, half-herbaceous or suffruticose, hairy or tomentose plants, with alternate, lanceolate or ovate (or sublinear) pubescent leaves. Racemes terminal, or lateral, erect. Alæ broader than the capsule. (Sp. 22–30).

22. P. pubiflora (Burch. ! Cat. No. 6205) ; suffruticose, rigid, villous ; leaves petiolate, spreading, ovato-lanceolate or lanceolate (the lower ones ovate), mucronate, laxly villous and ciliate, flat, thickish ; racemes terminal, few-flowered ; *pedicels hispid,* much shorter than the flower ; bracts ovate ; alæ ovate, mucronulate, *hairy, as well as the elliptic-ovate anterior sepals ;* keel amply crested, lat. petals oblong-obovate, upturned, with connivent sides ; capsule elliptic-oblong, sub-emarginate. *DC. Prod.* 1. *p.* 322.

HAB. S. Africa, *Masson ! Burchell !* Kaffer's Kraal, *Herb. Hooker !* (Herb. Banks ! Hook., Burch. !)

More or less hairy in all parts, especially on the sepals and alæ. *P. pubiflora* of E. & Z. ! No. 156 is only a hairy form of *P. bracteolata ;* its calyx is quite glabrous.

23. P. gracilipes (Harv.) ; root woody, stems suffruticose, tufted, hispidulous, filiform, much branched ; leaves alternate, elliptical or oblong, the uppermost linear, obtuse, tapering into a short petiole, puberulous or glabrescent, with recurved edges ; racemes lateral and terminal, 3–4 *flowered*, erect ; pedicels *as long as or longer than the flower*, clavate, erect ; bracts minute, obtuse, ovate, persistent ; alæ broadly *ovate, obtuse*, submucronulate, corolloid, ant. sepals oval, obtuse ; keel crested, lat. petals obcordate, *deeply cloven ;* capsule obcordate. *Drege No.* 7191, 7192.

Hab. Los Tafelberg, Tambukiland, 6–7000ft., *Drege !* (Herb. Hook., Sond., Benth.)

3–6 inches high, densely tufted, fastigiate. Racemes very lax and few-flowered, with long pedicels. Alæ whitish, with green veins. Resembles *S. serpentaria*, but differs in the alæ and petals, &c. Drege's 7192 from Klein Bruintgeshoogte 3–4000ft. has narrower leaves, and is nearly glabrous, but otherwise very similar. The pedicels are deflexed in fruit.

24. P. hispida (Burch. in DC. Prod. 1. p. 323) ; *whole plant softly villous ;* root woody ; stems numerous, ascending or decumbent, subsimple ; leaves alternate, shortly petiolate, the lower oval or ovate, the upper lanceolate or linear, *flat*, midribbed ; racemes terminal, *many-flowered, elongating ;* bracts ovate-acuminate, membranous, reflexed, persistent ; pedicels *shorter* than the flower ; alæ *obovate, very obtuse*, ant. sepals broadly elliptical, 3-veined ; keel small, amply crested, lat. petals spathulate ; capsule obcordate. *E. & Z. No.* 171. *P. lanata, E. Mey. ! and P. erubescens, E. Mey. ! (dwarf). Drege No.* 7190.

Var. *β*. **declinata** ; decumbent, less hairy, the lower leaves roundish or oval, the upper obtusely ovate. *P. declinata, E. Mey.*

Hab. Among shrubs on mountain sides in Uitenhage, Albany and Caffirland, *E. & Z. !* &c. Vanstaaden's Berg, Zurreberg and Bothasberge, *Drege!* var. *β*. between the Omsamcaba and Omtendo, *Drege !* (Herb. T.C.D., Hook., Sond).

Stems 8–12 inches long or more, numerous, spreading from a robust crown. The whole plant is clothed with long, silky, spreading hairs, which are sometimes very copious (as in *P. lanata*, E. Mey.) and sometimes shorter and fewer. Var. *β*. has much broader and more obtuse leaves than usual, but is connected with *α*. by many intermediate forms. The ant. sepals are twice as broad as the posterior. The flowers are small and greenish, perhaps flesh-colour when fresh. *P. erubescens*, E. Mey.! appears to be merely a starved form of this plant.

25. P. Ohlendorfiana (E. & Z. 170) ; *whole plant villous*, root woody ; stems numerous, filiform, subsimple ; leaves alternate, on short petioles, *ovate*, obtuse or subacute, thin, midribbed ; racemes lateral and terminal, elongating ; bracts small, ovate-subulate, membranous, persistent ; pedicels shorter than the flower ; alæ clawed, broadly *ovate, obtuse*, ant. sepals elliptical, obtuse, concave ; keel very short, amply crested, lat. petals obcordate, clawed ; capsule obcordate. *P. ovalis, E. Mey. ! P. tomentosa ? E. & Z. ! (non Thunb.)*

Hab. Mountains of Caffraria, *E. & Z. !* Katberg, 4–5000ft., Drege. Nov.–Dec. (Herb. T.C.D., Hook., Sond.)

The stout and woody root throws up a tuft of slender, threadlike stems, simple or branched. Leaves ½–¾ inch long, thin and flexible. Flowers pale, greenish or perhaps flesh coloured. Racemes few or many-flowered. The wings are longer than the keel. Very nearly allied to *P. hispida*, but with larger flowers, differently shaped alæ, and less variable foliage. *P. tomentosa*, E. & Z. ! according to an authentic fragment in Herb. Sond. seems to be a monstrous state of this species or of *P. hispida ;* it comes from Kat-River-berg.

Group 6. ASBESTINÆ. Small suffrutices, all (except *P. Bowkeræ*) with very thick, woody roots, and puberulous. Racemes lateral or terminal. *Alæ greenish, narrower than the capsule.*

26. P. Bowkeræ (Harv.); suffruticose, slender, diffuse, *glabrous;* branches *angular;* leaves scattered, *linear-lanceolate, flattish, erect, acute at both ends,* mucronate; racemes terminal and lateral, *lax, erect;* pedicels as long as the flower; bracts minute, ovate, subacute, persistent; alæ *ovate-acute,* tapering at base, ant. sepals oblong, obtuse or subacute; keel with a small, multifid crest, lat. petals oblong, *sharply and equally bilobed;* capsule obcordate, broader than the alæ.

HAB. Grassy hills and vales, Graaf Reynet, *Miss Bowker,* (Herb. Hooker.)

A very slender, slightly branched, elegant little plant, 6–12 inches high. The lowest leaves are broader than the rest. Flowers small, 2 lines long; the alæ greenish and veiny, the keel purple. Racemes 2–4 inches long, very lax. Intermediate in character between the *Asbestinæ* and *Virgatæ* Groups. Sent from the Cape, several years ago, by Mrs. Barber (formerly Miss Bowker).

27. P. illepida (E. Mey.! in Herb. Drege.); root woody; stems suffruticose, diffuse or trailing, branched; the branches striate, minutely downy or glabrous, angular; leaves scattered, *linear-lanceolate, acute at both ends, mucronate,* midribbed; racemes lateral, few flowered, *patent or somewhat reflexed;* pedicels clavate, as long as the flower, deflexed; bracts very minute, *deciduous;* alæ herbaceous, *narrow-oval,* acute or mucronulate; ant. sepals ovate, *acute;* keel with an ample, multifid crest, lat. petals *broadly spathulate;* capsule obcordate, broader than the alæ. *P. refracta, E. & Z.! No.* 178, *(not of Burch.)*

HAB. On Mountains. Adow and Van Staaden's Berg, Uitenhage, *E. & Z.!* (Herb. T.C.D., Hook., Sond.).

Stems 3–6 inches high, many from the same root. Whole plant often minutely downy. Flowers greenish, inconspicuous.

28. P. asbestina (Burch. Trav. 1, p. 543); pubescent; stems woody, diffuse or trailing, rigid, much and divaricately branched; leaves flat, petiolate, midribbed, thickish, thinly downy, lower obovate or roundish, retuse, upper oblong, tapering to the base, mucronulate; racemes lateral, 2–3 flowered; bracts deciduous; alæ herbaceous, narrow, *obliquely oblong, acuminate, tapering at base,* ant. sepals ovate, mucronulate; keel amply crested, lat. petals *oblong, obtuse;* capsule deeply notched, wider than the alæ. *DC. Prod.* 1. *p.* 323. *E. & Z.!* 184.

VAR. β. **rigens**; all the leaves linear, tapering to the base, with slightly revolute margins and a minute recurved mucro. *P. rigens, Burch.! Cat.* 1821. *DC. Prod.* 1. *p.* 323.

HAB. On sandy hills among the Zwarteberge and Winterhoeks Bergen, *E. & Z.!* Nieuwveldsberg, Beaufort, and in the Zuureberge, Uitenhage, *Drege!* Orange River, *Zeyher.* (Herb. T.C.D., Hook., Sond., Burch.!)

Stems widely spreading 6–12 inches long or more. All the younger parts pubescent. The lower leaves are frequently broad and very short; the upper gradually passing from oblong to linear. The flowers are small; the keel flesh colour; the alæ green, and very narrow, almost cymetar-shaped. Burchell's *P. rigens,* according to an authentic specimen, differs merely in having narrower leaves; it seems to be a starved variety.

29. P. serpentaria (E. & Z.! No. 181); root woody; stems diffuse or prostrate, puberulous; leaves *subsessile, elliptical or oblong*, obtuse, mucronulate, midribbed, with *reflexed* margins and a paler under side; racemes lateral, 2–8 flowered; pedicels not longer than the flower; bracts ovate, deciduous; alæ *sub-herbaceous, elliptical*, mucronate, with white edges, ant. sepals ovate, sharply mucronate, concave; keel with a multifid crest, lat. petals *broadly cuneate, truncate, on long claws;* capsule sharply notched, broader than long, with a wide, marginal wing. *P. lævigata, E. Mey.! in Herb. Drege.*

HAB. Stony places near the Fort of Chumiberg and at Fort Beaufort, Kat River, *E. & Z.!* Caffirland, and towards Natal, *Drege!* Natal, *Gueinzius! Sanderson!* Herb., T.C.D., Hook., Sond.)

Stems decumbent, 6–12 inches long or more, subsimple or much branched, slender. Flowers small, with green alæ, and purple or flesh coloured keel and petals: the petals large (in proportion) and standard-shaped. According to *Ecklon* and *Zeyher*, confirmed by *Dr. Pappe*, the root is a Caffir remedy for the bite of serpents, whence the specific name, and the colonial "*Kaffir Schlagen Wortel.*" It is nearly allied to *P. amatymbica*, from which it differs in habit; in its more obtuse leaves, broader and more elliptical alæ, &c.

30. P. amatymbica (E. & Z.! No. 182); root woody; stems half-herbaceous, *tufted, erect*, sub-simple, filiform, downy; leaves *lanceolate, acuminate; pungently mucronate*, the lower ones broader, with reflexed edges; racemes lateral, few-flowered, with a wavy rachis; bracts deciduous; alæ herbaceous, *ovato-lanceolate, acute, mucronate*, ant. sepals ovate, sharply acuminate; keel with a multipartite crest, ant. petals obovate or cuneate, truncate; capsule oval-oblong, emarginate, wider than the alæ. *P. acuminata, E.Mey.! in Herb. Drege.*

HAB. Mountains near Shiloh, Tambukiland, *E. & Z.! Drege!* Aapjies River and Macallisberg, *Burke!* Gekau, Caffraria, *Drege!* (Herb. T.C.D., Hook., Sond.).

Densely tufted, 3–6 inches high, with slender, wiry stems. The lower leaves are somewhat ovate; the midribs prominent. Flowers small, the alæ green and very narrow; keel and petals flesh coloured or purple. Allied to *P. asbestina*, but differing in the foliage and habit.

Group 7. PARVIFLORÆ. Minute, herbaceous species, with linear, scattered leaves. Racemes terminal, densely many-flowered, spiciform, with very minute flowers. Keel with a small, not much divided crest. (Sp. 31–32).

31. P. pallida (E. Mey.! in Herb. Drege); herbaceous, branched near the base, filiform; leaves few, linear, tapering at base, *obtuse;* racemes terminal, spike-like, elongating, many-flowered; pedicels much shorter than the minute flowers; bracts *oval, obtuse, persistent;* alæ broadly elliptical, veiny, very obtuse, ant. sepals obovate or obcordate; keel crested, *twice as long* as the spathulate petals; capsule *obcordate.*

HAB. Between Nat Voet and the Gareep, 1–1500 feet, *Drege!* (Herb. Hook., Benth.).

A small annual (?), 2–3 inches high, or perhaps more; nearly allied to the following, but differing in bracts, sepals, petals, &c. The leaves are sometimes oblong and mucronulate. Flowers nearly sessile, whitish or white. Crest of the keel small, and but little cut.

32. P. capillaris (E. Mey.! in Herb. Drege); herbaceous, erect, sub-simple, glabrous, with angular stems; leaves few and distant, *linear-subulate, acute*, or mucronate; racemes terminal, spike-like, elongating, many-flowered; pedicels much shorter than the minute flowers; bracts *subulate, deciduous;* alæ oval, obtuse, 3–nerved, ant. sepals narrow-oblong, subacute; keel subglobose, crested, *shorter* than the ovate-oblong petals; capsule *oval-orbicular.*

HAB. Between the Omsamculo and Omcomas, in boggy spots, under 500 feet. *Drege!* Macallisberg, *Burke and Zeyher!* (Herb. T.C.D., Hook., Sond.).

A very slender annual (?), 2–6 or 12 inches high, its virgate branches ending in a long dense raceme of minute, subsessile, whitish flowers. The keel has a fleshy crest cut into four or eight lobes.

APPENDIX.—Species unknown to me, or altogether doubtful.

33. P. tomentosa (Thunb. Fl. Cap. p. 556); "stem shrubby, erect, villous, branching, 2 feet high, or more; branches sub-opposite; leaves opposite, sessile, ovate, mucronate, entire, with revolute margins, glabrous on the upper, tomentose on the lower surface, crowded but spreading, an inch long; flowers sessile in the axils of the leaves throughout the whole length of the branches."—*Thunb., l. c.*

A very imperfect specimen, without flowers, exists in Herb. Holm.!

34. P. intermedia (DC. Prod. I. p. 322); "leaves oblong-linear, mucronate, with revolute margins; ramuli glabrous; bracts persistent, of equal length; pedicels something longer than the flower; alæ sub-cuspidate." *DC. l. c.*

Can this be our Var. β of *P. peduncularis?* or a var. of *P. bracteolata?*

35. P. Burmanni (DC. Prod. I. p. 322); "leaves linear, rather obtuse; branches velvetty; racemes supra-axillary; pedicels shorter than the flower; bracts deciduous." *DC. Burch. Cat.* 6437. *Burm. t.* 73, *f.* 4.

36. P. pungens (Burch. Cat. No. 1598. trav. p. 304); "leaves linear, subacute, narrow, few; branchlets divaricating, glaucous, rigid, spinous at the point; racemes 2–4-flowered." *DC. l. c.*

37. P. polyphylla (DC. Prod. I. p. 324); "leaves oblong, acute at each end, crowded, scabrous on the back and the edges; branches downy; pedicels axillary, one-flowered; alæ suborbicular." *DC. l. c.*

38. P. venulosa (Dietr.); "branches terete, smooth; leaves alternate, oblong, mucronulate, glabrous; bracts ovate, equal, persistent; alæ ovate, acute, veiny; lat. petals bifid, with nearly equal, acute lobes." *Walp. Repert.* I. *p.* 235, *No.* 38.

39. P. longifolia (Dietr.); "branches terete, hairy; leaves alternate, lanceolate, acute, glabrous; bracts lanceolate, obtuse, subequal, persistent; alæ oblong, apiculate; lat. petals bifid, inner segments very long and narrow, outer patent, obtuse." *Walp. l. c. No.* 39.

40. P. diffusa (Dietr.); seems to be one of the endless forms of *P. oppositifolia. See Walp. Repert.* I. *p.* 235. *No.* 41.

II. MUNDTIA, Kunth.

Sepals 5, the two lateral (*alæ*) much larger than the rest, winglike and coloured. *Petals* 3, united at base and attached to the staminal tube; the lower one keel shaped, enclosing the stamens, emarginate, with a multifid crest below its apex (in the Cape species); lateral petals small, oblong. *Stamens* 8, united into a split tube. *Style* compressed, thickened upwards, two-lobed, the posterior lobe horizontal, the anterior vertical. *Fruit*, a fleshy one or two-seeded *drupe*. *DC. Prod.* 1. *p.* 337.

Rigid glabrous shrubs, with scattered leaves, spinous-tipped branches, and axillary, solitary, pedicellate flowers. Two species are known, one a native of the Cape, the other of Brazil. The generic name is in honour of *Mundt*, a meritorious collector of S. African plants.

1. **M. spinosa** (DC. Prod. 1. p. 338); divaricately much branched; branches angular, furrowed, glabrous, spine-tipped; leaves either elliptical, obovate, cuneate or linear, obtuse or mucronulate, thick, glabrous; flowers minutely pedicellate; alæ sub-orbicular; petals nearly as long as the amply crested keel, linear or spathulate, obtuse. *Polygala spinosa, Linn. Thunb. Fl. Cap. p.* 556.

VAR. *α*. **latifolia**; robust, with short and leafy branches; leaves oval or obcordate. *M. tabularis, E. & Z.!* 229.

VAR. *β*. **angustifolia**; leaves linear-oblong, linear-cuneiform or sublanceolate; branches leafy, longish and somewhat virgate. *M. spinosa, E. & Z.!* 230. *M. albiflora, E. & Z.!* 231. *M. montana, E. & Z.!* 232. *M. glauca, E. & Z.! No.* 233. *Drege, No.* 7224, *a, c, and* 7255.

VAR. *γ*. **scoparia**; nearly leafless, or with a few oval-oblong, obtuse or subacute leaves; branches long and virgate, slender and deeply furrowed. *M. scoparia, E. & Z.!* 235. *Drege,* 7254, *δ*, 7256, *& Muraltia,* 3290.

HAB. Dry, stony places throughout the Colony. *γ*. Clanwilliam, E. & Z.! (Herb. T.C.D., *&c.*).

A much branched, rigid, furzelike, thorny bush, 2–3 feet high, varying in the shape and abundance of its leaves, and in the colour of its flowers, which are either red or white. The varieties above indicated pass one into the other. The most distinct looking is *γ*., which is more slender than the others and often nearly leafless; but sometimes sparsely leafy. *M. desertorum, E. & Z.! is Muraltia juniperifolia, DC.! (non E. & Z.)*.

III. MURALTIA, Neck.

Sepals 5, dry and membranous, subequal, the two lateral somewhat longer than the rest. *Petals* 3, united at base and attached to the staminal tube; the lower one hood-shaped, with a two-lobed crest below the apex; lateral petals oblong, free or cohering by their edges. *Stamens* 8, united into a split tube. *Capsule* membranous, compressed, with 4 horns or tubercles at its upper angles; rarely hornless. *DC. Prod.* 1. *p.* 335.

Small but rigid *shrubs* or *undershrubs*, peculiar to South Africa, with fascicled, or rarely scattered, rigid, mostly pungent-mucronate, very entire, narrow, subulate, linear, or rarely lanceolate or ovate small leaves; and axillary, solitary, subsessile or pedicellate bright-purple flowers, tribrateate at base. The generic name is in honour

of John von Muralt, a Swiss botanist. The genus naturally divides itself, as pointed out by Turczaninow, into two subgenera; the first (Eumuraltia) distinguished by subsessile flowers, and with few exceptions by fascicled leaves and long-horned capsules; the second (Psilocladus) by manifestly pedicillate flowers, solitary leaves, and emarginate or merely 4-tubercled capsules.

ARTIFICIAL ANALYSIS OF THE SPECIES.

Sub-genus 1. Eumuraltia. Fl. sessile, or nearly so. Sepals subequal. Leaves *mostly* fascicled.

A. Ovatifoliæ: Leaves broad at base, ovate or ovate-acuminate (1–3).

Leaves glabrous ... ... (2) **Cliffortiæfolia.**
Leaves ciliate. Sepals acute. Capsule horned ... (1) **serpylloides.**
Leaves villous or tomentose. Sepals obtuse. Caps. hornless ... ... (3) **squarrosa.**

B. Subulifoliæ. Leaves from a broad base subulate or linear mucronate (4–21).

* Leaves straight, *very erect,* (scattered or subfasciculate) (21) **filiformis.**
** Leaves straight (not hook-pointed), patent or reflexed.
Leaves mostly solitary ... ... (20) **stipulacea.**
Leaves fascicled :
† Leaves *subulate,* pungent-mucronate :
rounded at back : { petals oblong, subacute ... (14) **longicuspis.**
{ petals subulate-acuminate ... (13) **acipetala.**
keeled at back.
Slender; with *acicular* leaves { leaves glabrous ... (16) **asparagifolia.**
{ leaves tomentose (17) **acicularis.**
Robust, with broadly triangular leaves.
Petals spoon-shaped, as long as the keel ... (9) **Heisteria.**
Petals linear. Leaves 3–4 times as long as broad ... ... (12) **conferta.**
Pet. linear. Leaves 6–12 times as long as broad :
Sepals oval or ovate, mucronate.
Leaves subequal in each fascicle ... ... (7) **ericæfolia.**
Lowest leaf in fascicle very long ... ... (8) **ononidifolia.**
Sepals lanceolate, acuminate ... ... (15) **satureoides.**
†† Leaves *linear,* blunt; the youngest mucronulate.
Robust, much branched shrubs :
Petals longer than keel, spathulate or spoon-shaped ... ... (10) **macropetala.**
Petals shorter than keel, linear, flat ... ... (11) **calycina.**
Slender, trailing or decumbent ... ... (19) **divaricata.**
*** Leaves arched backwards, or strongly hooked at point.
Leaves broadish, channelled, strongly { pubescent ... (4) **rubeacea.**
hooked ... ... { glabrous ... (5) **macrocarpa.**
Leaves slender, recurved, but { a robust shrub ... (6) **laricifolia.**
not hooked ... ... { slender and trailing ... (18) **tenuifolia.**

C. Diversifoliæ. Leaves of two forms on the same branch; the lowest narrow, linear or subulate; the upper broader, more or less lanceolate, midribbed. (22–28).

* Erect shrubs, with rodlike branches, and pale green foliage; glabrescent or downy :
Leaves crowded, mostly clavate or spathulate. Capsule with subulate horns ... ... (22) **mixta.**
Leaves laxly set, mostly very slender, linear subulate. Caps. with filiform horns ... ... (23) **macroceras.**

Lower leaves lax, linear; upper imbricate, lanceolate.
Caps. with filiform horns... ... ... (24) **alopecuroides.**

** Low growing or diffuse shrubs, hairy and ciliate; leaves full green, squarrose.
Capsule hornless ... ... (25) **anthospermifolia.**
Capsule with slender horns:
tufts of leaves densely crowded; leaves squarrose, straight-pointed ... ... (26) **phylicoides.**
tufts laxly set: upper leaves strongly hooked at point ... ... (27) **Candollei.**

** Depressed, but robust, divaricately much branched; leaves glabrescent, dotted ... ... (28) **Pappeana.**

D. Lancifoliæ. Leaves uniformly tapering much to the base; in outline either lanceolate, ovato-lanceolate, obovate, or clavate (29–41).

* Leaves lanci-subulate, lanceolate or ovato-lanceolate.
Branches hispid; leaves ciliate with long white hairs:
Leaves lanci-subulate, crowded. Capsule hornless (29) **ciliaris.**
Leaves lanci-subulate, laxly set. Capsule horned (30) **incompta.**
Leaves lanceolate, thin and flat, midribbed ... ... (34) **pilosa.**
Branches downy or glabrescent; leaves softly hairy or quite glabrous.
A trailing, slender, villous or tomentose undershrub (31) **diffusa.**
A slender, glabrous, erect, tufted suffrutex ... ... (35) **lancifolia.**
Robust, diffusely branched; leaves scattered, linear-lanceolate ... ... (36) **Dregei.**
Robust, divaricate; leaves ovato-lanceolate or ovate oblong ... ... (39) **thymifolia.**

** Leaves linear-clavate, or obovate, blunt, mucronulate.
Leaves softly downy or velvetty, with short pubescence ... ... (33) **pubescens.**
Leaves glabrous or nearly so:
Leaves clavate, subterete, punctulate ... ... (37) **depressa.**
Leaves broadly linear, flattish, midribbed ... ... (38) **dumosa.**
Leaves obovate, flat, thick and fleshy ... ... (40) **obovata.**
Leaves ciliate, midribbed and veiny, broadly obovate (41) **reticulata.**

Sub-genus 2. PSILOCLADUS. Flowers manifestly stalked. Sepals very unequal. Leaves scattered.

E. Acutifoliæ. Leaves acute; either subulate, linear, lanceolate or ovate (42–46).

A robust, much branched, rigid shrub ... ... (42) **juniperifolia.**
Suffruticose, diffuse; leaves subulate, keeled ... ... (43) **pauciflora.**
Suffruticose, diffuse; leaves flat, ribbed (stems sharply angular).
Leaves cordate-acuminate, three-nerved ... ... (44) **trinervia.**
Leaves linear-acuminate, three-nerved ... ... (45) **Beiliana.**
Leaves linear, acute, one-nerved ... ... (46) **angulosa.**

F. Obtusifoliæ. Leaves blunt, or (at least) not pungent. (47–51).

Leaves oval or subrotund, convex, petiolate ... ... (47) **carnosa.**
Leaves oblongo-lanceolate, flat, ribbed, sessile ... (48) **crassifolia.**
Leaves linear or linear oblong.
Divaricately much branched, robust and woody... (49) **rigida.**
Slender, virgate, suffruticose: leaves channelled ... (50) **striata.**
Slender, virgate; leaves subacute, furrowed above (51) **leptorhiza.**

Sub-genus 1. EUMURALTIA. Flowers sessile, or on minute *pedicels shorter* than the subtending bracts. *Sepals* of nearly equal length.

Capsule almost always horned (except in *M. squarrosa, anthospermifolia* and *ciliaris*). *Leaves* very generally fascicled. (Sp. 1–41.)

* Group 1. OVATIFOLIÆ. Leaves ovate or ovate-acuminate, not tapering at base. (Sp. 1–3.)

1. **M. serpylloides** (DC. Prod. 1. p. 335); diffuse, much branched, the branches hairy; leaves fascicled, ovate or ovato-lanceolate, taper-pointed and mucronate, flat, midribbed, the *recurved margin and midrib hispido-ciliate;* flowers sessile; sepals lanceolate, *acute or acuminate;* lat. petals linear, obtuse, rather shorter than the keel; ovary densely hairy; capsule ciliate and hispid, longer than the erect, *subulate horns. M. marifolia,* E. & Z. No. 217.

HAB. Sandy hills among bushes, in the Cape District. Clasenbosch & Constantia, *E. & Z.!* Cape flats, *W.H.H.* &c. Rondebosch, *Drege!* (Herb. T.C.D., Hook., Sond.)

A spreading, small, and rather slender shrub, with *thin* leaves, somewhat concave on the *lower* surface, elegantly ciliate on the ribs and margin. The subtending leaf of the fascicle is always broadest, and generally typically ovate, but in some specimens becomes somewhat lanceolate. The calyx is long in proportion to the small, pale-coloured corolla.

2. **M. Cliffortiæfolia** (Eck. and Zey. 188); robust, rigid, much branched, the branches pubescent; leaves sub-fasciculate, densely crowded, ovate, acuminate, rigid and pungently attenuato-mucronate, patent, *glabrous, concave above,* ribbed and *keeled;* flowers sessile; sepals elliptic-oblong, *obtuse,* ciliolate; lat. petals linear, subacute, straight; capsule (fide *Meisn.*) "obliquely ovate, with 4 *subulate horns* of its own length." *Meisn. in Hook. Lond. Journ.* 1. *p.* 475.

HAB. Among shrubs, and in sandy places, near Gaurits River, George, *E. & Z!* *Krauss,* (Herb. Sond. T.C.D.)

A very distinct species, among the largest and most robust, with leaves, as Meisner well observes, resembling those of an *Epacris.* They are think and hard, but not fleshy, nearly ½ inch long, and 2–3 lines broad, and of a pale colour when dry. Each has generally 2 or 3 much smaller leaves in its axil. The flowers are small, pale, and hidden among the leaves.

3. **M. squarrosa** (DC. Prod. 1. p. 335); robust, woody, branches curved, pubescent or tomentose; leaves fascicled, dense, ovate-acuminate, rigid, *villous* or *tomentose,* ciliated, *recurved, concave, keeled,* taper-pointed and pungent-mucronate; fl. sessile; sepals ¼ of corolla, elliptic-oblong, *obtuse* or subacute; petals shorter than the keel, linear, tapering to an acute point; capsule pubescent, *obcordate, hornless,* one loculus frequently abortive. *Polygala squarrosa, Thunb. Fl. Cap. p.* 558. *Muraltia alopecuroides, E & Z. No.* 201. *(non DC.) & M. ruscifolia E. & Z.* 189. *Drege, No.* 7220, 7221.

VAR. β. **subulata**; leaves broadly subulate, much acuminate. *Zeyher,* 1951. *Drege,* 7222.

HAB. South Africa, *Thunberg, Masson,* In Karroo plains and near sea shore, in the district of Uitenhage, *E. & Z.* George and Uitenhage. *Drege!* (Herb. T.C.D., Hook. Sond.)

A very rigid, coarse growing shrub, 2–3 feet high, thickly covered with pungent leaves, and mostly very hairy. Flowers conspicuous, bright purple. *E. & Z.'s.* No. 201. exactly agrees with Thunberg's specimen of *M. squarrosa,* in Herb. Holm.;

their No. 189 has the tufts of leaves rather less dense. The essential character lies in the *capsule*, which is very unlike that of any other species. VAR. β., founded on a specimen from Kakerlak valley, differs in having much narrower and more tapering leaves; Drege's 7322 is a much starved state, but both have the peculiar capsule.

** Group 2. SUBULIFOLIÆ. Leaves subulate or linear-mucronate, from a broad base; *not at all narrowed* or *very slightly narrower* towards the base. (Sp. 4–21.)

4. M. rubeacea (E. & Z. En. No. 200); erect, densely branched, hairy; branches hispid; leaves fascicled, crowded, rigid, *linear-uncinate, strongly hooked*, concave above, keeled below, mucronate, *scabrous, pubescent, ciliated*; fl. sessile, sepals ⅔ of corolla, elliptic-oblong, acute, mucronate, ciliolate; petals oblong, obtuse, capsule . . . . . ? *M. metalasiæfolia, E. & Z.* 199. *Drege, No.* 7215.

HAB. Mountains of Hott. Holland, near Palmiet River; and near Klyn River, Caledon, *E & Z. Pappe!* (Herb. T.C.D., Benth., Sond.)

A very dense little bush, 8–12 inches high, rough and hairy. Flowers of average size, purple. I cannot distinguish *E. & Z's. M. metalasiæfolia* by any tangible character. Known from the following by its hairiness.

5. M. macrocarpa (Lehm. MSS.); rigid, divaricately much branched; branches rough, the younger twigs downy; leaves fascicled, rigid, *linear-uncinate, strongly hooked*, pungent mucronate, channelled above, keeled below, scabrous or serrulate at the edge, *glabrous*, flowers sessile, sepals ⅔ of corolla, broadly elliptical, mucronulate, ciliolate; lat.-petals erect, broadly oblong, obtuse; capsule glabrous, oval, with 4 divergent *broadly subulate* horns of its own length. *E. & Z. No.* 198. *M. hamata, E. Mey.*

HAB. In the Nieuweveldt, Beaufort, *E. & Z. Drege*, also (fide Drege) on the Schneuwberg. (Herb., T.C.D., Hook., Sond.)

A very rigid, depressed, squarrose, robust, *and nearly glabrous* shrub, 6-12 inches high. Leaves many in a fascicle, very rigid. Flowers of average size. Capsule 2 lines broad, flat. I have compared the original specimen marked by *Lehmann* with *E. Meyer's M. hamata* and find them closely to agree. There are old specimens, probably from *Thom or Mundt*, in Herb. *Hooker*. This plant is closely allied to *M. rubeacea*, from which its nearly glabrous stems and foliage chiefly distinguish it.

6. M. laricifolia (*E. & Z.* No. 209); shrubby, much branched; branches downy; leaves fascicled, sub-equal, linear-subulate, pungent, *squarrose or recurved*, slightly narrowed to the base, flattish above, keeled, rough edged; flowers sub-sessile; bracts ovate; sepals *lanceolate, acute, or acuminate*; petals as long as the keel, broadly linear, tapering to an obtuse point; capsule . . . . ?

HAB. Grassy Fields, near Boschesman's River, and Oliphant's Hoek, Uitenhage, *E. & Z.* (Herb. T.C.D. Sond., Benth.)

A rigid scraggy shrub, with spreading slender branches. Leaves ½ inch long, very narrow, and not very dense, 4–6 in each tuft. Sepals ⅔ of corolla. Flowers often crowded near the ends of the branches.

7. M. ericæfolia (DC. ? Prod. 1. p. 336); robust, shrubby, much branched; branches pubescent; leaves fascicled, crowded, *sub-equal*, patent, subulate, pungent, straight (or somewhat curved), concave above, keeled, *glabrous*, rough-edged; fl. sessile; sepals broadly *elliptical, concave, mucronulate*, ciliolate. ribbed; petals broadly linear, *obtuse or subacute*;

capsule ovate, longer than the divergent, *subulate* horns. *M. juniperifolia, E. & Z. No.* 196 *(non DC.) Zeyher,* 1950. *Drege,* 7218, 7223.

VAR. β. **curvifolia**; leaves recurved, but not hooked. *M. Sprengelioides, E. & Z.* No. 197.

HAB. Mountains of Uitenhage, (Krakakamma and Van Staaden's Berg), and the Langekloof, *E. & Z.* Zwarteberg, *Drege.* (Herb. T.C.D., Hook., Sond.)

A rigid furze-like bush, closely allied to the following, from which it differs more in aspect than by precise character. *M. juniperifolia* DC. is a very different plant, and belongs to the section "Psilocladus." Whether this be the true "*M. ericæfolia,*" I cannot say; but it agrees fairly with *De Candolle's* and *Thunberg's* descriptions, and is of common occurrence, and therefore likely to have been described by early writers.

8. M. ononidifolia (E. & Z.! No. 195); robust, divaricately much branched, branches pubescent; leaves fascicled, crowded, *very unequal,* horizontally patent, *the primary leaf of each tuft* 2–4 *times as long as the rest,* subulate, triquetrous, flattish above, keeled, rough and ciliolate at the edge, pungent; fl. sessile; sepals ovate or ovato-lanceolate, mucronate; petals broadly linear, obtuse; capsule ovate, *longer* than the sharply subulate horns. *M. heterophylla, E. Mey.! in Herb. Drege.*

HAB. District of Caledon, *E. & Z.!* George, *Drege!* Grahamstown, *T. W.* (Herb. T.C.D., Hook., Sond.).

A very rigid, furzelike bush, with spine-tipped leaves bristling in all directions; the lowest leaf of each tuft 1 inch long. The young leaves are ciliate with rigid bristles; the older rough-edged. Meisner (Lond. Journ. 1. p. 474) refers "*M. heterophylla,*" *E. Mey.!* to *M. linophylla. E. & Z.*; but the specimens we have seen belong to the present species. Probably Drege distributed two plants under one name.

9. M. Heisteria (DC. Prod. 1. p. 335); robust; branches pubescent; leaves fascicled, *linear-subulate, keeled,* patent or reflexed, pungent, rough-edged and sub-ciliate; fl. subsessile; sepals lanceolate, acute; petals *spoon-shaped,* nearly as long as the keel; capsule *shorter* than its subulate horns. *Pol. Heisteria, Linn. Thunb. Cap. p.* 557. *Bot. Mag. t.* 340. *M. conferta, E. M.! in Pl. Drege. Drege, No.* 7216, 7219.

HAB. Common throughout the Colony, in dry rocky places; by road sides, &c. (Herb. T.C.D., &c.).

A very rigid shrub, 1–3 feet high or more, densely branched; the branches spreading, virgate, with reddish bark. Leaves 3–4 lines long, thick, nearly triquetrous; sometimes downy. Flowers vivid purple, conspicuous, thickly studded along the twigs.

10. M. macropetala (Harv.); robust, divaricately branched; branches pubescent, often spine-tipped; leaves in sub-distant tufts, linear, squarrose, thickish, *semi-terete, blunt, with a short pungent mucro,* the primary leaf of each tuft much *shorter* and more acute than the rest; fl. sessile; sepals oblong-lanceolate, acute; petals *longer than the keel and crest, arched, spathulate,* concave, obtuse; capsule with 4 subulate horns of its own length. *Drege, No.* 7226, 7227! *M. stipulacea, Andr. Bot. Rep. t.* 363 *(good!).*

HAB. Herb. Banks, 1771.—Laauwskloof, near Groonekloof, 1000 ft. *Drege!* Mountains of Caledon, *Pappe,* (54*a*). (Herb. T.C.D., Benth., Sond.).

A woody, erect shrub, 1–2 feet high, with spreading, stiff, straight branches, the young ones woolly, the older downy. Leaves few in each tuft, $\frac{1}{3}$, $\frac{1}{2}$, $\frac{3}{4}$ inch long, furrowed on the upper, rounded on the lower surface; the subtending leaf often a mere scale. Flowers bright purple, as large as those of *M. Heisteria.* Andrews' figure, above quoted, well represents our plant, which is probably the *M. stipulacea* of many gardens and herbaria; but is quite different from the plant so named by Thunberg.

11. M. calycina (Harv.); robust, with woolly branches; leaves fascicled, *linear, very obtuse,* thick and fleshy, concave above, keeled below, thinly hairy and ciliate, the young ones woolly, pointless, the primary leaf broader than the rest; fl. subsessile; calyx *nearly as long as the corolla, coloured;* sepals broadly ovato-lanceolate, acute or acuminate; petals *shorter* than the keel, *linear, obtuse, flat;* ovary hispid, horned; capsule . . . . . . ? *Drege, No.* 7228!

HAB. Between Cape L'Aguillas and Potberg, on limestone hills, 500 feet. *Drege!* (Herb. Sond.!).

A rigid, woody, branching shrub, woolly in the younger parts. Leaves, many in each tuft, the subtending one shorter and broader than the rest; all fringed with woolly hairs and more or less pubescent. Calyces remarkably long, and bright purple or pink. Flowers deep purple. Apparently a well-marked species.

12. M. conferta (DC. ? Prod. 1. p. 335); robust, densely branched; branches downy; leaves tufted, *crowded, short, patent or recurved,* oblong-linear or somewhat lanceolate, thick and rigid, mucronate, concave above, keeled below, rough at the edge, glabrous; fl. sessile; sepals scarcely half as long as corolla, *elliptic-oblong, acute or sub-acute;* petals rather shorter than the keel, broadly linear, obtusely *deltoid* or narrowed at point; capsule shorter than the *slender* horns. *M. conferta, E. & Z.! No.* 211; *and M. Burmanni. E. & Z.!* 191. *Pappe!* (49 & 38).

VAR. β. **gracilis**; branches more slender and virgate; fascicles of leaves sub-distant; leaves bluntly keeled; flowers bright purple. *M. ericæfolia, Pappe, No.* 46.

HAB. Brede River, Swell., and at Saldanha Bay, *E. & Z.!* Swellendam, *Thom!* Mouth of Berg River and Pot River, *Dr. Pappe!* β. on the Cape Flats, *Dr. Pappe!* (Herb. T.C.D., Hook., Sond.).

A strong-growing, much branched, erect shrub, 1–2 feet high, more or less densely leafy. Leaves varying in shape from nearly linear to oblong-lanceolate, but *never much* narrowed at base, 2–3, rarely 4 lines long, flattish or hollow above. *M. Burmanni, E. & Z.* and *Pappe* (49) has longer, more patent, and more exactly linear and less crowded leaves. Our var. β. is much more slender than usual, and with more distant leaves, and almost intermediate with *M. divaricata* in appearance; but its essential characters seem to bring it under our species. I am quite ignorant whether ours be the plant of D'C., which I only know by his brief description.

13. M. acipetala (Harv.); suffruticose, diffuse, with filiform, hairy branches; leaves fascicled, short, linear, pungent-mucronate, thick and hard, channelled above, round backed, glabrous, with downy edges; fl. sessile; sepals oblong, subacute; petals shorter than the amply-lobed keel, *subulate-attenuate, tapering to a sharp point;* capsule glabrous, as long as the *broadly-subulate* horns.

HAB. Roadsides near the path leading up Table Mt., *Magillivray!* Simon's Bay, *C. Wright!* (Herb. Hook., T.C.D.).

A small, procumbent, many stemmed undershrub, with sub-simple branches and short leaves. Its chief character is found in the very acute, taper-pointed lateral petals.

14. M. longicuspis (Turcz., animadv. No. 2823); shrubby, diffuse, with virgate, hairy branches ; leaves fascicled, closely set, subulate (from a broad base), *attenuate-mucronate*, channelled above, round-backed, glabrous or downy, ciliolate ; fl. sessile ; sepals oblong or *oblongo-lanceolate, acute ;* petals nearly as long as the keel, oblong, tapering to a sub-acute point ; capsules with four *slender* horns of its own length. *M. mixta, E. M.! (pro parte) in Pl. Drege !* also, *Drege, No.* 7238 ! *M. affinis, Harv. in Herb. T.C.D.*

HAB. Table Mt. near the summit, *Pappe!* Dutoitskloof, 2–3000 ft. *Drege!* (Herb. T.C.D., Hook., Benth.).

A small shrubby or suffruticose plant, with many virgate, simple or branching stems, widely spreading or sub-erect, 1–1½ feet long. Leaves ⅓ inch long, exactly subulate, with a very long sharp mucro. Flowers small.

15. **M. satureoides** (Burch. Cat. 5617) ; shrubby, divaricately much branched ; twigs downy ; leaves fascicled, densely set, short, thick and rigid, *linear-subulate*, mucronate, very patent, flattish above, keeled below, rough at the margin, glabrous ; fl. (small) sessile ; sepals oblong-lanceolate, acuminate and mucronate, *unequal*, the alæ nearly as long as the petals, which are much shorter than the keel, broadly linear and blunt ; capsule with four subulate horns of its own length. *DC. Prod.* 1. *p.* 336. *E. & Z.! No.* 218. *M. ericæfolia, E. Mey.! in Herb. Drege.*

HAB. Mountains round Capetown, *E & Z.! Dr. Pappe, Drege, &c.* (Herb. T.C.D., &c.).

This has the habit of *M. conferta*, but is much smaller and more slender, with much narrower and more awl-shaped leaves, smaller and paler flowers, and a very different calyx and capsule.

16. **M. asparagifolia** (E. & Z.! No. 220) ; divaricately much branched, rigid, with downy branches ; *leaves in dense, starry fascicles, needle-shaped*, straight, taper-pointed, glabrous or roughish, widely-spreading ; fl. (*minute*) sessile ; sepals *linear-lanceolate, acuminate ;* petals broadly linear obtuse, shorter than the amply-lobed keel ; capsule . . . . . ? *Drege, No.* 7251 !

HAB. Near Mr. Brink's farm, "Laatstegift," Hott. Holl., *E. & Z.!* Drakensteenberg, *Drege!* (Herb. T.C.D., Hook., Sond.).

A minute, depressed, but very rigid shrub, 3–8 inches high, with divergent branches ; readily known by its very slender, star-tufted, needle-leaves, like those of *Asparagus capensis*, bristling in all directions. *Drege's* specimens are rather stouter and more scabrous than *Ecklon's*, but otherwise similar. The capsule is said to have "setaceous horns, shorter than itself."

17. **M. acicularis** (Harv.) ; root woody ; stems *densely tufted*, filiform, suffruticose, villous ; leaves fascicled, crowded, *needle-shaped*, pungent, *villoso-tomentose*, erecto-patent ; fl. sessile ; sepals lanceolate, acuminate and pungent, equalling the short, broadly linear-obtuse petals ; ovary-glabrous, horned ; capsule . . . . . ? *Drege, No.* 7252.

HAB. Winterhoeksberg, Tulbagh, 2–3000 feet, *Drege!* (Herb. Sond., T.C.D., Hook.).

Root thick and woody, throwing up a multitude of very slender, sub-simple stems, 3–6 inches high. Stems and leaves clothed with curled hairs ; the leaves extremely narrow, close together, erect or the points patent.

18. **M. tenuifolia** (DC. Prod. 1. p. 336) ; suffruticose, or shrubby, *diffuse or procumbent*, with slender, downy branches ; leaves in sub-

distant fascicles, linear-subulate, *very narrow, squarrose or recurved,* semi-terete, mucronate, *glabrescent ;* fl. (small) sessile or sub-sessile ; sepals not half the length of corolla, oblong, acute, mucronate ; petals nearly as long as keel, linear, obtuse or subacute ; capsule shorter than the slender horns.

VAR. α. **major** ; flowers quite sessile ; petals much shorter than the keel and very obtuse. *M. micrantha, E. & Z.!* 222. *Pappe,* 51.

VAR. β. **intermedia** ; flowers on short stalks ; petals nearly equalling the keel, obtuse or subacute. *Sparmann ! in Herb. Holm.*

VAR. γ. **minor** ; flowers subsessile and smaller ; petals nearly as long as keel, taper-pointed ; whole plant very slender. *M. tenuifolia, E. & Z.! No.* 219. *Drege, No.* 7253. *Pappe,* 50.

HAB. Var. α. Mountains near Zonderende R., Sw., *E. & Z.!* Var. γ. same locality ; and near Gnadendahl, *E. & Z.! Drege!* Var. β. C.B.S., *Sparmann!* (Herb. T.C.D., Holm., Sond., Benth.).

Closely resembling *M. diffusa* in habit and in the size of the flower ; but the leaves are more decidedly tufted, very much more slender, without obvious midrib, but concave and bluntly keeled, and their points are often hooked backwards. The sepals are less acuminate, and the pubescence less copious. The three varieties above noted chiefly differ in size ; var. β. (which is founded on an old specimen in Herb. *Holm.*) is larger and stouter than either α or γ ; but in *character* approaches nearest to γ, in *aspect* to α.

19. **M. divaricata** (E. & Z.! No. 223) ; suffruticose, slender, *procumbent,* pubescent ; leaves in sub-distant fascicles, patent, linear-filiform, *terete, fleshy, and somewhat narrowed towards the base,* mucronate, the younger ones downy ; fl. (small) sessile ; sepals elliptical, mucronate ; petals broadly linear, upturned, sub-acute ; capsule shorter than the slender horns. *M. ericæfolia, E. & Z.!* 224. *Drege,* 7237 !

VAR. β. **obtusifolia** ; leaves obtuse or mucronulate, linear-clavate. *M. laxa, E. & Z.!* 228 *(non DC.).*

HAB. Among sub-alpine shrubs, near the Waterfall, Tulbagh, *E. & Z.!* Var. β. on Mountains in the Onder-bokkeveld, *E. & Z.!* (Herb. T.C.D., Hook., Sond.).

A trailing plant, 1–2 feet long, with filiform, sub-simple branches, appressedly hairy when young, roughish or smoothish afterwards. Leaves thicker, more fleshy (nearly terete) than in *M. tenuifolia,* to which this is nearly allied. Crest of the keel very ample. *E. & Z.'s M. laxa* scarcely differs, except in having more distinctly club-shaped and obtuse leaves, but on the same root mucronate and pointless leaves occur.

20. **M. stipulacea** (D.C.? Prod. 1. p. 336) ; suffruticose, slender, branched from the base ; branches virgate, downy ; leaves *distant, linear-subulate, patent,* rigid, mucronate, flattish above, round-backed, rough-edged, the lowermost sub-fasciculate, the *uppermost solitary ;* fl. sessile ; sepals ovate-elliptical, acute ; petals as long as the keel, linear, sub-acute ; capsule with subulate horns of its own length. *Pol. stipulacea, Linn. fide Thunb.! in Herb. Holm. Thunb. Cap. p.* 558.

Hab. Houtniqua, *Thunberg!* C.B.S., *Thom?* (Herb. Hook.! Holm.!)

Root thick and woody ; stems numerous, a foot long, spreading, sub-simple. Leaves ¾ inch long, very narrow, incurved or recurved, the lower ones with 2–3 small axillary leaves, the upper solitary and sub-distant. Flowers purple, nearly half as long as the leaf in whose axil they are set. Calyx more than half the length of corolla. This plant has been much misunderstood. I describe from *Thunberg's* original specimen in Herb. Holm., with which others, probably from *Dr. Thom*, in Herb. Hooker, perfectly agree. The plant does not occur in any recent collection that I have seen.

21. M. filiformis (Thunb. Fl. Cap. p. 558); erect or ascending, shrubby, slender, many stemmed; branches *virgate*, downy or glabrous; leaves either *solitary or with axillary tufts, very erect, straight* and rigid, subulate, taper-pointed and pungent, channelled above, ribbed or keeled below, the younger ones fringed with woolly hairs; fl. sessile; sepals lanceolate or ovato-lanceolate, acute or acuminate; petals shorter than the amply lobed keel, broadly linear, obtuse; capsules with subulate horns of its own length. *E. & Z.! Nos.* 215 *and* 216! *M. linophylla and M. virgata, DC. Prod. p.* 336. *Pol. micrantha, Andr. Bot. Rep. t.* 424. *Drege, No.* 7230, 7231.

HAB. Rocky or sandy situations, among shrubs. Very common on the Cape Flats, and throughout the Cape district. (Herb. T.C.D., Hook., Sond.).

When growing in moist, sandy places, this little shrub is erect and rod-like, 1–2 feet high, with many simple stems, and then answers to *M. virgata, Burch.*; when found in dry, stony places it is shorter, more diffuse and branching, and often scrubby, and then seems to be *M. linophylla, Burch. DC.* points to a distinction in the sepals; but I find no fixed limits to this character in the very numerous and varied specimens I have examined. *M. virgata, E. & Z.!* is quite different, and appears to be a state of *M. alopecuroides*, in which the filiform leaves are in excess; almost uniting that species with *M. macroceras.*

*** Group 3. DIVERSIFOLIÆ. Leaves of two forms on the same stem or branch; the lower leaves narrow, *linear or subulate;* the upper at least (and especially *those subtending the flowers*), *broader, more or less lanceolate, and mid-ribbed.* (Sp. 22–28).

22. M. mixta (Linn.); robust, virgate, with downy branches; leaves fascicled, and densely imbricated, the lower ones linear, the rest *lineari-clavate or spathulate*, attenuate at base, mucronate, flat above, somewhat keeled below, *glabrous;* fl. sessile; calyx not one-third of corolla; sepals elliptical, obtuse, scarcely mucronulate; petals nearly as long as keel, broadly linear, obtuse; capsule glabrous, oblong, with *subulate* horns of its own length. *E. & Z.! No.* 207. *DC. Prod.* 1. *p.* 336.? *M. depressa, E. Mey.! (non Burch.).*

HAB. Cape Flats, near Duicker's Valley, *E. & Z.!* Rondebosch, *Drege!* near Kuyl's Riv., *Dr. Pappe* (49). (Herb. T.C.D,, Hook. Sond.)

A strong-growing, erect shrub, 1–2 feet high, with rod-like branches, densely clothed with leaves; sometimes so densely as to hide the branches. The denuded branches are covered with raised tubercles. The leaves, when dry, are of a pale, yellowish green. Very closely allied to *M. alopecuroides. M. Kraussiana, Meisn. in Hook., Lond. Journ., vol.* 1. *p.* 473, which I have not seen, appears, from description, to belong to our plant.

23. M. macroceras (DC. Prod. 1. p. 336); shrubby, slender, erect, with virgate, minutely downy branches; leaves fascicled, laxly set, *linear-subulate, very narrow, squarrose, or recurved*, pungent, glabrous or downy, the upper and floral leaves more or less lanceolate, flat, and mid-ribbed; fl. sessile; sepals one-third of corolla, linear-oblong, acute or mucronate; petals nearly as long as keel, broadly subulate, sub-acute; capsule *shorter than the very slender horns. Burch.! Cat. No.* 3985. *E. & Z! No.* 208. *Drege,* 7242. *Zey. No.* 74.

HAB. Mts. of Swellendam and Uitenhage, E. & Z.! &c. (Herb, T.C.D., Hook., Sond.).

Scarcely to be kept distinct from those varieties of *M. alopecuroides*, in which the filiform leaves are in excess, and the lanceolate few and scattered. The flowers are more scattered and the whole plant more slender and more glabrous than ordinary *M. alopecuroides*, but of the same pale colour. *Burchell's* original specimen is even less glabrous than those we describe.

24. M. alopecuroides (DC. Prod. I. p. 335); erect or diffuse, with virgate, downy branches; leaves fascicled, the lower ones linear-subulate and semi-terete, or squarrose and linear, with a small pungent mucro, *upper very densely imbricated, lanceolate, acuminate, flat, mid-ribbed*, mucronulate, ciliate, pubescent or glabrous; fl. sessile, crowded towards the ends of the branches; calyx one-third of corolla, sepals oval-oblong, mucronulate; petals as long as keel, broadly linear, tapering to an obtuse point; capsule rather *shorter than the slender filiform horns. Pol. alopecuroides, Linn., Thunb., Fl. Cap. p.* 557. *Bot. Mag. t.* 1006. *M. squarrosa, E. & Z.!* 213; *and M. Burchellii*, 212; *and M. virgata, E. & Z.!* 210. *Drege, No.* 7232–7235, 7244, 7247–7250!

VAR. α. **latifolia**; linear leaves few and confined to the bases of the branches, which are densely imbricated throughout nearly their whole length with broadly lanceolate-acuminate and pungent leaves. *Drege,* 7249, 7232, &c.

VAR. β. **typica**; linear leaves in sub-distant tufts, occupying two thirds of the branch; lanceolate leaves forming a densely imbricated oblong spike at the end of the branch. *M. squarrosa, E. & Z.!* 213.

VAR. γ. **filifolia**; linear leaves distributed over the greater part of the branch; lanceolate few and only subtending the upper flowers. *M. virgata, E. & Z.! No.* 210.

HAB. In rocky, subalpine places, throughout the Colony. (Herb. T.C.D. &c.)

An erect or spreading woody shrub, 1–2 feet high, with rodlike branches. It varies considerably in different places, but all its ordinary forms may be reduced to the above varieties. Usually the upper or lanceolate leaves are much acuminate and mucronate, but in some varieties they are barely acute; they vary also greatly in pubescence. The colour of the foliage is a pale yellow green.

25. M. anthospermifolia; (E. & Z. No. 204.); suffruticose, erect or diffuse; branches pubescent; leaves densely set, subfacisculate, erectopatent, lanceolate-acuminate, flat above, ribbed below, *the younger ones hispid and ciliate with long white hairs*, tapering to a recurved, pungent point; lowest leaves somewhat filiform; flowers (small) sessile; sepals short, elliptical, obtuse; petals very short, linear, truncate and emarginate; ovary hairy and capsule *hornless.*

HAB. Mountains near Mr. Joubert's farm, Kannaland, Swell, *E. & Z.!* Devil's bush, *Dr. Pappe* No. 62[a]). (Herb. Sond., T.C.D.)

Allied on the one hand to *M. phylicoides*, which it resembles in foliage; and on the other to *M. ciliaris*, of which it has the hornless capsule, but not the leaves. It may possibly be merely a broad leaved form of *M. ciliaris.*

26. M. phylicoides (Thunb.! DC. Prod. 1 p. 337); dwarf, shrubby, densely ramose, *with hispid branches;* leaves very densely crowded, fasciculate, squarrose, the lower linear-subulate, the upper lanceolate, acuminate, broad and narrow intermixed, *hispid and ciliate, flat, not* *obviously ribbed;* fl. sessile; sepals ovato-lanceolate, acute; petals hatchet-shaped, with an ovate, acute limb; ovary downy, horned; capsule hairy, *with long slender horns. Polygala phylicoides, Thunb.! Cap. p.* 558. *M. Thunbergii, E. & Z.!* 203 *(pro parte). M. ciliaris, Sieb. No.* 16 *and No.* 38. *M. aspalatha, DC. Prod.* 1. *p.* 336. *E. & Z. No.* 202.

HAB. About Table Mt. and Constantia, *E. & Z.! W.H.H. Dr. Pappe* (63) Simon's Bay, *C. Wright!* (Herb. T.C.D., Sond., Holm.!).

6–12 inches high, with subcorymbose, densely leafy and generally very hairy branches. The habit is similar to that of *M. ciliaris*, but the foliage, petals and capsule are abundantly different. It is often mistaken in herbaria for *M. ciliaris*. A specimen from Thunberg, in Herb. Holm., establishes the above reference.

27. ? M. Candollei (E. & Z. No. 214.); suffruticose, slender, hispid; branches virgate, *hispid;* leaves fascicled, the lower ones linear-uncinate, semiterete, squarrose, mucronate, hairy and ciliate, in subdistant tufts; upper densely imbricate, *broadly lanceolate, hook-pointed, flat, midribbed, hairy and ciliate,* with long, spreading, white hairs; fl. sessile among the upper leaves; sepals $\frac{1}{3}$ of corolla, oval-oblong, mucronulate; petals equal to keel, linear, obtuse; capsule ?—*M. Candolleana, Meisn. in Hook. Lond. Journ.* 1 *p.* 471 ?

HAB. Zwarteberg, near Caledon, *E. & Z.!* (Herb. Sond.)

Of this plant I only know the specimen in Herb. Sonder, collected by Ecklon. Whether it be the same as that described by Meisner, who makes no reference to Ecklon, I cannot say. To me the plant I have seen looks like a very hairy form of *M. alopecuroides*, with remarkably uncinate leaves.

28. ? M. Pappeana (Harv.); robust, woody, divaricately much branched, with villous twigs; lower leaves *linear-spathulate, thick and fleshy, round backed, flat above, dotted, obtuse* and pointless or minutely mucronate; upper and floral leaves lanceolate, acutely mucronate, the younger ones villous; fl. sessile; sepals broadly oval, very obtuse, $\frac{1}{2}$ as long as corolla; petals shorter than keel, broadly linear, blunt; capsule . . . . . . ?

HAB. Breede River, Swellendam, *Dr. Pappe* (68). Oct. (Herb. T.C.D.)

A strong-growing but dwarf bush with the aspect of *M. depressa*, but differing in foliage. Having as yet seen but a single imperfect specimen, I propose the species with much hesitation. The younger branches are but just starting, and when fully grown would probably assume a different character.

**** Group 4. LANCIFOLIÆ. Leaves *uniformly tapering much to the base;* in outline either *fusiform, lanceolate, ovato-lanceolate, obovate,* or *clavate.* (Sp. 29–41.)

29. M. ciliaris (DC. Prod. 1. p. 336.); shrubby, divaricately much branched, the branches patently hairy; leaves densely crowded, patent and squarrose, fascicled, *lanci-subulate,* tapering to the base, but very narrow, channelled and keeled, *hispid and ciliate with long hairs,* taper-pointed and pungent; fl. sessile; sepals elliptic or ovate, bluntish, mucronulate; petals linear-oblong, truncate and emarginate; ovary roundish, hairy; *capsule without horns. M. fasciculata, E. & Z.!* 205. *Zey.* 1944.

Hab. Mountains near Swellendam, *E. & Z.!* Pappe! Voormansbosch, *Zeyher!* Grahamstown, *C. J. F. Bunbury!* (Herb. T.C.D., Hook., Sond.)

A small, multifid, corymbose shrub, very thickly covered with leaves, and very hairy. Flowers conspicuous, brightly coloured, with an amply crested keel. This looks like *M. phylicoides*, but besides the different foliage, may always be distinguished by the petals and capsule. E. & Z. confounded these plants in the specimens distributed to different herbaria.

30. M. incompta (E. Mey!)! shrubby, divaricately much branched, the branches patently hairy; leaves in sub-distant tufts, lanci-subulate, and very narrow, semi-terete, *hispid and ciliate with long hairs*, pungent; fl. sessile (small); sepals ovate-oblong, mucronate; petals linear-oblong, acute or sub-acute; capsule hispid, ovate, as long as its *subulate horns. M. ciliaris, E. & Z.! 206. (non D.C.)* Drege, *No. 7245 (pro parte).*

Hab. Paarlberg & Dutoit's Kloof, *Drege!* Mts. near Waterfall, Tulbagh, *E. & Z. Pappe*, (66.) (Herb. T.C.D., Hook., Sond.)

A much branched, slender, hairy or downy, small shrub, diffuse or procumbent. Tufts of leaves separate, at short intervals. Allied to *M. tenuifolia & M. divaricata* as well as to the species of the present group. It would be desirable to see more specimens.

31. M. diffusa (Burch. Cat. 916, & 354); suffruticose, *slender, diffuse, or prostrate*; leaves sub-fasciculate or scattered, lanci-subulate or *lanceolate, flat, ribbed*, patent, squarrose, taper-pointed, and sharply mucronate, downy or villous or tomentose; fl. (minute) sessile; sepals *lanceolate-acuminate*, mucronate; petals linear-oblong, obtuse; capsule with four long slender horns. *DC. Prod.* 1. *p.* 336. *E. & Z.* 221. *M. micrantha, Sieb!* 241, *& M. squarrosa, Sieb!* 17 *&* 44.

Hab. On dry hills round Capetown, Wynberg, and Camps Bay; common. Witsenberg, *Zeyher!* 68. (Herb. T.C.D., Hook., Sond., &c.)

A widely spreading, much branched, nearly prostrate species, with subdistant leaves, varying from broadly lanceolate to very narrow, almost subulate; varying also much in the pubescence. It resembles *M. tenuifolia* in general habit, but differs in leaves and floral characters. A specimen from *Thunberg* (1774) in Herb. Holm. marked "*Pol. mixta*," belongs to this species.

32. M. Zeyheri (Turcz. Animadv. No. 2814); "branches pubescent; leaves lanceolate, acute, mucronate, glabrous, or softly hairy on the lower side at the rib and margin, solitary or tufted, and then the outer leaves smaller; fl. sub-sessile; petals half as long as the keel, gibbose at base." *Turcz.*

Of this plant I can say nothing further.

33. M. pubescens (DC. Prod. 1. p. 336); *clothed in all parts with short, horizontally patent, soft hairs*; stems numerous, diffuse or prostrate, filiform, not much branched; leaves crowded, fascicled, *linear-oblong or clavate*, tapering to the base, *obtuse*, minutely mucronulate, thick, flattish, obscurely ribbed; fl. sessile, sepals not ⅓ of corolla, oblong, obtuse, mucronulate, ciliolate; petals equalling the keel, broadly linear, obtuse, separate; capsule . . . . ? *M. rupestris, Pappe, MS. No.* 71.

Hab. Crevices of rocks, Junkershoek Mt., Stellenbosch, *Dr. Alexander Prior. Dr. Pappe.* (Herb., Alex.-Pr. T.C.D.)

Root thick and woody, throwing up many trailing stems, 3–6 inches long, equably downy, and well covered with softly downy leaves. Flowers pale flesh-coloured, except the top of keel, which is dark purple. Our specimens seem to agree well with *De Candolle's* short diagnosis.

34. M. pilosa (DC. Prod. I. p. 337 ?); branches hispido-pubescent; leaves fascicled, densely set, *lanceolate, attenuate-mucronate, rigid, thin & flat, midribbed, ciliate on margin and midrib, with long white hairs;* fl. sessile; sepals $\frac{1}{4}$ of corolla, narrow-oblong, acute and mucronate, ciliolate; petals shorter than keel, broadly linear, obtuse or emarginate; ovary villous, horned; capsule . . . . ? *M. pilosa. E. Mey. in Herb. Drege! Pol. pilosa, Poir ?*

HAB. Caffraria, at the Omsamwabo, 1-2000 ft., *Drege.* (Herb. Benth.)

Of this I have seen but a single branch, in Herb. *Bentham.* It is nearly related to our *M. lancifolia,* but has the pubescence of *M. serpylloides.*

35. M. lancifolia (Harv.); root woody; stems numerous, slender, suffruticose, the younger ones minutely pubescent; leaves fascicled, *lanceolate, acuminate, pungent, thin and flat, midribbed,* rough-edged, *glabrous;* fl. sub-sessile, sepals *linear-lanceolate,* acuminate; petals shorter than keel, broadly linear, obtuse or emarginate; ovary glabrous, horned; capsule . . . . . ? *M. conferta, Hochst.! Pl. Kraus.* 253. *M. tenuifolia, Meisn. in Hook. Lond. Jour.* I. *p.* 474.

HAB. Port Natal, on the Tafelberg, *Krauss! Sanderson!* (Herb. T.C.D., Hook.)

Root thick and woody, throwing up many tufted stems of unknown length. The specimens seen appear to have been burned-over the previous year, and therefore are probably not characteristic of the ordinary plant. The leaves are of thinner substance than in most, strongly ribbed on both surfaces and veiny when dry.

36. M. Dregei (Harv.); *Shrubby, diffusely much branched,* with glabrous or scaberulous branches; leaves scattered, patent, *linear-lanceolate, acute* and mucronulate, flat, midribbed, rough-edged; fl. sessile; sepals $\frac{1}{3}$ of corolla, ovate-oblong, mucronate, ribbed, ciliolate; petals *linear-oblong,* obtuse or sub-acute, shorter than the keel; ovary glabrous, horned; capsule . . . . . ? *M. laxa, E. Mey. in Herb. Drege (non DC.)*

HAB. Between Groenekloof and Saldanha Bay, under 500ft. *Drege!* (Herb. T.C.D., Hook., Sond.)

A procumbent or straggling bush with thickly set, but not fascicled leaves, $\frac{1}{2}$–$\frac{3}{4}$ inch long and a line wide, tapering to both ends. It is allied to *M. thymifolia,* but the leaves are much longer and narrower and the whole plant glabrous.

37. M. depressa (Burch! Cat. No. 6264.); *shrubby, depressed,* densely branched, with minutely downy or glabrous branches; leaves tufted, short, *linear-clavate, tapering to the base, obtuse,* mucronulate, *fleshy, subterete,* punctulate, glabrous or slightly downy; fl. sessile; sepals elliptic-oblong or lanceolate, mucronulate; petals broadly subulate, subacute, scarcely equalling the keel; ovary glabrous, tubercled; capsule . . . . . ? *DC. Prod.* I. *p.* 336. *Zeyher,* 1945.

HAB. C.B.S., *Burchell!* Calcareous spots, between Zwartkops and Koega Rivers, *Zeyher!* (Herb. T.C.D., Hook., Sond.)

A rigid, scraggy, robust, small and nearly glabrous shrub. The leaves are remarkably obtuse; the younger ones minutely mucronate and channelled, the older, convex at both sides. Fl. small and pale. Zeyher's specimens are rather larger than Burchell's, with longer leaves and blunter sepals. The species is nearly allied to *M. dumosa.*

38. M. dumosa (DC. Prod. I. 337 ; fide E. Mey !) ; robust, divaricately much branched, with downy or glabrescent, often spine-tipped branches ; leaves fascicled, *broadly linear*, narrowed towards base, *very obtuse* or with a recurved mucron, thick and rigid, flat above, midribbed below, glabrous, the young ones downy ; fl. sessile ; sepals $\frac{1}{3}$ of corolla, *elliptical*, *obtuse* or mucronulate ; petals nearly as long as the keel, linear, sub-acute ; ovary glabrous ; capsule with subulate horns. *Drege* 7229. *M. rubicunda, Turcz ! M. stipulacea, E. & Z. No.* 194.

HAB. Blauweberg, near Groenekloof ; on the Giftberg, and Witberg, 5000ft. *Drege !* Lion's Mt. near Cape Town, *E. & Z. ! Dr. Pappe* (48) (Herb. T.C.D., Hook., Sond.)

A rigid, coarse, scrubby bush with the habit of *M. thymifolia*, to which it is very nearly allied, and from which it differs in the more densely-tufted, narrower and more obtuse leaves ; and in the smaller and blunter sepals. E. & Z.'s "Pol. 5. 117. 11," on which *P. rubicunda Turcz !* is founded, seems to me identical with Drege's plant.

39. M. thymifolia (Thunb !) ; robust, divaricately branched, with pubescent, often spine-tipped branches ; leaves scattered or sub-fasciculate, thick and rigid, *ovate-oblong or ovato-lanceolate, or lanceolate*, flat, midribbed, *acute*, with a short pungent mucro, rough-edged, the younger ones downy ; fl. sessile ; sepals ovato-lanceolate, acute and mucronate ; petals as long as the keel, broadly linear, obtuse ; capsule as long as the broadly subulate horns. *E. & Z. ! No.* 193, *also M. serpylloides, E. & Z !* 190 *(non DC.) M. brevifolia, DC. Prod.* I. *p.*335. *(non E. & Z. !) Pol. thymifolia, Thunb. ! Cap. p.* 557.

VAR. **β. aspera** ; leaves strigoso-pubescent, scabrid. *M. aspera, Lehm. ! in E & Z. ! E. M. No.* 192.

HAB. Dry hills round Cape Town, *E. & Z. ! Drege ! Dr. Pappe !* &c., extending northwards to Saldanha Bay. (Herb. T.C.D., Hook., Sond.)

A very rigid, spreading shrub, 1–2 feet high. Leaves commonly thinly scattered, sometimes crowded, variable in form and insertion. Flowers pale. This is the true *P. thymifolia*, Thunb. according to original specimens ; and seems also to be *M. brevifolia, DC.* The common or normal state has been distributed by E & Z. as "*M. serpylloides.*" Their *M. thymifolia* is a more slender variety. *M. aspera Lehm. !* according to the original specimen in Herb. Eckl., differs merely in being rougher and more hairy.

40. M. obovata (DC. prod. I. p. 337.) ; robust, shrubby, rigid, divaricately much branched, *glabrous ;* leaves approximate, solitary or subfascicled, *obovate, thick and fleshy*, flat, *obtuse*, with a minute, reflexed point ; fl. sessile ; sepals ovate, acute ; petals as long as keel, linear, subacute ; capsule as long as the subulate horns. *M. brevifolia, E. & Z. !* 225. *(non DC.)*

HAB. Sandy places near Saldanha Bay, *E. & Z. !* Between S. Helena Bay and Oliphant's River, *Drege !* (Herb. T.C.D., Hook., Sond.)

A small, but stout and very rigid, spreading shrub, glabrous on all parts except the very young leaves and twigs, which are microscopically downy. Leaves $\frac{1}{3}$ inch long, 1–1$\frac{1}{2}$ lines wide, very blunt. Flowers resembling those of *M. thymifolia*, to which species this is nearly allied.

41. M. reticulata (Harv.) ; shrubby, depressed, divaricately much branched, with downy, often spine-tipped branches ; leaves scattered or tufted, *broadly obovate or elliptical, flat, pubescent and ciliate, mid-*

ribbed and veiny, with a minute, reflexed point; fl. (minute) sessile; sepals linear-oblong, obtuse; petals as long as the keel, linear, obtuse; capsule broader than long, ciliolate, with short, flat, subulate horns. *M. obovata, E. & Z.! 226. (non DC.)*

HAB. Stony, mountain situations round Hercules' fountain, Clanwilliam, *E. & Z.!* (Herb., T.C.D., Hook., Sond.)

A scrubby, seemingly prostrate shrub, with very patent branches, the older ones hardening at the points into rigid spines. On the principal branches the leaves are rosulate; on the shoots scattered. They are $\frac{1}{3}$ inch long, 1–1½ lines wide, of thin substance and not rigid, conspicuously veiny. Flowers among the smallest in the genus, hidden under the broad leaves. The capsule is peculiarly broad and the horns short.

Sub-genus 2. PSILOCLADUS. Flowers manifestly pedicellate, the pedicel always longer than the subtending bracts. *Sepals* unequal, the alæ longest. *Capsule* either emarginate, or 4-tubercled at top, not horned. Leaves scattered. (42–51.)

***** Group 5. ACUTIFOLIÆ. Leaves acute; subulate, linear, lanceolate or ovate. (42–46.)

42. M. juniperifolia (DC. Prod. 1. 336; non E. & Z.) *shrubby, robust,* much branched, glabrous; leaves scattered, short, *subulate, triquetrous,* flat above, keeled, *patent,* straight and rigid, pungent, rough-edged; pedicel twice as long as the minute roundish bracts; sepals elliptical, obtuse, the alæ twice as long as the others; petals as long as the keel, broadly linear, obtuse, recurved; capsule ovate, emarginate, hornless. *M. Sprengeliodes, Burch.! Cat.* 4957. *Mundia desertorum, E. & Z.! No.* 234.

HAB. South Africa, *Masson!* in Herb. Banks. Desert places in the Langekloof and Zwarteberg, *E. & Z.!* (Herb. T.C.D., Hook., Sond.)

A very densely branched and rigid shrub, 6–12–15 inches high, nearly glabrous. Leaves $\frac{1}{4}$ inch long, very hard and sharp. It is strange that E. & Z., whose specimens are plentifully in fruit, should have mistaken this well-marked species for a Mundtia. I have compared an authentic specimen of Burch. 4957, with the original from Masson in Herb. Banks, and can detect no specific difference.

43. M. pauciflora (DC. Prod. 1. p. 337.); suffruticose, slender, glabrous, with *filiform* branches; leaves few and distant, scattered, *subulate,* pungent, *flat above, with a medial furrow, triquetrous, keeled;* pedicels about as long as the leaves; sepals elliptic-oblong, obtuse or acute, the alæ twice as long as the rest, petals longer than the amply-lobed keel, spathulate, entire or immarginate; capsule elliptic-oblong, acutely bidentate. *Pol. pauciflora, Thunb.! Fl. Cap. p.* 558. *E. & Z.! En.* 186.

VAR. β. **Zeyheri**; sepals acute and mucronate; teeth of the capsule divergent.

HAB. Mts. near Tulbagh; and at Puspasvalley, Swell., *E. & Z.!* Dutoitskloof, *Drege!* (Herb. T.C.D., Hook., Sond.)

A very slender, filiform, much branched, half-woody plant, with a few very narrow leaves and long stalked minute flowers. The capsule is twice or thrice as long as broad, sharply bifid, the lobes either connivent or divergent.

44. M. trinervia (DC. Prod. 1. p. 335); suffruticose, robust, diffuse, with sharply angular, rough branches; leaves scattered, sessile, *cordate-acuminate, acute, flat, three-nerved and netted* on the under side, glabrous, rigid, scabrido-denticulate; pedicels rather shorter than the leaves; sepals oblong, obtuse, alæ longest; petals equalling the amply lobed keel, broadly linear, obtuse; capsule pendulous, sharply bifid. *Pol. trinervia, Thunb. ! Fl. Cap. p.* 559. *Drege, No.* 7213!

HAB. South Africa, *Thunberg! Masson!* Groenekloof, *Drege!* (Herb. Holm., Banks, Sond., Benth.)

A remarkable plant, apparently rare; not found by Eck. & Zey. The leaves are half inch long, 3–4 lines wide, in shape somewhat resembling those of *Hallia cordata*, but shorter in proportion to their breadth. The larger ones are sometimes 5-nerved. *Drege's* specimens are smaller and less luxuriant than *Thunberg's*, but otherwise the same.

45. M. Beiliana (Harv.); suffruticose, slender, diffuse, with angular, glabrous branches; leaves scattered, sessile, *linear-acuminate and mucronate, flat, obtuse at base, three-nerved,* glabrous, rigid, scabrido-denticulate; pedicels shorter than the leaves; sepals oblong, obtuse, the alæ longest and broadest; petals longer than the amply lobed keel, broadly linear, obtuse; capsule pendulous, oblong, sharply bifid. *Pol. Beiliana, E. & Z. !* 179 *& P. Muraltioides, E. & Z. !* 183. *Drege,* 7214. *Zeyher,* 1749. *M. Sprengelioides, Turcz. Anim. No.* 2839 *(non Burch.)*

HAB. Near the Zwartkops' River, Uitenhage; and the Winterhoek, Tulbagh, *E. & Z. ! Drege!* Between R. Zonderende & Assagaiskloof, *Zeyher.!* (Herb. T.C.D., Hook., Sond.)

Spreading over the soil, ascending or suberect, 1–2 feet long, much branched, the younger twigs two-edged, the older 3–4 angled, all the angles sharp. Lowest leaves ¼ inch, upper ½–1 inch long, 1–1½ lines wide. Nearly related to *M. trinervia,* but more slender, with much narrower and differently shaped leaves.

46. M. angulosa (Spreng. fide Turcz. Anim. No. 2836); suffruticose, slender, diffuse, with angular and furrowed, rather rough branches; leaves scattered, sessile, *linear, acute, narrowed at base, flat, one-nerved,* glabrous, rigid; pedicels shorter than the leaves; sepals ovate-oblong, acute, mucronate, the alæ longest; petals equalling the amply lobed keel, spathulate; capsule . . . . . ? *M. striata, E. & Z. ! No.* 227 *(non DC.) M. linifolia, Harv. ! in Herb. Sond.)*

HAB. Mts. near Piquinierskloof, Clanw., *E. & Z. !* (Herb. Sond.)

Very near *M. Beiliana,* from which it differs in having narrower, strictly linear leaves and spathulate petals. I had formerly named it *M. linifolia,* but it seems to agree sufficiently with Turczaninow's diagnosis, above quoted. His specimen came from Zeyher, but without reference to the *M. striata* of the Enumeratio.

******* Group 6. OBTUSIFOLIÆ. Leaves blunt, not pungent. (Sp. 47–51).

47. M. carnosa (E. Mey.! in Herb. Drege); shrubby, glabrous, and somewhat viscid, with round branches; leaves *petiolate,* scattered, thick and fleshy, *oval or subrotund, convex* on both surfaces, smooth, muticous; pedicels very short, cernuous; sepals very obtuse, elliptic-ovate, alæ broadly oval, twice as long as the rest; petals shorter than the keel, linear, obtuse; capsule . . . . . ?

HAB. Zwarteberg, near Vrolyk, in moist, rocky places 4–5000 ft., *Drege!* (Herb. T.C.D., Hook., Sond.)

A rigid shrub, a foot or more in height, more like a *Mundtia* than one of the present genus. The leaves are distinctly petioled, 2 lines long, 1 line wide, very thick and blunt: they spring from tubercular cicatrices. The ovary is oblong, without horns or tubercles; fruit unknown.

48. M. crassifolia (Harv.); shrubby, much branched, diffuse, glabrous, with angular branches; leaves alternate, thick and rigid, *oblongo-lanceolate*, mucronulate, *flat*, midribbed, with rounded edges; pedicels shorter than the leaves; sepals elliptic-oblong, obtuse, the alæ twice as long as the rest; petals equalling the keel, linear-spathulate, obtuse; capsule . . . ? *Pol. polyphylla, E. & Z.! No.* 185. *(non DC.)*

HAB. Summit of the Winterhoeks Berg, Tulbagh, *E. & Z.!* (Herb. T.C.D., Hook., Sond.)

A small but stout shrub, 6–12 inches high, spreading, densely covered with leathery leaves $\frac{1}{3}$ inch long and 1 line wide; the shape varying from nearly linear to lanceolate and elliptic-lanceolate. The whole plant is perfectly glabrous.—Distributed by *E. & Z.* as "*Pol. polyphylla, DC.*," but does not agree with De Candolle's diagnosis.

49. M. rigida (E. Mey.! in Herb. Drege); *robust*, woody, *divaricately much branched*, with glabrous, pungent branches; leaves few and small, scattered, sessile, *linear-oblong*, *flat on both sides*, but thick and fleshy, obtuse, with a minute point; pedicels shorter than the leaves; sepals elliptic-oblong, very obtuse, the alæ much longer than the rest, petals longer than the keel, linear, channelled, obtuse; capsule oblong, pendulous, bidentate.

HAB. Kamiesberg, and between Pedroskloof and Leliefontein, *Drege!* (Herb. T.C.D., Hook., Sond.)

A very scraggy, coarse, densely branched and rigid shrub; the branches spreading at right angles. Leaves 2–3 lines long. Flowers pale, small.

50. M. striata (DC., ex Thunb. Fl. Cap. p. 559); erect, *slender*, *with rod-like, angular and furrowed branches*, glabrous or nearly so; leaves few and distant, scattered, erect, short, *linear*, *obtuse*, sub-mucronulate, *channelled above*, round-backed, glabrous; pedicels solitary or in pairs, cernuous, shorter than the leaves, twice as long as the minute bracts; sepals *broadly elliptical*, very obtuse, the alæ twice as long as the rest; petals longer than the keel, linear, obtuse; capsule pendulous, oblong-oval, sharply bidentate. *M. brevicornu, E. Mey.! in Herb. Drege (an DC.?) M. viminea, Turcz. Anim. No.* 2837. *Pol. striata, Thunb. l. c.*

HAB. On the Cape Flats, *Dr. Pappe* (47). Groenekloof & Dassenberg; and between Bergvalley & Langevalley, *Drege!* Piquetsberg, *Zeyher!* (71). (Herb. T.C.D. Hook., Sond).

A slender, tall, naked suffrutex, $1\frac{1}{2}$–2 feet high, sparingly branched. Leaves $\frac{1}{4}$–$\frac{1}{3}$ inch long, and as far apart, either close-pressed or somewhat patent. Flowers small and pale, axillary, or clustered in small, sub-umbellate fascicles at the ends of the branches. This agrees very well with *Thunberg's* description, and being a plant of the neighbourhood of the Cape, is likely to have been found by early explorers. De C.'s *M. brevicornu* is either badly described, or is something different.

51. M. leptorhiza (Turcz. Anim. No. 2838); erect, *slender*, *with rod-like, angular and furrowed, glabrous branches;* leaves scattered, distant, narrow-linear, *mucronate, keeled, flat above, with a narrow, medial furrow*, erect, glabrous; pedicels solitary or in pairs, twice as long as the minute bracts; sepals *linear-oblong*, very obtuse, strongly ribbed, the alæ rather longest; petals equalling the keel, linear, obtuse; capsule pendulous, sharply bifid, the lobes acuminate. *Zeyher, No.* 1948! *M. sulcata, Harv. in Herb. Hook.!*

HAB. Grootvadersbosch, Swell., *Zeyher! Pappe* (60). (Herb. T.C.D., Hook., Sond.) Very closely allied to *M. striata*, but more slender, with narrower and acute leaves, sharply furrowed down the middle; narrower sepals, and more deeply bifid, almost horned capsules.

APPENDIX.—Doubtful or insufficiently known species.

52. M. brevicornu (DC. Prod. I. p. 337); "leaves linear, obtuse, glabrous, tubercled at base on the outside, scattered; branches virgate, glabrous; flowers axillary, sessile; capsule with very short horns."

53. M. Burmanni (DC. l. c.); "leaves ovato-linear, appressed, straight." *Pol. ericoides, Burm. Fl. Cap. Prod. p.* 20.

54. M. Poiretii (DC. l. c.); "leaves subulate, glabrous, as well as the virgate branches; flowers axillary, sessile, crowded." *P. ericoides, Poir. Dict.* 5. *p.* 497.

55. M. ? laxa (DC. l. c.); "leaves lanceolate, solitary; flowers racemose." *Pol. laxa, Thunb. Cap. p.* 559.

56. M. fasciculata (DC. l. c.); "leaves very slender, subciliate; branches fascicled; fl. sessile, solitary or in pairs." *Pol. fasciculata, Poir. l. c.*

57. M. parviflora (DC. l. c.); "leaves setaceous, smoothish, minute, stipulate ?; branches slender; fl. axillary, solitary, sessile." *Pol. parviflora, Poir. l. c.*

58. M. humilis (DC. l. c.); "leaves lanceolate-linear, erect, scattered; stem short, much branched; fl. axillary, sessile, solitary." *Pol. humilis, Lod. Cab. Bot. t.* 420.

59. M. micrantha (Dietr.); "branches slender, glabrous; leaves scattered, linear-subulate, pungent, glabrous; fl. axillary, sessile." *Walp. Repert.* I. *p.* 244.

ORDER XIV. FRANKENIACEÆ, St. Hil.

(BY W. H. HARVEY.)

(Frankeniaceæ, St. Hil.—DC. Prod. I. p. 349. Endl. Gen. No. cxcii. Lindl. Veg. Kingd. No. cxvii.)

Flowers regular. *Calyx* tubular, 4–5 toothed, ribbed, persistent, hardening after flowering. *Petals* 4–5, hypogynous, with long claws, twisted-imbricate in æstivation, deciduous. *Stamens* generally 6, or 5, hypogynous, alternating with the petals; filaments subulate; anthers extrorse, versatile. *Ovary* free, unilocular, with 3–4 parietal placentæ and numerous ovules; style single, filiform. *Capsule* enclosed in the calyx, many seeded, splitting. *Seeds* minute, albuminous, with an axile embryo.

Small, heathlike perennials or undershrubs, inhabiting salt-marshes or sea coasts. Stems much branched, jointed and knotted; leaves opposite, alternate or whorled, or densely tufted, very entire, and often linear, with revolute margins. *Flowers* solitary or cymose, sessile, fugacious, pink or purple.

A small Order, of which less than thirty species are known, but these are scattered along the shores of the four quarters of the world. They resemble in habit some of the Caryophylleæ, from which their placentation and the structure of their seeds separate them. In foliage they look like small heaths; whence the English name "Sea-heath." Their true affinities are not very clearly ascertained, but they are generally considered to be related to Violaceæ and Turneraceæ. Their properties and uses are unimportant. *Beatsonia portulacifolia*, a sea coast plant of St. Helena, is said to have been once used in that island as a substitute for Tea.

I. FRANKENIA, Linn.

Calyx tubular, 4–5-fid, ribbed and furrowed. *Petals* 4–5, hypogynous, with long claws. *Stamens* 6, rarely 5, hypogynous. *Ovary* sessile, one-celled, with 3–5 parietal placentæ; ovules few, style filiform; stigma tripartite. *DC. Prod.* 1. *p.* 349. *Endl. Gen. No.* 5053.

Small, perennial, herbaceous or suffruticose plants, with diffuse, rarely erect, wiry stems; small, heathlike, opposite or alternate, frequently fascicled leaves; and sessile flowers in terminal cymes. Petals purple or pink, very delicate and soon withering; calyces hard and dry, remaining as a permanent case to the capsule. There are several species (though not nearly so many as have been *made* by botanists), inhabiting salt marshes and sea-coasts in most parts of the temperate zones, north and south. In England they are called Sea-Heaths. The botanical name is given in honour of *J. Franken*, formerly Professor of Medicine at Upsala, Sweden.

1. **F. capitata** (Webb & Berth. Fl. Canar 1. p. 131. t. 16); stems woody, prostrate or ascending, much branched, glabrous or hairy; leaves *linear*, glabrous, *with strongly revolute margins;* petioles short, ciliated, petals much longer than the glabrous, puberulous or hispid calyx.

VAR. *α*, **lævis**; stems and calyces quite glabrous. *F. lævis, Linn. DC. Prod.* 1. *p.* 349. *E. Bot. t.* 205. *E. & Z.* 241.

VAR. *β*. **hebecaulon**; stems velvetty; calyces subglabrous. *Lowe. Mad. p.* 48.

VAR. *γ*. **Nothria**; leaves broader and less revolute; stems pubescent or hispid; calyces glabrous or downy. *F. Nothria, Thunb. Fl. Cap. p.* 295. *DC. l. c. E. & Z.! No.* 239.

VAR. *δ*. **Krebsii**; stems pubescent or hispid; calyces hispid. *F. Krebsii, Ch. & Schl. Lin.* 1. *p.* 36. *E. & Z.! F. hispida ? DC. l. c.*

HAB. In salt marshes, &c., from Capetown to Uitenhage; not uncommon. (Herb. T.C.D., Hook.)

Very variable in the amount and nature of the pubescence, and somewhat in the breadth of the leaves. The above 4 varieties occur at the Cape, and three of them (*lævis, Nothria & Krebsii*) have usually been held for species; but after a careful examination of many specimens, in the Dublin & Hookerian Herbaria, and from various parts of the world, I find it impossible to determine where one variety begins and another ends. Several other reputed species might be added to the list of synonyms. The least variable character is found in the linear, strongly revolute leaves; but this is less marked in *Nothria* than in the other varieties.

2. F. pulverulenta (Linn.); stems herbaceous, diffuse or prostrate, more or less hispid or pubescent; leaves *broadly oval* or *obovate*, flattish, pale and powdery-puberulous on the lower surface, the margins *slightly reflexed*; petioles ciliated; calyces glabrous, sepals linear, obtuse. *DC. Prod.* 1. *p.* 349. *E. & Z. No.* 237. *Lam. Ill. t.* 262. *f.* 3. *E. Bot. t.* 2222. *F. pulverulenta & F. nodiflora, of Drege's Coll.*

HAB. In salt marshes, by the seaside; and on the Great Karroo, Beaufort; and in Namaqualand, near the Orange river. Zwartkops' river, Uitenhage, *C. Zeyher!* (Herb. T.C.D., Hook.)

Stems weak and straggling, less woody than in *F. capitata*, from which this is easily known by its broad, flat leaves, and by the smaller flowers. The underside of the leaves is generally minutely downy. In *Drege's* specimens marked "*nodiflora*" the surface is smooth, but minutely dotted. The pubescence of the branches varies greatly in different specimens.

3. F. nodiflora (Lam. Ill. t. 262. f. 4); "leaves ovate, glabrous, the petiole not ciliated; stems prostrate, glabrous, as well as the calyces." *DC. Prod.* 1. *p.* 349. *Burch. Cat. No.* 513.

HAB. South Africa. (Unknown to me: possibly a mere variety of *F. pulverulenta.*)

ORDER XV. ELATINACEÆ.

(By W. H. HARVEY.)

(Elatinaceæ, Lindl. Veg. Kingd. No. 181. Elatineæ, Cambess. Endl. Gen. No. 219.)

Flowers minute, regular. *Sepals* 3–4–5, separate, acute, imbricate. *Petals* as many as the sepals and alternate with them. *Stamens* hypogynous, as many or twice as many as the petals; filaments subulate, free; anthers introrse, 2-celled. *Ovary* free, 3–4–5-locular, with axile placentæ and numerous ovules; styles as many as the loculi; stigmas capitate. *Capsule* dry, splitting, the valves falling away from the persistent column. *Seeds* many, cylindrical, straight or curved, exalbuminous.

Small, often minute, herbs or undershrubs, growing in wet places; with opposite, entire or denticulate leaves, and interpetiolar membranaceous stipules. *Flowers* axillary, solitary or tufted or in lax cymes.

A small group of minute marsh plants, found in both hemispheres. They were formerly included in Caryophylleæ, from which they differ in the structure of the fruit, the capitate stigmas, and the exalbuminous seeds. They are, perhaps, more nearly related to Hypericineæ, but differ in the structure of the seed vessel and the presence of stipules. A relationship with Lythraceæ has also been pointed out; and Lindley places them among the Rutaceous Orders. They are of no particular use; and are too insignificant for cultivation.

I. BERGIA. Linn.

Sepals 5, separate. *Petals* 5, hypogynous. *Stamens* 10, those opposite the petals shortest and sometimes deficient. *Ovary* 5-celled; styles 5, separate; ovules numerous. *Capsule* 5-celled, 5-valved, many seeded, *DC. Prod.* 1, *p.* 390. *Endl. Gen.* 5476.

Herbs or small suffrutices, found in moist places throughout the tropics and warmer parts of the temperate zones. Leaves opposite, simple, entire or toothed, often with shortened leafy branchlets in the axils, and then densely crowded. Stipules membranous or bristle-pointed. Flowers minute, axillary, sessile or stalked, solitary or in cymes. This genus, which is scarcely distinguishable from *Elatine*, is named in honour of *Peter Jonas Bergius*, an early explorer of S. African Botany and author of a Cape Flora, published at Stockholm, 1767.

1. **B. glomerata** (Linn. f. Suppl. 243); stems *prostrate*, woody, quadrangular, distichously much branched; leaves *crowded*, *obovate*, toothed or nearly entire; stipules *ovate*, *laciniate*; flowers axillary, solitary, nearly *sessile*; sepals oblong, acute, ciliated; petals spathulate-linear, obtuse. *DC. Prod.* 1. *p.* 390.

HAB. In wet places near the Zwartkops River, Uitenhage, *Thunberg*, *E. & Z.!* (Herb. T.C.D., Hook.).

Perennial. Stems many from a thick root, 1–2 feet long, rough, furrowed, densely set with sub-opposite, spreading, distichous branches. Leaves ¼ inch long, closely set, with tufts of smaller leaves in their axils, glabrous, Stipules membranous, inciso-pinnatifid. Flowers bracteate; sepals and petals narrowed to the base. Stamens 10. Styles 5, divergent; capsule 5-angled and 5-celled.

2. **B. decumbens** (Planch. in Herb. Hook.); stem suffruticose, diffuse, setulose or glabrous, oppositely branched; leaves *remote*, *lanceolate*, sessile, *sharply serrate*, acute; stipules *setaceo-subulate*, *ciliated*; cymes axillary, few flowered, much shorter than the leaves; pedicels longer than the flowers; sepals acuminate and mucronate; petals obovate-oblong, sub-obtuse. *Harv. Thes. Cap. t.* 24.

HAB. Macallisberg, *Burke!* (Herb. Hook., T.C.D.).

Stems many from the same crown, decumbent or ascending, terete, the younger ones 4-angled, sparsely pubescent, 1–2 feet long; branches 1–2 inches apart. Leaves 1–1½ inches long, 1–2 lines broad. Stipules rigid, erect. Cymes imperfect, reduced to verticilasters, 3–6 or more flowers in each. Sepals concave and sharply keeled, with scarious edges; petals as long as the sepals, obtuse. Stamens 10, the 5 opposite the petals smaller with narrower filaments.

3. **B. anagalloides** (E. Mey! MS.); *glabrous*, suffruticose at base; stems herbaceous, procumbent, oppositely branched; leaves remote, *oblong or elliptical*, tapering at base into a short *petiole*, *denticulate* towards the obtuse apex; stipules (*small*) linear-subulate, *entire*; flowers solitary, axillary, on filiform *pedicels much longer than the leaves*; sepals *elliptical*, *obtuse* or mucronulate, albo-marginate; petals longer than the sepals, obovate, *obtuse*, entire. *Fenzl. in An. Wien. Mus.* 1. *p.* 344. *Walp. Repert.* 2. *p.* 786.

HAB. Orange River, on sand hills at the right bank, near Verleptram, *Drege!* (Herb. Sond.).

Root thick and woody. Stems several from the crown, decumbent or prostrate, somewhat woody at base, 8–12 inches long, terete or sub-compressed, the nodes about an inch apart. Branches slender, filiform. Leaves ½ inch long. Pedicels 1 inch long or more. Flowers purplish-rose-colour, about the size of those of *Anagallis arvensis*. Sepals with a broad white margin. Stipules much less conspicuous than in the other species.

4. **B. polyantha** (Sond. Linn. 23. p. 17); herbaceous, glabrous, many stemmed; stems 4-angled; leaves opposite, oblong, sub-acute, narrowed at base, sessile, entire or sub-denticulate; stipules lanceolate; flowers axillary, solitary, *on long pedicels, crowded towards the ends of the branches*,

bracteate; calyx 5-parted, sepals *acuminate*, with scarious margins petals *shorter* than the sepals, *ovate, acute. Sond. l. c. Lancretia humifusa, Pl. in Herb. Hook.*

HAB. Rhinosterkopf, by the Vaal River and Sand River Hills, *Burke and Zeyher!* 540. (Herb. Hook., Sond.).

A small *annual*, with many short, decumbent stems, spreading from the crown. Stems 1–2 inches long. Leaves ½ inch long, 2–3 lines wide, the lowermost larger, the upper gradually dwindling into bracts. Stipules a line long, those of the lower leaves entire, of the upper serrate or ciliate. Petals pale rose coloured. *Sond. l. c.*

ORDER XVI. HYPERICINEÆ, Chois.

(By W. SONDER.)

(Hyperica, Juss. Gen. 254. Hypericineæ, Chois. DC. Prod. I. p. 541. Endl. Gen. No. 218. Hypericaceæ, Lindl. Veg. Kingd. No. 146.).

Flowers regular. *Sepals* 5, rarely 4, united at base, persistent, strongly imbricate, two external to the others. *Petals* alternate with the sepals, unequal-sided, spirally twisted in æstivation, bordered with black dots! *Stamens* mostly indefinite, polyadelphous; filaments united in 3–5 parcels, filiform; anthers 2-celled. *Ovary* free, unilocular, or imperfectly 3–5-locular, with numerous ovules; styles 3–5, separate or connate at base. *Fruit* a dry or fleshy capsule, uni- or pluri-locular; seeds exalbuminous.

Trees, shrubs or herbs, with resinous juices; opposite or whorled, often quadrangular branches, and opposite, very entire, pellucidly-dotted, penninerved, simple, exstipulate leaves. *Flowers* terminal or axillary, in cymes or panicles, yellow, rarely red or white, showy, but not fragrant.

Dispersed throughout the tropics and temperate zones of both hemispheres, but much more abundant in the northern temperate zone. About 300 species are known. This family may be considered as the representative, in temperate climates, of the tropical order GUTTIFERÆ—whose gummy properties are possessed by its species in a weak degree. None are of much value, but many are cultivated for ornament. In the earlier medicine their repute as febrifuges and vulneraries was greater, and the *Hypericum* itself was dedicated to S. John.

I. HYPERICUM. Linn.

Calyx 5-parted, the sepals equal, or the two outer largest, imbricate. *Petals* 5, hypogynous, twisted-imbricate in æstivation. *Stamens* numerous, polyadelphous, in 3–5 parcels. *Ovary* sessile, unilocular, or imperfectly 3–5-locular, with many-ovuled, parietal placentæ. *Capsule* dry and dehiscent, (or rarely fleshy and indehiscent). *Seeds* numerous, very rarely definite or solitary. *DC. Prod.* I. *p.* 543. *Endl. Gen.* 5464.

Shrubs, under-shrubs, or herbs diffused throughout the temperate and subtropical zones of both hemispheres, but much more common in the northern than the southern hemisphere. Leaves opposite, either petiolate or sessile or amplexicaul, very entire or serrulate, very generally pellucid-dotted, exstipulate. Flowers yellow, cymose or panicled, the petals margined with black dots. These plants were called ὑπηρικον, by Dioscorides.

1. **H. aethiopicum** (Thunb. Cap. p. 439.); glabrous; stem herbaceous, erect, or decumbent at base, terete; leaves *ovate, obtuse* at base, sessile, *subamplexicaul*, pellucid-dotted, with a revolute margin; pa-

nicle terminal, dichotomous; *sepals lanceolate, very acute, glandularly toothed;* styles 3, erecto-patent nearly as long as the (half ripe) capsule. *DC. Prod.* 1. *p.* 552. *E. & Z.* 415. *Zey!* 149.

VAR. β. **glaucescens**; stems dwarfer, erect; leaves glaucescent, closely set, longer than the internodes, black-dotted.

HAB. Grassy hills. Robbeberg, Houtniquas, *Thunberg.* Uitenhage, Caffraria and Port Natal, *E. & Z.! Drege, Krauss, Sanderson,* var. β. at Aapjes River and Macallisberg, *Zeyher.* (Herb. Thunb. T.C.D., Sond., Hook.)

An under-shrub, 6–12 inches high; the stem and the erect opposite, filiform, virgate branches ferruginous. Leaves ½ inch long, 4 lines wide, often smaller on the branchlets, with a wide base, perfectly sessile, paler on the lower surface, penninerved. Panicle cymose. Sepals 2 lines long. Petals oblong, yellow, striolate, twice as long as the calyx. *Stamens* numerous, rather shorter than the petals; the anthers as well as the sepals and petals dotted with black. Specimens collected by *Dr. Gueinzius* and *Mr. Sanderson* at Port Natal have larger and broader leaves than usual, but do not otherwise differ from the common variety.

2. H. humifusum (Linn.); glabrous; stem *prostrate,* filiform, *terete or obsoletely two-edged;* leaves *elliptic-oblong,* pellucid-dotted, flat; *sepals oblong, obtuse;* petals once and half as long as the calyx. *E. Bot. t.* 1226. *DC. Prod.* 1. *p.* 549.

HAB. Cape (no locality indicated, *Drege,* 7530. (Herb. Sond.)

Stems 6–12 inches long. Cauline leaves 4–5 lines long. Flowers pale yellow, solitary, terminal and in the forks of the branches. A common European plant. The Cape specimens are taller than usual, but not otherwise different.

3. H. Lalandii (Chois. in DC. Prod. 1. p. 550.); glabrous; stem herbaceous, erect or decumbent at base, simple or somewhat branched, 4-*angled;* leaves *lanceolate, acute* or sub-obtuse, with a revolute margin, pellucid-dotted, appressed to the stem, sub-amplexicaul; panicle terminal, few-flowered, dichotomous; *sepals linear-lanceolate, very entire. E. & Z.* 416.

VAR. α. **lanceolata**; leaves lanceolate, or linear-lanceolate; petals as long as the calyx.

VAR. β. **latifolia**; leaves ovate, sub-acute; petals as long as the calyx.

VAR. γ. **macropetala**; leaves lanceolate; petals twice as long as the calyx. *H. Lalandii, Drege, c.*

HAB. Slopes of the Zuureberge, and Zwartehoogdens, near Grahamstown, Alb.; Vanstaadensberg and Krakakamma, Uit., *E. & Z.!* Dutoitskloof, Galgebosch, and Bosjeman's River, *Drege!* Port Natal, *Drege, Gueinzuis, Krauss,* &c. Aapjes River, *Zeyher,* 150. β. Uitenhage, *Zeyher.* (Herb. Hook., T.C.D., Sond.)

An under-shrub, 6–18 inches high. Stems many from the same root, mostly straightly erect and simple, acutely 4-angled, pale green, purplish below. Leaves subcoriaceous, one-nerved, few-veined, closely set in the lower part of the stem, as long as the internodes, or longer, more distant on the upper half and shorter than the internodes, very variable in size, in the smaller specimens 3–4 lines long and 1 line wide, in the larger ½–1 inch long;—in γ. broadly ovate, 6 lines long and 4 lines wide, like those of *H. æthiopicum.* Cyme commonly 3–5-flowered, rarely many-flowered. Cal 2 lines long. Pet. yellow. Stamens numerous. Styles 3, erect. Capsule glabrous.

"*Hypericum verticillatum,*" Thunb. *Fl. Cap. p.* 439, is *Linum quadrifolium,* L., according to the specimens preserved in Herb. Thunb.!

ORDER XVII. TAMARICINEÆ. Desv.

(By W. H. HARVEY.)

(Tamaricineæ, Desv. DC. Prod. 3. p. 95. Endl. Gen. No. ccxxi. Tamaricaceæ, Lindl. Veg. Kingd. N. cxviii.)

Flowers minute, regular. *Calyx* 4–5 parted, persistent, imbricate. *Petals* 4–5, imbricated, marcescent. *Stamens* hypogynous, as many or twice as many as the petals; filaments monadelphous; anthers two-celled, incumbent. *Ovary* free, unilocular, with parietal or basal placentæ, and numerous ovules; styles mostly 3, free or confluent. *Capsule* dry, splitting into valves; seeds numerous, comose, without albumen.

Shrubs or suffruticose plants, inhabiting sea-shores or salt plains. Leaves minute, scalelike, imbricated, sessile. Flowers racemose or spiked, minute, crowded, white or pink.

A very small Order, chiefly natives of the northern temperate zone of the Old World, especially along the shores of the Mediterranean, and in the salt plains of Central Asia. A single species only occurs in Southern Africa. These plants are of doubtful affinity, and were formerly placed, in Calycifloræ, near *Saxifragaceæ;* but appear to be rather a reduced form of the Hypericoid-group. An affinity with *Salicaceæ* has also been asserted. Their officinal uses are unimportant; the bark is astringent and bitter. As ubstance resembling manna is obtained, in Arabia Petræa, from *T. mannifera.*

TAMARIX, Linn.

Sepals 4–5, unequal, imbricated. *Petals* 4–5, hypogynous, equal. *Stamens* 4–10, rising from the glandular margin of a fleshy, hypogynous disc; filaments separate at base. *Styles* usually 3. *Ovary* one-celled, with parietal placentæ; ovules numerous. *Capsule* 3–, rarely 2–4-valved, many seeded; seeds with a terminal tuft of silky hairs. *Endl. Gen. No.* 5484. *DC. Prod.* 3. *p.* 95.

Shrubs or small trees, growing in desert places, generally where there is much salt in the ground; as, along the sea shores of Middle and Southern Europe, in North Africa, and in the salt plains of Central Asia. Leaves very minute, often reduced to mere sheathing scales. Flowers small, pink or white, conspicuous by their abundance, in terminal spicate racemes; disagreeably scented. The European species smells like a pigstye. "Named from the *Tamarici,* a people who inhabited the banks of the Tamaris, now *Tambria,* in Spain, where the Tamarisk abounds." *Hook. Br. Fl.*

1. **T. articulata** (Vahl, Symb. 2. p. 48. t. 32); glabrous, glaucous; branchlets seemingly articulate; leaves very minute, sheathing, mucronulate; spikes lateral and terminal, lax, few flowered; flowers 5-cleft, subsessile; capsule 3-valved. *DC. Prod.* 1. *p.* 96. *T. orientalis, Forsk.—Eck. & Zey. !* 2150. *T. usneoides, E. Mey. in Pl. Drege.*

HAB. Barren places in Namaqualand, *Eck. & Zey. ! Drege !* (Herb. T.C.D.)

A much branched bush or small tree, 20–30 feet high; the younger branches and twigs marked at the nodes with a circular leafscar, so as to appear jointed. *Leaves* reduced to a mere sheath, closely investing the branch, with or without a minute lateral point. The flowering branches are repeatedly divided, or panicled, every branchlet ending in a few-flowered imperfect raceme or spike. Sepals deltoid. Petals oval. Stamens protruding. Capsule in our specimens (from *Drege)* certainly three, not 4-valved. Ecklon & Zeyher gathered this plant without flowers or fruit. Some of *Drege's* specimens are in flower; others bear slender, cone-like bodies, formed of imbricated and enlarged leaves, probably the result of insect punctures.

According to Eck. & Zey. the Hottentots call this plant "*Daweep,*" and the Boers "Abiquasgeelhout." The species is found also in North Africa, Arabia, and Persia.

ORDER XVIII. CARYOPHYLLEÆ, Juss.

(By W. SONDER.)

(Caryophylleæ, Juss. Gen. 299. DC. Prod. I. p. 388. Endl. Gen. No. ccvii.; and part of Portulaceæ, Endl. ccvi. Caryophyllaceæ, Lindl. Veg. Kingd No. clxxxviii., and Illecebraceæ, No. clxxxix.)

Flowers regular, sometimes apetalous. *Calyx* free, persistent, 4-5 cleft or parted, with imbricated æstivation. *Petals* as many as the calyx lobes (rarely indefinite) or none, hypogynous or slightly perigynous. *Stamens* as many or twice as many as the calyx lobes, rarely fewer or more numerous, hypogynous or slightly perigynous; filaments slender, free, or connate at base. *Ovary* free, sessile or stipitate, unilocular, or imperfectly or completely 3–5 locular; ovules one or many, on cords rising from the base of the ovary or from axile placentæ; styles or filiform stigmas as many as the carpels. *Fruit* dry, either a one-seeded *utricle* or a many-seeded, dehiscent capsule; very rarely baccate. *Seeds* reniform or lenticular, with floury albumen. *Embryo* excentric, curved round the margin of the seed; the radicle next the hilum.

Herbs, rarely suffrutices, of small size, dichotomously or trichotomously, rarely irregularly branched; usually with swollen nodes. Leaves opposite, scattered or densely tufted and rosulate, quite entire, usually narrow, often awl-shaped or grass-like. Stipules either none, or scarious; sometimes copious, lacero-multifid. *Flowers* bisexual, in cymes, fascicles or umbels; very rarely axillary and sessile.

As here understood this is a large and much diversified Order, dispersed over most Parts of the globe, but greatly more abundant in the northern temperate zone. Upwards of 1200 species are known. Many occur in high latitudes, and are among the last flowering plants seen in the arctic regions, or on the borders of perpetual snow, within the temperate or tropical zones. The first two of the following sub-orders (see table), *Sileneæ* and *Alsineæ,* may be regarded as *typical;* the others either as *aberrant* members of this Order, or as types of separate, but closely allied Orders. In the *typical* groups the leaves are opposite, exstipulate, and connate at base; the stems are distinctly jointed; and the capsule is unilocular, with a free central placenta. The *Paronychiaceæ* differ by the constant presence of the membranous stipules, and the irregular insertion of the leaves; but through *Lepigonum* and *Spergula* are connected with *Alsineæ*. The *Mollugineæ* and *Polpodeæ* differ rather more widely; by their plurilocular ovary, the position of their stamens, and their want of true petals; but in vegetation, in their stipules, and in general aspect they harmonize so completely with *Paronychiaceæ,* and so little with *Portulaceæ* (to which Order they are referred by Endlicher and Fenzl), that we venture to retain them. For this, however, Dr. Harvey is alone responsible; and at his suggestion *Acrosanthes* and *Scleranthus* are also omitted, on account of their strikingly perigynous stamens, and general habit. All the Sub-Orders are very closely allied to *Mesembryaceæ* and *Portulaceæ,* on the one hand; and to *Chenopodiaceæ* and *Amarantaceæ* on the other; and, in a strictly natural arrangement, all these Orders would form a natural subclass or "Alliance." But, in the present work, having adopted the principles of the De Candollean system, we are compelled to place them in different sub-divisions.

TABLE OF THE SOUTH AFRICAN GENERA.

Sub-order 1. SILENEÆ. *Calyx* tubular, 5-toothed. *Petals* and *Stamens* hypogynous, on the summit of a more or less evident *gynophore* (or stipes of the ovary). *Ovary* unilocular, many-ovuled. *Capsule* many-seeded. *Leaves* opposite, *exstipulate.*

I. **Dianthus.**—*Calyx* bracteate at base. Styles 2.
II. **Silene.**—*Calyx* naked at base, 5-toothed. Styles 3.
III. **Agrostemma.**—*Calyx* naked at base, with leafy lobes. Styles 5.

Sub-order 2. ALSINEÆ. *Calyx* 5-parted. *Petals* and *Stamens* hypogynous or sub-perigynous, without obvious gynophore. *Ovary* unilocular, many-ovuled. *Capsule* many-seeded. *Leaves* opposite, *without stipules.*

IV. **Stellaria.**—Styles 3.
V. **Cerastium.**—Styles 4–5.

Sub-order 3. PARONYCHIACEÆ. *Calyx* 5-parted. *Petals* 5. *Stamens* sub-perigynous, opposite the calyx lobes, when of the same number. *Ovary* unilocular; one or many-ovuled. *Leaves* opposite, alternate, or fascicled, *always furnished with membranous stipules.*

* *Ovary* 1–2 *ovuled. Fruit indehiscent, one-seeded.*

VI. **Corrigiola.**—*Calyx* 5-parted. *Petals* roundish. *Stamens* 5. *Leaves* alternate.
VII. **Herniaria.**—*Calyx* 5-cleft. *Petals* filiform. *St.* 5 or fewer. *Leaves* opposite, hairy.
VIII. **Pollichia.**—*Calyx*-tube urceolate, becoming fleshy and juicy in fruit, 5-toothed. *Stamen* 1–2. *Leaves* whorled.

** *Ovary many-ovuled. Capsule* 3–5 *valved, many-seeded.*

IX. **Polycarpon.**—*Sepals* sharply keeled, and mucronate. *Petals* entire. *Style* trifid.
X. **Polycarpæa.**—*Sepals* membranous, not keeled. *Pet.* entire. *St.* 5. *Style* trifid.
XI. **Lepigonum.**—*Sepals* green, not keeled. *Pet.* entire. *St.* 10 or fewer. *Styles* 3.
XII. **Spergula.**—*Sepals* green, not keeled. *Pet.* entire. *St.* 5–10. *Styles* 5.
XIII. **Drymaria.**—*Sepals* green. *Petals* deeply bifid or quadrifid.

Sub-order 4. MOLLUGINEÆ. *Calyx* 5-parted or cleft. *Petals* none, or (*in Orygia* and *Glinus,*) numerous, linear. *Stamens* alternating with the sepals, if of the same number; sometimes indefinite. *Ovary* 3–5-celled; the cells with one or many ovules. *Capsule* loculicidal, the valves septiferous in the middle. *Leaves alternate or spuriously whorled, aggregate or rosulate; stipules generally present, membranous.*

* *Petals linear or spathulate, indefinitely numerous.*

XIV. **Orygia.**—*Sepals* cuspidate, unequal. *Seeds* 2-eared at base. *Flowers* in *stalked* cymes.
XV. **Glinus.**—*Sepals* ovate, subacute. *Seeds* simple at base. *Flowers* in *sessile* umbellules or fascicles.

** *Petals none. Capsule many-seeded.*

XVI. **Mollugo.**—*Calyx* 5-parted. *Stigmas* linear. *Stipules* obsolete.
XVII. **Pharnaceum.**—*Calyx* 5-parted. *Stigmas* obovate, fleshy. *Stipules* indefinitely numerous, scarious, *fimbriato-lacerate.*
XVIII. **Hypertelis.**—*Flower* as in *Pharnaceum. Stipules* adnate with the sheathing base of the petiole, tooth-like, *not lacerate.*
XX. **Cœlanthium.**—*Calyx* urceolate or campanulate, semi-quinquefid. *Stipules* fimbriato-lacerate.

*** *Petals none. Ovules one in each cell. Seeds solitary.*

XIX. **Psammotropha.**—*Flower* of *Pharnaceum. Stipules* none or lacerate.

Sub-order 5. POLPODEÆ. *Calyx* 4-parted; segments petaloid, *fimbriato-lacerate. Petals* none. *Stamens* 4, alternate with the calyx lobes, hypogynous. *Capsule* obcordate, compressed, 2-celled, 2-valved; seeds solitary. *Leaves imbricate.*

XXI. **Polpoda.**—(Character of the Sub-order.)

Sub-order 6. ADENOGRAMMEÆ. *Calyx* 5-parted. *Petals* none. *Stamens* 5, alternate with the calyx lobes. *Ovary* unilocular, with a solitary ovule. *Style* filiform; stigma capitate. *Utricle* indehiscent, one-seeded. *Leaves whorled; stipules obsolete; flowers in axillary umbellules.*

XXII. **Adenogramma.**—(Character of the Sub-order.)

Sub-order I. SILENEÆ. (Gen. I.–III.)

I. DIANTHUS, L.

Calyx tubular, 5-toothed, with two or more imbricated bracts at base. *Petals* 5, with long claws. *Stamens* 10. *Ovary* unilocular, with many ovules; *styles* 2, filiform, stigmatose on their inner face. *Capsule* cylindrical or oblong, opening by 4 teeth. *Endl. Gen.* 5244. *DC. Prod.* 1. *p.* 355.

Herbs or small suffrutices, common in Europe and Central Asia, rare in North America and South Africa. Leaves opposite, connate, generally grass-like; flowers terminal, solitary or in paniculate cymes, or densely fascicled. Pinks, and Carnations and "Sweet William" are well known garden examples of this genus. All the Cape species are perennial. The name Dianthus is derived from διος, *divine* and ανθος *a flower*; in allusion to the fragrance of the Carnation.

* *Stem simple, one-flowered* (Sp. 1–2).

1. D. cæspitosus (Thunb. Prod. p. 81); densely tufted; stems erect, simple, one-flowered, *glabrous;* radical leaves crowded, linear-setaceous, sub-triangular, keeled, *scabrous;* upper cauline scale-like; bracts 4–6, *lanceolate,* 2–3 *times shorter than the* (1½–2 *inch long*) *tube;* petals toothed or slightly cut, the lamina obovate. *Thunb. Fl. Cap. p.* 393. *DC. Prod.* 1. *p.* 363. *E. & Z. No.* 243, *ex parte. D. albens, Jacq. Herb.—DC. l. c. non Ait. D. angulatus, E. Mey!*

HAB. In the Karroo, near Gauritz River, George, and in sandy places, Cannaland, *E & Z.!* Lange Kloof, *Drege!* Dec. (Herb. Thunb. Vind. Sond.)

Root thick, throwing up many tufts. Rad. leaves 1½–3 inches long, rigid, ½ line wide. Stems very rarely with one branchlet, 6–12 inches long, straightly erect, with linear 2–6-line long, leaves. Calyx much longer than in any other species; its bracts sometimes 8, pale-edged; its teeth 3 lines long, acuminate. Petals flesh-coloured, the lamina ½ inch long, 2–3 lines wide.

2. D. scaber (Thunb. Prod. p. 81); stems erect or decumbent at base, simple, one-flowered, *scabrid;* radical leaves aggregate, linear-subulate, sub-triangular, keeled, scabrid; upper cauline scale-like; bracts 4, *ovato-lanceolate, twice or thrice as short as the* (*inch-long*) *tube;* petals toothed or lacerate, the lamina broadly obovate, short. *Thunb. Fl. Cap. p.* 393. *D. micropetalus, Ser., DC. Prod.* 1. *p.* 359. *D. prostratus, E. & Z.!* 248, *ex parte. D. scaber, var. graminifolia, E. Mey. and D. micropetalus, E. Mey. D. ramentaceus, Fenzl.* (a larger and slightly branched variety.)

Var. β. glabratus; leaves sub-glabrous. *D. crenatus, E. Mey.!*

HAB. Grassy fields in Uitenhage, *E. & Z.!* Highest ranges of Wittberg, 5–6000 ft.

and at Klipplaat River, *Drege!* Gt. Fish River, *Burke*, Sandfontein and Zeekoe Valley, *Zeyher*, 80. Springbokkeel and Kamos, Boschjesman's land, *Zey. No.* 26. Orange River and in Ceded Territory, *E. & Z.* β. at Graham's Town, *Zey.* 82. and Dutoit's Kloof, *Drege.* (Herb. Thunb., Vind., Hook., Sond., T.C.D.)

Herbaceous, many-stemmed, 6–12 inches high. Stems filiform, simple, rarely with a single branchlet. Rad. leaves rigid, uncial; upper cauline, 2–3 lines long. Bracts acute, glabrous, pale-edged. Lamina of petals about 2 lines long, broadly obovate, sometimes shortly toothed, sometimes fimbriate. Best known from the preceding by its much shorter calyx tube.

** *Stems paniculately branched. Petals entire or toothed.* (Sp. 3–6.)

3. D. incurvus (Thunb. Fl. Cap. p. 392); stems *curved, sub-erect,* branched above, glabrous; *lower leaves long,* linear setaceous, sub-trigonous, *glabrous,* upper scale-like; bracts 4, *ovate-acute,* mucronulate, 2–3 times shorter than the ($\frac{1}{2}$–$\frac{3}{4}$ inch long) tube; calyx teeth ovate, acute, ciliolate; petals entire or sub-denticulate, the lamina obovate. *E. & Z.!* 242. *D. albens, Ait. Hort. Kew. Ed.* 1. *vol.* 2. *p.* 90. *Willd. Sp.* 2. 677. *Lindl. Bot. Reg. III.* 256. *in not. D. Burchellii, Ser., DC. Prod.* 1. *p.* 359. *D. albens, E. Mey.!*

Hab. On the hills below the east side of Table Mt., in the Cape Flats, and near Simon's Bay. Also at the Zondag and Zwartkops Rivers. (Herb. Thunb. Hook., T.C.D., Vind., Lehm., Sond.)

Root long, branching, many-stemmed. Stems 6–12 inches high. Lower leaves crowded, 1–2 inches long, $\frac{1}{2}$ line wide. Flowers white or pale rosy, flesh-coloured, or purple: according to Thunberg cernuous. Bracts short and broad, membrane-edged. Lamina of the petals 4 lines long, 2 lines wide.

4. D. holopetalus (Turcz. Animadv. 1. p. 99. excl. syn. D. albens E.& Z.); stems *erect,* branched, glabrous or minutely downy; leaves serrulato-scabrid, sub-5-nerved, flat, the lower ones *linear-elongate,* the cauline gradually shorter; bracts 4–6, *lanceolate-acuminate,* membrane-edged, twice shorter than the (half-inch long) tube; petals oblong, *entire. D. crenatus, E. & Z.! col. postr.*

Hab. Skurfdeberg and Riebeckskasteel, *Zey. No.* 78! Puspas Valley, and at Tulbagh, *E. & Z.!* Nov. Dec. (Herb. Vind., T.C.D., Sond.)

1$\frac{1}{2}$–2 feet high, straightish. Lower leaves 3–4 inches long, 1 line wide. Bracts aristato-acuminate, with a wide margin subciliate at base. Calyx teeth 1 line long. Lam. of petals 3 lines long, a line wide, obtuse, not acute. Known from *D. crenatus,* Thunb. by the twice shorter calyx; from *D. incurvus* by the straight stem, flat leaves and thinner and more acuminate bracts.

5. D. crenatus (Thunb. Prod. p. 81); stems erect, much branched, terete, glabrous; branches compressedly angular, one-flowered; lower leaves long, linear-lanceolate, 3 nerved, *glabrous or roughish near the top,* upper gradually shorter; bracts about 6, ovato-lanceolate, about twice as short as the (inch-long) tube; calyx teeth lanceolate, ciliolate; petals *crenato-dentate,* the lamina broadly ovate. *Thunb. Cap. p.* 392. *DC. l. c. D. cæspitosus, prostratus, et scaber, E. & Z.! ex parte. D. prostratus, Hochst.! D. inæqualis, E. Mey! Zey.* 1958.

Hab. Hills between Swellendam and Howhoek, *Thunb.* Zwartkop's river and Grahamstown, *E. & Z.!* Coega R., *Drege.* Port Natal, *Krauss.* Nov. Dec. (Herb. Thunb., Vind., T.C.D., Sond.)

1–1$\frac{1}{2}$ feet high. Lower leaves 2–3 inches long, 1$\frac{1}{2}$–2 lines wide; upper rameal 6–2 lines long. Calyx somewhat striate; bracts ovate, acuminate. Lamina of the petals 5 lines long, 3 lines wide, very shortly toothed. It varies with a subsimple stem.

6. **D. Kamisbergensis** (Sond.); stems slender, erect, branched at top, terete, glabrous; radical leaves *linear-setaceous, with revolute margins*, scabrid, cauline gradually shorter; bracts 4, *broadly ovate*, mucronate, membrane-edged, thrice as short as the (half-inch long) tube; petals small, obovate, entire or denticulate. *D. micropetalus, E. & Z.!* 244., *non. Ser.*

HAB. Mountain sides. Kamisberge, Cannaland, in sandy places, *E. & Z.!* (Herb. T.C.D., Sond.)

1–1½ feet high, very slender, having at the summit a few longish, one-flowered branches. Rad. leaves 2–3 inches long, ½ line wide; upper cauline 3–2 lines long. Calyx 4–6 lines long, the teeth 1 line long. Claws of the petals not longer than the calyx, lamina 1½–2 lines long.

*** *Stems branched. Petals deeply digitate or pinnatifid.* (Sp. 7–9).

7. **D. Zeyheri** (Sond.); glabrous, *glaucous;* stem *straight*, terete, branched, branches angular, leaves rigid, *7–9-nerved, scabrid-edged*, the lower cauline oblongo-lanceolate, intermediate lanceolate-acuminate, and uppermost shorter and narrower; bracts 4–6, ovate, awned, 4 times shorter than the (1½ inch long) tube; petals broadly obovate, fimbriated. *D. albens, Turcz., non Ait.*

HAB. Macallisberg, Nov. Dec. *Burke and Zey.*, 79. (Herb. Hook., T.C.D., Sond.)

1½–2 feet high. Lower leaves 1½–2 inches long, 2–3 lines wide, very acute; intermediate equalling the internodes; upper 1 line wide. Branches one-flowered. Calyx pale, with acuminate teeth. Claws of the petals longer than the calyx, the lamina white, 6 lines long and 5–6 lines wide, the fringe 2 lines long. Known from the following by the straighter habit, glaucous colour, more leathery leaves, larger petals and longer fringe.

8. **D. prostratus** (Jacq. Schoenb. 3. t. 271.); *green*, glabrous; *sterile stems prostrate*, the flowering erect, paniculate; lower leaves lanceolate, or sublinear, *three-nerved, with a smooth margin*, scabrid at the point, upper scalelike; bracts 4–6, ovato-lanceolate, 3–4 times shorter than the (1½ inch long) tube; petals *obovate*, fimbriated. *DC. l. c. p.* 364. *D. crenatus, Bot. Reg. t.* 256. *D. prostratus, E. M.! in Hb. Drege, ex parte. D. inæqualis, E. M. b.*

HAB. Buffelrivier and at Enon, *Drege!* Port Natal, *Gueinzius, Plant.* (Herb. Vind., T.C.D., Sond.)

Three feet high. Stem terete, the branches angularly compressed. Lower cauline leaves generally broadest, oblong-lanceolate, interm. linear-acuminate, 1–1½ inches long, uppermost 3–2 lines long. Calyx very delicately striate, the teeth 2 lines long. Petals flesh-coloured or white, the claws exceeding the calyx, the limb sometimes shortly fimbriate, sometimes deeply multifid. It resembles *D. crenatus*, from which it only differs by the longer calyx and fimbriate petals.

9. **D. pectinatus** (E. Mey. in Herb. Drege); stems erect, branched, glabrous; rad. leaves linear, elongate, *scabrid-edged*, uppermost cauline, scalelike; bracts 4–6, lanceolate, cuspidate-acuminate, thrice as short as the (inch long) tubes; petals oblong, *pectinato-multifid. D. crenatus, E. & Z.!* 246. (a var. with broader leaves) *and D. albens E. & Z. No.* 247. (with narrower leaves).

HAB. Mountain places. Puspas Valley, and near Tulbagh, in Kaireberg,

Clanw., *E. & Z.!* Namaqualand, *v. Schlicht.* Skurfdeberg and 24-Rivers, *Zey.!* 77. Macallisberg, *Zey.* 81. Onderbokkeveld, *Drege.* (Herb. T.C.D., Sond.)

Stems 1-2 or more feet high. Rad. leaves 2-4 inches long, 1-2 lines wide, rigid, 3-nerved. Bracts very acute, membrane-edged. Claws of the petals longer than the calyx, the lamina cleft in capillary ribbons nearly to the middle.—Nearest to *D. holopetalus*, Turcz.; but differs in the longer calyx and multifid petals. From *D. scaber* it is known by a more branching habit, longer radical leaves, and narrower, multifid petals.

II. SILENE, Linn.

Calyx tubular, 5-toothed, naked at base. *Petals* 5, with long claws, the limb entire, bifid or multifid. *Stamens* 10. *Ovary* more or less 3-5-locular at base, unilocular above; styles 3 (very rarely 2-5) filiform, stigmatose on their inner face. *Capsule* membranous, opening by 6 teeth. *Endl. Gen.* 5248. *DC. Prod.* 1. *p.* 367.

A vast genus, dispersed over the globe. The species are either annual, perennial, or suffrutescent; leaves opposite or whorled, broad or narrow, very frequently pubescent or viscidly hairy and glandular. Flowers cymose, or panicled; the minor divisions of the inflorescence often unilateral. The name is said to be derived from σιαλον, *saliva;* in allusion to the viscidity of many. The English name "*Catchfly*" also alludes to this clamminess.

SECT. 1. **Elisanthe,** *Fenzl.* *Flowers* solitary, terminal, or in dichotomous cymes. *Capsule* unilocular at base. (Sp. 1—7.)

1. S. bellidioides (Sond.); pubescent, scabrid; stem erect, somewhat viscid, branched at top; radical leaves *spathulate, obtuse, apiculate,* narrowed into a long petiole, cauline obovate or oblong, gradually smaller; peduncles few, erect, the lower long; fruiting calyces *oblong,* the thecaphore ¼ as long as capsule. *S. bellidifolia, E. & Z.!* 261, *non Jacq. nec Thunb. S. bellidifol. var stricta, Fenzl.*

HAB. In fields, near the Zwartkops River, *E. & Z.! Zey. No.* 1964 *ex parte,* near Port Elizabeth, *Drege.* Knysna, *Krauss.* (Herb. Vind., Sond.)

1-2 feet high. Rad. leaves thickish, 3-4 inches long, including the petiole, 4-6 lines wide, hook-pointed; the lower cauline 1 inch, upper 6-4 lines long, 2-1 line wide. Peduncles 2-5. Calyx nearly an inch long, 10-ribbed, pubescent, with lanceolate-subulate teeth, 2-lines long. Petals white, bifid. Teeth of the capsule short: thecaphore 2 lines long. Allied to the following.

2. S. capensis (Ott. DC. Prod. 1. p. 379.); viscoso-pubescent; stem erect, with spreading branches; radical leaves *oblong-acute;* narrowed into the petiole; cauline sessile, lanceolate or linear-lanceolate, gradually smaller; flowers erect, trichotomously panicled; calyces elongate, the fruiting ones *clavate; thecaphore at least half as long as the capsule. E. & Z.!* 259. *Zey. No.* 1964 *ex parte.*

HAB. Berg River, Cape; Brackfontein, Clanw.; Zwartkops Riv. Uit.; Grahamstown, and in Caffraria, *E. & Z.!* Knysna, *Pappe.* Gift Berg and Nieuweveldsberg 3-4000f. *Drege!* Slaaykraal, *Burke.* Port Natal, *Miss Owen! Sanderson!* July-Oct. (Herb. Vind., Hook., T.C.D., Sond. &c.)

Stem 2-3 feet high, leafy, cymoso-paniculate. Rad. leaves 4-6 inches long, 6-12 lines wide; the lower cauline 2 inches long, upper an inch or less. Calyx 14 lines long, in flower 2-3 lines wide, 10 ribbed, veiny, with linear-subulate teeth. *Petals* large, white, bifid, with obtuse lobes; the claws longer than the calyx. Capsule ovate-oblong, 7 lines long; the thecaphore variable in length. It varies with stem and leaves clothed with long spreading hairs, or rough with a short pubescence.

3. S. undulata (Ait. Kew. 2. p. 96); viscidly pubescent; stem erect or decumbent at base, branching; leaves lanceolate, *undulate*, the radical spathulate, acute, petiolate; cymes few flowered or paniculate; flowers erect; fruiting calyces *clavato-cylindrical, the thecaphore ⅓ of capsule. S. noctiflora, Thunb. Cap. p.* 394. *E. & Z. No.* 260. *Hb. Un. It. No* 757. *S. Thunbergii, E. Mey! non E. & Z. S. diurniflora, Kunz. in Linn.* 17. *p.* 578.

HAB. Hills round Table Mountain, on both sides and at the Paarl, and elsewhere in Cape and Stellenbosch, *E. & Z., W.H.H.* &c. Mts. near Liefde *(S. bellid. var. foliosa, Fenzl.)* and in Dutoitskloof *(S. ornata var. florida, Fenzl.)* and *S. undutata?* 2962, *Drege.* Jul.–Oct. (Herb. Holm., Vind., T.C.D., Sond.)

Stem either sub-simple, and six inches high; or trichotomously branched and 1–2 feet high, with alternate spreading branches. Cauline leaves sessile, the lower 2 inches long, oblong-lanceolate or oblong, acute, upper uncial, acuminate, with curled margins. Calyx 8–10 lines long, 10 ribbed, venulose, teeth subulate, 1½–2 lines long. Petals white or reddish, their claws longer than the calyx, lamina bifid, the lobes denticulate: corona very short. Capsule ovate; thecaphore variable. Known from *S. noctiflora*, L., which it greatly resembles, by the undulate leaves, the longer calyx, and chiefly by the thecaphore, which is very short or scarcely any in *S. noctiflora*. It differs from the following by its white flowers.

4. S. ornata (Ait. Kew. 2. p. 96); *viscoso-pubescent;* stem erect, branching; *leaves lanceolate, flat,* the radical petiolate; flowers erect, dichotomously cymose or paniculate; fruiting calyces *elliptic-oblong;* thecaphore ¼ as long as the capsule. *Bot. Mag. t.* 382. *DC. l. c. p.* 379. *E. & Z. !* 262. *S. bellidifolia, Thunb. Cap. p.* 394. *S. Meyeri, Fenzl.*

HAB. Sandy places on Cape Flats, near Doornhoogde and Riet valley, and near the hot springs, Caledon, Sept. Oct., *Thunberg, E. & Z.! Drege,* &c. (Herb. Holm., T.C.D., Sond.)

In size, habit and foliage this is very like the preceding, but the leaves are not undulate and the flowers are never white. Calyx 8 lines long, 10 ribbed and veiny, with lanceolate teeth. Petals deep blood-red, semi-bifid, with broad, denticulate lobes; the claws not longer than the calyx. Corona small. *S. ornata var. stricta,* Fenzl., in Herb. Drege, judging from a very imperfect specimen, certainly does not belong to this species.

5. S. Eckloniana (Sond.); stems *ascending,* slightly branched above, leafy, *villous;* leaves *coriaceous,* spathulate, obtuse, sub-retuse, *with the margin revolute, ciliated,* glabrous above, hairy on the midrib below, the radical larger; flowers terminal and axillary, cymose and sub-paniculate, with erect peduncles; fruiting calyces *oblong,* thecaphore ¼ of capsule.

HAB. Sea shore near Cape Recief, Feb. 1830, *Dr. Ecklon!* (Herb. Sond.)

Stems many from the same root, prostrate, terete, a span long, not glandular but clothed with long, spreading, soft hairs. Leaves fleshy-coriaceous, spathulate or obovato-spathulate, the midrib and veins impressed on the upper, raised on the lower side; the radical crowded, 2–3 inches long, 6–9 lines wide at the apex, 2–3 lines wide at base; lower cauline uncial, upper semi-uncial. Cal. 8–10 lines long, villoso-pubescent, 10 ribbed, with lanceolate teeth. Petals not seen. Capsule 8 lines long, resembling that of *S. bellidioides.* Seeds black, punctato-rugulose.

6. S. caffra (Fenzl in Pl. Drege); stem *erect, quite glabrous,* somewhat branched; lower *cauline leaves lanceolate-oblong, acute,* glabrous on both sides, *ciliolate,* the lowermost cilia longer; upper leaves gradually shorter; the terminal cymoso-paniculate pendulcles and the calyces pubescent; fruiting calyces oblongo-clavate, thecaphore $\frac{1}{3}$ of capsule.

HAB. Caffraria, *Drege.* No. 5342 (Herb. Vind.)

Lower part wanting. Stem (in the solitary specimen seen) 2 feet long, straight, sub-simple, terete below, angular above. Lower leaves narrowed at base, 2–3 inches long, 4 lines wide, equalling the internodes, upper gradually smaller, narrowed from a broader base, thrice shorter than the internodes. Branches terminated by 3-flowered cymes: the lateral peduncles uncial. Calyx inch long, 10 striate, with subulate teeth. Petals white, the claws longer than the calyx, lamina bifid, coronate at base. Capsule equalling the calyx tube.

7. **S. Mundtiana** (E. & Z.! 252); downy; stem *procumbent at base, weak, dichotomously much branched,* leafy; leaves narrow-spathulate, acute, ciliate at base, the uppermost smaller, lanceolate; flowers about 3 on each ramulus, the lower flower sub-remote, the two terminal paired, unequally pedunculate; fruiting calyx clavate; thecaphore equalling the (immature) capsule.

HAB. Among shrubs. Paardekop, near Plettenberg Bay, *Mundt.!* (Herb. Sond.)

Root perennial, many stemmed. Stems slender, terete, a span long, knee bent; branches erect, filiform, covered with a very short, reversed down. Leaves patulous, puberulous, evidently ciliate at base, 6–8 lines long, 1–1½ line wide, the uppermost sub-remote, half the size; lower mostly dry. Flowers pedunculate; peduncles of the terminal flowers unequal, one 4–5 lines, the other scarcely 2 lines long. Calyx 5 lines long, 10 nerved, with ovate, obtuse teeth. Petals white (or flesh-coloured?) cleft to the middle, coronate. Ripe fruit wanting. Doubtfully referred to this section.

SECT. 2. STACHYOMORPHA, Otth.—*Flowers* racemoso-spicate, rarely sub-solitary. *Capsule* trilocular at base. (Sp. 8–13.)

* *petals undivided, or sub-emarginate* (Sp. 8.)

8. **S. gallica** (Linn. Sp. 595); viscoso-pubescent, branching; leaves oblong, the lower obovate; flowers alternate, racemose; calyx hispid tubular, in fruit ovate; thecaphore much shorter than the calyx. *Thunb. Cap. p.* 393. *E. & Z.!* 258. *Hb. Un. It. No.* 759. *S. micropetala, DC. l. c. p.* 372. *S. lusitanica, L.*

β. **quinquevulnera** (Koch, Syn. Fl. Germ. ed. 2. p. 109); petals bright red on the disc, white at the edges. *S. quinquevulnera,* Linn. *E. Bot. t.* 86.

γ. **anglica** (Koch, l. c.); stem more branching and spreading. *S. anglica, Linn. E. Bot. t.* 1178.

HAB. A troublesome weed in cultivated ground, introduced from Europe. (Herb. Vind., Holm., T.C.D., Sond.)

Annual, erect or spreading, 6–12 inches high. Leaves acute, ½–1 inch long. Peduncles very short. Flowers white or red; in *β.* dark red. Calyx 4 lines long, ribbed. Petals entire, toothed or emarginate. Capsule subsessile, more evidently pedicellate in the lower flowers. This is the "*Gunpowder-weed*" of the Colonists; its black seeds resembling gunpowder.

** *petals deeply bifid* (Sp. 9—13.)

9. **S. clandestina** (Jacq. Coll. Suppl. 3. t. 3. f. 3); pubescent; stem branched; leaves linear or linear-lanceolate, the lower ones linear-spathulate; flowers laxly racemose or sub-solitary; fruiting calyces *oblongo-*

clavate; thecaphore 2–3 times shorter than the capsule. DC. l. c. p. 376. *E. & Z.!* 250. *S. linifolia, Willd. En. p.* 473. *DC. l. c. S. cernua, Thunb.! Cap. p.* 394. *Hb. Un. It. No* 760. *S. recta, Bartl. Lin.* 7. *p.* 623. *E. & Z.!* 249. *S. Constantia, E. & Z.!* 251.

VAR. α. **minor**; stem slender, generally much branched from the base; the cauline leaves linear; flowers sub-solitary or few in a raceme.

VAR. β. **major**; stem taller, branching; cauline leaves linear-lanceolate; flowers more numerous, racemose, cernuous before opening.

HAB. Sandy places and on hills round Capetown; at Berg River; on Assagaiskloof and at Malabarshoogde, and Ebenezer, &c.; Zwartkops River, Uit. *E. & Z.! Drege! Pappe, &c.; Zey. No.* 83, 1960, *and* 1961. (Herb. Holm., Jacq., Lehm., T.C.D., Sond.).

Annual, filiform, 3–12 inches high, covered with a short pubescence, sub-simple or branched from the base; branches erecto-patent. Leaves in the dwarfer specimens 6 lines, in the larger 1–2 inches long, ½–1 line wide, scabrid, ciliate at the margin, chiefly at base. Branches in the lesser specimens 1–2 flowered, in the larger, 3–6 flowered, racemose, secund. Calyx 5 lines long, downy, 10-nerved, nerves raised, teeth obtuse. Petals, according to Thunberg, white; judging by the dried specimens, rosy; lamina, short, bifid, coronate. Capsule 4 lines long, with short, recurved teeth.

10. S. Burchellii (Otth., DC. Prod. 1. p. 374); pubescent, roughish; stem simple or branched, leafy, denudate above; *the leaves lanceolate or linear-lanceolate,* the lower broader; flowers racemose, the fruiting calyces *very long, clavate; thecaphore as long as the capsule or longer.*

VAR. α. **angustifolia**; leaves lin.-lanceolate or linear. *S. cernua, Bartl. E. & Z.!* 256, *non Thunb. S. cernua, Drege, ex parte.*

VAR. β. **pilosellæfolia**; stems decumbent at base; lower leaves crowded, spathulato-lanceolate, acute or mucronulate, upper linear-lanceolate. *S. pilosellæfolia, Cham. & Schl. Linn.* 1. *p.* 41.

VAR. γ. **latifolia**; stem erect, leafy to the middle; lower leaves oblong, acute or oblongo-lanceolate, rough; upper lanceolate.

HAB. Sandy places on the Cape Flats. Zwarteberg, Caledon, *Zey.! No.* 1957. Voormansbosch, Swell., *Mundt.* Sondag's River and on the Gareep, *Drege!* Stony hills at Thaba Unca, *Zey. No.* 84. β. Grahamstown; and at the Zwartkops River, and near Adow, Uit., *E. & Z.!* Plettenberg's Bay, *Mundt.* γ. Shady places at the Crocodile River, *Zey.! No.*84. Aug.–Dec. (Herb. Hook., Vind., T.C.D., Sond.).

Root perennial, many-stemmed. Stems a foot or more in height, erect or decumbent. Leaves in α, uncial, a line wide; in β. twice or thrice as wide, 1–1½ inch long. Flowering raceme cernuous, fruiting erect, 6–12 flowered. Peduncles ½–2 lines long, the bracts lanceolate. Calyx 9–10 lines long, with coloured striæ, hairy, the teeth obtuse. Petals flesh-coloured, rather longer than the calyx, bifid, coronate. Capsule ovoid, 5 lines long. Known from the preceding, to which it is allied, by its perennial root, stems not filiform, wider leaves, and longer calyx and thecaphore. *S.acuta,* E. Mey.! in Herb. Drege, seems to belong to var. γ. It differs merely in the fewer flowered racemes and smaller flowers. The only specimen we have seen of it was gathered between Omtendo and Omsamculo, Caffr.

11. S. Thunbergiana (E. & Z.! 253, excl. syn.); *pubescent;* stem erect, much branched, leafy, naked at top, scabrid; leaves *obovate, obtuse, mucronulate,* the upper smaller, often narrower; flowers racemose; *bracts lanceolate;* calyces minutely pubescent, in flower cylindrical, in fruit longish-clavate; thecaphore as long as the *ovate* capsule. *S. crassifolia, Bartl., Linn.* 7. *p.* 623, *non L. nec Thunb. Herb. Un. It. No.* 758.

HAB. On the ascent of Table Mt., and on the Lion's Mt., Capetown, *Brehm, E. & Z.! Drege. Pappe, W.H.H., &c.* (Herb. Hook., T.C.D., Sond.).

Perennial, 1–2 feet high, with spreading branches, below covered with spreading hairs, above shortly pubescent. Leaves ½–1 inch long, 4–6 lines wide, pubescent on both sides, the upper ones remote. Raceme secund, 4–6 flowered; pedicels 1–3 lines long, the lowest longest. Calyx 8 lines long, shortly pubescent on the nerves, the teeth obtuse, 1 line long. Petals carneous or purple, the lamina bifid, coronate. Caps. 4 lines long. Differs from the two following in the taller and more slender stem; green leaves, not fleshy; more slender, and not funnel-shaped calyx, and by the fruit; from *S. obtusifolia*, Willd., to which it comes nearest, in the stem naked and scabrous near the summit, the longer and narrower calyx and longer thecaphore.

12. S. crassifolia (Linn. Sp. 597); *hairy;* stem procumbent, branched, leafy; *leaves suborbicular or obovate, fleshy*, hairy; raceme secund; *bracts ovate;* calyx *shaggy*, the flowering funnel-shaped; thecaphore *half as long as the bell-shaped capsule. Thunb. Fl. Cap. p.* 393. *E. & Z.!* 255.

HAB. Sands near the sea-shore. Rietvalley, *E. & Z.! W.H.H., Gueinzius.* Sea-shore at Blueberg, *Pappe!* Oct. Nov. (Herb. Vind., T.C.D., Sond.).

Yellowish when dry. Root many stemmed. Stems 1–2 feet long, from a decumbent base erect, thickened at the nodes. Leaves obtuse, ½–1 inch long, ½ inch wide or wider, densely hairy on both sides. Raceme 6–12 flowered; fl. short-stalked. Calyx 7 lines long, 10-nerved, hairy, with long, jointed hairs, the teeth short, obtuse. Petals purple, the lamina short, bifid. Ripe capsule suburceolate.

13. S. primulæflora (E. & Z.! 254); *pubescence short, reversed;* stem ascending, slightly branched, leafy; leaves *roundish-spathulate*, mucronulate, thick, scabrid; raceme secund; bracts *ovato-lanceolate;* calyx rigidly *pubescent*, the flowering funnel-shaped; thecaphore as long as the campanulate capsule. *S. colorata var. hirta, Fenzl, in Hb. Drege. P. Thunbergiana, Krauss.*

VAR. *β.* **ciliata**; leaves glabrate, ciliate at base. *S. col. var. ciliata, Fenzl.*

HAB. Sea-shore. Mouth of Zwartkops River and Algoa Bay, *Sparmann, E. & Z.! Zey. No.* 1963. *Drege,* 3559 *a.* Sea-shore between Omtendo and Omsamculo, *Drege,* 3559 *b.* Zitzekamma, *Krauss.* (Herb. Vind., T.C.D., Lehm., Sond.).

Differs from the preceding by the short, reversed pubescence of the stem; the mucronate leaves, and the calyces shortly pubescent (not shaggy) and the longer carpophore. Stem jointed, fistular. Leaves 1–1½ inch long, 6–9 lines wide; the largest 2 inches long, the upper smaller. Raceme 4–6 fl., at length pendulous. Cal. obconical, 8 lines long, nearly 10-nerved, with short teeth. Petals purple or rosy. Capsule 5 lines long and nearly 4 lines wide. *S. obtusifolia, Willd.*, (*S. colorata, Schousb.*) is very like this; but differs in the whole pubescence being soft, the leaves thinner, the calyces shorter, but chiefly in the thecaphore being scarcely half the length of the capsule; the capsule oblong-ovate, and only 2 lines wide.

III. AGROSTEMMA, L.

Calyx tubular, coriaceous, 5-toothed; the teeth long and leafy. *Petals* 5, clawed, not coronate. *Stamens* 10. *Styles* 5. *Capsule* unilocular, opening by 5 teeth.

An annual corn-field plant, with opposite leaves and large purple flowers. The name *αγρου στεμμα, crown of the field,* is appropriately given to this ornamental weed.

1. A. Githago (Lin.); silky, erect; leaves lanceolate-linear; segments of the calyx twice as long as the corolla. *E. Bot. t.* 741. *Schkuhr, t.*

121. *Githago segetum, Desf. Lychnis Githago, Lam. Encycl.* III. *p.* 643.

Hab. In corn fields; introduced from Europe.
One to two feet high. Flowers purple, handsome.

Sub-Order II. Alsineæ. (Gen. iv.-v.)

IV. STELLARIA, L.

Calyx 4–5 parted. *Petals* 4–5, bifid or bipartite. *Stamens* 8–10, or rarely fewer. *Styles* 3, filiform. *Capsule* unilocular, many-seeded, opening by 6 teeth. *Endl. Gen.* 5240. *DC. Prod.* 1. *p.* 396.

Cosmopolitan herbs and weeds of cultivation. Leaves opposite, broad or narrow, generally glabrous. Cymes dichotomous. Flowers small and white, on long peduncles. Name from *stella, a star,* from the star-like flowers.

1. **S. media** (Villars, Delph. III. p. 615); stems diffuse, ascending or erect, with a line of hairs along one side; leaves with ciliate petioles, ovate, acute, quite glabrous, the upper ones sessile; flowers solitary, axillary and terminal; pedicels long, pubescent; petals as long as the calyx or shorter. *E. Bot. t.* 537. *Alsine media, Lin. Schk. t.* 85. *E. & Z.!* 264.

Hab. A weed in cultivated ground, everywhere; introduced from Europe.—"*Chickweed.*"
Annual. Stems terete, 3–12 inches long. Petals sometimes minute or altogether deficient. Stamens 3–4, often 5, rarely 10.

V. CERASTIUM, L.

Calyx 4–5-parted. *Petals* 4–5, bifid. *Stamens* 4–10. *Stigmas* 4–5. *Capsule* unilocular, many-seeded, opening by 8–10 teeth.

Annual or perennial, cosmopolitan weeds and small herbaceous plants, generally hairy or silky; sometimes clammy. Stems dichotomous. Flowers cymoid, small and white.

* *Macropetala; petals considerably longer than the calyx. Root perennial.* (Sp. 1–2.)

1. **C. Dregeanum** (Fenzl, Ann. Wien. Mus. 1. p. 341); *glandularly pubescent, with a short, dense down;* stems flaccidly procumbent, branching; the flowering ones erect, simple; *lower leaves oval or elliptical, obtuse,* narrowed into a petiole, upper oblong, or oblongo-lanceolate, acute, without axillary tufts; cyme repeatedly forked, the branches patent; bracts leafy, ovato-lanceolate; fruiting pedicels twice as long as the calyx, cernuous; sepals ovato-lanceolate; petals broadly obcordate, ⅓ *longer than the calyx. Cer. brachycarpum, E. Mey. in Hb. Drege.*

Hab. Katberg and Lostafelberg, 5000–5600 f., *Drege.* (Herb. Vind., Sond.).
Biennial or perennial, pale green. Flowering stems 8–12 inches long. Lower leaves ½ inch long, cauline as long or longer. Sepals 3½ lines long. Stamens 10, twice as short as petals. Ripe capsule wanting.

2. **C. arabidis** (E. Mey. in Hb. Drege); *hairy, with long, simple, white hairs;* stems abbreviate, procumbent, scaly with the rudiments of leaves, the flowering ones erect, simple, glandularly viscid above; lower leaves crowded, *oblong, and oblongo-lanceolate,* obtuse, narrowed at base, the

cauline narrower, without axillary fascicles; cyme repeatedly forked; bracts leafy, oblong-linear; fruiting pedicels twice as long as the calyx; sepals ovato-lanceolate; petals obcordate, *one-half longer than the calyx. Fenzl, l. c. p.* 340.

VAR. β. **glutinosum**; stems weaker, shortly glanduloso-pilose from the base. *C. arvense, E. & Z.!* 268, *non Linn.*

HAB. On the Wittberg, 7–8000 f., *Drege!* β. High Mts. at Klipplaats River, Tambukiland, and at Kat River, *E. & Z.!* Oct.–Jan. (Herb. Vind., Sond.).

Perennial, with the *primâ facie* habit of *Arabis hirsuta*, straight, leafy, a span high. Lower leaves 1 inch long, 3 lines wide, upper sub-remote, ½ inch long. Flowers nearly as in *C. Dregei*, but the petals longer. Ripe capsule not seen. β has shorter lower leaves, and more slender flowering stems, erect or ascending, 6–9 inches long, covered with a short pubescence; perhaps a distinct species.

** *Micropetala; petals not exceeding the calyx. Root annual.* (Sp. 3–4.)

3. C. capense (Sond.); annual, simple or branched, pubescent at base, towards the apex, glandularly viscid, with short pubescence; lower leaves *obovate* or *oblong*, obtuse, upper smaller, oblong, acute; cyme *lax;* bracts herbaceous; fruiting pedicels cernuous, 1½–2.ce as long as the calyx; sepals lanceolate, acuminate, *scarious at the margin and the nude apex*; petals ⅓ shorter than the calyx; st. 10; the scarcely curved capsule twice as long as the sepals; seeds brown, densely tuberculated. *S. semidecandrum, pentandrum, and vulgatum, E. & Z.!* 265, 266, 267. *Zey.* 1965, *b.*

HAB. Sandy places below Table Mt. and at the summit. Also near Caledon Baths, at the Zwartkops River, and in Adow, *E. & Z.!* (Herb. Sond.).

Annual, 3–6 inches high, with the habit of *C. vulgatum. L.*, (*C. triviale, Link.*) and *C. glutinosum*, Fries. From the first it differs by its annual root, and all the bracts herbaceous; from *C. glutinosum* (*pumilum*, Curt., Lond., 2. t. 92), which it much resembles, by the pubescence on the lower stem, with patent, white, not glandular hairs, the lower leaves larger, all the bracts herbaceous, not the upper ones scarious at margin, the petals smaller and the seeds twice as large, brown, more densely covered with more raised tubercles. From *C. semidecandrum* it differs by its more robust habit, bracts never semi-scarious, larger calyces, shorter petals, and different seeds.

4. C. viscosum (Linn. Sp. p. 627. excl. syn.); annual, clothed with a soft, glandular or non-glandular pubescence; stem erect or ascending, branched; *leaves subrotund and oval*, the lower tapering into a petiole; cymes many-flowered, *with glomerate branches, all the bracts herbaceous and the calyces bearded at tip;* fruiting pedicels as long as the calyx or shorter. *E. & Z.!* 269. *C. vulgatum, L. Herb., non Sp. pl. E. Bot. t.* 789. *E. & Z.* 267 *ex parte. C. glomeratum, Thuill. Fl. Dan. t.* 1921. *C. ovale, Pers. Syn.* 1. *p.* 521.

HAB. Cape Flats and round Table Mt., in sandy places, *E. & Z.!* Groenekloof and the Paarl, *Drege!* (Herb. Sond., T.C.D.).

6–12 inches high, yellowish-green, the branches mostly ascending. Leaves larger than in the preceding. Panicle cymoid, crowded, at length more effuse and open. Fruit stalks erect or inclined. Petals as long as the calyx; capsule twice as long.

Sub-Order III. Paronychiaceæ. (Gen. vi.–xiii.)

VI. CORRIGIOLA, L.

Calyx 5-partite, the segments very blunt, albo-marginate. *Petals* 5, perigynous, as long as the calyx, roundish, entire. *Stamens* 5, on an obsolete perigynous ring, alternate with the petals. *Ovary* unilocular, with a solitary pendulous ovule. *Style* tripartite. *Nut* 3-angled, covered by the persistent calyx. *Endl. Gen.* 5197. *DC. Prod.* 3. *p.* 366.

Prostrate, glaucous, littoral herbs. Leaves alternate, with small membranous stipules. Flowers minute, crowded, greenish. Name, a diminution of *corrigia*, a *strap* or *thong*.

1. **C. litoralis** (L. Sp. 388); *corymbs leafy;* flowers pedicellate; cauline leaves linear-cuneate; *root annual. Schkuhr, t.* 85. *E. Bot. t.* 668. *C. capensis, Thunb. Fl. Cap. p.* 272. *C. telephifolia, Zey.!* 2502, *ex parte.*

Hab. Wet, sandy places, and in cultivated ground, throughout the Colony. (Herb. T.C.D., Sond.).

Stems filiform, weak, branching, glabrous, 6–12 inches long. Cauline leaves ½ inch or less; radical spathulate, larger. Stipules silvery. Flowers ½ line long. Petals white.

2. **C. telephifolia** (Pour. Act. toul. 3. p. 316); *corymbs leafless;* flowers pedicellate; cauline leaves obovate or oblong, narrowed at base; *root perennial. DC. Prod.* 3. *p.* 367. *E. & Z.* 1835.

Hab. Roadsides, near the Jetty, Capetown, *E & Z.*, and at the River Zonderende, *Zey.* 2502, *ex parte.* Sept.-Oct. (Herb. Sond.).

Closely resembling the preceding, but differing, besides the above characters, by the flowers and fruits twice the size and thicker leaves.

VII. HERNIARIA, Tourn.

Calyx green, persistent (unchanged), deeply 5-cleft; the tube cup-shaped, segments ovate. *Petals* 5, filiform. *Stamens* 5, or fewer, on a fleshy, perigynous disc. *Ovary* hidden in the calyx-tube, free, unilocular, with a single erect ovule; style short, bipartite. *Utricle* indehiscent, covered by the calyx. *Endl. Gen.* 5198. *DC. l. c. p.* 367.

Minute, prostrate or diffuse, herbaceous plants, of the old world, villous, pubescent, or rarely glabrous, densely branched. Leaves small, oval or oblong, opposite or alternate with membranous stipules. Flowers very small, green, densely clustered. Named from this plant having been formerly used in the cure of *Hernia.*

1. **H. hirsuta** (Linn. Sp. p. 317); herbaceous, prostrate, much branched, hairy; leaves oval-oblong, or ovate; clusters sessile, few flowered. *E. Bot. t.* 1379. *H. lenticulata, Thunb. Cap.* 245, *non Lin.* (the densely hairy variety). *H. capensis, Bartl. Lin.* 7. *p.* 624. *E. & Z.!* 1805 *and* 1806. *H. virescens, Salzm. in DC. Prod.* 3. *p.* 367 (the green, ciliated variety).

Hab. Sandy places and roadsides, fromCapeTown to Graaf Reynet, *E. & Z.! Drege.* Klipfontein and Assagaiskloof, *Zey.!* 2487. Thaba Unka, *Burke & Zey.* 611. (Herb. Sond., Hook., T.C.D.)

Perennial, many-stemmed; the stems 2–3 inches to a foot in length; the branches filiform; the whole plant covered with a short, generally greyish, down. Leaves fleshy, entire, 2 lines long, hairy at both sides, or the margin alone ciliate. Flowers minute, in clusters shorter than the leaves.

VIII. **POLLICHIA,** Soland.

Calyx-tube urceolate, becoming succulent in fruit; *limb* 5-toothed; the *throat* closed by 5 scale-like glands, alternating with the teeth. *Petals* 5, or none, very small, outside the glands. *Stamens* one or two, inserted at the throat of the calyx. *Ovary* unilocular, with two basal ovules. *Style* filiform, elongate, bifid. *Utricle* one-seeded, concealed in the succulent calyx-tube. *Endl. Gen.* 5208. *DC. Prod.* 3. *p.* 377.

Only one species known: a suffruticose, diffusely branched plant, with pseudo-verticillate, linear-lanceolate leaves and membranous stipules, crowded at the nodes. Flowers very minute, glomerate. Name in honour of *J. A. Pollich, M.D.*, author of a history of plants of the Palatinate.

1. **P. campestris** (Ait. Hort. Kew. 1. p. 5); *Sm. Spicil.* 1. *t.* 1. *DC. Prod.* 3. *p.* 377. *E. & Z.!* 1807.

HAB. Among shrubs, &c., Uitenhage and Albany, frequent, *E. & Z.! &c.* Winterveld and Nieuweveld, 3-4000 f., *Drege.* Also at Strandfontein and Matjesfontein. Port Natal, *Krauss.* (Herb. T.C.D., Sond., &c.).

Suffruticose, 1-2 feet high, with virgate, spreading, ramulose branches, pubescent or glabrous. Leaves 3-8 lines long, 1 line wide, linear-lanceolate. Stipules small, whitish-scarious.

IX. **POLYCARPON,** Loeffl.

Calyx 5-parted; the *sepals* herbaceous, membrane-edged, compressed, keeled and mucronate. *Petals* 5, emarginate or entire. *Stamens* 3-5, sub-hypogynous. *Ovary* unilocular, many-ovuled; *style* trifid. *Capsule* 3-valved, many-seeded. *Endl. Gen.* 5212. *DC. Prod.* 3. *p.* 376.

Small, glabrous, many-stemmed annuals or perennials, natives of the warmer parts of the temperate zone. Leaves opposite or whorled. Stipules membranaceous. Flowers small, white, in densely much-branched, dichotomous cymes. Name from *πολυ, much,* and *καρπος, fruit;* the English name is "*All-seed.*"

1. **P. tetraphyllum** (Lin. f. suppl. p. 116); leaves obovate, opposite, the cauline quaternate; flowers paniculate; petals emarginate, shorter than the calyx. *E. & Z.! No.* 1813. *Mollugo tetraphylla, Lin. Sp. p.* 89. *Barr. Ic. t.* 534. *E. Bot. t.* 1031.

HAB. Roadsides and waste places throughout the Colony. All the year round.

A small, decumbent, dichotomous annual, much-branched from the base. Stipules silvery. Sepals white-edged.

X. **POLYCARPÆA,** Linn.

Calyx 5-parted; the sepals semi-scarious or scarious, concave, nerveless, not keeled, entire, muticous. *Petals* 5, sub-hypogynous, entire or bidentate. *Stamens* 5, hypogynous. *Ovary* unilocular, many-ovuled; style trifid. *Capsule* 3-valved, many-seeded. *Endl. Gen.* 5216. *DC. Prod.* 3. *p.* 373.

Small, much-branched herbs, common throughout the tropical and sub-tropical regions of both hemispheres. Leaves opposite or pseudo-verticillate, linear or oval, or spathulate. Stipules membranous and silvery, shining, crowded at the nodes. Flowers numerous in fasciculate or corymbose cymes. Name, of the same derivation as *Polycarpon.*

1. **P. corymbosa** (Lam. Ill. n. 2798); stem erect, herbaceous, tomen-

6

tose, branched; leaves about 6 in a whorl, linear, awned; cymes corymbose, rather loose; calyces scarious, acuminate. *DC. l. c. p.* 374,—var. stricta. *Fenzl, in Hb. Drege;* stems straightish, branches erect, not divaricate.

HAB. At the River Omblas, near Port Natal, *Drege, Rev. Mr. Hewetson!* (Herb. Sond., T.C.D.).

1–1½ feet high; the stems and branches terete, clothed with a short, dense, whitish tomentum, at length glabrescent. Internodes about 1 inch long. Leaves 4, 6, or 8 together, linear, aristate-acuminate, 6–8 lines long, ½–¾ line wide, glabrous, or nearly so. Stipules silvery, ovato-lanceolate, lacerate at the point, and toothed at the margin, 2–3 lines long. Corymbs dense, an inch broad. Flowers white, scarious, 1½ lines long.

XI. LEPIGONUM, Fries.

Calyx 5-parted; *sepals* herbaceous (thickish), nerveless or obtusely keeled, equal, muticous. *Petals* 5, with short claws, entire, sub-hypogynous. *Stamens* 10 (or by abortion 5–3), on a sub-hypogynous ring. *Ovary* unilocular, many-ovuled; *style* 3–5-parted. *Capsule* 3–5-valved, many-seeded. *Spergularia, Endl. Gen.* 5218. *Arenariæ sp. Auct. vet.*

Cosmopolitan herbs, growing near the sea-shore or on salt ground. Leaves opposite or fasciculate, awl-shaped, fleshy. Stipules membranous. Flowers rosy or white, in racemiform cymes, or in the forks of the branches, pedicellate: the stalks generally reflexed after flowering. Name from λεπις, *a scale*, and γονυ, a *knee* or *node;* from the scale-like stipules at the nodes.

1. **L. rubrum** (Fries. Wahl. Fl. Gothob. p. 45); *annual or biennial;* leaves linear-filiform, somewhat fleshy, *flat* on both sides; stems prostrate and ascending, branched; branches racemose; pedicels reflexed after flowering; sepals lanceolate, obtuse, membrane-edged; *seeds 3-cornered-obovate, not winged, minutely tubercled. Kindberg, Syn. Lepig. p.* 5. *Koch. Syn. p.* 121.

VAR. α. **campestris**; glabrous or glandularly puberulent; leaves slender-filiform, mostly curved; stipules lanceolate-acuminate. *Arenaria rubra, α, campestris, L. sp. p.* 606. *E. Bot. t.* 852. *Stipularia rubra, Haw. Spergularia rubra, Fenzl. Spergula rubra, Bartl. E. & Z.! No.* 1817.

VAR. β. **pinguis**; glabrous, or glandularly puberulent; stems suberect, often straight; leaves thickish, filiform, scarcely curved; stipules broader; seeds larger. *Spergula marina, Bartl. E. & Z.! No.* 1815. *Lep. capense, Schrad. Kindl. l. c.*

HAB. Sandy and stony places round Table and Lion Mts.; and about Tulbagh. β. on the shore, Algoa Bay; at the Zwartkops River, and near the Gareep, *E. & Z.! Drege!* (Herb. Sond., T.C.D.).

3 inches to a span long. Leaves 4–8 lines long. Stipules triangular-ovate, silvery. Cymes racemiform, long or short; peduncles subtended by leaves, the fruiting ones longest. Calyx 1½–2 lines long. Petals rosy, as long as the calyx. Capsule equalling the calyx or longer.

2. **L. medium** (Fries. Nov. Fl. Suec. Mant. 3. p. 33); *perennial,* glabrous; leaves linear-filiform, sub-muticous, fleshy, *convex above and below;* stems prostrate and ascending, branched; branches racemose, the pedicels reflexed after flowering; sepals lanceolate, obtuse, nerveless,

membrane-edged; *seeds triangular-obovate, compressed, smooth*, wingless or a few of them somewhat winged, with a thickened margin. *Kindb. l. c. p.* 13. *Koch. Syn. p.* 121. *A. rubra, β. marina, L. sp.* 668. *E. Bot. t.* 958. *Spergula media, E. & Z. No.* 1816, *ex parte. Spergularia media, a, Fenzl, Hb. Drege.*

Hab. Zwartkop's River, *E. & Z.!* At the Gareep, *Drege.* Winterhoek, *Krauss.* (Herb. Vind, Sond.)

Very like the last, especially like var. *β.* but generally larger and more robust. Calyx 2 lines long. Petals pale-purple or rosy-white.

3. L. marginatum (Koch, Syn. p. 121); perennial; leaves linear-filiform, sub-muticous, fleshy, semi-cylindrical; stems prostrate, ascending, branched; branches racemose; peduncles reflexed after flowering; sepals lanceolate, obtuse, nerveless, membrane-edged; seeds *subrotund, plano-compressed, smooth, with a broadish, marginal wing. Arenaria media, Lin. sp.* 606. *excl. syn. Alsine marina, Wahl. Sperg. media, Bartl. E. & Z. No.* 1816. *Lepigonum marinum, Kindb. l. c. p.* 12.

Var. *α.* **glabra**; altogether glabrous.

Var. *β.* **glandulosa**; glandularly pubescent. *Aren. glandulosa, Jacq. Hort. Schoenb.* III. *t.* 355. *Spergularia glandulosa, Fenzl. pl. Drege.*

Hab. Everywhere in salt, damp ground near the sea-shore, throughout the Colony.

Generally more robust than the preceding, often glandularly downy, at length glabrescent. Sepals 2½–3 lines long. Flowers purple or whitish. Seeds brownish with a broad, white margin.

XII. SPERGULA, L.

Calyx 5-parted; sepals oval, herbaceous, membrane-edged. *Petals* 5, with short claws, ovate, sub-hypogynous. *Stamens* 5–10, on a sub-hypogynous ring. *Ovary* unilocular, many ovuled; *styles* 5, distinct, alternate with the sepals. *Capsule* 5-valved, many-seeded; the valves opposite the sepals. *Endl. Gen.* 5219.

Small weeds and herbs of the temperate zones. Leaves subulate or filiform, fleshy, crowded about the swollen nodes. Stipules scarious. Flowers in racemiform cymes; the pedicels deflexed after flowering. Name from *spargo*, to *scatter;* because the species are widely dispersed.

1. S. arvensis (Linn.); leaves linear-subulate, whorled, stipulate at base; flowers decandrous; seeds spheroid, hispid, with a narrow margin, black. *DC. Prod.* I. *p.* 394. *Lam. Ill. t.* 392. *fig.* I. *E. & Z.! No.* 1814.

Hab. A weed in cultivated ground; introduced from Europe.

XIII. DRYMARIA, Willd.

Calyx 5-parted. *Petals* 5, deeply 2–4-lobed or partite. *Stamens* 5, or fewer, sub-hypogynous. *Ovary* unilocular, many ovuled. *Style* filiform, trifid. *Capsule* membranous, 3-valved, many-seeded. *Endl. Gen.* 5220. *DC. Prod.* I. *p.* 395.

Herbaceous plants of tropical and sub-tropical regions. Stems slender; leaves opposite, cordate or sub-rotund, or lanceolate and linear; stipules setaceous; flowers small, white, in terminal, effuse cymes or panicles. Name from δρυμος, a *forest;* place of growth.

1. D. cordata (Willd. Hb., ex Roem. & Sch. 5. p. 406); stem and leaves glabrous, shortly petioled, ovato-subrotund, acute or mucronate, rounded at base, or obsoletely cordate; peduncles dichotomous, many-flowered; calyx glabrous, longer than the petals; ovary 7–10 ovuled. *DC. Prod.* 1. *p.* 395. *Cham. & Schl. Linn.* 1. *p.* 47. *Holosteum cordatum, L.*

HAB. Woods near the Hanglip, Aug. 1821., *Mundt and Maire.* (Herb. r. Berol).

A weak-growing, diffuse plant, a foot or more in length; stems and branches filiform, sometimes minutely puberulent. Leaves 6–8 lines long and wide, petiole 2 lines. Stipules small. Peduncles terminal, cymoid; flowers solitary in the forks, lateral peduncles elongate, 2–3 flowered or again cymoid. Calyx 2 lines long, with acuminate segments. A common tropical species. The Cape specimens, preserved in the Berlin Herbarium, quite resemble the American forms, distributed by Sieber.

Sub-Order IV. MOLLUGINEÆ. (Gen. xiv.–xx.)

XIV. ORYGIA, Forsk.

Calyx 5-parted, sepals of unequal length. *Petals* (*parastemons*) 15–30, spathulate-linear or oval, shorter than the calyx, at length confluent into a fleshy cup. *Stamens* 12–40, in the bottom of the calyx, some free, some united at base; filaments subulate-triangular; anthers versatile, oblong. *Ovary* globose, 5-celled, many-ovuled. *Stigmata* 5. *Capsule* dry, roundish, 5-angled, 5-furrowed, 5-celled, loculicidally 5-valved. *Seeds* many, on ascending funiculi, reniform, black, concentrically furrowed, eared at the hilum. *Endl. Gen.* 5184.

A suffruticose, glabrous, glaucous, much-branched, diffuse plant, found also in Arabia Felix and in the East Indies. Stem and branches angular. Leaves fleshy, orbicular, obovate or elliptical, very entire or retuse, muticous or mucronulate, alternate, petiolate. Cymes divaricate, opposite the leaves. Flowers pedicellate, at length reflexed; calyces at tips and margins purplish. Name, altered from the Arabic name of this plant, *Hörudjrudj.*

1. O. decumbens (Forsk. descr. p. 103). *DC. Prod.* 3. *p.* 455. *Fenzl. l. c. Portulaca decumbens, Vahl. Symb.* 1. *p.* 33. *Talinum decumbens, Willd. Sp.* 2. *p.* 864. *Axonotechium trianthemoides, Fenzl, l. c. p.* 355. *Telephium laxiflorum, Burch. DC. Prod.* 3. *p.* 366.

HAB. Stony hills near Sondag Riv., 1500–2000 f., *Drege!* Albany, *Mrs. Barber!* Rhinoster Kopf. *Burke!* (Herb. T.C.D., Sond.).

Largest leaves 1–2 inches, smallest 4–6 lines long; petiole 2–4 lines long, membrane-edged. Cymes sessile or pedunculate, 1–3 inches long, unequally forked, bracteolate, the middle pedicel one-flowered. Sepals 2–3 lines long, equalling the capsule.

XV. GLINUS. Loefl.

Calyx 5-parted. *Petals (parastemons)* none or indefinite, sub-perigynous, shorter than the calyx, very narrow, 2–3-forked or setaceous. *Stamens* 3–20 in the base of the calyx, free or united in parcels; filaments subulate; anthers versatile, oblong. *Ovary* ovate, 3–5-celled, many-ovuled. *Stigmata* 3–5. *Capsule* membranous, ovoid, 3–5-angled, 3–5-furrowed, loculicidally 3–5-valved. *Seeds* numerous, fixed to ascending, circumflexed funiculi, reniform, smooth or tubercled, with an entire umbilical strophiole. *Endl. Gen. No.* 5185.

Annual or suffruticose, branched, prostrate, glabrous, or stellato-pubescent, weed-like plants, found in all tropical and sub-tropical countries. Leaves alternate or pseudo-verticillate, entire or denticulate. Flowers in clusters or umbels, opposite the leaves, sessile and crowded, or solitary at the nodes, and on long pedicels. The name is γλῖνος, an old name for the maple, arbitrarily affixed to this genus.

1. G. lotoides (Lin. sp. p. 663); *tomentose,* with simple and stellate pubescence; stems diffuse, branching; leaves whorled, unequally obovate, tapering into a petiole; flowers whorled, *the pedicels shorter than the calyx;* stamens 3–12. *Fenzl, An. Wein. Mus.* 1. *p.* 357. *Mollugo hirta, Thunb. Cap. p.* 120. *D.C. Prod.* 1. *p.* 391. *E. & Z.! No.* 1818.

VAR. *α.* **candida** (Fenzl); whole plant densely albo-tomentose.

VAR. *β.* **pubescens** (Fenzl); less tomentose, greenish.

HAB. Cape, *Thunberg!* In sandy places, Oliphant's River, *E. & Z.!* Hills near George, *Drege!* Dreifontein, Oliphant's River and Aapje's River, *Zeyher,* 612. (Herb. T.C.D., Sond.).

Annual, branched from the base. Stems ½–1 foot long, somewhat woody at the base, terete, as thick as a crow's quill. Leaves obovate, obtuse or shortly acute, the radical rosulate, fugacious, on longish petioles, cauline very unequal, sub-undulate at the margin, ribbed and veined, lamina 3–8 lines long, narrowed into a petiole half that length. Flowers 2–8, umbellate, axillary. Sepals oblongo-lanceolate, 2–2½ lines long, inflexed at point, mucronulate. Petals often none. Capsule shorter than the calyx. Seeds very minute, brownish, muricato-tuberculate.

2. G. Mollugo (Fenzl, l. c. p. 359); *herbaceous, glabrous;* branches downy at the point; stems dichotomous; leaves whorled, unequal, obovate-oblong or lanceolate, very entire or denticulate at the apex; flowers axillary, whorled, *pedicels* 2–4 *times longer than the calyx;* sepals oblong, pointless; stamens 3-10.

VAR. **Natalensis** (Sond.); stem erect, leaves inch long, ovate-oblong or oblong, acute, tapering into the petiole, serrulate; flowers on longish pedicels; sepals acute; stamens generally eight. *Mollugo serrulata, Sond. Linn.* 23. 1. *p.* 15.

HAB. Port Natal, *Gueinzius!* (Herb. Sond., T.C.D.).

Annual, a foot long, with erecto-patent branches. Internodes 1½–2 inches long. Leaves in fours, the larger one inch long, 4 lines wide, narrowed into a petiole 2–3 lines long. Peduncles 4–6, about 4 lines long. Sepals 2 lines long, membrane edged. *St.* hypogynous, filaments sub-unequal, broader at base; anthers sub-emarginate at base. Ovary 3-furrowed, 3-celled; ovules on short funiculi. Stigmata 3, linear-terete, papillose. Capsule 1½ line long, obtuse, not inflated in the middle, 3-celled, 3-valved. Seeds very small, blackish brown, minutely muriculate; the strophiole white.

XVI. MOLLUGO. L.

Calyx 5-parted. *Petals* none. *Stamens* 3-5, rarely 6-10, hypogynous, those of the outer row alternating with the sepals, or when fewer than 5, opposite the ovarian-dissepiments. *Anthers* minute, globose. *Hypogynous-disc* none. *Style* scarcely any; *stigmata* 3, linear. *Capsule* thinly membranous, 3-celled, bluntly 3-angled, 3-furrowed, 3-valved, loculicidal; valves septiferous in the middle, the cells many-seeded. *Seeds* globose, not caruncRulate at the hilum. *Endl. Gen.* 5186. *Fenzl, l. c.* 2. *p.* 246.

Small, slender, procumbent, dichotomously branched annuals, natives of the tropics and warmer temperate zones, chiefly of the old world. Leaves linear or spathulate, whorled round the nodes. Stipules obsolete, fugacious. Cymes dichotomous, racemiform, axillary; or sessile or pedunculate umbellules at the nodes. Named from the resemblance in habit to *Galium Mollugo*, supposed to be the *Mollugo* of the ancients.

1. **M. Cerviana** (Seringe); quite glabrous, glaucous; stems filiform, straight, erecto-patent or diffuse; radical leaves rosulate, obovate, spathulate, linear, obtuse; cauline remote, narrow-linear; umbellules 3-5 flowered; pedicels longer than the calyx; stamens 5; seeds shining, small, very minutely reticulated.

VAR. α. **linearis** (Fenzl,); radical leaves narrow-linear. *M. Cerviana, DC. Prod.* 1. *p.* 392. *Pharnaceum Cerviana, Lin. Sp.* 1. *p.* 388.

VAR. β. **spathulæfolia** (Fenzl,); radical leaves cuneato-linear, obovate, and widely spathulate, mucronulate, pointless or retuse. *Pharm. Cerviana, Lam. Ill. t.* 214. *f.* 2.

VAR. γ. **pygmæa**; stems subsimple, ½–1 inch long, diffuse; branches spreading; radical leaves narrow-spathulate, obtuse, cauline little narrower; pedicels spreading; seeds very minute.

HAB. Sandy places at Kamos, *Zey.!* 622. β Breederiverspont, *Zeyher.* Between Driekoppen and Bloedrivier, 2000–3000 f. *Drege.* Gamkerivier, *Burke and Zey.!* 613. (Herb. T.C.D., Sond.).

Root slender. Stems 1–6 inches long, yellow-green. Radical leaves 2–8 lines long, ⅓–1 line wide, somewhat fleshy, one-nerved; cauline whorled, unequal. Pedicels capillary, 3–6 lines long. Sepals oval, membranous, white, with a green nerve, 1 line long. Capsule equalling the calyx. Seeds brown, 3-angled-pyriform.

XVII. PHARNACEUM, Linn.

Calyx 5-parted; sepals very obtuse, petaloid within and at the margin. *Petals* none. *Stamens* 5, perigynous, in the base of the calyx and alternate with the sepals; anthers linear-oblong, rarely sub-globose. *Hypogynous-disc* cup-shaped, 3–5-fid, rarely obsolete. *Style* none; *stigmata* 3, obovate, fleshy, crest-like, coloured or white, rarely terete. *Capsule* membranous, roundish-three-angled below, three-angled above, 3-furrowed, trilocular, 3-valved, loculicidal; seeds 4–8 in each cell, globoso-lenticular, acutely margined, or sub-globose with a raised dorsal line, not strophiolate. *Endl. Gen.* 5187. *Fenzl, l. c.* 2. *p.* 246. *Ginginsia, DC.* 3. *p.* 362.

Small, slender suffrutices or herbs, rarely annual, natives of South Africa. Leaves setaceous, filiform or linear, rarely lanceolate, or elliptical, mostly ending in a bristle; the cauline leaves of the perennial species alternate, crowded, the upper ones in dense brush-like fascicles; of the annual, whorled round the nodes. Stipules scarious, cut into a capillary fringe. Cymes simple or compound, forked, racemiform. Named from Pharnaces, King of Pontus.

SECT. 1. Branching suffrutices, with long, rarely short and crowded, woody branches, leafy throughout, comose at the summit. Leaves alternate, often fasciculiferous in the axils. Peduncles scape-like, cymose at the summit (Sp. 1–6).

1. **P. trigonum** (E. & Z.! 1831); suffruticose, small, rigid; branches *short*, appressedly leafy; leaves closely fasciculate, *needle-shaped*, *sub-*

trigonous, aristate-mucronate, persistent; stipules *setaceo-fimbriate,* at length *falling off;* peduncles scape-like, cymose, 3–5-flowered, 3–6 times as long as the leaves; seeds globoso-lenticular, margined, very smooth. *Fenzl, l. c. p.* 248.

Hab. Calcareous and clayey hills between the Koega and Zondag Rivers. *E. & Z.!* March. (Herb. Sond.).

2 inches high. Root perpendicular. Stems very short, bi-trichotomous from the base; the lower branches 3–4 lines. the upper 2–1 line long. Leaves close-pressed, the uppermost spreading, $1\frac{1}{2}$–$2\frac{1}{2}$ lines long, furrowed at back, glabrous. Stipules very minute, 2–3 times shorter than the leaves. Cymes subumbellate, on a shining peduncle, $\frac{1}{2}$–1 inch long, pedicels very slender, bracteolate and stipulate at base. Sepals oval-oblong, 1 line long. Stamens $\frac{1}{2}$ as long as the calyx. Capsule as long as calyx.

2. P. microphyllum (Lin. f. Sup. p. 185); suffruticose, erect; branches *sub-dichotomous, virgate,* leafy; leaves crowded, *ovato-terete, obtuse; stipules lanato-fimbriate, the fringe very copious and silky, intangled in woolly glomerules at the axils;* peduncles scape-like, cymose, many-flowered; seeds globoso-lenticular, acutely margined, black-brown, very smooth. *Thunb.! Fl. Cap. p.* 272. *Mollugo microphylla, Ser. DC. Prod.* 1. *p.* 392. *Ginginsia microphylla, DC. l. c.* 3. *p.* 368.

Hab. On the shore, at the end of Verloren Valley, *Thunberg!* (Herb. Thunb.).

This resembles the following in all characters, and only differs in having shorter, thicker, fleshy leaves, 1 line long, $\frac{1}{2}$ line wide. Peduncles an inch long. Flowers of *P. lanatum;* of which this may be a mere variety.

3. P. lanatum (Bartl. Linn. 7. p. 625); suffruticose, erect; branches *dichotomous, virgate;* leaves closely or laxly set, *filiform, pointless; stipules lanato-fimbriate, the fringe very copious, silky, intangled in woolly glomerules at the axils;* peduncles scape-like, cymose, many-flowered; seeds globose, lenticular, margined, very smooth. *Fenzl, l. c. excl. syn. P. microphylli. P. lanatum and microphyllum, E. & Z.* 1824, 1823.

Hab. In sandy ground. Cape Flats, near Rietvalley, at Saldanha Bay, at Hott. Holl., and at Brackfontein, *E & Z.!* Nieuweveld, *Zeyher,* 2496. (Herb. Sond.).

A suffrutex 1–$1\frac{1}{2}$ feet high. Branches 2–4 inches long, straight or bent, slender, woolly with the fringe of stipules. Leaves terete, filiform, obtuse, glabrous, furrowed beneath, spreading, 3-12 lines long; those of the fascicles shorter, unequal. Stipules white, dry, divided into long and very slender curled fringe. Peduncles 2–6 inches long; cyme rarely few-flowered, simple or proliferous. Pedicels spreading, glabrous, at base bracteated by the cushion of stipules, or whorl of leaves. Sepals white-edged, 2 lines long. Capsule longer than the calyx.

4. P. incanum (Lin. Sp. p. 389); suffruticose, erect, or from the base squarrose and diffuse; branches closely leafy, leaves more or less densely set, *setaceo-filiform, mucronate,* or awned; stipules fimbriato-lacerate, the segments capillary, elongate, *not interwoven in glomerules;* peduncles scape-like, cymose, many-flowered; seeds globose-lenticular, acutely margined, very smooth. *Fenzl, l. c. p.* 249. *Thunb. Fl. Cap. p.* 273. *E. & Z. No.* 1821. *P. confertum, E. & Z.!* 1822. *Hb. Un. No.* 626. *Sieb. No.* 210. *Ginginsia conferta, DC. l. c. p.* 363. *Bot. Mag. t.* 1883. *Lam. Ill. t.* 214. *f.* 3. *G. elongata, D.C. l. c.*

Hab. In sandy and stony places near Capetown; also in Swellendam, Worcester, Uitenhage, and on the Picquet and Camiesberge Mts. *Thunb., Sieber, &c., E. & Z.! Zeyher, No.* 2497, 2498, 2499. (Herb. T.C.D., Lehm., Sond.)

$\frac{1}{2}$–1$\frac{1}{2}$ feet high; branches 1–2 inches, or 3–6 inches long, slender, comose at the end. Leaves fleshy, glabrous, furrowed below, rigid, patent or recurved, in some specimens 2–4lines, in others 6–12 lines long. Stipules silvery, the sub-simple fimbrils never curled. Peduncles solitary or several from the terminal tuft of leaves, 1–16 inches long. Cyme 2–3-chotomous, the branches often racemose, spreading; the pedicels at base bracteated by a tuft of stipules or leaves. Sepals with wide margins. Anthers orange. Capsule ovoid-triangular, as long as the calyx or longer. Only to be known from the preceding by its stipules.

5. **P. reflexum** (E. & Z.! No. 1825); suffruticose, erect, di-tri-chotomous or densely branched, branches naked at base; leavesthickish or slender, filiform, muticous or mucronulate, scattered or crowded, *recurvo-patent, never crowded in a tuft at the end of the branches;* stipules at length deciduous, *pectinato-fimbriate,* the shreds very slender, curled, *but not interwoven in dense glomerules;* peduncles scape-like, cymose, many-flowered; seeds globoso-lenticular, very finely *granulated. Fenzl, l. c. p.* 251. *P. albens, L. fil. Suppl.* 186? *Thunb. Cap. p.* 274? *Ginginsia aurantia, DC. l. c. P. lineare, Andr. Bot. Rep. t.* 326.

HAB. In stony places of the Karroo, near Gauritz River, Swell., *E. & Z.!* Kritsemberg, *Lichtenstein.* Wupperthall and Silverfontein, Modderfontein and elsewhere in Namaqualand, *Drege!* Sep.–Nov. (Herb. T.C.D., Lehm., Sond.).

3–12 inches long, like the preceding, but more robust, Leaves unequal, fleshy, roundish, 2–6 lines to nearly 1 inch long, $\frac{1}{3}$–$\frac{2}{3}$ line wide. Stipules minute, silky. Peduncles short or long, terminated by a tuft of stipules or whorl of leaves. Cymes 2–3-chotomous or racemiform, simple or proliferous, forming a compound umbel. Sepals oblong or oval, 1–2$\frac{1}{2}$ lines long, yellow or white-margined. Stigmata purplish-orange or golden. Capsule of the preceding, but the seeds have a more evident margin and are very minutely granulated.

6. **P. detonsum** (Fenzl, l. c. p. 253); suffruticulose, squarrose; *branches woody, short, leafy;* leaves crowded, elongate-filiform, aristato-mucronate, rather straight; stipules subulate-setaceous, *simple, not fimbriate;* seeds globoso-lenticular, very smooth, shining. *P. patens, E. & Z.! No.* 1820, *ex parte.*

HAB. In sandy and stony fields at Konabshoogde, and Kat River, *Ecklon.* Rhinoster River, and at the Gareep, *Zey. Paron. No.* 3. Klipplaats River, *Drege.* (Herb. Sond.).

6 inches high, with a long, thick, perpendicular root; branches several, $\frac{1}{2}$–1$\frac{1}{2}$ inch long, slender, clothed at base with the rigid remains of leaves. Leaves $\frac{1}{2}$–1 inch long, rigid, subterete, somewhat crowded at the ends of the branches. Stipules greenish white, straight, not cut. Peduncles 2–3 times longer than the branches, tipped with a whorl of leaves, and bearing a simple or proliferous cyme. Sepals 1–1$\frac{1}{2}$ lines long, with white or yellow margins. Capsule rather longer than the sepals. Seeds black. Very near *P. dichotomum,* but the stem is branched, the branches woody, the stipules simple and the seeds different.

Sect. 2.—Suffrutices or perennial or biennial (scarcely annual) herbs, with an obsolete or very short, simple, or shortly branched, persistent stem *(caudex),* crowned with a tuft of linear-filiform leaves, and annually emitting scapelike, knee-bent, simple or branching stems. Cauline leaves whorled. (Sp. 7—11.)

7. **P. lineare** (Lin. f. Sup. p. 1825); caudex obsolete; annual-stems herbaceous, elongate, diffusely branched; leaves whorled at the nodes, *linear-filiform, pointless,* sub-terete, fleshy, unequal; stipules at length fugacious; sepals oval-oblong, with wide margins; *hypogynous-disc*

none; seeds globoso-lenticular, very smooth, shining. *Thunb. Fl. Cap. p.* 274. *Fenzl, l. c. p.* 253. *E. & Z.! No.* 1827. *Mollugo linearis, Ser. in DC. Prod* 1. *p.* 392.

HAB. Sandy places on the Cape Flats, Zwartland, and Krakakamma, *Thunberg, E. & Z., Drege &c.* Verchevally and 24 Rivers, *Zey.!* 623 (Herb. T.C.D., Sond.)

Aspect of *Spergula arvensis.* Stems several, 1–2 feet high, very smooth, the internodes 1–5 inches long, pale green, with minute, scarious, fimbriate stipules and a whorl of leaves at each node. Leaves 4–10 in a whorl, unequal, obtuse, ½–1½ inch long, ½–1 line wide, spreading, afterwards reflexed. Cymes terminal and axillary, pedunculate, dichotomously racemiform, mostly effuse, 4–6 inches long, including the peduncle. Sepals 2–3½ lines long, petaloid, with a green rib; anthers golden. Stigmata thick, orange or crimson, at length pale and whitish. Capsule ovoid, oblong, scarcely exceeding the calyx. Seeds black.

8. **P. dichotomum** (Lin. f. suppl. p. 186); caudex short, simple or divided; annual stems slender, erect or spreading, simple or branched, whorled with leaves at the nodes; *leaves linear-filiform, mucronulate;* stipules ciliato-lacerate; *hypog.-disc tripartite;* seeds globoso-lenticular, *minutely granulated,* shining. *Thunb. Fl. Cap. p.* 274. *Fenzl, l. c. p.* 254. *Mollugo dichotoma, DC. l. c.*

VAR. *α.* **linearis**; lowest leaves crowded in a tuft, 1–2 inches long, cuneato-linear ⅓–1 line wide at the point; cauline spreading, not reflexed, stipules few and soon falling off. *P. fluviale, E. & Z.! No.* 1828.

VAR. *β.* **filifolia** (Fenzl); all the leaves narrow-filiform, scarcely ⅓ line wide, the cauline spreading, not reflexed; stipules fimbriate, clustered at the nodes but not glomerated. *Ging. brevicaulis DC.* 3. *p.* 363. *Pharm. patens, E. & Z.! No.* 1820, *partly.*

VAR. *γ.* **barbata** (Fenzl); leaves linear-filiform; the cauline at length reflexed; stipules very copious, conglomerated, fimbriato-barbate.

HAB. Var *α*, stony ground, Zwartkops River, *E. & Z.! Zey.* 2493a. *β*, in the same place and at Quaggas Vlakte, at Adow, and Sondags River, *E. & Z. Zey.* 2493b. In the district of George, *Drege, Mundt.* &c. Cape Flats, *W.H.H.* *γ.* Between the Koussie and Gariep, Kl. Namaqualand, 1500–2000f. *Drege.* (Herb. T.C.D., Hook., Sond.)

A many-stemmed plant, 3–8–18 inches long. Caudex scarcely any, or ½–1 inch long, multifid. Stems, except in *γ.*, with few internodes. Leaves in *α*, 4–12 lines long; in *γ*, 3–6 lines. Stipules scarious, shining. Cymes racemiform, axillary and terminal, simple or proliferous, panicled; pedicels capillary, deflexed after flowering. Sepals oval, 1–1½ lines long. Capsule rather longer than the calyx. Seeds black. Very small specimens of the first year constitute *P. brevicaule, Bartl. Linn.* 7. *p.* 625. *E. & Z. No.* 1819., which is hardly distinguishable from *P. subtile, E. Mey.* (No. 14.)

9. **P. Zeyheri** (Sond.); suffruticulose; caudex very short, multifid, *densely clothed with lacerated, silky stipules;* stems numerous, scape-like, emerging from the dense tuft of leaves, terminating in a single whorl of leaves; *lower leaves linear-spathulate, acute, mucronate aristate,* flat, one-nerved, fleshy, upper narrow-linear, aristate; stipules minute, fimbriate; cymes racemiform, elongate; peduncles capillary, obsoletely stipulate at base; sepals elliptical, petaloid with a green dorsal rib; anthers oblong; seeds globoso-lenticular, smooth. *Zey. No.* 614. (220.)

HAB. Macallisberg, *Burke and Zeyher!* (Herb. Hook., Sond., T.C.D.)

5–6 inches long. Root long and hard. Caudex, with the branches ½ inch long;

stipules imbricating the branches, about 2 lines long, oblong, cleft to the middle into setaceous laciniæ. Lower leaves ½ inch long, narrowed at base, one line wide; upper terminating the shining stem, 3–4 lines long and ½ line wide. Cymes terminal, or rarely issuing from the lower leaves, 1–3 inches long; pedicels 4–6 lines. Sepals a line long. Stamens 5. Capsule equalling the calyx. Habit of *Mollugo Cerviana.*

10. P. croceum (E. Mey. ! in Hb. Drege); biennial or annual; caudex none or obsolete, undivided; *stems very many, scape-like,* straight, issuing from a tuft of radical leaves; leaves fleshy, cuneate-filiform, and filiform, *not furrowed,* long-awned; cymes 2–3-chotomously racemiform, simple or shortly proliferous; *anthers linear;* hypogynous disc fleshy, 3-parted; seeds globoso-lenticular, very smooth. *Fenzl, l. c.* 255.

VAR. α. **crassifolia** (Fenzl); radical leaves uncial, 1 line wide.

VAR. β. **tenuifolia** (Fenzl); rad. leaves semi-uncial, ⅓ line wide.

HAB. Var. α, Silverfontein; and β, Namaqualand, *Drege!* Sep. (Herb. T.C.D., Sond.)

Root slender. Radical leaves crowded in a very dense tuft. Stems scape-like, simple, 3–4 times longer than the radical leaves, ending in a whorl of short leaves. Stipules very short, scarious, lacerate, nearly hidden among the radical leaves; chiefly visible on the nodes of the inflorescence. Cymes 1–2 inches long. Pedicels erect, spreading. Anthers orange. Stigmata reddish-saffron colour. Capsule equalling the calyx.

11. P. gracile (Fenzl, l. c. p. 256); biennial or annual; caudex none; stems numerous, scape-like, issuing from a tuft of radical leaves; leaves tereti-filiform, awned; cymes trichotomously racemiform, simple; *hypogynous disc membranous, white, obsolete, often reduced to a single lobe; anthers elliptical;* seeds sub-globose, *marked with an obsolete dorsal line,* not margined, opaque, brown.

HAB. On the Great Karroo, 2–3000f. *Drege!* (Herb. Sond.)

Very like the preceding, but much more slender, with longer racemes, capillary pedicels and flowers half the size. Leaves 8–12 lines long, ¼ line wide, tipped with a hair. Stems 2–2½ inches long, very slender. Stipules obsolete on the nodes of the inflorescence. Cyme 2 inches long; racemes capillary, with no ramuli. Pedicels 4–6 lines long, spreading. Sepals 1 line long or shorter, white-edged. Stigmata white. Capsule equalling calyx. Seeds under a strong lens, minutely granulate.

Sect. 2. Annual or perennial herbs, without *caudex.* Leaves flat, oval, lanceolate or linear-lanceolate, in remote whorls. (Sp. 12–14).

12. P. serpyllifolium (Lin. f. Suppl. p 186); perennial, small, with a very short stem branched from the base; branches many, diffuse, very slender, filiform, subdichotomous, densely bearded at the nodes with silky, curled stipules; *leaves whorled, ovate, acute or obtuse,* mostly bristle-pointed, narrowed at base; peduncles 1-*flowered,* axillary and terminal; capsule globose; seeds sub-globose, reticulate. *Thunb. Fl. Cap. p.* 275. *Fenzl, l. c. p.* 257. *Mollugo serpyllifolia, DC. Prod.* 1. *p.* 391.

HAB. Rock-fissures of Bockland Mt., *Thunberg!* Blauweberg and near Ezelsbank, 3–5000 f. *Drege!* Oct.–Dec. (Herb. Thunb., Sond.)

An herb with the aspect of *Herniaria;* differing from all its congeners by the form of the leaves. Root perpendicular, simple, hard. Stem divided into several short trunks, from which spring many capillary, glabrous, fragile branches, 1–3 inches long, prostrate, densely cæspitose, simple or dichotomous. Internodes 2–3 lines long. Leaves 3–7 in a whorl, unequal, narrowed into a short or longish petiole, the lamina 1 line long, ½–¾ line wide. Stipules very minute, shorter or sometimes longer than the leaves. Peduncles mostly one-flowered, rarely two-flowered; pedicels capillary, bearded at the nodes, 4–12 lines long. Calyx globose, 1 line long. Stamens 5. Hypog.-disc 5-lobed. Caps. membranous, white, as long as calyx; cells 2–5 seeded. Seeds sub-compressed, brown.

13. P. distichum (Thunb. Fl. Cap. p. 275, excl. syn. Willd.); perennial; stems decumbent, simple or dichotomous, *elongate;* leaves about six in a whorl, *lanceolate,* sub-acute; peduncles axillary and terminal, *racemose or paniculate,* longer than the leaf; capsule sub-globose; seeds lenticular, *very smooth,* black. *E. & Z. ! No.* 1827. *Fenzl, l. c.* 258.

HAB. Among shrubs. Krakakamma, *E. &. Z. ! Zey, No.* 2492. At the Koussie River, Kl. Namaqualand, and on hills by the sea side between Omcomas and Port Natal, *Drege.* (Herb., Sond.).

Aspect of *Gallium Mollugo.* Stems many from the crown, 1–2 feet long, glabrous, slender, flexuous, decumbent at base, then ascending, rarely trichotomous. Internodes 1–3 inches long. Leaves 5–10 in a whorl, sub-equal, inch long, 1–2 lines wide, in some specimens only half that size, muticous or mucronulate, narrowed at base, spreading, at length reflexed. Stipules very shortly fimbriate, white. Peduncles axillary, racemose; pedicels remote, 4–6 lines long, capillary; peduncles terminal, cymose or paniculate. Calyx 1 line long. St. 5. Hypog. disc 5-lobed. Capsule rather longer than the calyx, many-seeded. Seeds shining.

14. P. subtile (E. Mey. in Hb. Drege); *annual;* stems one or several, from a rosette of leaves, *very slender,* simple or forked; *radical leaves oval or oblong, fugacious,* succeeded by *rosulate, linear-lanceolate* ones; cauline narrow-linear, whorled; stipules obsolete, fugacious; cymes 2–3-chotomous, racemose, divaricate, *sessile, hypogynous-disc obsolete,* tripartite; capsule globose; seeds globoso-lenticular, reticulate, black. *Fenzl, l. c. p.* 259.

HAB. Rocky places. Paarlberg, 1500–2000f. *Drege.* Between Capetown and the Salt River, *Bergius.* Sep.–Oct. (Herb. Sond.)

A very small plant, resembling *Mollugo Cerviana.* Root slender. Stems 1–2 inches high, ending in a few-leaved whorl, and a sessile cyme. Rad. leaves sub-petiolate, 4–12 lines long, ½–1 line wide; cauline narrower and shorter. Stipules very minute. Cymes racemiform, divaricate, the racemes inch long, capillary, 3–7 flowered; pedicels 1–3 lines long, deflexed after flowering. Calyx globose, ½–¾ lines long; sepals fleshy, oval. Anthers globose. Capsule many-seeded, scarcely longer than the calyx.

SECT. 4. Diffuse, sub-dichotomous annuals. Leaves subulate, fasciculately whorled at the nodes. Flowers sub-sessile. (Sp. 15.).

15. P. scleranthoides (Sond.); annual, glabrous; stems diffuse, or erect, slender, sub-dichotomous; leaves verticillately aggregate at the nodes, subulate, mucronate; stipules setaceo-lacerate; flowers axillary, aggregate, very shortly pedunculate; sepals mucronate, twice as short as the capsule; seeds lenticular, margined, very smooth.

HAB. Springbokkeel, *Zey. No.* 617. March. (Herb. Sond.)

Root simple, perpendicular, white. Stems numerous, pallid, 2–3 inches long, simple, or branched from the base, the branches 1–2, the nodes 3–5 lines apart. Radical and cauline leaves equal, all 4–6 lines long, ¼ line wide, cuspidate with a short mucro. Stipules scarious, a line long, from a broad base, narrowly lacerate. Flowers seemingly sub-sessile, but really pedicellate and cymose; the intermediate flower sub-sessile, the lateral 2-4, evidently stalked. Calyx 1 line long, 5 parted; sepals elliptical, mucronate, with a green dorsal nerve. Hypog. disc none. St. 5, shorter than the sepals; the anthers sub-globose. Stigmata 3, short, recurved. Capsule membranous, prismatic-3-angled, glabrous, 3-celled, loculicidally dehiscent, cells many-seeded. Seeds minute, lenticular, sub-emarginate, very smooth, shining, not strophiolate. Habit of a *Scleranthus;* or rather of *Minuartia* or *Loefflingia.*

APPENDIX. Doubtful species.

16. P. teretifolium (Thunb. Fl. Cap. p. 274.); caudex fruticose, erect, divaricately branched; leaves filiform, mucronate, ½ line long; pedicels shorter than the leaves. *Fenzl, l. c. p.* 259. *Moll. teretifolia, Ser. in DC. Prod.* 1. *p.* 393. *Ging. teretifolia DC. Prod.* 3. *p.* 363.

HAB. Near Witte Klipp, *Thunberg.* Oct.

" Stem suffruticose, terete, ash-coloured, entirely glabrous, a foot high. Branches opposite, divaricate, slightly divided. Rameal leaves whorled, terete, subulate, entire, spreading, ½ line long. Umbels lateral, simple, pedunculate. Peduncles capillary, shorter than the leaf." *Thunb. l. c.*

According to Fenzl it is allied to *P. incanum* or *P. reflexum,* and perhaps a variety of one or other. To me, however, it seems, by the description, not to belong to the present genus, but to *Adenogramma,* and to be near *A. rigida,* if not the same.

XVIII. HYPERTELIS, E. Mey.

Calyx 5-parted, the sepals very obtuse, membrane-edged, coloured. *Petals* none. *Stamens* 3–5, or 12–16 in two rows, the 5 outer alternating with the sepals, shorter, the inner connate at base and longer; or 20–30, in 3–5 parcels opposite the dissepiments of the ovary, connate at base, alternating with a few free stamens. *Anthers* oblong. *Hypogynous-disc* none. *Style* none; *stigmata* 3–5, fleshy, recurved, white. *Capsule* ellipsoid, 3–5 furrowed, 3–5 celled, 3–5 valved, loculicidally dehiscent, the valves septiferous in the middle: cells many-seeded. *Seeds* lenticular, or pyriform, very smooth, shining, not strophiolate. *Endl. No.* 5188. *Fenzl, l. c.* 2. *p.* 261.

Small, decumbent suffrutices or annual or perennial herbs, natives of South Africa. Leaves glaucous, filiform, sub-terete, pointless, fleshy, dilated at base into a stipulary, amplexicaul, truncate, unidentate sheath; all either alternate and crowded, or remotely whorled. Umbels simple, on long peduncles, rarely shorter than the leaves. Name not explained: ὑπερ, Τηλις: but why?

* *Suffruticose: stamens* 12 *or more.* (Sp. 1–2.)

1. H. verrucosa (Fenzl, l. c. p. 262.); suffruticulose, with a very short stem, and short branches; leaves *alternate, densely set, or fasciculately crowded;* stipulary sheaths broad; peduncles 4–8 flowered, 4–6 times longer than the leaves; peduncles, pedicels and calyces verrucose above; stamens 11–16, in two rows; seeds pyriform. *Pharnaceum verrucosum, E. & Z. No.* 1826. *P. salsaloides, Burch. Trav.* 1. *p.* 286.

VAR. β. **lævigata**; stem elongate, branching; peduncles and pedicels smooth; sepals warted.

HAB. Stony hills between Sackriver and Kopjesfontein, *Burchell.* At the Gareep, near Verleptram, and at the Gamke River, near Jackalsfontein, *Drege.* Gauritz River, Swell., *E.* & *Z.* Gamke, *Zey.* 615. Macallisberg and Bitterfontein, *Zey.*, *No.* 2842. *β.* Sondag's rivier, near Blauwekraus, Graaf Reynet and Albany, *Drege.* (Herb. T.C.D., Lehm., Sond.)

Root woody. Stem, branches and ramuli whitish. Branches 1–3 inches long; twice longer in *β*, leafy, glabrous. Leaves terete filiform, 4–10 lines long, ½–1 line wide. Stipulary sheath hyaline, with taper-pointed teeth. Peduncles straight, umbelliferous, axillary and terminal, 2–5 inches long. Umbels quite simple; pedicels 4–10 lines long, divergent after flowering; sepals 1½–2½ lines long. Capsule equalling the calyx. Seeds sub-compressed, black.

2. H. spergulacea (E. Mey. in Hb. Drege); suffruticose, stems ascending; lower leaves crowded, linear wedge-shaped; *cauline remotely whorled*, narrow-linear; stipulary sheaths short; peduncles, pedicels and calyces *quite smooth;* stamens 20–30; most of them united at base into 3–5 parcels, a few free; seeds lenticular.

Hab. Stony places near Verleptram, Gariep, *Drege!* (Herb. Sond.)

Habit of *Spergula* or of *Pharnaceum lineare.* A glabrous, glaucous suffrutex, about a foot high. Root whitish. Leaves at the base of stem fleshy, 6–8 lines long, ½–1 line wide. Stems simple or forked, branches filiform, rigid. Leaves 5–10 in a whorl, at the tumid nodes, unequal, 4–8 lines long. Peduncles 1–2½ inches long, sub-umbellate; pedicels 6–12 lines long, very slender. Sepals oval-elliptic, 2–3 lines wide. Capsule as long as calyx. Seeds shining, black. Easily known from the preceding by its long stems, remotely whorled (not crowded) leaves, and absence of tubercles.

*** Annuals or biennials; stamens* 3–5. (Sp. 3–4).

3. H. arenicola (Sond.); annual or biennial, quite glabrous, glaucous; stems prostrate, branched; leaves opposite or tufted, fleshy, linear-terete, obtuse, attenuate at base, radical none; stipulary sheaths wide, with subulato-acuminate teeth; *peduncles axillary* 3–4 *times shorter than the leaf,* 2–3 flowered; *pedicels equalling the peduncle;* sepals very smooth; *stamens* 3; seeds reniform, minutely roughened.

Hab. On moist, sandy places at Greenpoint, Capetown, *Zeyher, No.* 619, *W.H.H.* Jan.–Mar. (Herb. T.C.D., Sond.)

2–3 inches high, leafy. Root sub-simple. Stems numerous, dichotomous, somewhat knee-bent. Leaves erect, ½–1 inch long ½–1 line wide, longer than the internodes. Peduncles erect, pedicels longer than the calyx, bracteate at base. Disc fleshy. Sepals elliptic, sub-petaloid, 1 line long. Stamens on the disc: anthers subglobose. Stigmata 3 sessile. Capsule 3-celled, 3-valved, loculicidal; cells many-seeded. Seeds blackish.

4. H. Bowkeriana (Sond.); biennial, quite glabrous, glaucous; stems prostrate, branching; leaves alternate, fasciculate, fleshy, linear-terete, sub-acute; stipulary sheathes broad, with acuminate teeth; *peduncles, pedicels and calyces sparsely tuberculate* or smooth; peduncles axillary, *once or twice as long as the leaf,* 2–4 flowered; pedicels *much shorter than the peduncle;* stamens 5; seeds 3-angled-pyriform, very smooth.

Hab. Albany, *Miss Bowker* (now *Mrs. Barber).* (Herb., Hook.)

A leafy herb, very like *H. verrucosa*, but differing in the prostrate stems, shorter peduncles, and few stamens. Branches whitish. Leaves ½–1 inch long, ½ line wide. Stipules white. Peduncles 1–1½ inches long; pedicels 2–4 lines long. Sepals 1 line, in fruit 1½ line long, with a wide, white margin. Anthers sub-globose. Stigmata sessile. Capsule globose, as long as the calyx or a little longer. Seeds shining, brown.

XIX. PSAMMOTROPHA. E. & Z.

Calyx 5-parted; sepals oval, with petaloid margins. *Petals* none. *Stamens* 5, alternate with the sepals; anthers globose. *Hypogynous-disc* none. *Style* very short; stigmata 3–5, filiform. *Capsule* 3–5 angled, sub-globose, 3–5 celled, 3–5 valved, loculicidal; valves septiferous in the middle; cells one-seeded. Seeds globulose, granulate, not strophiolate. *Endl. Gen.* 5189. *Fenzl. l. c.* 2. *p.* 263.

Small suffrutices or much-branched perennials; stipulate or exstipulate. Branches bent at the nodes, and verticillately leaved; sometimes inordinately ramulose, quadrifariously imbricated throughout with densely crowded, rigid leaves, so as to be acutely tetragonal. Flowers minute, on lateral or terminal articulate peduncles, in simple umbellules, sessile at the nodes or crowded about the forks. Name ψαμμος, *sand* and τροφος, *a nurse*.

1. **P. quadrangularis** (Fenzl, l. c. p. 264.); fruticulose, flexuously much-branched; leaves sessile, subulate (½–3 lines long) mucronulate, or aristate, rigid, with thick margins, exstipulate, *those of the branches very closely set in four ranks;* of the peduncles whorled, equalling the lateral, sessile umbels. *Pharnaceum quadrangulare, Linn. f. Suppl. p.* 185. *Thunb. Fl. Cap. p.* 275. *E. & Z.! No.* 1832. *Mollugo quadrangularis, DC. Prod* 1. *p.* 393.

VAR. α. **mucronata** (Fenzl); leaves triangular-subulate, white edged, acute and mucronulate.

VAR. β. **subulifolia** (Fenzl); leaves subulate-setaceous, awned, one-coloured.

HAB. Koude Bokkeveld, on the mts. near Verkeerde Valley and elsewhere, *Thunberg*. Stony places near Ezelsfontein and in Onderbokkeveld, 2500–4000f., *Drege*. β Swellendam, *Drege*. Heerelogement, *Ecklon*. Oct.–Dec. (Herb. T.C.D., Sond.)

Aspect of *Andromeda hypnoides;* very smooth, a span long, erect or diffuse; the primary branches leafless, or covered with dead leaves, the ramuli ½–2 inches long, densely leafy. Leaves rigid, in α, ½–2 lines long, ⅓–½ line wide at base; in β, 1–3 lines long, ¼ line wide, pale green or yellow. Peduncles terminal, mostly solitary, very slender, ½–1 inch long, terminated with a whorl of leaves and 3–7 pedicels about an inch long, nodoso-articulate and laterally umbelliferous. Flowers minute, semiverticillate at the nodes of the pedicels; pedicels 1–2 lines long, persistent. Calyx ½ line long, globose. Capsule sharply 3–5 angled, equalling or exceeding the calyx. Seeds brown, not glossy.

2. **P. androsacea** (Fenzl, l. c.); suffruticose, di-tri-chotomous; branches filiform, knee-bent; leaves exstipulate, lanceolate, subattenuate at base, aristato-mucronate, *very densely tufted at the nodes and at the ends of the branches;* peduncles rising from the tuft, one or several, elongate, capillary, bearing sessile umbels at their nodes. *Pharnaceum mucronatum, Thunb. Fl. Cap. p.* 275. *Psam. mucronata, Fenzl, l. c.*

VAR. α, **marginata** (Fenzl,); leaves linear, rigid, keeled; keel and margins thickened, white.

VAR. β, **enervis** (Fenzl); leaves linear or lanceolate, membranous, almost nerveless, of one colour.

HAB. In grassy places. α, Blesbokvlakte, 4000f.; β. on the Katberg, 3500f.; and in stony places, Omsamcaba, 500–1000f. *Drege*. Caffraria, *E. & Z.* Tafelberg, Natal, *Krauss*. Nov.–Jan. (Herb. T.C.D., Sond.)

3–8 inches long, with the habit of *Androsace lactea,* or rather, of *Stylidium bulbiferum.* Stem very short, many-branched; the branches prostrate, repeatedly proliferous, the internodes ½–2 inches long. Leaves densely rosulate at the nodes and apices, 2–5 lines long, ½ line wide, glaucous-green. Peduncles straight, shining, 2–3 inches long, simple or dichotomous, with many lateral umbellules, one over another near the apex. Umbellules opposite a semiverticil of leaves; pedicels 1 line long, equalling the leaves, minutely bracteolate at base. Calyx ½ line long. Capsule 3–5 angled. Seeds rust-coloured, not glossy.

3. P. myriantha (Sond.); biennial or annual (??), *stemless;* radical leaves very densely rosulate, elongate, narrow-linear, awned, one-nerved; scape one or several, thickish, rather longer than the leaves, ending in a whorl of leaves; *peduncles rising from the whorl, 2–3-chotomous, much branched, spreading widely;* ramuli bearing lateral umbelules opposite semi-whorls of leaves. *Zey. No. 616.*

HAB. Macallisberg, *Burke & Zeyher.* (Herb. Hook.! T.C.D., Sond.)

12 inches high, including the panicle. Root sub-simple, perpendicular. Tuft of radical leaves very large, densely many-leaved. Leaves 2–4 inches long, ½ line wide, rigid, with a strong nerve, tipped with an awn 1 line long, glabrous, and smooth. Common peduncle or scape one or several, 4–5 inches long; the middle ones much thicker than the outer; all shining, pale or foxy, ending in a whorl of 8–20 leaves, about ½ inch long. Partial peduncles elongate, much divided, bearing umbels at the forks and nodes. Fl. minute. Pedicels a line long. Calyx globose, sepals sub-petāloid. St. 5. Style 5 fid. Capsule 5 angled, rather exceeding the calyx. Seeds globose, brown, granulated.

4. P. parvifolia (E. & Z. No. 1833.); herbaceous, perennial, *prostrate,* many-stemmed; *stems filiform, much branched;* leaves *small,* oval, elliptical, oblong, or somewhat linear, pointless or mucronulate, with thick margins, whorled, spreading or reflexed; *flowers sessile, densely crowded at the nodes. Fenzl, l. c. p. 266. Pharn. marginatum, Thunb. Cap. p. 275. Moll. marginata, DC. Prod. 1. p. 392.*

HAB. Sandy hills near Witteklipp, *Thunberg.* Zwartkop's River, *E. & Z.! Drege, Zey.* 2501. (Herb. Sond., T.C.D.)

Aspect of *Valantia muralis.* Stems many, from a thick root, elongate, 3–10 inches long, weak, somewhat angular, whitish, oppositely branched. Internodes of stem and branches ½–1 inch long; on the flowering ramuli much shorter. Leaves 1–2 lines long, 6–15 in a whorl, rigid, stellately patent and reflexed, with thickened edges. Stipules scarious, ciliately lacerated, crowded, ½ the length of the leaves. Flowers minute; the glomerules 3–10 flowered. Sepals ½ line long. Capsule 2–4 celled, very small, globose, angled, 2–4 seeded, sometimes abortively one-seeded. Seed globulose, granulated.

XX. COELANTHIUM, E. Mey.

Calyx funnel-shaped or bell-shaped, semi-quinquefid; the segments petaloid, obtuse. *Petals* none. *Stamens* 5, inserted between the calyx lobes, very short; anthers sagittate, erect, longer than the filaments. *Disc* none. *Stigmata* 3, roundish, fleshy. *Capsule* oblong, 3-angled, 3-celled, many-seeded, loculicidally 3-valved; the valves septiferous. Seeds globose, sub-compressed, not strophiolate. *Endl.* 5190. *Fenzl, l. c.* 2. *p.* 276.

Glabrous annuals, with scapelike stems springing from a tuft of radical leaves, 2–3-chotomous above, and breaking up into raceme-like cymes. Radical leaves obovate, oval or lanceolate; cauline filiform, whorled. Stipules fimbriato-lacerate.—Name, *κοιλος, hollow* and *ανθος, a flower.*

1. **C. grandiflorum** (E. Mey. in Hb. Drege); radical leaves *ovate-elliptical* and lanceolate, fleshy, on long petioles; cauline filiform, aristate; stems *simple*, straight, ending in cymes; stipules *crowded, rigid;* segments of the *funnel-shaped* ($2\frac{1}{2}$–3 lines long) calyx obovate, broadly membranaceous, flat; capsule prismatic, as long as the calyx. *Fenzl, l. c.*

HAB. Karoo-plains between Goedman's kraal and Kaus, Kl. Namaqualand, 2000f. *Drege!* (Herb. T.C.D., Sond.)

Root filiform. Rad. leaves 8–12 lines long, $1\frac{1}{2}$–3 lines wide. Stipules shining, 2–3 lines long. Stems numerous, 2–4 inches long, scapelike, with a terminal whorl of leaves, whence the cymes arise. Cauline leaves unequal, 4–6 lines long, filiform, with setaceous stipules. Cymes bifurcate, racemiform, leafless; pedicels 3–4 lines long, flexuous, stipulate at the thickened forks. Seeds brown, reticulated.

C. parviflorum (Fenzl, l. p. 268.); radical leaves *narrow-spathulate* and lanceolate, tapering into a petiole; cauline filiform, pointless or mucronulate; stipules setaceous, *rather loose;* stems erecto-patent, *panicled above*, the branches cymigerous; segments of the *campanulate* (1–2 lines long) calyx ovate, rounded-obtuse, membranaceous at the apex and slightly inflexed margins; capsule subtrigonous, equalling the calyx. *Pharnaceum ? semiquinquefidum, Hook. Ic. Pl. t.* 83. *Zey. No.* 621

HAB. Cape Flats, *Ecklon, Drege! W.H.H., Zeyher.* (Herb., T.C.D., Sond.).

Root long. Stems numerous, terete, 4–6 inches long; the primary branches opposite, the rest dichotomous. Internodes 1–2 inches long. Radical leaves 1–$1\frac{1}{2}$ inch long, 1–2 lines wide; cauline much narrower and shorter. Stipules very thin. Cymes bifurcate, racemiform, leafless; pedicels 1–2 lines long, bracteate at base. Calyx 1–$1\frac{1}{2}$, rarely 2 lines long, subturbinate. Known from the preceding by the dichotomous, panicled stem and patent branches, longer and narrower radical leaves and smaller flowers.

Sub-Order V. POLPODEÆ (Gen. XXI.)

XXI. POLPODA, Presl.

Calyx corolloid, 4-parted; the sepals (snow-white) fringed and lacerate, imbricated at base with 3–4 hard-margined, basally fimbriated bracts. *Petals* none. *Stamens* 4, hypogynous, alternate with the sepals; filaments exserted, anther-cells linear, divaricate at base. *Style* bipartite; the branches filiform, erect, stigmatose. *Capsule* broadly obcordate, 2-celled, compressed at right angles to the dissepiment, loculicidally bivalve; the valves septiferous. *Seeds* solitary, globoso-reniform, granulate, black, opaque. *Endl. No.* 5194. *Fenzl, l. c. p.* 300. *Blepharolepis, Nees.*

A small, diffusely-branched suffrutex; the branches very densely imbricated with minute leaves and sessile flowers, the latter forming a long, terminal, cylindrical, leafy, spike. Leaves minute, linear, alternate, sessile, fleshy, with hard margins, channelled, with recurved points, in the lower part, on each side, bordered with very broad, membranous, triangular-ovate, fringed, stipulary-laminæ. Flowers solitary or 2–3 together, closely invested with leaves, bracts and stipulary fringes densely spicate. Name, πολπωδης, *fringe.*

1. **P. capensis** (Presl., Symbol. 1. tab. 1.); *Blepharolepis Eckloniana, Nees ab. E. in Lindl. Nat. Syst. p.* 442.

HAB. On the hills round Capetown, at both sides of Table Mountain, and at Campsbay, *Ecklon, Sieber. W.H.H. Pappe*, &c. Piquetberg, *Zey. No.* 1450. *Drege*, 8262. Oct. Dec. (Herb. T.C.D., Sond.)

6–18 inches high, more or less branched : with the aspect of a *Lycopodium* or of a *Stoebe* or *Seriphium*. The minute flowers, hidden among the imbricated leaves, easily escape notice.

Sub-Order VI. ADENOGRAMMEÆ (Gen. XXII.)

XXII. ADENOGRAMMA, Reichb.

Calyx 5-parted, ovoid or globose, sepals coloured at the margin and on the inside. *Petals* none. *Stamens* 5, alternating with the sepals, connate at base into a membranous, hypogynous ring. *Anthers* oval, versatile. *Ovary* one-celled, with a single ovule attached to a basal ascending seed-cord. *Style* simple ; stigma capitate. *Utricle* indehiscent, conical, straight, compressed or lenticular, obliquely acuminate, smooth or papillated. *Seed* ovoid, with a membranous coat. *Endl., No.* 5195. *Fenzl, l. c. p.* 274. *Steudelia, Presl.*

Suffrutices or annual herbaceous plants ; slender, dichotomous, diffuse and glabrous. Leaves whorled, those of the perennial branches imbricate, obovate, lanceolate, linear or filiform, very obsoletely stipulate. Flowers small, sessile or on short pedicels, disposed in sessile, axillary umbels.

Sect. 1 : SUFFRUTICOSE ; stem and branches woody ; branchlets densely leafy ; leaves imbricate ; floriferous branches panicled, with whorls of leaves at the nodes. (Sp. 1.).

1. **A. rigida** (Sond.) ; suffruticose ; stem di-tri-chotomous ; branchlets leafy ; leaves crowded, from a dilated, imbricating base elongate-subulate, pungently mucronate, sub-recurved, without stipules ; flowering branches elongate, panicled, whorled with stipulate, lanceolate or subulate leaves at the distant nodes, those of the upper nodes shorter ; stipules setaceous, deciduous ; pedicels few, axillary, equalling the leaves, one-flowered. *Pharnaceum rigidum, Bartl. in Lin.* 7. *p.* 626. *Psammotropha rigida, Fenzl, l. c. p.* 266.

HAB. Sandy flats below Tigerberg, near Riet Valley, *Ecklon, W.H.H.* Dec. (Herb. T.C.D., Sond.)

Stem woody, as thick as a pigeon's quill, prostrate, many times divided ; branches 1-1½ inches long, their bases clothed with persistent leaves ; ramuli inch long, leafy. Leaves filiform-subulate, furrowed above, ½–1 inch long. Flowering branches terete, 6–12 inches long, with swollen nodes, the internodes 3–1½ inches long, branchlets sub-horizontally patent, elongate, the lowest ones ramulose, the uppermost short and very slender. Pedicels about 1 line long, capillary. Flowers ½ line long. Sepals white-edged. Capsule shining, one-seeded.

Sect. 2. DIFFUSE SUFFRUTICES, with long, much divided, filiform branches ; all the leaves whorled, either obovate, elliptical, lanceolate or linear. (Sp. 2–4).

2. **A. sylvatica** (Fenzl, l. c. p. 275.) ; leaves *obovate, oval or oblong*, aristate, *thinly membranous, withering, but long persistent* ; stipules *capillary*, fugacious. *Steudelia sylvatica, E. & Z.! No.* 1809.

HAB. In woods at the River Zonder Ende and near Gnadendhal ; also near Caledon's Baths, *E. & Z.!* Breede river, *Zey!* 2488. Near Gnadendhal, at Dutoit's Kloof, and at Vanstaadensberg, *Drege.* Oct. Jan. (Herb. T.C.D., Sond.)

Aspect of a *Galium.* Root thick, perpendicular, 3–6 inches long. Stems many, 1–3 feet long, filiform, weak, spreading or scandent, dichotomous or somewhat whorled, very smooth; the branchlets capillary. Lower internodes 2–4 inches long, upper shorter. Leaves 3–8 in a whorl, stellately patent, at length reflexed, bright green with a pale awn, 4–6 lines long, 1–3 lines wide, the uppermost narrower and shorter. Stipules setaceous, ½ line long. Flowers axillary, 3–7 together; pedicels as long as the leaves or longer. Calyx globose, ¾–1 line long, with obtuse sepals. Ovary ovoid, sub-compressed; style oblique.

3. A. diffusa (Fenzl, l. c. p. 275.); leaves *oblongo-or linear-lanceolate,* mucronate, *fleshy, rigid, deciduous;* stipules *obsolete* or none. *Pharn. diffusum, Bartl. Linn. 7. p. 625. P. Lichtensteinianum, Roem. and Sch. (excl. syn.) Mollugo Lichtensteiniana, DC. Prod. 1. p. 393. excl. syn. Steudelia diffusa, E. & Z.! No. 1811. Herb. Un. It. 627. ex parte.*

HAB. Sandy and stony places on the sides of Table Mountain, towards the North and East, *E. & Z.! Krauss, Drege, W.H.H.* Swellendam, *Mundt.* Klipfontein, *Zeyher. No.* 618, 2490. Aug.–Nov. (Herb. T.C.D., Sond.)

Very like the preceding, but the stems are slender and somewhat rigid, ascending at the points, and the leaves are smaller, fleshy and quickly deciduous. Stems 10–18 inches long. Leaves 3–8 in a whorl, 2–2½ lines long ½–¾ line wide, the upper ones smaller. Stipules capillary, minute, only obvious among the uppermost leaves. Flowers axillary, 2–7 together, pedicellate; pedicels longer or shorter than the leaves. Capsules sub-globose, shortly mucronulate, obsoletely rugulose, shining, chesnut-coloured.

4. A. lampocarpa (E. Mey. in Hb. Drege); leaves *very narrow-linear,* sub-carinate, hook-pointed, *deciduous;* stipules scarcely any. *Fenzl, l. c. p. 276. Steudelia capillaris, E. & Z.! No.* 1810.

HAB. Among shrubs on the mountains near Dutoitskloof, 1500–2000f. *Drege.* Mountain sides near Riv. Zonderende, *E. & Z.!* (Herb. T.C.D., Sond.).

A diffuse, much-branched suffrutex, closely related to *A. diffusa,* but differing in the slender stems, capillary branches, less fleshy linear leaves, 2–3 lines long, ⅓–½ line wide, prominently nerved below and hooked at the point. Flowers on pedicels longer or shorter than the leaves. Sepals ½–¾ line long, petaloid. Capsule shining.

Sect. 3. ANNUALS; simple or many-stemmed, diffusely branched, many-flowered; leaves linear-subulate, or subulate-setaceous. (Sp. 5–7).

5. A. physocalyx (Fenzl, l. c. p. 276.); leaves subulate-setaceous; flowers sub-sessile; calyx *inflated and bladdery,* angled by the spreading margins of the ovate sepals; capsule ovate, cuspidate, two-edged, very smooth, enclosed in the persistent calyx.

HAB. On the flats between the Breede and Zonderende Rivers, *Drege!* Sandy places at Breederiver's pont, *Zeyher, No.* 2500. Oct. (Herb. T.C.D., Sond.).

A weak, glaucous herb, 1–4 inches long, very smooth, with 3–5 erect, filiform stems; the lateral ones diffuse, dichotomous. Radical leaves crowded, cauline whorled, 3–4 lines long, mucronulate, incurved, the uppermost smaller. Stipules obsolete, capillary. Flowers axillary on the branchlets, on very short pedicels, at length, with the pedicels, deciduous. Calyx in flower greenish, 1 line; in fruit 2 lines long, brownish, the outer sepals broadly ovate, the inner 2 boat-shaped, narrower, clasping the shining black capsule.

6. A. galioides (Fenzl, l. c. p. 277); leaves linear-subulate, or filiform;

pedicels at length as long as the flowers or longer; *calyx close-pressed on the ovoid or lenticular, obliquely acuminate, punctate capsule, persistent. Pharn. Mollugo, Lin. Sp. p.* 389. *Berg. Fl. Cap. p.* 79. *Pluck. t.* 331. *f.* 4. *P. glomeratum, L. fil. Sp. p.* 185. *Thunb. Fl. Cap. p.* 273. *Moll. glomerata, DC. Prod.* 1. *p.* 392.

VAR. α. **planifolia** (Fenzl); leaves linear-subulate, flat above; capsule very minutely punctate, the acumination twice as short as the lower, swollen portion. *St. galioides, Presl. Symb. t.* 2. *E. & Z. No.* 1808. *Un. It.* 628.

VAR. β. **teretifolia** (Fenzl); leaves fleshy, linear-filiform, sub-terete; capsule coarsely punctate, the acumination nearly equalling the lower swollen portion, or half as long.

HAB. Cape Flats, and on mountain sides round Capetown, common. Zwartland; and near the Zwartkop's River and Assagaiskloof, *Thunberg. E. & Z.! &c. Zey.! No.* 620 *and* 2489. July–Nov. (Herb. Thunb., T.C.D., Sond.).

Glaucous green, 2 inches to 1½ feet long, very smooth, or rarely minutely downy below, with the aspect of a *Galium.* Stems ascending or diffuse, branched from the base; branches alternate, opposite or whorled, filiform; branchlets spreading, bearing the flowers. Lower internodes 2–3 inches long, upper shorter. Leaves 3–6 lines long, ½ line wide, 3–12 in a whorl; the uppermost very small, at length reflexed. Flowers copious, crowded in the axils, umbellate; pedicels unequal, ½–2 lines long. Sepals oblong, obtuse, white-edged, ½ line long. Capsule brown.

7. **A. Mollugo** (Reichb. Icon. Bot. Exot. 2. 3. t. 109); leaves linear-subulate or filiform; flowers of the umbellules pedicellate, mixed with a few sessile ones; *sepals deciduous in fruit;* capsule *globose at base,* depressed, *produced into a long, straightish, two-edged beak,* once and a-half longer, black, shining, *granulate* on the disc, and margined with raised, golden-yellow papillæ. *Fenzl, l. c. p.* 278. *E. & Z.! No.* 1812. *Sieb. Fl. Cap. No.* 261.

HAB. Sandy places in the Cape districts, as at Green Point and Muysenberg, *E. & Z.!* River Zonderende, Swell., *Sieber, E. & Z.* Zwartland, *Drege.!* (Herb. Sond.).

A slender plant; in habit, ramification and leaves closely resembling the last species, but smaller (2–4 inches long) more rigid, with fewer leaves; the upper ones smaller. Leaves either pointless or mucronate. Flowers in size, disposition and colour as in the preceding. Sepals oblong, obtuse, white-edged. Capsule black and shining, nearly 1 line long.

ORDER XIX. PHYTOLACCEÆ. R. Br.

(By W. H. HARVEY.)

Phytolacceæ, R. Br., Endl. Gen., No. ccviii. Phytolaccaceæ, Lindl. Veg. Kingd., No. cxciii. DC. Prod. vol. 13. part ii. p. 2.)

Flowers regular, often *apetalous. Calyx* 4-5-parted, persistent, its segments generally with membranous margins, imbricated. *Petals* 4-5, or none, fugacious. *Stamens* sub-hypogynous, as many as the calyx lobes and alternate with them, or irregularly more numerous; filaments free or connate at base; anthers erect or incumbent, bilocular. *Ovary* of two or several (rarely of one) apocarpous or connate, one-ovuled carpels; styles as many as the carpels. *Fruit* dry or baccate, breaking up into single-seeded nuts or utricules. *Seed* (as in Caryophylleæ).

Herbs or shrubs, rarely trees, with alternate (v. rarely opposite) simple, very entire leaves, with or without stipules. Flowers in spikes, racemes or cymes, of small size.

Natives of the tropics and the warmer temperate zone, more frequent in America than in the Old World. About 80 species are known. The affinities are obviously with the Caryophylloid group of Orders, from which the present is chiefly known by the structure of the ovary. *Gieseckia,* in habit, much resembles some Mollugineæ. *Limeum* and *Semonvillea* are African types, but found in North as well as South-Africa.

TABLE OF THE SOUTH-AFRICAN GENERA.

Tribe 1. LIMEÆ. *Fruit* of two, plano-convex, dry, hard carpels.

I. **Semonvillea.**—*Carpels* orbicular, flattened, with a marginal wing.
II. **Limeum.**—*Carpels* hemispherical, wingless.

Tribe 2. GIESEKIEÆ. *Fruit* of several, separate or confluent carpels.

III. **Giesekia.**—*Carpels* 3-4-5, warted or crested, hard and dry.
IV. **Phytolacca.**—*Carpels* 5-12, fleshy or juicy, separate or united into a many-lobed fruit.

I. SEMONVILLEA. Gay.

Flowers perfect. *Sepals* 5, separate, herbaceous, with membranous edges. *Petals* 5 or none, clawed. *Stamens* 5–7, hypogynous, the dilated filaments slightly connected at base. *Ovary* compressed, formed of two plano-convex carpels, united by their flat side; styles 2, filiform; stigmas sub-capitate. *Fruit* orbicular, dry, formed of two, separable, one-seeded, indehiscent, plano-convex cocci, winged round the margin. *Endl. Gen.* 5259. *DC. Prod.* 13. 2. *p.* 19.

Slender, branching, glabrous, annuals, natives of Senegal and of S. Africa. The leaves are very narrow, slightly fleshy without obvious veins, laxly scattered on the branches. Bracts 3 under each flower, and one under each short pedicel. Flowers minute. Named in honour of *M. Semonville,* a French Botanist.

1. S. fenestrata (Fenzl, Nov. Stirp. Dec. Mus. Vind. 5. p. 42); stem much-branched, erect or diffuse, glabrous; leaves very long, narrow-linear, mucronate, narrowed at base; carpels rather longer than the calyx, the wings entire and sub-continuous below, transparent between the nerves (as if glazed). *Moq. in DC. Prod. l. c. p.* 20. *Hook. Ic. t.* 587. *Ditroche furcata, E. Mey.!*

HAB. In little Namaqualand, *Drege!* Vaal and Caledon Rivers, *Burke and Zeyher!* (Herb. T.C.D., Hook., Sond.)

Root small and fibrous. Stems 12–18 inches high; the flowering branches irregularly forked. Leaves 2 inches long or more, not ½ line in width, distant, somewhat glaucous. Flowers in a very diffuse, branching cyme, subsessile. Sepals ovate, oblong, acute, 3-nerved, green in the middle, with a wide, white border. Petals yellowish, broadly elliptic-ovate. St. 7, with broad, ciliated filaments. Carpels tuberculated on the back, the tubercles radiating from the centre, and continuous with the dark-coloured nerves of the membranous wing.

II. LIMEUM. Linn.

Flowers perfect. Sepals 5, connate at base, herbaceous, with membranous edges. *Petals* 3-5, or none, clawed. *Stamens* 7 (rarely 5-8-10), hypogynous, united at base or free. *Ovary* sub-globose, formed of two hemispherical carpels, united by their flat side; styles 2, terminal, slender. *Fruit* dry, formed of two separable, one-seeded, indehiscent, hemispherical, wingless, dorsally-pitted or echinulate *cocci*. *Seed* vertical, with a peripheric embryo. *Endl. Gen.* 5258. *Moq. DC. Prod.* 13. 2. *p.* 19.

Small, herbaceous or woody, prostrate or erect, many-stemmed, perennial or annual plants, natives of S. Africa, and of the tropics of N. Africa. Leaves simple and very entire, narrow, somewhat fleshy, often glaucous, glabrous or glandularly hairy and viscid. Inflorescence cymoid, terminal or opposite the leaves. Flowers small, greenish-white or white. Bracts 3 under each flower, and one under the pedicel. *Limeum* was the ancient name of a poisonous plant, and derived from *λοιμος*, *poison*. These small weeds are acrid poisons.

Sub-genus 1. LIMEASTRUM. Flowers with *petals*. Inflorescence terminal, or sometimes lateral, loosely cymoid, pedunculated. (Sp. 1–6.)

1. L. Africanum (Burm. Prod. Fl. Cap. p. 11); stems herbaceous, prostrate, sub-simple, angular, glabrous, green; leaves petiolate, tapering and slightly clasping at base, *oblong or lanceolate-linear*, acute or obtuse, or mucronate, glabrous; cymes terminal, pedunculate, *loosely corymbose*, much branched; sepals ovate, acute or acuminate, *nerveless*; fruits pitted and furrowed, much shorter than the sepals. *Moq. in DC. Prod.* 13. 2. *p.* 21. *E. & Z.! No.* 1836.; *and L. litorale, E. & Z.! No.* 1837. *Moq. l. c.*

HAB. Common round Capetown, on the Lion Mt., &c. Tigerberg, and on the shore near Saldanha Bay, Wupperthal, and at Kaus, 3–4000 ft., Namaqualand, *Drege!* (Herb. T.C.D., Hook., Sond.).

Root thick, deeply descending. Stems many from the crown, 12–14 inches long, spreading in a circle, the ends upturned and flowering. Leaves $\frac{1}{2}$–1$\frac{1}{2}$ inch long. slightly fleshy, very variable in form. Sepals with broad, silvery-white margins, Cymes 2–2$\frac{1}{2}$ inches across, flat-topped.

2. L. canescens (E. Mey. MS.); "stems herbaceous, procumbent, branching, angular, glabrous, green; leaves petiolate, *oblong-linear*, *acute*, muticous or mucronulate, glabrous, *covered with a glaucous rime, the lower leaves obovate;* cymes terminal, pedunculate, corymbose, many-flowered; sepals ovate, acuminate, nerveless; fruits . . . . . ?" *Moq. in DC. Prod. l. c. p.* 21.

HAB. S. Africa, *Drege* 6229. Lange Valley, *Zeyher* 631, *ex parte*. (Herb. Sond.).

"Greenish-canescent, drying darker. Branches 1–1$\frac{1}{2}$ foot long, slender. Leaves 4–9 lines long, including the petiole of 1–2 lines, 1$\frac{1}{2}$–2 lines wide, tapering to each end. Cymes 5–7 lines wide, with crowded flowers. Sepals with white borders." *Moq. l. c.* I translate Moquin's diagnosis for this species, which I only know by a specimen in Herb. Sond., from Zeyher; by whom it was mixed with *L. capense*.

3. L. capense ((Thunb. Prod. p. 68); stems woody, erect or diffuse, angular, glabrous, dull green; leaves petiolate, tapering and somewhat clasping at base, *elliptical or obovate, obtuse, mucronulate*, glabrous; cymes terminal or lateral, on short peduncles, *densely* many-flowered, corymbose; sepals broadly oval, *obtuse, mucronulate, one-nerved;* fruits pitted and furrowed, as long as the sepals. *Thunb. Fl. Cap. p.* 342. *E. & Z! No.* 1838. *Moq. in DC. Prod. l. c. p.* 21. *Zey. No.* 2505, 630, 631, *ex parte. L. telephioides, E. Mey.! Moq. l. c.*

VAR. *β.* **intermedium**; leaves narrow-oblong, or oval, subacute;

cymes loosely corymbose; fruits as long as the obtuse sepals. *L. telephioides, Moq. in Herb. Hook. non E. Mey.!*

HAB. Dry spots. Adow; and near Zondag and Zwartkops Rivers, *E. & Z.!* Somerset, *Dr. Atherstone!* Between Eenkoker and Bitterfontein, Hartveld, *Zeyher!* (Herb. Hook., T.C.D., Sond., Benth.).

Root woody. Stems several, 3–6 inches high, sparingly branched. Leaves ½ inch long, 2–4 lines wide, mostly broadly obovate and very obtuse, with a minute, reflexed point. Flowers smaller than in *L. Africanum*, the sepals blunter and the cymes less spreading. Var. *β.* is nearly intermediate between *L. capense* and *L. Africanum*, having the calyx of the former, and the foliage and loose cymes of the latter species. By Moquin it is referred to *L. telephioides*, E. Mey.; but Drege's original specimens have the broadly obovate leaves and dense cymes of the common *L. capense*.

4. **L. Æthiopicum** (Burm. Prod. p. 11); stems diffuse, woody, angular and striate, minutely scaberulous or glabrous, pale or greenish; leaves on short clasping petioles, *linear or linear-lanceolate*, obtuse or acute, often mucronulate, glabrous; cymes terminal and lateral, on short peduncles, or sub-sessile, dense, capitato-corymbose, many-flowered; sepals *broadly oval, obtuse*, mucronulate, one-nerved; fruit as long as the sepals, pitted and furrowed. *Thunb. Cap. p.* 343. *E. & Z. No.* 1840. *Moq. l. c. p.* 22. *L. fluviale, E. & Z.! No.* 1839. *Moq. l. c. Zey.* 627, 628, 629, 2503, 2506.

HAB.. Sand hills and river banks in the eastern districts, Gauritz River, the Langekloof, Graaf-Reynet, and near the Zwartkops R., *E. & Z.!* Fish River and Geelbeck River; and Bitterfontein, *Burke and Zeyher*. (Herb. T.C.D., Hook., Sond.).

Nearly related to *L. capense*, from which it chiefly differs in its more woody stems, much narrower and more linear leaves, more prostrate growth and pale bark. *L. fluviale*, E. & Z.! is scarcely separable from the common form; the characters assigned to it are very variable.

5. **L. pauciflorum** (Moq. in DC. Prod. 13. 2. p. 23); perennial; stems woody, prostrate or decumbent, roundish, *glandularly pubescent and viscid;* leaves petiolate, broadly obovate, obtuse, thick and leathery, undulate, viscoso-puberulous; cymes shortly pedunculate, lateral, few-flowered, sub-simple; sepals herbaceous, *broadly lanceolate, acute;* fruits shorter than the calyx, deeply pitted and furrowed. *Moq.! in Herb. Hook. and l. c.*

HAB. Cape of Good Hope. (Herb. Hook.).

A very depressed, half woody, much branched, rigid perennial, minutely downy and viscid in every part. Flowers green, the sepals with scarcely any border. Cyme 3–6-flowered.

6. **L. viscosum** (Fenzl, Nov. Stirp. l. c.); *annual;* stems herbaceous, prostrate, widely spreading and branching, somewhat angular, *glandularly pubescent and viscid;* leaves on longish petioles, obovate or oblong, obtuse, retuse or mucronulate, pubescent; cymes on short, lateral peduncles, densely sub-corymbose, several-flowered; sepals *broadly elliptical,* obtuse, herbaceous; fruits as long as the calyx, pitted and ridged. *Moq. in DC. Prod. l. c. p.* 23.

HAB. Port Natal, *Krauss*, 135; *Gueinzius*. Springbokkeel, *Zeyher* 632. (Herb. Hook., Sond., T.C.D.).

A prostrate, much-branched annual, more or less pubescent and glandular. Leaves variable in form, sometimes almost obcordate, sometimes linear-oblong, ½–1½ inches long, 3–4 lines wide. Cymes 10–20-flowered; calyces viscid, green, with a very narrow dull-white border.

Sub-genus 2. DICARPÆA. *Petals* none. Inflorescence axillary or opposite the leaves, sessile, glomerulate. (Sp. 7–8.)

7. L. glomeratum (E. & Z.! En. No. 1841); stems diffuse or procumbent, divaricately branched, herbaceous, glandularly pubescent and viscid; leaves on longish petioles, elliptic-oblong or broadly linear, tapering at base, very obtuse, the petiole, mid-rib and margin minutely glandularly pubescent; cymes lateral, subsessile, dense, globose, 3–12-flowered; sepals ovate, sub-acute, herbaceous, nerveless; fruits as long as the sepals, pitted. *Moq. in DC. Prod. l. c. p.* 24; *also L. Meyeri, Fenzl. Moq. l. c. p.* 24. *Zey. No.* 626.

HAB. Siloh, near Klipplaat River, Caffraria, *E. & Z.!* Cape, *Drege!* Caledon River, *Burke and Zey.!* (Herb. Hook., T.C.D., Sond.).

Widely spreading and branching, with the aspect of *L. viscosum*, but the leaves are proportionably longer and less obovate, and the cymes nearly or quite sessile. Petals *generally* wanting. Moquin (*in DC. Prod, l. c.*) retains *L. Meyeri*, Fenzl, founded on the *L. glomeratum*, Drege, as distinct from the original plant of Eck. & Zey.; calling the first "*herbaceous*," the latter "*suffruticose*." I have examined and compared *Ecklon's* and *Drege's* original specimens in Herb. Sond.; and *Burke and Zeyher's* in Herb. Hook., and T.C.D., but failed to find any tangible character to separate them. All are herbaceous, and very probably *annual*, judging by the roots of such specimens as have them.

8. L. linifolium (Fenzl, Monogr. p. 342); *annual, glabrous;* stems diffuse, branching, angular and striate; leaves sub-sessile, narrow-linear, elongate, tapering at base, obtuse or sub-acute, mucronulate; cymes opposite the leaves, sessile, dense, sub-globose, several-flowered; sepals ovate, acuminate, nerveless, *albo-marginate;* fruits as long as the sepals, *bristling with sharp tubercles. Moq. in DC. Prod.* 13. 2. *p.* 24. *Dicarpæa linifolia, Presl. Drege,* 2631.

HAB. S. Africa, *Drege!* Caledon River, *Burke!* (Herb. Hook., Benth., Sond.).

A very distinct species, with very narrow leaves, an inch or more in length, and from ½ line to 1 line in breadth. The sepals have a wide, white border and the fruits differ from those of the other species.

III. GIESEKIA. Linn.

Flowers bisexual. *Sepals* 5, herbaceous, with membranous edges, sometimes coloured. *Petals* none. *Stamens* 5–15, hypogynous, separate, alternating singly or in parcels of two or three with the sepals; filaments expanded at the base, subulate. *Carpels* 5 (sometimes 3-4), sessile on a small torus, separate; ovules solitary, erect; styles 3-5, continuous with the inner angle of the carpel. *Fruit* lodged in the persistent calyx, consisting of 3-5, one-seeded, warted or crested, dry cocci; embryo peripheric. *Endl. Gen.* 5261. *Moq. in DC. Prod.* 13. 2. *p.* 26.

Small, annual, or rarely perennial weeds, growing in sandy soil and near river banks. Stems slender, spreading, forked; leaves opposite or alternate, entire, somewhat fleshy, generally pale beneath and dotted with hard, immersed points. Flowers minute, green or reddish-purple, in simple or compound cymes. Fruits dry and hard, nut-like. Named in honour of P. D. Gieseke, a German botanist, who published figures of plants in 1777.

1. G. pharnaceoides (Linn.); annual; stems diffuse or prostrate, striate, branched; leaves on short, membrane-edged petioles, elliptic oblong or lanceolate, obtuse, pale beneath; cymes sessile or shortly pedunculate, rather dense, many-flowered; flowers nearly as long as the pedicels; sepals concave, sub-acute, three-nerved; *stamens five*. *Moq. in DC. l. c. p.* 27.

HAB. Near the Caledon, Crocodile, and Vaal Rivers, *Drege!* *Burke and Zeyher*. (Herb. T.C.D., Hook., Sond.).

Root slender, with few fibres. Stems several, 10–12 inches long, spreading over the soil. Leaves variable in shape; in the Cape specimens 1 inch long and ¼ inch wide, tapering at base, and ending in a blunt point. The sepals are boat-shaped, pale green, with three dark-green or reddish veins. The cymes are not always sessile; but sessile and pedunculate occur on the same root, and I find it impossible, so far as S. African forms go, to separate "*G. linearifolia*" of authors from the common variety. In Herb. Hooker the sheet marked "*G. linearifolia*," by Moquin himself, bears two plants; one, *Limeum glomeratum*, from Muddy River, the other a *Giesekia*, with *sessile* glomerules, from Vaal River, *Zey.!* 624, *ex parte*.

2. G. pentadecandra (E. Mey. MS.); annual; stems diffuse, forked, angular and dotted; leaves on short channelled petioles, obovate, elliptic-oblong or linear, tapering at base, obtuse, pale below; cymes pedunculate, somewhat umbellate, lax, few-flowered; flowers much shorter than the pedicels, purplish; sepals concave, three-nerved; *stamens fifteen*. *Moq. in DC. Prod. l. c p.* 28.

HAB. South Africa, *Drege!* *Zeyher!* Natal, *Miss Owen*. (Herb. Hook., T.C.D.)

A small annual; branches 6–8 inches long. Leaves very variable in form and size. The cymes are more lax than in *G. pharnaceoides*, from which this species is easily known by its numerous stamens.

3. G. Miltus (Fenzl,); "*perennial, shrubby;* leaves subsessile, oblong, obtuse, *very red on both sides;* cymes subsessile or pedunculate, umbellate, 5-15-flowered; flowers shorter than the pedicels, very red." *Moq. in DC. Prod.* 13. 2. *p.* 28. *Miltus Africana, Lour.*

HAB. Dry places near the Orange River, *Drege*.

"Stems 4 feet long, prostrate, slender, glabrous. Leaves crowded, minute. Stamens 12. The whole plant strongly tinged with red." *Moq. l. c.* I am unacquainted with this remarkable plant.

IV. PHYTOLACCA, Tourn.

Flowers mostly bisexual. *Sepals* 5, petaloid, or herbaceous, with membranous edges. *Petals* none. *Stamens* 5-30, hypogynous, separate, on a fleshy disc; the five outer ones alternating with the sepals. *Carpels* 5-12, nearly separate, or cohering by the ventral sutures; styles as many as the carpels, continuous with the inner angles; stigmata decurrent on the face of the style. *Ovules* solitary, basal. *Fruit*, fleshy or juicy; embryo peripheric. *Endl. Gen.* 5262. *DC. Prod.* 13. 2. *p.* 31. *Pircunia, Moq. in DC. Prod. l. c. p.* 29.

Herbaceous, shrubby or arborescent plants, common to the warmer zones of both hemispheres. Stems erect or climbing. Leaves alternate, petiolate, penninerved, entire, with a narrow, discoloured margin. Flowers white or greenish, small, in axillary or terminal racemes or spikes. The roots of some are large and violently purgative, and the foliage is acrid and poisonous. Yet the young shoots of *P. decandra* (Poke or Pocan) are eaten as asparagus in N. America. A tincture of the berries is recommended in cases of chronic rheumatism. Named from φυτον, a *plant*, and "lacca," the colour called *lake;* because the berries contain a red dye. The S. African species belong to Moquin's genus *Pircunia.*

1. **P. stricta** (Hoffm. in Com. Goett. 12. p. 27. t. 3); stem shrubby at base, *ascending or erect,* furrowed, glabrous; leaves *lanceolate,* obtuse; mucronate, narrowed at base, and decurrent into a *short* petiole, roughish; racemes pedunculate, the rachis furrowed; glabrous or rough; sepals herbaceous, white-edged, spreading; stamens 7-8; carpels 6-7. *P. heptandra, Retz. Pircunia stricta, Moq. in DC. l. c. p.* 30.

HAB. Between the Fish River and Fort Beaufort, *Drege!* Orange River, *Drege! Burke!* Caledon River, *Burke and Zeyher!* Caffraria, *Mrs. Barber!* (Herb. Hook., T.C.D.).

Stems 1–2 feet high, sub-erect. Leaves varying much in comparative length and breadth, 3–4 inches long and ½ inch wide, or 2 inches long and 1 inch wide, sometimes ovato-lanceolate or oval. Peduncle long or short: the pedicels equalling the greenish flowers. Fruit blackish-purple, succulent; the calyx not reflexed.

2. **P. Abyssinica** (Hoffm. in Com. Goett. 12. p. 28. f. 2); stem shrubby, *climbing,* terete; leaves on long petioles, ovate or ovate-elliptical, *obtuse at base,* acute and recurvo-mucronate at the apex; racemes much longer than the leaves, the rachis *pubescent;* pedicels longer than the flowers; *calyx reflexed;* stamens 5-12; carpels 5-8. *P. dodecandra, Herit. Strip. t.* 69. *Pircunia Abyssinica, Moq. in DC. l. c.*

HAB. Between Omsamculo and Omcomas, *Drege!* Near Port Natal, *Mr. Sanderson.* (Herb. T.C.D., Hook.).

Stem 10–20 feet high, climbing and scrambling over rocks. Leaves on long and slender petioles, 2 3 inches long, 1½–2 inches broad, varying from ovate to nearly elliptical, but always acute. Peduncles short or obsolete, raceme several inches long, densely flowered. Calyces strongly reflexed soon after the opening of the flower. A native of Abyssinia, Madagascar and the Sandwich Islands.

ORDER XX. MALVACEÆ. Juss.

(By W. H. HARVEY.)

(Malvaceæ, Juss. Gen. 271. D.C. Prod. 1. p. 429 Endl. Gen. No. ccix. Lindl. Veg. Kingd. No. cxxx.).

Flowers regular. *Calyx* 5, rarely 3–4 cleft, with valvate æstivation, usually furnished at base with involucral bracts. *Petals* 5, with twisted æstivation, diliquescent, usually attached to the base of the staminal column. *Stamens* indefinite, united into a tube, whose dilated base covers over the ovary; anthers reniform, terminal, one-celled. *Ovary* of 5 or many *carpels,* whorled round a common axis, and either separable or united into a plurilocular capsule; ovules one or several, axile; styles as many as the carpels. *Fruit* dry or fleshy, dehiscent or indehiscent; seeds with little or no albumen, with a curved embryo, and leafy, plaited cotyledons.

Trees, shrubs or *herbs,* very generally with stellate pubescence. *Leaves* alternate, simple, palmately nerved and often palmately lobed or parted, stipulate. *Flowers* mostly large and showy, variously disposed.

A considerable tropical and sub-tropical Order, with outlying species in the temperate zones. The tropical species are generally ligneous, and many of them even arborescent; those of the temperate zones are herbaceous or suffruticose. Upwards of 1000 species are known. The properties of the Order are mucilaginous and innocuous, and several are in local use in affections of the throat, &c. Others, as the *Gombo* of the West Indies, yield fruits which are an ingredient in soups, or boiled as a vegetable. The inner bark of most species is exceedingly tough and strong, and an excellent material for making ropes, or for converting into strong packing-paper. The *Paritium tiliaceum* of Natal might be profitably employed for these purposes.

TABLE OF THE SOUTH AFRICAN GENERA.

Tribe 1. MALVEÆ. *Carpels* separable, disposed in a ring round a central axis. *Stigmata* as many as the carpels.

* *Styles filiform, stigmatose along the upper side. Calyx with an involucel.*

I. **Althæa.**—Involucel 6–9-leaved.
II. **Malva.**—Involucel of 3 leaves.

** *Stigmata terminal, capitellate. Calyx with an involucel.*

III. **Malvastrum.**—Involucel 3-leaved. Ovules solitary.
IV. **Sphæralcea.**—Involucel 3-leaved. Ovules 2-3 in each carpel.
V. **Sphæroma.**—Involucel trifid. Ovules 2–3 in each carpel.

*** *Stigmata terminal, capitellate. Calyx without an involucel.*

VI. **Sida.**—Carpels one-seeded.
VII. **Abutilon.**—Carpels 3–12 seeded.

Tribe 2. URENEÆ. *Carpels* 5, separable. *Stigmata* capitellate, twice as many as the carpels (10).

VIII. **Pavonia.**—Involucel 5-20 leaved.

Tribe 3. HIBISCEÆ. *Carpels* united into a many-celled capsule. *Stigmata* capitate, as many as the carpels. Staminal column naked and 5-toothed at the apex, bearing stamens along its external surface.

IX. **Hibiscus.**—Involucel many-leaved.
X. **Paritium.**—Involucel cup-shaped, many-toothed or cleft.

I. ALTHÆA. Linn.

Involucel monophyllous, 6–9 cleft. *Staminal column* bearing anthers at the multifid summit. *Ovary* of many carpels, whorled round a columnar torus; *styles* as many as the carpels, stigmatose along the inner face. *Fruit* of many reniform, dry, one-seeded, indehiscent, separable carpels. *Endl. Gen.* 5270. *DC. Prod.* 1. *p.* 436.

Annual or perennial herbs, natives of the old world, and chiefly of the northern hemisphere. Leaves lobed or deeply parted. Flowers axillary or in terminal racemes: purple, or white. The common "*Marsh Mallow*" of England is the type of this genus. It is named from *αλθω, to cure;* in allusion to the emollient properties of these plants.

1. **A. Ludwigii** (Linn.); stems hispid; leaves *on long hispid petioles, nearly glabrous,* deeply 5-lobed, the lobes cuneate, crenato-dentate; stipules deltoid, obtuse; pedicels axillary, single-flowered; *involucel* 8-9 *cleft;* calyx very hairy. *DC. Prod.* 1. *p.* 437. *Cav. Ic. t.* 423. *A. gariepensis, E. Mey.!*

HAB, Between Verleptram and the mouth of Orange River, *Drege!* (Herb. T.C.D., Hook.).

A diffuse, many-stemmed annual, more or less hispid. Flower stalks often clustered, much shorter than the petiole; flowers white. Common to the South of Europe, North Africa and Sub-tropical Asia.

2. A. Burchellii (DC. Prod. I. p. 438); "stem erect, hairy; leaves *cordate*, somewhat 5-lobed, coarsely toothed, *velvetty;* pedicels axillary, one-flowered, hairy, scarcely shorter than the petiole; *involucel 5-cleft.*" *DC. Urena pilosa, Burch. Cat.* 2557.

HAB. South Africa, *Burchell.* (Unknown to us.)

The plant distributed by *E. & Z.*, under this name is an undoubted *Pavonia (P. mollis, E. Mey.)*

II. MALVA. Linn.

Involucel 3-leaved, persistent. *Staminal-column* bearing anthers at the multifid summit. *Ovary* of many carpels, whorled round a central torus; styles as many as the carpels, stigmatose along their inner face. *Fruit* of many dry, one-seeded, hard-shelled, pointless, indehiscent carpels, separating at maturity from the axis; seed filling up the cavity. *Endl. Gen.* 5271. *DC. Prod.* I. *p.* 430. *Gray, Gen. Vol.* 2. *p.* 49. *t.* 116.

Herbs, natives of the temperate parts of the old world, with roundish or lobed, palmate-nerved leaves, and axillary, purple or rosy flowers. Several are weeds in cultivated ground, "mallows by the hedges," and as such are dispersed over the globe. All have emollient and mucilaginous properties. The name is an old one derived from μαλακη, *soft.*

1. M. parviflora (Linn.); procumbent or prostrate; leaves on long petioles, reniform, obtusely 5-7 lobed, crenate; flowers clustered in the axils, subsessile, the petals scarcely longer than the calyx; carpels 10-12, sub-glabrous, cancellated or netted, the margins raised, sharp, and denticulate. *DC. Prod.* I. *p.* 433. *M. rotundifolia, E. & Z.! (non L.); M. microcarpa, E. & Z.!* 298, *and M. flexuosa, E. & Z.!* 299. *M. pusilla, E. Bot. t.* 241.

HAB. Roadsides and waste places; a weed, introduced from Europe. (Herb. Hook., Sond.).

Flowers pale flesh colour, small. All the Cape specimens I have seen have the transversely ribbed and furrowed carpels and sharp ridges, characteristic of *M. parviflora.* The pubescence and size of leaves vary much.

III. MALVASTRUM. A. Gray.

Involucel 3-leaved (sometimes wanting). *Styles* as many as the carpels; *the stigmata terminal capitellate.* Other characters the same as in MALVA. *A. Gray, Gen.* 2. *p.* 59. *t.* 121, 122.

Herbs or small shrubs, with alternate, stipulate, generally lobed leaves and axillary or racemose flowers, which in the Cape species are purple, rose-red, or red and white, The genus comprises many American species with orange or flame-coloured flowers, natives chiefly of the warmer temperate zones, north and south. It is known from *Malva* by the capitellate stigmas, and frequently by the sub-dehiscent carpels. All the Cape species were formerly referred to *Malva.* They are difficult to characterize, and I fear that some of those here adopted may not prove permanently distinct; and perhaps, in some cases, I have mistaken the plants intended by previous authors. The materials in the Herbaria at my command are scanty and unsatisfactory. The generic name is an alteration from *Malva.*

Group 1. FRUTICOSA. *Virgate Shrubs, evidently ligneous.* (Sp. 1-9.)

1. M. fragrans (Gray and Harv.); frutescent, virgate; the branches, petioles and leaves *minutely glandularly pubescent* and thinly sprinkled with long, patent hairs; leaves *cordate at base, 5-7-lobed*, the lobes short, obtuse, *sub-equal*, or the middle one longest, crenate; stipules ovate, small; peduncles one-flowered, axillary, longer than the petiole; involucral leaves lanceolate, shorter than the acuminate, villoso-ciliate calyx lobes; carpels glabrous, corrugated. *Malva fragrans, Jacq. Hort. Vind.* 3. *t.* 33. *DC. Prod.* 1. *p.* 434. *(non E. & Z. nec Bot. Reg.) Cav. Diss. t.* 23. *f.* 3.

HAB. Cape of Good Hope, *Cavanilles, Dr. Pappe!* (Herb. Hook.).

Stem 6 feet high, or more, shrubby below, herbaceous above; the whole plant slightly viscid, with a balsamic odour. Leaves on long petioles, 3 inches long and broad, the lower ones distinctly 5-lobed, the upper 3-lobed or 3-5-angled. The garden specimens (Hort. Kew! Hort. Monsp.!) are minutely glandular, with a few scattered hairs, in other respects glabrous; a wild one from Dr. Pappe is minutely velvetty. Whether it be a species, or merely a luxuriant state of *M. capense*, β, I cannot determine. The leaves are very much larger and less toothed than in other Cape species. Flowers purple.

2. M. Capense (Gray and Harv.); shrubby, virgate, *slightly viscid, minutely downy or glandularly pubescent*, rarely glabrescent, *the younger parts villous;* leaves ovate-oblong, somewhat 3-lobed and angled, the middle lobe longest, the lateral roundish or obsolete, unequally toothed; stipules short, patent; peduncles axillary, 1-2-flowered, with pedicels longer than the petioles; invol. leaflets lanceolate or ovato-lanceolate, shorter than the ovate-acuminate, ciliated calyx lobes. *Thunb. Cap. p.* 551.

VAR. α. **glabrescens**; puberulous or glabrescent, the glandular hairs minute. *M. Capensis, Cav. Dis. t.* 24. *f.* 3. *DC. Prod.* 1. *p.* 434. *E. & Z.* 280.

VAR β. **balsamicum**; glandular hairs copious and much longer than in var α. *M. balsamica, Jacq. DC. l. c.*. *M. fragrans, Bot. Mag. (non Jacq.) E. & Z.* 281. *Drege*, 7317, 7319.

HAB. Near the Zwartkops River, Oliphant's Hoek, and Boschesman's River, Uitenhage, and near Grahamstown, Albany, *E. & Z.!* and others. Simon's Bay, *C. Wright!* (Herb. T.C.D., Hook., Sond.).

A tall, slender bush, 3–4 feet high, somewhat clammy to the touch, with a strong balsamic scent. Flowers purple. The involucral leaves vary much in size. Carpels glabrous, slightly corrugated.

3. M. calycinum (Gray & Harv.); shrubby, virgate, *downy or tomentose with short stellate hairs*, the younger parts villous; leaves cordate or ovate, somewhat 3-lobed, crenate or toothed, 3–5-nerved; stipules ovate, patent; flowers axillary or in axillary pseudo-racemes, pedicels longer than the petioles; invol. leaflets *broadly ovate*, acute, *nearly as long* as the ovate-acuminate, tomentose and ciliated calyx lobes. *Malva calycina, Thunb., Cap. p.* 550. *Cav. Diss. t.* 22. *f.* 4. *DC. Prod.* 1. *p.* 434. *E. & Z.!* 277. *Bot. Reg. t.* 297. *M. amoena, Bot. Mag. t.* 1998. *M. retusa, E. & Z.!* 278.

HAB. Cape, *Villette, W.H.H.* Saldanha Bay; Zwarteberg, Caledon, and Brede River, Swell., *E. & Z.!* (Herb. T.C.D., Hook., Sond.).

A much branched, virgate shrub, more densely tomentose and less glandular than *M. capense*, with shorter petioles, less lobulate leaves and longer and much broader involucral leaflets. Fl. rosy purple or crimson. The *M. amoena* of Bot. Mag. seems to be a mere garden variety. *M. amoena* E. & Z.! from Lions Mt., belongs to *M. calycinum;* their other specimens to *M. Capense* and *M. strictum*.

4. **M. strictum** (Gray & Harv.); shrubby, *straight* and erect, *the whole plant rough with short, rigid, stellate hairs;* leaves ovate, obtuse or sub-acute, somewhat 3-lobed, crenate or denticulate, the middle lobe longer, lateral lobes often obsolete; stipules ovate or oblong; peduncles axillary, one-flowered; invol. leaflets *narrow-linear*, sub-remote, *much shorter* than the cuspidate calyx-lobes; carpels *glabrous. Malva stricta, Jacq. Hort. Schoenb.* 3. *t.* 294. *DC. Prod.* 1. *p.* 434. *M. amoena, E. & Z.! ex parte.*

HAB. Hill sides, Uitenhage, *Zeyher!* 1967., *E. & Z.* (Herb. T.C.D., Sond.).

3–4 feet high, very rigid, virgate, harsh to the touch. Nearly related to *M. Capense* and also to *M. asperrimum*, but much rougher than the former, and its pubescence much shorter and less stellate, and its leaves less deeply cut than in the latter. Flowers purplish-rosy. Our specimens pretty nearly agree with Jacquin's figure.

5. **M. asperrimum** (Gray & Harv.); shrubby, virgate, *the whole plant densely covered with rigid, harsh, stellate bristles;* leaves ovate-oblong, *somewhat 5-lobed,* undulate and toothed, *corrugated above,* prominently nerved below, the middle lobe much the longest and acute, the lateral lobes obtuse or obsolete; stipules broadly subulate or lanceolate; peduncles axillary, 2–4-flowered; invol. leaflets narrow-linear, shorter than the deltoid-acute calyx lobes; *carpels covered with stellate bristles. Malva asperrima, Jacq. Schoenb.* 2. *t.* 139. *DC. Prod.* 1. *p.* 434.

VAR. β. **stellatum**; leaves narrower, more deeply lobed, the lobes inciso-dentate. *M. stellata, Thunb. fide E. & Z.* 286. *M. corymbosa, E. Mey.? M. bryoniæfolia, Drege! (non DC.).*

HAB. Blaauwberg, 3000–4000 f. *Drege!* Near Heerelogement, Clanw., *E. & Z.!* (Herb. T.C.D., Hook. Sond.).

3–4 feet high, erect; the older parts finally becoming glabrescent; but all the younger parts excessively rough with densely stellate hairs. Lower leaves 3 inches long and two broad, upper about half this size. Flowers purplish. Staminal column densely bristled. I have seen but a frustule of *M. corymbosa*, E.M., doubtfully referred to, above.

6. **M. grossulariæfolium** (Gray & Harv.); shrubby, densely stellato-pubescent, *velvetty,* the stellate hairs very short, but stiff; leaves *roundish-ovate or flabelliform,* shortly 3–5 lobed, undulate, crenato-dentate, 3–5-nerved; stipules broadly subulate; peduncles axillary, 1–3-flowered; invol. leaflets *narrow-linear* or lanceolate, shorter than the deltoid-cuspidate calyx lobes; carpels rounded, *glabrous. Malva grossulariæfolia, Cav. Diss. t.* 24. *f.* 2. *DC. Prod.* 1. *p.* 434? *E. & Z.!* 284. *M. amoena, Drege! (non. Sims.) M. deflexa, Turcz. Mosc.* 1858. *p.* 186.

VAR. β. **parvifolium**; leaves smaller and denser. *M. bryonifolia, Drege! also No.* 7322, 7324.

HAB. In the Karroo. George, and Graaf Reynet, *E. & Z.* β. Hexrivier, *Zeyher!* Nieuwefeldsberg, *Drege!* (Herb. Sond., Hook.)

A tall, virgate, densely stellate shrub, but *soft* to the touch; the leaves having a velvetty down in addition to the stellate hairs. Leaves 1–1½ inches in diameter. Carpels scarcely corrugated, or quite even. Drege's 7324 is less velvetty than the others, but not otherwise different. It comes from Gauritz River.

7. **M. bryonifolium** (Gray & Harv.); shrubby, *densely lanoso-tomentose, and covered with long, soft, stellate hairs;* leaves ovate-oblong, deeply 3-lobed, undulate and corrugated above, prominently nerved be-

low, the margin crenato-dentate, the middle lobe longest; stipules linear-subulate; peduncles 2–4-flowered, shorter than the leaves; involucral leaflets *narrow-linear*, shorter than the deltoid-acuminate, *very woolly* calyx lobes; carpels *covered with starry scales. Malva bryonifolia, Linn ?–fide E. & Z.* 285!

HAB. Oliphant's river's Bad, Clanw., *Ecklon!* Namaqualand, *Zeyher!* (Herb. Sond.)

A woody shrub, much more tomentose than any other Cape species; the whole plant pale, through the abundance of yellowish, densely crowded, stellate, but not rigid hairs. Stipules longer and narrower than in most. Flower-stalks short, corymbose. According to E. & Z. this is *M. plicata*, Thunb.

8. **M. virgatum** (Gray & Harv.); *thinly stellato-pubescent* and villous, or *sub-glabrous*, shrubby; leaves (small) on short petioles, *oblong or broadly linear*, tapering to the base, simple or trifid, sharply serrate, midribbed, *nearly glabrous;* stipules lanceolate; pedicels axillary, one-flowered; involucel below the base of the calyx, its leaflets narrow-lanceolate; calyx segments deltoid-acuminate, thinly stellate, and fringed with woolly hairs; carpels glabrous. *Malva virgata, Cav. Ic.* t. 18. f. 2. and t. 24. f. 1. non E. & Z. *DC. Prod.* 1. *p.* 434.

VAR. *α*. **Dilleniana**; leaves more or less 3-lobed. *M. Dilleniana, E. & Z.* 282.

VAR. *β*. **angustifolia**; more glabrous, with narrow-cuneate leaves. *Drege* 7320!

VAR. *γ*. ? **oblongifolia**; leaves oblong, scarcely narrowed at base. *Drege* 7318, *b!*

HAB. Mountain-sides, at the Langekloof, George, *E. & Z.!* Oliphant's River, *Dr. Gill*, S. Africa, *Drege!* (Herb. Sond., Hook.).

A slender, virgate, nearly glabrous shrub, with a few stellate hairs on the leaves and calyx. Leaves $\frac{3}{4}$–1 inch long, $\frac{1}{4}$–$\frac{1}{2}$ inch wide, sometimes deeply 3-lobed. Involucre sometimes remote. Drege's 7318 b, in Herb. Hook. seems certainly of this species, but 7318, *a*, in Herb. Sond., looks different. *M. virgata*, E. & Z.! is *M. albens*, readily known by its glabrous staminal column.

9. **M. tridactylites** (Gray & Harv.); somewhat viscid, *glabrous or downy;* leaves *obovate-cuneate, tapering very much to the base, subsessile,* 3-toothed or trifid or tripartite, the segments patent, toothed; stipules ovate, erect; peduncles axillary, 1–2-flowered, slender; invol. leafllets linear or lanceolate, much shorter than the deltoid calyx lobes; carpels round-backed, glabrous, rugulose. *Malva tridactylites, DC. Prod.* 1. 434. *E. & Z.* 292. *Thunb.! Cap.* 551; *also M. oxyacanthoides, Hornm. non E. & Z.*

VAR. *α*. **glabra**; glabrous and viscid. *M. stricta, E. & Z* 283! *non Jacq. M. reflexa, Andr. Rep. t.* 135.

VAR. *β*. **puberula**; minutely downy, the calyx stellate-hairy. *E. & Z.* 292!

VAR. *γ*. **stellulata**; stems, and under sides of leaves, thinly stellato-pubescent.

HAB. In dry, desert places. Swellendam, George and Graaf Reynet, *E. & Z.! Drege!* &c. (Herb. Hook., Sond., T.C.D.)

A rigid, divaricately branched shrub, variable in its pubescence, as may be expected in a plant of the Karroo, where the annual rains are so capricious: but readily known from other Cape species by its deeply cut and wedge-shaped leaves. The rosy purple flowers are on short or long stalks. Staminal column very hairy.

Group 2. SUFFRUTICOSA : *spreading or decumbent, half-herbaceous species.* (Sp. 10–15).

10. M. setosum (Harv.); suffrutescent; *stems, peduncles, leaves and calyces setose, with long, simple or tufted hairs;* leaves obovate-cuneate, tapering to the base, deeply 3-lobed, the lobes inciso-dentate; stipules oblong, falcate, spreading; peduncles axillary, 1–2 flowered; invol. leaves linear, shorter than the deltoid-acuminate calyx-lobes; carpels glabrous, somewhat rugulose. *Malva 7321, Drege!*

HAB. Piquetberg, near Groenekloof, under 1000f. *Drege!* (Herb. Hook., Sond.)

Herbaceous, scarcely at all woody, remarkably hispid, with horizontally patent, yellowish hairs. The leaves resemble in form those of *M. tridactylites*; the stipules are longer, narrower and curved, and the calyx-lobes much more taper pointed than in that species. Flowers rosy purple; staminal column *very hairy.*

11. M. albens (Harv.); sub-herbaceous; stems and petioles *strigose, with appressed, stellate hairs;* leaves cuneate at base, ovate, deeply 3-lobed, the lobes obtuse, inciso-serrate near the point, with scattered simple hairs on the upper, and stellate hairs on the lower surface; stipules oblong, subfalcate; peduncles axillary, 1–3 flowered; invol. leaflets linear, obtuse, shorter than the ovate-acuminate, stellato-strigose calyx lobes; *staminal-column quite glabrous;* carpels wrinkled, glabrous. *Malva albens, and M. grossulariæfolia, E. Mey.! Herb. Drege. M. virgata, E. & Z.!* ! 291 ! *non DC.*

HAB. Zwartland, *Dr. Pappe!* Riebeckskasteel, on hills under 1000f., and at Groenekloof, Sept.–Nov. *Drege!* (Herb. Hook., Sond., T.C.D.)

Stems scarcely branched, speading or decumbent, covered, but not thickly, with close pressed, stiff, stellate hairs. Leaves sub-distant, 1 inch long and broad. Flower stalks short, 1–2 together; flowers flesh coloured.—Known from allied species by the glabrous staminal column.

12. M. divaricatum (Gray & Harv.); suffruticose, *divaricately branched,* thinly stellato-pubescent; leaves petiolate, *obtuse at base,* plaited, deeply 3–lobed, *the lobes cuneate, inciso*-crenate, stellato-pubescent especially on the lower surface, the ribs prominent below; stipules ovate or oblong, acute; pedicels axillary, one-flowered; invol. leaflets narrow-linear or spathulate, shorter than the taper-pointed calyx-lobes; *carpels densely setoso-pubescent,* rugulose. *Malva divaricata, Andr. Rep. t.* 182. *DC. Prod.* 1. *p.* 434. *E. & Z.* 293, *non Drege; also M. oxyacanthoides E. & Z.!* 291, *non Hornm. M. microphylla, E. Mey!*

HAB. In the great Karroo, near Gauritz River, George; and between Zwarteruggens and Graaf Reynet, *E. & Z.!* Uitglugt and Nieweveld, *Drege!* (Herb. Sond.)

Slightly ligneous at base, herbaceous upwards. Leaves ½ inch or somewhat more in length, the upper surface sometimes glabrous or nearly so. *M. microphylla,* Drege! seems to be a very stunted form, more glabrous than usual, with subsessile flowers. Flowers either white, with a purple spot on each petal, or suffused rosy-purple.

13. M. racemosum (Harv.); sub-herbaceous; stems, petioles, and both surfaces of the leaves, strigose with appressed, stellate hairs;

leaves obtuse at base, deeply 3-lobed, undulate, the lobes broadly cuneate, crenate or incised, plaited; stipules ovate-oblong, acute; *flowers in the axils of the upper (depauperated) leaves, forming a pseudo-raceme,* peduncles 2–3 flowered; invol. leaflets *broadly ovate,* stellato-pubescent, shorter than the ovate-cuspidate, tomentose or stellate calyx-lobes; staminal column *laxly hispid;* styles elongate; carpels *glabrous. Malva racemosa, E. Mey.! in Herb. Drege.*

HAB. Between Grasberg river and the Waterfall, Tulbagh, *Drege!* (Herb. Hook., Sond.)

A nearly herbaceous plant, said to be 2 feet high. I have only seen fragments of what appear to have been diffuse or procumbent branches. The invol. leaflets are remarkably broad, the rosy flowers crowded about the ends of the branches. Leaves 1 inch long and broad.

14. M. procumbens (Harv.); procumbent, sparingly hairy; leaves *on long petioles, cordate at base,* obtusely 3-lobed, the lobes spreading, middle one longest, crenate, rough with scattered simple hairs on the upper, and stellate hairs on the lower surface; stipules short, reflexed; pedicels axillary, 1–2 flowered, *shorter than the petiole;* invol. leaflets *narrow linear,* obtuse; calyx lobes deltoid, acuminate; carpels 9–10, smooth and convex.

HAB. Gamke River, *Burke & Zeyher!* (Herb. Hook., T.C.D.)

Stems scarcely suffruticose, many from the same crown, widely spreading on the ground, alternately branched. Leaves about an inch long and broad, on petioles of their own length. Pubescence scanty and mostly stellate. Flowers pale rose-colour; the staminal column hairy.

15. M. dissectum (Harv.); suffruticose, procumbent, *nearly glabrous;* leaves petiolate, triangular, tripartite, *the middle lobe bipinnatifid, the lateral lobes pinnatifid,* the segments *linear,* fleshy, sub-acute; stipules ovate-oblong; pedicels axillary, about as long as the petiole, one-flowered; invol. leaflets linear, shorter than the acuminate calyx lobes; carpels ? *Malva asperrima, E. & Z.* 295! *non Jacq.*

HAB. In sandy, desert places near Kochman's Kloof, Swell., *E. & Z.!* (Herb. Sond.)

Readily known from other Cape species by its multifid leaves, which resemble those of a *Grielum.* The linear pinnæ and pinnules are scarcely a line broad; the whole leaf is ¾ inch long, on a petiole of more than half that length. Staminal column glabrous, or nearly so. Calyx-lobes ciliate.—Perhaps a *Sphæralcea?*

Group 3. DUBIÆ:—*Species imperfectly known.*

16. Malva rugosa (Desrous.); "pedicels 1-flowered, solitary, longer than the petiole; invol. leaflets ovate, acute; leaves subcordate, pinnatifido-sinuate, ridged." *DC. Prod.* 1. *p.* 434.

17. Malva retusa (Cav. Diss. t. 21. f. 1.); "pedicels solitary, longer than the petiole; invol. leaflets lanceolate; leaves oblong, very obtuse, 3-lobed, toothed, tomentose." *DC. l. c. p.* 434.

18. Malva anomala (Link. & Otto. Ic. Pl. select. t. 22.); "leaves tomentose, 3-lobed, middle lobe produced, cut; involucre connate with the calyx." *Walp. Repert.* 1. *p.* 294.

IV. SPHÆRALCEA, S. Hil.

Involucel of 3 (or 2) narrow leaflets. *Staminal-column* antheriferous at the multifid summit. *Stigmata* capitellate, as many as the carpels. *Ovules* 2–3 in each carpel. *Fruit* sub-globose, of many compressed, two-valved, dehiscing, 2–3-seeded carpels, united in a ring round a central torus, from which they slowly separate. *Gray*, Gen. Vol. 2. p. 69. t. 127. Endl. Gen. 5272.

Herbaceous or shrubby plants, chiefly natives of North and South America, and most numerous in Central America. There are at least two, and perhaps three Cape species, formerly referred to *Malva*, from which genus the present is known by its capitellate stigmas and 2–3-ovuled carpels. Leaves often deeply cut. Involucel frequently of setaceous, deciduous leaflets, but well developed and persistent in the Cape species. Flowers flesh-coloured or purple, never yellow. Name, *σφαιρα*, *a globe*, and *alcea*, an ancient name of the mallow.

1. S. elegans (Don, Dict. 1. p. 465); sub-herbaceous, decumbent, stellato-lanuginous; leaves deeply three-lobed, or tripartite, the lobes cuneate, inciso-pinnatifid, undulate, stellato-tomentose, bluntly toothed; stipules ovate; peduncles axillary, one-flowered; *invol. leaflets linear, obtuse, shorter than the bellshaped, densely tomentose calyx, whose segments are ovate-acuminate;* fruit globose, stellato-pubescent, of 20–25 carpels; seeds glabrous. *Malva elegans, Cav. Diss. t.* 16. *f.* 1. *DC. Prod.* 1. *p.* 435. *E.&Z. !* 289; *also M. anomala*, E. & Z.! 290., *M. venosa, E. & Z. !* 287, *and M. rugosa, E. & Z.* 288, *non DC. M. striata, E. Mey. !*

HAB. Mountains near Tulbagh; River Zonderende, Assagaiskloof and between Kochman's kloof and Gauritz River, *E. & Z. !* Near Ezelsfontein and on the Roodeberg, *Drege !* (Herb. T.C.D., Hook., Sond.)

Root and bases of stems woody. Stems numerous, procumbent or spreading, 2 feet long or more, not much branched. Leaves rather distant, on petioles of their own length. Flowers from the axils of the upper leaves, usually on short, simple stalks, large, pale, with dark purple veins. Calyx cleft midway. Sometimes the whole plant is white with dense, soft, stellate hairs; sometimes more glabrous. Ovules 3 in each carpel; stigmas minutely capitellate.

2. S. Dregeana (Harv.); suffruticose, patently hairy and lanuginose, with long, sub-simple hairs; leaves oblong, deeply three-lobed, undulate, the terminal lobe longest, inciso-pinnatifid, villoso-lanuginous, bluntly-toothed; stipules ovate, acute; peduncles axillary, one-flowered, *invol. leaflets adnate with the calyx-tube, broadly oblong-lanceolate, calloso-denticulate, longer than the villous calyx, whose narrow segments are much acuminate;* ovary *glabrous*, the carpels 2-3 ovuled. *Drege*, 7323! *Anisodontea Dregeana, Presl, Bot. Bem. p.* 18.

HAB. Piquetberg, on stony and rocky table-land, 1–2000 f. *Drege !* (Herb. Hook., Sond.).

Resembling *S. elegans*, but readily known by its very different calyx and involucel. The calyx lobes are nearly lanceolate, and the involucral leaves spring from the summit of the calyx-tube, and look like an outer row of sepals. The flowers are smaller than in *S. elegans*.

Malva divaricata, E. Mey.! in Herb. Drege, not of Andrews, appears, by a very bad specimen in Herb. Sond., to belong to an undescribed species of *Sphæralcea*. The stems are slender and glabrescent; the leaves oblong, deeply 3-lobed, inciso-pinnatifid and stellato-tomentose; the stipules ovate; peduncles axillary, one-flowered, longer than the leaves; invol. leaflets adnate, oblongo-lanceolate and longer than the acute calyx lobes, and carpels glabrous. Flowers deep purple. Though having many characters, as in *S. Dregeana*, the general aspect is very different. It was found between Driekoppen and Bloedrivier, 2500–3000 f., in April.

V. SPHÆROMA, Harv. (non DC., nec Schl.).

Involucel monophyllous, persistent, 3-cleft. Other characters as in *Sphæralcea.* (*Lavateræ Sp. DC. et auct.*)

Suffruticose, or shrubby S. African plants. Leaves alternate, entire or lobed, toothed, canescent below, stellato-tomentose. Peduncles axillary, 1–2-flowered, jointed. *Flowers* rosy purple. The plants referred here differ from *Lavatera*, in which they have been generally placed, in having capitellate stigmata and 2–3 ovuled carpels. From *Sphæralcea*, to which they are much more nearly allied, they differ in habit and in the *monophyllous* involucre; a character the same as that which separates *Lavatera* from *Malva*. I employ the rejected name *Sphæroma* (from σφαιρα, a *globe*,) once given to the genus now called *Sphæralcea.*

1. S. Julii (Harv.); *erect;* branches tomentose, and hispid with scattered, stellate hairs; leaves *on long petioles, 5-lobed*, the middle lobe longest, unequally crenate, scabrous on the upper, tomentose, stellato-pubescent and canescent on the lower surface; peduncles from the axils of the upper leaves, 2–3-flowered, densely stellato-pubescent, as are also the calyx and *the deeply and sharply cleft involucel. Lavatera Julii, Burch. Cat.* 2664. *DC. Prod.* 1. *p.* 438. *L. biflora, E. Mey.!*

Hab. Banks of Orange River, *Drege, Zeyher!* Caledon River, *Burke!* (Herb. T.C.D., Hook., Sond.)

A tall, coarse growing, half-woody plant, all the younger parts densely tomentose and hispid; the stems and upper surfaces of the leaves afterwards becoming nearly naked. Flowers purple, 1–1½ inches across. Involucel very deeply parted, but its leaflets are confluent at base,

2. S. prostrata (Harv.); diffuse or "prostrate" (*fide Drege*); branches stellato-tomentose, becoming sub-glabrous; leaves *on short petioles, oblong, obtusely 3-lobed*, the middle lobe *much the longest*, the lateral short or obsolete, crenate, pubescent on the upper, stellato-tomentose, and canescent on the lower surface; peduncles from the axils of the upper leaves, stellato-pubescent, as are the calyx and the *obtusely trifid involucel. Lavatera prostrata, E. Mey.!*

Var. β. **mollis**; leaves softly canescent, with very short pubescence, the lateral lobes obsolete. *Drege*, 7325.

Hab. Witbergen and rocky heights at Krai-rivier, 4–5000 f. *Drege!* Orange River, *Burke and Zeyher!* (Herb. T.C.D., Hook., Sond.).

Said to be prostrate, which we should hardly have guessed from the specimens before us. The leaves are often very imperfectly lobulate at base. *Flowers* purple. *Invol.* less deeply cut than in the preceding. In var β. the pubescence is much softer and shorter, and the leaves are smaller.

VI. SIDA. L.

Involucel none. *Staminal column* antheriferous at its multifid summit. *Ovary* of 5–15 carpels, united in a circle round a central torus, one-ovuled; ovule *pendulous. Styles* 5–15; stigmata capitellate. Fruit of 5-15 one-seeded carpels, dehiscent at the summit and at length separating from the axis; seed 3-cornered, suspended. *Endl. Gen.* 5289. *DC. Prod.* 1. *p.* 459, *in part. A. Gray, Gen.* 2. *p.* 61. *t.* 123.

An immense tropical and sub-tropical genus, very variable in aspect. The S. African species are few and mostly of exotic origin. Leaves entire or lobed, often penninerved. Pedicels jointed below the summit, solitary, or several together. Flowers small, yellow, or orange, or white; rarely purple. Name, used by Theophrastus and the early botanists for a plant of this family.

1. S. triloba (Cav. Diss. 1. t. 1. f. 2. and t, 131. f. 1.); suffruticose, glabrous or nearly so; leaves on long petioles, cordate, *3-lobed, the upper ones tripartite*, segments lanceolate, acute, coarsely-toothed; stipules short, subulate; pedicels axillary filiform, *much longer than the leaves;* calyx-lobes ovate, acute; carpels 7-8, glabrous, pointless. *Jacq. Schoenb. t.* 142. *DC. Prod.* 1. *p.* 466. *E. & Z.!* 319. *S. triloba and S. ternata, Thunb. Sida,* 7320, *Drege.*

HAB. Frequent in margins of woods and waste places in Uitenhage and Albany, and in Caffirland. (Herb. T.C.D., Hook., Sond.).

Root perennial. Stems woody at base, slender, diffuse, sub-simple or branched. Lower leaves cordate or reniform, less deeply cut than the upper, with blunt segments. Flowers pale yellow, small.

2. S. longipes (E. Mey.! in Herb. Drege); shrubby, the branches and petioles stellato-pubescent; leaves *ovate or oblong*, acute or obtuse, unequally serrate, penninerved, minutely stellato-pubescent, green or canescent below; stipules subulate; flowers axillary, *on slender pedicels much longer than the leaves;* calyx downy, with deltoid, acute, one-nerved segments; *carpels* 7-8, *pubescent, pointless. S. longipes and S. spinosa, E. Mey.! S. capensis, E. & Z., non Cav.*

HAB. Uitenhage and Caffirland, *Drege! E. & Z.!* Natal, *Dr. Sutherland.* (Herb. T.C.D.).

Very like some varieties of *S. rhombifolia*, and only to be essentially distinguished by the carpels. The pedicels are 3-4 inches long; the leaves an inch, on short petioles.

3. S. rhombifolia (Linn.); shrubby, branches and petioles stellato-pubescent; leaves rhomboid or oblong-oval, cuneate at base, crenate toothed, penninerved, glabrescent above, canescent on the lower side stipules filiform, deciduous; flowers axillary, solitary or clustered, on short or long pedicels; calyx scaberulous, with deltoid, acute, nerved segments; *carpels* 9-10, *smooth, minutely bicuspidate. DC. Prod.* 1. *p.* 462. *E. & Z.* 317. *Sida capensis, E. Mey.! Cav. Ic. t.* 12. *f.* 3.

HAB. Eastern districts and Port Natal, *E. & Z.! Drege, Gueinzius,* &c. (Herb. T.C.D.).

A common weed throughout the tropics and warmer temperate zones of both hemispheres; very variable in size and form of leaf, pedicel, &c. The flower stalks are *usually* short, but sometimes nearly as long as in *S. longipes.*

4. S. spinosa (Linn.); annual, stellato-pubescent; leaves ovate oblong or lanceolate, obtuse, unequally toothed, penninerved, *the petiole armed at base with a hooked spur;* stipules linear; flowers axillary, pedicellate; calyx villous, with deltoid, acute, nerved segments; *carpels 5, rugose, bicuspidate DC. Prod.* 1. 460.

HAB. Vahl River, *Burke!* (Herb. Hook.).

Resembles *S. rhombifolia* in aspect, but readily known by the rigid hook at the base of the leaf-stalk. A common weed in tropical countries.

5. S. cordifolia (Linn.); shrubby, *branches and petioles villous;* leaves *ovate or cordate,* obtuse, penninerved, crenate, velvetty, canescent below; stipules filiform, deciduous; flowers axillary, solitary or clustered, on short pedicels; calyx *tomentose,* with deltoid, acute, nerved segments; *carpels* 9-10, *wrinkled and netted, dorsally armed with two long, rigid bristles. DC. Prod.* 1. *p.* 464. *Sida velutina, E. Mey.!*

HAB. Port Natal, *Drege! Miss Owen, Gueinzius, &c.* (Herb. T.C.D., Hook., Sond.)
Flowers small, buff-coloured or orange, either on axillary pedicels or oftener crowded on short, axillary branchlets, with or without leaves. Common throughout the tropics.

VII. ABUTILON. Tourn.

Involucel none. *Ovary* of 5-20 or more carpels, closely united in a circle round a central torus; ovules 3-9 in each carpel. *Stigmas* capitate. *Fruit* of numerous follicular, membranous, 3-6 seeded carpels, dehiscing by the ventral suture and sub-persistent. *Endl. Gen.* 5292. *A. Gray, Gen. Vol.* 2. *p.* 65. *t.* 125, 126.

Herbs, shrubby plants or tall shrubs, common throughout the tropics, the whole plant generally clothed with soft, velvetty hairs, sometimes also tomentose and stellate-hairy. Leaves on long petioles, cordate, sub-entire or angular. Flowers axillary, yellow or orange. Distinguished from *Sida* as well by habit, as by the numerous ovules and seeds. The name is of unknown origin or meaning.

1. A. Sonneratianum (Cav. Diss. t. 6. f. 4); stems and petioles velvetty and villous; leaves cordate, acuminate, or 3-angled, repand, or unequally crenate, velvetty on both surfaces, canescent on the lower; stipules setaceous; pedicels longer than the leaves; calyx segments oblong-ovate, acuminate; *carpels* 9-10, *truncate, obtuse, stellato-pubescent. DC. Prod.* 1. *p.* 470. *E. & Z.* 318. *S. Asiatica, Thunb. Cap. p.* 548. *Drege,* 7332, 7333.

VAR. β. **prostrata**; small, prostrate, leaves obtusely 3-angled, quite entire. *S. prostrata, E. Mey.*

HAB. Woods and waste places, in the eastern districts, Caffraria and Natal. (Herb. T.C.D.)
The whole plant is softly velvetty; the younger stems and branches are also clothed with long, soft, spreading, deciduous hairs. Stem 1-2 feet high. Flowers yellow, an inch across. Carpels inflated, either quite obtuse, or slightly mucronate. A native also of tropical Asia, and often confounded with the following.

2. A. indicum (G. Don); stems and petioles more or less velvetty and villous; leaves cordate, acuminate, 3-5 angled or somewhat lobed, repand or crenate, soft, glabrescent on the upper, velvetty and canescent on the lower surface; stipules setaceous, reflexed, deciduous; peduncles axillary, one-flowered, longer than the petioles; calyx lobes ovate, mucronulate, velvetty and canescent, *carpels* 10-20, *or more, truncate, their upper angles acuminate or cuspidate, stellato-pubescent. Sida indica Linn. DC. Prod.* 1. *p.* 471. *S. populifolia, Lam. Anoda holosericea, E. Mey.!*

HAB. Caffraria and Port Natal, *Dr. Grant.* (Herb. Hook., T.C.D.)
Except in the *cuspidate* carpels, if this be a constant character, I scarcely see how this differs from the preceding. The pubescence, in our specimens, is less velvetty and more hairy than in *S. Sonneratianum,* and the leaves more decidedly lobed. I

have seen other scraps of *Abutila* from Caffirland which look different from either form, but are too imperfect for description. *A. indicum* is found throughout the tropics of both hemispheres.

VIII. PAVONIA. Cav.

Involucel 5-15 leaved, persistent. *Staminal column* naked and 5-toothed at the apex, bearing stamens on its external surface, below the summit. *Ovary* of 5 carpels united round a central axis; styles confluent below, 10-cleft above; stigmata ten, capitate. *Fruit* of 5 indehiscent, one-seeded carpels. *Endl. Gen. No.* 5275. *A. Gray, Gen.* 2. *p.* 75. *t.* 130.

Shrubs, suffrutices or herbs, chiefly tropical and sub-tropical. A large genus, variable in habit, chiefly American, with a few out-lying Asiatic and African species. Flowers yellow, white or red, often handsome. Name in honour of Don José Pavon, a Spanish botanist, and one of the authors of the *Flora Peruviana.*

Group 1. Lebretonia. *Involucel of* 5, *rarely of* 6 *leaflets.* (Sp. 1–3.)

1. P. macrophylla (E. Mey.!); stem and branches hispid with long, patent, stiff hairs; leaves on long hairy petioles, deeply cordate at base, 5-*angled and somewhat* 5-*lobed,* with shallow, rounded interspaces, coarsely crenato-dentate, laxly villous on the upper, stellato-pubescent on the lower surface; stipules linear; peduncles axillary, elongate, one-flowered; *invol. of* 5-6 *broadly ovate, fringed, connate leaflets;* calyx membranous, villous. *Pentameris macrophylla, E. Mey.! Pavonia Kraussiana, Hochst. in Walp. Rep.* 5. 90. *P. acuminata, Pl. Schimp.! P. crenata, and Urena mollis. Hochst. Pl. Schimp.*

Hab. Eastern districts and Port Natal, *Zeyher! Drege! Krauss!* Sand River and Macallisberg, *Burke & Zeyher!* (Herb. T.C.D., Hook.)

Herbaceous, tall and free growing, more hairy than tomentose. Leaves 2–3 inches long, equally broad, and either angled or lobed, the terminal lobe acuminate. The upper leaves are sometimes ovate, acuminate. Flowers flesh-coloured or rosy. Invol. leaflets more or less united at base, and in one of Burke's specimens united into a monophyllous, 5-cleft involucel! This plant is common to the warmer parts of the colony and to the country north and east of the frontier. It occurs also in Abyssinia, whence we have it under three different names! *Anoda cordifolia, E. Mey.!* in Herb. Drege, is another of its aliases.

2. P. microphylla (E. Mey.!); stems softly pubescent; leaves on long petioles, cordate, 3-*lobed, the lobes again crenato-lobulate, very obtuse,* puberulous on the upper, softly hairy on the lower surface; stipules subulate; peduncles axillary, one-flowered; *invol of* 5 *obovate, obtuse, leaflets;* calyx membranous and veiny.

Hab. In thickets near Port Natal, *Drege! Gueinzius!* (Herb, Hook., T.C.D., Sond.).

Herbaceous, slender, with spreading branches. Leaves distant, scarcely an inch long and broad, resembling, in shape those of *Veronica hederifolia.* Flowers yellow.

3. P. mollis (E. Mey.!); stem and branches hairy-tomentose; leavee on long, villous petioles, sub-cordate at base, with 5 short, deltoid, acuts lobes and rounded interspaces, coarsely and unequally toothed, 5-7-nerved, puberulous on the upper, velvetty and canescent on the lower surface; stipules setaceous; flowers axillary or in axillary, leafy racemes; *invol. of* 5-6 *linear leaflets;* calyx villous, diaphanous, its segments 3-ribbed. *Althæa Burchellii, E. & Z.! No.* 300.

HAB. About the Kat River settlement, *E & Z.!* S. Africa, *Drege!* Albany, *Mrs. F. W. Barber!* Natal, *Gueinzius!* (Herb. Hook., T.C.D., Sond.).

Herbaceous, tall and branching, softly hairy, with velvetty leaves, 3 inches in diameter. Flowers in our specimens on axillary, short racemose branches, with depauperated leaves ½–¾ inch long. Flowers white or rosy, on short pedicels.

Group 2. EUPAVONIA. *Involucel of* 10-14 *leaflets.*

4. P. præmorsa (Willd. Sp 3. p. 833); young branches tomentose; leaves petiolate, broadly obovate or flabelliform, truncate, obtusely dentate, 3-5-nerved, scabrous on the upper, canescent on the lower surface; stipules setaceous; pedicels axillary, 1-flowered, longer than the leaves; invol. of 12-14 very narrow, linear leaflets; calyces canescent, segments ovate, acute. *DC. Prod.* 1. *p.* 444. *E. & Z.* 302. *Jacq. Ic. Rar. t.* 141. *Cav. Diss.* 3. *t.* 41. *f.* 1. *Bot. Mag. t.* 436.

HAB. Bushy hills in Uitenhage and Albany, common. (Herb. T.C.D., Hook., Sond.).

A moderate sized shrub, with patent, straight, rod-like branches. Leaves about an inch long and broad, very constant in their form and pubescence; petiole much shorter than the lamina. Flowers bright yellow with a dark centre. Carpels downy, with a dorsal ridge. Ecklon and Zeyher's var. β. has leaves thrice as large as usual, but otherwise similar.

IX. HIBISCUS. Linn.

Involucel 5-20 leaved. *Petals* expanded. *Staminal column* 5-toothed at the apex, bearing stamens on its external surface. *Ovary* 5-celled; ovules numerous; *style* 5-cleft; *stigmata* 5, capitate. *Capsule* 5-celled, surrounded by the persistent calyx, 5-valved, loculicidal; seeds few or many in each loculus. *Endl. Gen.* 5277. *DC. Prod.* 1. *p.* 446 *(in part). A. Gray, Gen.* 2. *p.* 81. *t.* 133.

Trees, shrubs, or herbs, sometimes small annuals, common throughout the tropics and warmer parts of the temperate zones. Leaves alternate, entire or lobed, sometimes multi-partite, glabrous, or variously pubescent or stellato-hispid. Stipules lateral. Flowers generally large and showy, red, crimson, yellow, or more rarely, white; often with a dark centre and velvetty or silky surface. Stigmata often crimson, and much produced beyond the staminal column. Name from the Greek *ἱβίσκος*, an ancient name given to the marsh-mallow.

* *Involucel* 5-*leaved.* (Sp. 1–2.)

1. H. calycinus (Willd. Sp. 3. p. 817); shrubby; branches, petioles, and peduncles thinly *stellato-pubescent;* leaves on long petioles, roundish-cordate, *obtusely* 3-5-*angled,* crenate 5-7-nerved, velvetty and sprinkled with hairs; stipules subulate; peduncles axillary, shorter than the petiole; invol. of 5, *broadly-spathulate, cuspidate and bristle-pointed* many-nerved leaflets, about as long as the calyx; capsule ovate-acuminate, tomentose. *DC. Prod.* 1. *p.* 448. *H. calyphyllus, Cav. Diss. t.* 140. *H. ficarius, E. Mey.! H. borbonicus, Link?*

HAB. S. Africa, *Drege!* Caffraria, *Dr. Gill!* Mooje River, *Burke and Zey.!* Natal, *Gueinzius.* (Herb. T.C.D., Hook., Sond.).

A shrub, with slender, sub-simple branches and sub-distant leaves, pretty constant in shape and pubescence. The invol. leaflets are remarkable in form, and in their pungent points; and serve to distinguish this species from *H. Ludwigii,* its nearest ally. Cavanille's figure, except that it shows an infantile corolla, is very good. The mature flowers are large, yellow, with a dark-red centre.

2. H. Ludwigii (E. & Z. No. 312); shrubby; the slender branches and petioles *softly pubescent;* leaves petiolate, cordate at base, 5-angled, or with 5 *shallow, deltoid, acute lobes,* the middle one longest, crenate, rough with scattered hairs; peduncles axillary, shorter than the petiole; densely setose; invol. of 5, *broadly lanceolate,* many-nerved leaflets, longer than the calyx; capsule ovate-acuminate, densely setose. *H. calycularis, E. Mey.!*

HAB. Forest of Krakakamma, and at the Kat River, *E. & Z.! Drege!* (Herb. Sond.).

A tall, slightly branched, slender shrub, sometimes (according to Ecklon) rising above the tops of the surrounding trees, and making them gay with the abundance of its golden-yellow flowers. It bears the name of the late Baron v. Ludwig, of Capetown, a munificent patron of botany and horticulture, whose beautiful garden of Ludwigsburg, now obliterated, was, for many years, one of the attractions of Capetown.

** *Involucel* 8–10–14 *or many-leaved. Seeds glabrous. Root perennial; stems shrubby, suffruticose or herbaceous.* (Sp. 3–12.)

3. H. diversifolius (Jacq. Ic. Rar. t. 551); herbaceous or suffruticose; branches, petioles and nerves of the leaves *armed with sharp-pointed, hard tubercles;* lower leaves deeply 3–5-lobed, unequally toothed, upper elliptic-oblong or lanceolate, toothed or jagged; flowers in terminal or axillary racemes, the *pedicels very short and bristly;* invol. of many, rigid, subulate leaflets, shorter than the densely setose, tapering calyx-lobes. *DC. Prod.* 1. *p.* 449. *Bot. Reg. t.* 381. *H. ficulneus, Cav. Dis. t.* 51. *f.* 2. *E. & Z.!* 313. *H. macularis, E. Mey.!*

HAB. Near the Zwartkops River, Uitenhage, *E. & Z.! Drege!* (Herb. T.C.D., Hook., Sond.).

A stout, coarse-growing plant, 6–8 feet high, very harsh, rough and prickly. Branches ending in pseudo-racemes of large, yellow, dark-centred flowers. *H. macularis, E. Mey.!* according to the single specimen we have seen, seems only to differ in having a *gland* in the middle of each calyx lobe; but traces of a similar gland may often be detected in the common form.

4. H. ricinifolius (E. Mey !); herbaceous, laxly pubescent; *stem and petioles prickly;* leaves on long petioles, palmately 5–7 lobed, the lobes acuminate, sharply and unequally serrate, glabrescent, sprinkled with trifid hairs and prickles; *peduncles elongate, jointed below the flower, patently pubescent;* invol. of 9–10 narrow-linear, ciliate leaflets, rather shorter than the ovate-acuminate, thin, hispidulous calyx-lobes; capsule sub-globose, laxly setose; seeds glabrous. *H. ricinoides, Garke!*

HAB. Near Port Natal, *Drege!* (Herb. Sond.)

A tall, herbaceous plant, set with a few small prickles and thinly sprinkled with soft hairs. Flowers 2 inches across, yellow, with a dark purple centre.—Of this I have only seen a single specimen. It may be merely a glabrescent form of *H. vitifolius,* with more deeply divided leaves than usual.

5. H. persicifolius (E. & Z. No. 305.); stem shrubby, glabrescent, *armed with scattered, sharp tubercles;* petioles hispid; leaves *linear-lanceolate,* tapering to the base, sub-distantly callous-toothed, midribbed and penninerved, *nearly glabrous;* stipules setaceous; *peduncles shorter than the leaves;* invol. of 10-12 subulate leaflets, shorter than the ovato-lanceolate, acuminate, coriaceous calyx-lobes, which, as well as the peduncle, are densely clothed with golden-yellow echinate tubercles.

HAB. In the Great Karroo, between Graaf Reynet and Uitenhage, *E. & Z.!* (Herb. Sond.)

Apparently a branching shrub, nearly glabrous except on the petioles, peduncles and calyx, which latter are thinly covered with prominent, glandular-bristly-tubercles, like miniature hedgehogs. *Leaves* 3–4 inches long, ½ inch broad. *Flowers* said to be purplish. I have only seen a single, imperfect specimen.

6. H. mutabilis (Linn.); stellato-tomentose, *without prickles;* leaves on long petioles, subcordate at base, acutely 5–7-angled or somewhat lobed, unequally toothed; peduncles from the axils of the upper leaves *elongate, jointed below the flower;* invol. of 8–10 linear-leaflets, half as long as the ovate-acuminate, 5-nerved calyx-lobes; corolla expanded. *DC. Prod.* 1. *p.* 452. *Bot. Reg. t.* 589.

HAB. South Africa, (probably Natal,) *Drege!* (Herb. T.C.D., Hook.)

A large shrub or small tree, more or less tomentose. Leaves 3–4 inches long, and broad. Flowers crowded round the ends of the branches, large and handsome; *white* when they first open in the morning, *reddish* at noon, and *bright-crimson* towards evening. It is a native of India and China, and has long been cultivated in English hot-houses. Its South African habitat is questionable.

7. H. physaloides (Guill. & Per. Fl. Sen. p. 52.); *herbaceous,* tall; stem, petioles and peduncles tomentose, and hispid with long, patent, sub-simple hairs; leaves on long petioles, cordate at base, 5-angled or 5-lobed, *the lobes deltoid-acuminate or cuspidate,* crenate, velvetty on the upper, tomentose and somewhat hispid on the lower surface; peduncles elongate, axillary or in a terminal pseudo-raceme; invol. of 10 filiform, hairy, patent or reflexed, curved leaflets half as long as the shaggy calyx.; capsules densely hispid; seeds glabrous. *H. ascendens, Don, Gard. Dict.* 1. *p.* 482. *H. heterotrichus, E. Mey! H. variabilis, Garke.*

HAB. South Africa, *Drege!* Port Natal, *Gueinzius! Krauss!* 346. Macalisberg, *Burke!* (Herb. T.C.D., Hook., Sond.)

About 3–6 feet high, the branches above bare of leaves, or with depauperated foliage, and ending in a long pseudo-raceme of large, showy flowers. Petals deep red at base, yellow in the upper part. A native also of tropical Africa and the Canary Islands.

8. H. cordatus (Harv.); herbaceous, tall; stem, petioles and peduncles softly pubescent and sprinkled with stellate hairs; leaves on long petioles, oblong-ovate, cordate at base, 3–5-lobulate, *the lateral lobes short and obtuse,* the middle one ovate and acute, all crenato-dentate, with velvetty and stellate pubescence; peduncles axillary, as long as the petioles, jointed above the middle; invol. of 10–12 minute filiform leaflets, very much shorter than the semi-quinquefid, stellato-pubescent calyx; capsules acuminate, hispid; seeds glabrous.

HAB. Near the Vaal River, and at Macallisberg, *Burke & Zeyher!* (Herb. Hook. Sond.)

Two or 3 feet high, with the foliage of an *Abutilon.* It is nearly related to *H. physaloides,* but differs in the short and obtuse leaf-lobes, the less setose pubescence, shorter invol. leaflets and less shaggy calyces and capsules. I have only seen unopened flowerbuds and ripe capsules.

9. H. urens (Linn.); herbaceous, *procumbent,* densely stellato-tomentose and hispid; leaves on long petioles, *reniform,* 7–9-lobulate, *the lobes very short, obtuse* and unequally crenate, the ribs and veins prominent on the lower surface; flowers axillary, several crowded together, *on very short pedicels;* invol. of 10–12 linear, tomentose, acute leaflets half as long as the bell-shaped, shortly 5-lobed calyx. *DC. Prod.* 1. *p.* 447. E. & Z.! 309. *H. cucurbitinus, Burch? Trav.* 1. *p.* 278, *Cat. No.* 1481.

HAB. Karroo, between Beaufort and Graaf Reynet, *E. & Z.! Mrs. F. W. Barber!* Brack Rivier, *Burke!* (Herb. T.C.D., Hook., Sond.)

A strong growing, shaggy plant, very rough in all parts. Flowers deep crimson, handsome, opening throughout the whole summer.

10. H. pedunculatus (Cav. Dis. t. 66. f. 2.); *shrubby,* laxly stellato-hispid, rough; branches slender; leaves on short or longish petioles, 3–5-*lobed, the lobes obtuse,* the middle one longest, unequally toothed; peduncles axillary, *many times longer than the leaves,* one-flowered; invol. of 8–10 linear-lanceolate leaflets, equalling or exceeding the cuspidate calyx-lobes; capsule pubescent, seeds smooth. *DC. Prod.* 1. 446. *E. & Z.* 310. *Bot. Reg. t.* 231. *Thunb. Cap. p.* 549. *H. Kraussianus, Buching. in Walp. Rep.* 5. *p.* 91.

HAB. In woods and on hill sides. Galgebosche, *Thunberg, Drege!* Uitenhage, *E. & Z.!* Port Natal, *Dr. Sutherland.* (Herb. T.C.D., Hook., Sond.)

A graceful shrub, 2–4 feet high, with erect branches and distant leaves. The leaves are usually as above described, but the lobes are sometimes shallow or obsolete. The long flowerstalk is a constant mark. The flowers are deep rosy red, 1½ inch long, campanulate and slightly nodding.

11. H. Meyeri (Harv.); shrubby, slender; branches straight, *strigose with appressed stellate bristles;* leaves (small) on short petioles, 3-nerved, *ovate-oblong, obtuse,* unequally toothed and somewhat 3-lobed, the lateral lobes very short, *appressedly stellato-pubescent;* stipules setaceous, rigid; peduncles axillary, longer than the leaves, jointed below the flower; invol. of about 10 linear-subulate pubescent leaflets, shorter than the deltoid-acuminate calyx lobes; capsule puberulous, seeds smooth. *H. microphyllus, E. Mey.! (non Vahl.).*

HAB In the Valley of the River Omblas, *Drege!* Natal, *Gueinzius!* (Herb. Sond.)

A slender shrub, with rodlike branches and a very close-pressed and short, but rigid pubescence. Leaves ½–¾ inch long; petioles ¼ inch. Peduncles about twice as long as the leaves; flowers bright crimson, 1½ inch across, with narrow, cuneate petals. The leaves in the specimens seen can scarcely be called 3-lobed, but there is a tendency to such division; and, notwithstanding the short petioles, the habit reminds us of some depauperated forms of *H. pedunculatus.*

12. H. leiospermus (Harv.); *dwarf, erect, the whole plant, except the petals densely setose with stiff, spreading, stellate hairs;* leaves oval-oblong, 5-nerved, entire or 2–3-toothed near the point, on short petioles; stipules subulate; flowers axillary, pedunculate; involucre of about 10 subulate leaflets, nearly as long as the acuminate calyx-lobes; *capsules and seeds glabrous.*

HAB. South Africa, *Burke!* (Herb. Hook.)

Stems woody, about 6 inches high, divided near the base into several erect, simple branches. Pubescence very rough, rigid and copious. Flowers 1¼ inch diameter, crimson or rosy. Seeds quite naked. In habit resembling one of the following section.

*** *Involucel* 8–12 *leaved. Seeds woolly, hairy, silky or pubescent. Root perennial; stem shrubby or fruticulose.* (Sp. 13–18.)

13. H. æthiopicus (Linn.); *dwarf*, stellato-hispid; stems curved, decumbent, subsimple; leaves longer than their petioles, *elliptic-obovate, elliptical or oblong, sub-truncate, 5–7-toothed or sub-entire,* 3–5-nerved, stellately hispid, especially on the ribs and at the margin; stipules narrow-linear, herbaceous; peduncles axillary, mostly longer than the leaf, hispid and jointed below the flower; invol. of 10–12 narrow-lanceolate or subulate, ciliate leaflets, shorter than the acuminate, laxly setose and ciliate calyx-lobes; capsules nearly glabrous, the seeds *minutely pubescent. DC. Prod.* 1. *p.* 451. *E. & Z.!* 303. *Cav. Dis. t.* 61. *f.* 1. *Thunb. Fl. Cap.* 548.

VAR. β. **ovatus**; caulescent, leaves elliptical or elliptic-ovate, subentire, laxly hispid or glabrescent. *H. ovatus, E. & Z.!* 304. *Cav. Dis. t.* 50. *f.* 3? *Drege!* 9511.

VAR. γ. **helvolus**; caulescent; flowers orange; leaves broadly elliptical or oblong, hispid; invol. leaflets often lanceolate. *H. helvolus, H. ellipticus, and H. propinquus, E. Mey.! H. asperifolius E. & Z.!* 306.

VAR. δ. **diversifolius**; leaves varying from obovate to ovate, lanceolate or linear, on the same stem; smaller and more glabrous than the other varieties. *H. ovatus,* β. *angustifolius, E. & Z.!* 304, *ex parte. Drege,* 7327!

HAB. On grassy hills throughout the Colony, and extending to Port Natal. (Herb. T.C.D., Hook., Sond.)

A small suffruticose perennial, 3–6 inches, or rarely a foot high, and mostly procumbent; varying greatly in pubescence and in the form of its leaves. Of the above quoted synonyms Meyer's *H. helvolus* is the most unlike the ordinary form, but through *H. ellipticus* and *propinquus* it is brought near *H. ovatus, E. & Z.*, and so passes into the common *æthiopicus*. Var. δ, is from dry ground and much depauperated. Flowers (except in var. γ.) bright yellow, with a dark centre.

14. H. malacospermus (E. Mey.); dwarf, stellato-hispid; stem curved, sub-simple; *leaves sub-sessile, very long and narrow,* linear or linear-lanceolate, entire, or remotely toothed and cut, prominently 3–5-ribbed, the ribs and margins stellato-hispid; stipules linear-subulate, *rigid*, spreading; peduncles axillary, elongate, hispid and jointed below the flower; invol. of 10–12 subulate, strongly 3-ribbed ciliate leaflets, more than half as long as the setose, acuminate calyx-lobes; capsules nearly glabrous, *seeds densely albo-tomentose. H. malacospermus and H. tridentatus, E. Mey.! H. microcarpus, Garke. Kosteletskya malacosperma, Turcz! Bull. Mosc.* 1858. *p.* 192.

VAR. α. **luteus**; flowers yellow.

VAR. β. **purpureus**; flowers purple.

HAB. Caffraria and towards Port Natal, *Drege! Miss Owen!* Modder-river and Macallisberg, *Burke & Zeyher!* (Herb. T.C.D., Hook., Sond.).

Nearly allied to *H. æthiopicus*, but a larger, stronger and more rigid plant; the stems 12–14 inches high, and leaves 3–5 inches long, about ¼ inch wide, sometimes quite entire, but frequently jagged at the sides or trifid near the point. The specimens with purple flowers are not otherwise different from those with yellow. Seeds much more densely and softly pubescent than in *H. æthiopicus*. I do not know why Turczaninow has referred this plant to *Kosteletskya*; of which it does not possess the *solitary* seeds.

15. H. pusillus (Thunb. Cap. p. 550.); dwarf, hispid or glabrescent; stems decumbent, sub-simple; leaves on short petioles, *polymorphous* (ovate, or tricuspidate, or 3-lobed, or tripartite) *sharply and coarsely serrate*, rigid, nearly glabrous, with prominent nerves and veins, coarsely reticulate; stipules setaceo-subulate, spreading; peduncles axillary, longer than the leaves, jointed below the flower; involucre of about 10 subulate leaflets, half as long as the acuminate, stellato-pubescent calyx-lobes; *capsules glabrous, seeds woolly. Harv. Thes. t. 73. H. gossypinus, E. & Z.! 307, non Thunb. H. serratus, E. Mey.!*

HAB. Near the Zwartkops and Sondag rivers, Uit. *E. & Z.!* Graaf Reynet and Somerset, *Mrs. F. W. Barber!* Macallisberg, *Burke!* (Herb. T.C.D., Hook., Sond.).

Root thick and woody, throwing up several stems 6–12 inches long or more; the larger specimens branching. Leaves 1–1½ inch long, extremely variable in shape and cutting, but always thick and rigid and sharply toothed; rarely stellato-pubescent, mostly sub-glabrous. The flowers are rather large, crimson or deep rosy-purple, rarely varying to orange or yellow, and appear throughout the summer.

16. H. atromarginatus (E. & Z.! No. 308); laxly hispid or glabrescent; branches slender, virgate; leaves petiolate, *3–5 parted, with narrow, linear-lanceolate or cuneate, entire or jagged or pinnatifid lobes*; stipules setaceous, stiff; peduncles axillary, many times longer than the leaves, one-flowered; invol. of 10-12 subulate, rigid, pungent, leaflets, half as long as the linear-lanceolate, acuminate calyx-lobes; *seeds clothed with long, silky hairs. H. lasiospermus, E. Mey. H. macrocalyx, Garke!*

HAB. Caffirland *E. & Z.! Drege!* Crocodile River and Macallisberg, *Burke.* Somerset, *Mrs. F. W. Barber!* Natal, *Miss Owen!* (Herb. T.C.D., Hook., Sond.)

A graceful suffrutex, with distant, digitate leaves and large flowers on long peduncles. The petals are either yellow, or purple, or pale with dark purple bases. E. & Z. say that the leaves have black margins, whence the specific name; this character may be obvious on the living plant, but is not so on our dried specimens.

17. H. gossypinus (Thunb. Cap. p. 549.); *shrubby*, the rod-like branches and short petioles *densely setose with red-brown, rigid hairs*; leaves *ovate or elliptical, obtuse, serrulate*, pubescent on the upper, tomentose on the lower surface; stipules setaceous; peduncles axillary, longer than the leaves, jointed and swollen below the flower, densely tomentose; invol. of 10–12 filiform, hairy leaflets, shorter than the lanceolate, rufo-tomentose calyx lobes; capsule obtuse; *seeds clothed with long hairs. DC. Prod. 1 p. 453. H. ferrugineus, E. & Z.! an Cav.? H. fuscus, Garke!*

HAB. Near the Luris River, *Thunberg!* Kat River, *E. & Z.!* South Africa, *Drege!* Port Natal, *Gueinzius! Sanderson!* (Herb. T.C.D., Hook., Sond.)

A tall shrub, with long, straight, simple branches; the pubescence generally dark red-brown. Flowers small, yellow; the petals externally stellate-hairy.

18. H. leptocalyx (Sond. in Linn. 23. p. 17.); suffrutescent; stems slender, *diffuse, branching*, hispid with long patent hairs, leaves petiolate, *triangular-hastulate or subcordate*, repand or sub-crenulate, obtuse, stellato-hispid, especially on the lower surface; stipules filiform; peduncles axillary, solitary, as long as the leaves; invol. of 10-12 narrow-linear, piloso-ciliate leaflets *twice as long as the ovate-acuminate calyx lobes;* capsule *depressed, transversely rigid and furrowed, glabrous, the seeds puberulous.*

HAB. Port Natal, *Gueinzius!* (Herb. Hook., Sond.)

Stems many from the same crown, spreading, alternately branched. Leaves scarcely an inch long, on petioles of their own length. Flowers small, reddish buff colour (" rosy" *Sond.*). Very like *Pavonia odorata*, Willd., but besides differing generically, it wants the glandularly viscid pubescence of that species.

**** *Involucel many leaved. Root annual.* (Sp. 19–22.)

19. H. Trionum (Linn.); annual, hairy; leaves *polymorphous*, mostly deeply 3–5 lobed or partite, the segments coarsely toothed and incised, the lower leaves (sometimes all the leaves) ovate or oblong, less densely cut or simply toothed; peduncles axillary; involucels of about 10 narrow-linear, hairy and ciliate leaflets; calyces *5-angled, inflated, becoming bladdery after flowering*, the segments broadly ovate, blunt, with many-hispid nerves; capsules many-seeded, hispid, seeds smooth. DC. Prod. 1. p. 453. Cav. Diss. t. 64. f. 1. 2. 3. Bot. Mag. t. 209.

VAR. α. **hispidus**; lower leaves deeply 3-lobed, upper 3-5 parted, incised. *DC. l. c. E. & Z.! ex parte!*

VAR. β. **ternatus**; leaves mostly tripartite.

VAR. γ. **cordifolius**; lower leaves roundish-cordate, upper tripartite *H. pusillus, E. & Z.!* 314. *non Thunb.*

VAR. δ. **hastæfolius**; leaves hastate, 3-lobed, the middle lobe very long. *H. hastæfolius, E. Mey!*

VAR. ε. **cordatus**; all the leaves roundish cordate, obtuse, dentate. *H. uniflorus, E. Mey!*

VAR. ζ. **lanceolatus**; all the leaves ovato-lanceolate, serrate. *H. physodes, E. Mey!*

HAB. In waste, garden ground and in rich, damp soil throughout the colony, Caffraria and Port Natal, common. (Herb. T.C.D., Sond., &c.).

A common sub-tropical weed, with handsome primrose-coloured or buff flowers, with a dark-brown, velvetty centre. The species is best known by its bladdery, many-ribbed calyx.

20 H. cannabinus (Linn.); annual, nearly glabrous; *stem and petioles prickly;* leaves on long petioles, palmately 5-7-parted, the segments lanceolate, acute, serrulate, the middle nerve with a gland at base; flowers (large) *subsessile*, axillary; involucel of 9-10 subulate, bristly leaflets half as long as the calyx lobes, which are deltoid-acuminate, bristly or prickly, *and each having a large medial gland;* capsule globose, acuminate, very hairy, seeds glabrous. *DC. Prod.* 1. *p.* 450. *W. & A. Prod. p.* 50. *Roxb. Corom. t.* 190. *Cav. Diss. t.* 52. *f.* 1.

HAB. Between Omsamculo and Omcomas, and towards Port Natal, *Drege!* (Herb. Sond.).

4–6 feet high, with foliage like that of *Hemp*, and large white flowers, with a dark purple centre. Much cultivated in India for its *leaves*, which are used as a vegetable, and for its fibre, from which an inferior kind of hemp is prepared.

21. H. furcatus (Roxb. Cat. p. 31); annual; the branches, peduncles, petioles, and nerves of the leaves *armed with hooked tubercles;* leaves deeply 5-lobed, the upper ones 3-lobed, nearly glabrous, unequally toothed; *stipules lanceolate;* peduncles axillary, one-flowered; invol. of *many spathulate leaflets, each furnished in front with an inflexed, subulate appendix;* calyx lobes deltoid-acuminate, with a densely setose mid-rib. *DC. Prod.* 1. *p.* 449. *H. hamatus, E. Mey.!*

HAB. Port Natal, under 500 f., *Drege!* (Herb. Hook.).

Drege's specimens that we have seen are very imperfect, but unmistakeably referable to this common Indian species. Flowers yellow.

22. S. Surattensis (Linn); annual; the branches, peduncles and petioles armed with hooked tubercles; leaves deeply 5-lobed, with lanceolate, toothed lobes; *stipules broadly semi-cordate, clasping the stem;* peduncles axillary, one-flowered; invol. of many spathulate leaflets, each furnished in front with an inflexed, subulate appendix; calyx-lobes deltoid, acuminate, with a densely setose mid-rib. *DC. Prod.* 1. *p.* 449. *Bot. Mag. t.* 1356. *H. hypoglossum, E. Mey.!*

HAB. Port Natal, *Drege! Sanderson! Dr. Grant!* (Herb. Hook.).

Chiefly distinguished from *H. furcatus* by the shape and size of the stipules. *Flowers* yellow with a deep red or purple centre. The curious form of involucral leaflets which marks these two species may be referred to what is called, "*deduplication,*" and its occurrence in Malvaceæ, where the stamens (as Dr. Gray has ably shown) are developed in a similar way, is not without significance.

X. PARITIUM. A. Juss.

Involucel monophyllous, 10-12 cleft or toothed. *Petals* and *Staminal tube* as in *Hibiscus. Ovary* 5-celled, each cell partially divided into two by a spurious, parietal dissepiment; *ovules* numerous. *Capsule* 5-celled, the cells imperfectly bilocular. *Endl. Gen.* 5283. *Hibisci Sp., DC. et Auct.*

Tropical trees or shrubs, with cordate, entire or lobed leaves, glandular on the under surface. *Stipules* broad, ovate. *Flowers* yellow, with dark centre. The inner bark affords a strong fibre, suitable for many textile purposes, for cordage and paper. Cuba "*bast*" is the product of one of this genus. The name *Paritium* is an alteration of *Pariti*, the native name of these plants in India, fide *Rheede, Hort. Mal.* 1. *t.* 30.

1. P. tiliaceum (S. Hil. Fl. Braz. 1. p. 198); a tree; leaves on long petioles, roundish-cordate, with a sudden acumination, quite entire or crenulate, 5–7-nerved, glabrous on the upper, velvetty and canescent on the lower surface; stipules ovate, deciduous; flowers in terminal cymes; involucel 10-toothed. *Hibiscus tiliaceus, Linn. DC. Prod.* 1. *p.* 454. *Cav. Diss.* 3. *t.* 55. *Bot. Reg. t.* 232. *Wight, Ic. t.* 7.

HAB. Sea-shores. Mouth of the Omsamcaba, *Drege!* Port Natal, *T. Williamson*, &c. (Herb. T.C.D., Hook.).

An umbrageous tree, 20–30 feet high, with a round head, and leaves not unlike those of the lime tree in form, but of a thick, leathery substance, pale and softly velvetty beneath. Flowers either from the axils of the terminal leaves, or in cymoid few-flowered panicles, yellow, with a dark, purplish-brown centre.

ORDER XXI. STERCULIACEÆ, Vent.

(By W. H. HARVEY.)

(Sterculiaceæ, Vent. Malm. 2. 91. Endl. Gen. No. ccx. Lindl. Veg. Kingd. No. cxxvi. Bombaceæ, DC. Prod. 1. p. 475.)

Flowers mostly regular, frequently unisexual and apetalous. *Calyx* 5–7-fid, with valvate æstivation. *Petals* 5 or none, convolute. *Stamens* indefinite, monadelphous; anthers two-celled, opening outwards. *Ovary* of 3–5 carpels, separate or united in a plurilocular pistil; ovules numerous or definite, axile; styles more or less confluent. *Fruit* various; a capsule or berry, or of several follicles, dehiscent or indehiscent. *Seeds* with fleshy albumen or exalbuminous, and a straight or curved embryo, with plaited or flat cotyledons.

Trees, often of vast size, rarely *shrubs*, glabrous or with stellate pubescence. *Leaves* alternate, simple or digitately compound, with deciduous stipules or exstipulate. *Flowers* in some large and conspicuous; in others small and green.

Nearly related, in character, to the Malvaceæ, from which these plants are technically distinguished by the bilocular anthers, opening outwards. The habit is, however very different. The Order is almost exclusively tropical, with the exception of a few Sterculias found in New Holland, and the solitary South African representative. In the tropics the *Adansonia* (or *Baobab*), the *Silk Cotton Tree* (*Bombax Ceiba*), and others are among the grandest objects of the vegetable world. They are more remarkable for beauty than for useful products. The *Durian*, the most delicious but stinking fruit of the Malay peninsula, is yielded by a tree of this Order, *Durio Zibethinus*.

I. STERCULIA, L.

Flowers unisexual, without petals. *Calyx* coloured, campanulate or tubular, 5–7 cleft or parted; the segments leathery. *Males: staminal-tube* shorter or longer than the calyx, solid, bearing anthers at the 5–10-toothed extremity; anthers adnate, extrorse, 2-celled. *Female: staminal-tube* adnate to the carpophore, the abortive anthers surrounding the base of the ovary. *Ovary* stipitate, of 5 connivent or partially connate carpels: *styles* more or less united; *ovules* numerous. *Follicles* 5 or fewer, leathery or ligneous, few or many-seeded. *Endl. Gen. No.* 5320. *DC. Prod.* 1. *p.* 481.

Tropical trees, common to both hemispheres, but most numerous in Asia and Africa. Leaves alternate, petiolate, simple or digitately compound. Stipules deciduous. Panicles or racemes axillary or sub-terminal; flowers red, yellow, or particoloured, pubescent or woolly.

1. S. Alexandri (Harv.); leaves digitately compound, quinate; leaflets oblong, obtuse, mucronulate, acute at base, glabrous, reticulate; racemes few-flowered, shorter than the petiole; calyx campanulate, 5–7-cleft, downy on both surfaces. *Harv. Proc. Dub. Un. Zool. & Bot. Assn.* 1. *p.* 140. *t.* 15. *Harv. Thes. Cap. p.* 1. *t.* 3.

HAB. In a narrow mountain valley, near Uitenhage, *Dr. Alexander Prior!* (Herb. Prior., T.C.D., Hook.).

A small tree, with a thick stem, and soft spongy wood. Leaves crowded near the ends of the branches, the petioles 3–4 inches long; the leaflets somewhat longer and ¾–1 inch broad, closely reticulated with veinlets, very obtuse, with a minute or obsolete mucron, towards the base narrowed or cuneate. Allied to *S. fœtida*, but differing in the obtuse not acuminate leaflets; in the short not elongate racemes; and in the downy not tomentose inner surface of the calyx. No one seems to have met with it but Dr. Alexander Prior, who found but a solitary tree, in a narrow kloof, somewhere among the Van Staaden Mts.; a locality rich in interesting plants, and probably still concealing other novelties.

ORDER XXII. BYTTNERIACEÆ, R. Br.

(By W. H. HARVEY.)

(Byttneriaceæ, R. Br. DC. Prod. I. p. 481. Endl. Gen. No. ccxl. Lindl. Veg. Kingd. No. cxxvii.)

Flowers regular. *Calyx* 4-5 cleft or parted, with valvate æstivation. *Petals* as many as the calyx lobes, clawed, with twisted (or induplicate-valvate) æstivation, or none. *Stamens* hypogynous, sometimes as many as the petals and opposite to them, sometimes twice or several times as many, some barren and opposite the sepals; filaments united at base or into a cup or tube; anthers two-celled, introrse. *Ovary* of 4–10 (rarely of one) carpels united round a central column; ovules in pairs, axile; style single, stigmas as many as the carpels. *Fruit* capsular, rarely indehiscent. *Seeds* albuminous, with a straight or curved embryo.

Trees, shrubs, or *undershrubs,* rarely *herbaceous,* with stellate pubescence. *Leaves* alternate, simple, penninerved or palmate-nerved, entire or pinnatifid, stipulate. Flowers variously arranged.

Natives of the tropics of both hemispheres, extending into the warmer parts of the temperate zone, and particularly abundant in Australia and South Africa. This Order is nearly allied to Sterculiaceæ, from which it chiefly differs in its introrse, not extrorse, anthers. From Malvaceæ it is known by the definite stamens, and bilocular anthers; and from Tiliaceæ by the monadelphous stamens. All these Orders are separated by minor characters, and might, without violence, be united. The most important product of the present group is *Cacao,* from which chocolate is prepared, and which is yielded by several American trees and shrubs of the genus *Theobroma.*

TABLE OF THE SOUTH AFRICAN GENERA.

Sub-order 1. HERMANNIEÆ. *Stamens* 5, opposite the petals, slightly connate at base; no barren stamens.

I. **Waltheria.**—*Ovary* unequal-sided, unilocular; style lateral. *Capsule* one-celled, one-seeded.

II. **Hermannia.**—*Ovary* 5-celled. *Filaments* flat, broadly linear, oblong, or obovate. *Capsule* 5-celled, many-seeded.

III. **Mahernia.**—*Ovary* 5-celled. *Filaments* abruptly dilated in the middle (or shaped like a cross). *Capsule* 5-celled, many-seeded.

Sub-order 2. DOMBEYACEÆ. *Stamens* 15–40, monadelphous at base, rarely all perfect; usually the 5 which are opposite the petals sterile, strap-shaped.

IV. **Dombeya.**—*Petals* marcescent. *Fertile-stamens* 10-15.

V. **Melhania.**—*Petals* deciduous. *Fertile-stamens* 5.

I. **WALTHERIA**, Linn.

Calyx campanulate, 5-cleft, 10-nerved, with or without a 3-leaved, lateral involucel. *Petals* 5, attached to the base of the stamen-tube, oblong, with slender claws. *Stamens* 5, opposite the petals; filaments united below into a tube; anthers oblong, erect, 2-celled. *Ovary* unequal-sided, unilocular; ovules 2, one over the other. *Style* one, lateral; the stigma multifid. *Capsule* unilocular, two-valved, one-seeded. *Endl. Gen.* 5336. *DC. Prod.* 1. *p.* 492.

Herbs, shrubs, or small trees, common throughout the tropics of both hemispheres, and straggling into the warmer temperate zone. *W. indica*, our only species, is an extremely variable plant, being sometimes almost herbaceous and sometimes nearly arborescent. Under various forms and names it is found in all hot countries. The name is in honour of *Augustin F. Walther*, a Saxon botanist of the last century, the owner of a botanical garden at Leipsic, and author of several botanical works,

1. W. indica (Linn.); shrubby, densely clothed with stellate and simple hairs; leaves ovate, oblong, or ovato-lanceolate, petiolate, plaited, unequally serrate, penninerved, tomentose; stipules subulate, withering; heads of flowers axillary, sessile or pedunculate, dense. *DC. Prod.* 1. *p.* 493; *also D. Americana, L., DC. l. c., and several other reputed species, see W. & A. Prod. Fl. Ind. p.* 67.

HAB. Macállisberg, *Zeyher!* (Herb. Sond., T.C D., Hook.).

A more or less tomentose, erect or prostrate, branching shrub, very variable in aspect and in the form of the leaves, which are 1–2 inches long, and very roughly hairy. Flowers small, in dense clusters, not quite sessile in our specimens. When the heads are manifestly pedunculate, it becomes *W. Americana*, Auct.

II. **HERMANNIA**, Linn.

Calyx campanulate, 5-cleft, exinvolucrate, often inflated. *Petals* 5, hypogynous, with hollow claws, spirally twisted in æstivation. *Stamens* 5, opposite the petals; filaments connate at base, flat, oblong or obovate. *Ovary* shortly stipitate, 5-celled; styles coalescing, separable. *Capsule* coriaceous, 5-celled, 5-valved, many-seeded, simple or crested at the summit. *Endl. Gen.* 5340. *DC. Prod.* 1. *p.* 493.

Small shrubs and undershrubs, almost all South African; a very few from North Africa and from *Mexico!* Pubescence stellate or woolly, rarely glandular, copious or scanty. Leaves alternate, either entire, toothed, or pinnatifid, sinuated or lacerated, often plaited; stipules petiolar, leafy or small, very rarely absent. Peduncles axillary or sub-terminal, often in pseudo-racemes or panicles, 1–3-flowered. Flowers numerous, yellow or orange, rarely cream-colour, or reddish; often sweetly-scented. Name in honour of *Paul Hermann*, a botanical traveller, afterwards Professor of Botany at Leyden; died in 1695. The species are subject to much variation, and perhaps many which are here retained may eventually be cut down, when fuller materials are consulted.

ANALYSIS OF THE SPECIES.

Sub-genus 1. EUHERMANNIA. Capsule rounded or 5-umboned at summit (*not horned or crested*). (Sp. 1–65.)

Group 1. **Althæoideæ.** Diffuse or decumbent *undershrubs, or herbs.* Leaves ovate or oblong, on longish petioles, plaited, crenate or rarely incised. Pubescence stellate and tomentose, copious or scanty. Flowering branches naked, or with a few small leaves, ending in a lax raceme or panicle. (Sp. 1–10.)

* *Calyx inflated; furnished with ciliated appendages* ... (1) **comosa.**

** *Calyx inflated, tomentose or villous:—*

Densely tomentose; stipules broad; petals with narrow, woolly-edged claws... ... ... (2) **althæoides.**

Stellato-canescent; stipules narrow-lanceolate; petals with broad claws, stellate at back... ... ... (3) **leucophylla.**

Thinly pubescent; stipules and bracts broad; claws of petals ciliate ... ... (4) **decumbens.**

Stellato-pubescent; leaves green above; stipules ovate (7) **macrophylla.**

*** *Calyx campanulate or cup-shaped:—*

Procumbent, glabrescent; stipules ovate; capsule short (5) **procumbens.**

Procumbent, stell.-pubescent; stipules cordate; capsule 3–4 times as long as the calyx ... ... (8) **prismatocarpa.**

Erect or diffuse; albo-tomentose; stipules subulate; capsule short ... ... (6) **candicans.**

Erect, sub-dichotomous; branches, glabrescent; stipules triangular; filaments linear-oblong ... ... (9) **scordifolia.**

Erect, sub-dichotomous; branches hispid and scabrous; stip. ovato-lanceolate; filaments obovate ... ... (10) **patula.**

Group 2. **Cuneifoliæ.** Much branched, rigid shrubs, erect or spreading. Leaves on short petioles, *cuneate and entire at base,* toothed towards the point, oblong, obovate or elliptical, plaited. Pubescence copious, stellate, or velvetty or scaly. Flowering branches short, leafy, *loosely panniculate or racemose.* (Sp. 11–18.)

* *Pubescence tomentose or coarsely stellate:—*

Leaves elliptic-oblong, very densely albo-tomentose and stellate; stipules subulate ... ... (11) **disermæfolia.**

Leaves sub-rotund, stell.-pubescent; stipules ovate... (12) **disticha.**

Leaves cuneate; stipules subulate; anthers short, obtuse... ... ... (13) **rigida.**

Leaves cuneate or obovate, stell.-pubescent; stipules ovate-cuspidate; calyx albo-tomentose ... ... 14) **cuneifolia.**

Leaves cuneate, albo-tomentose; stip. subulate; calyx minutely stellulate ... ... (15) **desertorum.**

Leaves cuneate, undulate, stell.-pubescent; flowers *very small,* panicled ... ... (16) **alnifolia.**

** *Pubescence velvetty and canescent, equally spread:* flowers panicled ... ... (17) **holosericea.**

*** *Pubescence squamulose and silvery;* stem divaricately branched ... ... (18) **pallens.**

Group. 3. **Scaberrimæ.** Suffruticose or shrubby. Leaves on short petioles, or sessile, oblong, coarsely crenate or sinuate, rarely sub-entire. Stipules leaf-like, mostly amplexicaul. Pubescence copious and *very rough; the stiff, spreading, stellate hairs not uniformly spread,* but separated by bald interspaces. Peduncles loosely racemose. (Sp. 19–25.)

* *Leaves white on the under surface:*

Leaves plaited, unequally toothed ... ... (19) **hirsuta.**

Leaves inciso-sinuate, or pinnatifid ... ... (20) **dryadifolia.**

** *Leaves green on both sides:—*

Suffruticose; fl. aggregated multibracteate ... ... (21) **orophila.**

Suffruticose; leaves short-stalked, oval-oblong, subentire, stipules cordate; flowers racemose ... ... (24) **latifolia.**

Shrubby; leaves short-stalked, cuneate, 3–5-toothed; stip. ovato-lanceolate: fl. racemoso-paniculate ... (22) **scabra.**
Shrubby; leaves sessile, oblongo-lanceolate, sub-entire; stipules cordate; fl. racemose ... ... (23) **stipulacea.**
Shrubby; leaves sub-sessile, imbricate, oval-oblong; stipules and bracts subulate; fl. racemose ... ... (25) **decipiens.**

Group 4. **Glomeratæ.** Much branched, rigid shrubs. Leaves on short petioles, oblong or ovate, entire or crenate. Pubescence copious, *closely set, and uniformly spread*, roughly stellate. Flowering branches leafy; *the peduncles very short, densely crowded in a terminal cluster.* (Sp. 26–29.)

* *Leaves quite entire or nearly so:—*
Leaves cordate-ovate; stip. obliquely decurrent, deltoid ... ... (26) **cordifolia.**
Leaves oblong or oval; st. subulate or lanceolate; calyx membranous and sub-inflated ... . (27) **salvifolia.**
Leaves roundish-oval; st. subulate; calyx thick, 5-ribbed, stellato-scabrous... ... ... (28) **chrysophylla.**

** *Leaves sub-rotund, crenate or cut,* plaited and undulate; stipules cordate, amplexicaul... ... ... (29) **glomerata.**

Group 5. **Flammeæ.** *Virgate* shrubs and under-shrubs. Leaves sessile or sub-sessile, cuneate, lanceolate or linear. Pubescence *scanty*, never copious, but variable in quantity, thinly stellate; the hairs short or appressed. Peduncles short, in a terminal raceme. (Sp. 30–37.)

* *Either quite glabrous, or the branches minutely stellulate.*
Leaves *coarsely toothed* above the middle... ... ... (30) **denudata.**
Leaves *entire;* linear-lanceolate or spathulate:—
Stipules much shorter than the leaves ... ... (31) **scoparia.**
Stipules equalling the flat leaves; branches glabrous (32) **linifolia.**
Stipules equalling the channelled leaves; branches scabrous ... ... (33) **filifolia.**

** *Branches and leaves more or less stellulate or hispid.*
Calyx semi-5-fid, the lobes deltoid; leaves thinly stellulate ... ... (34) **flammea.**
Calyx deeply 5-parted, the lobes narrow; leaves densely pubescent ... ... (35) **flammula.**
Calyx campanulate, 5-toothed, teeth short, taper-pointed ... ... (36) **Joubertiana.**
Calyx *inflated*, prominently 5-angled, *glabrous and glossy* ... ... (37) **angularis.**

Group 6. **Velutinæ.** *Canescent shrubs* (rarely suffrutices) *covered with very short and close, velvetty or powdery pubescence.* Leaves shortly petiolate or sessile, entire or denticulate. Stipules leaf-like. Peduncles short, either crowded or racemose. (Sp. 38-48.)

* *Calyx inflated, globose or urceolate.*
Leaves petiolate, obovate or oblong, obtuse;
serrate near the point, thinly pubescent ... .. (38) **hyssopifolia.**
entire, softly villous ... ... (41) **suavis.**
Leaves cuneate; stipules spathulate; calyx scaly ... (39) **ternifolia.**
Leaves cuneate; stip. oval or oblong; calyx densely stellulate ... ... (40) **trifoliata.**
Leaves ovato-lanceolate; stip. broadly lanceolate, nerved ... ... (43) **mucronulata.**

** *Calyx campanulate or turbinate, 5-cleft, 5-angled.*

Pubescence squamulose; the scales close pressed ... (42) **diversistipula.**
Pubescence softly velvetty:—
Stipules narrow-subulate, scarcely longer than the petiole ... ... (45) **lavandulæfolia.**
Stipules oblong or lanceolate, $\frac{1}{3}$–$\frac{1}{2}$ as long as leaf:
Calyx stellato-tomentose, campanulate ... ... (44) **velutina.**
Calyx obconic, the teeth triangular-acuminate ... (45) **Cavanillesiana.**
Calyx obconic, strongly ribbed and furrowed ... (47) **sulcata.**
Calyx downy, with short, distant, deltoid teeth (48) **gracilis.**

Group 7. **Lateriflorœ.** Peduncles *axillary,* distributed along the leafy branches, *not obviously racemose.* Leaves undivided. Pubescence stellate or tomentose. (Sp. 49-56.)

* *Petals longer than the stamens.*

Leaves rough and wrinkled or plaited, crenate toothed or lobed:
Stems nearly simple; leaves sub-sessile ... ... (49) **Sandersoni.**
Much branched; leaves on long petioles:—
Calyx globose; filaments obovate ... ... (50) **candidissima.**
Calyx campanulate; fil. linear ... ... (51) **floribunda.**
Leaves flat, quite entire or nearly so:—
Stipules none; calyx densely tomentose ... ... (55) **exstipulata.**
Stipules small, triangular; calyx thinly woolly ... (56) **Gariepina.**

** *Petals much shorter than the connivent, taper-pointed anthers.*

Erect, glandular; stipules ovate; petals obovate, equalling the calyx ... ... (52) **boraginiflora.**
Erect, not glandular; stip. subulate; petals *cucullate* and truncate, very small ... ... (53) **micropetala.**
Procumbent, canescent; stip. subulate; petals spoon-shaped, very small ... ... (54) **brachypetala.**

Group 8. **Pinnatifidæ.** Leaves deeply pinnatifid, bi-pinnatifid or multifid. Habit various. (Sp. 57-65.)

* *Calyx campanulate:—*

Leaves covered or sprinkled with flat, silvery scales:
pinnatifid; peduncles 2-flowered ... ... (57) **pulverata.**
bi-pinnatifid; peduncles 1-flowered ... ... (58) **argentea.**
Leaves stellato-pubescent; *petals entire:*
Branches stellulate; leaf-lobes narrow-linear (59) **tenuifolia.**
Branches glabrous; leaf-lobes cuneate; fl. buds acuminate ... ... (60) **paucifolia.**
Branches canescent; leaf-lobes cuneate; fl. buds ovate ... ... (61) **chrysanthemifolia**
Leaves stellato-pubescent; *petals deeply inciso-pinnatifid* ... ... (65) **incisa.**

** *Calyx globose and inflated:—*

Leaves bi-pinnatifid, with linear, obtuse lobes ... (62) **abrotanoides.**
Leaves palmati-partite and pinnatifid; ped. 1-flowered (63) **multifida.**
Leaves palmati-partite and pinnatifid; ped. 2–4-flowered ... ... (64) **halicacaba.**

Sub-genus 2. Acicarpus. Capsule truncate, crowned with 5-10 divergent, simple or plumose horns. (Sp. 66-70.)

* *Stems and branches divaricately much branched:—*

Leaves cuneate, 3–5-toothed ... ... (67) **spinosa.**
Leaves narrow-linear, entire ... ... (68) **linearifolia.**

** *Stems and branches virgate, slender*:—

Canescent; leaves linear-cuneate, entire or 2-3-toothed (66) **trifurcata.**
Stellulate; leaves linear-lanceolate, acute, entire ... (69) **filipes.**
Glabrous; leaves cuneate; petals thrice as long as calyx (70) **stricta.**

Sub-genus 1. Euhermannia.

Capsule rounded or 5-umboned at the summit (not horned or crested). (Sp. 1-65.)

Group 1. Althæoideæ. Diffuse or decumbent undershrubs or herbs. Leaves ovate or oblong, on longish petioles, plaited, crenate, or rarely incised. Pubescence stellate and tomentose, copious or scanty. Flowering branches naked, or with a few small leaves, ending in a lax raceme or panicle. (Sp. 1–10.)

1. H. comosa (Burch. Cat. 1683); "leaves tomentose, ovate, sinuato-dentate; stipules lanceolate-linear; peduncles 2-flowered; calyx *inflated, crinito-comose with ciliated appendices* (calycibus inflatis appendicibus ciliatis crinito-comosis)." *DC. Prod* 1. *p.* 493.

Hab. S. Africa, *Burchell*. (Unknown to me.)

2. H. althæifolia (Linn.); diffuse or decumbent, *densely tomentose and patently villous;* lower leaves on long petioles, ovate or sub-cordate, the upper oblong, rhomboid, or obovate, plaited or rugose, sub-entire or sinuato-dentate or crenate, 5–7-nerved, softly tomentose on both sides, sub-canescent; stipules *leafy, 3–5-nerved, ovate or ovato-lanceolate;* peduncles 2–3-flowered, longer than the upper leaves; bracts *linear or lanceolate,* as long as the pedicels; calyx tomentose and pilose, inflated, shortly 5-fid, the teeth deltoid-acute; petals twice as long as the calyx, *their narrow claws fringed with long white hairs,* longer than the roundish limb; ovary oblong, sub-sessile, densely hispid. *Jacq. Hort. Schoenb. t.* 213. *H. althæifolia and H. plicata, DC. Prod.* 1. 493. *Cav. Diss. t.* 179. *f.* 2. *Bot. Mag. t.* 307. *H. aurea, Jacq. l. c. t.* 214. *E. & Z.!* 323, *H. diversifolia, E. & Z.!* 324. *H. disermæfolia, E. & Z.!* 325, *non Jacq. H. fragrans, Link. Drege,* 7300–7302. *Zeyher,* 2000.

Hab. Common on the hills round Capetown, Cape district, Clanwilliam, Beaufort, Swellendam, and Graaf Reynet, *E. & Z.! Drege!* (Herb. T.C.D., Hook., Sond.).

I cannot certainly distinguish "*H. plicata,*" Willd., from the older *H. althæifolia.* It is said to have more cordato-ovate and hairy leaves, and ovate stipules; but leaves and stipules vary on the same specimen, and on specimens from the same locality, and I am persuaded that the more numerous the specimens examined, the less distinct will the two forms be found to be. *H. diversifolia, E. & Z.!* was founded on imperfect and half-starved specimens. *H. plicata, E. & Z.! non Willd.* is *H. candicans, β. H. plicata, Drege!* is *H. leucophylla.* Flowers large, bright yellow.

3. H. leucophylla (Presl, Bot. Bem. p. 20); diffuse or decumbent, *stellato-tomentose and canescent;* lower leaves on long petioles, cordate at base, ovate-oblong, sinuate and crenate, plaited and rugose, 3-5-7-nerved, stellato-pubescent on both sides, canescent on the lower; *stipules narrow-lanceolate, acute;* peduncles 2-flowered, short or longish, the bracts *subulate;* calyx *thinly tomentose* and stellato-pubescent, inflated, membranous, shortly 5-fid, the teeth deltoid; petals not much exceeding the calyx, *their broad claws stellato-pubescent at back*, much longer than the roundish limb ; ovary pedicellate, broader than long, stellato-pubescent. *H. plicata, E. Mey.! in Herb. Drege.*

HAB. Winterveld and Nieuweveld, *Drege!* (Herb. T.C.D., Hook., Sond.).

Very similar to *H. althæifolia*, of which it has the habit and foliage; but the stipules and bracts are smaller, narrower and more acuminate; the pubescence is more stellate and less tomentose, without intermixture of long, white hairs; the petals have much broader claws and smaller limbs, and are stellato-pubescent at back, instead of being fringed with long white hairs; the ovary is shorter, and evidently stipitate. These characters, if constant, may define the species.

4. H. decumbens (Willd.! Suppl.); decumbent, suffruticose, more or less *fasciculato-pubescent or glabrescent;* leaves petiolate, elliptic-oblong or ovate, obtuse, either sub-entire, crenate or coarsely toothed, sparsely stellate or glabrescent on the upper, more densely stellato-pilose on the lower surface ; *stipules ovate or ovato-lanceolate*, ciliate or sub-glabrous ; *flowers in a pedunculate, terminal, lax raceme*, pedicels short, very hispid; *bracts broadly ovate or ovato-lanceolate, glabrescent;* calyx somewhat inflated, semi-quinquefid, hispid or villous ; lobes deltoid, acute ; petals twice as long as the calyx, *the ciliate claw* longer than the limb ; filaments narrow-spathulate, longer than the anthers ; ovary ellipsoid, pubescent. *DC. Prod.* 1. *p.* 494. *E. & Z.!* 328; *also H. pratensis, E. & Z.!* 327; *and H. collina, E. & Z.!* 326. *H. argyrata, Presl! Drege*, 7303, 7304.

VAR. α. **hispida**; densely stellato-hispid, especially on the young branches and peduncles. *Willd.! Herb. Ber.* 12,327. *H. decumbens, E.&Z.!*

VAR. β. **argyrata**; stellato-canescent, with very broad bracts.

VAR. γ. **collina**; the leaves glabrescent, sub-entire ~~or~~ crenulate. *H. collina and H. pratensis, E. & Z.!*

HAB. Sandy flats. Riedvalley, Cape; Langehoogde and near the Pot and Klynrivers, Caledon, *E. & Z.!* Between Cape L'Agulhas and the Potberg, *Drege!* Kalebaskraal, *Zeyher!* (Herb. Willd., T.C.D., Hook., Sond.).

A smaller, rougher and greener plant than *H. athæifolia*, with shorter calyces, a more racemose inflorescence, and very broad, sub-glabrous bracts. All the above forms run into each other.

5. H. procumbens (Cav. Diss. t. 177. f. 2.); *half-herbaceous, procumbent, glabrescent;* the branches long, sub-simple, ascending, *sparsely-stellulate;* leaves sub-distant, petiolate, *oblong or obovate, obtuse, coarsely toothed*, cuneate at base, *glabrous or thinly stellulate on the under-side;* stipules *ovate*, shorter than the petiole ; peduncles racemose, drooping, short, one flowered ; bracts minute ; calyx *campanulate, nearly glabrous*, shortly 5-fid, the lobes ovate, acute ; petals twice as long as the calyx, the claw as long as the limb ; filaments cuneate ; much longer than the anthers ; ovary obovoid, pubescent. *DC. Prod.* 1. *p.* 495. *E. & Z.* 329, *ex parte.*

HAB. Common among the *Euphorbia Caput Medusæ* at Greenpoint, Capetown, in places where water lies in winter. (Herb. T.C.D., Hook., Sond.)

Stems several from a woody crown, trailing on the ground and generally quite simple, 1–2 feet long. Young leaves thinly stellulate, older glabrous or nearly so. Flowers orange-yellow, in a secund, terminal raceme. Some of E. & Z.'s specimens under their No. 329 belong to *Mahernia vesicaria;* these seem to be what

Presl proposed to call *H. leucanthemoidea*. I cannot guess what the *H. Zeyheriana*, Presl may be.

6. H. candicans (Ait. Hort. Kew. 2. p. 412.); suffruticose, *erect or diffuse;* branches albo-tomentose or stellato-pubescent, rarely glabrescent; leaves on long petioles, elliptical or ovate-oblong, obtuse, subentire or undulato-crenate, flat or somewhat plaited, *albo-tomentose on one or both sides; stipules subulate or narrow lanceolate*, much shorter than the petiole; peduncles racemoso-paniculate, 1–3 flowered, bracteate near the summit; bracts *subulate* or lanceolate; flower-buds tomentose; calyx cup-shaped, albo-tomentose or pubescent, angular, semi-5-fid, the lobes deltoid-acuminate; petals twice as long as the calyx, the glabrous claw longer than the oval-oblong limb; filaments obovate, equalling the hastate anthers; ovary albo-tomentose. *Jacq. Hort. Schoenb. t.* 117. *DC. Prod.* 1. *p.* 494. *H. præmorsa, Wendl., non E. & Z. Drege* 7297, 7299.

VAR. *α.* **incana**; leaves albo-tomentose on both sides; stem canescent, sub-erect. *H. incana, E. & Z. !* 333. *H. althæoides, E. & Z. !* 334, *ex parte. H. mollis, E. & Z. !* 336. *Willd. Enum.* 692. *DC. l. c. Zeyher*, 112.

VAR. *β.* **discolor**; leaves green and stellato-pubescent above, albo-tomentose below. *H. candicans, E. & Z. !* 327. *H. plicata E. & Z. !* 341. *H. nemorosa, E. & Z. !* 335. *H. discolor, Otto ! and Dietr.*

VAR. *γ.* **fistulosa**; stems fistular, decumbent, glabrescent; leaves minutely stellulate on the upper, canescent on the lower surface; flowers in pedunculate racemes. *H. fistulosa, E. & Z. !*

VAR. *δ.* **depressa**; prostrate, robust, but dwarf; leaves smaller, canescent below.

HAB. Frequent in Uitenhage and Albany. George, *E. & Z. !* Schneeberg and various places in Caffirland, *Drege !* Somerset, *Mrs. F. W. Barber !* var. *γ.* near the Berg River, *E. & Z. ! δ*, near the Zwartkops, *Zeyher !* (Herb. T.C.D.. Hook., Sond., &c.).

Very variable in size and pubescence, but tolerably constant in other characters. I have no hesitation in uniting, under one head, the various specimens of E. & Z. above quoted. Var *γ.* is the most glabrous, and seems to have grown in a very moist station, whence also its fistular stem; *δ*, on the other hand is starved with drought. *H. incana*, Thunb. ! referred here by E. & Z., belongs to *H. holosericea.* Willdenow's *H. mollis*, described from garden specimens, resembles some of the more canescent wild specimens.

7. H. macrophylla (Turcz. Bull. Mosc. 1858. p. 216.); "stem decumbent, fistular, glabrescent, branched; branches, petioles and peduncles stellately hairy; leaves petiolate, ovate, or ovate-oblong, obtuse, somewhat incised, the lobules crenate; the upper leaves entire at base, sub-acute, green and sparsely stellate above, densely stellato-tomentose and silvery below; stipules *ovate, acute*, caducous; racemes axillary, compound, many-flowered, longer than the leaf; pedicels unilateral, cernuous, bracteolate; calyx *inflated.*"

HAB. South Africa, *Zeyher*, fide Turcz. l. c. (unknown to me).

This seems, judging by its stipules, to be some form of *H. althæifolia.*

8. H. prismatocarpa (E. Mey ! in Herb. Drege); suffruticose, procumbent; branches sparsely stellato-pilose; leaves on longish petioles, ovate-oblong, rugose and sub-plicate, unequally eroso-denticulate, glabrescent on the upper, stellato-canescent on the lower surface; *stipules cordate-ovate, amplexicaul, cuspidate, much shorter than the petiole; bracts ovate;* calyx *campanulate*, 5-angled, thickly stellulate, 5-toothed, the teeth short, deltoid, with wide interspaces; petals twice as long as the calyx, the claw longer than the limb; filaments obovate, somewhat longer than the cuspidate anthers; *capsule prismatic, 5-angled and 5-crested, 3-4 times as long as the calyx. H. hirsuta, E. & Z.!* 339, *non Schrad.*

HAB. Riebeckskasteel, *Drege!* Cape flats, *W.H.H.* Koeberg, *E. & Z.!* (Herb. T.C.D., Hook., Sond.)

A trailing or decumbent, slightly branched, half-herbaceous plant; the branches leafy below, naked above, and ending in a compound pseudo-raceme. Fl. yellow. It much resembles *H. althæifolia*, but is less tomentose, has a different calyx, and is readily known when in fruit, by its *long*, sharply 5-angled capsules.

9. H. scordifolia (Jacq.! Schoenb. t. 120.); suffrutescent, *erect, sub-dichotomous;* branches *naked*, or with few leaves, *nearly glabrous;* leaves petiolate, ovate-oblong, crenato-dentate, or inciso-lobulate, the lobules crenate, minutely stellulate on both surfaces; stipules *triangular, minute;* peduncles laxly panicled, 1-2 flowered, bracteolate; calyx campanulate, puberulous, 5-fid, the segments deltoid-acuminate; petals not much exceeding the calyx, the claw as long as the limb; filaments *broadly linear, equalling the cuspidate anthers;* capsule short, 5-angled, pubescent. *DC. l. c. p.* 494. *H. paniculata, E. Mey.!*

VAR. β. **integriuscula**; leaves crenate, not lobed. *Drege*, 7298. *Zey.* 117.

HAB. The Gariep, near Verleptram, on the river banks and in stony bottoms, *Drege!* β. Ebenezer, Oliphants R. *Drege!* Brandenberg, *Zeyher!* C.B.S., *Forsyth!* (Herb. Jacq., Hook., Benth., Sond.)

A very straggling species, with a few leaves below; the flowering branches nearly or quite bare of leaves; the flowers small, few and distant, growing from the axils of depauperated leaves. It resembles *H. patula*, but is much less hairy and its leaves are differently shaped. The leaves vary much in incision in the present species. Our specimens with crenulate leaves exactly agree with Jacquin's originals in *Herb. Vind.*

10. H. patula (Harv.); suffrutescent, *erect, sub-dichotomous;* branches *hispidulous and scabrid;* leaves on short petioles, *cuneate-obovate, dentate towards the apex*, sparsely stellato-hispid on both surfaces; stipules *ovato-lanceolate, equalling the petiole;* peduncles racemose-paniculate, 1-2-flowered, bracteolate; bracts lanceolate; calyx campanulate, sparsely hispid, semi-5-fid, the segments deltoid-acuminate, plumoso-ciliate; petals twice as long as the calyx, the ciliate claws as long as the limb; filaments *obovate, much longer than the hastate anthers;* ovary ovoid, hispid. *H. scordifolia, E. & Z.!* 344; *also H. denudata E. & Z.!* 344, and *Mahernia odorata, E. & Z.!* 405. *Drege*, 7294.

HAB. Mountain sides near Tulbagh, *E. & Z.!* Paarlberg, *Drege!* (Herb. T.C.D., Hook., Sond.)

A very diffuse, forking, roughly, but not densely pubescent species, allied to *H. scordifolia*. Branches flexuous. Leaves scattered, 1 inch long, not $\frac{1}{8}$ inch wide at top, tapering at base; the upper ones small, narrow and entire. Flowers orange.

Calyx-lobes remarkably acuminate, the flower-buds cuspidate. Capsule scarcely longer than the calyx, deeply 5-lobed, stellato-hispid.

Group 2. CUNEIFOLIÆ. Much branched, rigid shrubs, erect or spreading. Leaves on short petioles, *cuneate and entire at base*, toothed towards the point, oblong, obovate, or elliptical, plaited. Pubescence copious, stellate or somewhat velvetty or scaly. Flowering branches short, leafy, *loosely paniculate or racemose.* (Sp. 11-18.)

11. H. disermæfolia (Jacq ? Schoenb. t. 121.); shrubby, *densely albo-tomentulose;* leaves petiolate, elliptic-oblong, or oblongo-lanceolate, obtuse or sub-acute, dentate or crenate, undulate, plaited and somewhat rugulose, *albo-tomentose on both sides with minute fascicled hairs;* stipules subulate deciduous; peduncles racemose, 1–2-flowered; bracts subulate, calyx cup-shaped, shortly 5-fid, the teeth broadly deltoid, densely albo-tomentose; petals glabrous, the claw as long as the limb; filaments linear-spathulate, longer than the anthers; ovary albo-tomentose. *DC. l. c. p.* 494. *H. bryonifolia, E. & Z.!*

HAB. Moderfontein, 1500–2000f. *Drege!* Lislap, *Burke & Zeyher!* (Herb. T.C.D., Hook., Sond.)

A tall shrub, white-hairy in all parts; the hairs of the tomentum very short, but densely faciculate-stellate, the tufts looking under a pocket lens like microscopic hedgehogs. Leaves in our specimens seldom more than an inch long, ½ inch broad, thick in substance. Flowers yellow, in a loose raceme. Anthers acute, but not acuminate. I follow E. Meyer in referring this plant to Jacquin's.

12. H. disticha (Schr. & Wendl.! Sert. Han. t. 10.); shrubby, much branched; *young branches setose, older scabrous;* leaves on short petioles, *sub-rotund* or *obovate,* broadly cuneate at base, very obtuse, eroso-dentate, rugose, thinly hispid on the upper, densely stellato-hispid on the under side; stipules *ovate-acuminate;* peduncles racemose-paniculate, 1–2-flowered, the bracts ovate, cuspidate; calyx campanulate, *with piliferous glands on the ribs and margin, otherwise glabrous,* 5-fid, the lobes triangular, acuminate; petals about twice as long as the calyx, the claw rather longer than the limb; filaments oblong or obovate, longer than the acute anthers; ovary ovoid, tomentose. *DC. Prod.* 1. *p.* 494. *H. rotundifolia, Jacq.! Schoenb. t.* 118.

HAB. South Africa; locality uncertain, (Herb. Hook., Benth., Wendl.! Jacq.!)

A stout shrub, 3 feet high, diffusely branched, hispid in most parts with scattered or tufted hairs, rising from gland-like tubercles. Flowers small, yellow, in few-flowered panicles. Not found by E. & Z., or Drege. We describe from an old, probably garden specimen; compared with Wendl. and Jacq.'s originals.

13. H. rigida (Harv.) *suffruticose, rigid,* erect; branches patent, stellato-pubescent and scabrid; leaves petiolate, cuneate, round-topped, crenulate, plaited, densely *stellato-canescent on both sides;* stipules broadly subulate, *nerved;* peduncles racemose, distant, one-flowered, short; calyx campanulate, wide-mouthed, pubescent, semi-5-fid, the lobes deltoid-acuminate; petals twice as long; filaments obovate, *longer than the obtuse anthers;* capsule scarcely longer than the calyx, pubescent.

HAB. Namaqualand, Zeyher! 1126. (Herb. Sond.)

Allied to *H. candicans* and *H. patula,* as well as to *H. cuneifolia,* but different from any of these. It is a tall, straggling plant, remarkably rigid when dry, with a rough, but not a copious pubescence. The leaves are strongly plaited on the upper surface, the ribs very prominent below; they are ½ inch long, ⅓ inch wide; the petioles ¼ inch long. Flowers yellow.

14. H. cuneifolia (Jacq. ! Schoenb. t. 124.); *shrubby, much branched;* branches stellato-pubescent; leaves on short petioles, cuneate or obovato-cuneate, obtuse, crenato-dentate toward the point, stellato-pubescent on both sides; stipules *ovate-cuspidate* or broadly subulate; peduncles racemoso-paniculate, 2-flowered, spreading; bracts ovate-acute; calyx cup-shaped, *albo-tomentose,* 5-fid, the lobes deltoid; petals twice as long, the claws equalling the limb; filaments obovate, longer than the anthers; ovary oblong, tomentose. *DC. Prod.* I. *p.* 495. *E. & Z.* 342, and *H. præmorsa, E. & Z.!* 338. *H. alnifolia, Drege! H. multiflora, Jacq. ! Schoenb. t.* 128.

HAB. About Capetown, &c. on mountain sides, *E. & Z.!* &c. common. (Herb. T.C.D., Sond.)

A rigid, much-branched, scrubby, sometimes prostrate bush, much resembling *H. alnifolia* in foliage; but with flowers twice as large as in that species. I cannot distinguish *H. multiflora,* Jacq. by any tangible characters. Flowers bright yellow.

15. H. desertorum (E. & Z. 382.); shrubby, *divaricately much branched; young twigs thinly tomentose;* leaves on short petioles, broadly cuneate, plaited, coarsely toothed near the rounded apex, *albo-tomentose on both sides;* stipules *subulate,* equalling the petioles; peduncles racemose, short; calyx campanulate, angularly ventricose, *minutely and sparsely puberulous,* 5-toothed, the teeth distant, cuspidate; petals twice as long as the calyx, the claw equalling the limb; filaments obovate, as long as the taper-pointed anthers; ovary 5-lobed, pubescent.

HAB. Desert plains near Graaf Reynet, *E. & Z.!* Gamke River, *Burke & Zeyher!* 123. (Herb. Sond., Hook.)

A scraggy, small bush with woody branches and few leaves and flowers. It is closely allied to *H. cuneifolia,* but has canescent leaves and young branches and a dissimilar calyx. It has the habit of some forms of *H. pallens,* but a very different pubescence.

16. H. alnifolia (Linn.); shrubby, much branched; branches stellato-pubescent; leaves on short petioles, broadly cuneate or obovate, *undulate and plaited,* subtruncate, and crenato-dentate toward the point, stellato-pubescent on both sides; stipules ovate cuspidate; flowers *(small)* racemoso-paniculate, peduncles 2–3-flowered, short; calyx cup-shaped, 5-angled, sparsely stellate, 5-fid, with deltoid-acuminate lobes; petals twice as long as the calyx, the ciliate claw as long as the limb; filaments obovate, as long as the cuspidate anthers; ovary albo-tomentose, ovoid. *Jacq. Schoenb. t.* 291. *Cav. Diss. t.* 179. *f* 1. *Bot. Mag. t.* 229. *Zey.* 1985. *H. alnifolia* and *H. multiflora, E. & Z. H. hirsuta, Drege.*

HAB. Frequent on hill sides, near Capetown. Hott. Holland; Swellendam, &c. (Herb. T.C.D., Hook., Sond.)

Either erect, spreading or prostrate; when starved, having leaves ¼–⅓ inch long and broad, when in good soil ½–¾ inch long. Whole plant thinly stellulate; the upper surfaces of old leaves glabrescent. Flowers of small size, but in great abundance; bright yellow. Chiefly known from *H. cuneifolia* by its much smaller flowers and more branching inflorescence.

17. H. holosericea (Jacq.! Schoenb. t. 292.); shrubby, much branched; *branches velvetty-canescent;* leaves on short petioles, elliptical or obovate, somewhat cuneate at base, rounded and crenulate at the apex, penni-nerved, *velvetty-canescent on both sides;* stipules lanceolate, as long as the petioles; flowers *(small)* in drooping, unilateral, branched racemes; peduncles 2 or more together, short, 2–4-flowered; calyx obconic, velvetty, 5-toothed, teeth deltoid-acuminate; petals twice as long as the calyx, the claw exceeding the limb, nearly glabrous; filaments oblong, as long as the anthers; ovary ellipsoidal, tomentose. *D.C. Prod.* 1. *p.* 495. *H. incana, Thunb. Fl. Cap. p.* 505.

HAB. Dry, rocky ground in Uitenhage and Albany, common. E. & Z.! &c. (Herb. T.C.D., Hook., Sond.)

A much-branched shrub, resembling *H. alnifolia*, but differing in pubescence, &c. Leaves $\frac{1}{2}$–$\frac{3}{4}$ inch, rarely 1 inch long, varying in shape from exactly elliptical to obovate-oblong and cuneate. Flowers abundant, but small, yellow.

18. H. pallens (E. & Z.! 378.); shrubby, divaricately much branched, *stellato-squamulose and canescent;* leaves petiolate, obcordate or obovate, entire or crenulate near the apex, one-nerved, *squamulose on both sides;* stipules ovate-oblong, acute or acuminate, shorter than the petioles; peduncles racemose, 1–2-flowered; bracts short, ovate; calyx cup-shaped, *squamulose,* 5-toothed, the teeth broadly deltoid, acute; petals twice as long as the calyx, pubescent, the wide claw equalling the oval limb; ovary stellato-squamose, oblong and 5-umbonate. *H. flammea, var. Drege! H. secundiflora, E. & Z.!* 380.

VAR. β. **glabrescens**; leaves glabrescent, very sparingly scaly. *H. multiflora, Drege!*

HAB. Between the Coega and Sondag Riviers, Uitenhage; and at Potrivier, Langehoogde and Bontjeskraal, Caledon, *E. & Z.!* Orange River, *Burke & Zeyher!* var. β. Albany, *Mrs. Barber.* (Herb. T.C.D., Hook., Sond.)

A much-branched, scraggy, very rigid, small bush, more or less densely covered with close-pressed, flat, silvery, star-shaped scales. Flowers yellow. Filaments broadly linear, longer than the anthers. *H. secundiflora,* E. & Z.! is more diffuse and slender, as if "drawn up" among bushes. *H. triphylla,* E. & Z.! 379, by a fragment in Herb. Sond., is merely the present species, with sub-fasciculate leaves. Our var. β. may possibly be *H. multiflora,* Jacq.: it certainly agrees with his remark, "*florum tanto numero onustus ut totus inde flavescat.*"

Group 3. SCABERRIMÆ. Suffruticose or shubby. Leaves on short petioles, oblong, coarsely crenate or sinuate, rarely sub-entire. Stipules leaf-like, mostly amplexicaul. Pubescence copious and *very rough, the stiff, spreading, stellate hairs not uniformly spread,* but separated by bald interspaces. Peduncles loosely racemose. (Sp. 19–25.)

19. H. hirsuta (Scrad! & Wendl.! Sert. Han. t. 4); suffruticose, slender, diffuse; branches hispid with spreading, white bristles; leaves petiolate, *oblongo-obovate, obtuse, plicate, eroso-dentate,* ciliate, sparsely hairy on the upper side, *albo-tomentose* and villous below; stipules cordato-cuspidate, amplexicaul; peduncles racemoso-paniculate; filiform, longer than the leaves, 1–2-flowered, bracteate near the summit; bracts lanceolate; calyx campanulate, villous, 5-toothed, the teeth distant, deltoid-acute; fil. obovate, longer than the anthers. *DC. Prod.* 1. *p.* 495. *non E. & Z. H. muricata, E. & Z.!* 343, *and H. verrucosa,* 340? *H. scabra, Jacq. Schoenb. t.* 127, *non Cav.*

HAB. South Africa; *cult.* Swellendam? *E. & Z.!* (Herb. R., Ber., Sond., Benth.) Our description is taken from a garden specimen, which agrees perfectly with the figure above quoted, and also with Schrader's original in Herb. Berol. Ecklon's *H. muricata,* of which I have seen only a very bad specimen, belongs either to this species or to *H. dryadifolia. H. verrucosa,* E. & Z., from the Karroo, between Beaufort and Graaf Reynet, seems very similar; to judge by a fragment in Herb. Sond.

20. H. dryadifolia (Harv.); suffruticose, slender, diffuse; branches hispid with spreading, white bristles; leaves petiolate sub-distant, oblong, obtuse, *inciso-sinuate or pinnatifid,* ciliate, sparsely hispid or glabrescent above, *albo-tomentose* below; stipules amplexicaul, cordato-cuspidate; peduncles racemose-paniculate, filiform, longer than the leaves, 2-flowered near the summit; bracts ovato-lanceolate; calyx campanulate, pubescent, the 5 teeth deltoid-acuminate, with rounded interspaces; petals twice as long as the calyx, the claw longer than the limb; filaments cuneate, longer than the anthers: ovary ovoid, pubescent. *Mahernia dryadiphylla, E. & Z.!* 400.

HAB. Mountain sides, at Brackfontein, Clanw. *E. & Z.!* (Herb. T.C.D., Sond.) Very near *H. hirsuta,* but with more deeply divided, sinuate (not toothed) leaves; perhaps a mere variety. Its stamens are clearly those of a *Hermannia,* and so is the habit.

21. H. orophila (E. & Z.! 369); suffruticose, erect; branches scabrous with stellato-piliferous tubercles; leaves *tapering at base into a short petiole* or *sub-sessile,* cuneate, truncate, toothed at the extremity, *green,* ciliate and stellato-scabrous on both sides; stipules lanceolate-subulate; peduncles *aggregate at the ends of the branches, short;* bracts *numerous, subulate,* longer than the pedicels; calyx 5-angled, *inflated, thin,* stellato-pubescent, semi-5-fid, segments ovate, mucronulate; petals not much exceeding the calyx, the ciliate claw shorter than the broadly oblong limb; filaments linear, longer than the obtuse anthers; ovary downy.

HAB. Mountain sides. Hott. Holl., and at the Zwarteberg, and Klynriver's-berg *E. & Z.!* Swellendam, *Dr. Pappe!* (Herb. Sond.)
If the specimens are mature, this is a small plant, 3–6 inches high, sub-simple, densely leafy and very rough. The flowers are said to be "purplish-violet, with reddish calyces." It requires further examination.

22. H. scabra (Cav. Diss. t. 182. f. 2, non Jacq.); *shrubby,* erect; branches hispid with spreading bristles, very rough; leaves small, *sub-sessile, cuneate, truncate and 3–5 toothed* at the extremity, *green,* ciliate, stellato-scabrous on both sides; stipules ovato-lanceolate, patent; peduncles *racemoso-paniculate, filiform,* bracteate near the summit; 1–2 flowered, very scabrous; calyx campanulate, stellato-pubescent, semi-5-fid, the segments triangular, acute; petals twice as long as the calyx, the albo-ciliate claw equalling the rounded limb; filaments cuneate, longer than the anthers; capsule stellulate, shorter than the calyx.

DC. Prod. I. *p.* 495? *E. & Z.!* 350, *non Drege.* *H. biflora, E. & Z.!* 351.

HAB. Tulbagh, *E. & Z!* Riebeck's Casteel, Bergvalley and Langevalei, *Zeyher!* (Herb. Sond.)

An erect, shrubby plant, 1–2 feet high, extremely rough with rigid, scarcely stellate hairs. Leaves turning dark in drying. Flowers orange-yellow. *H. angularis*, Drege! (non Jacq.), judging from a fragment, seems to belong to the present species.

23. H. stipulacea (Lehm! in Eck. & Zey. En. 340.); shrubby, erect; branches densely hirsute; leaves *sessile, oblong or sub-lanceolate, entire* or denticulate at the point, stellato-hispid on both sides; stipules *leafy, ⅔ as long as leaf, cordate-acuminate,* ciliate and stellato-hispid; peduncles racemose, short, 1–2 flowered; bracts ovato-lanceolate; calyx campanulate, densely stellato-tomentose, 5-fid, the segment deltoid-acuminate; petals about twice as long as calyx, the hispid and ciliate claw equalling the ovate limb; fil. linear-cuneiform, much longer than the anthers; ovary ovoid, pubescent at the angles. *Drege,* 7276.

HAB. Near the mouth of the Zwartkops, Uitenhage, *E. & Z.!* near the mouth of the Orange River, *Drege!* (Herb. Hook., Sond.)

Very rough in all parts with tufted, patent hairs, but only tomentose on the *calyx.* Leaves crowded, erect, not ½ inch long, about ⅓ longer than the leaf-like stipules. Flowers orange. Chiefly known from *H. scabra* by the sessile leaves and large stipules.

24. H. latifolia (E. & Z.! 348.; non Jacq.); suffruticose, erect; branches virgate, stellato-pilose; leaves *on short petioles elliptic-oblong,* obtuse at each end, sub-entire or undulate and denticulate at point, sparsely stellato-pilose, specially on the ribs and margin; stipules leafy, broadly cordate, acuminate, ¼ as long as leaf, 3-nerved and hispid; peduncles laxly recemose, longer than the leaf, 2-flowered, densely hispid; bracts ovate, acuminate; calyx campanulate, densely stellato-tomentose, the segments deltoid-acuminate; petals scarcely twice as long as calyx.

HAB. Winterhoek's Mt., Uitenhage, *E. & Z.!* (Herb. Sond.)

Of this I have only seen a single branch. It is very close to *H. stipulacea,* but the leaves are evidently petiolate and much larger; 1–1½ inches long, ½ inch broad. The flowers are similar, but rather larger.

25. H. decipiens (E. Mey.!); shrubby; branches scabrid; leaves *subsessile,* densely set, erect, *elliptic-oblong or sub-acute, 3-toothed* at the truncate extremity, stellulato-scabrid on both sides; stipules and bracts *subulate;* peduncles shortly racemose, 2-flowered, very short; calyx campanulate, densely *stellato-tomentose,* the teeth deltoid, petals somewhat longer than the calyx, the *woolly-edged* claw longer than the limb; fil. narrow-linear, longer than the anthers; ovary obovoid, *tomentose.* *H. decipiens, E. M. & Drege* 7277.

HAB. Zwarteberg, *Drege!* (Herb. Hook., Sond.)

A small, but woody and branching shrub. Leaves crowded and imbricate, not ½ inch long, scarcely ¼ wide. Pubescence short, but rough. Flowers rather small, orange. [Perhaps more properly referable to the "Flammeæ" group.]

Group. 4. GLOMERATÆ. Much-branched, rigid shrubs. Leaves on

short petioles, oblong or ovate, entire or crenate. Pubescence copious, closely set, roughly stellate. Flowering branches leafy, *peduncles very short, densely aggregated in a terminal cluster.* (Sp. 26–29.).

26. H. cordifolia (Harv.); shrubby, robust; branches virgate, stellato-scabrid; leaves on short petioles, *cordate* or *cordato-ovate*, sub-acute or obtuse, rugose, prominently veined below, undulato-crenulate, densely stellato-tomentose on both sides, thick; stipules *obliquely decurrent, broadly deltoid,* shorter than the petioles; peduncles sub-terminal, aggregate; calyx urceolate, stellato-tomentose, the segments triangular, acute; petals broadly obovate, glabrous, the claw as long as the limb; fil. obovate, as long as the acuminate anthers.

HAB. Piquetberg, *Zeyher!* 111. (Herb. Benth., Hook., Sond.).

Apparently a large and woody shrub, with the habit of *Malvastrum strictum*, but leaves resembling those of *Sida cordifolia.* Leaves 1–1½ inch long, very thick like coarse cloth to the touch. The stipules are remarkable for their oblique insertion, and are broader than long. The specimens seen are scarcely in full flower; a single open flower in Herb. Benth. supplied the above character.

27. H. salvifolia (Linn.); shrubby, spreading and much-branched; branches densely stellato-hispid; leaves on short petioles, *linear-oblong or elliptical, quite entire,* densely stellato-pubescent on both sides, thickish; stipules acute, deciduous, either subulate or lanceolate; peduncles crowded at the ends of the branches, very short, flowers *surrounded by numerous linear-subulate bracts;* calyx oblong, 5-angled, *membranous and subinflated,* stellato-villous, the teeth short, deltoid, connivent; petals somewhat longer than the calyx, revolute, the claw *much shorter* than the oblongo-spathulate limb; fil. linear, longer than the anthers; ovary stellato-pubescent. *H. salvifolia, H. micans and H. involucrata, DC. Prod.* 1. *p.* 494. *Cav. Ic. t.* 180. *f.* 2. *& t.* 177. *f.* 1. *H. latifolia Jac. Schoenb. t.* 119.

VAR. α. **oblonga;** leaves linear-oblong or oblongo-lanceolate, sub-acute, 4–6 times as long as broad. *H. salvifolia, Auct. vetust. non E. & Z.*

VAR. β. **ovalis;** leaves exactly oval, obtuse, twice as long as broad; stipules subulate or linear-lanceolate. *H. micans, Drege! E. & Z.!* 347. *H. involucrata, E. & Z.!* 354, and *H. chrysophylla,* 346, *ex parte. Drege,* 7280. *Zey.!* 1993.

VAR. γ. **grandistipula;** leaves oval; stipules leafy, broadly lanceolate.

HAB. Winterhoek's berg, Uit., *E. & Z.!* Paarl., and Aasvogelberg, *Drege!* Various other localities throughout the Colony. (Herb. T.C.D., Hook., Sond., Banks.)

A much-branched, very rough shrub. The vars. α and β are very distinct in the shape of the leaves, but identical in other respects. Flowers small orange, the the tips of the petals rolling back. According to Herb. Banks. this is the true *H. salvifolia,* L., a species frequently confounded with others. I cannot distinguish *H. involucrata,* Cav. by any certain character. E. & Z.'s "*H. involucrata*" is made up, partly of our plant and partly (perhaps chiefly?) of *H. suavis,* Presl. The stipules in the present species vary much in breadth, and in var. γ. are almost leafy.

28. H. chrysophylla (E. & Z.! 346, ex parte); erect, shrubby; branches densely stellato-hispid; leaves on short petioles, roundish or elliptical, quite entire, densely stellate and very scabrous on both sides, thickish; stipules narrow-subulate, deciduous; peduncles in dense, terminal racemes; bracts subulate; calyx campanulate, *strongly 5-ribbed, stellato-scabrous, the lobes triangular-cuspidate, erect; petals twice as long as the calyx, pubescent, the claw equalling the oval limb;* fil. broadly linear, longer than the anthers; ovary stellato-pubescent.

HAB. Mts. between Langekloof and Winterhoek, *E. & Z.!* (Herb. Sond., Benth.)

This has quite the aspect of the broad-leaved vars. of *H. salvifolia*, but very different calyx and petals, and more orange flowers. E. & Z. partly confounded it with that species.

29. H. conglomerata (E. & Z. 352); shrubby, *divaricately much-branched;* branches stellato-hispid or scabrous; leaves petiolate, cuneate-ovate or sub-rotund, very obtuse, *crenate or incised, plaited and undulate,* glabrescent on the upper, fasciculate-hairy on the lower side; stipules *amplexicaul, cordate-cuspidate,* pilose; flowers crowded at the ends of the branches; bracts subulate; calyx somewhat inflated, densely albo-villous and hairy, the lobes broadly subulate, acute; petals not longer than the calyx, the claw about as long as the oval limb; fil. linear, much longer than the anthers; ovary sub-globose, pubescent. *H. glomerata, E. Mey.!* in Herb. Drege.

HAB. Karroo Plains. Near Sondag and Zwartkops Rivers, Uit., *E. & Z.!* Zuureberg, *Drege!* (Herb. T.C.D., Hook., Sond.)

A very spreading, hispid shrub, 1–2 feet high. Leaves ½–¾ inch long on petioles of half that length. Flowers small and yellow, sub-capitate; the calyces covered with long, white hairs.

Group 5. FLAMMEÆ. Virgate shrubs and under-shrubs. Leaves sessile or sub-sessile, either cuneate, lanceolate or linear. Pubescence *scanty,* never copious, but variable in quantity, thinly stellate; the hairs very short or appressed. Peduncles short, in a terminal raceme. (Sp. 30–37).

30. H. denudata (Linn.); frutescent, *quite glabrous;* leaves on short petioles or sub-sessile, *lanceolate, acute, coarsely toothed above the middle,* mid-ribbed; stipules semi-hastate, or ovato-cuspidate, leafy, ¼ as long as leaf; peduncles racemoso-paniculate; 1–2–3-flowered, with ovate-acuminate bracts; calyx campanulate, glabrous, the segments triangular-acuminate; petals about twice as long as the calyx, the obconic claw shorter than the roundish limb; fil. *obovate,* longer than the anthers; *ovary glabrous,* short, pentagonal; capsule scarcely longer than the calyx. *DC. Prod.* 1. *p.* 495. *Cav. Diss. t.* 181. *f.* 1. *Jacq. Schoenb. t.* 122. *H. quercifolia, E. & Z.!* 383, *ex parte majori.*

HAB. South Africa, *Zeyher!* 115. (Herb. Sond., Hook., Benth.)

2–3 feet high, not much branched, straggling, not strongly woody; branches curved. All parts of this plant are quite glabrous. Leaves 1–2 inches long, ½ inch wide. Flowers small, yellow. E. & Z.'s 383, according to a frustule in Herb. Sond. seems to belong to *H. patula.* Zeyhers' specimens (115) are excellent, perfectly agreeing with the plant of Linnæus.

31. H. scoparia (Harv.); *suffruticose, decumbent;* branches erect, sparsely hispid or glabrous; leaves sub-sessile, tapering at base, scattered, the lower ones linear-cuneate, often 3-toothed at the point, *upper linear or linear-lanceolate, acute, quite entire,* ciliate or glabrous;

stipules linear-lanceolate, not half as long as the leaf; peduncles racemose, mostly one-flowered, bracteate near the apex, bracts 2–3, subulate; calyx campanulate, nearly glabrous, the segments deltoid-cuspidate, ciliate; petals twice as long as the calyx, obovate, the claw as long as the limb; fil. *obovate*, longer than the taper anthers; ovary stipitate, ovoid, *stellulate on the angles.* *Mahernia scoparia, E. & Z. !* 404. *H. filifolia, Drege ! non Cav.*

HAB. C.B.S., *Masson !* Ried Valley and in Zwartland, *E. & Z !* Cape Flats, *Dr. Pappe!* Between Groenekloof and Saldanha Bay, *Drege!* (Herb. T.C.D., Sond., Banks.).

Stems 1–1½ feet long, several from the same crown, trailing and nearly glabrous. Leaves, save a few of the lowest ones, linear, an inch long, 1 line wide. Fl. bright yellow, pendulous, in secund racemes. In Herb. Banks is a specimen marked "*H. filifolia,*" Lin. f.; and possibly it may be Linnæus's plant. It certainly is not that of Cavanilles. Though referred by E. & Z. to *Mahernia*, its stamens are those of the most typical Hermannia.

32. H. linifolia (Linn.?); diffuse, suffruticose; branches ascending, *glabrous;* leaves *sessile, tufted or scattered, linear-spathulate,* sub-acute, tapering at base, flat, very entire, glabrous; *stipules leaf-like, as long as the leaves,* lanceolate, one-nerved, flat; peduncles laxly racemose, one-flowered, bracts subulate; calyx glabrescent, deeply 5-cleft, segments patent, lanceolate, acute; petals scarcely longer than the calyx, glabrous, the claw *shorter* than the broadly oblong limb; fil. *linear-oblong,* equalling the anthers; ovary stellato-pubescent. *E. & Z.!* 371. *Drege,* 7285. *DC. Prod.* 1. *p.* 495?

HAB. Karroo districts of George, Beaufort and Graaf Reynet, *E. & Z.!* Between Zwarteberg and Kendo, 2–3000 f. *Drege!* (Herb. Sond., Benth.).

Nearly allied to *H. filifolia,* but more glabrous, with thinner and flatter leaves, less woody stems, and somewhat different flowers. Corolla strongly convolute, dark-fulvous or reddish orange. Whether this be the little known and imperfectly described *H. linifolia, L.,* I cannot tell.

33. H. filifolia (Linn.?); erect, shrubby; branches virgate, scabrid; leaves sessile, *fascicled, narrow-linear, acute, thickish, very entire, glabrous;* stipules like the leaves and equally long, subulate or lanceolate, one-nerved and somewhat keeled; peduncles racemoso-paniculate, 2-flowered, bracts filiform; calyx stellato-scabrid, somewhat 5-angled, 5-fid, the segments ovate-oblong, acuminate; petals not much exceeding the calyx, the ciliate claw much shorter than the broadly-oblong limb; fil. oblong, longer than the obtuse anthers; ovary tomentose; capsule 5-lobed and umboned, stellato-pubescent. *DC. Prod.* 1. *p.* 495. *Cav. Diss. t.* 180. *f.* 3. *E. & Z. !* 372. *Zeyher,* 2013, 2004, 2010? 126. *Dregе,* 7288.

VAR. β. **passerinoides**; branches and calyces more densely stellulate; leaves shorter and fl. smaller. *H. passerinoides, E. & Z. !* 373. *Zey.* 2008.

HAB. About Port Elizabeth and on the Karroo, near Graaf Reynet, *E. & Z.* Near Grahamstown; on the Winterberg and at Gamke River, *Zeyher!* β. Riv. Zonderende, *E. & Z.!* (Herb. Sond.)

An erect shrub, 1–2 feet high, turning black in drying. Leaves ½–1 inch long, not a line wide, channelled above, rounded below, *mostly* glabrous. Stipules sometimes longer and broader than the leaves. I have no doubt that this is Cavanilles' plant;

his description is excellent; but Jacquin's (*Hort. Schoenb. t.* 123.) may be different. *H. passerinoides, E. & Z.* scarcely differs by a constant character; but almost connects the present species with some varieties of *H. flammea!*

34. H. flammea (Jacq. Schoenb. t. 129); erect or diffuse, shrubby; branches virgate, stellato-scabrid; leaves *scattered* or fascicled, sub-sessile, rigid, *dimorphous, either linear-spathulate and acute,* or *cuneate, truncate and 3–4-toothed* or sub-entire, *glabrous or thinly stellulate* on one or both sides; stipules leafy, linear-acute or lanceolate, one-nerved, shorter than the leaf; peduncles laxly racemose, short, 1–2-flowered; bracts subulate; calyx campanulate, 5–angled, thinly stellato-pubescent, *semi-5-fid, lobes broadly triangular, acute;* petals about twice as long as the calyx, the ciliate claw shorter than the limb; fil. linear-oblong, much longer than the obtuse anthers; ovary pubescent; capsule oblong, umboned, stellulate. *DC. Prod.* 1. *p.* 495. *E. & Z.!* 365–367, *and partly* 368, *and* 370.

VAR. α. **Jacquini**; leaves cuneate, 3–5-toothed, glabrous above, stellulate below; calyx lobes spreading. *Drege,* 7271, 7273, 7292.

VAR. β. **polymorpha**; leaves cuneate or linear-spathulate, often tufted, glabrescent; calyx lobes erect or conniving. *H. polymorpha, E. & Z.!* 366. *Drege,* 7293.

VAR. γ. **falcata**; leaves pubescent, cuneate, 3–5-toothed; stipules linear, sub-falcate. *H. falcata, E. & Z.!* 365. *Drege,* 7305, *Zey.* 2005.

HAB. Frequent on hills and mountain sides throughout the Colony. Uitenhage and Albany, common. Also about Muysenberg and Simon's Bay. (Herb. T.C.D., Hook., Sond.).

Generally turns black in drying. Very variable in the shape, breadth and size of the leaves, and the amount of pubescence; but more constant in its flowers. Usually an erect, slightly branched shrub, 1–2 feet high. Nearly allied to *H. filifolia.*

35. H. flammula (Harv.); erect, shrubby; branches stellulate; leaves sub-sessile, linear-cuneate, three-toothed or entire, *densely stellato-pubescent on both sides;* stipules linear or lin.-subulate, leafy, not half as long as leaf; peduncles somewhat racemose; one-flowered, as long as leaves; calyx campanulate, angular, stellate, *deeply 5-parted, the lobes narrow-triangular, acute;* petals twice as long as the calyx, *the hispid claw much shorter than the limb;* filaments linear-oblong, longer than the obtuse anthers; capsule pubescent, umboned. *H. trifurca, E. & Z.!* 370, *ex parte; non Linn. Zey.* 2009. *Drege,* 7287.

HAB. Hartebeeste River, Caledon, and Vormansbosch, *Zeyher!* Swellendam, *Thom!* (Herb. Hook., Sond.).

Very near *H. flammea,* but with smaller, more densely pubescent leaves, a more deeply parted calyx with narrower and more pointed segments, and smaller flowers. It does not turn black in drying. It was formerly cultivated at Kew, and old garden specimens, but without name, are preserved in Herb. Hook. E. & Z., confounded under their "*H. trifurca,*" the present plant, with *H. flammea* and *H. angularis* and the following.

36. H. Joubertiana (Harv.); shrubby, erect; branches stellulate; virgate; leaves narrow-cuneate or spathulate, the lower ones 3-toothed, *glabrescent or sparsely stellulate;* stipules ½ as long as leaves, linear- oblong, sub-acute; peduncles *racemoso-fasciculate,* one-flowered, pubescent; calyx campanulate, sparsely stellate, 5-*toothed, the teeth short, triangular,* *taper-pointed;* petals more than twice as long as the calyx, *the ciliate claw nearly equalling the obovate limb;* fil. linear-oblong, much longer than the obtuse anthers; capsule stellulate, umboned.

HAB. Caledon, *Miss Joubert!* (Herb. Sond.).

It differs from *H. flammea* and *H. flammula* chiefly in the very short calyx, with taper-pointed segments, the longer and narrower petals and smaller flowers, and somewhat clustered inflorescence.

37. H. angularis (Jacq. Schönb. t. 126); erect, shrubby; branches stellato-scabrid, virgate; leaves sub-fasciculate, sub-sessile, *rigid, cuneate or linear-spathulate, truncate,* mostly 3–5-toothed, glabrescent above, stellato-scabrid below; stipules leafy, linear-acute, ciliate, one-nerved; peduncles laxly racemose, short, 1–2-flowered; bracts linear; *calyx inflated-pyramidal, prominently 5-angled, glabrous and shining, deeply 5-fid, the segments broadly ovate, acute, ciliolate; petals scarcely longer than the calyx,* the claws much shorter than the limb; filaments linear-oblong, scarcely longer than the obtuse anthers. *DC. Prod.* 1. *p.* 495. *H. trifurca, E. & Z.! ex parte; non Linn. nec. Jacq.*

HAB. Hills about Stellenbosch; Hoow Hoek Pass; Hemel and Aarde, &c., *E. & Z.* Caledon, *Dr. Alexander Prior.* (Herb. Sond., T.C.D).

A virgate shrub, 1–2 feet high, turning black in drying. Leaves ¾ inch long, 1–4 lines wide. Near *H. flammea,* from which it is readily known by its 5-angled, glabrous calyx. *H. angularis, E. & Z.* is only *H. hyssopifolia.* Jacquin's figure, except in having such large leaves, well represents our plant.

Group 6. VELUTINÆ. *Canescent shrubs* (rarely under-shrubs) *covered with very short and close, velvetty or powdery pubescence.* Leaves shortly petiolate or sessile, entire or denticulate. Stipules leafy. Peduncles short, crowded or racemose. (Sp. 38–48.)

38. H. hyssopifolia (Linn.); shrubby, erect; branches virgate, stellato-pubescent; leaves petiolate, *obovate or lanceolate, obtuse* or sub-acute, *serrulate towards the apex, puberulous and sub-canescent* on both sides; stipules *dimorphous, the lower small and narrow, the upper large, leafy, ovate or ovato-lanceolate or oblong, acute, one-nerved;* peduncles sub-terminal, 2–3-flowered, short; bracts lanceolate, deciduous; calyx globose, inflated, *villous,* 5-angled, the broadly ovate, mucronulate lobes connivent; petals scarcely longer than the calyx, *revolute, the claw shorter than the narrow limb;* fil. linear, much longer than the anthers; ovary oblong, stellulate. *DC. Prod.* 1. *p.* 494. *Thunb. Cap. p.* 504. *E. & Z.* 355. *Cav. Diss. t.* 181. *f.* 3. *H. angularis, E. & Z.!* 356, *non Jacq.*

HAB. Hills round Capetown, common; and generally throughout the Colony to Uitenhage and Albany. (Herb. T.C.D., Hook., Sond.).

Not much branched, 2–3 feet high, with long, simple, erect branches. Pubescence generally pale, not always so. Leaves commonly obovate and very obtuse; but varying as above. Calyx bladdery. Petals pale yellow.

39. H. ternifolia (Presl, Bot. Bem. p. 21.); shrubby, diffuse, slender; branches minutely scurfy; leaves tapering at base into a short petiole, *cuneate, crenulate at the blunt extremity,* stellulato-canescent on both surfaces; stipules leafy, *spathulate, obtuse;* peduncles sub-racemose, short, one-flowered; calyx inflated, globose, *squamulose, the teeth short, deltoid;* petals strongly convolute, *the glabrous claw longer than the obovate limb;*

fil. narrow-linear, much longer than the obtuse anthers; ovary obovate, 5-angled, pubescent. *H. triphylla, Drege! non Cav.*

HAB. Between Groenekloof and Saldanha Bay, *Drege!* (Herb. T.C.D., Hook., Sond.).

A small, slender shrublet, canescent in all parts with very white, short, tufted hairs. The foliage is like that of *H. trifoliata*, but the inflorescence and calyces are very different. Stipules more than half as long as the leaves.

40. H. trifoliata (Linn.); shrubby, erect; branches virgate, *stellulate;* leaves crowded, sub-sessile, cuneate, truncate or emarginate, entire or crenulate at the point, stellato-canescent on both sides; stipules leaf-like and nearly as long as the leaf, *broadly oblong or elliptical, obtuse;* flowers *densely crowded* at end of branches, on short 1–2-flowered peduncles; calyx globose, 5-angled, somewhat inflated, *5-fid, densely stellulate, the lobes ovate, acute;* petals strongly convolute, the claw *shorter* than the broadly cuneate, spreading limb; fil. oblong, longer than the sub-acute anthers. *DC. Prod.* 1. *p.* 494, *non E. & Z. Cav. Diss. t.* 182. *f.* 2. *H. imbricata, E. & Z.* 381.

HAB. Between Cape L'Agulhas and Potberg, *Drege!* Paardekop, near Plettenburg Bay, *E. & Z.!* (Herb. T.C.D., Hook., Sond.).

A densely leafy, strongly woody, but small shrub. Leaves imbricating, $\frac{1}{2}$–1 inch long, tapering much at base; the stipules $\frac{2}{3}$–$\frac{3}{4}$ as long, called "lateral leaves" by Linnæus. Pubescence short and close, often yellowish, velvetty. Corolla deep orange.

41. H. suavis (Presl, Bot. Bem. p. 21); shrubby, erect, much branched; branches *tomentose;* leaves *obovate or obcordate, round-topped, entire, softly villous* and stellato-canescent on both sides; stipules leaf-like and nearly equalling the leaves, oblong or elliptical, obtuse; fl. crowded towards the ends of the branches, sub-sessile; bracts linear; calyx urceolate, stellate and *villous, 5-toothed,* the teeth deltoid, connivent; petals *revolute,* the claw shorter than the narrow, strap-shaped limb; filaments oblong, longer than the small, obtuse anthers; ovary stellate and tomentose, hispid at top. *H. involucrata, E. & Z.! ex parte. Drege,* 7268.

HAB. Between Coega and Sondag Rivers; and Cradockstaad, near Port Elizabeth, &c., *E & Z.!* Small Deel, *Burke!* (Herb. T.C.D., Hook., Sond.).

Near *H. trifoliata,* under which name it occurs in Herbaria; but differing in foliage, pubescence, and petals, &c. Also allied to *H. hyssopifolia.* Flowers pale yellow or cream-coloured.

42. H. diversistipula (Presl, Bot. Bem. p. 21); fruticose, the branches *minutely scurfy;* leaves sub-sessile, cuneate or obcordate, entire or 2–3-toothed, *canescent, and stellato-squamulose* on both sides; stipules leaf-like, linear-oblong, obtuse, nearly as long as the leaves; peduncles from the axils of the upper leaves, 1–2-flowered; calyx *5-angled, deeply 5-cleft, squamulose, the lobes ovate, sub-acute;* petals not much longer than the calyx, the ciliate claw shorter than the limb; fil. oval-oblong, longer than the sub-acute anthers. *Drege,* 7274. *Zey.* 2014.

VAR. β. **nana**; dwarf; flowering branches naked, racemulose above.

HAB. S. Africa, *Niven!* Assagaiskloof, *Zeyher!* Breede River, *Drege!* β, Storm Valley, near Breede River poort, *Zeyher!* (Herb. Hook., Sond.).

Very like *H. trifoliata* in foliage, but with pubescence of a different structure. Here the surface is clothed with minute, close-pressed, flat scales, stellulate at the margin.

43. H. mucronulata (Turcz. Bull. Mosc. 1858. p. 217); shrubby, erect, canescent; branches virgate; leaves *oblongo- or ovato-lanceolate, acute, quite entire*, one-nerved, somewhat veiny below, softly velvetty on both sides; stipules *leaf-like, petiolate, broadly lanceolate, one-nerved;* flowers crowded at the ends of the branches; bracts linear; calyx stellato-scabrid, *inflated*, 5-toothed, the teeth short, deltoid, connivent; petals not much longer than the calyx, the claw *shorter* than the narrow limb; fil. broadly-oblong, rather longer than the anthers; ovary canescent. *H. salviæfolia, E. & Z.!* 353. *Drege*, 7267. *H. salicifolia, Harv. MSS.*

HAB. Mountain sides, Vanstaadensberg, *E. & Z.! Drege!* (Herb. T.C.D., Hook., Sond.)

A virgate shrub, 2–3 feet high, softly velvetty. Very near *H. velutina;* but the stipules are much larger and more leaf-like, nearly as long as the leaves, and the calyx is *inflated* and much rougher.

44. H. velutina (DC. Prod. 1. p. 495); shrubby, erect, canescent; leaves distinctly petiolate, *oblong or sub-lanceolate, acute*, sub-acute, or mucronulate, quite entire, softly velvetty on both sides; stipules *leaf-like, linear-oblong or lanceolate, one-nerved, acute, half as long as the leaf;* peduncles racemulose, short, 2–3-flowered, bracts narrow linear; calyx *campanulate*, stellato-tomentose, semi-5-fid, segments triangular, acute; petals ⅓ longer than calyx, the claw shorter than the ovate-oblong limb; fil. broadly obovate, shorter than the anthers; ovary oblong, canescent. *E. & Z.!* 357. *H. lavandulæfolia, Drege !*

HAB. Among shrubs. Uitenhage and Albany, frequent, *E. & Z.! &c.* (Herb. T.C.D., Hook., Sond.).

2–3 feet high. Very like *H. lavandulæfolia*, but the leaves are more decidedly petiolate, and the stipules very different; the petals and filaments are also dissimilar. Drege's 7282, probably belongs here.

45. H. lavandulæfolia (Linn.); shrubby, erect, canescent; leaves tapering at base into a very short petiole, oblong or sub-lanceolate, acute or mucronulate, entire or denticulate near the apex, softly velvetty on both sides; *stipules narrow-subulate, somewhat longer than the petiole;* flowers racemulose, peduncles 2–3-flowered, bracts subulate; calyx *turbinate*, velvetty or stellato-tomentose, semi-5-fid, 5-angled, the lobes triangular, acute; petals twice as long as the calyx, the claw longer than the roundish limb; fil. oblong or obovate, longer than the anthers; ovary oblong, canescent. *DC. Prod.* 1. *p.* 495. *Thunb. Cap. p.* 501. *Jacq. Schoenb. t.* 215. *Cav. Diss. t.* 180. *f.* 1. *E. & Z.!* 360. *H. odorata, Ait ! fide Herb. Banks. Drege*, 7281, 7283.

HAB. Cape Flats, near Salt River; and on the Kars River, Caledon, *E & Z.* (Herb. T.C.D., Hook., Sond., Thunb.).

2–3 feet high, erect, and much branched, woody. Very like *H. velutina*, but differing in stipules, &c. The leaves in our specimens are much smaller and narrower than in Jacquin's figure, otherwise very good.

46. H. Cavanillesiana (E. & Z.! 361); erect, shrubby, canescent; leaves *narrow-obovate or spathulate*, tapering at base into a short petiole, acute or recurvo-mucronulate, softly velvetty on both sides, quite entire; stipules *leaf-like, one-nerved, linear-acute*, ½ as long as the leaves; flowers racemose, bracts subulate; calyx *obconic*, 5-fid, the teeth triangular-acuminate; petals twice as long as the calyx, the claw shorter than the oblong limb; fil. *linear-oblong*, longer than the obtuse anthers. *Drege*, 7272.

HAB. Between Assagaiskloof and Breede River, *E. & Z.!* Caledon, *Zeyher!* Gauritz River, *Drege!* (Herb. T.C.D., Hook., Sond.).

Intermediate in character between *H. velutina and H. lavandulæfolia;* but with smaller flowers than either, and in longer racemes. Petals dark, ruddy-orange. E. & Z.! refer, perhaps correctly, to *Cav. Diss. t.* 180. *f.* 1. *x.*; but that figure does not clearly show the stipules.

47. H. sulcata (Harv.); shrubby, *diffuse*, canescent; leaves *narrow-obovate, tapering at base, obtuse or mucronulate*, entire or crenulate, softly velvetty on both sides; stipules leaf-like, linear, acute, ⅓ of leaf; flowers (*small*) *paniculato-racemose*, partial peduncles 2–3-flowered, bracts subulate, small; calyx *obconico-prismatic, ribbed and furrowed, stellato-tomentose*, 5-fid, the lobes triangular, petals twice as long as the calyx, the claw much shorter than the spathulate limb; ovary oblong, pubescent. *H. odorata, E. & Z.!* 359, *non Ait.*

HAB. Clayey hills at Sondag River, and near Port Elizabeth, *E & Z.!* Algoa Bay, *Forbes!* (Herb. Hook., Sond.).

A small, but woody and scrubby bush, much branched and densely covered with leaves. Flowers pale yellow, in dense, many-flowered panicles. The calyx is remarkably narrow and strongly 10-ribbed. It is nearest to *H. Cavanillesiana*, but has broader and more obovate leaves, and different flowers and inflorescence.

48. H. gracilis (E. & Z.! 358); shrubby, the young branches minutely scurfy; leaves on short petioles, sub-fasciculate, obovate-oblong or oblongo-lanceolate, tapering at base, obtuse or acute, the lower denticulate, the upper entire, all velvetty and canescent on both sides; stipules leafy, *broadly linear, sub-acute, one-nerved;* inflorescence . . . . . ? calyx campanulate, *downy, 5-toothed, the teeth deltoid, acute, with wide, rounded interspaces;* petals twice as long.

HAB. Karroo, Graaf Reynet, *E. & Z.!* (Herb. Sond.).

Of this I have examined the original (very incomplete) specimen in Herb. Ecklon. It has nearly the foliage of *H. velutina*, but differs in the calyx from that and all the allied species. Flowers yellow, of moderate size. Stipules ⅔ as long as the leaves.

Group 7. LATERIFLORÆ. Peduncles *axillary*, distributed along the branches, *not obviously racemose.* Leaves undivided. Pubescence stellate or tomentose. (Sp. 49–56.)

49. H. Sandersoni (Harv.); *suffruticose*, erect, many-stemmed; stems *sub-simple*, stellato-strigose and tomentose; leaves *shortly petiolate or subsessile*, broadly elliptical or obovate, rugose, *crenato-dentate*, stellato-hispid on the upper, stellato-tomentose and discoloured on the lower side; stipules *ovate* or *ovato-lanceolate*, equalling the petiole; peduncles axillary (or sub-terminal) rather long, very hispid, 1–2-flowered, bracteolate in the middle; calyx campanulate, short, angular, hispid and rough, the lobes deltoid-acuminate; petals twice as long as the calyx, *the claw shorter than the broadly oblong limb;* filaments *linear*, about as long as the acuminate anthers; capsule umboned, hispid, scarcely longer than broad.

HAB. Port Natal, *Mr. Sanderson!* (Herb. Hook., T.C.D.).

Root very thick and woody. Stems 6–8 inches high, with few, sub-distant leaves,

1–1½ inch long, ½–¾ inch wide. Flowers very rough. Petals strongly convolute, reddish-orange.(?) Not nearly related to any other species.

50. H. candidissima (Spreng., fide E. & Z.! 331); shrubby, much branched, canescent; branches albo-tomentose and stellato-hispid; leaves *on long petioles, ovate or cordate-ovate, obtuse, sinuate or lobulate, crenate, plaited and undulate, densely stellato-tomentose on both sides;* stipules *lanceolate-linear,* obtuse; peduncles axillary, 1–2-flowered, short; bracts linear, small; calyx *sub-globose,* densely stellate and tomentose, the teeth deltoid, acuminate; petals about twice as long as the calyx, pubescent and ciliate, the claw longer than the oval limb; fil. *broadly obovate,* scarcely longer than the hastate anthers; ovary pedicellate, ovoid, sparsely pubescent. *H. vestita, E. Mey. in Herb. Drege.*

HAB. Karroo, between Beaufort and Graaf Reynet, *E. & Z.! Drege!* Vlekport River, *Zeyher!* Graaf Reynet, *Mrs. F. W. Barber!* (Herb. T.C.D., Hook., Sond.).

A bushy shrub, wholly covered with rough, pungent, densely fascicled white hairs, "which" (says Mrs. Barber) "sting something like a nettle." Leaves as broad as long, rarely exceeding ½ inch. In luxuriant specimens the lower leaves become somewhat lobed. Drege's 7298, from between Zwarteberg and Kendo, scarcely differs, except that the leaves are flat, not plaited, the stipules rather broader, and the habit less woody.

51. H. floribunda (Harv.); shrubby, erect; branches virgate, roughly stellato-pubescent; leaves *on longish petioles, ovate or sub-cordate, obtuse, repand or crenate, stellato-pubescent on the upper, tomentose on the lower side;* stipules *subulate,* scabrous; peduncles axillary, *two or more from the same axil,* short, 1–2 or more flowered; calyx *campanulate,* semi-5-fid, stellato-pubescent, the lobes deltoid-acuminate; petals scarcely exceeding the calyx, the broad claw equalling the oval limb; fil. *narrow-linear,* longer than the acute anthers; ovary canescent; capsule tomentose, as long as the calyx.

HAB. Vaal Rivier, *Burke and Zeyher.* Jan. (Herb. Hook., Sond., T.C.D.).

More allied to *H. candidissima* than to any other, but far less hairy, with different inflorescence, calyx, anthers, &c. Flowering specimens, gathered in Jan., are in Herb. Hook.; the other specimens seen, collected in the same locality, in May, have withered leaves and are in fruit.

52. H. boraginiflora (Hook. Ic. Pl. t. 597); suffruticose, erect, stellato-pubescent and *glandular;* branches scabrous; leaves *on short petioles,* oblong or obovate, flat, dentate, thinly pubescent above, more thickly stellulate below; stipules minute, *ovate, acute;* flower-stalks axillary, 1-flowered, as long as the leaves, minutely bracteate above the middle; calyx campanulate, laxly stellate, deeply 5-fid, the lobes subulate; petals *scarcely as long as the calyx, obovate;* fil. obovato-spathulate, *much shorter than the taper-pointed, connivent and exserted anthers;* ovary obovoid, pubescent.

HAB. Macallisberg, *Burke and Zeyher.* (Herb. Hook, T.C.D.).

1–2 feet high, rather thinly pubescent; the branches leafy below, flowering towards the ends, the upper leaves gradually smaller and narrower. Lower leaves 1 inch long. The long, ciliate, tapering, exserted anthers, in this and the two following species, stand close together in a cone, like those of a *Borage.*

53. H. micropetala (Harv.); suffruticose, *erect,* stellato-pubescent; leaves on short petioles, the lower oblong or obovate, toothed, the upper smaller, more lanceolate, entire, all stellato-pubescent, specially on the lower side; stipules minute, *subulate;* flower-stalks axillary, one-flowered, longer than the leaf, minutely bracteolate beyond the middle; calyx *densely tomentose and stellulate*, deeply 5-parted, the lobes subulate; petals *shorter than the filaments, cucullate and truncate;* fil. obovate, much shorter than the connivent and exserted anthers; ovary oblong, stellato-pubescent.

HAB. Delagoa Bay, *Forbes!* (Herb. Hook.)

Nearly related to *H. boraginiflora*, but not glandular; and differing in calyx, petals and ovary. Also near the following, but with a different habit, smaller, less uniform and not canescent leaves; a more hairy calyx and shorter and differently shaped petals. Though not found within the geographical limits of our Flora, I am unwilling to omit a species so nearly allied to others of this section.

54. H. brachypetala (Harv.); suffruticose, *diffuse or procumbent, stellato-canescent;* branches tomentose; leaves on short petioles, *oblong*, obtuse, flat, denticulate, *velvetty and canescent*, especially on the lower side; stipules subulate; fl. stalks axillary, deflexed, as long as the leaf, bracteate near the summit, bracts subulate; calyx *stellate and villous*, deeply 5-parted, the lobes subulate; petals *much shorter than the calyx, spoon-shaped;* filaments narrow-obovate, much shorter than the connivent and exserted anthers; ovary ovoid, tomentose; capsule truncate, as long as the calyx.

HAB. Macallisberg, *Burke & Zeyher!* Zooloo Country, *Miss Owen!* (Herb. Hook., Sond., T.C.D.)

More canescent than either of the preceding, with larger and more uniform leaves, and flowers distributed along the branches, not confined to their upper half. The pedicels are remarkably patent, with nodding flowers, and are gradually deflexed in fruit.

55. H. exstipulata (E. Mey. in Hb. Drege); much branched, *shrubby, canescent;* leaves on short petioles, oblong, obtuse, *very entire*, flat, one-nerved below, densely stellato-canescent on both sides; *stipules none;* flower-stalks axillary, one-flowered, jointed in the middle, hispid above the joint, *ebracteate;* flowers nodding; calyx campanulate, *densely tomentose and stellato-hispid*, deeply 5-parted, the lobes triangular, acuminate; petals scarcely longer than the calyx, the obconic claw much shorter than the broadly oval, spoon-shaped limb; filaments spathulate, scarcely as long as the acuminate anthers; ovary ovoid, stellulate.

HAB. Hills and flats near the mouth of the Orange River, *Drege!* (Herb. T.C.D., Hook., Sond.).

The whole plant is densely clothed with very short, closely set, stellate, whitish hairs. Leaves seldom ½ inch long, not ¼ inch wide. The pubescence on the upper half of flower-stalk and on the calyx is much coarser than on other parts. Flowers orange yellow. Filaments intermediate in character between those of *Hermannia* and *Mahernia*, being narrow for ⅔ their length and then suddenly widened; but there is no tubercle or hispid thickening of the back.

56. H. Gariepina (E. & Z.! 384); *divaricately much-branched*, shrubby, canescent; leaves on short petioles, oblong, obtuse, quite entire or 3-toothed, flat, one-nerved, densely stellato-*tomentose* on both sides; *stipules minute, triangular, deciduous;* fl. stalks axillary, one fl., jointed, hispid above the joint, ebracteate; fl. nodding; calyx campanulate, *membranous, thinly lanuginous and stellate*, deeply 5-cleft, the lobes triangular-acuminate; petals twice as long as the calyx, the obconic claw much shorter than the broadly oval limb; fil. cuneate, cordate at top, rather shorter than the tapering anthers; ovary 5-umboned, stellato-tomentose; capsule 5-pointed. *H. racemosa, E. Mey.!*

HAB. Boschesman's Land, *E. & Z.!* Silverfontein, *Drege!* (Herb. T.C.D., Hook., Sond.).

A small densely branched bush, more tomentose than *H. exstipulata*, with larger flowers, a different ovary and capsule and evident stipules. The petals, according to E. & Z., are pale violet. The filaments are intermediate with those of a *Mahernia;* and the deeply umboned capsule connects with our section *Acicarpus.*

Group 8. PINNATIFIDÆ. Leaves deeply pinnatifid, bi-pinnatifid or multifid. Habit various. (Sp. 57–65.)

57. H. pulverata (Andr. Rep. t. 161.); suffruticose, *squamuloso-canescent*; branches leafy below, naked above; leaves on longish petioles, deeply inciso-pinnatifid or bi-pinnatifid, the lobes cuneate, lobed, *both surfaces sprinkled or covered with silvery, stellulate appressed scales;* stipules ovato-lanceolate; peduncles racemoso-paniculate, elongate, 2-flowered; calyx campanulate, 5-cleft, canescent, the lobes deltoid, acute; petals glabrous, twice as long as calyx, the claw shorter than the limb; fil. oblong, longer than the obtuse anthers; ovary obovoid, pubescent. *H. pulverulenta, DC. Prod.* 1. *p.* 496. *E. & Z.!* 375; *and H. argentea, E. & Z.!* 376.

HAB. Karroo districts of George and Uitenhage, *E. & Z.!* Sondag River, *Drege!* Albany, *Mrs. F. W. Barber!* (Herb., T.C.D., Hook., Sond.)

Stems woody at base, sub-herbaceous upwards, 12–18 inches high, paniculately branched. Leaves and petioles each 1 inch long, deeply cut. Flowers orange-yellow.

58. H. argentea (Sm. in Rees. Encycl.); "leaves bi-pinnatifid, covered with stellate scales, the lobes decurrent; peduncles racemose, one-flowered." *DC. l. c. p.* 496.

I cannot say how this differs from the preceding. *H. argentea*, E. & Z.! is not distinguishable from *H. pulverata.*

59. H. tenuifolia (Bot. Mag. t. 1348); erect, suffruticose, slender, sub-dichotomous; branches *thinly stellulate;* leaves *sub-sessile*, deeply pinnatifid, *the lobes narrow-linear*, acute, glabrous above, sparsely stellulate below; stipules short, broadly ovate, amplexicaul; peduncles racemoso-paniculate, elongate, bracteate near the summit, 2-flowered; bracts ovate; fl. buds *acuminate;* calyx campanulate, glabrous, the lobes deltoid-cuspidate; petals twice as long, the narrow, downy claw as long as the ovate limb; fil. broadly obovate, longer than the hastate anthers; ovary obovoid, pubescent. *DC. Prod.* 1. *p.* 496. *E. & Z.* 374; *and H. coronopifolia, E. & Z.!* 377; *also Mahernia pinnata, E. & Z.! non Linn.*

HAB. Hills near Caledon and Somerset, *E. & Z.!* Hartebeeste River, *Zeyher*, 2001. Between the Breede River and Bokkeveld, and Hex River, *Drege!* (Herb. Hook., Sond.).

Stems 12–18 inches high, leafy below, bare above, paniculately branched. Leaves ½–⅔ inch long, channelled above, somewhat fleshy, each lobe tipped with a stiff hair; as are also the calyx segments. Fl. orange-yellow.

60. H. paucifolia (Turcz. Bull. Mosc. 1858, p. 218); erect, suffruticose, slender, dichotomo-paniculate; branches zigzag, naked above, *glabrous;* leaves *on long petioles,* oblong, inciso-pinnatifid or sub-bipinnatifid, *the lobes cuneate, lobed or cleft,* glabrescent above, stellato-pubescent and paler below; stipules broadly ovate, short; peduncles racemoso-paniculate, 2-flowered; fl. buds *acuminate;* calyx campanulate, angular, puberulous and glandular, the segments short, deltoid-acuminate; petals twice as long as calyx, the claw rather shorter than the roundish limb; fil. narrow-cuneate, as long as the tapering anthers; ovary pubescent. *H. dissecta, Harv.*

HAB. Bitterfontein, *Zeyher!* 118. (Herb. Hook., Sond., Benth.).

Stems 12–18 inches high. It resembles *H. tenuifolia,* but the leaves are more compound, the stems glabrous, and the anthers taper-pointed, &c. It differs from the following in pubescence, the shape of the flower-buds, &c.

61. H. chrysanthemifolia (E. Mey.! in Herb. Drege); suffruticose, diffuse, leafy below; branches naked, paniculate, *pubescent and canescent;* leaves petiolate, inciso-pinnatifid or sub-bipinnatifid, the lobes cuneate, obtusely lobulate or cleft, *stellulate above, canescent below;* stipules ovate, short; peduncles racemoso-paniculate, *tomentose,* 2-flowered; fl. buds *ovate;* calyx campanulate, stellato-canescent, the segments deltoid, acute; petals twice as long; filaments broadly oblong, longer than the acute anthers; ovary *albo-tomentose.*

HAB. Kaus Mts., near Goedemanskraal, Rustbank and Kookfontein, *Drege!* (Herb. Hook., Benth.).

Our specimens are small and imperfect. Flowers yellow. It differs from the last chiefly in pubescence, flower-buds and anthers.

62. H. abrotanoides (Schrad.); scarcely suffruticose, slender, *stellato-tomentose, procumbent;* stems angular, stellulate; leaves petiolate, *bi-pinnatifid, the lobules linear, obtuse,* stellato-canescent, especially on the lower side; stipules *subulate;* peduncles axillary, 2-flowered, shorter than the leaves, bibracteolate; calyx globose, inflated and tomentose, with 5 short, deltoid-cuspidate teeth; petals not twice as long, erect, convolute, the claw longer than the limb; filaments broadly obovate.

HAB. C.B.S. (Herb. Wendland!)

Of this I have seen but a single branch. The flowers are bright yellow. It differs from the following, apparently, in the incision of the leaves.

63. H. multifida (DC. Prod. I. p. 493); "leaves canescent, *palmati-partite,* the lobes pinnati-partite, lobules linear, entire, somewhat channelled; pedicels *one-flowered,* shorter than the leaves; calyx inflated, puberulous." *DC. Burch. Cat.* 1627.

HAB. South Africa, *Burchell.*

64. H. halicacaba (DC. Prod I. p. 493); "leaves whitish on the lower, punctate on the upper surface, the lower ones palmately 5-parted, upper 3-parted, lobes pinnatifid, the middle longer; *peduncles 2–4-flowered,* shorter than the leaves; calyx inflated, puberulous." *DC. Burch. Cat.* 2020.

HAB. South Africa, *Burchell.*

65. H. incisa (Willd.! Sp. 3. 599); suffruticose; stems slender, patently hispid; leaves sub-sessile, deeply pinnatifid, the segments patent, linear, acute, ciliate; stipules lanceolate, ciliate; peduncles racemose, pedicels one-flowered; calyx sub-glabrous, *divided nearly to the base, the lobes lanceolate; petals deeply inciso-pinnatifid. DC. Prod.* 1. *p.* 496.

HAB. Uncertain. (Herb. Willd.!)

I describe from the original specimen in Willdenow's Herb.; but consider this curious plant to be a garden monstrosity. If a genuine species, none can be more distinct.

Sub-genus 2. ACICARPUS.

Capsule truncate, crowned with 5–10, divergent, simple or plumose horns. (Sp. 66-70.)

66. H. trifurcata (Linn.); shrubby, *virgate, canescent;* branches downy or glabrescent; leaves on short petioles, *linear-cuneate, truncate,* 2–3-toothed, or entire and obtuse or sub-acute, *canescent and pulverulent on both sides*; stipules minute, subulate; peduncles racemose, one-flowered, pendulous, shorter than the leaf; calyx campanulate, villous, semi-5-fid, veiny, the lobes ovate, acute; petals twice as long, obovate; fil. obovate, shorter than the tapering, ciliate anthers; capsule plumoso-villose, truncate, ten-horned. *DC. Prod.* 1. *p.* 495. *Cav. Diss. t.* 178. *f.* 2. *Jacq. Schoenb. t.* 125. *Thunb.! Cap. p.* 503. *H. bicornis, E. & Z.!* 385, *and Mahernia hilaris, E. & Z.!* 389. *M. incana, E. & Z.!* 388.

HAB. Cape, *Masson!* Near Saldanha Bay; along the Berg River and near Stellenbosch, and at Brackfontein, *E. & Z.!* Silverfontein, and near Groenekloof, *Drege!* (Herb. T.C.D., Hook., Sond., Banks., Thunb.).

2–3 feet high, pallid and more or less velvetty with very short hairs. Flowers violet or purplish, in terminal pseudo-racemes. *H. trifurcata* of E. & Z.! is *H. angularis*, Jacq., for the most part.

67. H. spinosa (E. Mey.!); shrubby, *divaricately much branched,* the branches glabrous; leaves (small) petiolate, scattered, cuneate, sub-truncate, coarsely 3–5-toothed, *minutely stellulate on both sides;* stipules minute, subulate, deciduous; peduncles axillary, as long as the leaves, jointed, one-flowered, *the lower half persistent and hardening into blunt spreading spines;* calyx glabrescent, veiny, *deeply 5-fid, the segments broadly subulate;* petals narrow-obovate, not much longer than the calyx; fil. linear-cuneate, shorter than the tapering anthers; ovary crested; capsules villous, with ten divergent horns. *Mahernia spinosa, Burch. in DC. Prod.* 1. *p.* 497? *Mahernia, No.* 88, *Herb. Eck.! Zey.* 124.

HAB. Gamke River, *Burke and Zeyher*. Nieuweveld, *Drege*. (Herb., T.C.D., Hook., Sond.).

A small, scraggy bush, 6–12 inches high, with pale bark, and few leaves. Leaves ¼ inch long. Fl. small, purplish? The spines formed of old peduncles, are ¼–½ inch long. I cannot say whether or not this be *Mahernia spinosa* of Burchell; but our plant is a *Hermannia*.

68. H. linearifolia (Harv.); shrubby, *divaricate;* branches glabrous or viscidulous; leaves fascicled or scattered, *narrow-linear or linear-lanceolate, tapering at base, obtuse, entire, glabrous and slightly viscid;* stipules subulate; peduncles racemose, cernuous, shorter than the leaf, one-flowered; calyx semi-5-fid, glabrous, glandular, the lobes broadly subulate; petals obovate, flat, not much longer than the calyx; filaments spathulate, rather shorter than the tapering anthers; ovary top-shaped; capsule obconic, crowned with ten divergent horns. *H. linifolia, E. Mey.! non Linn.*

HAB. Winterveld and Nieuweveld, *Drege!* Orange River, *Burke & Zeyher!* Somerset, *Mrs. F. W. Barber!* (Herb. Hook., Sond., T.C.D.).

A stout woody, much-branched, scrubby bush, with densely leafy divergent branches. Leaves a line broad, nearly an inch long, mostly tufted. Flowers purplish.

69. **H. filipes** (Harv.); *slender, suffruticose;* branches stellato-pubescent; leaves on short petioles, sub-distant, linear-lanceolate, acute at each end, quite entire, nearly glabrous, or sparsely stellulate; stipules minute; *peduncles axillary, filiform, longer than the leaves, jointed, bracteate, one-flowered,* nearly glabrous; calyx campanulate, deeply 5-parted, glabrescent, segments from a broad base subulate; petals scarcely longer than the calyx; fil. much shorter than the taper-pointed anthers; capsule glabrescent except the angles, *shortly 5-horned.*

HAB. Zooloo Country, *Miss Owen!* (Herb. T.C.D.).

Very distinct from any described species. I have only seen a single specimen, evidently a young one. The root is scarcely fibrous; the stem unbranched, 6–8 inches high, the leaves 1½ inch long, 1–2 lines wide, nearly an inch apart; the peduncles 1¾–2 inches long and very slender, quite naked and strictly one-flowered.

70. **H. stricta** (Harv.); shrubby, slender, *minutely viscid, nearly glabrous;* branches *virgate,* straight; leaves sub-distant, on short petioles, cuneate-obovate, truncate, toothed above, glabrous, viscidulous; stipules minute, subulate; peduncles racemose, one-flowered, jointed and minutely bracteolated below the flower; calyx sparsely hairy and ciliate, deeply 5-parted, segments broadly subulate, mid-ribbed; *petals thrice as long as the calyx, broadly obovate, expanded; fil. broadly spathulate, much longer than the oblong, blunt anthers;* capsule with ten, plumoso-villous horns. *Mahernia stricta, E. Mey.! Turcz. Mosc.* 1858, *p.* 222.

HAB. Between Natvoet and the Gareep, *Drege!* (Herb. T.C.D., Hook., Sond.).

Readily known by its large, *expanded* flowers, which are very like those of *Mahernia grandiflora,* or of an *Oxalis.* The stamens and inflorescence are those of a *Hermannia.* It seems to be a small, slender and slightly branched shrub, more or less viscid, and nearly glabrous, save for a few scattered, stellate hairs.

APPENDIX.—*Doubtful, or little known species.*

H. triphylla (Cav. Dis. 6. t. 178. f. 3, non Linn); "leaves roughish, tripartite, the parts cuneate, truncate and toothed at the apex, the intermediate petiolate; stipules lanceolato-subulate." *DC. Prod.* 1. *p.* 494.

H. glandulosa (Link, Enum. 2. p. 179); "leaves oval, unequally crenate, sub-pubescent; stipules ovate, acute, often incised; stem glandularly hairy." *DC. l. c.*

H. melochioides (Burch. Cat. 2957); "leaves glabrous, ovate, unequally toothed; stipules ovate-acuminate; pedicels one-flowered, shorter than the leaf; filaments filiform-linear." *DC. l. c.*

H. bryonifolia (Burch. Cat. 2131); "leaves roughly stellulate, cordate-ovate, unequally toothed; stipules linear-lanceolate; pedicels one-flowered, patent, nodding." *DC. l. c.*

H. coronopifolia (Link, En. 2. p. 180); "leaves linear, pinnatifid, fleshy, glabrescent; stem pubescent." *DC. l. c. p.* 496.

H. hispidula (Rchb. Ic. Cult. t. 69); "calyx cup-shaped; flowers laxly panicled; leaves lanceolate, toothed beyond the middle, acute, hispidulous, as well as the branches." *Walp. Repert.* 1. *p.* 346.

H. discolor (Otto, & Dietr. Gart. 8, 314); "branches hairy; leaves oblong, obtuse, sub-cordate, undulato-crenate, rugose, hairy and green above, glauco-tomentose below; stipules subulate; fl. term., sub-racemose; peduncle short, aggregate, one-flowered; calyx campanulate." *Walp. l. c.*

H. venosa (Bartl. in Ott. & Dietr. l. c. 315); "hairy; leaves ovato-sub-rotund, toothed, undulate, veiny, rugose; stipules ovate, much acuminate, the point toothed; ped. axillary, solitary; fl. aggregated at the apex of the peduncle, involucred with bracts; calyx sub-urceolate." *Walp. l. c.*

H. glauca (Hort. Herrenb. Ott. & Dietr. p. 330); "branches glabrous; leaves shortly petiolate, lanceolate, acute, serrate above, glabrous, flat, glaucous; stipules ovate, entire; fl. terminal, sub-racemose; ped. axillary, 2-flowered." *Walp. l. c. p.* 347.

H. leucanthemifolia (Ott. & Dietr. l. c.); "branches glabrescent; leaves shortly petiolate, lanceolate, toothed above, glabrous, smooth; stip. obliquely cordate, sub-entire; fl. terminal, panicled; calyces 5-angled, campanulate." *Walp. l. c. p.* 347.

H. cluytiæfolia (Otto & Dietr. l. c. 332); "branches tomentulose; leaves short-stalked, oblong and lanceolate, mucronulate, quite entire, holosericeous, whitish; stipules large, lanceolate, holosericeous; peduncles axillary, 3-fl.; calyx campanulate, angulate." *Walp. l. c.*

III. MAHERNIA, Linn.

Stamens 5, opposite the petals; their filaments suddenly dilated and mostly tuberculated in the middle, or somewhat cruciform. Other characters, as in *Hermannia*. *Endl. Gen.* 5341. *DC. Prod.* 1. *p.* 496.

Small shrubs or under-shrubs, or perennial herbs, almost all natives of S. Africa; a few from North Africa. Leaves alternate, frequently deeply cut or pinnatifid, rarely entire. Stipules petiolar. Peduncles mostly two flowered, terminal or opposite the leaves; pedicels slender, bracteate at base. Flowers nodding, red, orange, violet or yellow. The name is an anagram of *Hermannia*, from which genus this differs, as well by the artificial character in the filaments, as by the natural one of *terminal*, not *axillary*, inflorescence. I have grouped the species chiefly by differences in the incision and pubescence of the leaves.

ANALYSIS OF THE SPECIES.

Group 1. **Verticillatæ.** Leaves spuriously whorled; that is, the stipules large, resembling the leaves, and parted to the base into two or three leaf-like segments. (Sp. 1–6.)

* *Erect or sub-erect:—*

Softly hairy and woolly. Calyx inflated, 5-toothed... (1) **grandistipula.**
Rough, or thinly pubescent. Calyx deeply 5-parted (2) **heterophylla.**

** *Procumbent:—*

Bracts linear, leaf-like, separate ... ... (3) **verticillata.**
Bracts connate into an amplexicaul involucre;
Leaves, stems and calyces densely glandular ... (5) **humifusa.**
Leaves glabrous, *pinnatifid* ... ... (4) **diffusa.**
Leaves thinly stellate, *bi-pinnatipartite*... ... ... (6) **multifida.**

Group 2. **Pinnatifidæ.** Stipules simple. Leaves alternate, deeply inciso-pinnatifid, *the lobes linear or oblong.* (Sp. 7–14.)

* *Erect or sub-erect:—*

Stellato-pubescent. *Capsule* bladdery, much inflated (7) **vesicaria.**
Stellato-pubescent. *Capsule* not inflated ... ... (8) **incisa.**
Glabrous; branches exuding a gummy substance ... (9) **pulchella.**

** *Procumbent or very diffuse.*

Bracts separate; stipules linear, entire ... ... (11) **sisymbriifolia**
Bracts connate; stipules ovate;
Branches scabrous, leaves stellulate ... ... (10) **pilosula.**
Glabrous, or glandular; leaf-lobes flat and broad... (12) **bipinnata.**
Pubescent; leaf-lobes very narrow, channelled ... (13) **anthemifolia.**
Stellulate and canescent; peduncles very short ... (14) **marginata.**

Group 3. **Lacerifoliæ.** Leaves unequally inciso-lacerate, or crenato-lobulate; not regularly pinnatifid, nor equally toothed. (Sp. 15–19.)

Capsule 2–3 times as long as the calyx:—
Sub-erect, suffruticose, glabrescent or minutely glandular ... ... (15) **coccocarpa.**
Trailing, herbaceous, scabrous and glandular ... ... (16) **scabra.**
Capsule not much longer than the calyx:
Petioles much shorter than the leaves,
Hispid and scabrous; bracts distinct ... ... (17) **rutila.**
Glabrous, stellulate, or glandular; bracts connate ... (18) **erodioides.**
Petioles long; pubescence whitish ... ... (19) **nana.**

Group 4. **Dentatæ.** Leaves linear, ovate, sub-rotund, obovate or oblong, equally toothed or sub-entire, sub-glabrous (not tomentose). (Sp. 20–28)

Bracts more or less connate:—
Stems filiform; leaves sub-rotund, distant ... ... (20) **gracilis.**
Stems shrubby or woody, rigid:—
Leaves broadly elliptical or sub-rotund, close set ... (21) **Linnæoides.**
Leaves oval or oblong, glabrous ... ... (22) **ovalis.**
Stems suffruticose or sub-herbaceous; leaves oblong.
Scabrous; leaves eroso-crenate, margined with hispid glands ... ... (23) **veronicæfolia.**
Pubescent; leaves toothed; stipules variously cut ... (24) **Abyssinica.**
Thinly stellulate and viscid; stipules entire; leaves toothed ... ... (25) **stellulata.**
Bracts distinct not connate:—
Leaves linear-lanceolate; peduncles very short ... ... (26) **linearis.**
Leaves oblong or oval; flowers very small ... ... (27) **parviflora.**
Leaves cuneate-obovate; flowers very large... ... ... (28) **grandiflora.**

Group 5. **Tomentosæ.** Leaves oblong, ovate or cordate, albo-tomentose or woolly. (Sp. 29–33.)

Leaves broadly elliptic or ovate-oblong, cordate at base:—
Petals twice as long as calyx; ovary tomentose... ... (29) **chrysantha.**
Petals scarcely longer than calyx; ovary pubescent ... (30) **betonicæfolia.**
Leaves broadly ovate-oblong, velvetty, rounded at base; the calyx teeth acuminate, as long as the petals ... (31) **vestita.**
Leaves narrow-oblong or linear, cuneate at base:—
Procumbent; stipules and bracts deeply incised... ... (32) **oblongifolia.**
Sub-erect; stipules and bracts subulate ... ... (33) **tomentosa.**

Group 1. VERTICILLATÆ. (Sp. 1–6.)

1. M. grandistipula (Buching.); suffruticose, sub-erect, pilose or villous; leaves shortly petiolate, oblong or linear-oblong, obtuse, toothed or sub-entire, sprinkled with long, soft, tufted hairs, membranous; *stipules leafy, palmately 3–4-parted, the segments lanceolate;* peduncles shorter than the leaves, 2-flowered; bracts palmatifid; *calyx globose, inflated, villous, 5-toothed, the teeth short, deltoid;* petals scarcely longer than the calyx, pubescent, obtuse; ovary obconic, canescent. *Herm. decumbens, Drege, non Willd.*

HAB. Port Natal, *Krauss, No.* 175, *Sanderson!* Between Gekau and Basche, *Drege!* (Herb. T.C.D., Hook.)

Stems about 1 foot high, branching below, sub-erect, woody at base, covered with long, white, soft, stellate hairs. Leaves 1–1½ inches long, rarely ½ inch broad, green, chiefly hairy on the nerves of the lower surface, rarely sub-entire. Stipules longer than the petioles, quite like leaves, deeply cut. Calyx becoming bladdery in fruit. Filaments cruciform, with a very hairy, transverse tubercle above the middle; anthers short. A remarkably distinct species, easily known by its stipules and inflated calyces.

2. M. heterophylla (Cav. Diss. 6. p. 324. t. 178. f. 1.); *stems erect, flexuous, branching,* scabrous; leaves glabrous, scaberulous or pubescent, spuriously whorled, one petiolate, cuneate-oblong, 3–4-toothed at the point, the rest (leaf-like stipules) sessile, linear-oblong, entire; peduncles longer than the leaves, 2-flowered; bracts *broadly ovate, connate into a bifid-hood; calyx pilose, deeply 5-fid,* the segments ovate, acute; petals short-clawed, broadly obovate; ovary ovoid, stipitate, pubescent. *DC. Prod.* 1. *p.* 496. *E. & Z.! No.* 386.

VAR. β. **Namaquensis**; more glabrous, with shorter petioles, and narrower and more cuneate leaves. *M. Namaquensis, E. Mey! Turcz. Mosc.* 1858. *p.* 219.

VAR. γ. **pubescens**; densely pubescent in all parts, *Hb. Banks.*

HAB. Near the Berg River and at Saldanha Bay, *E. & Z.!* var. β. between Kaus, Natvoet and Doornport, 1–2000f. *Drege!* var. γ. Cape, *Masson!* (Herb. Hook., Sond., Banks.)

Erect and shrubby, variable in pubescence. This species is allied to *M. diffusa*, but differs in habit and foliage. Cavanille's figure is characteristic. The hood-shaped involucre is remarkably large.

3. M. verticillata (Linn.); *procumbent,* the branches ascending or erect, scabrous; leaves spuriously whorled, linear, simple or cut, tapering at base, acute, glabrous or scabrous, ciliolate; stipules leafy, numerous, resembling the leaves; peduncles filiform, one or two-flowered; *bracts linear, elongate, distinct;* calyx campanulate, semi-5-fid, lobes deltoid, acute; petals broadly obovate, flattish, twice as long as the calyx; ovary obovate, puberulous. *DC. Prod.* 1. *p.* 496., *non Cav. E. & Z.! No.* 446, *ex parte.*

HAB. Hills about Groenekloof, *Pappe!* Simonsbay, *M'Gillivray.* (Herb. Hook., Sond.).

Stems trailing, 2 feet long or more, with many erect, simple branches, 4–6 inches high, branches and leaves generally scabrous, sometimes very rough. Stipules quite like the leaves, ¾ inch long, ½ line wide. Always to be known from *M. diffusa* which it resembles, by its separate (not connate) bracts, and usually by its more simple leaves.

4. M. diffusa (Jacq. Schoenb. 2. t. 201); diffuse; the branches ascending, scabrous; leaves cuneate at base, sub-sessile, *pinnatifid*, glabrous, the lobes linear, sub-acute; stipules leafy, 3 or 4 in an imperfect whorl, lanceolate, midribbed; peduncles filiform, 2-flowered, bracteate near the summit; *bracts ovate, connate into a bifid hood;* calyx campanulate, deeply cut, the lobes ovate, sub-acute, ciliolate; petals broadly obovate, nearly flat, twice as long as the calyx; ovary obovate, pubescent. *DC. Prod.* 1. *p.* 496. *E. & Z.! No.* 403.

VAR. β. **simplicifolia**; leaves varying to simple, and sparingly pinnatifid on the same branch. *M. verticillata, Auct. plur. ex parte. Cav. Ic. t.* 176. *f.* 1. *(quoad iconem) E. & Z.! No.* 406, *ex parte.*

HAB. Hills round Capetown, extending to Hott. Holland, Blauweberg, and Jackal's Valley, *Zeyher!* (Herb. T.C.D., Hook., Sond.).

Stems trailing, rough with raised, hard-points. Leaves and leafy stipules spuriously whorled. Nearly allied to *M. verticillata,* with which it is very frequently confounded in Herbaria, and chiefly to be known by its more constantly pinnatifid leaves, more deeply cut calyces, and specially by the *connate* bracts.

5. M. humifusa (E. & Z.! No. 402); procumbent, *all parts densely glanduloso-pubescent;* leaves on longish petioles, oblong, deeply incised or sub-pinnatifid, the lobes oblong, obtuse; stipules leafy, 2–3-parted (spuriously whorled), the segments broadly lanceolate, sometimes cloven; peduncles 2-flowered; bracts broadly ovate, connate; calyx campanulate, hairy or glandular, semi-5-fid, the segments ovate-oblong, sub-acute; ovary roundish, pubescent; capsule as long as calyx. *Herm. procumbens, E. Mey.! (non Cav.), and Herm. Drege, No.* 7308, *ex parte.*

HAB. Sand hills near the mouth of the Berg River and at Saldanha Bay, *E. & Z.!* Paarlberg, and between Groenekloof and Saldanha Bay, *Drege!* (Herb. Sond., Hook.)

Chiefly known from *M. diffusa* by the abundant glandular pubescence, more luxuriant growth, larger involucres, &c. The corolla is brownish or purplish red, nodding, of average size. Stipules usually 3–4-parted to the base, sometimes single and incised.

6. M. multifida (E. Mey.! in Hb. Drege); procumbent, suffruticose, *villoso-canescent;* stems filiform, with scattered, stellate hairs; leaves *bi-pinnati-partite*, the lobes linear, obtuse, thinly stellato-pubescent; *stipules leafy, incised or palmatifid;* peduncles 2-flowered; bracts oblong, connate at base; flower-buds acute; calyx campanulate, stellato-canescent, the segments broadly subulate, acute; petals ovato-lanceolate, sub-acute; ovary oblong, stellato-pubescent.

HAB. Between Pedroskloof and Leliefontein, *Drege!* (Herb.Sond., Hook., T.C.D.)
Nearly allied to *M. diffusa*, but with more decompound leaves, different calyces and stellato-canescent pubescence.

Group 2. PINNATIFIDÆ. (Sp. 7–14).

7. M. vesicaria (DC. Prod. 1. p. 497); diffuse, stems short, sub-erect, stellato-pubescent; leaves sub-sessile, cuneate-obovate, entire in the lower half, incised or toothed above, the lobes oblong, acute, sparsely stellato-pubescent on both sides; stipules lanceolate; peduncles *sub-terminal*, 2–4-flowered, or cymoso-paniculate, stellato-hispid; bracts linear-oblong, distinct; calyx deeply 5-parted, pubescent, the lobes ovate-oblong, acute; petals cucullate, pubescent; anthers ovate-oblong, sub-obtuse; *capsule inflated and bladdery*, 5*-lobed. Herm. vesicaria, Cav. Diss.* 6. *t.* 181. *f.* 2. *H. procumbens, E. & Z. ! ex parte, non Cav. H. procumbens and M. ovata, E. Mey. ! Hb. Drege. M. incisa, E. & Z. ! No.* 401, *non Jacq.*

HAB. Hills round Capetown, common *W.H.H.* Cape Flats, *Pappe!* summit of Table Mt., *E. & Z.!* Groenekloof, Paarl and Drakenstein, *Drege!* (Herb. T.C.D., Hook., Sond.).

Root thick and woody. Stems 8–12 inches high. Leaves scarcely an inch long, on very short petioles or sub-sessile, always cuneate at base, variously cut, generally hispidulous and turning black in drying. The inflorescence is nearly that of Hermannia, but the filaments are those of *Mahernia*. The large bladdery fruit, nearly ¾ inch in diameter, is the most obvious character.

8. M. incisa (Jacq. ! Schoenb. t. 54); *shrubby, erect; branches rough with simple and fascicled hairs;* leaves shortly petiolate, oblongo-lanceolate, deeply inciso-pinnatifid, *pubescent on one or both sides;* stipules linear-lanceolate, acute; peduncles sub-terminal, 2-flowered, bracts connate; calyx deeply 5-cleft, pubescent, the segments lanceolate, acute; anthers short, ovate-oblong, obtuse. *DC. Prod.* 1. *p.* 496.

HAB. Uncertain. Cult. in European gardens. (Herb. Jacq. !).

Said to be a shrub, 2 feet high, erect and much branched. The flowers are rather small, white or yellowish. The character here given is chiefly taken from Jacquin, whose specimens, in Herb. Vind., have neither flower nor fruit. The nearest wild plant to this, known to us, is a specimen in Herb. Sonder, found by Zeyher at Groenekloof, but it does not perfectly coincide with Jacquin's description.

9. M. pulchella (Cav. Diss. 6. t. 177. f. 3); *erect, suffruticose, quite glabrous*, or nearly so, the branches resinous; leaves petiolate, narrow-oblong, inciso-pinnatifid, the entire lobes and the sinuses very obtuse; stipules small, ovate; peduncles shorter than the leaves, 2-flowered; bracts connate, hoodshaped, incised; calyx semi-5-fid, resinous-dotted, glabrous, the segments deltoid, acute, half as long as the obovate petals; ovary obovate, stellato-pubescent. *DC. Prod.* 1. *p.* 496. *M. biserrata, E. & Z ! non Cav. M. diffusa, E. Mey. ! non L. Hermannia,* 2316, and 7316 *Drege. Herm. pulchella, Linn. Thunb. Fl. Cap. p.* 507. *M. glandulosa, Presl. Turcz. Mosc.* 1858, *p.* 220.

HAB. Cape *Masson!* in Hb. Banks. Karroo, near Gauritz River, *E. & Z.! No.* 399. Silverfontein, Zondag River, Nieuwefeldsberg and the Winterfeld, *Drege!* Wolfkopf, *Burke!* Albany, *Mrs. F. W. Barber!* (Herb. T.C.D., Hook., Sond.)

An erect, wiry undershrub, 8–12 inches high, branching, glabrous in every part, but generally resinous and clammy. The young stems are sometimes microscopically puberulent. Leaves polished 1–1½ inch long, ¼ inch wide. Flowers purple, campanulate. Burchell's *M. vernicata* (not of E. & Z.!) seems to come very close to this, if it be different.

10. M. pilosula (Harv.); diffuse, much branched; branches scaberulous and hispid with scattered, stellate hairs; leaves deeply pinnatifid or bi-pinnatifid, the lobes linear, sub-acute, glabrous on the upper, sparsely hairy on the the lower side; stipules ovato-lanceolate, acuminate; peduncles filiform, 1–2-flowered; bracts connate into an incised hood; calyx semi-5-fid, puberulous, the lobes ovate-oblong, sub-acute; petals twice as long as the calyx; ovary obovoid, stipitate, pubescent. *M. myrrhifolia, E. & Z.! No.* 407, *non Thunb. Drege, No.* 7307!

VAR. β. **latifolia**; leaves broader, inciso-pinnatifid; the lobes entire, cultrate.

HAB. Sandy ground. Zwartland, *E. & Z.!* near the Paarl and at Groenekloof, *Drege! Pappe!* var. β. Groenekloof, *Zeyher!* (Herb. Hook., Sond.)

Root woody and thick. Stems several, elongate, with filiform branches. Leaves 1 inch long, deeply cut, all parts except the upper surfaces of the leaves hispidulous. Flowers campanulate, ⅓ inch long. E. & Z.! refer their plant to *H. myrrhifolia*, Thunb., which, according to a specimen in Herb. Holm., is a *Hermannia.* Var. β. agrees tolerably with Jacquin's figure of *M. glabrata,* but is by no means glabrous. It differs from the typical form of the species chiefly in having less compound and somewhat broader leaves.

11. M. sisymbriifolia (Turcz. Mosc. 1858. p. 221); suffruticose, slender, glandularly puberulous and stellato-hispid in all parts; leaves inciso-pinnatifid, the lobes few, distant, entire, linear, obtuse; *stipules linear oblong, obtuse, entire;* peduncles longer than the leaves, 2-flowered; *bracts like the stipules, separate;* calyx hispid, deeply 5-fid, segments ovate-oblong, acute, half as long as the narrow-obovate petals; filaments very narrow, with obsolete tubercles; anthers ovate, acute; capsule sub-sessile, inflated, roundish, 5-lobed, broader than long, stellato-pubescent. *Drege, No.* 7306! *M. brachycarpa, Harv. MS.*

HAB. Blaawberg, 2–3000f. *Drege!* (Herb. T.C.D., Hook., Sond.)

A slender species, resembling *M. pilosula* in miniature, but with different pubescence, stipules and bracts, and less compound leaves, &c. Capsules 2 lines in diameter, the calyx small, deflexed. Flowers very small.

12. M. bipinnata (Linn.); procumbent, suffruticose, nearly glabrous; stems filiform, glabrous or scaberulous; *leaves pinnatifid, or bi-pinnatifid, the lobes linear, flat, obtuse, glabrous, or nearly so;* stipules ovate-acute, amplexicaul, short, entire or cut; peduncles 2 flowered, the bracts semi-confluent; calyx campanulate, the segments broadly subulate, gradually acuminate, ciliolate; petals oval-oblong; ovary ellipsoidal, glandular; style hispidulous. *DC. Prod.* 1. *p.* 496. *Cav. Diss. t.* 176. *f.* 2. *Drège,* 7315, 7316. *E. & Z.! No.* 409. *M. jacobæafolia, Turcz. Mosc.* 1858, *p.* 220.

VAR. β. **acutifolia**; leaf-lobes acute.

VAR. γ. **glandulosa**; glandularly pubescent in all parts; leaves simply pinnatifid.

HAB. Caledon River and Wolfkop, *Zeyher!* Schneuwberg, *Drege!* Albany, *Mrs. Barber!* Between Kochman's Kloof and Gauritz R., *E. & Z.!* β. Gamke River, *Zeyher.* γ. Port Natal, *Miss Owen!* (Herb. T.C.D., Hook., Sond.)

Stems trailing, with ascending or sub-erect branches. Leaves very generally bi-pinnatifid, but sometimes on the same root, simply pinnatifid, now and then sprinkled with minute, tubercular asperities, which also occur on stem, peduncles and calyx. Peduncles as long as, or much longer than the leaves. E. & Z.'s specimen in Herb. Sond. has more obtuse flower-buds and broader and shorter calyx-lobes, but otherwise it closely agrees with the more common variety. The incision of the stipules, which De Candolle attributes to his *M. resedæfolia*, seems to be a very variable character. Ecklon's "*M. pinnata*" is *Hermannia tenuifolia*. Our var. γ. may possibly be a species.

13. M. anthemifolia (Harv.); suffruticose, diffuse or procumbent; *stems pubescent;* leaves bi-pinnatifid, puberulous, *the lobes very narrow, linear, obtuse, channelled;* stipules ovate, amplexicaul, cut; peduncles short, 2-flowered, with connate bracts; calyx pubescent, deeply 5-fid, its lobes broadly subulate, acute; petals ovate-oblong; ovary stellato-pubescent, obovate.

HAB. S. Africa, *Zeyher!* (Herb. Sond.)

Very near *M. bipinnata*, but the leaves are more finely decompound, the lobes are much narrower and manifestly channelled, and the peduncles shorter. The habit is rather different, and the pubescence more copious.

14. M. marginata (Turcz. Mosc. Bull. 1858. p. 221); suffruticose, decumbent, many-stemmed, small, *thinly clothed with short, white, substellate hairs;* leaves petiolate, *ovate, obtuse, inciso-pinnatifid,* the lobes obtuse, sub-canescent on both sides; stipules ovate, acute; peduncles scarcely equalling the leaves, spreading, 2-flowered; bracts ovate, acute, connate; calyx deeply 5-fid, the lobes ovate-oblong, acute, half as long as the petals; anthers longer than the filaments; capsule sub-globose, pubescent. *M. geranioides, Harv. in Herb.*

HAB. Springbokkeel, *Zeyher! No.* 133. (Herb. Sond., Hook., Benth.)

Root woody and thick. Stems cæspitose, 3–6-inches long, slender and sub-simple, spreading from the crown. Leaves ½–1 inch long, deeply cut. Very like the more pinnatifid forms of *M. erodioides*, but much more hairy, and with a different hairiness, with shorter peduncles, patent or deflexed in fruit, and shorter, nearly globular capsules. It is also allied to *M. nana*, of which it has the pubescence, but differs in foliage and flower-stalks, &c.

Group 3. LACERIFOLIÆ. (Sp. 15–19).

15. M. coccocarpa (E. & Z. No. 397); suffruticose or woody, much branched, *erect or diffuse, glandularly pubescent and sub-glabrous;* leaves petiolate, oblong or linear, *deeply incised or pinnatifid, the lobes obtuse,* entire, or sparingly toothed, glandular or glabrous, or thinly stellulate; stipules amplexicaul, ovate-acuminate, much shorter than the petiole; peduncles filiform, much longer than the leaves, the bracts connate; calyx-segments *broadly subulate, acute,* more than half as long as the obovate petals; ovary elliptic-oblong; *capsule thrice as long as the calyx. Drege,* 7312.

VAR. β. **ustulata**; scrubby, rigid; leaves depauperated and more glabrous; flower stalks shorter.

HAB. Konabhoogde, *E. & Z.!* Somerset, *Dr. Atherstone, Mrs. Barber,* Grassy Hills, beyond Fish River, *Mr. Bunbury.* β. Assagaisbush, *Burke.* (Herb. T.C.D., Hook., Sond.)

A densely-tufted, branching, sub-erect, suffrutex, 6–12 inches high, sometimes densely glandular, sometimes nearly glabrous. Leaves 1–1½ inches long, very variable in shape and incision. The broader leaved forms resemble *M. erodioides.* Flowers purple. Capsule ¾–1 inch long.

16. M. scabra (E. & Z. ! No. 398); *diffuse;* branches, petioles and peduncles *rough with capitellate hairs;* leaves ovate-oblong, obtuse, *lacero-pinnatifid,* the lobes short and broad, variously toothed; upper surface glabrous, or sparsely stellulate; stipules amplexicaul, broadly-ovate, deeply toothed or cloven; peduncles very long, the bracts *connate into a hoodshaped involucre;* pedicels hispid; calyx puberulent and scabrid, semi-5-fid, *the lobes ovate sub-acute;* capsule oblong obtuse, *pubescent,* 2–3 times as long as thc calyx. *M. lacera, E. Mey.! in Hb. Drege, also Drege, No.* 7307 *α.*

HAB. Between the Zwartkops and Zondag's River, *E. & Z.! Drege!* Blauweberg, *Zeyher!* (Herb. T.C.D., Hook., Sond.)

A slender, trailing plant, varying in scabridity; sometimes very rough all over with pin-headed hairs, and sprinkled with stellate hairs. Leaves on longish petioles, 1 inch long, irregularly pinnatifid. Peduncles 4–5 times as long as the leaves. Flowers reddish. Very like the following, from which it differs chiefly in the involucre.

17. M. rutila (Jacq. Schoenb. t. 263); "stems procumbent, hairy, scabrous; leaves on short petioles, oblong, obtuse, lacero-pinnatifid, and unequally lobed and serrated, plaited, rough on both surfaces; stipules lanceolate, acute, sometimes serrated; peduncles scarcely longer than the leaves, *the bracts lanceolate, acute, small;* calyx semi-5-fid, hairy, *the lobes lanceolate, acute;* ovary obovate." *Jacq. l. c. abbrev.*

HAB. Formerly cultivated in Schoenbrun Gardens, Vienna.

The above description is compiled from Jacquin. His figure shows a plant very like *M. scabra,* but differing in stipules, peduncles and separate bracts.

18. M. erodioides (Burch. Cat. 1491); suffruticose, decumbent and many-stemmed; stems filiform *glabrous, or sparsely stellate, or viscoso-pubescent;* leaves shortly petiolate, glabrous, or nearly so, oval-oblong, or cordate-ovate, obtuse, inciso-crenate or inciso-pinnatifid, with obtuse lobes, thin; stipules ovate, short, amplexicaul; peduncles much longer than the leaves; bracts ovate-acute, connate; calyx glabrous or glandularly pubescent, its lobes *deltoid-acuminate,* half as long as the petals; capsule longer than the calyx, *nearly glabrous. DC. Prod* 1. *p.* 496. *E. & Z.! No.* 394. *Drege, No.* 7309, 7313. *Herm. multicaulis, E. Mey! in Hb. Drege. Turcz. Mosc. p.* 220.

VAR. *α.* **glabra**; glabrous, except a few close-pressed, stellate hairs on the ribs and veins of the leaves.

Var. *β.* **viscidula**; stems, peduncles and calyx glanduloso-pubescent.

VAR. *γ.* **latifolia**; leaves cordate-ovate, crenate, not much cut.

HAB. South Africa, *Burchell.* Karroo plains in many places, throughout the colony and on the Schneuweberg, Kamdebosberg and other mountain ranges, *E. & Z.! Drege!* var. *γ.*, Sand River, *Burke & Zeyher!* (Herb. T.C.D., Hook., Sond.)

Root thick and woody. Stems very slender, 2–8 inches long, not much branched. Leaves ½–1 inch long, more or less deeply incised, sometimes sparsely stellate beneath. Peduncles 2 inches long, bracteate beyond the middle. Flowers ¼ inch long. Filaments with a nearly glabrous, cordate tubercle; anthers taper-pointed, nearly equalling the petals. Our var. *γ.* has leaves 1½ inch long, cordate at base, much less deeply crenate than usual; with narrower petals and a more lunate tubercle on the filaments. The two forms seem to run into one another.

19. M. nana (E. & Z.! No. 395); minute, suffruticose, *thinly covered with very short, white, stellate hairs;* leaves on long petioles, elliptical or sub-rotund, inciso-crenate; stipules ovate, amplexicaul; peduncles much longer than the leaves, the bracts oblong-acute; calyx campanulate, the lobes broadly subulate, half as long as the petals. *M. linnœoides, E. Mey! non DC.*

HAB. Karroo, near Gaaup, Beaufort, *E. & Z.!* Desert near the Gamke River, *Zeyher, No.* 134. Between Dweka and Zwartbulletje, *Drege!* (Herb. Sond., Benth.)

Root woody. Stems slender, 2–4 inches long, slightly branched. Leaves scarcely ½ inch long; petioles nearly as long. Allied to *M. erodioides,* but with shorter, broader, rounder, less cut leaves, and different pubescence.

Group 4. DENTATÆ. (Sp. 20-28).

20. M. gracilis (Harv.); stems trailing, *very slender,* sub-simple, nearly glabrous; leaves remote, *sub-rotund,* 5–7-crenato-lobulate, sparsely stellate on both sides; stipules amplexicaul, ovate-acuminate, shorter than the petioles; peduncles much longer than the leaves; bracts connate at base, subulate; calyx glabrous, deeply 5-fid, the lobes broadly subulate, acute, ⅔ as long as petals; anthers taper-pointed, much longer than the filaments; ovary pubescent.

HAB. Zooloo Country, *Miss Owen.* (Herb. T.C.D.)

Of this well-marked and pretty little plant I have as yet only seen a single specimen, which I should probably have referred to *M. Linnœoides,* Burch. had not a very different plant, equally answering to De Candolle's diagnosis, been so named by Eck. & Zey. The present is one of the slenderest of the genus, with leaves like those of *Veronica hederœfolia,* but smaller, ¼ inch in diameter. Peduncles more than an inch long, flowers 2 lines long.

21. M. Linnæoides (Burch. Cat. No. 1878; fide E. & Z.); shrubby, procumbent, glabrescent or stellato-pubescent; leaves on short petioles, stellato-pubescent on one or both sides, *broadly elliptical or roundish,* very obtuse, crenato-dentate stipules; *ovate-acuminate,* equalling the petioles; peduncles glandularly pubescent and viscid, much longer than the leaves; bracts ovate-acuminate, connate beyond the middle; calyx campanulate, semi-5-fid, densely glandular, with deltoid, acute segments more than half as long as the petals; ovary turbinate, densely pubescent. *DC. Prod.* 1. *p.* 497.

Var. *α,* **glabrescens**; upper surface of the leaves nearly glabrous. *M. Linnœoides, E. & Z.!* 391.

VAR. *β.* **hispidula**; more hairy in all parts, both surfaces of the leaves stellato-pubescent. *Herm. rotundata, E. Mey!*

HAB. Bothasberg, Grahamstown, *E. & Z.!* near the Fish River, *Drege!* (Herb. T.C.D., Hook., Sond.)

Stems spreading widely, a foot long or more, much branched, woody and robust. Leaves densely set, ½–¾ inch long, mostly elliptical, sometimes orbicular, varying much in hairiness. The young branches, peduncles and calyces are thickly clothed with gland-tipped hairs.

22. M. ovalis (Harv.); shrubby, sub-erect or diffuse, nearly glabrous, viscidulous; branches with a few scattered, appressed, stellate hairs; leaves on short petioles, *quite glabrous, oval or oblong oval,* obtuse, cuneate at base, serrato-dentate; stipules ovate-acuminate, equalling the petioles; peduncles much longer than the leaves, roughish; bracts connate beyond the middle or cucullate; calyx campanulate, either glabrous or viscoso-puberulous, or tomentose, with deltoid-acute lobes half as long as the broadly oval petals; ovary oblong, albo-tomentose. *M. vernicata, E. & Z.! 393 (non Burch.)*

VAR. β. **cucullata**; bracts confluent into a cup-like cucullus, *Drege, No.* 7311. *M. saccifera, Turcz. Mosc.* 1858. *p.* 219.

VAR. γ. **tomentosa**; bracts cucullate; calyx tomentose. *M. tomentosa, Wendl.*

HAB. Fields near the Zwartkops R. *E. & Z.!* Hassagais' Kloof, *Zeyher!* (Herb. Sond., T.C.D.)

Distinctly shrubby, sub-erect or spreading, a foot high, branched; glabrous in every part, except occasionally a woolliness on the calyx and a few scattered stellate hairs on the young branches. The whole plant shines, as if varnished. Leaves ¾–1 inch long. The bracts vary in degree of coherence, and are sometimes altogether confluent, as in var. β.

23. M. veronicæfolia (E. & Z.! 392); suffruticose, decumbent, *thinly sprinkled with stipitate, stellate hairs;* stems filiform; leaves distant, shortly petiolate, glabrous above, sparsely stellate at the margin and on the lower surface, *oblong, obtuse, eroso-crenate,* glanduloso-ciliolate; stipules ovato-cuspidate, longer than the petioles, ciliolate; peduncles longer than the leaves; bracts connate beyond the middle, ovate-acuminate; calyx campanulate, glandular and stellate, with deltoid-acuminate lobes half as long as the broadly oval petals; ovary stellato-pubescent; capsule oblong, tuberculate.

HAB. Limestone hills in Uitenhage and Albany, *E. & Z.!* (Herb. Hook., Sond., T.C.D.)

Stems many, 1–2 feet long, branching. Stems, margins, and underside of leaves, peduncles, and calyces, are sprinkled with glandshaped tubercles crowned with a starry tuft of short, white, deciduous hairs; the glandular base is persistent. Leaves 1–1½ inches long, ⅓ inch wide, the petioles scarcely ¼ inch long. Peduncles 2 inches long. Flowers ⅓ inch.

24. M. abyssinica (Hochst! Pl. Schimp. No. 320); suffruticose, decumbent, thinly stellato-pubescent; stems filiform; leaves shortly petiolate, *oblong, coarsely toothed,* obtuse at both ends, or sub-acute, *the younger ones fasciculate-pubescent,* the older glabrescent; *stipules cut or palmatifid,* shorter than the petioles; peduncles shorter than the leaves; bracts connate below; calyx deeply 5–cleft, with hairy, ovate-acute lobes ⅔ as long as the narrow-obovate petals; capsule oblong, stellato-pubescent or glabrous. *M. dentata, Harv. in Herb.*

HAB. Vaalrivier, *Burke & Zeyher, No.* 120. (Herb. Hook., T.C.D., Sond.)

Allied to *M. veronicæfolia,* but less woody, more hairy, and with a different structure of hair; with smaller flowers, narrower petioles, shorter peduncles and divided stipules. The peduncles very rarely exceed the leaves in length. Flowers small, yellow. I cannot distinguish the South African from the original Abyssinian specimens.

25. M. stellulata (Harv.); suffruticose, decumbent; the filiform stems, branches and peduncles viscoso-pubescent and thinly stellulate; leaves *linear-oblong*, obtuse at both ends, dentate, sprinkled on both sides with stellate hairs; *stipules broadly ovate, equalling the petioles; viscidulous, quite entire;* peduncles about equalling the leaves; bracts connate; calyx deeply 5-cleft, with broadly subulate, viscoso-pubescent and thinly stellate lobes not much shorter than the obovate petals; ovary obovate, glandular.

HAB. Near Grahamstown, *Herb. Hooker!*

This looks very like *M. abyssinica*, but has very different stipules. The elegantly stellate, scattered, appressed hairs, and glandular pubescence afford further characters. The leaves also are less coarsely-toothed, and the flowers seem to have been reddish.

26. M. linearis (Harv.); shrubby, branching, sub-erect, glabrous, or nearly so; leaves on very short petioles, *linear-lanceolate, very narrow*, sub-acute, either quite entire, or remotely or obsoletely denticulate; stipules short, deltoid-subulate; peduncles much shorter than the leaves, *bracts distinct*, oblong-acute; calyx glabrous, deeply 5-fid, with broadly subulate, acuminate lobes, half as long as the oblong, sub-acute petals; ovary sub-globose, 5-lobed; style hispid. *Herm.* 7310, *Drege!*

HAB. Somerset. Zwartkey, on mountain plains, 4000f. *Drege!* (Herb. Hook., Sond.)

A small branching shrub, nearly allied to *M. pulchella*, from which it chiefly differs in the longer, narrower and sub-entire leaves. Leaves 1–1½ inch long $\frac{1}{10}$–$\frac{1}{8}$ inch broad, quite glabrous; the stems and branches slightly scaberulous. Stipules either triangular-acute or awl-shaped.

27. M. parviflora (E. & Z.! 396); decumbent, thinly stellulate; stems filiform; leaves on short petioles, cuneate at base, *oblong or oval, crenate, obtuse*, glabrescent or minutely stellulate; stipules ovate, acuminate; peduncles equalling the leaves, bracts separate, lanceolate; *flowers very small;* calyx campanulate, semi-5-fid, with deltoid segments; ovary sub-globose, pubescent. *Herm. parviflora, E. Mey.! and H. diffusa, E. Mey.! in Hb. Drege.*

HAB. Near the Zwartkey River, *E. & Z.!* Katriver, Stormberg and Witberg, 5–6000f. *Drege!* Herb. Hook., Sond.)

Stems 1–1½ feet long, spreading over the soil, with short erect branchlets. Whole plant minutely stellato-puberulent. Leaves ¼–¾ inch long, sometimes rather deeply cut. Fl. reddish purple, $\frac{1}{10}$ inch long.

28. M. grandiflora (Burch. Cat. 2333); shrubby, divaricately branched, glandular or glabrous; leaves cuneate-obovate, inciso-dentate or sub-pinnatifid, with obtuse lobes, scattered or tufted; stipules small, lanceolate or ovate; peduncles elongate 2-flowered, with oblong-acute, ternate bracts; pedicels pubescent; calyx obconic, deeply cleft, pubescent, with broadly subulate lobes; *petals 4–5 times as long as the calyx, spreading, obovate, tapering at base;* anthers short, ovate-acute; ovary obovate, stellato-pubescent. *DC. Prod.* 1. 497. *Herm. grandiflora, Hort. Kew.*

VAR. *α*. **Burchellii**; viscoso-pubescent; leaves stellulate, obovato-lanceolate, toothed; stipules ovato-lanceolate. *M. grandiflora, Burch.! in Bot. Reg. t.* 224.

VAR. β. **glabrata**; leaves obovato-cuneate, deeply cut, glabrous, as well as the branches and peduncles; stipules ovate. *M. grandiflora, Drege! M. oxalidiflora, E. & Z.! 387. (non Burch.).*

HAB. Karroo Districts, *Masson!* Plains north of Litakun, *Burchell.* Nieuweveld, Beaufort, *Drege!* Gamke R., *Zeyher!* (Herb. Hook., Banks, Benth., Sond.)

A branching shrub, 1–2 feet high, the young parts sprinkled with minute, stipitate glands; sometimes almost glabrous. Leaves rather rigid, varying from cuneate to almost lanceolate, variably incised. Flowers large, funnel-shaped, pendulous, brick-red, sweetly scented, nearly ¾ inch across. Stamens very short; the anthers small, ovate, shortly cuspidate; filaments with a narrow wing and obcordate, hispid, expansion below the summit. *Mahernia stricta, E. Mey.!* has the filaments and inflorescence of *Hermannia,* under which genus it will be found: in its mere corolla it closely resembles *M. grandiflora.*

Group 5. TOMENTOSÆ. (Sp. 29–33).

29. M. chrysantha (Planch. in Hb. Hook.); suffruticose, decumbent, albo-tomentose; leaves petiolate, elliptic-oblong, obtuse, subcordate at base, crenulate, corrugated, and at first pubescent, but growing glabrous on the upper side, albo-tomentose with prominent nerves and veins on the lower; stipules membranous, broadly ovate, acute, cut; peduncles elongate, *with incised bracts;* calyx *turbinate, woolly,* semiquinquefid, with deltoid-acuminate lobes; *petals twice as long as the calyx,* stellato-pubescent, with a narrow cucullate claw, and ovate limb; ovary obovate, *densely tomentose. Turcz. Mosc.* 1858. *p.* 219. *Melhania chrysantha, E. Mey.! Herm. geniculata, E. & Z.!* 321.

HAB. On the Zuureberg and between the Keiskamma and Buffalo River, *Drege!* Dornkopf, *Burke!* (Herb. T.C.D., Hook., Sond.).

Larger and more woody than *M. betonicæfolia,* with longer leaves and flower-stalks, longer and narrower petals, and more acuminate and more woolly calyx-lobes, but in other respects so similar that we suspect it is merely a very luxuriant variety, and retain the species chiefly in deference to the opinion of other botanists.

30. M. betonicæfolia (E. & Z.! 320); suffruticose, decumbent, albo-tomentose; leaves petiolate, elliptic-oblong, very obtuse, cordate at base, crenulate, corrugated and at first velvetty, afterwards glabrescent above, albo-tomentose, with prominent nerves and veins below; stipules membranous, broadly ovate, lacero-dentate; peduncles equalling the leaves, *with subulate, distinct bracts;* calyx *swollen,* tomentose, campanulate, semi-5-fid, with deltoid lobes; petals stellato-pubescent, *scarcely longer than the calyx,* cucullate, round-topped; ovary egg-shaped, *pubescent;* style glabrous. *Mahernia cordata, E. Mey.! in Hb. Drege.*

HAB. Stormberg 5–6000f, *Drege!* Thaba Unka and near the Vaal River, *Zeyher & Burke!* (Herb. Hook., Sond., T.C.D.)

Stems half herbaceous, not much branched, 8–12 inches long. Stems, petioles, peduncles, calyx and the under side of leaves densely covered with white, stellate, soft hairs. Leaves 1½ inch long, ¾ inch broad, sometimes rounded, but mostly cordate at base, the upper side with depressed venation and raised areolæ, and the mature ones glabrous, except for a few scattered stellate hairs. Stamens with a large, cordate hispid tubercle. Calyx lobes not acuminate,

31. M. vestita (Harv.); suffruticose, sub-erect, tomentose; stem robust, flexuous; leaves petiolate, *ovato-oblong, obtuse, rounded at base,* crenulate, both surfaces densely stellato-pubescent, somewhat velvetty, penni-nerved below; stipules ovate-oblong, unequal-sided, foliaceous, mostly entire; peduncles *shorter than the leaves, with subulate, distinct bracts;* calyx campanulate, *very deeply 5-fid, with ovato-subulate, acuminate lobes,* as long as the stellato-pubescent, ovate-limbed petals; ovary sub-globose, stellato-pubescent.

HAB. Macallisberg, *Burke and Zeyher!* (Herb. Hook.)

Stems 8–12 inches long, robust, somewhat zig-zag and and paniculately branched, the lateral branches floriferous, with smaller leaves. Leaves 1½ inches long, 1 inch wide, thickish, soft and velvetty. This is nearly related to *M. betonicæfolia,* but the leaves are not cordate at base, nor corrugated; the stipules are different, the peduncles shorter, and the calyx much more deeply cleft, with much narrower and more acuminate segments. The stem is more robust, more woody and erect.

32. M. oblongifolia (Harv.); suffruticose, decumbent; young branches tomentose, the older pubescent or glabrescent; leaves *cuneate at base, narrow* and *elongate-oblong, obtuse,* repando-denticulate, especially in the upper half, more or less tomentose, the older pubescent; *stipules* (withering) *ovato-lanceolate, acuminate, deeply incised;* peduncles scarcely as long as the leaves, tomentose, 2–3-flowered, *with incised bracts;* calyx piloso-tomentose, deeply cut, with broadly subulate segments half as long as the narrow-obovate petals; petals obovate pubescent. *Herm. humilis, E. Mey. in Hb. Drege.*

HAB. Between Caledon, and Thaba Unka and Siebenfontein, *Burke and Zeyher!* 108, 114. Between Sternsbergspruit and Kraunberg, 4500f. *Drege!* (Herb. Hook., Sond., T.C.D.).

A more tomentose plant than *M. abyssinica,* with much larger and longer and less toothed leaves, much attenuated at base, an albo-tomentose calyx and larger (red or purple ?) flowers. Leaves 2 inches long, ½ inch wide, of thin texture; stipules membranous, many ribbed. *Zeyher's* No. 114 is more tomentose than usual, but not otherwise different.

33. M. tomentosa (Turcz. Mosc. 1858., p. 218); suffruticose, suberect and sub-simple, albo-tomentose; leaves on short petioles, *broadly linear,* cuneate at base, coriaceous, entire, denticulate near the apex, obtuse or sub-acute, canescent on both surfaces, penni-nerved on the lower; stipules *subulate,* equalling the petioles; peduncles scarcely as long as the leaves, *with subulate bracts;* calyx turbinate, deeply cut, with broadly subulate, acute lobes; petals not much longer than the calyx, pubescent on the claw only, with an ovate glabrous limb; *fil. narrow-linear, hispid on the outside;* ovary oblong, canescent. *M. angustifolia, Harv. in Herb.*

HAB. Macallisberg, *Burke & Zeyher!* (Herb. Hook., T.C.D., Sond.).

Root woody. Stems tufted, 3–4 inches high, sub-simple and erect. Leaves 2 inches long or more, ¼ inch wide, nearly linear. Fl. stalks 1–1½ inches long, mostly one-flowered. About the size of *Melhania Burchellii,* and much resembling it. Its stamens are scarcely those of a *Mahernia,* but neither are they those of a *Hermannia;* and the habit is that of the present genus.

APPENDIX. Little known and doubtful species.

M. resedæfolia (Burch. Cat. 2280); "leaves glabrous, pinnati-partite, the lobes linear, entire; stipules deeply trifid." *DC. Prod.* 1. *p.* 496.

M. seselifolia (DC. prod. I. p. 496); "leaves pinnati-partite, lobes filiform, entire, acute, the lowest and the stems setoso-hispid, the branches virgate, glabrous above; peduncles elongate, 1–2-flowered." *DC. l. c.*

M. vernicata (Burch. Cat. 1461., Trav. I. p. 278) *DC. l. c. p.* 496, seems to be the same as *M. pulchella*, (Cav. (?)

M. glabrata (Cav. Diss. 6. p. 326. t. 200. f. I.); "leaves thinly stellato-puberulous, lanceolate, pinnatifido-dentate, the teeth undivided; peduncles 2-flowered, elongate." *DC. l. c. p.* 497. *Jacq. Schoenb. t.* 53. *Rchb. Exot. t.* 66. *Herm. glabrata, Linn. f. Suppl.* 301.

M. oxalidiflora (Burch. trav. I. p. 295; Cat. 1536); "leaves glabrous, pinnatifid, the lobes entire, sub-acute; stipules ovate, acute; branches erect, scaberulous; peduncles 1–2-flowered, longer than the leaves. Allied to *M. glabrata*, but the leaves are more deeply cut and the flowers twice the size. *DC. l. c.*

M. biserrata (Cav. Diss. 6. p. 326. t. 200. f. 2); leaves glabrous, oblong-lanceolate, doubly toothed; peduncles 2-flowered, as long as the leaves. *Herm. biserrata, Linn. f. Suppl.* 302. *DC. l. c. p.* 497.

M. violacea (Burch. Cat. 3098); "leaves ovate, obtuse, toothed, glabrous; peduncles 1–3-flowered, opposite the leaves, and longer than them." Habit of *Melochia pyramidata. DC. l. c.*

M. spinosa (Burch. trav. I. p. 279); may be *Herm. spinosa* (see page 205.

M. fragrans (Rchb. Ic. Exot. t. 65); "leaves linear-lanceolate, acutely inciso-pinnatifid; bracts connate at base, acuminate, ciliate." *Walp. Rep.* I. *p.* 347.

Sub-order II. DOMBEYACEÆ.

IV. DOMBEYA. Cav.

Involucel 3-leaved, unilateral, deciduous, *sometimes wanting. Calyx* 5-parted, persistent. *Petals* 5, hypogynous, obovate, unequal sided, convolute in æstivation, at length scarious and persistent. *Stamens* 15–20, slightly connate at base; 5 sterile, strap-shaped or filiform, alternate with the petals; the rest shorter, 2–3 opposite each petal, antheriferous; anthers oblong, erect, slitting. *Ovary* sessile, 3–5-celled; ovules 2–4 in each cell, collateral; *style* simple 3–5-cleft, with revolute arms, stigmatose above. *Capsule* coriaceous, 3–5-celled, septicidal. *Endl. Gen.* 5346, *and Xeropetalum, Del. Endl.* 5346. *DC. Prod.* I. *p.* 498.

Shrubs and small trees found in the tropical and sub-tropical parts of Africa and of the African Islands, Madagascar and Mauritius; rare in tropical Asia. Leaves alternate, cordate, or lobed. Flowers rosy or white, often bursting forth on leafless branches, in umbels or corymbs. *Xeropetalum*, Del., merely differs in having no involucel; but in the following species this character, though found in all, varies so greatly in each, as to show that it is of secondary importance. The name is in honour of Joseph Dombey, a botanical traveller in Chili and Peru.

1. **D. Dregeana** (Sond. in Linn. 23. p. 18); branches terete, glabrous; leaves on long, pubescent petioles, *cordate, acute or acuminate,* angled or even 3-lobed, toothed, minutely stellato-pubescent, 5–7-ribbed; peduncles filiform, axillary, longer than the leaves; umbels 2–4-flowered, with hairy pedicels; *involucral leaflets broadly ovate,* acuminate, and, as well as the lanceolate, reflexed sepals, tomentose; the 5 sterile stamens spathulate, as long as the 5-fid style; ovary globose, tomentose. *Leeuwenhoekia tiliacea, E. Mey.! (ex parte) Xeropetalum tiliaceum, Endl.*

HAB. In the districts of Uitenhage and Albany, and in Caffirland, *Drege! Mrs. F. W. Barber, Mr. Brownlee, &c.* (Herb., T.C.D., Hook., Sond.)

A shrub or small tree, well-covered with aspen-like leaves and showy flowers, on long peduncles, springing from the axils of the uppermost leaves. The flowers are upwards of an inch across (rosy?), and the petals wither without falling off, remaining expanded.

2. **D. Natalensis** (Sond. in Linn. 23. p. 18); branches terete, glabrous; leaves on long petioles, *cordate, acute* or acuminate, somewhat angular, toothed, minutely stellato-pubescent, 5–7-ribbed; peduncles filiform, axillary, longer than the leaves, umbels 4–6–8-flowered, with canescent pedicels; *invol. leaflets narrow-subulate,* and, as well as the lanceolate, reflexed sepals, canescent; the 5 sterile stamens clavate, as long as the 5-cleft style; ovary globose, tomentose. *Leeuwenhoekia tiliacea, E. Mey.! ex parte.*

HAB. Port Natal, *Gueinzius! Drege!* (Herb. T.C.D., Sond.).

Closely resembling *D. Dregeana,* from which it is best known by its narrow subulate involucral leaflets. The flower buds are more fusiform and less tomentose than in *D. Dregeana,* and the umbels have usually more rays.

3. **D. rotundifolia** (Harv.); young branches stellato-tomentose, older glabrous; leaves petiolate, *sub-orbicular, very obtuse,* repando-crenulate, stellato-pubescent on both sides, becoming glabrous above, 5–7-nerved and netted on the lower side; peduncles one or several, tomentose, mostly forked, each arm bearing an umbel of 6–12 flowers; *invol. leaflets narrow-linear,* obtuse, shorter than the bud, deciduous and as well as the lanceolate reflexed sepals, tomentose; the 5 sterile stamens linear clavate, longer than the 3–5-fid style; ovary globose, tomentose. *Xeropetalum rotundifolium, Hochst. Sond in Linn. 23. p. 18.*

HAB. Port Natal, *Gueinzius,* &c. Macallisberg, *Burke & Zeyher!* (Herb. T.C.D., Hook., Sond.)

A divaricately branched, rigid shrub. Leaves 1½–2 inches broad, somewhat cordate at base, strongly netted below and very obtuse. Flowers crowded round the ends of the naked branches in many-flowered umbels; the peduncles long or short. Petals white? The involucel is constantly found on the young buds, but soon falls away. The style is sometimes trifid.

V. MELHANIA, Forsk.

Involucel 3-leaved, sub-unilateral. *Calyx* 5-parted, persistent. *Petals* 5, hypogynous, obovate, unequal sided, convolute in æstivation, deciduous. *Stamens* 10, connate at base; 5 sterile, strap-shaped, opposite the petals; 5 alternate fertile, shorter; anthers sagittate, erect, slitting.

Ovary 5-celled; ovules several in each cell; style 5-fid at summit. *Capsule* 5-celled, loculicidal. *Endl. Gen. No.* 5348. *DC. Prod.* I. *p.* 499.

The S. African species of this genus are small, half woody, slightly branched, tomentose or softly hairy plants, with the aspect of *Hermannia*, but with larger flowers than commonly occur in that genus. Their leaves are ovate or linear-oblong, obtuse at base. The peduncles are axillary, 1–2-flowered. Stipules narrow-subulate. In tropical Africa and Asia there are shrubby or arborescent kinds, resembling *Dombeya* in aspect. The generic name is taken from *Mt. Melhan*, in Arabia, where one of the original species is found wild.

* *Erect, with broadly cordate involucral leaflets.* (Sp. 1–2.)

1. **M. didyma** (E. & Z. No. 410); erect, tomentose; leaves petiolate, *ovate-oblong*, rounded at base, obtuse or sub-acute, *serrate or sub-entire*, softly pubescent on the upper, canescent and tomentose on the lower side; peduncles axillary, much longer than the petiole, forked, 2-flowered; invol. leaflets broadly cordate, acuminate, cuspidate, tomentose, longer than the densely hirsute calyx. *M. leucantha, E. Mey.! in Hb. Drege.*

HAB. Winterberg and Zuureberg Mts. *E. & Z.!* At Enon, and on the Kei, in mountainous places, *Drege!* Natal, *Krauss, No.* 217! (Herb. T.C.D., Hook., Sond.)

1–2 feet high, simple or branched near the root. Leaves 2–3 inches long, 1 inch broad, sometimes nearly entire, but generally sharply toothed; the under side of the young leaves rusty, of the older pale, with red dots. This is the largest and strongest of the S. African species.

2. **M. linearifolia** (Sond. in Linn. 23. p. 18.); erect, stellato-tomentose; leaves *linear, obtuse at each end, entire,* obsoletely denticulate at the extremity, softly pubescent on the upper, tomentoso-canescent and rufous-dotted on the under side; peduncles much longer than the petiole, forked, two-flowered; invol. leaflets broadly cordate, acuminate, cuspidate, tomentose, equalling the densely hirsute, rufous calyx.

Hab. Port Natal, *Gueinzius!* (Herb. Sond.)

Stem a foot high or more, the whole plant rust-coloured. Nearly allied to *M. didyma*, but the leaves are much narrower, not perceptibly ovate at base, and quite entire, save 2–3 minute denticles at the tip.

** *Diffuse or prostrate, with narrow invol. leaflets.* (Sp. 3–4.)

3. **M. prostrata** (DC. Prod. I. p. 499); diffuse or prostrate, canescent, *thinly tomentose;* leaves broadly linear, obtuse at both ends, mucronulate, entire or denticulate at the tip only, *the younger minutely pubescent, the older glabrous on the upper*, canescent, thinly tomentose and rufous-dotted on the lower side; peduncles 2–3 times as long as the petiole, one-flowered; invol. leaflets narrow, *ovate-acuminate*, cuspidate, equalling the tomentose, rufescent calyx. *Burch. Cat. No.* 2153. *Zey. No.* 140.

HAB. S. Africa, *Burchell.* Rhinosterkopf, *Zeyher!* (Herb. Hook., Sond.)

This differs from *M. linearifolia* in its thinner and whiter pubescence, and its much narrower involucral bracts, which are *ovate*, not *cordate* at base. The one flowered peduncle is an uncertain character.

4. **M. Burchellii** (DC. Prod. I. p. 499); diffuse, *velvetty and canescent;* leaves oblong-linear, rounded at base, tapering to a sub-acute point, denticulate, *velvetty on both surfaces*, rufous-dotted on the lower; peduncles equalling the petioles, one or two-flowered; invol. leaflets *narrow, sublanceolate*, shorter than the lanceolate, albo-tomentose calyx lobes.

Hab. S. Africa, *Burchell*, Cat. 2417. Aapjes River, *Zeyher!* Zooloo Country, *Miss Owen!* (Herb. Hook., Sond., T.C.D.)

The smallest of the Cape species. It differs from *M. prostrata* by its velvetty pubescence, copious in all parts, its shorter flower-stalks and much narrower bracts and calyx lobes.

Order XXIII. TILIACEÆ, Juss.

(By W. H. Harvey.)

(Tiliaceæ, Juss. Gen. 290. D.C. Prod. 1. p. 503. Endl. Gen. No. ccxii. Lindl. Veg. Kingd. No. cxxxi. Elæocarpeæ, Juss., DC. Prod. 1. p. 519.)

Sepals 4–5, separate or united, valvate in æstivation, *Petals* 4–5 (or none), clawed, deciduous, entire or multifid, imbricate in the bud. *Stamens* rarely definite, hypogynous; filaments filiform, separate or connate at base; anthers 2-celled, introrse. *Pollen* smooth. *Ovary* free, sessile or on a columnar torus (*gynophore*), of 2–10 carpels, and as many cells; styles single; stigmas 2–10; ovules definite or indefinite. *Fruit* dry or succulent, frequently hispid or prickly, dehiscent or indehiscent. *Seeds* albuminous, with an axile embryo.

Trees, shrubs, or herbaceous plants, with simple, branched or stellate hairs. Leaves mostly alternate, and petiolate, simple, penni-nerved or palmately nerved, entire or lobed. Stipules deciduous. Flowers rarely unisexual, solitary, racemose or corymbose, naked or bracteate.

Chiefly distinguished from Byttneriaceæ by the free, and usually indefinite stamens; from Sterculiaceæ by the introrse anthers; and from Malvaceæ by the bilocular anthers and smooth pollen. All these Orders are closely allied, and agree in having tough, fibrous bark, suitable for textile purposes; and innocuous, mucilaginous qualities. The typical genus, *Tilia* (the *Lime* or *Linden*), furnishes from its bark, the material of the Russian mats of commerce. Some of the *Grewiæ* have edible fruits. About 400 species are known, natives chiefly of tropical and sub-tropical countries.

TABLE OF THE SOUTH AFRICAN GENERA.

* *Flower 4-parted.*

I. **Sparmannia.**—Outer-stamens moniliform, without anthers.

** *Flower 5-parted.*

II. **Grewia.**—*Fruit* fleshy, of 1–4 hard-shelled, 1–2 seeded *drupes.*

III. **Triumfetta.**—*Capsule* globose, covered with hooked or straight, sharp-pointed bristles, few-seeded.

IV. **Corchorus.**—*Capsule* pod-like, 2–5-valved, splitting, many-seeded.

I. SPARMANNIA, Thunb.

Sepals 4, lanceolate, pointless, deciduous. *Petals* 4, hypogynous, obovate, spreading horizontally. *Stamens* indefinitely numerous; the *outer* sterile, moniliform; the *inner* fertile, with nodose filaments. *Ovary* sessile, 4–5-celled; ovules numerous; style columnar, stigma 5-toothed. *Capsule* globose or oblong, 4–5-celled, 4–5-valved, loculicidal, covered with rigid bristles. *Endl. Gen. No.* 5369. *DC. Prod.* 1. *p.* 503.

Large, handsome, free-growing shrubs or small trees, natives of Africa. Only three species are yet known; two of them S. African, and one from Abyssinia. The leaves are alternate, softly tomentose, stellato-pubescent, on long petioles, 5–7-angled or lobed, crenate or serrate, palmately nerved. Stipules lateral, small, persistent. Flowers in terminal umbels, white. Named in honour of Dr. Andrew Sparmann, a Swede, who travelled in S. Africa and afterwards accompanied Captain Cook in his second voyage.

1. **S. Africana** (Linn. f. Sup. p. 265); arborescent; branches terete, patently hairy; leaves on long petioles, *cordate-acuminate 5-7-angled, unequally toothed, softly hairy on both sides,* 7-9-ribbed below; stipules subulate; peduncles elongate, many-flowered; invol. bracts subulate, acute; sepals white, membranous, lanceolate; petals obovate; barren filaments numerous; capsules sub-globose, 5-celled. *DC. Prod.* 1. *p.* 503. *Vent. Malm. t.* 78. *Bot. Mag. t.* 726. *E. & Z.! No.* 411. *Thunb. Cap. p.* 432.

HAB. In moist woods, district of George, *E. & Z.! Drege!* &c. Cult. in England. (Herb. T.C.D., Hook., Sond.)

A quick-growing arborescent shrub, 10–20 feet high, with spongy wood and half herbaceous branches, the whole plant thickly covered with long, soft, spreading hairs. Leaves 5–6 inches long, and 3–4 broad, pale green. Flowers conspicuous; the sepals and petals white; the barren filaments yellow, with a purple tip; the fertile ones purple.

2. **S. palmata** (E. Mey. in Herb. Drege); shrubby; the branches terete, virgate, stellato-pubescent; leaves on long petioles, *deeply 5-7-lobed, the lobes much acuminated, inciso-sinuate, and unequally-toothed, minutely stellato-pubescent on both sides,* prominently 5-7-nerved below; stipules subulate, deciduous; peduncles sub-terminal, elongate, densely 12 or more-flowered; invol. bracts subulate, acute; sepals linear-oblong, coriaceous, purplish within; petals narrow-obovate; barren filaments few; capsules elliptic-oblong, 4-celled. *Urena ricinocarpa, E. & Z.!* 301.

HAB. Between the Omtata and Omsamwubo, 1000–2000 f. *Drege!* Sources of the Kat River and Makasana River, *E. & Z.!* (Herb. Sond.)

A virgate shrub, with soft, spongy wood and half herbaceous branches, much smaller in all parts than S. Africana; with smaller and more densely clustered, purplish flowers; deeply parted leaves and 4-celled capsules. It is closely allied to *S. Abyssinica*, but differs in pubescence, and in the shape and section of the leaves. The hairs, in *S. Abyssinica*, are long, simple, patent and soft, and the leaf-lobes are more acuminate, and less pinnatifid.

II. GREWIA, Linn.

Sepals 5, linear, leathery, coloured within, deciduous. *Petals* 5, each with a nectariferous gland or pit, at the base, within; inserted at the base of a columnar *torus*, which supports the stamens and ovary. *Stamens* numerous; filaments filiform; anthers roundish. *Ovary* 2-4-celled; style simple. *Drupe* four-lobed (or 1-2-lobed), containing 1-4 hard-shelled, 1-2-celled, 1-2-seeded pyrena. *Endl. Gen.* 5376. *DC. Prod.* 1. *p.* 508.

Trees and shrubs, natives of the tropical and sub-tropical parts of Asia and Africa. Leaves alternate, undivided, entire or serrulate, 3–7-nerved at base, often pale on the under side. Stipules lateral. Pubescence stellate. Flowers purple or yellow, in axillary or terminal cymules or solitary; sometimes umbellulate. Petals generally shorter than the sepals, sometimes very short, or altogether wanting. Named in honour of Nehemiah Grew, M.D., of London, author of a famous work on the anatomy of plants, published in 1682, and containing very elaborate illustrations of vegetable structure, wonderfully accurate, considering the imperfect condition of the microscope in those early times.

1. **G. occidentalis** (Linn.); young twigs puberulous, older glabrous; leaves petiolate, *ovate*, obtuse or acute or acuminate, *crenate, glabrous*, 3-nerved at base; peduncles opposite the leaves or terminal, forked, 3-6-flowered; flower-buds *roundish-oval, obtuse;* sepals lanceolate, canescent, *nerveless*, longer than the petals; ovary villous; drupe depressed, sparingly hispid. *DC. Prod.* I. *p.* 511. *E. & Z.! No.* 412. *G. obtusifolia, E. & Z.! No.* 412. *G. trinervis, E. Mey.! and Drege, No.* 7265.

HAB. In mountain ravines, from Capetown to Port Natal, common. (Herb. T.C.D., &c.)

A middle sized tree, nearly glabrous, except on the young shoots and about the inflorescence. Flowers purple. Leaves 1–2 inches long, very variable in shape, sometimes sub-rotund and very obtuse, sometimes much acuminate.

2. **G. caffra** (Meisn. in Hook. Lond. Journ. 2. p. 53); young twigs and leaves scabrous, older glabrous; leaves on short petioles, *ovate-oblong*, acutely *mucronate, finely serrulate, glabrous*, 3-nerved at base; peduncles opposite the leaves, forked, 2-6-flowered; fl.-buds *linear-oblong, swollen at base;* sepals one-nerved, longer than the petals, canescent; ovary scaly; drupe globose, glabrous, yellow.

HAB. Port Natal, *Krauss! Gueinzius! T. Williamson!* Delagoa Bay, *Forbes!* (Herb. T.C.D., Hook.).

Very much resembling *G. officinalis*, but well characterized by its sharply serrulate (not crenate) leaves and oblong (not sub-globose) flower-buds and by the very dissimilar drupe. Flowers purple.

3. **G. obtusifolia** (Willd. Hort. Berol. p. 566); "leaves oblong-elliptical, obtuse at both ends, *hairy*, acutely and unequally serrate." *DC. Prod.* I. *p.* 512.

HAB. South Africa, *Willdenow*.

We are unacquainted with this species.

4. **G. flava** (DC. Hort. Monsp. p. 113); branches and twigs rigid, *canescent;* leaves on short petioles, *exactly elliptical, very obtuse*, crenulate, *glabrous above, canescent below*, 3-nerved at base; peduncles one-flowered; fl.-buds oblong; *sepals 3-nerved*, longer than the bifid petals, canescent; drupes bilobed, black, hispid and furrowed. *DC. Prod.* I. *p.* 509.

HAB. Beyond the Orange River, *Burchell!* Woods near Uitenhage, *E. & Z.!* (Herb. T.C.D., &c.)

A rigid, divaricately much branched, canescent shrub, with small, exactly oval leaves and yellow flowers. The berries are eaten by the country-folk. Leaves ½ inch long, ¼ inch wide.

5. **G. cana** (Sond. in Linn. 23. p. 20); twigs densely *tomentose and canescent;* leaves on short petioles, *oblong or linear-oblong*, obtuse, finely serrulate, 3-nerved at base, velvetty and *canescent on both surfaces;* peduncles solitary, one-flowered; fl.-buds sub-globose, tomentose; sepals 3-nerved, longer than the yellow petals; ovary villous; drupe depressed, black, thinly hispid. *Sond. l. c.*

HAB. Macallisberg, *Burke and Zeyher!* (Herb. Hook., T.C.D., Sond.).

A much and divaricately branched shrub. Nearly allied to *G. flava*, but with longer and more oblong, downy leaves, which are serrulate (not crenulate). The fruits are eaten by the Bechuanas, who also make a sort of beer from them.

6. G. hermannioides (Harv.); twigs densely *stellato-tomentose, rufescent;* leaves on short petioles, *elliptical or oblong, obtuse,* calloso-serrulate, 3-nerved at base, velvetty on the upper, *stellato-tomentose on the under side;* peduncles axillary, solitary, hispid, forked, 2-flowered; fl. buds globose, tomentose; sepals 3-nerved, longer than the petals; drupes globose, hispid.

HAB. Macallisberg, *Burke!* (Herb. Hook.)

Near *G. cana*, but with very different pubescence. Leaves 1 inch long, ½ inch broad. Only one specimen seen.

7. G. monticola (Sond. in Linn. 23. p. 20); twigs densely *tomentose, with reddish, stellate hairs;* leaves *almost sessile, unequal-sided, half-cordate at base,* oval-oblong, acute, unequally calloso-serrate, minutely pubescent or glabrate above, densely albo-tomentose below; peduncles axillary, forked, 2-3-flowered; fl.-buds globose, tomentose; sepals 3-nerved, longer than the petals; ovary villous; drupe didymous, shining, *glabrous.*

HAB. Macallisberg, *Burke and Zeyher!* (Herb. Hook., T.C.D., Sond.)

A much branched, spreading shrub; all the younger parts clothed with rusty-red, stellate hairs. Readily known from *G. discolor*, which it much resembles, by the unequally sided leaves.

8. G. discolor (Fresen, in Mus. Senkenb. 2. p. 159); *twigs densely tomentose, with reddish, stellate hairs;* leaves on short petioles, *equal-sided,* elliptic-oblong, acute, somewhat cuneate and 3-nerved at base, serrulate, minutely pubescent above, albo-tomentose below; peduncles two or more from the same axil, forked 2-3-flowered; fl.-buds globose or oval, tomentose; sepals narrow, shorter than the petals; ovary densely hairy; drupe? *Sond. in Linn.* 23. *p.* 21.

HAB. Macallisberg, *Zeyher*, 146. (Herb., Sond.).

A large shrub or small tree. Old branches glabrous, with an ashen bark; younger fulvous and densely tomentose. Leaves 1-1½ inch long and ½ inch broad. I have not seen the original plant of Fresenius (a native of Abyssinia), and adopt from Dr. Sonder the above name for the S. African specimen here described. It closely resembles *G. monticola*, but the leaves are narrower, equal-sided at base and on longer petioles.

9. G. hispida (Harv.); twigs *roughly stellate;* leaves sub-sessile, *ovate, obtuse, coarsely toothed,* 3-nerved at base, veiny, *roughly stellato-pubescent and green on both sides;* peduncles few-flowered, shaggy; fl.-buds subglobose, hispid; sepals scarcely longer than the petals; ovary very hispid.

HAB. Port Natal, *Sanderson*, 33! (Herb. T.C.D.).

Seemingly a small shrub. The older branches are naked, with a rough, dark-coloured bark. Leaves ¾-⅞ inch long, ½ inch wide, roundish or elliptical, equal-sided, unequally toothed; the veins prominent on the under side. Peduncles much shorter than the leaves, scale of the petals and gynophore crowned with dense, white bristles; ovary similarly setose. Very unlike any other Cape species.

10. G. lasiocarpa (E. Mey.! in Herb. Drege); scandent; twigs and petioles *very hispid,* with spreading, stellate hairs; leaves petiolate, *roundish-ovate or cordate,* obtuse, unequally dentato-crenate, *3-5-nerved at base, reticulate and harsh* on the upper, softly pubescent on the lower side; peduncles solitary, axillary, longer than the petioles, forked, hispid, 2-4-flowered; fl.-buds globose, *rufo-tomentose;* sepals 3-nerved, longer than the petals; ovary hirsute; drupes hispid.

HAB. Between the Omtata and Port Natal, in several places, *Drege!* Natal, *Gueinzius!* Trans-Kei Country, *Mr. Bowker!* (Herb. Hook., T.C.D., Sond.).

A climbing shrub. Leaves much larger and broader than in any other S. African species, 2½–3½ inches long, 2–2½ inches broad, minutely netted and (when dry) finely wrinkled on the upper surface, which is hispidulous in the younger, and smooth in the older leaves. Allied to *G. Asiatica,* and still more nearly to *G. villosa,* W. & A. but differing from both in its flowers, the character of pubescence, and rugulose reticulation of leaves.

III. TRIUMFETTA, Linn.

Sepals 5, linear, membranous, coloured, fornicate and dorsally mucronate or horned, deciduous. *Petals* 5, without glands, inserted at the base of a short columnar torus, which supports the stamens and ovary. *Stamens* definite, 5-30; filaments thread-like; anthers roundish. *Ovary* 2-5-celled; the cells divided by a false, parietal, vertical dissepiment; ovules in pairs; style filiform; stigma 2-5-lobed. *Capsule* sub-globose, covered with straight or hooked prickles, 2-5-celled; cells 1-2-seeded. *Endl. Gen.* 5372. *DC. Prod.* 1. *p.* 506.

Shrubs, suffrutices or herbs, dispersed through tropical and sub-tropical countries; many of them weeds in tilled ground, and rapidly disseminated by their bur-like capsules which stick to passing quadrupeds, &c. The leaves, in the same species, vary greatly in shape; they are entire or lobed, serrate, and many-nerved, with glands frequently on the serratures, at the under side. Flowers small, yellow or orange, solitary or clustered, on simple or branched stalks. Name in honour of J. B. Triumfetti, an Italian botanist of the seventeenth century, and author of several botanical works.

1. T. rhomboidea (Jacq. Am. p. 147. t. 90); herbaceous, tall, diffusely much branched, *stellulato-pubescent;* lower leaves 3–5-lobed, upper rhomboid, acuminate, cuneate at base, the uppermost small and lanceolate, all minutely pubescent on both sides, unequally serrulate, 5-nerved; peduncles clustered, axillary, 3-flowered, very short; sepals hispid, mucronulate; stamens 10–15, glabrous; capsules albo-tomentose, their prickles glabrous above, hooked, and ending in a white point. *DC. Prod.* 1. 507. *T. glandulosa, Lam.? T. velutina, Vahl.? T. riparia, Hochst. Pl. Kr. No.* 56. *T. diversifolia and T. angulata, E. Mey.!*

HAB. About Port Natal, *Mr. Hewetson, Krauss, Drege, Sutherland.* (Herb. T.C.D., Hook., Sond.)

Stem 3–4 feet high, panicled, variable in pubescence and foliage. This plant is probably originally West Indian, but is now spread over most tropical countries. Probably many of the reputed *local* species ought to be reduced to it.

2. T. pilosa (Roth, Nov. Sp. p. 223); herbaceous, tall, erect, *densely tomentose with stellate hairs;* lower leaves 3-lobed, upper ovate-oblong, acuminate, the very uppermost small and lanceolate, all stellately hispidulous on the upper, and densely tomentose and hispid on the lower side, unequally serrate, subcordate and 5-nerved at base; peduncles several together, very short, 3-flowered; sepals hispid; stamens 10, glabrous; capsules hispid, their prickles hairy below, glabrous above, hooked, and ending in a white point. *DC. Prod.* 1. *p.* 506. *W. & Arn. Prod. p.* 74. *T. tomentosa, E. Mey.!*

HAB. South Africa, *Drege!* Natal, *Dr. Sutherland.* (Herb. Hook., T.C.D., Sond.).

Stems 4–6 feet high, strong and half woody at base. Lower leaves 3–5 inches long, 1–3 inches wide, obscurely 3-lobed or angled; upper much smaller, ovate or lanceolate; the serratures alternately large and small. Our specimens quite agree with the East Indian plant. Bojer's *T. tomentosa,* from the Mauritius, is very similar; but in that the setæ of the capsule are straight. Is the character constant?

3. T. effusa (E. Mey. in Hb. Drege); herbaceous, tall, diffusely branched, *rough with rigid, stellate hairs;* lower leaves . . . . ? *upper tricuspidate,* rounded at base, acuminate, all stellately pubescent and rough on both sides, unequally toothed and jagged, 5-nerved; peduncles clustered, bracteate above the middle, 1–3-flowered, *longer than the flower; pedicels half as long as the tomentose sepals;* stamens 10, glabrous; capsules hispid, their prickles ciliate at base, glabrous above, hook-pointed.

HAB. Natal country, *Drege! T. Williamson!* (Herb. T.C.D., Hook., Sond.).

Closely allied to *T. Vahlii,* and perhaps not distinct. The inflorescence is much more lax, and the peduncles longer than in the allied species. The leaves resemble those of *T. angulata,* Lam.

4. T. trichocarpa (Sond.! in Linn. 23. p. 19); erect, *shrubby,* hispid with stellate hairs; leaves *subsessile, elliptic-oblong,* sub-acute, toothed, stellato-pubescent on both sides, 3-nerved at base, reticulately veined below; stipules subulate, deciduous; peduncles 3-flowered; sepals fornicate, hispid; petals spathulate; stamens 15–16, glabrous; ovary 2-celled; capsules *densely setose, the prickles very long, straight, patently hairy (not hook-pointed.) Harv. Thes. Cap. t.* 52.

HAB. Macallisberg and near Vaal River, *Burke & Zeyher!* Zooloo Country, *Miss Owen!* (Herb. Hook., Sond., T.C.D).

A rigid, erect shrub, hairy in all parts with coarse, but not thickly set stellate hairs; branches very erect. Leaves 1 inch long, ½ inch wide, mostly oblong-oval, not lobed, but unequally toothed. Petiole one line long. Peduncles shorter than the leaves, slender, hairy. Sepals linear, hooded at point and long-horned behind. Style minutely bifid. Capsule sub-globose, ¼ inch in diameter, bristling like a porcupine, with *straight,* very hairy, and long setæ.

IV. CORCHORUS, Linn.

Sepals 4–5, ovate or lanceolate, unequal, deciduous. *Petals* 4–5, obovate, clawed, hypogynous. *Stamens* numerous, mostly indefinite. *Ovary* sessile, or shortly stipitate, 2–5-celled; ovules very numerous; style short; stigmas 2–5. *Capsule* podlike or roundish, 2–5 celled, 2–5-valved, loculicidal, many-seeded. *Endl. Gen.* 5371. *DC. Prod.* 1. *p.* 504.

Herbaceous, suffruticose or shrubby plants, common throughout the tropics. Leaves alternate, serrate; the teeth sometimes produced into long, setaceous points. Stipules lateral. Flowers yellow, on short, simple or branched peduncles, axillary or opposite the leaves. *C. olitorius* and *C. capsularis* are used in the East as pot-herbs, and are also extensively cultivated for their fibrous bark, from which the "*jute,*" now largely imported into England, as a cheap substitute for hemp, is procured. "Gunny-bags" are made of it. The generic name is the Greek κορχορος,

an old word applied to this or some similar plant, from κορεω, to purge; because the qualities are laxative.

1. C. trilocularis (Linn.); *annual;* leaves ovate-oblong or lanceolate, obtuse or sub-acute, *serrulate;* peduncles opposite the leaves, *equalling the petiole,* 1–2-flowered; capsules slender, linear, elongate, 3–4-angled, 3–4-celled, *scabrous,* tapering to a blunt point. *DC. Prod.* 1. *p.* 504. *Jacq. Hort. Vind. t.* 173. *C. asplenifolius, E. Mey.! non Burch.*

Hab. Near Port Natal, *Drege! Gueinzius! Sanderson!* (Herb. Hook., T.C.D., Sond.).

An erect or diffuse pubescent or hispid annual, 1–3 feet high. Leaves 1–2 inches long, variable in breadth and pubescence; the lowermost serratures frequently prolonged into bristles. Capsule often 4-celled, rough with minute, sharp points. A common weed in tropical Africa and Asia.

2. C. asplenifolius (Burch. Cat. 1737., Voy. 1. p. 400); *perennial, woody at base;* stems *prostrate,* herbaceous, hispid; leaves ovate-oblong, or oblong, *sparsely hispid, dentato-crenate;* peduncles opposite the leaves, *obsolete,* 3–4-flowered; pedicels as long as the *ovate, setose flower-buds;* capsules linear, cylindrical, *hispid,* attenuated to a blunt point. *DC. Prod.* 1. *p.* 505.

Hab. Near Saltponds beyond the Orange River, *Burchell.* Near Grahamstown? *Dr. Atherstone!* (Herb. Hook.)

Stems many from the same crown, widely spreading, flexuous, with patent branches. Leaves distichous, an inch apart, 1–1½ inches long, the shorter ones subovate, the rest ovate-oblong or linear oblong, all rough with stiff bristles.

3. C. serræfolius (Burch. Cat. 1962., Voy. 1. p. 537); perennial, *woody at base;* stems *prostrate,* herbaceous, villous, or glabrescent; leaves linear or linear oblong, *glabrous,* coarsely toothed; peduncles opp. the leaves, obsolete, pedicels shorter than the *obovate, acute, glabrous flowerbuds;* capsules linear, *twisted,* 6-valved, *rough with sharp points. DC. Prod.* 1. *p.* 504.

Hab. Plains beyond the Orange River, *Burchell.* Thaba Unka, *Burke & Zeyher!* (Herb. Hook., T.C.D.).

This resembles *C. asplenifolius,* but has narrower, more linear, and more strongly serrate leaves, and is nearly glabrous. The capsules are muricated, not setose.

Order XXIV. HIPPOCRATEACEÆ. Juss.

(By W. H. Harvey.)

(Hippocraticeæ, Juss. Ann. Mus. 18. 483. Hippocrateaceæ, Kunth. DC. Prod. 1. p. 567. Endl. Gen. ccxxxvii. Lindl. Veg. Kingd. No. ccxxiv.)

Flowers perfect, regular, small and greenish white. *Calyx* free, 5-parted, with imbricate æstivation. *Petals* alternate with the sepals, sessile, inserted on the margin of a fleshy disc. *Stamens* 3, within the disc, hypogynous and monadelphous; anther cells frequently confluent. *Ovary* immersed in the disc, trilocular; ovules definite or indefinite, anatropal; style short, stigma simple or 3-lobed. *Fruit* either capsular or fleshy; *seeds* exalbuminous.

Small trees or shrubs, with opposite *(rarely alternate)*, simple, entire or serrulate, coriaceous leaves. Stipules small, soon falling off. Flowers in axillary cymes or panicles, inconspicuous. A small Order, containing less than a hundred species, inhabiting the warmer zones, but most numerous in South America. Their properties are unimportant; the seeds of some, and the pulpy fruit of others are edible. The affinities are with *Celastraceæ* on the one part, from which the strictly hypogynous stamens divide this Order, and with *Euphorbiaceæ* on the other, from which the bisexual flowers distinguish it.

I. SALACIA, Linn.

Calyx deeply 4–5-parted. *Petals* 4–5, inserted round a crenate perigynous disc. *Stamens* 3, strictly hypogynous; filaments flattened, recurved; anthers short, 2-celled, adnate, opening outward by a longitudinal slit. *Ovary* half sunk in the disc, trilocular; style short; stigma 3-lobed; ovules two or more in each cell. *Fruit* fleshy, often by suppression, one-seeded. *Endl. Gen.* 5702. *DC. Prod.* 1. *p.* 571.

Trees or shrubs, with the aspect of *Cassine* or *Celastrus*, common within the tropics; a few straggling into the temperate zones. Leaves thick, rigid, entire or serrulate, glabrous or nearly so, almost always opposite. Flowers minute, greenish or whitish, in axillary cymes or corymbs. Named from *Salacia*, the wife of Neptune.

1. S. Zeyheri (Planch. ! in Herb. Hook.); leaves *opposite*, shortly petiolate, elliptical or obovate, with thickened margins, crenulate or repando-dentate; cymes axillary, few-flowered; flowers four-parted, the ovate-oblong, obtuse petals thrice as long as the roundish sepals. *Rhamnus Zeyheri, Spreng ! in Herb. Zey.*

HAB. Eastern Districts, *Zeyher !* (Herb. Hook.)

A glabrous shrub or small tree. Leaves 1 inch long, ½ inch wide, obtuse, with a minute tubercle in each indentation of the margin. Flowers greenish, 4-parted in the specimens examined. Petals persistent. Stamens 3, distant, within the margin of a crenulate, fleshy disc; filaments curved backwards; anthers very short. Ovary trilocular.

2. S. Kraussii (Hochst. in Flora, xxvii. 1. 306); leaves *alternate*, subsessile, elliptical or oblong-lanceolate, glossy above, reticulate below, repando-denticulate; fascicles subumbellate, axillary, several-flowered; flowers 5-parted, the buds globose; sepals very unequal, much shorter than the petals. *Walp. Rep.* 5. *p.* 147. *S. alternifolia, Hochst. ! in Pl. Krauss. No.* 348. *Sond. in Linn.* 23. *p.* 25. *Diplesthes Kraussii, Harv. Lond. Journ.* 1. *p.* 19.

HAB. Margins of woods, near Port Natal, *Dr. Krauss! Dr. Pappe ! Gueinzius !* (Herb. T.C.D., Hook., Sond.)

A glabrous shrub. Leaves very variable in shape, sometimes broadly and shortly obovate or oval, 1½ inch long, sometimes almost lanceolate, 2½ inches long, always obtuse, remotely denticulate, with very shallow indentations. Pedicels numerous, simple, ½ inch long, rising from axillary tubercles, imperfectly umbellate; flower, greenish. The genus *Diplesthes*, formerly proposed by me for this plant, is untenable.

ORDER XXV. MALPIGHIACEÆ, Juss.

(By W. SONDER.)

(Juss. Gen. 252., DC. Prod. 1. p. 577., Endl. Gen. ccxxviii. Lindl. Veg. Kingd. No. cxxxix.)

Flowers perfect, regular. *Calyx* 5-parted, equal, persistent, mostly with conspicuous glands at the base of one or more of the segments externally. *Petals* 5, hypogynous, clawed, spreading; the lamina concave and often jagged, imbricate in æstivation. *Stamens* twice as many as the petals, those opposite the petals sometimes abortive or wanting; filaments connate at base; anthers erect or incumbent, introrse, longitudinally slitting. *Ovary* free, 3–2-celled; ovules solitary, pendulous; *styles* 3, distinct or confluent; stigmata simple. *Fruit* various, either fleshy, or woody and 1–3-winged, the carpels cohering or separate; seeds exalbuminous; embryo commonly conduplicate.

Trees or shrubs, often climbers; with opposite, rarely alternate, petiolate, simple, penninerved, very entire or rarely toothed or lobed, often stipulate leaves. Pubescence various, often copious, silky and shining. Flowers in corymbs or racemes, terminal or lateral: petals red or yellow, rarely white or blue.

A considerable Order, chiefly tropical; about $\frac{1}{10}$ of the species are natives of the Old World, the rest American and chiefly from the Southern Continent. Only 4 species have yet been found in South Africa.

TABLE OF THE SOUTH AFRICAN GENERA.

I. **Acridocarpus.**—*Petals* sub-entire. *Ovary* 3-lobed. *Fruits* with dorsal wing.
II. **Triaspis.**—*Petals* fringed. *Ovary* 6-winged. *Fruits* with a marginal wing.

I. ACRIDOCARPUS, Guill. & Per.

Calyx 5-parted. *Petals* 5, longer than the calyx, clawed, sub-entire unequal, glabrous. *Stamens* 10, fertile; filaments short; anthers large, cordato-lanceolate. *Carpels* 3, united in a sharply 3-lobed ovary; *styles* 2–3, elongate, divergent, flattened, acute. *Nuts (samaræ)* 2–3, confluent at base, each above expanded into a straight or oblique vertical wing, thickened along its upper margin. *A. de Juss. Mon. p.* 228. *Endl. Gen. No.* 5576.

African trees or shrubs, sometimes climbers. Leaves alternate or opposite, entire, glabrous or silky, exstipulate. Racemes terminal and lateral; flowers yellow. Name *ακρις, a locust,* and *καρπος, fruit.*

* *Flowers racemose.*

1. **A. Natalitius** (Juss. l. c. p. 232); branches glabrous, the younger tomentose; *leaves alternate, oblong-obovate, obtuse, glabrous,* leathery; racemes terminal, simple, elongate, the rachis and pedicels rusty-tomentose; bracteoles subulate, without glands; ovary distylous; wings of the fruit broad, obliquely obovate, glabrescent, veiny with nerves. *Harv. Thes. Cap. t.* 19. *Banisteria Kraussiana, Hochst. in Pl. Krauss. No.* 261.

HAB. Margins of woods, Port Natal, *Krauss, Gueinzius, &c.* Oct. (Herb. Sond., Hook., T.C.D.)

Branches erect, terete. Leaves mostly alternate, some nearly opposite, 3–5 inches long, 1–2 inches wide, the upper smaller, shining on the upper surface, paler beneath; the petiole 3–4 lines long, with two glands at its summit. Racemes 6 inches long, or longer, pyramidal; lower pedicels uncial, spreading. Calyx lobes obtuse, 1 line long. Petals 5–6 lines long, unequal. Styles as long as the petals. Wings of the fruit an inch long, 7–8 lines wide, upturned.

** *Flowers umbellate.* (Sp. 2–3.)

2. A. Galphimiæfolius (Juss. l. c. p. 237); branches tomentose, at length glabrous; leaves opposite, short, ovate, mucronate, *glaucous below and sparingly piliferous;* petiole bi-glandular below the summit; umbels four-flowered, terminating the branches and ramuli; ovary and fruit tristylous. *A. pruriens, β. lævigatus, Sond. in Linn. 23. p. 22.*

HAB. Algoa Bay (quere, Delagoa Bay?), *Forbes* in Hb. Lindl. Port Natal, *Gueinzius.* (Herb. Sond.)

Branches voluble; branchlets opposite, divaricate. Leaves 9–18 lines long, 4–7 lines wide, green above. Petiole 2–4 lines long. Umbels 3–5-flowered; the peduncles at base furnished with minute, subulate bracts, somewhat silky, 4–6 lines long, bibracteate at the summit, and articulated with the thickened, 2 line long pedicels. Calyx lobes oblong, obtuse, 2 lines long, downy externally. Petals shortly clawed, obovate or sub-orbicular, 4–5 lines long, veiny. Styles long, unequal. Ovaries 3, hairy. Samaræ (fide Juss.) 3, on an enlarged, pyramidal receptacle, oblong, the sides rufo-sericeous, above and outwards produced into a puberulent wing, 5 lines long and 2½ wide. Perhaps a mere variety of the next species?

3. A. pruriens (Juss. l. c. p. 238); branches whitish, tomentose; leaves opposite, short, ovate, mucronulate, *silvery and silky on the lower side;* petioles bi-glandular above the middle; umbels 3–5-flowered, terminating the branches and ramuli; ovary and fruit tristylous. *Banisteria pruriens, E. Mey.! in Hb. Drege.*

HAB. Woods near Port Natal, common, *Drege, Gueinzius, &c.* Macallisberg, *Zeyher.* (Herb. Sond., Hook., T.C.D.)

Branches voluble. Leaves 1–1½ inches long, ½–1 inch wide, obtuse or sub-retuse, mucronate, the younger appressedly pubescent above, older glabrate, all silky tomentose on the under side. Petiole tomentose, 3–6 lines long. Umbels mostly ternate; peduncles silky, 1–1½ lines long, bibracteolate at the apex and articulated to the 2–3 line long pedicels. Calyx and petals as in the preceding. Styles long, unequal. Ovaries 3, hairy. Samaræ 3, silky pubescent, the wings appressedly downy, 5 lines long, 2½ lines wide, twice as long as the styles.

II. TRIASPIS, Burch.

Calyx short, 5-parted, without glands. *Petals* longer than the calyx, clawed, fringed. *Stamens* 10, fertile, unequal, connate at base and adnate to the stipe of the ovary. *Carpels* 3, expanded into a wing at each side, united into a 3-lobed, 6-winged, shortly stipitate ovary. *Styles* 3, glabrous, elongate, flat, acute. *Samaræ* 3, or fewer, winged at the margin, the wing shield-like, sometimes interrupted at the apex, commonly dorsally crested in the middle. *Juss. l. c. p.* 250. *Endl. Gen.* 5569.

Shrubs, often climbing, found in tropical, as well as in Southern Africa. Leaves opposite or sub-alternate, entire, glabrous or pilose, exstipulate. Flowers in axillary racemes or corymbs, or panicled, rose-coloured. Name from τρις, *three,* and ασπις, *a shield;* from the shield-like wings of the carpels.

1. T. hypericoides (Burch. Trav. 2. p. 280. Id. p. 290); leaves linear, glabrous on both sides; wings of the fruit rounded, semi-orbicular. *A. Juss. l. c. p. 505. t. 17.*

HAB. Kosi Fountain, lat. 27° 52′, *Burchell.*

Stem shrubby, 3–4 feet high, erect, branched; branches brown, glabrous, opposite, spreading. Leaves opposite, rarely sub-alternate, petiolate, lanceolate, linear, glaucous, very entire, 1–1½ inch long. Pedicels axillary, solitary, 3–6-flowered. Flowers rosy, inodorous. Sepals lanceolate, erect. Petals spreading, concave. Immature fruit, sprinkled with branched hairs. Fruits with a roundish wing.

ORDER XXVI. ERYTHROXYLEÆ, Kunth.

(BY W. SONDER.)

(Kunth, in Humb. Nov. Gen. 5. 175. DC. Prod. 1. p. 573. Endl. Gen. No. ccxxiv. Erythroxylaceæ, Lindl. Veg. Kingd. No. cxl.)

Flowers perfect, regular. *Calyx* free, persistent, 5-parted or cleft. *Petals* 5, hypogynous, equal, broad at base and each furnished in front with a plaited, bifid scale, imbricated in æstivation. *Stamens* 10, hypogynous; *filaments* flat, connate at base; *anthers* erect, 2-celled, slitting lengthwise. *Ovary* free, 2–3-celled, with solitary ovules; two cells abortive; *styles* 3, distinct or confluent; *stigmata* capitate. *Drupe* by abortion unilocular, and one-seeded; seed with little or no albumen, a straight embryo and superior radicle.

Shrubs or small trees, natives of the tropical and sub-tropical regions of both hemispheres, most abundant in America. Leaves alternate, rarely opposite, simple, penninerved or 3-nerved, entire, glabrous. Stipules intra-axillary, scale-like. Flowers axillary, solitary or clustered, small, whitish or greenish; the peduncles springing from imbricated scales.

This Order consists of a single genus, containing 70 or 80 species. It is closely allied to *Malpighiaceæ*, from which it is known by the want of glands on the calyx, the peculiar petals and the capitate stigmas. A Peruvian species (*Erythroxylon Coca*) is very largely used in S. America, especially by the miners, as a stimulant to the nervous system. Its leaves, mixed with chalk, are chewed, and effects similar to those resulting from the immoderate use of opium are produced.

I. ERYTHROXYLON, L.

Calyx 5-parted, 5-angled at base. *Styles* 3, distinct. *Endl. Gen. No. 5597, Sec. 1. DC. Prod. 1. p. 573.*

The generic name is derived from ερυθρος, *red,* and ξυλον, *wood;* several of the species have bright red wood.

1. E. caffrum (Sond. Linn. 23. p. 22); glabrous; branches terete, branchlets compressed; leaves on very short petioles, ovate or obovate-oblong, emarginate, *acute at base, coriaceous, netted with veins, shining; stipules rigid, very acute, sub-persistent, half as long as the petiole;* peduncles 2–4, on axillary depauperated ramuli, twice as long as the flowers; calyx-lobes triangular; petals obovate; stamens longer than the clavate styles.

HAB. Port Natal, *Gueinzius, T. Williamson.* (Herb. Sond., T.C.D.)

A shrub. Leaves 1–2 inch long, 8–10 lines wide, dark green above, paler below, netted on both sides with raised veins. Petioles channelled, 1–1½ lines long. Stipules ¼ line long. Peduncles 4 lines; petals 2 lines long. Stamens 10, fil. connate at base. Styles 3.

2. E. pictum (E. Mey.! in Herb. Drege); glabrous; branches terete, slightly angled by the decurrent petiole; branchlets compressed; leaves short-stalked, obovate-oblong or obovate, very obtuse, slightly emarginate, *obtuse at base, sub-coriaceous, opaque and nearly veinless* (or faintly veined) *on both surfaces, pale below; stipules soon deciduous, membranous, longer than the petiole;* peduncles about 2, axillary; calyx-lobes triangular; drupe oblong.

HAB. In the valley, near the river, between Omsamculo and Omcomas, towards Port Natal, *Drege*. April. Albany. *Mrs. F. W. Barber!* (Herb. Sond., T.C.D.)

Branches here and there tubercled. Leaves 2–2½ inches long, 11–14 lines wide, when dry (often) dark red on the upper side, with a very narrow white margin, without evident veins; below pale, the mid-rib purpurascent, the primary veins obsolete, secondary immersed. Petioles 2–3 lines; stipules 3 lines long. Peduncles 4 lines long. Calyx very short, exceeded by the staminal tube. Petals not seen. Drupe 4 lines long, blackish-blue. This differs from the preceding by its thicker branches and larger and neither glossy nor conspicuously netted leaves. Mrs. Barber's specimens (in Herb. T.C.D.) are less brightly coloured, with rather more evident nerves.

ORDER XXVII. OLACINEÆ, Mirb.

(By W. SONDER.)

(Mirb. Bull. Philom. (1813.) DC. Prod. 1. p. 531. Endl. Gen. No. ccxxiii. Benth. Linn. Trans. 18. p. 676. Olacaceæ, Lindl. Veg. Kingd. No. clxii. Olacaceæ, Icacinaceæ and Aptandraceæ, Miers.)

Flowers perfect, regular. *Calyx* small, free or partly adnate, truncate or toothed, unchanged or enlarged in fruit. *Petals* 4–6, hypogynous or sub-perigynous, with valvate æstivation, free, or united in pairs, or combined at base into a tube. *Stamens* as many or twice as many as the petals, inserted with them, coalescing or free; those opposite the petals often barren; anthers introrse, 2-celled, slitting lengthwise. *Ovary* seated on a small or thickened *torus*, which is sometimes concrete with the calyx, unilocular, or spuriously or incompletely 3–4-celled, or rarely excentrically 3-celled. *Ovules* 2–4, collateral, rarely solitary, pendulous from the summit of a free-central placenta, or attached to the sides of the ovary or spurious dissepiments; *style* simple; stigma truncate or lobed. *Drupe* subtended by the unaltered or enlarged calyx, one-celled and one-seeded (rarely 2–3-seeded). *Seed* with much albumen, and a small, axile or basal embryo.

Trees or shrubs, erect or climbing, unarmed or spiny, glabrous or thinly pubescent, natives of the tropics and sub-tropical regions of both hemispheres. Leaves alternate, simple, very entire, exstipulate, without glands. Flowers sometimes polygamous, small, either axillary and racemose, spiked or cymose, or terminal and cymoso-paniculate, rarely solitary. Bracts scaly, more often minute, rarely the younger imbricated; bracteoles small, connate in a cup or none. Only two South African genera are known, and they belong to different Sub-Orders.

Sub-Order I. OLACEÆ.

Ovary spuriously 3–4-celled at base, unilocular above, with a central placenta adhering below to the spurious dissepiments, free above. *Ovules* as many as the spurious cells, pendulous from the apex of the placenta. *Seed* spuriously erect. *Inflorescence axillary, racemose; the racemes rarely reduced to a single flower.*

I. XIMENIA, Plum.

Calyx free, unchanged in fruit. *Petals* 4–5, free, hairy at base, within. *Stamens* twice as many as the petals, all fertile, free, hypogynous. *Drupe* baccate, with a bony one-seeded nut. *Endl. Gen.* 5490. *DC. Prod.* 1. *p.* 533. *Benth. l. c.*

Tropical trees or shrubs, usually armed with axillary spines. Leaves coriaceous, ovate or lanceolate, entire. Peduncles axillary, one-flowered, or corymbose and many-flowered. Named after Fr. Ximenes, a Spaniard, author of an account of the medical plants and the animals of New Spain.

1. **X. caffra** (Sond. Linn. 23, p. 21); tomentose; branches spiny, reddish; leaves elliptical or oblong-elliptical, obtuse, emarginate, coriaceous, the adult glabrous above; peduncles axillary, solitary or in pairs, single-flowered; petals externally pubescent, densely bearded within; drupe oval, glabrous.

VAR. β.? **Natalensis**; branches and leaves glabrous; petioles, peduncles and calyces pubescent.

HAB. In woods at Macallisberg, *Zeyher, No.* 1847. β. at Port Natal, *Gueinzius.* Oct. Dec. (Herb. Sond.)

A shrub. Branches sub-divaricate, terete or angular, clothed with a short tomentum, more copious on the ramuli, and armed with short (3–4 lines long) spines. Leaves short-stalked, 1½ inch long, 10–12 lines wide, entire, one-nerved, the younger at each side clothed with brownish pubescence, the adult shining above. Peduncles 3 lines long. Calyx 4-fid. Petals 4, lanceolate, 3–3½ lines long, recurved at the points. Drupe nearly an inch long, fleshy, edible. Var. β. resembles *X. laurina*, DC., (*X. Americana, var. inermis*, Hochst.) but differs in its thick leaves and peduncles always one-flowered and pubescent.

Sub-Order II. ICACINEÆ.

Ovary completely unilocular, or excentrically and completely 3-locular. *Ovules* 2 in each cell, collaterally pendulous from the apex of a placenta adhering to one side of the ovary, lying one above the other (one funiculus being longer). *Style* excentrical. *Seed* pendulous. *Inflorescence cymose, or panicled, axillary or terminal.*

II. APODYTES, E. Mey.

Calyx free, 4–5-toothed, unchanged in fruit. *Petals* 4–5, glabrous within, united at base, among themselves and with the stamens. *Stamens* 5, all fertile. *Ovary* one-celled. *Fruit* ovato-reniform, subcompressed, oblique, with a fleshy protuberance on one side. *Benth. l. c. p.* 683.

Only one species known.

1. **A. dimidiata** (E. Mey.! in Hb. Drege); *Benth. l. c. tab.* 41. *Plectronia ventosa, Thunb.! in Herb. Holm. (non Auct.)*

HAB. In woods. Grootvadersbosch, *Zeyher, No.* 2024. Stellenbosch, *E. & Z.!*

Terebinth, No. 9. Bothasberg, Alb., *E. & Z.! Tereb. No.* 8. In Uitenhage and George, *E. & Z.! Ter. No.* 7., *Drege, Krauss.* Port Natal, *Drege, Krauss, &c.* Oct. April. (Herb. Sond., T.C.D., Hook.).

A tree, 20–50 feet high, with grey bark, and angular, pubescent branches. Leaves alternate, petiolate, ovate-elliptical, oblong, or ovate-acuminate, 1–2½ inches long, obtuse, very entire, rounded or cuneate at base, perennial, glabrous and shining on the upper, sparsely pubescent on the lower side, at length glabrous, and drying black. Panicles terminal, loosely branched, as long as the leaves or longer. Bracts minute, ovate or lanceolate. Flowers scarcely more than 1 line long. Calyx minutely downy. Corolla glabrous. Drupe baccate, 2 lines long, including the fleshy appendage, 4 lines wide. Seed ovate-reniform, compressed, with a thin seed coat; albumen abundant, fleshy, black.

ORDER XXVIII. SAPINDACEÆ. Juss.

(By W. SONDER.)

(Sapindi, Juss. Gen. 246. Sapindaceæ, Juss. An. Mus. 18. 476. DC. Prod. 1. p. 601. Endl. Gen. No. ccxxx. Lindl. Veg. Kingd. No. cxxxvi.)

Flowers often polygamous or unisexual. *Calyx* 4–5-parted, with imbricate æstivation, the sepals mostly unequal and partly connate. *Petals* 4–5 (or sometimes none), alternate with the sepals and often bearded, or furnished with scales on the inner face, imbricated in æstivation. *Disc* fleshy, free or adhering to the bottom of the calyx, interposed between the petals and stamens, regular or irregular. *Stamens* inserted within the disc, 8–10 or fewer, rarely 12–20, often excentrical; filaments filiform, free or connate at base; anthers 2-celled, introrse. *Ovary* free, 2–4-celled, with axile placentæ; ovules 1–3, rarely more numerous; *style* simple or 2–4-cleft. *Fruit* capsular or fleshy; seeds often arillate, exalbuminous; embryo usually curved or convolute, with incumbent or rarely accumbent cotyledons and an inferior radicle.

Trees, shrubs or herbaceous plants, natives of the warmer temperate and the tropical zones. Leaves alternate, rarely opposite, very often compound, ternate or pinnate or supra-decompound. Stipules none or small and caducous. Inflorescence racemose or paniculate.

A considerable Order much diversified in vegetation and in the structure of its flowers, but not readily divisible, except into sub-orders. We append to it two genera, *Ptæroxylon* and *Aitonia*, whose affinities are uncertain, but which cannot be very far removed. *Ptæroxylon* differs by its slightly imbricate, erect petels, its regular and isomeric stamens, winged seeds and accumbent cotyledons; in all but the last of these characters resembling *Cedrelaceæ*. *Aitonia*, which has been sometimes referred to *Meliaceæ*, sometimes to the Geranial-group of Orders, and sometimes to *Rutaceæ*, differs from *Sapindaceæ* chiefly by the insertion of its stamens, outside the disc, the collateral ovules and the slightly curved embryo. In external habit it is not unlike *Dodonæa*, and its bladdery capsules resemble those of *Cardiospermum* and *Erythrophysa*.

TABLE OF THE SOUTH AFRICAN GENERA.

* Flowers furnished with petals.

† Capsule bladdery.

I. **Cardiospermum.**—*Sepals* 4, unequal. *Herbaceous climbers.*

II. **Erythrophysa.**—*Calyx* bell-shaped, 5-lobed, coloured. *An erect shrub.*

†† Fruit fleshy or drupaceous. *Trees.*

III. **Schmidelia.**—*Flowers* 4-parted. *Ovary* 2-lobed. *Fruit* dicoccous.

IV. **Sapindus**—*Flowers* 5-parted. *Ovary* 3-lobed. *Fruit* tricoccous.
V. **Hippobromus.**—*Flowers* 5-parted. *Petals* naked. *Ovary* undivided. *Fruit* drupaceous.

** Flowers without petals.

VI. **Dodonæa.**—Capsule 2–3–4 winged, membranous. *Shrubs.*

Tribe 1. SAPINDEÆ. *Ovules* mostly solitary, rarely 2. Embryo curved, or rarely straight. (Gen. i.–v.)

I. CARDIOSPERMUM. L.

Sepals 4, two of them smaller. *Petals* 4, with internal, unequal appendages. Two *hypogynous-glands* between the petals and stamens. *Stamens* 8. *Styles* 3. *Fruit* inflated, of 3 membranous, dorsally-winged carpels, connate by their inner faces, valveless. *Seeds* globose, with a wide, cordate scar; cotyledons incumbent, large, transversely folded in the middle; radicle short, inferior, pointing to the hilum. *Endl. Gen. No.* 5598. *DC. Prod.* 1. *p.* 601.

Tropical and sub-tropical climbing, herbaceous plants, natives chiefly of the New World. Leaves alternate, biternate or supra-decompound, petiolate; leaflets toothed or cut, often dotted. Flowers sometimes dioecious, in axillary racemes or panicles, the common peduncle bearing near the summit a pair of simple tendrils, or abortive pedicels. The S. African species is found commonly throughout the tropics of both hemispheres. Name, *καρδια, the heart*, and *σπερμα, a seed;* from the heart-shaped hilum of the seed.

1. **C. Halicacaba** (Linn. Sp. 925); stem, petioles and leaves glabrous; leaves biternately cut, the segments petiolate, inciso-dentate; hypogynous glands rounded, short. *DC. l. c. Lam. Ill. t.* 317. *Rumph. Amb.* 6. *t.* 24. *f.* 2. *Bot. Mag. t.* 1049. *C. microspermum, E. Mey.! in Herb. Drege, non H.B.K.*

HAB. In sandy places near the Kei River, and near Omsamwubo, *Drege!* Umlaas River, Natal, *Krauss, Gueinzius.* (Herb. Sond., T.C.D., Hook., &c.)

Stem furrowed and angular. Leaves petiolate; the common petiole ½–1½ inches long, the partial shorter, unequal, intermediate longer; leaflets ovate or oblong, acuminate, glabrous or minutely puberulous at each side, sometimes on longer petiolules, 1–1½ inches long, the side leaflets usually smaller and less acuminate. Peduncles longer than the leaves, triparted at the apex, below which there are two tendrils; pedicels 3–4, which are 6–10 lines long in fruit. Flowers small. Fruit thinly downy, netted with veins, 8 lines long. and 10 wide. Seeds as large as peppercorn, black, with a white cordate spot.

II. ERYTHROPHYSA, E. M.

Flowers perfect. *Calyx* campanulate, sub-oblique, coloured, 5-lobed, the lobes obtuse, sub-unequal. *Petals* 4 (the place of the fifth vacant), inserted under the margin of a fleshy, cup-shaped *disc*, on long, linear-filiform, pilose claws; the limb oblong, obtuse, hooded at the base and furnished with a short, petaloid, toothed and crested, but beardless scale. *Stamens* 8, ascending, inserted in a tuft beneath a rostrate-acuminate, fleshy gland, at the side of the flower, where the fifth petal is deficient; filaments exserted, hairy; anthers oblong, 2-celled, dorsally inserted above the base, at length bifid at base and versatile. *Ovary* shortly stipitate, 3-angled, tapering into a short 3-sided style, 3-celled;

ovules 2, one above the other, axile, the lower pendulous, the upper ascending-erect. *Fruit* inflated, of 3 membranous, dorsally winged, valveless carpels, connate by their inner faces. *Seeds* solitary, globose, compressed, on very short seed-cords; testa purple, without aril; radicle next the hilum, inferior; cotyledons? *W. Arn. in Hook. Journ. Bot.* 3. *p.* 258.

A rigid, glabrous shrub, with alternate branches. Leaves impari-pinnate, crowded near the ends of the branches; the rachis interruptedly winged, the wings narrow-obovate oblong, cuneate; leaflets obovate, flat, undulate, very entire, mucronulate, acute at base, sub-sessile, flowers racemoso-corymbose, red. Name, ερυθρος, *red*, and φυσα, *a bag;* from the red, bladdery fruit.

1. E. undulata (E. Mey.! in Pl. Drege). *Erythrophila undulata, Arn. l. c.*

HAB. At Uitkomst, 2000–3000 f., and on the Mts. before Kamisberge and Kasparskloof, Elleboogfontein and Geelbecks Kraal, 3–4000 f. Aug. *Drege!* Namaqualand, *Backhouse!* (Herb. Sond., T.C.D.).

Branches rigid, with short, ashen, divaricated branchlets. Leaves an inch long; leaflets 3 lines long, 2 lines wide. Racemes corymbose, or panicled, abbreviate, leafy at base, the rachis 1–1½ inches long. Pedicels glandular. Calyx lobes downy within. Claws of the petals 2 lines long, lamina 4–5 lines. Ripe fruit 1½ inches long, 2 inches wide, obtusely emarginate at base, shortly stipitate, purplish, netted with veins. Only one seed seen, in which the cotyledons were not developed.

III. SCHMIDELIA, Linn.

Flowers often monœciously polygamous. *Calyx* 4-parted. *Petals* 4, bearded on the face. *Stamens* 8. *Ovary* didymous. *Fruit* of two fleshy, one-seeded carpels. *Endl. Gen. No.* 5605. *DC. Prod.* 1. *p.* 610.

Tropical and sub-tropical trees and shrubs. Leaves alternate, trifoliolate or simple; leaflets toothed, serrate, or sub-entire, often pellucid dotted. Flowers small, in axillary racemes. Named in honour of Casimir Christopher Schmiedel, once Professor at Erlange, and author of several botanical treatises, between 1751 and 1793.

* *Leaves trifoliolate* (Sp. 1–4).

1. S. melanocarpa (Arn. in Hook. Journ. 3. p. 153); branchlets glabrous or downy; leaves trifoliolate, leaflets petiolulate, oval, acuminate at each end, dentato-serrate, glabrous, bearded in the axils of the veins; peduncles divided into 2–3 subspicate racemes, thinly downy; *carpels black, sub-globose,* one often abortive. *Presl, Bot. Bem. p.* 40. *S. Africana, DC., ex Hochst. Rhus melanocarpa, E. Mey.! in Hb. Drege.*

HAB. Among shrubs, at Port Natal, *Drege! Gueinzius! Williamson!* (Herb. E. Mey., Sond., T.C.D.)

Branches and branchlets whitish. Petioles 1½–2 inches long, minutely downy, little channelled, the lateral petiolules 1–2 lines, intermediate 2–3 lines long. Leaflets 3–4 inches long, 1½–2 inches broad, somewhat leathery, when dry dark above, pale green below, with raised veins, the margin coarsely toothed or sub-serrate. Infl. longer or shorter than the leaves, its branches spreading, 1–3 inches long. Fl. minute, whitish, on pedicels ½–1 line long, solitary, in pairs or ternate. Calyx concave, ½ line long. Filaments hairy, exserted, with versatile anthers. Style bifid. Carpels 1 line long.

2. S. leucocarpa (Arnott, l. c.); branchlets glabrous or downy; leaflets petiolulate, *ovate-acuminate,* much tapering at base, *dentato-serrate,* *glabrous, bearded in the axils of the veins;* peduncles divided in 2–3 subspicate racemes, thinly downy; *carpels whitish, obovate,* one often abortive. *Presl, l. c. Rhus leucocarpa, E. Mey.! Hb. Drege.*

HAB. Among shrubs at the river near Omsamwubo, *Drege!* (Herb. E. Mey., Sond.)

Very like the preceding in the form of the leaves, the inflorescence and flowers, but differing in the leaves not turning dark in drying, and especially in the obovate white carpels, 1½ lines long. According to Arnott this is the true *S. Africana,* DC., to which opinion I cannot assent. Neither this nor the preceding agrees with the excellent figure of *S. Africana, in Rich. Tent. Fl. Abyss. t.* 27.

3. **S. Natalensis** (Sond.); branchlets *glabrous;* leaflets sessile, *oblongo-lanceolate, obtuse,* narrowed at base, *the margin revolute, remotely and obtusely toothed,* shining above, pale below, *quite glabrous on both sides;* peduncles divided in 2–3 spiked racemes, minutely downy; carpels sub-globose (immature) blackish. *S. erosa, Arn. l. c. Presl, l. c. excl. syn. Thunb. Rhus erosa, E. Mey.! non Thunb.*

HAB. Among shrubs near Omsamculo and Natal, *Drege! Plant!* April. (Herb. E. Mey.! Sond., T.C.D.)

Branches and branchlets as in the preceding. Common petiole shorter than the leaf. Leaflets more coriaceous than in the preceding specimen, 3 inches long, 8–12 lines wide, below netted with raised veins, regularly and shortly toothed, not *erose.* Flowers (scarcely larger), peduncles, calyx, petals, stamens, young carpels and styles as in *L. melanocarpa.*

4. **S. decipiens** (Arn. l. c.); branchlets and petioles *minutely downy; leaflets sessile, oblongo-lanceolate or obovate, narrowed at base, toothed near the apex,* with revolute margins, glabrous, paler below, and bearded in the axils of the veins; peduncles *undivided,* equalling the leaves; fl. spicato-racemose; carpels 2–1, rather large, obovate (when dry), reddish. *S. undulata and S. decipiens, Arn. l. c. Presl, l. c., excl. syn. Jacq. Rhus spicata, Thunb. Fl. Cap. p.* 265. *R. undulata and R. decipiens, E. Mey.!*

HAB. In woods. Zitzekamma and Krakakamma, Keyskamma and Oliphant's hoek, and towards Port Natal, *Drege! E. & Z.! Zey. No.* 2249. Dec.–Mar. (Herb. E. Mey., Sond., T.C.D.)

A smaller shrub than the preceding, with small leaves, and short, undivided racemes. Branches ash-coloured; branchlets whitish. Common petiole 1 inch long. Middle leaflet 1½–2 inches long, 7–10 lines wide; lateral smaller, all coriaceous, shining above, cuneate and very entire in the lower half, with a few larger or smaller teeth from the middle to the apex, obtuse or obtusely acuminate, mucronulate. Spikes including the peduncle 1–2 inches long. Flowers small, greenish or brownish, shining; their parts, &c., as in *S. melanocarpa.* Ovaries divaricate; style bifid. Carpels twice as large as in the other species, 2½–3 lines long, 2 lines wide. *Rhus undulata, Jacq. Schoenb. t.* 346, though very like this species, differs in its paniculate, dioecious, pentandrous flowers.

** *Leaves simple.* [*unifoliolate.*]

5. **S. Dregeana** (Sond.); branchlets, petioles and leaves glabrous; leaves simple, broadly ovate, acute, sinuato-dentate, pale below, netted with veins; peduncles solitary, undivided; flowers racemoso-spicate. *S. monophylla, Presl, l. c., non Hook. f. Rhus monophylla, E. Mey.! Hb. Drege.*

HAB. Among shrubs, near Omsamculo and Omcomas, *Drege!* March. (Herb. E. Mey., Sond.)

Leaves 3 inches long, 2 inches wide, articulate to the apex of an inch long petiole, dull green above, paler or livid below. Spikes, including the peduncle, 2–4 inches long. Flowers as in the other species.

IV. SAPINDUS. Linn.

Calyx unequally 5-parted or of 5-sepals, 2 outside 3. *Petals* 4, 5, 6, naked or bearded at the base inside, or furnished with a scale above the claw. *Stamens* 8–10 or rarely more, on a hypogynous disc. *Style* one; stigmata 3. *Fruit* of 3 fleshy or leathery, sub-globose carpels, connate at base, 1–2 often abortive. *Seed* erect, with or without aril; cotyledons thick, curved, the outer embracing the inner, which is transversely folded; radicle acute, inferior. *Endl. Gen.* 5610, also *Pappea, E. & Z.! En. p.* 53. *Endl. Gen.* 5635. *Harv. Gen. S. A. Pl. p.* 36.

Tropical and sub-tropical, unarmed trees with abruptly pinnate, or by abortion of one leaflet often impari-pinnate, more rarely simple, coriaceous leaves. Flowers polygamous, in axillary racemes or in terminal panicles. Name, a syncope of *Sapo-Indicus, Indian Soap;* the aril of the seeds of *S. Saponaria* is used in S. America as soap.

* *Leaves pinnate.* (Sp. 1–2.)

1. S. capensis (Sond.); branches and leaves glabrous; rachis not winged; leaves petiolate, abruptly pinnate, leaflets in 3–4 pairs, sessile, oblong, narrowed at base and apex, *obtusely toothed*, coriaceous, undulate, netted with veins, with minute pellucid dots; panicle axillary, shorter than the leaf; carpels 2-1, sub-globose, glabrous.

Hab. Langekloof, George, 2–4000 f., *E. & Z.! Drege, No.* 8266. Decr. (Herb. Sond.)

Branches ash-coloured. Leaves alternate. Rachis 2½–3 inches long, narrow. Leaflets 2–2½ inches long, 7–12 lines wide, cuneate, acuminate, sub-obtuse, paler below; the teeth obtuse, ½–1 line long. Panicle 1–2 inches long, minutely downy. Flowers unknown. Fruit short-stalked, fleshy; each carpel as big as a cherry, one-seeded. Seed erect, brownish purple, shining, larger than a pea.

2. S. oblongifolius (Sond.); diœcious; leaves petiolate, abruptly or unequally pinnate, leaflets 5-7 pairs, petiolulate, alternate or opposite, oblong, netted with veins, glabrous; panicle axillary, thinly pubescent, the flowers fascicled; stamens 15-16; carpels 3-1, fleshy, sub-globose, one-seeded, the younger yellow-tomentose, the older glabrous. *Rhus oblongifolia, E. Mey.! Hb. Drege. Prostea oblongif. Arn. in Hook. Lond. Journ. Vol.* 3. *Presl, l. c. Simaba lachnocarpa, Hochst. Pl. Krauss, No.* 113. *Sapindus capensis, Hochst. Flora,* 1843. *p.* 80, *excl. syn.*

Hab. Woods at Omsamculo and near Port Natal, *Drege, Krauss, Hewitson, and Williamson.* April. (Herb. T.C.D., Sond.).

A tree. Leaves 6–12 inches long. Leaflets 2–3½ inches long, 1–1½ inch wide, obtuse, commonly with a short, obtuse mucro, pale below, the lower ones on petiolules 2–3 lines long, the upper sub-sessile. Panicle fulvous-orange, ½–1 foot long, its primary branches spreading, branchlets short, and fl. minutely pedicellate. Sepals 5, concave, rounded-obtuse, fringed and ciliate, 1⅛ line long. Petals 5, shorter than the sepals, obtuse, fringed with woolly hairs, having at the base inside a fringed petaloid scale. Filaments densely hairy, shorter than the oblong, sub-apiculate anthers. Calyx and petals of the female flowers as in the male. Ovary not seen. Style hairy, with a glabrous point, curved. Ripe fruit of 3, 2, 1 carpels, each carpel obovate-globose, smaller than a cherry, indehiscent; seed erect, obovate-globose, exarillate, with leathery coat; cotyledons thick, curved, the outer embracing the transversely plaited inner one; radicle short, acute, inferior.

** *Leaves simple.*

3. S. Pappea (Sond.); branches and leaves glabrous; leaves simple, alternate, obovate-oblong, obtuse at base, unequal, with revolute margins; racemes crowded to the ends of the branches, the male racemes equalling the leaves, the female 3-8-flowered, shorter. *Pappea capensis, E. & Z.! En. p. 53. Hook. Ic. Pl. t. 352. Pappe, Fl. Med. p. 3. Sylv. Cap. p. 6. Kiggelaria integrifolia, E. Mey.! Hb. Drege, non Jacq.*

HAB. In woods, between the Zwartkop's and Coega River, Uit.; and in the Bothasberge, near the Fish River, Albany, *E. & Z.! Zey. No.* 151. At Silverfontein, Namaqualand, 2–3000 f., *Drege!* Sep.–Nov. (Herb. Hook., T.C.D., Sond.).

A tree 15–20 feet high, with hard wood. Branches spreading, ash-coloured. Leaves crowded at the ends of the branches, 1½–3 inches long, 6–10 lines wide, pale underneath. Racemes 1–2 inches long. Flowers minute. Calyx ½ line long. Stamens with woolly filaments, exserted. Ovary hairy, 3-celled, cells 1-ovuled. Fruit globose, fleshy, downy, red, about as large as a small cherry. The fruit, called "Wilde Preume, Oliepitten, Wilde Amandel, t'Kaambesje," is edible, and a bland oil is expressed from the seeds.

V. HIPPOBROMUS, E. & Z.

Flowers polygamo-diœcious. *Calyx* 5-sepalled, persistent, unequal. *Petals* 5, obovate, hypogynous, without scales, imbricate. *Male: stamens* 8, inserted within an annular disc; anthers oblong, basifixed. *Ovary* abortive, silky-tomentose. *Female: ovary* 3-celled, seated on a disc. *Style* short, thick; stigma capitate, trifid. 8 abortive stamens. *Pericarp* drupaceous, fleshy, of 1-3 pyrena. Seed pendulous, exalbuminous; cotyledons green, fleshy, conduplicate; radicle very short, near the hilum. *E. & Z. En. p. 151. Endl. Gen. No. 5637.*

A resiniferous tree, with abruptly pinnate leaves; and sub-opposite leaflets. Panicles axillary, short; flowers reddish, velvetty; the calyx lobes rounded-concave and, like the petals, ciliato-laciniate. Name, *ἵππος, a horse,* and *βρῶμος, a bad smell;* colonial name "*Paardepis.*"

1. H. alata (E. & Z.!); *Pappe, Sylv. Cap. Rhus alatum and R. pauciflorum, Thunb. Fl. Cap. p. 268. Weinmannia dioica, Spreng. Suppl. ad Syst. Veg. p. 18.*

HAB. In woods, Albany and Uitenhage; also in Caffraria, *Thunb., E. & Z., Drege, Krauss.* Jul.–Dec. (Herb. Lehm., Sond., T.C.D.).

A tree. Branches erecto-patent, grey, the younger ones yellowish-red, thinly downy. Leaves petiolate, the rachis winged, thinly tomentose. Leaflets sessile, in about 5 pairs, serrate or toothed, or quite entire, with reflexed margins, ovate-obtuse or obovate, cuneate at base, mostly oblique, coriaceous, glabrous or appressedly pilose on the mid-rib, green above, paler below, 1–1½ inches long, 6–8 lines wide; in some branches of larger size (var. *latifolia,* E. & Z.). Panicles dense, reddish, ½–1 inch long; bracts minute, at the divisions. Calyx silky, 1 line long. Petals glabrous, little longer. Stamens exserted. Fruit the size of a large pea.

Tribe 2. DODONÆACEÆ. *Ovules* 2-3 in each cell. *Embryo* spirally rolled.

VI. DODONÆA, L.

Flowers polygamous. *Calyx* 3-4-5-parted. *Petals* none. *Stamens* 5 or more; filaments very short; anthers erect, thick, 4-sided. *Ovary* sessile, 3-4-angled, 3-4-celled; ovules in pairs; style angular, central, 3-4-cleft. *Capsule* membranous, 2-3-4-winged. *Endl. Gen.* 5626.

Shrubs or small trees, tropical or sub-tropical, very numerous in Australia. Leaves alternate, exstipulate, simple or impari-pinnate, often exuding gummy secretions. Flowers small, green, generally in racemes, axillary or terminal. Named in honour of Rambert Dodoens, commonly called *Dodonæus*, a famous physician and author of a Historia Plantarum. He died in 1585.

1. D. viscosa (Linn. Mant. 238); *leaves obovate-oblong,* cuneate at base, viscid; flowers racemose, or racemoso-paniculate; fruits 2-3-winged, as long as the stalk or longer. *DC. Prod.* 1. *p.* 616. *A. Gray, Gen.* 182. *Plum. ed. Burm. t.* 247. *f.* 2. *Sloane, Hist. Jam.* 2. *t.* 162. *f.* 3. *D. Natalensis, Sond. Linn.* 23. *p.* 23.

HAB. Port Natal, *Gueinzius, Williamson.* (Herb. Sond., T.C.D.).

Branches terete, reddish; branchlets sub-compressed. Leaves sessile, much attenuate at base, very entire or sub-crenulate, penni-nerved, acute, 3–3½ inches long, 10–12 lines wide. Racemes panicled, half as long as the leaves, with spreading, viscid branchlets. Flowers small, hermaphrodite. Calyx 1 line long. Fruit, not quite mature, 4–5 lines long and wide. The Natal specimens differ from American ones by more acute leaves, the old ones scarcely viscid and somewhat smaller fruits.

2. D. Thunbergiana (E. & Z.! 419); *leaves lanceolate or linear-lanceolate,* narrowed at base, repandly sub-denticulate; flowers racemose; fruits 2-3-winged, as long as the pedicel or longer. *D. angustifolia, Thunb. Fl. Cap. p.* 383.

VAR. β. **linearis**; branchlets more crowded; leaves linear, acute or obtuse, mucronulate. *D. Mundtiana, E. & Z.!* 420. *D. linearis, E. Mey.!*

HAB. Mountain sides, at the Paarl, Piquetberg, Hex rivier, *Thunberg, Krauss, Drege,* 7532, *Zeyher* 152. Hassagaiskloof, *Zeyher* 2027. Tulbagh, and at Brackfontein, *E. & Z.* β. Hills near Kochman's Kloof, at Gauritz Rivier, *Mundt.* In Zwartruggens, near Uitenhage, and Graaf Reynet, *E. & Z.!* Modderfontein, and near Leliefontein, *Drege.* Aug.–Nov. (Herb. Sond., T.C.D.).

A shrub, 5–10 feet high, much branched, glabrous. Leaves somewhat viscid, 1½–2½ inches long, 3–4 lines broad; in β. 1–2 lines broad, mid-ribbed below. Racemes ½ inch, simple or branched, dense; peduncles lengthening in fruit, Flowers green, polygamous. Calyx obtuse, a line long. Stamens 8, rarely 7–6. Style trifid. Capsule 5–6 lines long and wide, resinous-shining. A decoction of the root is used as a purgative in fever. *Pappe, Fl. Cap. Med. Prod. p.* 4.

(GENERA WHOSE AFFINITY IS VERY UNCERTAIN.)

* PTÆROXYLEÆ, Sond.

(By W. SONDER.)

PTÆROXYLON, E. & Z.

Flowers polygamously diœcious. *Sepals* 4, short, obtuse. *Petals* 4, at first erect, slightly imbricate, then spreading, concave. *Hypogynous-disc* annular, glandular. *Stamens* 4, alternate with the petals; fil. glabrous; anthers oblong, fixed below the middle, bifid at base. *Ovary* compressed, obcordate, 2-celled; ovules solitary, affixed at the inner angle above the base. *Styles* 2, or connate in one; stigmas capitate. *Capsule* compressed, bi-lobed at the apex, cordate at base, 2-celled, 2-seeded, the cells internally dehiscing, at length separating, and suspended from a bi-partite axis. Seeds compressed, produced into a membranous wing; cotyledons flat, somewhat fleshy, accumbent to the inferior, cylindrical radicle. *E. & Z. En. p.* 54. *Endl. No.* 5636.

A tree, with abruptly or impari-pinnate leaves; leaflets unequal-sided, 5–8 pair, decreasing, very entire; racemes axillary, panicled, shorter than the leaves, crowded at the ends of the branches; fl. small, white or yellowish. Name, from *πταιρω, to sneeze, ξυλον, wood; sneeze-wood,* colonial name of the tree.

1. Pt. utile (E. & Z. l. c.); *Harv. Thes. Cap. t.* 17.

HAB. Hill sides near the Boschjesman's Rivier and in Adow and Coega, Uitenhage, *E. & Z., Zeyher, No.* 2025. *Drege* 6814; and in other parts of the Eastern districts, common. (Herb. Sond., T.C.D.)

A tree, 20–30 feet high, with a trunk 2–4 feet diameter. Leaves opposite. Rachis pubescent, sub-compressed, 3–5 inches long. Leaflets ovate-oblong, obtuse, or sub-retuse, mucronulate, oblique at base, coriaceous, veiny, 1 inch long, 4–6 lines wide or less. Panicles 1–2 inches long. Petals 2 lines long. Capsule 7–10 lines long. 5–6 lines wide. Seed 2 lines, or including the wing, 6–8 lines long. The wood, *Nieshout*, Sneezewood, is handsome, takes a fine polish, is strong, durable, and somewhat like mahogany, *Pappe, Silv. Cap. p.* 5.

** AITONIEÆ, Harv.

(By W. H. HARVEY.)

AITONIA, Linn. f.

Calyx short, deeply 4-parted, slightly imbricate in æstivation, deciduous. *Petals* 4, much longer than the calyx, erect, oblong, sessile, strongly imbricate and slightly convolute in æstivation. *Stamens* 8, hypogynous, monadelphous, exserted; filaments ascending, subulate, flat, their dilated bases united in a short tube; anthers incumbent, oblong, introrse, 2-celled, slitting longitudinally. *Hypogynous-disc* cup-shaped, crenulate, fleshy, within the stamens. *Ovary* sessile, 4-lobed, 4-celled; *ovules* 2 in each cell, collateral, axile, ascending, sub-sessile, semi-amphitropal; *style* filiform, exserted, *stigma* simple. *Capsule* inflated and membranous, deeply 4-lobed, and acutely 4-angled, 4-celled, loculicidal, the valves septiferous. *Seeds* 2–1, in each cell, reniform, with a thick, corrugated, leathery and rather loose testa; *embryo* green, incurved, exalbuminous, enclosed in a membranous integument, separable from the testa; radicle short, straight; cotyledons elliptic-oblong, leafy. *Endl. Gen. No.* 5548. *A. Juss. Mem. Mus.* 19. 186.

A shrub with alternate or fascicled, very entire, linear-oblong, obtuse, glabrous, one-nerved, coriaceous, evergreen, leaves; solitary, axillary, purplish, stalked flowers and bladdery capsules. It is named in honour of W. Aiton, Gardener to King George III., and author of the "*Hortus Kewensis.*"

1. A. capensis (Linn. f. Suppl. 303); *Ait. Hort. Kew. Ed.* 1. *Vol.* 3. *p.* 431. *Thunb. Fl. Cap. p.* 508. *E. & Z.! No.* 426. *Curt. Bot. Mag. t.*

173. *Cav. Diss. t.* 159. *Lam. Encycl. t.* 571. *Lodd. Cab. Bot. t.* 682. *Rchb. Exot. t.* 229. *Burm. Afr. t.* 21. *f.* 2. *Don, in Ed. N. Phil. Journ.* 1832, *p.* 242, *and* 1833, *p.* 262.

HAB. Karroo, near Gouds River and Slang River, *Thunberg.* Gauritz River, George; and between Uitenhage and Graaf Reynet, *E. & Z.!* Zwarteberg, 2-3000 f.; between Kaus, Natvoet, and Doornpoort, 1-2000 f. (Orange River Mouth); and near the Great Fish River, Albany, *Drege!* Winterhoek, *Dr. Pappe!* Woods near Uitenhage, *Dr. Alexander Prior!* (Herb. T.C.D., &c.)

A much branched, rigid shrub, 5-10 feet high, with long, virgate, leafy branches. Twigs angular, minutely downy. Leaves densely set, scattered, or fascicled on tubercle-shaped, abortive ramuli, linear, lin.-oblong, or spathulate, tapering at base into a minute petiole, 1-1½ inch long, 1½-3 lines wide; obtuse or sub-acute, glabrous, one-nerved, veinless, coriaceous. Pedicels axillary, shorter than the leaves, minutely puberulous, as are also the calyx and petals. Calyx 1½ lines long. Petals 5-6 lines long, 2-3 lines broad, purplish. Ovary pubescent. Capsule pink or purple, 1½ inches in diameter, of a thin, membranous substance, much inflated, deeply lobed and sharply angled.

ORDER XXIX. MELIACEÆ, Juss.

(By W. SONDER.)

(Meliæ, Juss. Gen. 263. Meliaceæ, Juss. Mem. Mus. 3. 436. DC. Prod. 1. p. 619. Ad. de Juss. Memoire Mus. vol. xix. Endl. Gen. No. ccxxv. Lindl. Veg. Kingd. No. clxxiii.)

Flowers perfect, or abortively unisexual, regular. *Calyx* short, 4-5-cleft or cup-shaped, with imbricate æstivation. *Petals* 4-5, longer than the calyx, sessile, with valvate or slightly imbricate æstivation. *Stamens* hypogynous, twice as many as the petals; *filaments* united into an entire or toothed tube; anthers either on the teeth of the tube, or sessile within the orifice. A hypogynous, fleshy, cup-like *disc* surrounds the ovary. *Ovary* free, pluri-locular, with axile placentæ; *ovules* in pairs, rarely solitary; *style* simple, stigma capitate or peltate, obsoletely lobed. *Fruit* a berry, capsule or drupe. *Seeds* with or without albumen, with a straight embryo and leafy or fleshy cotyledons.

Trees or shrubs. Leaves alternate or rarely sub-opposite; usually pinnate or bi-pinnate, rarely simple; the leaflets entire or cut or serrated. Stipules none. Flowers either panicled, corymbose, racemose or spiked, usually of small size.

This Order is chiefly tropical, a few species only straggling into the warmer parts of the temperate zone. Most are extremely bitter, astringent and acrid, and are either powerful medicines or poisons. All parts of the common Cape Lilac (*Melia Azederach*) are intensely bitter, with strongly purgative qualities; its extract is, in small doses, useful as a vermifuge; in larger it is violently poisonous.

ANALYSIS OF THE SOUTH AFRICAN GENERA.

I. **Turræa.**—*Leaves* simple. *Petals* very long, strap-shaped.
II. **Melia.**—*Leaves* bi-pinnate. *Fruit* drupaceous.
III. **Trichilia.**—*Leaves* simply pinnate. *Fruit* a dry, splitting capsule.
IV. **Ekebergia.**—*Leaves* simply pinnate. *Fruit* fleshy, indehiscent.

I. TURRÆA, Linn.

Calyx cup-shaped, 5-toothed. *Petals* 5, very long, strap-shaped, convolute in æstivation. *Stamens* 10, connate in a long tube, 10-toothed at the summit, the anthers inserted between the teeth. *Ovary* sessile, 5-10-20-celled; ovules in pairs, superposed on axile placentæ. *Style* 1; *stigma* thickened. *Capsule* 5-celled; cells 2-1 seeded, the valves septiferous. *Seeds* compressed; cotyledons flat; radicle superior. *Endl. Gen. No.* 5519. *Juss. Mem. Mus. xix. p.* 217. *t.* 12 *f.* 3. *DC. Prod.* 1. *p.* 620.

Trees or shrubs, found in Tropical Asia, Madagascar and South Africa. Leaves alternate, petiolate, simple. Flowers on short branchlets, somewhat tufted. Named in honour of Geo. Turra, an Italian botanist, professor at Padua, died in 1607.

1. T. obtusifolia (Hochst. in Flora, xxvii. 1. p. 296); branches and branchlets glabrous; *leaves obovate*, narrowed into a short petiole, entire or obtusely 3-lobed, with revolute margins, glabrous on both sides, paler beneath, the young ones pubescent on the mid-rib; *the axillary, subsolitary peduncles and the calyx glabrous;* petals elongate, ligulate, somewhat longer than the staminal tube; stigma exserted, *shortly cylindrical. T. lobata, E. Mey.! in Herb. Drege, non Lindl.*

HAB. In woods at Omsamcaba, *Drege!* Port Natal, *Krauss, Gueinzius, Sanderson!* Feb.–Jul. (Herb. T.C.D., Vind., Sond., Hook.)

A shrub, with a grey or reddish bark. Leaves 1–1¾ inches long, 4–10 lines wide, obovate or obversely lanceolate, in some specimens 3-lobed below the point, the terminal lobe largest. Peduncles 6–10 lines long. Calyx campanulate, shortly 5-toothed, 1 line long. Petals glabrous, 1–1½ inches long. Style slightly longer than the petals; stigma ½ line long.

2. T. heterophylla (Sm. in Rees Cycl. 36. No. 6); branches glabrous, branchlets thinly pubescent, leaves short-stalked, *ovate, acute or produced into a short obtuse point*, undivided or sub-trilobed, the younger pubescent above, silky beneath, the full-grown pale beneath, glabrous, villous on the nerves; flowers clustered at the ends of the branches, *the peduncles and calyces silky tomentose;* petals elongate ligulate, longer than the staminal tube; style much exserted, the stigma *depressed-globose. DC. Prod.* 1. *p.* 620. *T. floribunda, Hochst. l. c. p.* 297.

HAB. In woods, Omtata and Omsamwubo, *Drege.!* Port Natal, *Krauss, Gueinzius, Sanderson*, &c. (Herb. T.C.D., Hook., Vind., Sond.)

A shrub, with a grey or brown bark. Leaves falling away before the flowering season, ovate, obtuse at base, 2–3½ inches long, 12–16 lines wide, the nerves and veins very prominent on the lower side. Petioles tomentose, 2–3 lines long. Peduncles 3–8 lines long, crowded on the top of very short branchlets, which look like common peduncles, often very numerous, bracteolate at base. Calyx shortly and acutely 5-toothed ("downy," *Sm.* "hairy," *DC.).* Petals 10–14 lines long. Style much longer than the petals. Capsule woody, 10-furrowed and lobed, 4–5 lines long, 5–6 lines wide, 3–5-celled, loculicidal; cells 2-seeded, the valves septiferous in the middle. Seeds 3 lines long, incurved, purple, shining.

II. MELIA, Linn.

Calyx small, 5-fid. *Petals* 5, oblong-linear, spreading, convolute in æstivation. *Stamens* 10, the filaments connate into a tube 20-toothed at the summit, the anthers sessile within the throat of the tube. *Ovary* on a raised torus, 5-celled; *ovules* 2 in each cell; *style* filiform; *stigma* capitate, 5-angled. *Drupe* ovate, with a 5-furrowed and 5-celled, bony *stone;* cells one-seeded. *Endl. Gen. No.* 5520. *DC. Prod.* 1. *p.* 621.

Trees with bi-pinnate or decompound leaves, natives of the tropics of the Eastern hemisphere. Flowers paniculate, sweetly scented, lilac-coloured. Name, from μελια, the Greek name for the *Ash.*

1. M. Azederach (Lin. Sp. p. 550); leaves bipinnate; leaflets smooth, incised, subquinate. *E. & Z. ! En. No.* 424. *Herb. Un. It. No.* 509.

Cultivated in gardens throughout the Colony. "Cape Lilac," or Pride of China.

III. TRICHILIA, Linn.

Calyx short, 4–5-toothed or cleft. *Petals* 4–5, oblong, or ovate, subimbricate in æstivation. *Stamens* 8–10, the filaments sometimes subdistinct, sometimes united into a tube, antheriferous within. *Ovary* 2–4-celled; ovules 2 in each cell. *Style* short, continuous with the ovary; stigma capitate, 2–4-lobed. *Capsule* 2–4-celled, loculicidally 2–4-valved; valves septiferous in the middle; cells 1-2-seeded. *Seeds* with a fleshy arillus. *Endl. Gen. No.* 5541. *A. Juss. l. c. p.* 235. *DC. Prod.* 1. *p.* 622.

Tropical and sub-tropical trees or shrubs, natives chiefly of America, very few African. Leaves alternate, imparipinnate. Flowers in axillary panicles. Name, τριχα, *by threes;* the ovary and capsule are usually trilocular.

* *Staminal tube deeply ten-cleft; segments hairy within.*

1. T. Dregeana (E. Mey. in Hb. Drege); leaves impari-pinnate, leaflets in 2–5 pair, *lanceolate-oblong, acute or sub-acuminate,* glabrous on both sides, or strigilose on the nerves beneath; panicles axillary, short, divided from the base; petals tomentose on each side, four times as long as the *obtusely 5-fid calyx; staminal tube 10-cleft to the middle,* the segments very hairy within, antheriferous at the bifid apices; capsule globose, cells 2–1-seeded

VAR. β. **oblonga**; leaflets 2–3 pair, *broadly elliptic-oblong, obtuse. T. Dregeana β. oblonga, Harv. Thes. Cap. t.* 76.

HAB. In woods at Port Natal, *Drege! Gueinzius!* Var. β. from the same district, *Mr. Sanderson!* (Herb. T.C.D., Hook., Sond.)

Flowering branches thick, with a brownish-grey, rugged bark. Leaves large; the rachis somewhat furrowed, tomentose, 3–4 inches long, including the petiole; leaflets sub-opposite, shortly petiolulate, 3–4½ inches long, 1–1½ inches wide, shining above, paler beneath, midrib and veins prominent. In β. the leaflets are 2½–3½ inches wide, and 1½–2 inches long. Panicles 1–2 inches long, thinly downy. Calyx lobes 1 line long. Petals 5 lines long, erect, the margin slightly imbricating, afterwards spreading. Staminal tube glabrous on the outside, ¼ shorter than the petals; anthers oblong, glabrous. Style as long as the staminal tube, appressedly pubescent, glabrous near the top; stigma thick, angled, sub-globose. Capsule 8–10 lines in diameter, 3–4-celled, loculicidal, the valves woody. Seeds 2, collateral, 6 lines long, concealed in fleshy arillus; cotyledons thick, radicle short, immersed.

** *Staminal tube undivided.* (Sp. 2–3).

2. T. Ekebergia (E. Mey. in Hb. Drege); leaves impari-pinnate, leaflets 3–5 pair, *ovate, very acute,* mostly somewhat narrowed at the oblique base, glabrous on both sides; panicles axillary, *on long peduncles,* the branches bracteolate, cymose; petals tomentose at each side, thrice as long as the acutely 4–5-toothed calyx; staminal tube *undivided,* antheriferous at summit. *Ekebergia capensis, Hochst. Pl. Krauss. Ek. Meyeri, Presl, Bot. Bem. p.* 25.

HAB. Woods at Sitzekamma and Port Natal, *Drege, Krauss,* &c. (Herb. T.C.D., Sond.)

Branchlets nodose. Leaves crowded at the ends of the branches, a foot or more in length, including the flattened petioles; leaflets opposite, 2–3 inches long, 12–15 lines wide, ovate or ovate-oblong, the lower on very short petiolules, the terminal with a petiolule ½ inch long. Panicle glabrous, 3–4 inches long, on a compressed peduncle of the same length or longer; branches of the panicle 1–1½ inches long, loosely cymose, 7–15-flowered, with a minute, subulate bract under each pedicel. Flowers 4–5-fid. Calyx shortly campanulate. Petals two lines long, oblong, longer than the downy staminal tube. Anthers 8–10, oblong, glabrous. Style glabrous, equalling the staminal tube; stigma thick, obscurely lobed. This much resembles *Ekebergia capensis* from which it differs in the wider, more ovate leaflets, the longer and more slender panicles, the acutely-toothed calyx, the scarcely-imbricate petals and longer staminal tube.

3. **T. natalensis** (Sond. Linn. 23. p. 23); glabrous; leaves impari-pinnate, 2–4 pair; leaflets opposite, sessile, *lanceolate, acuminate, sub-falcate, serrate,* paler beneath; racemes crowded at the ends of the branches, simple, much shorter than the leaves; flowers short-stalked, 4-petalled, octandrous; stamens glabrous, connate in a short tube; stigma 4-lobed.

HAB. Port Natal, *Gueinzius!* (Herb. Sond.)

A tree. Leaves crowded at the end of the branches, alternate, petiolate; the rachis, including a 1–2-inch petiole, three inches long. Leaflets 2–3 inches long, 6–8 lines wide, the terminal petiolulate, all nearly equal-sided at base, penni-nerved, venulose. Racemes inch long, lax-flowered, glabrous, the pedicels solitary or in pairs, one line long, bracteolate. Calyx 4-fid, the lobes acute, one line long, with valvate æstivation. Petals ovate-oblong, ½ longer than the calyx, spreading, re-curved, glabrous. Ovary 2-celled; style short; fruit unknown. Perhaps not a true *Trichilia?* It resembles *Harpephyllum caffrum,* Bernh., but the leaves are serrate and the flowers not dioecious.

IV. EKEBERGIA, Sparm.

Calyx short, 4–5-fid to the middle, lobes obtuse, imbricate in æstiva-tion. *Petals* 4–5, scarcely longer than the calyx, elliptical or oblong, imbricate in æstivation. *Stamens* 10 (rarely 9), united in a campanulate 10-toothed, short tube; the teeth antheriferous. *Ovary* 4–5-celled, surrounded by an annular disc; ovules in pairs; style short, thick; stigma discoid, obsoletely lobed. *Berry* dry, globose, 4–5-celled, 5-seeded, or abortively 1–2-seeded; seeds exarillate. Embryo exalbu-minous, with fleshy cotyledons and a superior radicle. *Sparm. Act. Holm.* 1779, *p.* 282. *t.* 9. *Juss. l. c. p.* 233. *t.* 17. *Endl. Gen. No.* 5538.

A large tree with ashen bark. Leaves impari-pinnate, in 4–5 pair, the leaflets op-posite, sessile, very entire, oblong, acuminate at each end, coriaceous, glabrous, with revolute margins. Flowers in axillary panicles. The name is in honour of Charles Gustavus Ekeberg, Captain of the Swedish East Indiaman in which Sparmann sailed to China.

1. **E. capensis** (Sparm. l. c.); *Thunb. Fl. Cap. p.* 582. *E. & Z.! No.* 425. *Pappe, Sylv. Cap. p.* 6. *Trichilia capensis, Pers. Ench.* 1. *p.* 468. *Zey. No.* 2029.

HAB. Primitive woods in George, Uitenhage, Albany, and Caffraria, *Thunberg, E. & Z.!* &c. Aug. (Herb. T.C.D., Lehm., Sond.)

This is the Essenhout or Essenboom of the Colonists. Wood hard and white. Branches and branchlets nodose after the leaves have fallen, glabrous. Leaves crowded at the ends of the branches, alternate, petiolate, the petiole, including the channelled rachis, 4–8 inches long. Leaflets 2–3 inches long, ¾–1 inch wide, some smaller, all unequally narrowed at base, paler underneath. Panicles 1–3 inches long, loose, the common peduncles 1–2 inches long, compressed, smooth; pedicels cernuous, 1–2 lines long. Calyx 1 line long, with roundish, obtuse segments. Petals tomentose at both sides. Berry (when dry) blueish-black, as large as a small cherry, 1–2 seeded; seeds oblong, sub-compressed.

ORDER XXX. **AMPELIDEÆ**, Kunth.

(By W. H. HARVEY.)

(Vites, Juss. Gen. 267. Viniferæ, Juss. Mem. Mus. 3. 444. Ampelideæ, Kunth, in Humb. Nov. Gen 5. 223. DC. Prod. 1. p. 627. Endl. Gen. No. clxiv. Vitaceæ, Lindl. Veg. King. No. clx.)

Flowers usually perfect, regular, minute. *Calyx* small, cup-shaped. *Petals* 4–5, separate or connate, concave, with valvate æstivation. *Stamens* as many as the petals and opposite them; filaments short; anthers introrse, 2-celled. *Ovary* surrounded by a cuplike, fleshy disc, 2-celled (rarely 3–6-celled); ovules in pairs or solitary, ascending or erect. Style simple; stigma capitate. *Berry* 2–6-celled, pulpy; seeds bony, with much hard albumen, and a small basal embryo.

Shrubs, erect or more commonly trailing, or climbing by tendrils formed from abortive peduncles. Branches tumid at the nodes, often succulent. Lower leaves opposite, upper alternate, petiolate, rarely simple, generally palmately compound, sometimes pedate or pinnate. Peduncles opposite the leaves, simple or branched; pedicels umbellulate or cymose.

Natives of tropical and sub-tropical countries of both hemispheres. The grapevine (*Vitis vinifera*), the type of this Order, now cultivated throughout the warmer temperate zones, is originally from the Caucasian region, whence it has accompanied civilized mankind in all their wanderings. None of the wild vines of South Africa have esculent fruit. [*Leea crispa*, Lin., a reputed Cape species, is here omitted, as there is no valid evidence to show that it exists within the colony. It probably was introduced to British gardens from "Cape Coast," a locality which may have been confounded with "the Cape."]

TABLE OF THE SOUTH AFRICAN GENERA.

I. **Vitis.**—*Petals* connate into a hood. *Stigma* sessile.
II. **Cissus.**—*Petals* distinct. *Style* cylindrical, stigma sub-capitate.

I. VITIS, L.

Calyx short, cup-shaped, obsoletely 4–5-toothed. *Petals* 5, cohering by their edges into a cap-like (*calyptriform*) body, separating at base, and falling off in one piece. *Style* none; stigma sessile, depressed. *DC. Prod.* 1. p. 633. *Endl. Gen.* 4567.

Climbing shrubs, of which the common grape-vine is universally known. The solitary Cape species is admitted with doubt. *Vitis*, the classical name of the grapevine.

1. V. hispida (E. & Z.! 428); leaves deeply cordate at base, with 5 short, deltoid-acuminate, cuspidate lobes, sharply toothed and hispid on both surfaces; the middle lobe longest.

HAB. In woods. Oliphantshoek, near mouth of Boschesman's River, Uitenhage, *E. & Z.!* (Herb. Sond.)

This species requires confirmation. I have only seen a fragment, consisting of three leaves on a broken portion of a branch, without flower or fruit. It resembles *V. latifolia*, Roxb., but is roughly hairy and more sharply lobed. Perfect specimens are desired.

II. CISSUS, L.

Calyx cup-shaped, obsoletely 4–5-crenate. *Petals* 4–5, inserted outside a fleshy, hypogynous disc, concave, hooded at the apex, separate, deciduous. *Style* cylindrical; stigma capitate or simple. *Endl. Gen.* 4566. *DC. Prod.* 1 *p.* 627.

Climbing, rarely erect, shrubby or half shrubby plants, found throughout the tropics and warmer temperate zones of both hemispheres. Stems tumid at the nodes, often fragile, sometimes succulent, leaves simple, or variously compound, often pellucid dotted. Flowers minute, greenish, in clusters or panicles, opposite the leaves. Tendrils simple or branched. Name *κισσος*, the Greek name for the vine.

Group 1. SIMPLICIFOLIÆ. *Leaves* simple, entire or lobed. *Stipules* membranaceous, deciduous. (Sp. 1–3).

1. C. tetragona (Harv.); *puberulent;* stems *succulent, green, sharply quadrangular*, marginate; leaves petiolate, cordate at base, angularly sub-trilobed, the lobes short, ovate, distantly denticulate, minutely puberulent on both sides; flowers ?

HAB. Near Port Natal, *Mr. R. W. Plant.* (v. v. cult. in Hort. *Wilson Saunders*).

Stems herbaceous, fleshy, about $\frac{4}{10}$ inch square, green, minutely downy, sharply 4-angled, with a slightly elevated, discoloured (brownish-red) marginal line on the angles, and a more or less evident intermediate groove; the internodes 4–8 inches long. Petioles 1–1½ inch long, pubescent, the stipules small, ovate, soon withering. Leaves 3–5-nerved at base, carnose, pale green, 2–3 inches long and broad; the lower ones 3-lobed, the lateral lobes shallow and blunt, the middle longest and ovate; upper leaves somewhat triangular-hastate, with a tendency to be angularly-lobed, acute or obtuse; the margin with very minute denticulations, 2–3 lines asunder. Flowers not seen. Described from a living plant in the rich collection of *W. Wilson Saunders, Esq.* It is allied to *C. quadrangularis*, L., and others with similar habit, but appears to be distinct.

2. C. fragilis (E. Mey! in Hb. Drege); *glabrous;* stems half-herbaceous, weak, striate; leaves on long petioles, *angularly cordate acuminate*, mucronulate, distantly ciliato-denticulate, three-nerved at the base, and penni-nerved, thin and membranous; inflorescence loosely panicled, the lateral branches cymoid, few-flowered; pedicels much longer than the flowers.

HAB. In woods on the Omblas Hills, Natal, *Drege!* Mar.-Apr. (Herb. Hook., Benth., Sond.).

A straggling climber. Leaves green, thin, quite glabrous; the lateral veins prolonged into marginal ciliary processes. Fruit fleshy. Allied to *C. arguta*, H. f., but differing in several respects.

3. C. Capensis (Willd. Sp. 1. p. 655); young twigs and leaves rufopubescent, older glabrous; stems terete, striate; leaves on long petioles, broadly *cordato-reniform, very obtusely 5-angled*, repando-dentate, three-nerved at base and penni-nerved, obtuse, coriaceous; stipules ovate; inflorescence thyrsoid, densely tomentose, the peduncles elongate, with several lateral and one terminal *sub-capitate* cymes; flowers woolly. *DC. Prod.* 1. *p.* 629. *Vitis capensis, Thunb. Fl. Cap. p.* 212. *E. & Z.!* 427.

VAR. β. **Dregeana**; leaves *3-lobed or tri-partite*, the segments entire or repand. *Cissus*, 7526 *Drege! C. Dregeana, Bernh.! Flora*, 27. *p.* 297. *C. capensis*, β. *integrifolia, E. & Z.!*

HAB. In mountain ravines. Eastern side of Table Mountain, and in Uitenhage, *E. & Z.! Drege*, &c.

Leaves 2 inches long, and 3 broad, truncate or concave at base, between cordate and reniform, sinuous and obscurely toothed, or in β. deeply lobed and cut nearly to the base. All the young parts and the inflorescence are covered with short, red-brown, woolly hairs.

Group 2. EXSTIPULATÆ. *Leaves* without stipules, rigid, 1–3-5-foliolate; leaflets entire or bluntly toothed or sinuate. (Sp. 4-11.)

4. C. unifoliata (Harv.); branches angular, glabrous, or nearly so; leaves petiolate, *unifoliolate;* leaflet elliptical or ovate-oblong, obtuse or sub-acute, coriaceous, one-nerved, equal sided, quite entire, mucronulate, the younger leaves and twigs rusty-puberulous; stipules none; cymes shorter than the leaves, few-flowered; petals 4, deltoid.

HAB. Collected by *Zeyher*. (Herb. T.C.D., Sond.).

A much-branched, sub-erect or trailing shrub, densely leafy, the young parts rusty. The leaves in all the specimens seen are unifoliolate, with an imperfect joint at the summit of the petioles. In other respects the species is nearly related to *C. Thunbergii*. No habitat is given with the specimens, which are marked "*Z. n. N. n. E. Ampelid.* 2. 104. 3."

5. C. Thunbergii (E. & Z. 430); shrubby, much-branched; branches *reddish-brown*, pubescent; leaves *petiolate, ternate* (or quinate); leaflets thick and rigid, oblong or obovate, very obtuse, tapering at base, entire, glabrous above, *the younger rusty-pubescent beneath*, the lateral pair unequal-sided and shorter than the terminal; cymes terminal or lateral, pedunculate, divaricate, *rusty-pubescent;* petals 5-6, glabrous. *Rhus digitatum, Thunb., fide E. & Z.! Cissus ferruginea, E. Mey.! Drege* 7524, 7525.

HAB. Woods, throughout the eastern districts and Caffraria. (Herb. Hook., T.C.D., Sond.).

A rigid, sub-erect or trailing shrub, having all the younger parts, the under-sides of the leaves and the inflorescence clothed with foxy hairs. Cirrhi simple, opposite the leaves. Cymes generally shorter than the leaves. Drege's 7524 is a young shoot of this species with quinate leaves and hirsute petioles.

6. C. semiglabra (Sond.! in Linn. 23. p. 24); the young shoots and opening leaves fulvo-villous; leaves petiolate, digitate, *mostly 5-foliolate;* leaflets petiolulate, obovate-oblong, cuneate at base, obtuse, the outer ones unequal sided, all coriaceous, very entire and glabrous on both sides; *cymes shorter than the petiole*, rusty-tomentose, several-flowered; tendrils sub-simple.

HAB. Port Natal, *Gueinzius*, 98. (Herb. Sond.).

Too nearly related to *C. Thunbergii*. The upper leaves are sometimes trifoliolate, and *C. Thunbergii* has, occasionally, 5-foliolate leaves.

7. C. pauciflora (Burch. Cat. 3009); shrubby, much-branched; branches *ash-coloured*, pubescent, especially the younger; leaves *on very short petioles* or quite sessile, ternate; leaflets thick and rigid, oblong, obovate or cuneate, obtuse, or 2–3-crenate, *minutely appressed-pubescent*, especially beneath, *afterwards nearly glabrous*, the lateral unequal sided and shorter than the terminal; cymes pedunculate, 3-6-*flowered, glabrescent*; petals 5–6; berries glabrous. *DC. Prod.* 1. *p.* 630. *Drege* 7519, 7520, 7521.

VAR. α. **cirrhiflora**; leaflets very entire. *C. cirrhiflora, E. & Z. !* 432.

VAR. β. **tridentata**; leaflets repandly 2-3-toothed. *C. tridentata, E. & Z. !* 433.

HAB. Woods in Uitenhage and Albany, common. (Herb. T.C.D., Hook., Sond.)

A more slender plant than *C. Thunbergii*, with smaller leaflets and nearly sessile leaves. The two varieties run into one another.

8. C. dimidiata (E. & Z. ! 434); *silky and canescent;* stems shrubby, sub-erect, densely branched; branches canescent, twigs tomentose; *leaves sessile, ternate;* leaflets thick and rigid, silky on both sides, the middle one cuneate, truncate and very obtuse and repand at apex, *the lateral dimidiate, straight along the inner margin, semi-ovate and repando-dentate on the outer;* cymes few-flowered; petals ?

HAB. Rocky ground. Bothasberg, between Grahamstown and the Fisch Rivier, *E. & Z. !* Bushy dells near Grahamstown, *Mr. Bunbury*. (Herb. T.C.D., Sond.).

Very near *C. sericea*, and perhaps a mere variety. The leaves are, however, more uniformly sessile, smaller, and the leaflets of different form, and more constantly, though bluntly toothed.

9. C. sericea (E. & Z. ! 435); *silky and canescent;* stems shrubby, sub-erect, much and densely branched; branches erect, ash-coloured, canescent; leaves *shortly petiolate*, or sub-sessile, ternate; leaflets thick and rigid, silky on both sides, entire or repand, of nearly equal length, the medial elliptic-oblong, tapering at base, *the lateral linear-oblong, unequal-sided;* cymes sub-sessile, 3-6-flowered; petals hairy.

HAB. Among shrubs. Winterhoeksberg, Uit. *E. & Z. !* Herb. Hook., T.C.D., Sond.).

This seems to be an erect shrub. It is clothed in every part with minute, close-pressed, white hairs. In other respects it is nearly intermediate between *C. Thunbergii* and *C. pauciflora*.

10. C. cuneifolia (E. & Z. ! 431); shrubby, and much-branched; branches tuberculate, thinly pubescent; leaves *petiolate*, ternate, leaflets rigid, nearly glabrous or pubescent, *eroso-dentate* or sub-entire, penni-nerved, *the medial bluntly rhomboid or obovate*, or truncate, cuneate at base, the lateral somewhat *ovate*, unequal-sided, very obtuse and *toothed on the outer margin;* cymes short-stalked, 12-20-flowered, dense, with villous branches; petals glabrous. *C. inæquilatera, E. Mey. ! in Herb. Drege.*

HAB. Woods. Eastern Districts, Caffraria and Natal, *E. & Z. ! Drege! Krauss*, 132. (Herb. T.C.D., Hook., Sond.)

Strongly woody, climbing or trailing. Leaves larger and leaflets broader and more toothed than in the allied species, pale underneath, and in the young leaves thinly appressed-puberulent, with strongly marked nerves. Cirrhi simple. Medial leaflets 1–2 inches long, 1 inch broad, often flabelliform. Fruits glabrous. Petals mostly 5. *Drege's* 7522, in Hb. Sond. seems to be a sub-entire leaved variety of this species.

11. C. rhomboidea (E. Mey. !); young shoots and opening leaves fulvo-villous, otherwise nearly glabrous; leaves petiolate, ternate; leaflets petiolulate, netted beneath, the medial rhomboid, cuneate at base, *acute,* paucidentate above the middle, *with sharp, small teeth,* the lateral obliquely ovate, acute and few-toothed; cymes few-flowered, on short hairy peduncles; tendrils simple.

HAB. Near Port Natal, *Drege* ! Howison's Poort, *H. Hutton !* (Herb. T.C.D., Sond.).

Allied to *C. cuneifolia,* but the leaflets are more acute, the medial one acuminate, and the serratures cilii-form, more than tooth-like.

Group 3: DIGITATÆ. *Leaves* digitate, soft and membranous, the leaflets sharply serrate. *Stipules* membranous, deciduous. (Sp. 12-15).

12. C. cirrhosa (Pers. Ench. 1. p. 142); *glabrescent or hispid or pilose,* especially on the young parts; stems weak and straggling; leaves on longish petioles, quinate or ternate; leaflets sub-sessile, cuneate at base, obovate, acute, sharply serrate, thin, penni-nerved, glabrous or hispidulous on the nerves and veins; cymes on long peduncles, divaricately much-branched; style filiform; berries tomentose. *DC. Prod.* 1. *p.* 631. *Cissus quinata, Ait., DC. l. c. Vitis cirrhosa, Thunb. Fl. Cap. p.* 212. *E. & Z. !* 429. *Zey. !* 2030. *Drege !* 7518.

VAR. β, **glabra**; stems, leaves and inflorescence nearly glabrous. *Drege,* 7517,

HAB. In woods and by river banks, among bushes. Uitenhage and Albany, frequent. *E. & Z. !* &c. (Herb. T.C.D., Hook., Sond.)

A fragile climber, varying much in its pubescence, which is sometimes glandular. Leaflets somewhat succulent, usually 5, but sometimes 3 or 7. Var. β. is sometimes quite glabrous; but intermediately pubescent states occur.

13. C. lanigera (Harv. Thes. t. 65); *densely villoso-pubescent;* stems ligneous, angularly compressed, striate; leaves on longish petioles, quinate; leaflets sub-sessile, *obovate or obovato-lanceolate,* acute, sharply and doubly serrate, paler below, closely penni-nerved and *densely tomentose,* cymes pedunculate, diffuse; style filiform; berries tomentose. *Zey. !* 155.

HAB. Macallisberg, *Burke and Zeyher !* Also near Port Natal. *R. W. Plant.* (Herb. T.C.D., Hook., Sond.)

A strong growing, woody climber. Tendrils simple or nearly so. Leaflets 3–4 inches long, 2 inches wide, unequally serrate, with mucronulate teeth. Stipules ovato-lanceolate, oblique. This species is readily known from the preceding by its copious, white-woolly pubescence. It is the *Tambesi* of the Zooloos "who hold it in great esteem," according to Mr. Plant, "as a remedy for tooth-ache. The root is used, rubbed on the gums." *W. Saund. in litt.*

14. C. hypoleuca (Harv.); pubescent; stems weak, striate, minutely pubescent or glandular; leaves on long petioles, quinate, leaflets *petio-*

late, broadly lanceolate, acute or acuminate, serrate, green and *glabrous above, densely tomentose and pale beneath,* penni-nerved ; cymes pedunculate, diffuse, with patent tomentose branches , style filiform. *C. quinata, b., Drege !*

HAB. Omsamwubo, *Drege !* Port Natal, *Gueinzius !* (Herb. T.C.D , Hook, Sond.)

Allied to *C. cirrhosa*, but with more distinctly petiolulate, lanceolate leaflets, less strongly toothed (the teeth shallow and broad, but mucronate) and densely albo-tomentose underneath.

15 ? C. Sandersoni (Harv.) ; the weak succulent stems and the long petioles *tomentose ;* leaves 5–7-foliolate ; leaflets sub-sessile, *broadly obovato-trapeziform,* acute, *doubly inciso-serrate, scaberulous on both sides,* thin, penni-nerved, fringed with minute, gland-tipped hairs ; cymes ?

HAB. Transvaal, Natal, *Mr. Sanderson !* (Herb. Hook.)

Very unlike the other Cape species of this section, and remarkable for its large, membranous, strongly serrate and minutely glandular leaflets. Perfect specimens are, however, required to establish the species fully.

Group 5. PEDATÆ. *Leaves* pedate ; the lateral segments bi-tri-foliolate. (Sp. 16.)

16. C. bigemina (Harv.) ; stems slender, herbaceous, striate, glabrous ; leaves on long petioles, pedate, 5-foliolate, the petiole trifid, the medial branch unifoliolate, the lateral forked and bearing twin leaflets ; leaflets all petiolate, ovate or lanceolate, acuminate, coarsely toothed, thin and sub-glabrous, penni-nerved ; cymes pedunculate, few-flowered, not much branched, nearly glabrous ; style very short.

HAB. Port Natal, *Gueinzius.* (Herb. Hook.).

A slender, trailing and climbing plant, differing from all the other known Cape species by the composition of its leaves, which, though of 5 leaflets, are formed on a ternate plan ; the two lateral petiolules forking and bearing two leaflets, smaller and more ovate than the terminal or central one. The nerves and veins are hispidulous, and a few bristles are sprinkled over the leaflets which are otherwise glabrous.

Group 6. PINNATÆ. *Leaves* pinnate or bi-pinnate. (Sp. 17.).

17. C. orientalis (Lam. ill. t. 84. f. 2.) ; glabrous ; stems half herbaceous, weak, striate ; leaves bi-tri-pinnate, leaflets stalked, ovate, acute, sharply serrate ; stipules ovate, reflexed ; cymes on long peduncles, di-tri-chotomous and much-branched, corymbose ; pedicels much longer than the flowers ; styles filiform. *C. glabra, E. Mey.! in Hb. Drege.*

HAB. Between Omtata and Omsamwubo, 1000-2000f. *Drege !* Macalisberg, *Burke and Zeyher !* Feb. (Herb. T.C.D., Hook., Sond.).

A straggling climber. Leaves very compound, twice or thrice pinnate, the pinnules trifoliolate. Leaflets 1 inch long, sharply and deeply serrate ; the teeth mucronate. Cymes very lax, many-flowered, widely spreading and panicled. I venture to refer the S. African specimens to *C. orientalis*, of which, however, I have only seen imperfect specimens. If not absolutely indentical, ours is a very closely allied plant.

Drege's *Cissi*, No. 9514 and 7527, having (in Hb. Sond. !) neither flowers nor fruit, cannot be satisfactorily determined.

ORDER XXXI. GERANIACEÆ, DC.

(By W. H. HARVEY.)

(Gerania, Juss. Gen. 268. Geraniaceæ, DC. Prod. 1. p. 637. Endl. Gen. No. ccliv. Lindl. Veg. Kingd. No. clxxxvii.)

Flowers perfect, regular or irregular. *Sepals* 5, persistent, unequal, imbricate in æstivation. *Petals* 5 (rarely 4-2-1 or none), clawed, with convolute æstivation, caducous. *Stamens* monadelphous or polyadelphous, hypogynous, twice or thrice as many as the petals, some frequently abortive. *Ovary* of 5, uniovulate carpels, cohering round an awl-shaped or beak-like torus, to which the styles adhere; *stigmas* 5, filiform. *Fruit* of 5, membranous, one-seeded carpels, separating at maturity from the persistent, enlarged torus; seed exalbuminous; embryo curved and folded, with leafy cotyledons.

Herbaceous plants or shrubs; sometimes stemless, or with tuberi-form, underground stems. Branches tumid and easily broken at the joints. Leaves opposite or alternate, simple or variously cut, rarely compound, usually stipulate. Stipules free or adnate with the petiole. Peduncles opposite the leaves, or in the axils of dichotomous branches, rarely single-flowered, mostly either two-flowered or umbelled; the umbel involucrate. Petals showy, white, yellow, blue, red, purple or spotted.

A considerable Order, widely distributed, chiefly in the temperate Zones, perhaps three-fourths of the species being South African. The genus *Pelargonium* is one of those most charateristic of the Cape Flora, and a universal favourite with horticulturalists, who have wonderfully changed, by high culture and hybridization, the aspect of these flowers. None are particularly useful to mankind. They have astringent properties, especially in the roots; many have remarkably aromatic leaves, and some yield fragrant resins. The stems of the *Sarcocaulons* dry up into masses of resin. Some of the tuberous-rooted kinds may be eaten. No poisonous plants occur in the Order.

TABLE OF THE SOUTH AFRICAN GENERA.

I. **Monsonia.** *Stamens* 15, in 5 parcels of three each.
II. **Sarcocaulon.** *Stamens* 15, monadelphous.
III. **Geranium.** *Stamens* 10, monadelphous. *Flowers* regular.
IV. **Erodium.** *Stamens* 5. *Flowers* regular. *Calyx* equal at base.
V. **Pelargonium.** *Stamens* 7, or fewer. *Flowers* irregular; petals $\frac{2}{3}$, or the lower ones absent. *Calyx* prolonged at base into a tube.

I. MONSONIA, Lin. f.

Sepals equal at base, mucronate. *Petals* spreading equally, longer than the calyx. *Stamens* 15, connate at base, and spreading in 5 sets; each set formed of 3 stamens whose filaments cohere for half their length. *DC. Prod.* 1. *p.* 638. *Endl. Gen.* 6047.

Perennial or annual, herbaceous or suffruticose plants, with slender stems. Leaves alternate, simple, sub-entire or toothed, or deeply lobed and cut. Peduncles bracteate above the middle, one, two, or umbellately several-flowered. Named in honour of Lady Ann Monson, a lady of considerable botanical taste and acquirements.

Sect. 1. HOLOPETALUM. *Petals* obovate, entire or emarginate. *Leaves* ovate, oblong, or lanceolate, crenate or toothed. (Sp. 1–5.)

1. M. ovata (Cav. Diss. t. 113. f. 1.); *suffruticose*, diffuse; leaves on longish petioles, *ovate or cordate*, obtuse or acute, crenate; stipules pungent; peduncles filiform, bi-bracteate in the middle, one-flowered; petals entire. *DC. Prod.* 1. *p.* 638. *E. & Z. !* 439. *M. emarginata*, *L'Her. Ger. t.* 41. *E. & Z. !* 440. *M. præmorsa*, *E. Mey. !*

VAR. **biflora**; peduncles 2-flowered, 4-bracted. *E. & Z. !* 441.

HAB. Hill sides, among grass. Uitenhage and Albany, and in Caffraria, common. (Herb. T.C.D., Hook., Sond.)

Root-stock perennial. Stems many, procumbent, hairy and branched. Petioles longer than the lamina. The leaves vary from broadly cordate to ovate-oblong. Flowers an inch broad, streaked. Sepals aristate. *M. præmorsa*, E. Mey. has emarginate leaves, but is not otherwise different.

2. M. biflora (DC. Prod. 1. p. 638); *annual*, erect, much-branched; leaves on short petioles, *linear-oblong*, obtuse, cuneate at base, crenate; stipules pungent; peduncles short, 2–4 bracteate, generally 2-flowered; petals entire. *DC. Prod* 1. *p.* 638. *M. angustifolia*, *E. Mey. ! Hochst. Pl. Abyss !*

HAB. Dry hills of the Eastern and N. Eastern districts; Caffraria and Port Natal. (Herb. T.C.D., Hook., Sond.)

A much-branched, hairy annual, with narrower leaves, shorter petioles and peduncles, and smaller flowers than *M. ovata*. It is found also in Abyssinia.

3. M. Burkeana (Planch ! in Hb. Hook); *perennial*, suffruticose at base, diffuse; leaves on short petioles, *elliptic-oblong*, obtuse, rounded at base, crenate; stipules pungent; *peduncles short, pluri-bracteate*, 4–3-*flowered;* petals entire.

HAB. Macallisberg and Aapjes River, *Burke & Zeyher !* (Herb. T.C.D., Hook., Sond.)

Very like *M. biflora*, but evidently with a perennial root; shorter and more oval leaves, and more numerous flowers. The pubescence varies greatly.

4. M. attenuata (Harv.); *perennial;* stems sub-simple, *erect;* leaves on short petioles, *linear-lanceolate, attenuate, sharply serrate;* stipules setaceous, pungent; peduncles short, *one-flowered;* petals entire.

HAB. Mohlamba Range, Natal, 6–5000f. *Dr. Sutherland.* (Herb. Hook.)

Known by its very narrow and *sharply* serrate leaves. Leaves 2 inches long, not 2 lines wide. Flowers veiny, an inch across. Stems 6–12 inches long, whitish, villous.

5. M. umbellata (Harv.); *annual*, much-branched, diffuse; leaves on longish petioles, *ovate or cordate*, denticulate, plaited; *stipules membranous, lanceolate;* peduncles elongate, umbellately 5-6-flowered, pluri-bracteate; sepals with a reflexed mucro; petals narrow-cuneate, not much longer than the calyx, emarginate.

HAB. Bitterfontein, *Burke & Zeyher !* (Herb. Hook., Sond., T.C.D.).

A small patently hairy annual, looking like an *Erodium*. Leaves ½–¾ inch long, on petioles somewhat longer than the lamina. Flowers ½ inch across, much smaller than in any other species. This resembles small specimens of *M. Senegalensis*, but differs by its many-flowered peduncles, &c.

Sect. 2. ODONTOPETALUM. *Petals* coarsely toothed at the summit. *Leaves* lobed or multifid. (Sp. 6–8.)

6. M. lobata (Willd. 3. p. 718); suffruticose, sub-simple; leaves on long petioles, cordate at base, sub-rotund or triangular, *5-7-lobed, the lobes shallow, rounded,* crenate or toothed; peduncles elongate, one-flowered; petals coarsely toothed. *DC. Prod.* 1. *p.* 638. *Bot. Mag. t.* 385. *Sw. Ger. t.* 273. *E. & Z.! No.* 442. *M. filia, L. f. Cav. Diss. t.* 74. *f.* 2. *Ger. anemoides, Thunb. Fl. Cap. p.* 510.

HAB. Near the Berg River, and 24-Rivers, *Thunberg!* Tulbagh, *E. & Z.!* Caledon, *Dr. Thom!* (Herb. Hook., Benth., Sond.).

Readily known from *M. speciosa* by its broad, slightly cut leaves. The pubescence is very variable.

7. **M. speciosa** (Linn. f. Suppl. p. 342); suffruticose, sub-simple, glabrescent or villous; leaves on very long petioles, *5-parted, the segments bi-pinnatifid and cut into many linear lobes;* peduncles elongate, bracteate in the middle, one-flowered; petals coarsely toothed. *DC. Prod.* 1. *p.* 638. *Curt. Bot. Mag. t.* 73. *Cav. Dis. t.* 74. *f.* 1. *Sw. Ger. t.* 77. *E. & Z.!* 444, *and M. pilosa,* 433! *(non Willd.)*

HAB. In fields. Hott. Holland, the Paarl, Groenekloof, &c. *E. & Z.*, and others. (Herb. T.C.D., Hook., Sond.).

Stems tufted, slender. Leaves crowded, the petiole 2–3 times as long as the lamina. Fl. 2 inches broad, greenish and striped on the outside, white, with a red basal spot within. The pubescence is either scanty or copious, varying extremely in different specimens.

8. **M. pilosa** (Willd. En. 717); suffruticose, sub-simple, hispid and glandular; leaves on long petioles, *deeply 5-lobed, the lobes inciso-pinnatifid and toothed;* peduncles elongate, bracteate, viscoso-pubescent, one-flowered; petals coarsely toothed. *DC. Prod.* 1. *p.* 638. *Sw. Ger. t.* 199. *M. filia, Andr. Rep. t.* 276, *non Lin. f.*

HAB. Cape of Good Hope, *Thunberg! Forsyth!* (Herb. Benth., Holm.).

This is only to be known from *M. speciosa* by its broader and less divided leaves. The pubescence varies in both species.

II. SARCOCAULON, DC.

Sepals equal at base, mucronate. *Petals* spreading equally. *Stamens* 15, monadelphous at the base; filaments subulate, not cohering in parcels. *DC. Prod.* 1. *p.* 638.

Divaricately branched, fleshy or succulent, rigid shrubs, armed with spines formed out of persistent and hardened petioles. In a young state, some of these spinous petioles bear leaves, but the ordinary leaves are tufted or solitary, in the axils of the thorns. Peduncles one-flowered, bibracteate at base. The name is compounded of σαρκος, *flesh,* and καυλον, *a stem;* from the fleshy stems.

1. **S. Burmanni** (DC. Prod. 1. p. 638); leaves obovato-cuneate, *inciso-crenate,* glabrous or downy; petals twice as long as the mucronate sepals. *Burm. Afr. t.* 31. *Ger. spinosum, Cav. Diss. t.* 75. *f.* 2.

HAB. On dry, clayey hills or plains, Oliphant's River, *E. & Z.!* Bitterfontein, Gamke Rivier, and near Beaufort, *Zeyher!* Jackal's Fontein and Silverfontein, *Drege!* (Herb. T.C.D., Hook., Sond.).

Leaves ½–¾ inch long, fleshy, on short petioles. Flowers 1½–2 inches broad. Stamens monadelphous, 5 long, and 10 shorter.

2. **S. Patersoni** (DC. l. c.); leaves cuneate or obcordate, obtuse or mucronulate, *entire,* glabrous; petals not twice as long as *the obtuse, mu-*

cronulate sepals. Paters. Itin. t. 14. *S. Patersoni and S. L'Heretieri. E. & Z.!* 436, 437. *Mons. macilenta, E. Mey. and M. obcordata, ex parte, Drege,* 7515.

HAB. In very dry places. Eastern and N. Eastern districts, Hermanskraal, and by the Fish River; also near Beaufort, *E. & Z.! Mrs. F. W. Barber!* Gamke River, *Burke and Zeyher.* (Herb. T.C.D., Hook., Sond.)

Known from *S. Burmanni* by its entire leaves, and from *S. L'Heretieri* by the very minute point of the elliptical sepals. The flowers are smaller than in either.

3. S. L'Heretieri (DC. l. c.); leaves obovate or obcordate, acute or obtuse, entire, glabrous; *petals not much exceeding the cuspidate, attenuate sepals. M. spinosa, L'Her. t.* 42. *M. obcordata, ex parte, E. Mey.!*

HAB. Modderfontein, *Drege!* (Herb. Hook., T.C.D.).

Best known from *S. Patersoni*, which is often confounded with it, by the long points of the sepals. The petals are larger or smaller in different specimens.

III. GERANIUM, L.

Sepals equal at base. *Petals* spreading equally. *Stamens* 10, all perfect, the alternate longer. *Glands* at the base of the longer stamens. *DC. Prod.* 1. *p.* 639. *Endl. Gen.* 6046.

Herbaceous plants, rarely suffruticose; a few shrubby or arborescent. Leaves on long petioles, opposite or alternate, palmately lobed. Peduncles 1–2-flowered, opposite the leaves, or in the forking of the branches. Name *γεpανιον*, of the Greeks, from *γεpανος, a crane;* because the capsule with its beak resembles the head of a crane. The English name is *Crane's-bill.*

1. G. incanum (Linn. Sp. p. 957); suffruticose at base, diffuse, slender, appressedly silky; leaves *digitately* 5-7-*parted, the segments cut into many linear lobes,* pubescent on the upper, canescent on the lower surface; peduncles 2-flowered; sepals canescent; petals entire. *DC. Prod.* 1. *p.* 640. *Burm. Ger. t.* 1. *Cav. Diss. t.* 82. *f.* 1. *E. & Z.!* 445. *Thunb. Cap. p.* 511. *Zey.* 160. *Drege,* 7510.

HAB. Cape Flats; hills at the Eastern side of Table Mountain, and elsewhere in the Western districts, frequent. A variety with red flowers, at Van Staadensberg, Uit., *E & Z.!* Mohlamba Range, Natal, *Dr. Sutherland!* (Herb. T.C.D., Hook., Sond.).

Bases of the decumbent branches woody. All parts are covered with short, close-pressed, silky hairs, but the upper sides of the leaves are green. The sepals are either aristate, mucronate, or simply acute. The flowers, $\frac{1}{2}$ inch across, are white or a purplish-rosy. This is the "*Berg-thee*" of the colonists.

2. G. sericeum (Harv.); suffruticose, slender, decumbent; leaves *digitately* 5-*parted* or very deeply divided, the segments cuneate, incised, *both surfaces white, with appressed silky hairs;* the long, 2-flowered peduncle tomentose and canescent; petals emarginate. *Ger. incanum, E. Mey! non L.*

HAB. Wildschutsberg and Compasberg, 5–6000 f., *Drege! Zeyher!* (Herb. T.C.D., Hook., Sond.).

This resembles *G. canescens* in the shape of its leaves, but is not glandular, and every part of the plant is silvery-white, with copious, appressed silky hairs. Flowers red.

3. G. canescens (L'Her. t. 38); suffruticose at base, slender, diffuse; leaves *deeply* 3–5-*lobed,* the lobes cuneate and cut, *appressedly pubescent* *above, canescent beneath;* the long, 2-flowered peduncle, the calyces and carpels clothed with gland-tipped hairs; petals emarginate. *DC. Prod. l. c. p.* 640. *G. glandulosum, Lehm.! in E. & Z.!* 447. *Drege,* 7511.

HAB. Near Somerset, Hott. Holland, *E. & Z.!* At the Paarl, *Rev. Mr. Elliott.* (Herb. T.C.D., Sond.).

Leaves much less deeply divided and with broader lobes than in *G. incanum.* The stems, peduncles and petioles are thinly tomentose; the glandular hairs look like a small *Mucor.* Drege's 7511 is thrice the usual size, but not otherwise different.

4. G. ornithopodum (E. & Z.! 449); suffruticose at base, sub-erect; leaves deeply 5-lobed, the lobes cuneate and cut, *pubescent on the upper, villous and pale* on the lower side; branches, petioles and peduncles densely villous, with long, white, simple (or glandular !) hairs. *Drege,* 7513 *! also G. contortum, E. & Z.!* 450, *and G. flexuosum, E. Mey.*

HAB. Ceded territory, *E. & Z.!* Mountains near Grahamstown, *Zeyher!* (Herb. T.C.D., Hook., Sond.).

This is chiefly known from *G. canescens* by the villous or softly shaggy, patent, not appressed and silky pubescence. The glandular hairs are, I fear, no constant mark of either species. A specimen from Port Natal is almost intermediate between the two. *G. contortum,* E. & Z.! is scarcely worth distinguishing.

5. G. caffrum (E. & Z.! 448); suffruticose at base, diffuse, slender; branches angular; leaves *green on both sides,* digitately 3-5-parted, the segments lanceolate, inciso-pinnatifid, the upper surface sparsely and minutely appresso-puberulent, *lower glaucous* and glabrous, except along the strigilose nerves; the 2-flowered peduncles, calyces and carpels mostly glandular and pubescent; petals emarginate. *Drege,* 7512 ! *Zeyher,* 2038 !

HAB. Kat River, *E & Z.!* Zwartkops, *Zey.!* Zuureberg, Klipplaat River, and near Rhinosterkopf, Nieuweveld, *Drege!* Albany, *Mrs. F. W. Barber!* (Herb. T.C.D., Hook., Sond.).

The least hairy of the Cape species, though far from glabrous. The hairs are very minute, rigid and close-pressed. Flowers white. The glandular hairs vary in copiousness, in different specimens.

IV. ERODIUM, L'Her.

Sepals equal at base. *Petals* spreading equally. *Stamens* 5, perfect, bearing anthers; 5 sterile, subulate or obsolete. *Glands* at the base of the sterile stamens. *Endl. Gen. No.* 6045. *DC. Prod.* 1. *p.* 644.

Herbaceous plants, rarely suffruticose, with pinnate-parted, lobed or entire leaves and membranous stipules. Peduncles mostly umbellately many-flowered, rarely one-flowered. Name from *ερωδιος, a heron;* because of the long-beaked fruit. English name *Heron's bill.*

* *Perennial.* (Sp. 1–2.)

1. E. incarnatum (L'Her. Ger. t. 5); suffruticose at base, the stems angular, glabrous; leaves on very long petioles, scaberulous, the lowest *cordate, crenate, the rest digitately-tripartite, with deeply cut or 2-3-parted segments;* stipules subulate, ciliate; peduncles several-flowered; petals twice as long as the lanceolate, setose sepals. *DC. Prod.* 1. *p.* 648. *Cav. Diss. t.* 97. *f.* 3. *Sw. Ger. t.* 94. *Bot. Mag. t.* 261. *E. & Z.! No.* 454. *Zey.!* 2040.

HAB. Hott. Holl., near Palmiet River, *E. & Z.!* Zwarteberg and Caledon'sbad, *Zeyher!* Cult. in England, 1787. (Herb. T.C.D., Hook., Sond.).

A beautiful little plant, with multifid, much-lobed foliage, and bright crimson (or flesh-coloured) flowers. The petals are each marked with a dark and light band. The foliage and general aspect are those of *Pelarg. patulum*, β.; but the flowers are very different. Stipules and bracts rigidly ciliate, subulate.

2. **E. arduinum** (Willd. 3. p. 637); "stemless; leaves *cordate, 5-lobed*, crenate, obtuse; peduncles many-flowered." *DC. Prod.* 1. *p.* 648.

HAB. South Africa, *Burmann.*

Of this I know nothing. Perhaps a var. of the preceding.

** *Annual or biennial.*

3. **E. maritimum** (L'Her.): annual or biennial, diffuse, pubescent; leaves on long petioles, *cordate*, obtuse, pubescent, *crenate;* stipules subrotund; *peduncles* 1-3-*flowered, shorter than the leaves;* sepals oval, mucronate, *longer than the petals. DC. Prod.* 1. *p.* 648. *Thunb. Cap. p.* 511. *Cav. Diss. t.* 88. *f.* 1. *E. Bot. t.* 646.

HAB. Paardeberg, *Thunberg.*

Of this I have seen no Cape specimens. Leaves ½ inch long. The whole plant is much smaller than the following, which resembles it. E. & Z.'s "*E. maritimum,*" No. 452, belongs, according to Ecklon's original specimens in Herb. Sond., to *Pelarg. chamædryfolium.*

4. **E. malachoides** (Willd. 3. p. 639); annual or biennial, diffuse, hispidulous; lower leaves on long petioles, *cordate-ovate*, obtuse, pubescent, *unequally cut or lobed*, upper deeply 3-parted and jagged; stipules ovate, obtuse; *peduncles elongate, several flowered;* sepals oval, aristate, *equalling the petals. DC. Prod.* 1. *p.* 648. *Cav. Diss. t.* 91. *f.* 1.

HAB. Sands near Greenpoint, *W.H.H.* (Herb. T.C.D.).

A diffuse or prostrate annual, varying much in size and in the cutting of the leaves. Branches often 1–2 feet long, hispid. Beak of the fruit 1½ inches long. A littoral plant in the South of Europe, the Canary Islands, North Africa, and even in Peru. The Cape specimens are very similar to the European.

5. **E. moschatum** (Willd. Sp. 3. p. 631); biennial, procumbent; *leaves pinnati-partite*, segments petiolulate, ovate, obtuse, unequally toothed and cut, hispidulous; stipules broadly ovate, filmy; peduncles elongate, many-flowered, glandularly pubescent; sepals mucronate, shorter than the petals. *DC. Prod.* 1. *p.* 647. *E. & Z.!* 453. *Jacq. Vind. t.* 55. *Cav. Diss. t.* 94. *f.* 1. *E. Bot. t.* 902.

HAB. Common near cultivation. Introd. from Europe (Herb. T.C.D.).

A strong-growing plant, sometimes cut for fodder, and sometimes grown in gardens for its musky fragrance. The leaves are fern-like, 6–12 inches long, variably pubescent. It is common to Europe, N. Africa, and S. America.

V. PELARGONIUM. L'Her.

Calyx 5-parted, the uppermost segment produced at base into a slender, nectariferous tube, which is decurrent along the pedicel, and adnate to it. *Petals* 5, rarely but 4, or 2, more or less unequal. *Filaments* 10, unequal, monadelphous; 2–7 fertile, the rest without anthers. *DC. Prod.* 1. *p.* 649. *Endl. Gen. No.* 6048.

A large genus, very variable in habit, and almost exclusively South African. Flowers in umbels, rarely sub-solitary. The generic name is derived from πελαργος, *a stork;* in allusion to the long beak of the fruit. English, "*Stork's-bill*."

For convenience of study the species are grouped, according to what appear to be their natural affinities, into fifteen sections, depending on several characters, either of habit or floral structure. I have not found it easy to affix exact definitions to them, and must, therefore, recommend the following abstract to the careful study of the student. After he has become acquainted with one or two typical species of each section, he will find but little difficulty in associating others with them.

SYNOPSIS OF THE SECTIONS.

Sec. 1. HOAREA. *Stemless.* with *tuberous* roots. *Petals* 5 or 4. (Sp. 1–42.)

* *Leaves either all entire ; or entire and laciniate together*. (Sp. 1–17.)
** *Leaves three-lobed or tripartite.* (Sp. 18–24.)
*** *Leaves deeply pinnatifid or pinnati-partite.* (Sp. 25–42.)

Sec. 2. SEYMOURIA. *Stemless,* with *tuberous* roots. *Petals* only two. (Sp. 43-46.)

Sec. 3. POLYACTIUM. *Caulescent,* with *tuberous* roots. *Leaves* lobed, or pinnately-decompound. *Umbels* many-flowered. *Petals* sub-equal, obovate, *entire or fimbriato-lacerate.* (Sp. 47-66.)

* *Leaves sub-radical. Petals entire. Stipules ovate or cordate.* (Sp. 47–53.)
** *Leaves scattered on a simple or branched stem. Petals entire.* (Sp. 54–60.)
*** *Leaves sub-radical. Petals entire. Stipules subulate, rigid.* (Sp. 61–63.)
**** *Petals fimbriato-lacerate.* (Sp. 64–66.)

Sec. 4. OTIDIA. *Stem* succulent and knobby. *Leaves* fleshy, pinnately or bi-pinnately compound. *Petals* sub-equal, *the upper eared at base. Stamens* 5. (Sp. 67-72.)

Sec. 5. LIGULARIA. *Stem* either succulent or slender and branching. *Leaves* rarely entire; mostly much cut or pinnately decompound. *Petals* sub-unequal, spathulate, *the uppermost tapering at base. Stamens* 7. (Sp. 73-92.)

* *Stem short, undivided, armed with spine-like* (free) *stipules.* (Sp. 73–75.)
** *Stem armed with spine-like, persistent petioles. Stipules adnate.* (76.)
*** *Stem simple or branched. Stipules conspicuous, adnate to the petiole.* (77–81.)
**** *Stem slender, much branched. Stipules minute or obsolete, adnate.* (Sp. 82–85.)
***** *Stem slender, branching. Stipules free, ovate or subulate.* (Sp. 86–89.)

Sec. 6. JENKINSONIA. *Shrubby* or succulent. *Leaves* palmately nerved or lobed. Two upper *petals* on long claws, very much larger than the lower. *Stamens* 7. (Sp. 93-95.)

Sec. 7. MYRRHIDIUM. *Slender* suffruticose or annual. *Leaves* pinnatifid or pinnatisect. *Petals* 4 (rarely 5); two upper largest. *Calyx-segments* membranous, *strongly ribbed and mucronate or taper-pointed. Stamens* 5, rarely 7. (Sp. 96-101.)

Sec. 8. PERISTERA. *Herbaceous,* diffuse, annual or perennial. *Leaves* lobed or pinnatifid. *Flowers* minute. *Petals scarcely longer than the calyx.* [Habit of *Geranium* or *Erodium.* (Sp. 102-108.)

Sec. 9. CAMPYLIA. *Stem* short, sub-simple. *Leaves* on long petioles, undivided, entire or toothed. *Stipules* membranous. *Flowers* on long pedicels. Two upper *petals* broadly obovate, 3 lower narrow. *Fertile-stamens* 5; two of the *sterile* ones recurved. (Sp. 109-116.)

Sec. 10. DIBRACHYA. Much branched, with weak, jointed stems. *Leaves* peltate or cordate-lobed, *fleshy*. *Petals* obovate. *Stamens* 7, the two upper very short. (*Ivy-leaved.*) (Sp. 117-118.)

Sec. 11. EUMORPHA. Slender, suffruticose, or herbaceous. *Leaves* on long petioles, *palmately* 5-7-nerved, reniform, lobed or palmatifid. *Petals* unequal, the 2 upper broad. *Stamens* 7. (Sp. 119-125.)

* *Glabrous and glaucous. Flowers pedicellate.* (Sp. 119-122.)
** *Pubescent. Flowers sub-sessile.* (Sp. 123-125.)

Sec. 12. GLAUCOPHYLLUM. *Shrubby. Leaves* carnose, simple or ternately compound, *the lamina articulated to the petiole. Stamens* 7. (Sp. 126-130.)

Sec. 13. CICONIUM. *Shrubby*, with carnose branches. *Leaves* either obovate or cordate-reniform, palmately many-nerved, undivided. *Petals* all of one colour, *scarlet, pink or white. Stamens* 7, 2 upper very short. (Sp. 131-134.)

Sec. 14. CORTUSINA. *Caudex* short, thick and fleshy; branches (if present) slender and half herbaceous. *Leaves reniform or cordate, lobulate, on long petioles. Petals* sub-equal, two upper broadest. *Stamens* 6-7. (Sp. 135-141.)

Sec 15. PELARGIUM. Much branched shrubs or suffrutices, not fleshy. *Leaves* entire or lobed (never pinnati-partite). *Stipules* free. Inflorescence frequently panicled, the partial peduncles umbelled. Two upper petals longer and broader than the lower. *Stamens* 7. (Sp. 142-163.)

* *Leaves oval, ovate, cordate, or multangular.* (Sp. 142-145.)
** *Leaves bluntly 3-5-lobed, pubescent; not scabrous or viscid.* (Sp. 146-148.)
*** *Leaves scabrous, sharply three-lobed and serrated.* (Sp. 149-150).
**** *Leaves oblong, cordate at base, 5-7-lobed or pinnatifid, visc.-pubescent.* (Sp. 151-155.)
***** *Leaves deeply palmatifid or palmati-partite, fœtid.* (Sp. 156-162.)

Sec. 1. **HOAREA.** *Stemless* perennials, with tuberous or turnip-shaped root-stocks. *Leaves* annual, fascicled, all radical. *Scapes* simple or branched, leafless. *Flowers* umbellate. *Petals* 5 or rarely 4, unequal. *Fertile* stamens usually 5, long or short. *Hoarea and Dimacria, Sw. DC. l. c.* (Sp. 1-42.)

* *Leaves either all entire, sub-entire or crenate; or entire and laciniate on the same root.* (Sp. 1-17.)

1. **P. longifolium** (Jacq. Ic. Rar. t. 518); leaves *simple* (*or pinnatifido-laciniate*), ovate, oblong, or lanceolate, tapering to the base and apex,

acute or sub-obtuse, with immersed veins, glabrescent, softly pubescent, or puberulous, often ciliated; stipules linear-subulate, acuminate, pubescent; scapes branching, pubescent or villous, with villous bracts and pluri-flowered umbels; calyx *glanduloso-pubescent and thinly villous, the segments lanceolate, acute, gland-tipped, with very narrow, membranous margins. DC. Prod.* 1. *p.* 649. *Cav. Diss. t.* 102. *f.* 1. *P. lanceæfolium, Sw. Ger. t.* 387, *and P. auriculatum, Sw. t.* 395.

VAR. *α.* **virgineum**; leaves mostly entire; flowers flesh-coloured, or white and veiny, upper petals obovato-spathulate. *P. virgineum, Pers. DC. l. c. P. undulatum, Andr. Rep. t.* 317. *Hoareæ, Sp. E. & Z.! No.* 462, 463, 464, 469, 470. *Drege, No.* 9519, 7494.

VAR. *β.* **longiflorum**; leaves mostly entire; flowers primrose-colour, with dark lines or spots; petals very long and narrow-linear. *P. longiflorum, Jacq. Ic. t.* 521. *P. depressum, Jacq. Ic. t.* 520. *Sw. Ger. t.* 290. *Hoareæ, E. & Z.! No.* 466, 472. *P. otites, E. Mey.!*

VAR. *γ.* **laciniatum**; leaves entire and *inciso-pinnatifid;* flowers flesh-coloured or white, with dark lines or spots; upper petals obovato-spathulate. *P. laciniatum, Pers. DC. l. c. Andr. Rep. t.* 131, *t.* 204. *P. auriculatum, Willd., DC. l. c. p.* 651. *Sw. Ger. t.* 395. *P. ciliatum, Jacq. Ic. t.* 519. *Hoareæ, E. & Z.! No.* 473, 474, 478. *P. purpurascens, Pers.? DC. l. c.*

VAR.? *δ.* **ciliatum**; leaves ovate or ovato-lanceolate, acute or obtuse, not tapering at base, rigidly ciliolate, glabrous or pubescent. *P. ciliatum, L'Her. Ger. t.* 7. *DC. l. c. p.* 650. *P. auriculatum, E. Mey.! Herb. Drege. H. erythrophylla, E. & Z.!* 457. *Hoareæ, Zey.!* 175, 2043.

HAB. Dry clayey or sandy ground, chiefly in the western districts, common. (Herb. T.C.D., Hook., Sond.)

Very variable in the shape of the leaves, and the amount of the pubescence, which is always soft and silky; sometimes the leaves are quite glabrous. Var. δ. may perhaps be a species. The other synonyms quoted appear to me undistinguishable.

2. P. Meyeri (Harv.); leaves on rigidly ciliate petioles, ovate, obtuse, coriaceous, glabrous or nearly so, entire; stipules adnate, lin.-subulate, ciliate; scapes branched, pluri-flowered, with subulate, setulose bracts; calyx sub-sessile, the glandularly pubescent tube not twice as long as the lanceolate, *strigulose* segments; petals obovate. *P. ficaria, E. Mey.! in Hb. Drege, non Willd. P. ovalifolium, E. & Z.!* 461, *non Sw., also Hoareæ, E. & Z.!* 459, 460.

HAB. Between Eikenboom and Riebekskasteel, Cape, *Drege!* Tigerberg, Witsenberg and near Constantia, *E. & Z.!* (Herb. T.C.D., Hook., Sond.).

Known from all forms of *P. longifolium* by the stiff, appressed, short bristles on the calyx-lobes. The flowers are pale purple. Our specimens are in very bad order, and the description may need to be corrected. Can this be the *P. ciliatum, Cav. Diss. t.* 118. *f.* 2, (*P. parnassioides, L'Her.*) which is said to be different from the plant of L'Heretier.

3. P. angustifolium (Thunb.? Cap. p. 514, non L'Her.); leaves on long hairy petioles, lanceolate, acute at each end, *glabrous, except the ciliate margins and mid-rib;* stipules adnate, membranous; scapes *hispid,* umbellately branched, the partial umbels pluri-flowered, with subulate, shaggy bracts; calyx nearly sessile, *glandular and patently hairy,* the tube longer than the lanceolate, pale-edged segments. *DC.? Prod.* 1. *p.* 680. *E. & Z.! No.* 465.

HAB. Sandy fields at Zwartland, *Thunberg*. Steenberg, Cape, *E. & Z.!* (Herb. Sond.).

Described from a single leaf, and a scape in Herb. Ecklon. It seems to agree, except in the form of the leaf, which Thunberg calls "elliptical," with Thunberg's description, and differs from *P. longifolium* chiefly in pubescence. The hairs, wherever they occur, are rigid, very long and spreading. Zeyher's 172, from the Berg River (Herb. Hook., Sond.), though not quite the same, is nearer to this than to anything else.

4. **P. ensatum** (Thunb.? Cap. p. 515); leaves on long, hairy petioles, obovato-spathulate, sub-acute, much attenuated at base, *membranous, hispid on both sides;* stipules adnate, subulate, hispid; scapes pubescent, branched, the partial umbels many-flowered, with subulate, shaggy bracts; calyx sub-sessile, the tube pubescent, *thrice as long as the linear, obtuse, albo-marginate segments. DC. Prod.* 1. *p.* 680. *E. & Z.! No.* 467. *Drege,* 7495.

HAB. Mountain sides in the Langekloof, *E & Z.!* Piquetberg, *Drege!* (Herb. Sond.)

5. **P. crinitum** (Harv.); leaves on long, setose petioles, oval-oblong, sub-entire, *densely and rigidly hairy on both surfaces;* stipules subulate; scapes elongate, branched, patently hispid, the umbels few (4-6) flowered, with subulate, pilose bracts; calyx tube hispidulous, 3-4 times as long as the lanceolate, acute, membrane-edged, hispid sepals; stamens 5, elongate, declined; petals broadly ovate. *Pelarg. Drege,* 1290.

HAB. Dutoitskloof, 1-2000 f., *Drege!* (Herb. Hook., Benth., Sond.).

This has almost exactly the inflorescence of the following species, but the leaves are excessively hirsute on both surfaces, more so than in any other, except perhaps some forms of *P. hirsutum*, a species with very dissimilar flowers and sepals.

6. **P. oblongatum** (E. Mey.! in Herb. Drege); leaves (*not perfectly known*) on long, *glabrous* petioles, oblong-obovate, crenate, *glabrous*, subciliate (?); scapes patently hispid, branched; *umbels* 4-8-*flowered*, with ovato-lanceolate, setose bracts; calyx tubes sparsely hispid, 3-4 times as long as the lanceolate, acute, membranous edged, hispidulous segments; stamens 5, very long, declinate; petals broadly obovate.

HAB. Kaus Mountain, near Kookfontein, Namaqualand, 29° s., 3-4000 f., *Drege!* (Herb. T.C.D., Hook., Sond.)

On all the specimens seen, the leaves, which probably precede the flowers, are so far decayed that it is not easy to ascertain their proper shape. Our description may therefore require correction, The flowers are among the largest in this section, with remarkably long calyx-tubes and stamens, and cream-coloured petals, the upper ones with purple veins. It seems to be related to *P. Grenvilleæ, Andr.*

7. **P. ochroleucum** (Harv.); leaves on hispid petioles, *oblong, obtuse,* entire or lobulate or repand, broad at base, glabrous on the upper surface, ciliolate, sparsely pubescent beneath; stipules subulate, scarious; peduncles elongate, branched, *patently hispid;* umbels *densely many-flowered,* with lanceolate, villoso-canescent bracts; calyx tube twice as long as the bracts, villoso-canescent, as are also the oblong, mucronate, white-edged segments; petals scarcely twice as long as the cal.-lobes,

narrow-obovate. *P. reflexum, E. Mey.! non Pers. H. theiantha, E. & Z.!* 490, *ex parte.*

HAB. Near the Great Fish River, *Drege!* Karroo, near Gauritz River, *E. & Z.!* Kamanassie Hills, George, *Dr. Alexander Prior!* (Herb. T.C.D., Hook., Sond.)

I cannot find any description or figure agreeing with this plant, which seems distinctly characterized by its oblong, slightly carnose leaves, hispid scapes and very densely crowded, short flowers. The petals also are of a colour unusual in this section; the two upper being greenish yellow, the lower white. E. & Z's *H. theiantha* is a fictitious species, founded on the scape and root of this plant, combined with a leaf of some *Composite*, probably of a *Sphenogyne!* as appears by Ecklon's specimen in Herb. Sond.

8. P. moniliforme (E. Mey.! in Hb. Drege); root moniliform; leaves on pilose petioles, oblong or ovate, obtuse, tapering at base, *crenato-lobulate, pubescent or villous;* stipules subulate; *scapes elongate, quite simple,* patently pilose; bracts densely barbate, umbels many-flowered; calyx tube patently hispid, 5-6 times as long as the *linear, obtuse, white-edged, bearded segments. Zey. No.* 2067.

HAB. Silverfontein, Namaqualand, 2–3000 f., *Drege! Zeyher!* (Herb. T.C.D. Hook., Sond.)

Root a string of several roundish tubers, connected by a slender cord. Leaves several, thin and membranous, spreading. Scapes 12–18 inches high, quite simple. Umbel 12-20-flowered. Flowers white, the upper petals with purple blotches. The calyx-lobes resemble those of *P. hirsutum,* but the tube is much longer, and the leaves and scapes very different.

[*Simple leaved* Hoareæ, *unknown to me.* (Sp. 9–17.)]

9. P. punctatum (Willd. Sp. 3. p. 645); leaves broadly ovate, repand, glabrous, on softly puberulous petioles; stipules subulate; scapes branching, the partial umbels very densely many-flowered; flowers subsessile, the calyx tube thrice as long as the segments; upper petals linear-spathulate (pale yellow), spotted, the three lower linear, shorter; fertile stamens two. *DC. Prod.* 1. *p.* 650. *Andr. Rep. t.* 60.

HAB. Introduced to England from the Cape, 1794.

10. P. radicatum (Vent. Malm. t. 65); leaves oblongo-lanceolate or sub-elliptical acute, very entire, nearly glabrous, ciliated, on ciliate petioles; stipules adnate, subulate; umbel simple, many-flowered; pedicels much shorter than the softly pubescent calyx tube, which is four times as long as the lanceolate segments; petals spathulate, retuse (pale yellow); stamens five. *DC. l. c. p.* 650. *Sw. Ger. t.* 174. *Ger. ciliatum, Andr. Rep. t.* 247.

HAB. Introduced to England from the Cape, about 1802.

11. P. spathulatum (Andr. Rep. t. 152); leaves broadly lanceolate, tapering at base, glabrous, sub-ciliate; scapes puberulous, branched, partial umbels 5-8-flowered, petals linear-spathulate, (pale yellow) recurved or revolute; stamens five. *DC. l. c. p.* 650. *Andr. Rep. t.* 282, *(var. β. curviflorum.)*

HAB. Introduced to England, 1800.

12. P. oxaloides (Willd. Sp. 3. p. 642); "leaves oblong, sagittate, entire, fleshy, glabrous; umbel compound." *DC. l. c. Burm. Ger.* 71. *t.* 2. *Cav. Diss. t.* 97. *f.* 1.

13. P. chelidonium (Houtt. fl. syst. 8. t. 61. f. 1); leaves sub-rotund, truncate at base, acute, very entire, pubescent; umbel compound." *DC. l. c.*

14. P. velutinum (L'Her.); "stemless? leaves deeply cordate, very obtuse, undivided, crenato-subsinuate, canescent and tomentose on the lower surface; umbel compound." *DC. l. c. Burch. Cat.* 2828.

"Petals, when dry, linear, sub-undulate, very dark-coloured." (quere, is this the same as *P. Sibthorpiæfolium, Harv.?*)

15. P. bifolium (Willd. 3. p. 645); "leaves two only, cordate, sub-acute, sharply toothed; umbel simple." *DC. l. c. Burm. Afr. t.* 35. *f.* 1. *Cav. Diss. t.* 115. *f.* 3.

16. P. revolutum (Pers. Ench. 2. p. 226); leaves cordate, obtuse, nerved, entire, often auriculate at base; scapes branching, pubescent; bracts lanceolate, revolute; umbels several flowered; calyx tube about twice as long as the lanceolate sepals; petals linear-spathulate (rose-red). *Andr. Rep. t.* 354.

HAB. Introduced to England, 1800, *Niven.*

17. P. Grenvilleæ (Andr. Ger. cum ic.); stemless; leaves spathulate-ovate or obovate, coarsely crenate, villous; scapes very long, simple or branched; umbels many-flowered, with subulate bracts; calyx tube thrice as long as the lanceolate segments; upper petals obovate, emarginate, clawed, much longer than the lower; fertile stamens four, declinate. *Grenvillea conspicua, Sw. Ger. t.* 262. *f.* 2.

HAB. Namaqualand, whence it was sent to *Lord Grenville*, 1810.

A superb species, now lost to English gardens, and only known by Andrews' figure. The leaves resemble those of *Primula vulgaris* in shape and size. The scape is 12 inches high, as thick as a swan's quill; the flowers very numerous, of large size, flesh-coloured, the upper petals darkly lined at base and with a deep red central spot.

** *Leaves deeply 3-lobed, or 3-parted; or simple and tripartite on the same root.* (Sp. 18–24.)

18. P. heterophyllum (Jacq. Ic. Rar. t. 516); leaves on ciliated petioles, *diverse*, some oblong or ovate, entire, some three-lobed or 3-parted, the terminal lobe largest, *rigidly ciliate, glabrous* or nearly so; scapes branching, patently hispid; umbels pluriflowered, with subulate bracts; calyx pubescent and setose, the tube nearly twice as long as the lanceolate, acute lobes; petals narrow-spathulate. *DC. Prod.* 1. *p.* 651, *non E. & Z.*

HAB. Groenekloof, *Zeyher, No.* 171. (Herb. Vind., Hook., Benth., Sond.)

A small species with glabrous, fleshy leaves, about 1 inch long, and ¾ inch wide, bordered with rigid, short, swollen hairs; mostly trifid. All the petals long and narrow, the two upper blotched in the middle, 3 lower flesh-coloured. Stipules subulate, scarious. Uppermost stamen very short, the two lowermost much longer than the intermediate pair, as in *Dimacria*, Sw. Our specimens are badly dried, but agree tolerably with *Jacquin's* originals.

19. P. oxalidifolium (Pers. Ench. 2. p. 227); leaves on longish, *rigidly setose* petioles, tri-foliolate, *the leaflets petiolulate, broadly ovate or subrotund*, obtuse, ciliate, and sprinkled with rigid hairs; scapes puberulous and laxly setose, branched; umbels pluriflowered with lanceolate bracts; calyx pubescent and sparsely setose, the lobes broadly lanceolate, acuminate, half as long as the obovate, long-clawed petals. *DC. Prod.* 1. *p.* 651.? *Andr. Bot. Rep. t.* 300?

HAB. Cape, *Masson! Thom!* Between Breede River and Bokkeveld, *Drege!* Herb. T.C.D., Hook., Sond.)

I describe from *Drege's* specimens, doubtfully referred to the authorities cited. *Andrews* represents the umbels as densely many-flowered and the petals narrow and sulphur yellow, with dark spots on the upper ones. In *Drege's* the petals are broader, more obovate, rosy-purple and veiny. A specimen in Herb. Banks, (23) marked *P. auritum*, and another in Herb. Holm., marked "*P. proliferum var.*," belong to this.

20. P. triphyllum (Jacq.! Ic. Rar. t. 515); leaves on long, *puberulent* petioles, tripartite, nearly glabrous, *the segments obovate, obtuse, crenate;* scapes and calyces pubescent; calyx-lobes lanceolate, reflexed, albo-marginate; petals cuneate. *DC. Prod.* 1. *p.* 651.

HAB. Formerly cultivated in Hort. Schoenb. (Herb. Vind.)

Of this I have only seen a garden specimen, preserved in the Imperial Herbarium, Vienna.

21. P. nervifolium (Jacq.! Ic. Rar. t. 517); leaves on long, sparsely setulose petioles, tripartite, *rigidly ciliate*, glabrous, the segments elliptical or ovate, 5-7*-nerved, glaucous;* scapes branched, hispid, with pluriflowered umbels; calyx tube hispid, 4 *times as long* as the lanceolate, spreading, albo-marginate segments; petals cuneate. *DC. Prod.* 1. *p.* 651.

HAB. Formerly cultivated at Vienna. (Herb. Vind.)

Only known from *Jacquin's* figure and a garden specimen in the Vienna Herb. Its flowers resemble those of *P. oblongatum*, E. Mey.

22. P. reflexum (Pers. Ench. 2 p. 227); leaves on strigose petioles, *pinnately-tripartite, the middle lobe petiolate, deeply* 3*-fid, the lateral sessile unequally bilobed*, the lobules obovate, obtuse, strigoso-ciliate; scapes branching, setulose; umbel few-flowered; calyx puberulous and setulose, with lanceolate, acute, gland-tipped segments, half as long as the spathulate petals. *DC. l. c. Andr. Rep. t.* 224. *H. bijuga, E. & Z.!* 499.

HAB. Near Tulbagh, *E. & Z.!* (Herb. Sond.)

The solitary specimen I have seen looks so like the figure quoted that I venture to refer it as above. In Andrews' plant, however, the petals are white, with dark spots; in Ecklon's, purple, with lines.

23. P. attenuatum (Harv.); leaves on very long, strigose petioles, tripartite, *strigoso-pubescent, the segments linear, pinnatifid, the ultimate lobes long and narrow;* stipules adnate, subulate; scapes branched, calyx sub-sessile, the tube tomentose, 3–4 times as long as the lanceolate, *taper-pointed*, albo-marginate segments; petals linear, elongate, the upper emarginate. *Zeyher, No.* 169.

HAB. Uitkomst, Brandenberg, *C. Zeyher*. (Herb. Sond.)

Known by its leaves and especially by the taper-pointed, almost awned calyx segments, and the very long, narrow petals, which are thrice as long as the calyx lobes. Flowers white. Umbels pluriflowered with subulate bracts.

[*Triphyllous* Hoareæ *unknown to me.*]

24. P. trifidum (Willd.); "leaves tripartite, the segments linear-cuneiform, three-toothed at the summit; umbel simple." *Burm. Afr. t.* 35. *f.* 2. *Ger. trifidum, Cav. Diss. t.* 115. *f.* 1. Flowers blood-red.

*** *Leaves either pinnatifid, or pinnately parted, or pinnatisect; or with simple and pinnatifid leaves on the same root.* (Sp. 25–42).

25. P. barbatum (Jacq. Ic. Rar. t. 513); leaves petiolate, *decompound-pinnatifid*, the laciniæ laxly set, erecto-patent, narrow-linear 3-4-fid or pinnatifid, the margins often inflexed, and *the apices tipped with 3-4 bristles*; scapes branching; calyx tomentose-villous, the segments *lanceolate-acute*, with a narrow margin, and apical, bearded gland. *DC. Prod.* 1. *p.* 652., *non E. & Z.! G. proliferum, Cav. Diss. t.* 120. *f.* 3. *H. barbata, Sw. Ger. t.* 391 ? *E. & Z.! No.* 475, 476, 479. *Zey. No.* 168, 2041, 2344. *P. laciniatum, setosum, and limbatum, of Drege's plants.*

HAB. Rocky and sandy ground, in the Western Districts about Capetown, &c. (Herb. T.C.D., Hook., Sond.)

! I fear this is only a var. of *P. longifolium*, with constantly multipartite leaves. There is no difference in the flower, and the pubescence can scarcely be depended on. The present is sometimes nearly glabrous, sometimes densely pubescent. Some of Drege's specimens, marked "*barbatum*," belong to our plant, others to what appear a different species. *P. pennæformis*, E. & Z.! 477 (Herb. Sond.) is much larger, with broader and flatter leaf-segments, and white, or cream-coloured flowers, wanting the dark spot; it may be a distinct species.

26. P. hirsutum (Ait. Hort. Kew. 2. p. 417; leaves petiolate, *polymorphous* (simple, pinnatifid, bi-pinnatifid or almost pinnati-partite), *rigidly setose on both sides*; scapes branching, pubescent or villous; umbels many-flowered, with bearded bracts; calyx softly pubescent, and sparsely villous, the tube about twice as long as the *linear, subobtuse, broadly white-edged segments. DC. Prod.* 1. *p.* 652., *also P. melananthum, P. dioicum and P. atrum, l. c. p.* 653. *Ger. hirsutum, Burm. t.* 68. *f.* 2.

Var. *α.* **melananthum**; petals blackish-purple. *P. melananthum, Jacq.! Ic. Rar. t.* 514. *Sw. Ger. t.* 73. *Andr. Rep. t.* 209. *P. atrum, L'Her. Ger. t.* 44. *Sw. Ger. t.* 72. *Hoareæ, E. & Z.! No.* 481, 486, 487, 488, 489. *Drege, No.* 7489.

VAR. *β.* **carneum**; petals rosy, veiny or white. *E. & Z.! No.* 468, 482, 483, 484, 485, 494 ? *Drege, No.* 7490, *b.*, 7491, 7492.

HAB. Stony Hills and dry ground, throughout the Colony, frequent. Herb. T.C.D., Hook., Sond.)

Very inconstant in the shape of the leaves and colour of the petals, but all the forms agree in the rigid pubescence of the leaves, the plumoso-barbate bracts, and the linear, albo-marginate sepals. Drege's, 7490, *a*, (Hb. Benth.) belongs, perhaps, to *P. heterophyllum*.

27. P. campestris (E. & Z.! 480); leaves petiolate, *polymorphous* (ovate and entire, deeply inciso pinnatifid, or lobed), obtuse at base, *nearly glabrous on the upper*, pilose on the lower side, ciliated; stipules pilose; scapes short, simple or branched, pubescent and villous; bracts broadly lanceolate, silky; calyx softly pubescent, the *sepals lanceolate, acuminate-mucronate*, albo-marginate.

HAB. Grassy fields, near the Zwartekop's River, Uit. *E. & Z.!* (Herb. Sond., Benth.)

A small species. Leaves on short petioles, 1–1½ inches long, ¾ inch broad, mostly lobed, or pinnatifid, sometimes entire; varying in amount of pubescence. Flower stalks 3–4 inches high; umbels 8–10-flowered. Allied to *P. hirsutum*, but with acuminate petals and not so rigidly hispid.

28. P. roseum (Ait. Kew. Ed. 2. vol. 4, p. 161); leaves *on long, pubescent petioles*, sinuato-pinnatifid, obtuse, *tomentose*, the segments bluntly toothed; *scape long, simple; umbel densely many-flowered*, the bracts subulate, bearded; calyx tube tomentose, 3–5 times as long as the lanceolate segments; *upper petals emarginate, much longer than the lower. DC. Prod.* 1. *p.* 651. *Andr. Rep. t.* 173. *Sw. Ger. t.* 262.

HAB. Cult. in England, 1794, *Masson, Forsyth!* Oliphant's River, *Niven*. (Herb. Holm., Benth.)

My description is chiefly drawn from Sweet and Andrews, compared with a specimen from Forsyth (Hb. Benth.), and another in Hb. Holm. This fine, but little known species, is now lost in English gardens. The flowers are very numerous, and bright rosy-red.

29. P. pilosum (Pers. Ench. 2. p. 227); leaves on long, hispid petioles, *pinnate-partite*, pinnæ in 3-4 pair, sub-alternate, *deeply bi-trifid or many-lobed*, softly hairy, the lobes obtuse; scapes branching, pilose; umbel 4-5-flowered, with subulate, setose bracts; calyx tubes 6-8 times as long as the bracts, laxly setulose, as are also the broadly-lanceolate, gland-tipped segments, petals narrow. *DC. Prod.* 1. *p.* 652. *Andr. Bot. Rep. t.* 259.

HAB. In the Stormsvalley, R. Sonderende, *Zeyher! No.* 2046. (Herb. Hook., Sond.).

Seemingly a good species. The leaves are thin and membranous, and very softly hairy, with long, slender hairs. The petals in our specimens seem to have been white, with purple streaks. *Andrews* figures them a lake-red.

30. P. astragalifolium (Pers. Ench. 2. p. 227); stemless or nearly so; leaves on long, setose petioles, *pinnati-partite*, pinnæ *in many* pairs, *lanceolate-oblong* sub-acute, *simple or bi-trifoliolate*, hirsute on both sides; stipules subulate; scapes pubescent and villous, branching; umbels many-flowered, with villous bracts; calyx silky, the segments lanceolate, acute, gland-tipped, narrow-margined. *DC. Prod.* 1. *p.* 653. *Sw. Ger. t.* 103. *Cav. Diss t.* 104. *f.* 2. *P. foliolosum, DC. P. pinnatum, E. Mey.! ex parte.*

VAR. α. **minor**; smaller; leaflets somewhat oval and mostly simple. *Sw. Ger. t.* 103. *Zey.!* 2045. *H. pinnata, E. & Z.!* 492, *non DC.*

VAR. β. **foliolosum**; taller; leaflets numerous, more oblong and frequently bipartite. *Andr. Rep. t.* 311. *P. foliolosum, E. & Z.!* 496. *ex parte, Drege!* 7499. *Zey.!* 167, 2044. *H. lessertiæfolia, E. & Z.!* 495.

VAR. γ. **tri-foliolatum**; small; leaflets numerous, narrow-oblong, acute at each end, mostly tripartite. *Hoarea trifoliata E. & Z.!* 493.

HAB. Rocky and dry situations. About Capetown and Wynberg, and in Caledon and Worcester Districts. γ. near the Waterfall, *E. & Z.!* (Herb. T.C.D., Hook., Sond.)

Very near *P. pinnatum*, but a stronger plant, with thicker and more opaque leaves, more setose pubescence and narrower and sharper leaflets, disposed to become compound. The petals vary in breadth and colour in different specimens, and afford no valid characters.

31. P. pinnatum (Linn.); stemless; leaves on long villous petioles, pinnati-partite, pinnæ few or many, *ovate or sub-rotund*, sub-acute or obtuse, *thin*, alternate or sub-opposite, silky on both sides; scapes pubescent and villous, branching; umbels pluri-flowered, with villous bracts; calyx silky, the segments lanceolate, acute, gland-tipped, with a narrow margin. *Sw. Ger. t.* 46. *Cav. Ic. t.* 115. *f. H. astragalifolia, E. & Z. !* 497, *and No.* 498. *Drege,* 7500, 7501. *P. viciæfolium, DC. Prod.* 1. *p.* 653. *P. coronillæfolium, Andr. Rep. t.* 305. *DC. l. c.*

HAB. Near Capetown and Constantia, *E. & Z., W.H.H.* Nieuwekloof, *Drege!* (Herb. T.C.D., Hook., Sond.).

Nearly allied to *P. astragalæfolium*, but with thinner, broader and more obtuse leaflets and a more silky pubescence. Flowers white or flesh coloured, either veiny or with a dark spot on each upper petal. The citation of Eck. & Zey. and of Drege's collections must be cautiously followed, as their tickets, for both species, are misplaced in some herbaria.

32. P. carneum (Jacq. ! Ic. t. 512); leaves on long, *setulose* petioles, *bipinnately divided*, segments linear, obtuse, *setulose;* scapes simple, pluri-flowered; bracts lanceolate; calyx setulose, the tube 3-4 times as long as the lanceolate segments; petals spreading, the upper obovate. *DC. Prod.* 1. *p.* 654.

HAB. Formerly cult in Hort. Schoenb. (Herb. Vind. !).

Petals rosy white, with red veins. DC. quotes "*Cav. Dis. t.* 120. *f.* 1," but that figure ill accords with Jacquin's specimen, which has much larger flowers.

33. P. rapaceum (Jacq. Ic. rar. t. 510); leaves on long, hairy petioles, erect, linear (in outline) *bi-pinnati-partite*, pinnæ numerous, closely set, short, softly hairy, multifid, the segments linear or narrow-cuneate, more or less cut; stipules subulate, attenuate; scapes simple or branched, villous, the umbels densely many-flowered and bracts bearded; calyx pubescent, with lanceolate, acute, albomarginate, hair-pointed segments; *two upper petals narrow, reflexed,* 3 *lower broader, oblong, connivent, straight. DC. Prod.* 1. *p.* 651., *also P. nutans, and P. corydalifolium, DC. l. c.*

VAR. α. **selinum**; flowers rosy-white or flesh-coloured, the upper petals mottled at base. *G. selinum, Andr. Rep. t.* 239.

VAR. β. **luteum**; fl. pale sulphur yellow, the upper petals with a dark spot. *H. carinata, Sw. Ger. t.* 135. *Bot. Mag. t.* 1877. *E. & Z. ! No.* 502. *also* 500, 501. *Zey. !* 166.

VAR. γ. **corydalifolium**; fl. primrose-yellow, the upper petals red at base; pinnæ lacero-pinnatifid. *H. corydalifolia, Sw. Ger. t.* 18.

HAB. On dry, stony, mountain sides, in the Cape, Stellenbosch, and Swellendam districts. Frequent near Capetown. (Herb. T.C.D., Hook., Sond.).

This species is readily known by the connivent lower petals, which, with the reflexed upper petals, give the flower somewhat the look of a papilionaceous corolla. It chiefly varies in the degree of incision of the short, scarcely inch long, pinnæ, and in the colours of the petals. *P. fissifolium*, E. Mey. ! non Pers., belongs to our var. γ.

34. P. appendiculatum (Willd. Sp. 3. p. 651); nearly stemless; leaves on long, villous petioles, densely silky and villous, bi-pinnatipartite, the pinnæ cut into many narrow-linear, obtuse lobes; *stipules broadly ovate, or ear-shaped, rigid, patent, adnate and decurrent along the petiole;* scapes branched near the base, pilose; umbels many-flowered, with densely hairy bracts; calyx tube 8-10 times as long as the hispid, linear, obtuse sepals. *DC. Prod.* 1. *p.* 662. *Cav. Diss. t.* 121. *f.* 2, *Thunb. ! Fl. Cap. p.* 529. *E. & Z. !* 503.

HAB. Clayey soil, in the Langvalley, Worcester, *E. & Z. !* (Herb. T.C.D., Hook., Sond.).

The ear-shaped, broad and rigid stipules, and the silky, decompound leaves distinguish this plant from all allied species. *P. pulchellum* has somewhat similar stipules, but very different leaves.

[*Pinnate or pinnatifid-leaved* Hoareæ *unknown to me.* (Sp. 35–42.)]

35. P. fissifolium (Pers. ench. 2. p. 227); leaves pinnatisect, the segments bi-trifid and cut, naked; umbel simple, several-flowered; petals obtuse, all with deep red, oblong spots. *DC. l. c. p.* 652. *Andr. Rep. t.* 378.

HAB. Introduced to England, 1795.

36. P. setosum (Sw. Ger. t. 38); leaves pinnatisect, pubescent, pinnæ sub-opposite, bi-partite, the segments cuneate, 3-5-toothed, the teeth setose at the summit; scapes branched, the umbels several-flowered; calyx tube sub-sessile, twice as long as the limb; upper petals reflexed, lower sub-connivent. *DC. l. c. p.* 652.

HAB. Cult. in England, 1821.

37. P. bubonifolium (Pers. Ench. 2. p. 227); leaves glabrous, pinnatisect, segments few and distant, inciso-lobate, acute; scapes simple, about 5-flowered; petals linear-spathulate, emarginate, white, the two upper red-spotted at base. *DC. l. c. p.* 652. *Andr. Rep. t.* 328.

HAB. Introduced to England, 1800, *Niven.*

38. P. violæflorum (Sw. Ger. t. 123); sub-caulescent; leaves tripartite or pinnatisect, the segments oblongo-lanceolate, glabrous, very entire, ciliated, acuminate and tipped with a few rigid bristles, the lower deeply bifid; petioles hispid; scapes branched, the partial umbels several-flowered; petals (white) reflexed, the lower much the shortest; stamens 5. *DC. l. c. p.* 652.

HAB. Introduced to England, 1822, *Colvill.*

39. P. floribundum (Ait. Kew. Ed. 2. vol. 4. p. 163); leaves pinnati-partite, segments bifid, in several pair; scapes branched, the umbels several-flowered; petals spathulate, white, all with red streaks or spots. *DC. l. c. p.* 652. *Andr. Rep. t.* 420.

HAB. Introduced to England, 1795.

40. P. penniforme (Pers. Ench. 2. p. 227); leaves pinnati-partite, segments lanceolate-linear, acute; scapes branched, the umbels several-flowered; petals yellow, red-blotched at base, narrow-cuneate. *DC. l. c. p.* 652. *Ger. laciniatum, var. bicolor, Andr. Rep. t.* 269.

HAB. Intoduced 1800. Possibly a var. of *P. longifolium?*

41. P. centauroides (L'Her.); stemless; leaves hairy, pinnati-partite; the segments remote, cut or entire; umbel simple, densely many-flowered. *DC. l. c. p.* 652.

42. P. incrassatum (Bot. Mag. t. 761); leaves glabrous, deeply pinnatifid or pinnati-partite, somewhat fleshy, segments lobed, obtuse; scape pubescent, simple or branched, umbels several-flowered; calyx tube 3–4 times as long as the segments; upper petals obcordate, lower spathulate. *DC. l. c. p.* 654. *Ger. incrassatum, Andr. Rep. t.* 654.

HAB. Introduced to England, 1801.

Flowers of large size, (somewhat like those of *P. roseum*,) deep red, with dark streaks.

SECT. 2. **SEYMOURIA**. Stemless, tuberous rooted, resembling *Hoareæ*, but: *petals* two only, the three lower abortive. *Stamens* 5, sub-equal, straight, exserted. (Sp. 43–46.)

43. P. dipetalum (L'Her. Ger. t. 43); stemless; leaves *ovate, acute*, glabrous; umbel simple; flowers pentandrous, dipetalous." *DC. Prod.* 1. *p.* 650.

HAB. S. Africa.

Only known to me by the figure above quoted. The umbel is represented as 3-flowered, the petals strongly reflexed, obovate, streaked at base, and purplish.

44. P. asarifolium (Sw.); stem scarcely any; leaves on hairy petioles, *roundish-cordate, blunt*, entire, ciliate, glabrous and shining above, hairy underneath; stipules adnate, minute; scapes branched, villous, many-flowered, with subulate bracts; calyx tubes about as long as the reflexed segments; petals spathulate, emarginate, reflexed. *Seymouria asarifolia, Sw. Ger. t.* 206.

HAB. Intoduced to English gardens, 1822.

Flowers dark purple.

45. P. trifoliatum (Harv.); leaves on long hairy petioles, *tripartite*, hairy on both sides, the segments broadly cuneate, coarsely crenate; stipules adnate, membranous; scapes branched, slender, hispid; umbels few-flowered; calyx sub-sessile, the tube hispidulous, twice as long as the reflexed, lanceolate segments; petals spathulate; stamens four. *Drege*, 7497.

HAB. Klein Drakenstein, Stell., *Drege!* (Herb. Hook., Benth., Sond.).

The specimens are in poor condition, and our diagnosis may need correction.

46. P. Niveni (Harv.); leaves on long hairy petioles, *oblongo-lanceolate*, tapering at base, hairy on both sides; stipules adnate, subulate; scapes branched, slender, hispid; umbels pluri-flowered; calyx sub-sessile, the tube hispidulous and glandular, twice as long as the reflexed, lanceolate segments; petals spathulate; stamens four.

HAB. Elevated places in Sweetmilk valley, *Niven!* (Herb. Sond.)

Founded on a single, imperfect specimen, which much resembles *P. trifoliatum*, except in the shape of the leaves. It requires verification, and had it not belonged to the *dipetalous* section, I should scarcely have ventured to name it.

SECT. 3. **POLYACTIUM.** *Root* tuberous or incrassated. *Stem* succulent and nodose, often very short; flowering branches herbaceous, annual. *Leaves* on long petioles, lobed or pinnately-decompound. *Peduncles* elongate; umbel densely many-flowered, the pedicels much shorter than the calyx-tube. *Petals* 5, sub-equal, obovate, entire or fimbriato-lacerate. *Stamens* 5–7, one filament much broader than the rest. *Flowers* evening scented. (Sp. 47–66.)

* *Leaves crowded on a short stem, or radical. Petals entire. Stipules ovate or cordate, leafy or membranous, withering.* (Sp. 47–53.)

47. P. lobatum (Willd. Sp. 3. p. 650); stemless or nearly so; radical-leaves deeply cordate at base, digitately 3-5-nerved, and three-lobed or tri-partite, softly villous above, tomentose underneath, the margin inciso-crenate and doubly serrate; stipules sub-orbicular, acute, scarious; peduncles patently hairy, branched, bearing two or more long-stalked umbels; calyx segments softly pubescent, linear-oblong, obtuse, shorter than the obovate, *dull-coloured* petals. *DC. Prod.* 1. *p.* 662. *Cav. Diss. t.* 114. *Sw. Ger. t.* 51. *E. & Z.! No.* 504. *P. spondylifolium*, *E. & Z.!* 507. *non Sw. P. arenarium*, *E. & Z.!* 506.

HAB. Dry, clayey, sandy and rocky ground. Frequent near Capetown and through the Western districts. Winterhoeksberg, Uitenhage, *E. & Z.!* (Herb. T.C.D., Hook., Sond.).

Leaves often of large size, 6–12 inches broad, very soft, and when young canescent below, variable in shape and incision; sometimes 3-foliolate, the terminal leaflets petiolate, the lateral lobulate. Flowers aromatic in the evening, as in most of this section; the petals either dark brown with a pale margin, or dull yellow-brown.

48. P. heracleifolium (Lodd. Cab. t. 437); stems short and deflexed, herbaceous; leaves thickish, softly villous above, tomentose underneath, *oblong, deeply inciso-pinnatifid or somewhat pinnate*, serrated, the terminal segments very large, inciso-lobulate, and petiolate, the lateral small, simple or tripartite; stipules broadly cordate, acute, scarious; calyx segments pubescent, oblong, sub-acute, half as long as the obovate, *greenish-yellow* petals. *Sweet, Ger. t.* 211! *DC. Prod.* 1. *p.* 604? *P. geifolium*, *E. Mey.! P. hybridæfolium E. & Z.!* 509.

HAB. On the Gauritz Riv., George, *E. & Z.!* Between the French Hoek and and Donkerhoek, *Drege!* (Herb. T.C.D., Hook., Sond.).

Leaves 2–5 inches long, more oblong than those of *P. lobatum*, and more pinnatisect, in a somewhat panduriform manner, like those of *Geum rivale*. Peduncles elongate, pubescent, the umbel 10–12-flowered, with ovate-lanceolate bracts. *Drege's* specimens are very like *Sweet's* figure, above quoted.

49. P. pulverulentum (Colv. in Sw. Ger. t. 218); stem very short; leaves thickish, glabrescent, pubescent or villous, especially underneath, *cordate, obtuse, pinnate-lobulate*, and bluntly toothed, the lobules short and round; stipules cordate, acute, membranous; calyx *setulose*, the segments linear, acute, shorter than the obovate, brown-disked petals. *P. primulæforme, E. & Z.!* 505. *P. testaceum, E. Mey.! Drege*, 7507. *a. and b.*

VAR. β. **pedicellatum**; pedicels much longer than the bracts. *P. pedicellatum, Sw. Ger. t.* 250.

HAB. In grassy fields, of the Eastern Districts. Uitenhage and Caffirland, *E. & Z.!* Zuureberg and between the Gekau and Basche, *Drege!* Fish River, *Burke!* (Herb. T.C.D., Hook., Sond.).

This in somewhat resembles *P. lobatum*, but the leaves are much smaller, of thicker substance, quite simple, with no disposition to become tri-partite, more obtusely toothed, and much less hairy, sometimes nearly glabrous above, and only sparsely pubescent below. *Sweet* says "the leaves are covered with a powdery pubescence, quite white when young;" and he speaks also of an allied, cultivated species "with rounder and smoother leaves of a greasy appearance." This latter may probably be Meyer's "*testaceum*" which is more glabrous than Ecklon's plant; but mere pubescence is a fallacious character.

50. P. radulæfolium (E. & Z.! 510); stem short; leaves thickish and rather rigid, glabrescent or villoso-pubescent, *broadly ovate, deeply inciso-pinnatifid, the lobes cuneate, cut or multifid*, toothed; stipules subrotund, acute, scarious; calyx *setulose*, its segments linear, sub-acute, reflexed, half as long as the dull-coloured, obovate petals. *P. multiradiatum, E. Mey.! ex parte, non Wendl.*

HAB. Near the Zwartzkops River, Uit., *E. & Z.!* Albany, *Mrs. F. W. Barber.* (Herb. T.C.D., Hook., Sond.).

Near *P. heracleifolium*, but with more rigid and much more divided leaves, 2–4 inches long, ⅔ of their length wide. Peduncles long; the umbel pluri-flowered; with lanceolate bracts. Petals dull yellow-brown, darker in the middle. Flowers scented at night.

51. P. flavum (Ait. Hort. Kew. 2. p. 418); stem short, succulent; radical and lower leaves *4-pinnately decompound, the segments not decurrent, very narrow and hairy*, toothed and deeply cut; pubescence very copious, patent, white; stipules ovate, acute, scarious; calyx tube striate, setulose, segments lanceolate, reflexed, half as long as the narrow-obovate, thickish, dull-coloured petals. *DC. Prod.* 1. *p.* 662. *Sw. Ger. t.* 254. *Ger. daucifolium, Cav. Diss. t.* 120, *non E. & Z.! P. flavum, E. & Z.!* 514 *and P. coniophyllum, E. & Z.!* 515.

HAB. In sandy or clayey soil. Cape, Caledon and Clanwilliam Districts, Witsenberg, *Zeyher!* (Herb. T.C.D., Hook., Sond.)

Stemless, or with a short, decumbent or deflexed stem. Leaves 6–12 inches long, excessively divided, very hairy, the segments almost filiform. Peduncles long, umbel many-rayed, bracts ovato-lanceolate. Flowers greenish yellow, each with a dark centre, or very dark brown, with a pale border, sweetly aromatic at night. *Cavanille's* specific name "*daucifolium*" (carrot-leaved) is much more appropriate than that which the law of priority forces us to adopt. The leaves are much more finely divided than in any state of *P. triste.*

52. P. anethifolium (E. & Z.! 516); stems short, herbaceous; radical and lower leaves *tripinnately-decompound*, the segments not decurrent, very narrow, *sparsely pubescent, filiform, elongate;* stipules ovate, acute, scabrous; sepals lanceolate, reflexed, half as long as the thickish, narrow-obovate, dull-coloured petals. *P. callosum, E. Mey. ! P. peucedanifolium, E. & Z.* 518 ?

HAB. Groenekloof, *E. & Z. !* Piquetberg, *Drege !* (Herb. Hook., Sond.).

Very like *P. flavum* and perhaps a mere variety, with less finely divided and much more glabrous leaves. The flowers are deeply yellow, with dark brown centres and evening-scented. Peduncles long, patently pubescent; the umbel many-rayed, with ovato-lanceolate bracts. Meyer's "*P. Callosum*" is more glabrous, with longer ultimate leaf-segments, but otherwise very similar: and "*P. peucedanifolium, E. & Z.!*" from Tulbaghsberg, judging from a miserably imperfect specimen in Hb. Sond. is not different.

53. P. triste (Ait. Hort. Kew. 1. vol. 2. p. 418); stem short or scarcely any, deflexed, succulent; radical and lower leaves *bi-tri-pinnately-decompound, the segments decurrent*, toothed and laciniate, the teeth gland-tipped; pubescence copious, patent, pilose; stipules sub-rotund, mucronate, scarious; calyx tube striate, setulose, much longer than the pedicel, segments lanceolate, spreading or reflexed, half as long as the narrow-obovate, thickish, dull-coloured petals. *DC. Prod* 1. *p.* 662. *Bot. Mag. t.* 1641. *P. millefoliatum, Sw. t.* 220.

VAR. α. **daucifolium**; occasionally caulescent; leaves sub-tri-pinnatifid, their segments narrow. *Pol. triste, E. & Z. !* 512, *daucifolium, E. & Z.!* 513 *and P. multiradiatum, E. & Z. !* 521, *non Wendl.*

VAR. β. **filipendulifolium**; caulescent; leaves sub-bi-pinnatifid, their segments broader. *Ger. triste, Cav. Diss. t.* 107, *P. filipend. Sw. t.* 85, *E. & Z. !* 511. *P. papaverifolium, E. & Z.!* 508.

VAR γ. ? **laxatum**; leaves diffusely-decompound, 4-pinnate, the pinnæ petiolate, the ultimate segments distant, linear, decurrent, simple or pinnatifid; pubescence scanty. *P. laxatum, Harv. in Herb. Sond.*

HAB. Frequent in clayey soil, in Cape and Stellenbosch Districts. γ. at the 24 Rivers, *Zeyher !* (Herb. T.C.D., Hook., Sond.)

Often quite stemless, but sometimes with a longish, unbranched, ascending or knee-bent stem. Leaves 8–12 inches long (in γ. 2 feet long and broad), very much lobed and cut in an unequally pinnate order, the pairs of laciniæ unequal, short and long mixed. Flowers dull brownish yellow, with dark spots, or the petal dark brown with a pale border, very sweetly aromatic at night. Peduncles long, umbel many-flowered, with lanceolate bracts. Our var. γ., founded on a specimen in Hb. Sond., has enormously large and laxly-lobed leaves; the primary and secondary segments petiolulate. It probably grew under some modifying circumstances of soil or aspect.

** *Leaves scattered on an elongated, simple or branched, erect or ascending stem. Petals entire. Stipules ovate or cordate, membranous, withering.* (Sp. 54–60.)

54. P. multiradiatum (Wendl. ! Coll.); sub-caulescent; flowering branches herbaceous, slender, hairy, with swollen joints and long internodes; lower leaves pinnati-partite, hairy, the pinnæ petiolate, deeply pinnatifid, with cuneate, flat, incised segments, upper sub-bi-pinnatifid; stipules broadly cordate, withering; peduncles very long, 20–30-flowered; outer flowers sub-sessile, inner on long, villous pedicels; calyx tube glabrous, 2–3 times longer than the obtuse, villous segments; petals obovate, dark, with a pale border; fertile stamens five. *DC. l. c. p.* 655. *Sw. Ger. t.* 145.

HAB. Introduced to Europe 1820. (v. v. c.)

Root tuberous, very large. Stem short; the flowering branches leafy, 1–2 feet high. Leaves 10–12 inches long or more, decompound, with patent and sub-distant segments. Peduncles 8–10 inches long, villous. Bracts lanceolate, hairy. I have only seen cultivated specimens of this plant.

55. P. apiifolium (Jacq. f. Eccl. 1. t. 27); caudex thick and fleshy, glabrous; flowering branches herbaceous, slender, hairy, with long internodes; leaves *pinnati-partite, the pinnæ petiolate, pinnatifido-pinnate,* the segments cuneate, flat, laciniate, *glabrous and glaucous;* stipules small, triangular; peduncles simple; flowers sub-sessile, the calyx tube 3–4 times longer than the obtuse, reflexed segments; petals obovate, reflexed, dark, with a pale border. *DC. Prod.* 1. *p.* 662.

HAB. Cultivated in Europe, 1809.

Only known to me from Jacquin's figure and description, from which together I have drawn up the above diagnosis. Though placed by De Candolle widely apart from *P. multiradiatum,* it appears to me to be nearly allied. I cannot credit Jacquin's description of the stamens and calyx tube.

56. P. quinquevulnerum (Willd.! Enum. 703); suffruticose, sparingly branched; leaves *bi-pinnatifid, hairy and rough, the segments linear,* unequally cut and toothed; stipules broadly cordate, mucronate; calyx tubes setulous, equalling the pedicels, segments oblong, blunt; petals obovate, velvetty, *purple, with a pale edge. DC. Prod.* 1. *p.* 664. *Andr. Rep. t.* 114. *Sw. Ger. t.* 161.

HAB. Grown from seed sent from the Cape, 1796, *Andrews.* (Hb. Willd.)

Sweet supposes this to be a hybrid between *P. bicolor* and *P. triste.* The flowers are scentless.

57. P. bicolor (Ait. Hort. Kew. 2. p. 425); stems shrubby, succulent, sparingly branched, pubescent; leaves cordate at base, *pinnately-trifid or tripartite,* pubescent, *the lateral segments broadly cuneate,* bilobed and cut, the terminal trifid, toothed; stipules cordate, acute; calyx tubes *nearly sessile,* segments lanceolate, villous, reflexed; petals obovate, *purple, with a pale* border. *Jacq. Hort. Vind. t.* 39. *Cav. Diss. t.* 111. *Bot. Mag. t.* 201. *Sw. Ger. t.* 97. *DC. Prod.* 1. *p.* 664.

HAB. Cult. in England, since 1778. (v. v. c.)

Stem 1–2 feet high, exclusive of the annual flowering branches. Leaves on long petioles, 3–4 inches long, and nearly as broad, softly pubescent. A very old garden plant, but not sent by any recent collector from the Cape. Is it a garden hybrid?

58. P. sanguineum (Wendl. Coll. 2. f. 53); caudex shrubby, fleshy, nodose, glabrous; flowering branches herbaceous, slender, lax, with swollen nodes and long internodes, hairy; leaves *pinnati-partite, glabrous,* thickish, the pinnæ sessile, decurrent, laciniato-pinnatifid, concave, with acute lobes; stipules amplexicaul; flowers pedicellate; the calyx tube long or short; segments reflexed; petals *narrow obovate, scarlet. Jacq. f. Eccl. t.* 57. *Sw. Ger. t.* 76. *DC. Prod.* 1. *p.* 662.

HAB. Cultivated, but its history uncertain. *Sweet* supposes it may be a hybrid, between *P. multiradiatum* and *P. fulgidum,* having the leaves of the former and the flowers of the latter.

59. P. fulgidum (Willd. Sp. 3. 684); stem shrubby succulent; *densely pubescent;* leaves *pinnately 3-parted, silky on both sides,* the lateral segments broadly cuneate, 3-lobed and cut, the terminal oblong, pinnatifid and lobulate; stipules broadly cordate, acute; peduncles branched; calyx-tube much swollen at base and in the throat, thrice as long as the pedicel, the segments linear, obtuse, softly hairy; *petals obovate, scarlet, DC. Prod.* 1. *p.* 663. *Dill. Ellth. t.* 130. *f.* 137. *Cav. Diss. t.* 116. *f.* 2. *Sw. Ger. t.* 69. *E. & Z.!* 519. *Zey.!* 193. *G. tectum, Thunb.! Cap. p.* 525, *fide Herb. Holm.!*

HAB. Cult. in England, since 1732. Sandy ground: near Saldanha Bey, *E. & Z.!* Between Langvalley and Heerelogement, Clanw. *Drege! Zeyher!* (Herb. Hook. Sond., Banks.)

An old inhabitant of conservatories, though not now so frequently met with as formerly. It is a free flowerer, and one of the parents of many beautiful hybrids. Flowers brilliant scarlet, with black lines. Garden specimens are much more glabrous, with thinner leaves than wild ones.

60. P. gibbosum (Willd. Sp. 3. p. 684); stem shrubby, succulent, *much swollen* at the distant nodes; leaves *glaucous, and nearly glabrous, pinnati-partite,* the segments in 1–2 pair with a terminal, the lowest petiolate, *all broadly cuneate,* cut or lobed; stipules small, ovate-acuminate, membranous; pedicels very short; calyx-tube thrice as long as the villous and pubescent, linear-lanceolate segments; petals obovate, *greenish yellow. DC. Prod.* 1. *p.* 662. *Sw. Ger. t.* 61. *E. & Z.!* 520. *Cav. Diss. t.* 109. *f.* 1. *Burm. Afr. t.* 37. *f.* 2.

HAB. Introduced 1712 to England from the Cape. Riedvalley and False Bay, Cape, *E. & Z.!* Oliphant's River, *Zeyher!* (Herb. Benth., Sond.).

The "Gouty-Geranium" of English greenhouses. The nodes of the long, straggling stems are 5–6 inches apart, and conspicuously swollen; the bark is pale, smooth and glaucous. A few spreading hairs are scattered here and there; in other respects the plant is nearly smooth. It is a well marked and easily recognized species.

*** *Leaves crowded on a short stem, or sub-radical. Petals entire. Stipules subulate, acuminate, rigid.* (Sp. 61–63.)

61. P. aconitophyllum (E. & Z.! 517); stem scarcely any; radical leaves on long, angular, sparsely hairy petioles, broadly-ovate in outline, but deeply lobato-pinnatifid or pinnati-partite *(polymorphous),* the segments *cuneate or broadly lanceolate,* lobed or entire, acute, mucronate, strigose and often pilose; stipules broadly subulate, much acuminate, rigid; scapes much longer than the leaves, pilose; calyx softly pubescent, the segments lanceolate, reflexed, half as long as the obovate petals. *P. polymorphum, E. Mey.!*

HAB. Caffraria and Natal. Near the Kei, *E. & Z.!* Grassy fields between the Gekau and Bashe Riv. *Drege!* Coast land, near Natal, *Sutherland! Sanderson! Gueinzius!* and Macallisberg, *Burke and Zeyher!* (Herb. T.C.D., Hook., Sond.)

Very variable in the shape of the leaves, which are 2–4 inches long and about equally wide; sometimes with a broad, moderately-sized lamina, and sometimes cut down to the ribs into many narrow segments. Umbels 12–40-flowered. Petals dull yellow, without large spot.

62. P. Zeyheri (Harv.); radical leaves on long petioles *capillaceo-multi-partite,* bi-tri-pinnate, the pinnæ and pinnules petiolulate, filiform, very long and slender, channelled above, almost quite glabrous; stipules broadly subulate, much acuminate, rigid, pubescent; scape longer than the leaves, angular ; bracts, pedicels and calyx-tubes lanuginous, cal. segments lanceolate, half as long as the obovate petals.

HAB. Crocodile River and Macallisberg, *Burke & Zeyher !* (Herb. T.C.D., Hook., Sond.)

Stem an inch or two long, leafy at the crown. Leaves 8–16 inches long, the petiole as long ; the lamina cut into filiform shreds, sometimes setaceous, sometimes twice or thrice as thick, always very long and entire. The specimens, with slender leaf-lobes, are quite glabrous, the others have a few scattered hairs. These latter approach *P. anethifolium* in aspect, but differ decidedly in the stipules and other characters. The petals are pale, spotless, and of thinner substance than usual in this group. The broad-stamen is taper-pointed, or semi-lanceolate. I have pleasure in inscribing this remarkable species to the memory of its discoverer, the late estimable *Charles Zeyher* of Uitenhage, one of the ablest and most indefatigable explorers of the Botany of South Africa.

63. P. flabellifolium (Harv.) ; radical leaves on long, terete, hairy petioles, *broadly flabelliform, truncate at base, pilose, palmately many-lobed, and many-nerved,* the lobes sharply toothed, the nerves prominent, radiating, with closely netted, prominent veinlets; stipules broadly subulate, acuminate, rigid ; scapes very long and hairy ; umbel densely many-flowered, with lanceolate bracts ; calyx softly pubescent, the segments lanceolate, acute, half as long as the (dark red ?) obovate petals.

HAB. Trans Vaal and Stockspruit, Natal, *Mr. Sanderson !* (Herb. Hook.)

Leaves 5–10 inches long and broad, rigid. The strongly-netted and prominent venation distinguishes this from any of the broad-leaved forms of *P. aconitophyllum ;* the section of the leaf is also different, and the size much greater. The name "*flabellifolium*" is given by *Sweet* to a hybrid (*Sw. vol.* 5. *t.* 48) which must not be confounded with our species.

**** *Petals fimbriato-multifid.* (Sp. 64–66.)

64. P. schizopetalum (Sw. Ger. t. 232); stem short, succulent; radical leaves petiolate, oblong, obtuse, strigose on both sides, *inciso-pinnatifid,* the lobes cuneate, deeply toothed or cut at the apex ; stipules lanceolate, acuminate ; scapes much longer than the leaves, scabrous and hispid ; umbel pluri-flowered ; calyx setulose, with linear lanceolate segments; petals bi-partite, their segments fimbriato-multifid. *P. uitenhagense, E. & Z. !* 523.

HAB. Van Staadensberg, Uit., *E. & Z. !* (Herb. T.C.D., Hook., Sond.)

Stem often scarcely any. Petioles 3–6 inches long, supporting a deeply pinnatifid or almost pinnati parted lamina of the same length. Petals sub-equal, the two upper greenish-yellow, the lower brownish-purple, all cut into slender, forking shreds. This curious plant was cultivated in England by Mr. Colvill in 1821, but like many others of similar habit, has disappeared. Sweet says the flowers are unpleasantly scented.

65. P. amatymbicum (E. & Z. ! 522) ; stem short, succulent ; radical leaves petiolate, softly pubescent and sub-canescent beneath, *ovate-oblong, obtuse, glabrous above, pinnato-lobulate,* the lobules mostly broader than long, coarsely toothed at the apex ; stipules lanceolate acuminate ; scapes very much longer than the leaves, pilose, many-flowered ; calyx villoso-pubescent, with linear-lanceolate segments ; petals bi-partite, their segments fimbriato-multifid. *P. fimbriatum, E. Mey. !*

HAB. Mountain sides, Caffraria, *E. & Z. !* Katberg, *Drege !* (Herb. T.C.D., Hook., Sond.).

Very like *P. schizopetalum*, but the leaves are more ovate, much less deeply lobed, and the upper surface is glabrous and the margin purple. The flowers are rather larger and all the petals suffused with purple; the pubescence of the calyx is different, and the umbel has more rays.

66. P. caffrum (E. & Z.! 524); stem short and succulent; radical leaves on long petioles, somewhat digitately *pinnati-partite or sub-bipinnati-partite, the segments narrow-linear*, very entire, with revolute margins, sub-acute, sparsely villous, especially on the nerves; stipules lanceolate, acuminate; scapes much longer than the leaves, villous; umbel many-flowered, the pedicels longer than the bracts, slender, and, as well as the calyx, villous; petals bi-partite, their segments fimbriato-multifid.

HAB. Winterberg, Caffr., *E. & Z.!* Van Staadensberg, Uit., *Zeyher!* (Herb. Sond., Hook.)

Distinguished from the two preceding species by the narrow, linear segments of the multi-partite leaves. In an old specimen, without special locality, in Herb. Hook., the petals are very dark, almost black; in those collected by Zeyher they are brownish-yellow, with purple lines. The leaves occasionally vary to palmatifid, with broad lobes, glaucous below.

SECT. 4. **OTIDIA.** *Root* branching. *Stem* succulent and knobby. *Leaves* fleshy, pinnately or bi-pinnately divided. *Stipules* minute. *Petals* narrow, nearly equal, the two upper eared at base. *Fertile* stamens 5. (*Inflorescence often dichotomous; umbels few or pluri-flowered; fl. small.* (Sp. 67–72.)

67. P. carnosum (Ait. Hort. Kew. 2. p. 421); stem succulent, the branches swollen at the nodes; leaves shortly petiolate, oblong, *deeply pinnatifid*, somewhat fleshy, *puberulous or glabrous, the segments flat, cuneate*, sharply cut or sub-pinnatifid; stipules small, ovate, acute; peduncles long, branched, hispidulous; umbel many-flowered with short ovato-lanceolate bracts; pedicels patently setose, much longer than the calyx-tube, which is swollen at base, calyx segments *lanceolate, acute*, hispidulous. *DC. Prod.* 1. *p.* 655. *E. & Z.!* 525. *Dill. Elth. t.* 127. *fig.* 154. *Sw. Ger. t.* 98! *P. laxum, E. & Z.!* 527, *non Sw. P. dasycaulon, E. & Z.!* 526, *non Sims? P. crassicaule, E. Mey.! non L'Her.*

HAB. In dry or clayey places, eastern districts. Between Zwartkops and Sondag Rivers, Uitenhage, and near Fish River, Albany, *E. & Z.!* Between the Little and Great Fish River, *Drege!* Albany, *Mrs. F. W. Barber!* (Herb. T.C.D., Hook., Sond.)

Stem 1–2 feet high, clumsy, not much branched; the branches slender, with smaller leaves, or somewhat naked. Leaves 2–4 inches long, not cut quite to the mid-rib, the segments 2–4 lines wide. This plant was cultivated in England in 1724, and is still sometimes seen in collections. The flowers are generally white, and of small size, but Mrs. Barber's specimens seem to have been flame colour, the upper petals darker. *Sweet* says the leaves are "hairy on both sides," a character agreeing better with *P. ferulaceum.*

68. P. crithmifolium (Sm. Ic. pict. 1. t. 13); stem succulent, the branches swollen at the nodes; leaves on longish petioles, *sub-bipinnati-partite*, fleshy, glabrous or pubescent, *the segments narrow-linear, semi-terete, channelled*, inciso-pinnatifid; stipules minute, cordate; *flowering branches naked*, dichotomous and divaricate, downy; umbels 4–6-

flowered, with oblong, obtuse, small bracts; pedicels downy or glabrescent; longer than the calyx-tube, calyx-segments *linear-oblong, obtuse.* *DC. Prod.* 1. *p.* 655. *Sw. Ger. t.* 354. *P. paniculatum, Jacq. Schoenb. t.* 137. *P. laxum, Sw. Ger. t.* 196, *non E. & Z. Drege,* 7479 ? *P. dasyphyllum, E. Mey.! Jenkinsonia dichotoma, E. & Z.!* 543.

HAB. North-west districts. Between Kaussie and Silverfontein, *Drege!* Bitterfontein, *Zeyher!* Louisfontein, Clanw., *E. & Z.!* (Herb. T.C.D., Hook., Sond.)

A very straggling plant with panicled inflorescence and more fleshy, narrower and more divided foliage than in others of this section. *P. dasyphyllum, E. Mey.!* from Silverfontein, looks like a starved state, with simpler leaves and short cuneate and trifid leaf-lobes.

69. P. ceratophyllum (L'Her. Ger. t. 13); "stem shrubby, carnose, branched; leaves fleshy, pinnati-partite, the lobes linear, semi-terete, sub-canaliculate, entire or 3-toothed; umbels many-flowered." *DC. Prod.* 1. *p.* 653. *Bot. Mag. t.* 315.

HAB. South Africa. (Unknown to me.)

Probably a mere var. of *P. crithmifolium,* with simpler leaves.

70. P. dasycaulon (Bot. Mag. t. 2029); "stem shrubby, carnose, tuberculated; leaves fleshy, pinnati-partite, the lobes inciso-pinnatifid, sub-trifid; umbels about 3-flowered. *DC. l. c. p.* 655.

To judge by the figure in *Bot. Mag.* this and the preceding may be probably referred to *P. crithmifolium.*

71. P. ferulaceum (Willd. Sp. 3. p. 687); stem succulent, the branches swollen at the nodes; leaves on longish petioles, oblong, deeply *pinnatifid or pinnati-partite,* thickish, *pubescent on both sides,* the segments flat, cuneate, sharply incised or pinnatifid; stipules small, ovate-acuminate; flowering branches sub-dichotomous, the peduncles short, downy and the umbels 4–6-flowered, with short ovate bracts; pedicels downy, about as long as the swollen tube, cal.-segments *linear-oblong, white-margined,* downy, sub-acute. *DC. Prod.* 1. *p.* 654. *Burm. Afr. t.* 36. *f.* 1. *Cav. Diss. t.* 110, *f.* 2. *E. & Z.!* 528. *Drege,* 7481; *also P. carnosum,* β. *E. Mey.! P. Burmannianum, E. & Z.!* 529. *Zey.* 179.

VAR. β. **polycephalum**; downy or glabrescent, with broader and less deeply cut leaf-lobes, broader and more obtuse and paler bracts and stipules, and denser, sub-capitate umbels. *P. polycephalum, E. Mey.! in Herb. Drege.*

HAB. In sandy soil, north-west districts. Oliphant's River, Clanw., and Zwartland, Stell., *E. & Z.!* Silverfontein, *Drege!* both varieties. (Herb. T.C.D., Benth., Sond.)

Very like *P. carnosum* but more hairy and with much shorter and denser flowers; more paniculate branching; a softer and closer pubescence, and differently-shaped calyx-segments, &c. It is also a western, not eastern species.

72. P. alternans (Wendl. H. Herrenh. 1. p. 14. t. 9); stem succulent, much branched, the branches short and knobby; leaves petiolate, pinnati-partite, *pilose, the segments petiolate,* cuneate, trifid or 3–4-parted and toothed; stipules minute, ovate; peduncles short, *hispid,* with 3–4-flowered umbels and oblong, hairy bracts; pedicels *very short or obsolete,* the pubescent calyx-tube as long as or longer than the linear, obtuse segments. *DC. Prod.* 1. *p.* 655, *Sw. Ger. t.* 286. *Otidia microphylla, E. & Z.!* 531, *and O. corallina, E. & Z.!* 530.

HAB. Cultivated in England, 1791. Near Heerelogement and Kraus Valley, Clanw., *E. & Z.!* (Herb. Benth., Sond.)

This species is remarkable for its clumsy, many-knobbed stem, and small, white-hairy leaves. Though *E. & Z.'s* specimens are very imperfect, I have no hesitation in referring their *O. microphylla*, at least to this place; and *O. corallina* was founded on miserably defective scraps of a similar plant. Flowers small, white, with narrow petals.

Sec. 5. **LIGULARIA**. *Root* branching. *Stem* either succulent or suffruticose and slender. *Leaves* rarely entire, mostly incised, or decompound, pinnati-partite. *Pubescence* hairy, very rarely wanting. *Stipules* (except in *P. pulchellum*) small or obsolete. *Petals* sub-unequal, narrow spathulate, the two upper tapering at base or clawed. *Fertile* stamens 7. (Sp. 73-92.)

* *Stem short, thick, undivided, nodose, armed with persistent spine-like (free) stipules.* (Sp. 73-75.)

73. P. stipulaceum (Willd.! Sp. 3. p. 655); root-stock caulescent, moniliform, tuberous, armed with persistent stipules; leaves chiefly from the crown, on long filiform petioles, oblong, *cordate or hastate at base, somewhat 3-lobed,* inciso-dentate, ciliate, sparsely hispid; stipules subulate, free; scapes often branched, with few-flowered umbels and lanceolate bracts; calyx-tube sub-sessile, 4-5-times as long as the lanceolate, villous segments; petals broadly spathulate or obovate. *DC. Prod.* 1. *p.* 651. *Cav. Diss. t.* 122. *f.* 3. *Thunb. Fl. Cap. p.* 519. *P. pallens, Andr. Ger. cum Ic. Sw. Ger. t.* 148.

HAB. Near Hantam, *Thunberg.* Leliefontein, Namaqualand, 4-5000 f. *Drege!* (Herb. Willd., T.C.D., Hook., Sond).

Root-stock 3-4 inches long, half above ground; the upper portion succulent, with a hard, brown, rough bark, armed with hardened stipules and bases of petioles. Petioles 3-4 inches long; leaves 1½ inches, thin and membranous, glabrescent. Flower-stalks 8-12 inches high; umbels 2-3-flowered; calyx-tube nearly 2 inches long. Petals cream-coloured.

74. P. articulatum (Willd. Sp. 3. p. 655); "stem very short, armed with persistent stipules; leaves *reniform,* 5-lobed, villous, the lobes trifid; umbel simple, few-flowered. *Cav. Diss.* 4. *p.* 252. *t.* 122. *f.* 1. *DC Prod.* 1. *p.* 651. *Thunb. Fl. Cap. p.* 520.

HAB. Near Hantam, in the Roggeveld, *Thunberg.*

75. P. hystrix (Harv.); stem fleshy, moniliform or closely knobby, densely armed with persistent spine-like stipules, slightly branched; leaves from the crown, on slender, strigose and glandular petioles, *pinnati-partite, with very narrow, strigose, deeply incised or parted segments;* stipules rigid, subulate, free; peduncles densely glandular, simple; umbel pluri-flowered, with short subulate bracts; flowers sub-sessile, the glandular calyx-tube 4-5 times longer than the lanceolate, albo-marginate segments; petals small.

HAB. Collected by *Masson!* (Herb. Banks.)

Stem 6-8 inches high, ½ inch thick, consisting of a string of round knobs, bristling with rigid spines, formed of hardened stipules. The habit is nearly that of a *Sarcocaulon.*

** *Stem incrassated, armed with hardened, spine-like petioles. Stipules adnate.* (Sp. 76.)

76. P. crassipes (Harv.); stem short, carnose, thickly covered with the persistent, thickened and hardened bases of old petioles; leaves *on very thick, tapering petioles,* pinnati-partite, pubescent, ... ; stipules minute, adnate, subulate; peduncles elongate, branched or panicled, with many-flowered umbels and lanceolate bracts; flowers sub-sessile, the calyx-tube very long, glandular, 4-6-times longer than the oblong, obtuse, scabrous segments; petals dark purple, small.

HAB. Collected by *Masson!* (Herb. Banks.)

Leaves not perfect in the only specimen seen, decompound, once or twice pinnate. The hardened petioles, 1-2 inches long, spreading in all directions, are very characteristic.

*** *Stem simple, succulent, or branched and suffruticose. Stipules adnate to the petioles, deltoid or subulate, conspicuous.* (Sp. 77-81.)

77. P. pulchellum (Curt. Bot. Mag. t. 524); stem short and succulent; leaves on short, hairy petioles, oblong, silky, *inciso-pinnatifid, the lobes oblong, acute; stipules broadly ear-shaped,* acute, adnate, rigid, veiny, silky; scapes branched, pilose, with many-flowered umbels and lanceolate, silky bracts; outer flowers pedicellate, inner sub-sessile, the calyx-tube many times longer than the lanceolate, hairy sepals. *DC. Prod.* 1. *p.* 665. *Sw. Ger. t.* 31. *P. pictum, Andr. Rep. t.* 168? *DC. Prod. l. c.*

HAB. Cultivated in Europe, 1695. Kaus Mountain, Namaqualand, *Drege!* (Herb. Benth., Jacq.)

Stem scarcely branched, sometimes obsolete. Stipules remarkably broad and stiff, like those of *P. appendiculatum.* The petals are white, each with a large, deep red spot. Seemingly a rare species.

78. P. hirtum (Jacq.! Ic. Rar. t. 536); stem short, fleshy, villous, armed with the persistent bases of old leaves; leaves on villous petioles, *bi-tri-pinnate,* the segments narrow-linear, partite, *densely and softly hairy;* stipules narrow-subulate, adnate; peduncles patently hairy, scape-like, but often leaf bearing in the middle, branched; umbels 3-8-flowered, with villous bracts; calyx tube villoso-hispid, the segments oblong, white-edged, obtuse. *DC. Prod.* 1. *p.* 661. *Cav. Diss. t.* 117. *f.* 2. *Sw. Ger. t.* 113. *E. & Z.!* 540, *also* 538 *and* 539. *P. tenuifolium, L'Her. t.* 12. *DC. l. c.*

HAB. Mountain sides, in the western districts. Lion's Mountain; and at Saldanha Bay, and Brackfontein, *E. & Z.!* Near Groenekloof and at the Paarl, *Drege!* (Herb. Banks., T.C.D., Hook., Sond.)

Stem short, slightly branched, erect or decumbent. Leaves very finely divided, like those of a carrot. Flowers rather small, rosy purple, the two upper petals darker and spotted. The whole plant is very hairy. In the Stockholm Herb. is a specimen of this plant from Thunberg, marked "*P. abrotanifolium.*"

79. P. dissectum (E. & Z.! 536); stem short, *suffruticose,* sub-simple; leaves on very long, patently hairy petioles, *tripartite, the segments fastigiate, multifid,* sparsely pilose, their divisions narrow-linear, patent, acute and hair-tipped; stipules minute, deltoid, adnate. membranous;

peduncles scape-like, longer than the leaves, retrorsely strigose near the summit, villous below; umbels few-flowered, with small bracts; calyx-tube sub-sessile, retrorsely strigose and thinly pilose, 3–4 times as long as the lanceolate, acuminate segments; petals narrow, on long claws, the two upper longer, emarginate. *P. setosum, E. Mey.! in Hb. Drege, non Sw.*

HAB. Near Philipstown and Fort Beaufort, Kat River, *E. & Z.!* Kaffirland, *Drege!* also the Nieuweveldsberg, Beaufort, and between Blauberg and Tigerberg. Cape. Albany, *Mrs. F. W. Barber!* (Herb. T.C.D., Hook., Sond.)

Stem decumbent at base, the leafy portion erect, 2–3 inches long. Leaves crowded; the petiole 4–5 inches long, the multifid lamina 1 inch across. The habit is like that of *Monsonia speciosa* in miniature. Petals white. The 7 fertile stamens are sub-equal, and rather shorter than the calyx segments.

80. P. multifidum (Harv.); *shrubby, slender* (prostrate?), the branches rusty, angular, pubescent; leaves petiolate, sub-rotund in outline, *digitately 3-7-parted, the segments cuneate,* fastigiate, deeply inciso-lobulate, *both surfaces appressedly hairy;* stipules broadly subulate, adnate; peduncles filiform, persistent, few-flowered, with deciduous bracts; calyx-tube sub-sessile, pubescent, 4–5-times as long as the oblong, obtuse segments; petals spathulate, not much exceeding the calyx. *Drege,* 9460. *Zeyher!* 2054.

HAB. Between the Zwarteberg and River Zonderende, *Zeyher!* (Herb. Sond.)

A rigid, though slender, branching suffrutex, probably growing in dry ground. Leaves scarcely more than ½ inch in diameter, densely clothed with rigid, white hairs. Flowers in bad preservation. The petals seem to have been dull yellowish.

81. P. sericeum (E. Mey.!); *whole plant silky;* stem short, dichotomous, woody, imbricated with persistent leaf-bases; leaves densely sericeous, *canescent,* cuneate, deeply trifid, the lateral segments linear, obtuse, the medial cuneate and 3-lobed; stipules adnate, subulate, spreading; peduncles branched, the partial 1–3-flowered; bracts lanceolate, connate; sub-sessile, the tube several times longer than the oblong, blunt sepals; petals obovate.

HAB. Kaus Mountain, near Goedman's-kraal, 3–4000 f. *Drege!* (Herb. T.C.D., Hook., Sond.)

A densely tufted, small, but stoutly woody plant, with closely imbricated, *sub-sessile,* shining and silky leaves. Flower-stalks often very short, but branched; the involucres frequently one-flowered. Calyx-tube 1½ inches long. Flowers white.

**** *Stem slender, branching, suffruticose. Stipules adnate, very minute or obsolete.* (Sp. 82–85.)

82. P. abrotanifolium (Jacq. Schoenb. t. 136.); shrubby, slender, much branched, *velvetty-canescent;* leaves pulverulent, flabelliform, tripartite, *the lateral segments deeply 2-3-lobed, the terminal multifid, with linear, channelled, blunt lobes;* stipules tooth-like, adnate, obsolete; peduncles filiform, few-flowered; bracts minute, ovate; calyx-tube much longer than the pedicels, or the hairy segments; upper petals broader and shorter than the lower. *DC. Prod.* 1. *p.* 661. *Cav. Diss. t.* 117. *f.* 1. *Sw. Ger. t.* 351. *P. artemisiæfolium, E. Mey., non DC.*

HAB. North-western districts and Namaqualand. Nieuweveld, near Rhinosterkopf, Kenko, Nieuwe Hantam, &c., *Drege!* Africa's Hoogte, *Burke and Zeyher!*

Somerset, *Mrs. F. W. Barber!* Cultivated since 1796 in England. (Herb. T.C.D., Hook., Sond.)

A slender, strongly aromatic, canescent shrub, with leaves like those of "Southern Wood" (*Artemisia Abrotanum*). It chiefly differs from *P. incisum* in the petals and pubescence, and from *P. exstipulatum* in the multifid leaves. Flowers white, the two upper petals with a red spot, or rosy. Cavanille's figure is very bad.

83. P. incisum (Willd.! Sp. 3. p. 686); suffruticose, densely branched, the branches rough with the bases of old leaves; leaves shortly petiolate, *pubescent, tri-partite, the segments pinnatifid or multifid*, with linear, obtuse lobules; stipules minute, deltoid, slightly adnate; peduncles few-flowered, with oblong bracts; calyx tube sub-sessile, 4 times as long as the lanceolate, acute segments; *upper petals longer than the lower. DC. Prod.* 1. *p.* 661. *E. & Z.!* 533. *Sw. Ger. t.* 93. *P. canescens, E. & Z.!* 534. *Zey.!* 2086.

HAB. Clayey hills and river banks, in Swellendam and George. Hassagaiskloof, and near the Gauritz River, *E. & Z.!* (Herb. Willd., T.C.D., Hook., Sond.)

A much branched, leafy, pubescent or hairy bush, 1 foot or more in height. Leaves less than an inch across. Flowers small, pale; the petals narrow-linear or spathulate, the two upper with a dark red spot. The pubescence is not *velvetty*, as in the preceding, to which this species is allied.

84. P. exstipulatum (Ait. Kew. 2. p. 431); shrubby, *thinly canescent;* leaves velvetty (small), roundish-ovate, obtuse, truncate at base, three-lobed, the lobes cuneate, inciso-crenate, the lateral ones small; stipules very minute, adnate; peduncles elongate, few-flowered, with ovate, minute bracts; pedicels and calyces velvetty, the cal.-lobes lanceolate; petals spathulate, short. *DC. Prod.* 1. *p.* 678. *L'Her. Ger. t.* 35. *Cav. Diss. t.* 123. *f.* 1. *E. & Z.!* 637. *P. fragrans, Willd.! non Sw., nec Andr.*

HAB. Among shrubs. Near the Gauritz River, in the Karroo, and the Langekloof, *E. & Z.!* Introduced to England, 1779. (Herb. Sond., Willd.)

A slender shrub, velvetty and canescent in all parts, and commonly known in English green-houses as the "*Penny-royal scented.*" It is chiefly prized for its perfume, and small, neat foliage. Leaves scarcely exceeding ½ inch in diameter. The flowers are small and white.

85. P. ionidiflorum (E. & Z.! 532); shrubby at base, the branches slender, flexuous; leaves on *patently hairy petioles, oblong obtuse, cordate at base*, deeply inciso-pinnatifid, the lobes obtuse, crenato-dentate, the upper surface pubescent, the nerves on the lower surface ciliate; stipules very minute, tooth-like, adnate; peduncles branching, patently hairy; umbels 5–6-flowered, with short, glabrescent bracts; calyx-tube pedicellate or sub-sessile, slender, sparsely setulose, 4–5 times longer than the lanceolate, acute segments; petals narrow, the two upper longest. *P. cortusæfolium, E. Mey.! non L'Her.*

HAB. Rocky ground in Albany, *E. & Z.! Mrs. F. W. Barber*. Near Grahamstown, *Williamson!* (Herb. T.C.D., Hook., Sond.)

Stems thickish and woody below, the branches slender, 6-8 inches long. Leaves less than an inch long, or scarcely longer, ½–¾ inch broad, more or less deeply pinnatifid. The pubescence is sparse, but the hairs are long, white and silky. The flowers are small and purple, with darker streaks on the upper petals.

***** *Stem slender, branching, suffruticose. Stipules free, ovate or subulate. (Leaves in P. cardiophyllum undivided.)* (Sp. 86–89.)

86. P. ramosissimum (Willd. Sp. 3. p. 688); bushy, densely branched, the branches hairy; leaves on hairy petioles, *elongate, pinnati-partite*, the pinnæ very short, inciso-multifid or pinnatifid, scabrous and hairy, with linear lobules; stipules free, ovate, acuminate, membranous; peduncles alternately branched, the umbellules racemose, few-flowered, patent; bracts short, ovate; calyx scabrous, its segments linear, obtuse, very rough; petals narrow, with long claws. *DC. Prod.* 1. *p.* 661. *E. & Z.!* 537. *Burm. Afr. t.* 34. *f.* 2. *Zey.!* 2068 & 187.

HAB. Heathy ground near Beaufort, *E. & Z.!* Cradock, and in Namaqualand, *Burke and Zeyher!* (Herb. T.C.D., Hook., Sond.)

A small, bushy suffrutex, 1–2 feet high, hairy and rather viscous. Leaves very finely divided, 2–3 inches long and ½ inch wide. Old flower-stems often persistent, bearing several horizontally spreading peduncles.

87. P. tripartitum (Sw. Ger. t. 115); stem suffruticose, straggling; branches slender, brittle, pubescent; leaves on petioles longer than the lamina, *deeply trifid or tri-partite, the segments densely pubescent*, cuneate, inciso-trifid; stipules small, *ovate*, acute, membranous, free; peduncles spreading, pubescent; umbels 3–6-flowered, with ovate bracts; calyx-tube, scabrid, 4 times longer than the lanceolate, acute segments; petals narrow, the two upper longest. *DC. Prod.* 1. *p.* 661. *P. tripartitum, and P. fragile, Willd. P. trifidum, Jacq. Schoenb. t.* 134. *E. & Z.!* 535.

HAB. Cultivated, 1796. In the Karoo, between the Gauritz River and Langekloof, *E. & Z.!* Zuureberg, *Drege!* (Herb. T.C.D., Hook., Sond.)

A slender, half-shrubby, weak-growing plant, with flexuous branches and variously cut leaves. Petals cream-coloured, the two upper with a deep red medial stripe. Jacquin's figure, above quoted, exactly represents *Drege's* specimens.

88. P. Artemisiæfolium (DC. Prod. 1. p. 661.; fide E. & Z.); stem suffruticose, erect, sub-simple, glabrous; leaves on long petioles, *bi-pinnately-partite*, with *nearly glabrous*, linear-filiform, channelled, acute segments; stipules subulate, free; peduncles elongate, 2–3-flowered; bracts numerous, subulate; calyx tube on a longish pedicel, swollen at base, 2–3 times longer than the lanceolate, cuspidate, sparsely setulose segments; upper petals not twice as long as the sepals, lower spathulate. *E. & Z.!* 558. *Zey.!* 2066.

HAB. South Africa, *Dr. Grondale!* Mountain sides near Zonderende River, at Knoblanch, Swellendam, *E. & Z.!* (Herb. Sond., Hook.)

Placed by *E. & Z.!* in "*Myrrhidium*," but readily known from any of that section by the long pedicels to the calyx tube, the want of prominent ribs on the calyx segments, the subulate stipules and the erect habit. The leaves are usually as above described, but specimens sometimes occur with *ovate, inciso-lobulate* leaves; showing a tendency towards *P. cardiophyllum*.

89. P. cardiophyllum (Harv.); stem short, sub-simple, or with innovations; leaves on very long, nearly glabrous, channelled petioles, *rigid, cordate, crenato-dentate (rarely 3-lobed and laciniate)*, palmately 3–5-nerved, scaberulous or setulose on both sides; stipules subulate, attenuate, free; peduncles short, 2–3-flowered, with subulate bracts; pedicels elongate, exceeding the linear, curved, hispidulous calyx-tube, which is longer than the lanceolate segments. *Eumorphia elegans, E. & Z.!* 601, *non Sw. Zey.!* 2084.

VAR. *β.* **laciniatum**; leaves 3-lobed, the lobes laciniate and toothed.

HAB. Rocky mountain sides, above the Baths, and Baviansberg, Gnadenthal, *E. & Z.!* River Zonderende, *Zeyher!* β. in Herb. Sond., from *Zeyher!* (Herb. Hook., Sond.)

Stem scarcely any or 2–3 inches long, woody at base, the innovations herbaceous. Leaves crowded, the petioles 4–5 inches long, the lamina scarcely an inch. Except for the minute, scattered bristles the plant is glabrous. The lower petals are nearly as long as the upper, but are much narrower; all the petals are blotched, or suffused with rosy purple. Notwithstanding the very different foliage this species is closely allied to *P. artemisiæfolium*. The leaves, though generally cordate, are sometimes 3-lobed, and, as in var. β., *laciniate!*

[Ligulariæ *unknown to me.*]

90. P. minimum (Willd. Sp. p. 664); "stem scarcely suffruticose, erect; branches glabrous, herbaceous; leaves pinnatisect, the lower segments pinnatifid, upper entire and with linear-oblong lobules; umbels several-flowered." *DC. Prod.* 1. *p.* 660. *G. minimum, Cav. Diss. t.* 121. *f.* 3."

91. P. athamanthoides (L'Her. Ger. in ed. 47); "stem scarcely suffruticose, erect; branches sub-herbaceous, angular; leaves appressedly cinereo-tomentose, bi-pinnatisect, laciniate, the lobules linear-subulate, acute; umbels compound." *DC. l. c. p.* 660.

92. P. confusum (DC. Prod. 1. p. 661); "stem shrubby; leaves glaucous, glabrous, decompound, segments cuneate, inciso-dentate, the lowest divaricate; the many-leaved involucre and the calyx ciliated." *P. sanguineum, Willd., non Sw.* "Near P. abrotanifolium and P. tenuifolium (hirtum). DC. l. c."

Sect. 6. **JENKINSONIA**. *Stem* shrubby, or succulent and jointed. *Leaves* palmately nerved or lobed. *Stipules* rigid or membranous, persistent. *Petals* 4–5, *the two uppermost very much larger than the lower*, broadly obovate, emarginate or obcordate. Fertile stamens 7. (Sp. 93–95.)

93. P. quinatum (Bot. Mag. t. 547); stem shrubby, flexuous, minutely puberulous; leaves *on short petioles*, somewhat reniform, *deeply 5-lobed or 5-parted, the laciniæ bluntly 3-toothed or cleft, scurfy-puberulous;* stipules minute, deltoid, rigid, persistent; peduncles 1–2-flowered, patent, with 2–3 ovato-lanceolate, membranous bracts; pedicels very short, the calyx-tube scabrid, twice as long as the linear-lanceolate, acuminate, scabrid, white-edged and ciliate segments, the upper broader. *DC. Prod.* 1. *p.* 659. *Sw. Ger. t.* 79. *E. & Z.!* 541. *Ger. præmorsum, And. Rep. t.* 150. *Drege* 7476.

HAB. Cult. 1798. Oliphant's River, Clanw., *E. & Z.!* Namaqualand, *Drege!* (Herb. T.C.D., Hook., Sond.).

Stem much branched, distinctly jointed; the old parts shining, the younger branches, the leaves and peduncles minutely scurfy. Internodes an inch long; leaves about the same length. Flowers large, cream-coloured, with dark purple streaks, the two upper petals 1½ inch long, broadly obovate, emarginate, very much larger than the three lower, which are spathulate and white. Sepals ½ inch long or more. Style and 7 fertile stamens elongate, declined. *Drege*, 7476 (Herb. Sond.), partly consists of this species, partly of the barren stems of something else.

94. P. antidysentericum (E. & Z. 542); *root very large and turnip-shaped;* stems suffruticose (annually renewed?), slender, multangular and striate, spreading, much-branched; branches filiform; leaves on slender petioles, reniform, deeply 3-5-lobed, or parted, the segments incised or 3-lobed and toothed, softly pubescent; *stipules small, subulate, hardening into recurved thorns;* peduncles short, spreading, 2-3-flowered; pedicels twice as long as the calyx-tube, which is striate, scabrid, and about equalling the linear, acute segments. *Drege, No.* 7477!

HAB. Kamiesberge, Namaqualand, *E. & Z.!* Silverfontein, *Drege!* (Herb. Sond.) Roots, when full grown, the size of a man's head. Stems numerous, 2 feet long and more, weak and straggling; leaves and flowers small. "This plant is called *t'namie* by the aborigines, who boil the roots in milk and make use of them in dysentery." *Pappe.* I have not seen the flowers, which are said to be purplish; the style and fertile stamens longer than the calyx-segments. Does it belong to this section? The foliage and ramification, but not the stipules, resembles those of *P. patulum.*

95. P. tetragonum (L'Her. Ger. t. 22); stem *succulent,* branches sub-dichotomous, *obtusely* 4-3-angled, distinctly jointed, smooth; leaves few, on long, pilose petioles, *reniform-cordate, carnose, crenato-lobulate,* sparsely villous; stipules cordate, membranous, small, peduncles short, spreading, two-flowered, with 4 lanceolate bracts; flowers nearly sessile, the calyx-tube very long, sparsely setulose, the segments linear-lanceolate, white-edged, and ciliate, acute, the upper broader. *DC. Prod.* 1. *p.* 658. *Pl. Grass. t.* 96. *Bot. Mag. t.* 136. *Jacq. Ic. Rar. t.* 132. *Sw. Ger. t.* 99. *E. & Z.!* 544. *Chorisma flavescens, E. & Z.* 545.

HAB. Introduced to England, 1774, by *F. Masson.* Zoutpanshoogde, near the Zwartkops River, Uit., *E. & Z.!* also Swellendam, near the Gauritz River. (Herb. T.C.D., Hook., Sond.).

A remarkable species, readily known by its naked, square, succulent, often leafless stems. The few leaves are 1-1½ inch across. The flowers are of large size and generally 4-petalled, the two upper petals purplish and streaked, obovate, with long claws; the 2-3 lower not half as long, white and spathulate.

Sec. 7. **MYRRHIDIUM**. *Stems* slender, diffuse, biennial or suffruticose, rarely annual. *Stipules* foliaceous or membranous, withering. *Leaves* pinnatifid or pinnati-partite. *Petals* 4, rarely 5, the two uppermost obovate or cuneate, much larger than the lower, which are linear. *Sepals membranous, strongly ribbed and mucronate, or taper-pointed.* (Sp. 96-101.)

96. P. myrrhifolium (Ait. Kew. 1. vol. 2. p. 421); suffruticose at base or quite herbaceous, slender, diffuse, variously pubescent; branches angular; leaves *ovate-oblong, more or less deeply pinnatifid or bi-pinnatifid,* the segments glabrous or pubescent, or villous, linear or cuneate, flat or somewhat channelled, deeply cut or again pinnatifid; stipules broadly ovate, acuminate; peduncles 2-6-flowered; calyx-tube setulose, segments lanceolate, taper-pointed, strongly ribbed; upper petals much longer and broader than the lower.

VAR. α. **fruticosum**; *suffruticose, glabrescent;* leaves bi-pinnatifid, with narrow-linear *channelled* segments; peduncles 2-3-flowered; calyx-tube *very long;* flowers large, pale purple or white with purple streaks. *P. fruticosum, DC. p.* 661. *E. & Z.!* 557. *Cav. Diss. t.* 122. *f.* 2. *Zey.!* 2059.

VAR. β. **coriandrifolium**; herbaceous, *villoso-pubescent*, becoming subglabrous; leaves bi-pinnatifid, with narrow-linear *flattish* segments; calyx-tube *short;* flowers large, white, streaked with purple. *P. coriandrifolium, Jacq.! Ic. Rar. t.* 528. *DC. Prod.* 1. *p.* 657. *E. & Z.!* 556. *Sw. Ger. t.* 34.

VAR. γ. **athamanthoides**; herbaceous, *villous;* leaves bi-pinnatifid with linear-cuneate *flat* segments; peduncles 5–6-flowered; calyx-tube *mostly elongate;* flowers large, white with purple streaks. *P. athamanthides, L'Her. Ger. No.* 47. *DC. Prod.* 1. *p.* 660. *E. & Z.* 554.

VAR. δ. **intermedium**; flowers as in var. γ.; leaves as in var. ε. *P. athamanthoides, β. glabrescens, Harv. in Herb. Hook. and T.C.D.*

VAR. ε. **lacerum**; herbaceous, glabrescent; leaves bi-pinnatifid, with *linear-cuneate*, flat, deeply incised segments; calyx-tube *short;* flowers smaller, *rosy purple*, with dark streaks. *P. lacerum, Jacq. Ic. Rar. t.* 532. *DC. Prod.* 1. *p.* 657. *E. & Z.!* 555. *Zey.!* 2058. *Drege!* 7488.

VAR. ξ. **Synnoti**; leaves somewhat broader and less deeply cut than in var. ε; petals 4. *Jenk. Synnoti, Sw. t.* 342.

VAR. η. **pendulum**; *suffruticose*, with long pendulous branches, hairy; leaves bi-pinnatifid, with broad, cut segments; calyx-tube scarcely longer than the segments; fl. large, 4-petalled, rosy purple, with red streaks. *Jenk. pendula, Sw. Ger. t.* 188. *G. lacerum, Andr. Ger. cum ic.*

VAR. θ. **betonicum**; suffruticose at base, roughly pubescent; leaves more or less deeply *pinnatifid*, with broadly cuneate, coarsely toothed or lobulate segments; calyx-tube short; flowers rather small, with white, darkly streaked or rosy petals. *P. betonicum, Jacq. Ic. t.* 531. *Cav. Diss. t.* 118. *f.* 1. *E. & Z.!* 548. *P. bullatum, Jacq. Ic. t.* 530. *DC. Prod.* 1. *p.* 657. *E. & Z.!* 549. *Zey.* 2060. *P. myrrhifolium, DC. l. c. P. anemonifolium, Jacq.! Ic. Rar.* 535.

VAR. κ. **longicaule**; suffruticose at base, downy; leaves inciso-pinnatifid, the lobes broadly cuneate, inciso-dentate; calyx tube *long;* fl. rather large, white or rosy. *G. longicaule, Jacq.! Ic. Rar. t.* 533. *DC. Prod. l.* 657. *E. & Z.!* 550. *Zeyher*, 2057.

HAB. Common round Capetown and throughout the western districts; several of the vars. often growing together. Var. α. Zwarteberg, Cal., and Gauritz River, George, *E. & Z.!* Zonderende, *Zeyher!*

I have no hesitation in bringing together, under a single species, the forms here enumerated, and should not have retained several of them, even as varieties, had they not, by most botanists, been kept up as species. The greater the number of specimens examined, the less possible will it be found to fix limits between the forms.

97. P. multicaule (Jacq. Ic. Rar. t. 534); stem herbaceous, weak and straggling, angular, minutely scaberulous; leaves ovate-oblong, pinnatifid or pinnati-partite, *glabrous or nearly so*, with remote, narrow-cuneate, incised or sub-pinnatifid segments; stipules broadly ovate, acuminate; peduncles 5–10-flowered; calyx-tube setulose, equalling or exceeding the lanceolate, acuminate, ribbed and setulous segments; upper petals scarcely longer than the sepals. *DC. l. c.* 658. *E. & Z.!* 552–553. *Drege*, 7486, *α*, 7487. *Zey.* 2056.

HAB. Fields near the Zwartkop's River, Uit., and the Fish River, *E. & Z.!* &c. (Herb. Jacq., T.C.D., Benth., Sond.).

Chiefly known from *P. Myrrhifolium* by its more angular and glabrous stems, and the more distant and narrow, nearly glabrous leaf-segments. The calyx-tube rarely equals the sepals in length; the petals are rosy purple, with dark streaks. The flowers are sometimes very numerous. I retain this species, chiefly because it seems to be confined to the eastern districts, while *P. myrrhifolium* is quite a western plant. *Drege's* 7486, *a.* (Herb. Sond.) quite agrees with the ordinary plant; his 7486, *b*, (Herb. Hook.) has a calyx-tube thrice as long as the sepals.

98. P. urbanum (E. & Z. ! 546); stem herbaceous, weak and straggling, *tomentose;* leaves pinnatifid or bi-pinnatifid, pubescent or canescent on both surfaces, the lobes broadly linear or acute, crenate or bluntly lobulate; stipules broadly *ovato-cordate, obtuse or mucronulate, green, sub-persistent;* peduncles hairy, 3–5-flowered, with obtuse bracts; calyx-tube hispidulous, longer or shorter than the lanceolate, mucronate, ribbed and hispid sepals. *Drege,* 7485. *Zey.* 2089.

VAR. *α.* **pinnatifidum**; leaves pinnatifid, segments broadly cuneate, toothed; calyx-tube *shorter* than the sepals.

VAR. *β.* **bi-pinnatifidum**; leaves bi-pinnatifid, segments broadly linear, lobulate; calyx tube *longer* than the sepals. *P. anemonifolium, E. & Z.!* 551, *non Jacq.*

HAB. On sand hills and in dry places among shrubs. Var *α.* Near Uitenhage, *E. & Z.!* Port Elizabeth, *Zeyher! Drege!* Var. *β.* Near Cape L'Aguilhas, *E. & Z.!* Cape Flats, *Dr. Pappe!* (Herb. T.C.D., Hook., Sond.).

Well distinguished from *P. myrrhifolium* by its roundish-cordate, obtuse stipules and softer pubescence. The two varieties differ in the degree of leaf-incision and length of calyx-tube; both very variable characters. *P. anemonifolium, Jacq.!* according to the original in Herb. Vind.! is a var. of *P. myrrhifolium.*

99. P. candicans (Spreng. Syst. 3. p. 57); suffruticose, slender, diffusely branched; branches pubescent and hispid; leaves on long petioles, *canescent and silky on both sides, ovate-oblong, cordate at base,* more or less 3-lobed and crenulate, the basal lobes short, rounded, the middle lobe ovate, cuneate at base, sometimes lobulate; stipules ovate, acuminate; peduncles 2–3-flowered, with downy bracts; calyx-tube setulous, shorter than the lanceolate, acuminate ribbed and setulous sepals; upper petals not twice as long as the calyx. *E. & Z.!* 547. *Zey.!* 2061, 2062. *Drege,* 7484.

HAB. Between Swellendam and Kochman'skloof, *E. & Z.!* Assagaiskloof, and between the Zwarteberg and River Zonderende, *C. Zeyher.* Hills near the Karmelks River, *Drege!* (Herb. Sond., Hook., Benth.).

Stems weak, erect or decumbent, the upper branches with very long internodes, angular. Petioles much longer than the lamina, which is variously cut, sometimes nearly simple, sometimes 3-lobed, sometimes 3-parted and with the middle lobe deeply cut; always crenate and densely clothed with soft, shining, appressed, silky hairs. Upper petals much larger than the lower, with dark streaks. The foliage is quite unlike that of any other Cape species of this section, and most resembles that of *P. canariense.*

100. P. senecioides (L'Her. Ger. t. 11); *annual;* stem branching, zigzag, pubescent; leaves on long petioles, deltoid in outline, ter-quinately bi-pinnatifid, the lobes narrow-linear, incised, obtuse, pubescent; stipules minute, deltoid; peduncles opposite the leaves, patent, roughly pubescent; umbels few-flowered, with ovate, obtuse bracts; calyx-tube scabrous, the segments unequal, *the broader very blunt and strongly* 3–

5-ribbed, the narrower sub-acute, one-ribbed; petals obovate, sub-equal. *DC. Prod.* 1. *p.* 660. *Sw. Ger. t.* 327. *E. & Z.!* 559. *P. phellandrium, E. Mey.! Zey.* 183.

HAB. Cape District, at Tigerberg, &c., *E. & Z.!* Cape Flats, *Dr. Pappe!* Berg Valley and Langevalley, *Zeyher!* (Herb. T.C.D., Hook., Sond.).

A much-branched, diffuse annual, 1–2 feet high, with dense, roughish, glandular pubescence. The upper leaves are simply pinnatifid or merely incised and much smaller than the rest. The flowers are white, with red streaks on the outside, and the two upper petals, which are very broad, have a dark spot at the base and on the claw. Drege's 7482, seems to be a dwarf state of this species, with larger calyces than usual.

[*Species of* Myrrhidium *unknown to me.*]

101. P. caucalifolium (Jacq. Ic. Rar. t. 529); "stem herbaceous, biennial, hairy; leaves bi-pinnatifid, the lobes linear, glabrescent; peduncles one-flowered." *DC. Prod.* 1. *p.* 658. Petals white or flesh-coloured, with red veins.

Sec. 8. **PERISTERA.** *Stems* herbaceous, diffuse, slender, annual or perennial. *Leaves* lobed or pinnatifid. *Stipules* membranous. *Flowers minute! Petals* sub-equal, scarcely longer than the calyx-segments. *Fertile* stamens 5. (*These resemble* Gerania *and* Erodia *in habit.*) (Sp. 102–108.)

102. P. grossularioides (Ait. Hort. Kew. 1. vol. 2. p. 42); herbaceous, *glabrous, pubescent or villous,* many-stemmed; stems angular and furrowed; leaves on long petioles, roundish-cordate, crenate, more or less deeply 3–5-lobed or 3–5-parted, with cuneate, deeply cut lobes; stipules deltoid, acuminate; peduncles filiform, 3–8-flowered, with lanceolate bracts; calyx-tube funnel or trumpet-shaped, short or long, sepals *ovato-lanceolate, acuminate, nerved,* nearly as long as the minute petals. *P. grossularioides and P. anceps, DC. l. c.* 660. *E. & Z.!* 561–7. *Cav. Diss. t.* 119. *f.* 2.

VAR. α. **anceps**; nearly glabrous; the leaves roundish cordate, crenate or with shallow lobes. *Ger. anceps, Jacq. Coll.* 4. *t.* 22. *f.* 3. *P. nummulariæfolia, E. & Z.! Drege,* 7466, 6895.

VAR. β. **columbinum**; pilose or glabrescent; leaves multi-partite, with deeply incised segments. *P. columb. Jacq. Schoenb. t.* 133. *DC. l. c.* 654. *E. & Z.!* 567. *P. micropetalum, E. Mey.!*

VAR. γ. **pubescens**; whole plant pubescent or sub-canescent; leaves variously incised. *P. humifusum? and P. procumbens, DC. l. c. E. & Z.! Drege,* 7465, 7468. *Zey.!* 2065. *P. clandestinum, L'Her. Hook. f. Fl. Nov. Zeal.* 1. *p.* 41. *P. brevirostre, E. Mey.!*

VAR. δ. **iocastum**; whole plant sericeo-pilose; the pedicels of the flowers and the calyx-tube much longer and more slender than usual; flowers purplish or white, larger than in var. γ. *P. iocastum, E. & Z.! P. mutabilis, E. Mey.! Drege,* 7464, 7465.

HAB. In moist places throughout the Colony. Var. δ. in woody places, on the sides of Tulbagh Mt., Worcester, *E. & Z.!* (Herb. T.C.D., Hook., Sond.).

A very variable plant with minute, purple or white flowers. The above varieties appear to me to run, one into the other, and others might be given with intermediate characters.

103. P. althæoides (L'Her. Ger. t. 10); perennial, many-stemmed; stems herbaceous, slender, diffuse, terete, with very long internodes, canescent and downy; leaves often opposite, on very long petioles, *obtusely cordate*, somewhat 5–7-lobed, the lobes rounded, crenato-dentate, *silky above, densely albo-tomentose and veiny underneath;* stipules membranous, deltoid; peduncles filiform, 6–8-flowered, with lanceolate bracts and very short pedicels; *calyx densely villoso-tomentose*, the tube twice as long as the segments; flowers minute. *DC. l. c.* 655. *E. & Z.!* 560. *P. fragrans, E. Mey.! non Willd. Drege*, 7452, 7470. *Zey.!* 2063.

HAB. Kamiesberg, Namaqualand, *E. & Z.!* Uienvalei, *Drege!* also Wildchutzberg, Stormberg, and Witberg, Caffraria, *Drege.* (Herb., T.C.D., Benth., Sond.)

Apparently a distinct species, with soft, albo-tomentose leaves, like those of an *Abutilon* or *Althæa.* All parts white-hairy. Flowers like those of *P. grossularioides.* Zeyher's 2063, (from Zwarteberg) is less tomentose than usual, but is nearer this species than to anything else. The specimen in Herb. Banks, a garden one, is also less tomentose.

104. P. chamædryfolium (Jacq. Ic. Rar. t. 523); annual or biennial, many-stemmed, procumbent; stems angular, hairy; leaves on long, hairy petioles, *oblong, 3-lobed at base or pinnatifid*, obtuse, pubescent or hairy, denticulate; stipules membranous, *broadly subulate;* peduncles filiform, hairy, 4–8-flowered; pedicels longer than the hispid calyx-tube, *which is shorter than the oblong-acute sepals;* flowers small. *DC. Prod.* 1. *p.* 654. *Erodium maritimum, E. & Z.! No.* 452, *non aliorum.*

HAB. Sandy and dry hills, about Wynberg, &c., *W.H.H.* Simon's Bay. *C. Wright!* (Herb. Jacq., T.C.D., Hook., Sond.).

More robust than *P. grossularioides*, with more oblong leaves, larger flowers, especially a longer calyx, with broadly oblong sepals. Internodes distant. The leaves vary from exactly elliptic-oblong to 3-lobed, or even pinnatifid, on different parts of the same plant. The minute petals are a fine purple. E. & Z.! have confounded this plant with *Erodium maritimum.*

105. P. parvulum (DC. Prod. 1. p. 660); annual or biennial, herbaceous, *densely pubescent;* stems angular; leaves on long petioles, *oblong-cordate, obtuse, crenate, simple or inciso-pinnatifid, with cut lobes;* stipules *deltoid-acuminate;* peduncles filiform, 4–6-flowered; flowers minute, pedicellate with *a very short calyx-tube and elliptic-oblong, albo-marginate sepals*, equalling the petals. *P. nanum, L'Her. in Hb. Banks! P. humifusum, Willd.! P. procumbens, E. Mey.! Herb. Drege, non DC.; also Drege*, 7469.

HAB. Cape, *Masson!* Silverfontein, Namaqualand, and at Drakensteen, *Drege!* (Herb., Willd., Banks., Hook., T.C.D., Sond.)

Smaller than *P. chamædryfolium*, with more oblong leaves than *P. grossularioides*, from which it more particularly differs in the sepals. It has, as De Candolle remarks, the aspect of *Erodium maritimum*, but that the leaves are more deeply cut. Possibly it may be only a starved state of *P. chamædryfolium?*

106. P. fumarioides (L'Her.! in Herb. Banks); annual or biennial, densely pubescent and *canescent;* stems angular, diffuse; leaves on long petioles, *pinnati- or bi-pinnati-partite, the segments linear, entire or cut, obtuse;* stipules deltoid-acuminate; peduncles shorter than the petiole, many-flowered; flowers minute, pedicellate, with a very short or *obsolete* calyx-tube, and elliptic-oblong, obtuse, *white and silky* sepals, as long as the petals. *P. parvulum, E. & Z.!* 568, *non DC. P. columbinum, E. Mey.! non Jacq. Zey.!* 159, 161.

HAB. Karroo, between Beaufort and Graaf Reynet, *E. & Z.!* Bitterfontein, *Zeyher*. Nieuweveldsberg, Sneeuweberg, 6000 f.; and near the Gareep, 4200 f., *Drege!* Graaf Reynet, *Mrs. F. W. Barber*. (Herb. Banks, T.C.D., Hook., Sond.).

A straggling small plant with leaves like those of *Geranium incanum*. The petals are "bright crimson." The calyx-tube is sometimes nearly obsolete, when the plant almost becomes an *Erodium*, to which genus Zeyher referred his specimens.

[*Species of* Peristera *unknown to me.*]

107. P. clavatum (L'Her. ined.); "stems herbaceous, erect, puberulent; leaves reniform, obsoletely lobed, dentato-crenate; umbels many-flowered; calyx-tube clavate, twice as long as the acuminate segments." *DC. Prod.* 1. *p.* 660. (C.B.S., *Sonnerat.*)

108. P. distans (L'Her. ined.); "stems herbaceous, erect, 4-angled, glabrous; leaves remote, sub-rotund, 5-lobed, glabrous, the uppermost inciso-lacerate; umbels densely many-flowered, on very long peduncles." *DC. l. c. p.* 660. (C.B.S., *Sonnerat.*)

Sec. 9. **CAMPYLIA**. *Root* thickened, but scarcely tuberous. *Stems* short, suffruticose at base, simple or slightly divided. *Leaves* crowded, on long petioles, ovate, oblong, lanceolate or linear, toothed or lacerate. *Stipules* membranous, subulate, often bifid. *Peduncles* branched. *Flowers* on long pedicels. *Petals* unequal, the two upper shortly and broadly obovate, somewhat eared at base. *Fertile* stamens 5; two of the sterile ones frequently hooked. *Filaments mostly pubescent.* (Sp. 109–116.)

109. P. ovale (Burm. Cap. 19); *villoso-canescent and silky*; stem short, sub-simple, scaly and tomentose; leaves on long petioles, *elliptical, or oblong, or ovate, or sub-rotund,* obtuse or acute, crenate, penninerved, canescent or tomentose on both sides; stipules deltoideo-subulate, bifid, rufous; peduncles branched, the partial elongate, few or many-flowered; calyx-tube shorter than the pedicel; sepals villoso-canescent, elliptico-lanceolate, mucronate; staminal tube mostly hairy. *Ger. ovatum, Cav. Diss. t.* 103. *f.* 3. *P. eriostemon, Jacq. Schoenb. t.* 132. *P. trichostemon, Jacq. Ic. Rar. t.* 524. *P. blattarium, Jacq. Schoenb. t.* 131. *Sw. Ger. t.* 88. *P. verbasciflorum, DC. Sw. Ger. t.* 157. *P. holosericeum, Sw. Ger. t.* 75 *(hybrid). Ger. tomentosum, Andr. Rep. t.* 115. *Campylia cana, Sw. Ger. t.* 114. *P. scaposum, &c., DC. Prod.* 1. *p.* 656. *Ger. glaucum, Cav. Diss. t.* 103. *f.* 2, *non Andr.*

VAR. α. **blattarium**; leaves albo-tomentose, sub-rotund or oval; peduncles and calyces very villous. *E. & Z.!* 571. *Zey.!* 2072.

VAR. β. **veronicæfolium**; leaves lanceolate-oblong, silky, sub-canescent. *E. & Z.!* 572.

VAR. γ. **Dregei**; dwarf, thinly canescent, without villosity. *P. scaposum and P. canum, E. Mey.! in Hb. Drege.*

VAR. δ. **ovatum**; copiously villous; leaves mostly ovate, acute or obtuse, unequally toothed; flowers bright purple. *Camp. cana, E. & Z.!* 570. *Drege,* 7472! *P. ovale, L'Her. t.* 28. *DC. Prod.* 1. *p.* 666.

HAB. On dry hills and among bushes. Hassagaiskloof and Breede River, Swell., and Krum River, George, *E. & Z.!* Langekloof, *Drege!* Uitenhage, *E. & Z.! Drege!* Albany, *Dr. Stanger!* &c. (Herb. Jacq., T.C.D., Hook., Sond., Banks.).

Variable in size of leaf and flower, in the amount of pubescence, and partly in the form of leaf and sepals. Still, after carefully considering the figures and descriptions above quoted, and comparing with Jacquin's original specimens, and with many others from various collectors, I find it impossible to recognize, among the forms before me, more than a single species. The flowers vary from white, with dark blotches, to pale purple, and to deep purple with blotches. Burman's specific name *ovale* deserves to be restored, not merely for its antiquity, but for its appropriateness.

110. P. elegans (Willd. Sp. 3. p. 655); stem short, thick, sub-simple, scaly; leaves on long hairy petioles, *sub-rotund or oval, rigid*, coarsely toothed, *glabrous or sparsely hairy* above, the nerves strigose; stipules deltoid-acuminate, pubescent; peduncles branched, the partial long and villous 4–6-flowered; the long pedicels, short calyx-tube and lanceolate sepals villous. *DC. Prod.* 1. *p.* 666. *Andr. Rep. t.* 28. *Sw. Ger. t.* 36. *Camp. eriostemon, E. & Z.!* 569, *non Jacq.*

HAB. Early introduced to Europe. Near Cape L'Aguilhas, *E. & Z.!* (Herb. Hook., Benth., Sond.).

Nearly allied to *P. ovale*, especially to our var. γ, but the leaves are broader and rounder, more rigid, more decidedly toothed, and the pubescence is thin, and *hispid*, not silky. The stamens are downy. The flowers are large, a purplish rose-colour in our wild specimens; white, suffused with purple and streaked, in the cultivated.

111. P. Œnotheræ (Jacq. Ic. Rar. t. 525); *whole plant canescent;* stem short, sub-simple, scaly; leaves mostly shorter than the petioles, *lanceolate or linear-oblong*, cuneate or obtuse at base, toothed; stipules adnate, subulate, divergent; peduncles branched, the partial longer than the leaves, 3–4-flowered; pedicels not much longer than the calyx-tube, which is shorter than the broadly lanceolate, densely pubescent sepals. *DC. Prod.* 1. *p.* 656. *Camp. cartilaginea, E. & Z.!* 573. *P. elatum, E. Mey.! in Hb. Drege, non DC. Drege*, 7473.

HAB. Cult. 1796. Piquetberg, Stell., *E. & Z.!* Ezelsbank, 3–4000 f., and between Bergvallei and Langevallei, *Drege!* (Herb. Jacq.! T.C.D., Hook., Sond.).

Nearly related to *P. coronopifolium*, but the leaves are shorter, more equally toothed throughout, and not tapering much at base; the flowers also are smaller and the calyx-tube very short. Petals rosy purple. In Herb. Banks! is a specimen from Masson with longer and broader, more coarsely toothed leaves; 4–5-flowered umbels and longer pedicels; the general pubescence less canescent than usual.

112. P. coronopifolium (Jacq. Ic. Rar. t. 526); stem short, sub-simple, scaly; leaves *narrow-lanceolate or linear, tapering greatly at the base*, acute, toothed towards the upper extremity or entire, *thinly canescent*, with minute, appressed hairs; stipules subulate from a broad base, silky; peduncles branched, the partial shorter than the leaves, 2–3-flowered; pedicels four times as long as the short calyx-tube, whose pubescence, as well as that of the lanceolate sepals is reflexed; upper petals obovate. *DC. Prod.* 1. *p.* 656. *G. angustifolium, L'Her. in Hb. Banks.*

VAR. β. **lineare**; leaves narrow-linear, sub-entire, with sub-revolute margins. *Ger. lineare, L'Her. in Hb Banks. P. angustissimum, E. Mey.! Camp. staticephylla, E. & Z.!* 574.

HAB. Between Lang Valley and Oliphant's River; near Ezelbank; and on the Gift-berg, *Drege!* Var. β. Heerelogement, *E. & Z.!* Skurfdeberg, *Zeyher!* (Herb. Banks., T.C.D., Hook., Sond., Jacq.).

Readily known by its long, very narrow, tapering, scarcely canescent leaves. They vary a little in breadth and in the toothing of the margin. The flowers are a bright, violet purple.

113. P. tricolor (Cart. Bot. Mag. t. 240); stem *shrubby*, short, *branching, diffuse*, rough with the debris of old leaves; leaves on long, slender petioles, villoso-canescent, *lanceolate or oblong, inciso-dentate or lobed*, stipules subulate, adnate; peduncles branching, the partial 2–3-flowered; bracts subulate; flowers on long pedicels, the calyx-tube shorter than the lanceolate-acuminate, villous sepals. *DC. Prod.* 1. *p.* 657. *Sw. Ger. t.* 43. *E. & Z.!* 576. *P. violarium, Jacq. Ic. Rar. t.* 527. *P. elatum, Sw. Ger. t.* 96. *DC. l. c. E. & Z.!* 577.

Var. *β*. **concolor**; the lower petals rosy-purple. *P. capillare, E. & Z.! No.* 575. *Zey.!* 185. (vix. Cav. Diss. t. 97. f. 1.)

HAB. Early introduced to England. Near the Gauritz River, Swell., *E. & Z.!* Var. *β*. in the Winterhoek's berg, Tulbagh, *E. & Z.!* Witsenberg, *Zeyher*. (Herb. T.C.D., Hook., Sond.)

Stem divided near the base into several tufted, sub-simple branches, 3–12 inches long. Whole plant covered with soft, short, appressed silky hairs which are often white; young branches and peduncles often villous. The two upper petals are generally a very dark red, the 3 lower white; sometimes the upper are coloured at the base only, where there is always a darker spot, rough with small tubercles. Our var. *β*. seems merely to differ in the colour of the petals and the starved condition of stem and leaves. It scarcely answers to the *P. capillare*, which is only known by Burman's figure and description, copied by Cavanilles.

[*Species of* Campylia *unknown to me.*]

114. P. dichondræfolium (DC. Prod. 1. p. 656); "stem suffruticose, erect; leaves reniform, crenulate, canescent; peduncles 5-flowered; upper petals obovate, lower oblong. *Burch. Cat.* 3084.

115. P. carinatum (Sw. Ger. t. 21); "stem suffruticose, ascending; leaves ovate, unequally toothed and cut; *stipules carinate;* peduncles 2–4-flowered; upper petals oval, undulate, sub-emarginate." *DC. l. c. p.* 657."

116. P. capillare (Willd. 3. p. 660); "stem suffruticose, short; leaves lanceolate, deeply pinnatifid, pubescent; peduncles 2-flowered. *Ger. capillare, Cav. Diss. t.* 97. *f.* 1.

Sec. 10. **DIBRACHYA.** *Stem* succulent, weak and much branched. *Leaves* fleshy, peltate or cordate, 5-7-lobed. *Stipules* broadly ovate, membranous. *Flowers* pedicellate. *Petals* unequal, all obovate. *Fertile stamens* 7; the two uppermost very short. *Ivy-leaved.* (Sp. 117-118.)

117. P. peltatum (Ait. Hort. Kew. 2. p. 427); stem shrubby, branches angular, weak and straggling; leaves glabrous or pubescent, fleshy, *peltate*, radiately 5-nerved below, bluntly 5-angled or lobed, with very entire margins; peduncles elongate, 4-8-flowered; pedicels mostly shorter than the slender calyx-tube, which is 2-3 times longer than the acuminate, nerved segments; petals twice as long as the calyx. *DC. Prod.* 1. *p.* 666. *Cav. Diss. t.* 100. *f.* 1. *Bot. Mag. t.* 20. *P. scutatum, DC. l. c. Sw. Ger. t.* 95.

VAR. α. **glabrum**; glabrous or nearly so; flowers purplish-pink. *D. peltata, E. & Z.!* 578. *Zey.!* 2075.

VAR. β. **scutatum**; leaves and stems thinly hairy; calyx villous; flowers white, large, with purple streaks. *E. & Z.!* 579.

VAR. γ. **clypeatum**; whole plant pubescent. *D. clypeata, E. & Z.!* 580. *Drege*, 7459.

HAB. Among shrubs, in sheltered places, α, Swellendam, at the Breede and Gauritz Rivers, *E. & Z.!* β. Zwartekop's River, *E. & Z.!* Albany, and Caffirland, *Dr. Gill.* Port Natal, *Dr. Sutherland!* γ. Sondags River, *E. & Z.!* (Herb. T.C.D., Hook., Sond.).

All the above varieties grow intermixed and seem to run together. The pubescence is specially variable in quantity, and the flowers vary from white to red, and from large to small. Several garden varieties are cultivated as "*Ivy-leaved Geranium.*"

118. P. lateripes (L'Her. Ger. t. 24); "stem shrubby; branches fleshy, terete; leaves *cordate*, 5-lobed, sub-dentate, fleshy, glabrous; umbels many-flowered." *DC. l. c. p.* 666.

HAB. Cultivated in England, since 1807. (v.v.c.)

Very like the preceding, but here the petiole is inserted at the margin of the leaf, which is therefore "*cordate*," not "*peltate*." I have seen no wild specimens.

Sec. 11. **EUMORPHA**. *Stems* slender, suffruticose or herbaceous. *Leaves* on very long petioles, palmately 5-7-nerved, reniform, lobed or palmatifid. *Stipules* free, ovate or lanceolate. *Peduncles* elongate. *Petals* unequal, the two upper broad. *Fertile stamens* 7; the 3 upper shorter. (Sp. 119-125.)

** *Shrubby or suffruticose, glabrous and glaucous. Flowers pedicellate.* (Sp. 119-122.).

119. P. grandiflorum (Willd. Sp. 3. p. 674); shrubby, glabrous and glaucous; leaves on long petioles, palmately 5-7-nerved, deeply 5-7-lobed, the lobes coarsely toothed; stipules ovate, mucronate; peduncles about 3-flowered; *calyx-tube* 3-4 *times as long as the lanceolate segments, which are* ⅓ *as long as the petals. DC. Prod.* 1. *p.* 667, *non E. & Z.! Drege,* 7457. *Andr. Rep. t.* 12. *Sw. Ger. t.* 29.

HAB. Introduced to England, 1794, *Masson!* Giftberg, *Drege!* (Herb. Banks., Hook., Benth., Sond.)

Like *P. saniculæfolium*, but the leaves are not zoned, the flowers larger and white, and the calyx-tube very much longer.

120. P. variegatum (Willd. Sp. l. c.); "glabrous, glaucous; *leaves 3-5-lobed-palmati-partite*, the segments trifid, toothed; stipules ovate-cordate, acute; peduncles 2-flowered, *the calyx-tube* 5 *times as long as the segments, which are* ⅓ *as long as the petals." DC. l. c. p.* 667. *Ger. variegatum, Lin. Sup.* 305. *Cav. Diss. t.* 118. *f.* 3.

Is this more than a garden variety of the preceding?

121. P. saniculæfolium (Willd. Sp. 3. p. 673); *shrubby*, erect, glabrous and glaucous; leaves on long petioles, *mostly purple-zoned*, palmately 5-7-nerved, deeply 5-7-lobed, the lobes coarsely toothed; stipules ovate-oblong, mucronulate; peduncles branched, the partial ones

4-5-flowered; flowers on long pedicels, *the calyx-tube about as long as the lanceolate segments, which are half as long as the petals. DC. Prod.* 1. *p.* 668. *P. cortusæfolium, Jacq.! Ic. Rar. t.* 539. *G. tabulare, Cav. Diss. t.* 100. *f.* 2. *E. & Z.!* 602, 603, 604. *Drege,* 7456. *Zey.* 191. *P. fuscatum, Jacq.! Ic. t.* 540. *DC. Prod. l. c.*

HAB. About Table Mountain and Hott. Holl., also on Tulbagh Mt., *E. & Z.! W.H.H.*, &c. (Herb. Jacq., T.C.D., Hook., Sond.).

A much-branched bush with glaucous stems and foliage, either wholly glabrous, or with a few scattered bristles, especially about the calyx. Flowering branches mostly panicled. Upper petals obovate, much longer than the lower. Nearly allied to *P. grandiflorum*, but the flowers are much smaller and the calyx-tube shorter. Jacquin's *P. fuscatum*, of which I have examined the original specimens, does not materially differ.

122. P. patulum (Jacq. Ic. Rar. t. 541); *suffruticose*, flexuous, slender, glabrous and glaucous; leaves on long petioles, palmately 5-7-nerved, 3-5-lobed or parted, the lobes coarsely toothed or cut; stipules cordate, acuminate, entire or toothed, *ciliolate;* peduncles slender, 2-3-flowered; *pedicels shorter than the slender, curved calyx-tube, which is 2-3 times longer than the lanceolate sepals;* petals twice as long as the sepals. *DC. Prod.* 1. *p.* 668. *Eumorphia variegata, E. & Z.!* 606, *also* 605, 607, 608.

VAR. *α.* **latilobum**; leaves with 3-5 broad, shallow, toothed lobes. *Eum. marmorata, E. variegata and E. cataractæ, E. & Z.!*

VAR. *β.* **tenuilobum**; leaves 3-5-parted, with deeply incised and sharply toothed and cut segments. *Eum. tenuiloba, E. & Z.!* 607.

HAB. Mountain sides, western districts, Hott. Holl., Tulbaghsberg, Puspas Valley, Swell., and Gnadendahl, *E. & Z.!* (Herb. T.C.D., Hook., Sond.)

Much less shrubby, more slender, with smaller leaves and flowers than *P. saniculæfolium;* and certainly distinct though the differences seem small. Our two varieties chiefly differ in the degree of incision of the leaf, which varies in different specimens. Sometimes there are a few bristles on the calyx and nerves of the leaf.

[Doubtful species of this sub-section; described originally from garden specimens; very probably hybrids?]

P. albiflorum (Spin. Cat. 1818. p. 30); *DC. Prod.* 1. *p.* 667.

P. Curtisianum (Balb. & Spin. Cat.); *DC. Prod. l. c.*

P. Collæ (Spin. Cat. 1818. p. 31); *DC. l. c.*

P. Bellardii Spin. Cat. 1818. p. 30); *DC. l. c.*

P. hepaticæfolium (Spreng.); *DC. Prod.* 1. *p.* 668.

P. nobile (Hortul.); *DC. Prod.* 1. *p.* 668.

*** Scarcely suffruticose or herbaceous, pubescent. Flowers sub-sessile.* (Sp. 123–125.)

123. P. alchemilloides (Willd. 3. p. 656); perennial, diffuse, many-stemmed, pubescent; stems *herbaceous*, furrowed, leaves on very long petioles, 5-7-nerved, reniform, 5-7-lobed, the lobes rounded or cuneate, crenate or bluntly toothed; *stipules cordate-ovate,* acute; peduncles long, villous, 4-6-flowered; flowers sub-sessile, the calyx-tube many times longer than the narrow-lanceolate, striate segments. *DC. l. c. p.* 660. *Cav. Diss. t.* 98. *f.* 1.

VAR. *α.* **dentatum**; softly pubescent or villous; the leaves deeply lobed or sub-partite, with cuneate, bluntly toothed lobes. *P. alchemilloides, E. & Z.!* 592. *P. articulatum, E. & Z.!* 594. *P. geranioides, E. & Z.!* 595. *Zeyher*, 2076, 2077.

VAR. β. **aphanoides**; softly pubescent or villous; the leaves with shallow, rounded lobes, crenate or crenato-dentate. *P. aphanoides, E. & Z.!* 590. *P. dimidiatum, E. & Z.!* 591. *Drege*, 7462.

VAR. γ.? **ranunculifolium**; strigoso-pubescent, harsh; the leaves thicker than usual, deeply lobed, marked with a purplish circle, with toothed lobes. *P. ranunculophyllum, E. & Z.!* 593. *Drege*, 7458.

HAB. Hill sides among bushes, throughout the Colony and in Caffirland. Vars. α. and β. common. γ. at the Klipplaat and Zwart Key Rivers, Caffraria, *E & Z.! Drege!* (Herb. T.C.D., Hook., Sond.).

There is a short woody, perennial root-stock, from which spring several weak, annual stems 1–2 feet long or more, with distant nodes; the petioles are usually much longer than the lamina. Pubescence (except in γ.) copious and soft. The flowers are small, pale-rosy, with or without spots.

124. P. malvæfolium (Jacq. f. Eccl. t. 97); stems suffruticose, slender; branches weak and straggling, hairy, herbaceous, with distant nodes; leaves on long petioles, cordate at base, roundish, 5-lobed, hairy on both sides, with short, coarsely-toothed lobes; stipules small, ovate, acute, hairy, withering; peduncles long, 5–8-flowered; flowers sub-sessile, the slender calyx tube 3–4 times longer than the lanceolate, acute, hairy segments; petals *sub-equal, narrow obovate*, blood-red, with darker lines. *DC. Prod.* 1. *p.* 654, *P. lateritium and P. cynobastifolium, Willd.* ? *DC. l. c.*

Only known to me by Jacquin's figure and description. It seems to have very much of the habit of *P. alchemilloides*, especially of our var. γ.

125. P. tabulare (L'Her. Ger. t. 9); perennial, diffuse, many-stemmed, pilose; stems herbaceous, angular; leaves on long petioles 5–7-nerved, reniform, 5–7-lobed, the lobes rounded and crenate; *stipules lanceolate, rigidly ciliate;* peduncles long, glabrous, or sub-pilose, 4–6-flowered, with ciliate bracts; flowers sub-sessile, the calyx tubes much longer than the narrow-lanceolate, striate, often glandular segments. *DC. Prod.* 1. *p.* 660. *P. elongatum, Cav Diss. t.* 101. *f.* 3. *E. & Z.!* 602. *Drege*, 7463. *Ger. tabulare, Linn. Sp.* 947, *non Burm., nec Cav.*

HAB. Common about Table Mountain, &c., near Cape Town. (Herb. T.C.D., Hook., Sond.)

Very like *P. alchemilloides*, but readily known by its narrow and bristle-fringed stipules. The leaves are often, but not constantly, marked with a purplish circle or horse-shoe. Flowers small, pale, with narrow petals.

SECT. 12. **GLAUCOPHYLLUM**. *Shrubs. Leaves* carnose; *the lamina more or less perfectly articulated with the petiole*, simple or tri-partite. *Veins* inconspicuous. *Petals* unequal, the two upper broad. *Fertile stamens* 7. (Sp. 126–130.)

126. P. glaucum (L'Her. Ger. t. 29); suffruticose, brittle, glabrous and glaucous; leaves *unifoliolate*, leaflet articulated to the petiole, *lanceolate*, acute, quite entire (sometimes incised), somewhat fleshy; stipules subulate; peduncles short, 1–2-flowered; pedicels very short, the calyx tube 3–4 times as long as the lanceolate segments. *DC. Prod.* 1. *p.* 666. *Sw. Ger. t.* 57. *Andr. Ger. cum ic. G. lanceolatum, Cav. Diss. t.* 102. *f.* 2. *Bot. Mag. t.* 56. *Drege,* 9515. *G. cuspidatum, Willd.! P. diversifolium, Wendl.*

HAB. S. Africa, *Niven! Forsyth! Drege!* (Herb. Sond., Benth.).

Stem rather woody, erect, with few branches; the nodes distant and easily breaking asunder when dry. Leaves, in cultivated specimens, often irregularly laciniate. Flowers white, the two upper petals with a red blotch. The leaves and stipules vary in breadth, but not much in form. Willdenow's *P. cuspidatum!* (Herb. Berol., 12,447) is one of the many garden varieties with incised leaves.

127. P. lævigatum (Willd. 3. p. 685); suffruticose, brittle, glabrous, and glaucous; leaves *trifoliolate,* the leaflets articulated with the petiole, *narrow-lanceolate or linear,* fleshy, *either quite entire, incised, trifid, or subpinnatifid;* stipules *subulate;* peduncles short, 2-flowered; pedicels very short, the calyx tube 3-4 times as long as the lanceolate segments. *DC. Prod.* 1. *p.* 667. *Cav. Diss. t.* 121. *f.* 1. *E. & Z.* 609.

VAR. *β.* **oxyphyllum**; leaflets quite entire, *P. oxyphyllum, DC. l. c.*

HAB. Near Onzer, George, *Drege!* Langekloof, *E. & Z.!* var. *β.* Ezelbank, 4000f. *Drege!* (Herb. T.C.D., Hook., Sond.)

Readily known by its trifoliolate, fleshy, glaucous and mealy leaves, which, however, vary considerably in breadth and incision. I find it impossible to separate *P. oxyphyllum,* DC.; on the same branch there are often both simple and divided leaflets. Drege's specimens (our var. *β.*) are somewhat stronger and more mealy than usual.

128. P. divaricatum (Thunb.! Fl. Cap. p. 525); shrubby, slender, erect, much branched, *glabrous or roughly puberulous;* leaves on short petioles, *trifoliolate, the leaflets trifid,* with simple or tricuspidate, patent, glabrous or scabrous, fleshy, channelled, acute lobes; stipules *cordate,* acuminate; peduncles very short, 1–2-flowered; pedicels very short, the nearly glabrous calyx-tube 2-3 times longer than the hispid, lanceolate segments. *DC. Prod* 1. *p.* 682.

VAR. *α.* **glabrum**; glabrous or nearly so. *P. fruticosum. E. Mey.! in Hb. Drege, non Willd.*

VAR. *β.* **scabrum**; branches and leaves rough with short points. *Drege,* 7461.

HAB. South Africa, *Thunberg!* Zwarteberg, 4000f. and at Dreifontein, Mosselbay, *Drege!* Swellendam, *Dr Thom.!* (Herb. T.C.D., Hook., Sond.)

A much-branched, slender, small bush, distinctly woody below, half herbaceous above, having the habit of *P. abrotanifolium,* but differing in stipules and compound leaves, &c. I have examined an authentic specimen of Thunberg's in Herb. Holm. In Herb. Banks is a specimen collected by *Nelson,* marked in L'Heretier's writing, "*G. fruticosum, Cav. P. crithmifolium, L'Her.*"

129. P. ternatum (Jacq.! Ic. t. 544); suffruticose, erect, virgate, *scabrous and pubescent;* leaves on short petioles, trifoliolate, the leaflets *cuneate, concave,* somewhat fleshy, hispid and scabrous, *the lateral often trifid, the terminal deeply 3-lobed, all toothed at the truncate summit;* stipules ovato-lanceolate, sometimes cut; peduncles short, 1-2-flowered; pedicels shorter than the calyx tube, which equals or exceeds the lanceolate hispid segments. *DC. Prod.* 1. *p.* 678. *Cav. Diss. t.* 107. *f.* 2. *Bot. Mag. t.* 413. *Sw. Ger. t.* 165. *E. & Z.!* 639.

HAB. Cult. in England, 1789. Mountains near Mr. Joubert's farm, Cannaland, Swell., *E. & Z.!* (Herb. Banks., Sond., Jacq.)

A slender and brittle suffrutex, allied to *P. lævigatum*, but scabrous, with less fleshy and broader leaflets. Flowers pink, the lower petals narrow spathulate, upper obovate. The leaves vary much in pubescence and degree of incision. Nelson's specimen in Herb. Banks is much rougher than E. & Z.'s, and agrees better with Jacquin's figure.

130. P. spinosum (Willd. Sp. 3. p. 681); shrubby, divaricate, *glabrous* or nearly so; leaves articulated to *the long, divergent, rigid petioles*, 3–5-nerved, *broadly cuneate or triangular*, somewhat 3-fid, or 3-lobed, with toothed lobes; *the old petioles and the persistent subulate stipules hardening into spines;* peduncles 4–6-flowered; pedicels elongate, equalling the straight calyx tube, which exceeds the lanceolate, acuminate segments. *DC. Prod.* 1. *p.* 661.

HAB. Cult. 1796. Cape, *Forsyth!* Between Kaus, Natvoet and Doornpoort, Namaqualand, *Drege!* (Herb. T.C.D., Benth., Sond.).

A spreading shrub, remarkable for the 3–4 inch long, spinous, persistent petioles and spinous ¼–½ inch long stipules. Leaves 1 inch across, at first imperfectly, afterwards fully articulated with the petiole. Petals very narrow.

SECT. 13. **CICONIUM.** *Stem* shrubby, thick and succulent. *Leaves* obovate, or roundish-cordate or reniform, palmately many-nerved, crenate or lobulate. *Stipules* broadly ovate or cordate, free. *Umbels* few or many-flowered. *Petals all of one colour*, scarlet, pink or white, the two uppermost narrower than the lower. *Stamens* 7, the two uppermost very short. (Sp. 131–134).

131. P. acetosum (Ait. Hort. Kew. 2. p. 430); shrubby, the younger branches succulent, glabrous; leaves on short petioles, *glabrous*, somewhat fleshy, *obovate, obtuse, cuneate at base*, crenate; stipules obliquely ovate, small; peduncles few-flowered; pedicels very short, the calyx tube glandular, 3–4 times longer than the lanceolate segments; petals spathulate, sub-equal. *DC. Prod.* 1. *p.* 658. *Cav. Diss. t.* 104. *f.* 3. *Bot. Mag. t.* 103. *E. & Z.!* 587, *non Drege.*

HAB. Cult. in England 1724. Clayey places near Rivers. Uitenhage, *E. & Z.!* Somerset, *Mrs. F. W. Barber!* (Herb. T.C.D., Hook., Sond.).

A much-branched, but rather naked bush, with small, deciduous leaves and slender, pale flowers. The petals are at least twice as long as the calyx-segments. The leaves are acidulous, tasting like *sorrel.*

132. P. scandens (Ehr. Willd. ! Sp. 3. p. 666); shrubby, the branches flexuous; leaves petiolate, *roundish*, 5–7 *lobulate*, crenate, *glabrous, zoned ;* stipules broadly cordate; peduncles long, many-flowered; pedicels short, the calyx-tube thrice as long as the lanceolate, nearly glabrous segments; petals narrow linear. *DC. Prod.* 1. *p.* 658. *P. pumilum. Willd.? DC. l. c.*

HAB. Cult. in Europe. (Herb. Willd.!)

Probably a mere variety of *P. zonale.*

133. P. zonale (Willd. Sp. 3. p. 667); shrubby, the younger branches succulent, hispidulous; leaves on long petioles, roundish-cordate, glabrous or pubescent, *mostly with a dark horse-shoe mark above*, crenato-dentate, obsoletely many-lobed; stipules very broad, cordate-oblong;

peduncles long, many-flowered; flowers sub-sessile, the calyx tube glabrous or thinly pubescent, 4–5 times longer than the lanceolate segments; petals *narrow-cuneate or spathulate. DC. Prod.* 1. *p.* 659. *E. & Z.!* 585. *Cav. Diss. t.* 98. *f.* 2., *not good. Cic. densiflorum, E. & Z.!* 584. *P. lateritium, Willd.!*

VAR. *β.* **stenopetalum**; petals *very narrow linear. P. stenopetalum, Ehr. DC. l. c. p.* 658.

HAB. Cult. in England, 1710. Among shrubs and on hill sides in the Western Districts. Tulbaghsberg, near the Winterhoek and in the Langekloof, *E. & Z.!* Simon's Bay, *Wright!* (Herb. T.C.D., Hook., Sond.)

A large shrub, with juicy, green stems and thick leaves, usually, but not invariably, marked with a dark semi-circle, whence its name "*horse-shoe Geranium.*" The flowers vary from scarlet and crimson through all shades of red to pure white. *P. hybridum* and *P. monstrum*, old garden plants in England, appear to be varieties of this species; the first approaching *P. inquinans*.

134. P. inquinans (Ait. Hort. Kew. 2. p. 424); shrubby, the younger branches succulent, velvetty; leaves on long petioles, *orbicular-reniform, velvetty and somewhat viscoso-pubescent*, crenate, sub-undivided or obsoletely multi-lobulate; stipules broadly cordate; peduncles long, many-flowered; pedicels very short, the densely glandular and viscid calyx tube 3-4 times longer than the lanceolate segments; *petals broadly obovate. DC. Prod.* 1. *p.* 659. *Dill. Elth. fig.* 151. *Cav. Diss. t.* 106. *f.* 2. *E. & Z.!* 581, 583, 586. *Zeyher!* 2073, 2074. *P. cerinum, Sw. t.* 176. *E. & Z.!* 582. *Drege*, 7453, 7454.

HAB. Among shrubs and on hill sides. Eastern Districts, Uitenhage, Albany and Caffirland. (Herb. T.C.D., Hook., Sond.).

This is the parent of most of the "*scarlet Geraniums*" of English gardens, and has been cultivated since 1714. The flowers vary from intense scarlet to rose-colour and white. It is much softer and more viscid than *P. zonale*, without horse-shoe mark, and with broader and shorter petals.

SECT. 14. **CORTUSINA.** *Stem* caudiciform, short, thick and fleshy; branches (if any) slender and half herbaceous. *Leaves* on long petioles, reniform or cordate, crenate or lobulate, velvetty or pubescent. *Petals* sub-equal, obovate, the two upper broader. *Stamens* 6–7. (Sp. 135–141.).

135. P. echinatum (Curt. Bot. Mag. t. 309); stem fleshy, *armed with persistent, spine-like stipules;* leaves on long petioles, cordate-ovate, obtuse, somewhat 3-5-7-lobed, the lobes rounded, crenulate or bi-crenulate, pubescent above, albo-tomentose and nerved below; stipules subulate, rigid; peduncles elongate, branched, the partial 6-8-flowered; pedicels very short, the downy calyx-tube 5-6 times as long as the villous sepals; petals emarginate. *DC. Prod.* 1. *p.* 665. *Sw. Ger. t.* 54. *P. hamatum, Jacq. Schoenb. t.* 138. *Andr. Rep. t.* 158? *Zey.!* 2071.

HAB. Cult. 1795. North-Western Districts. Namaqualand, *Zeyher!* On the Roodeberg, 3500f., Kausberg, 4000f. and at Modderfontein, *Drege!* (Herb. T.C.D. Hook., Sond.).

This has the habit of *P. reniforme* and the leaves of *P. odoratissimum*, but differs from both in the spiny stipules. The flowers are mostly white, with a dark red spot on the upper petals; but are sometimes deep purple.

136. P. crassicaule (L'Her. Ger. t. 26); stem *short, thick and fleshy,* *smooth;* leaves from the crown, on long petioles, *ovate-reniform, with a cuneate attenuated base*, somewhat lobed, the lobes roundish and crenato-dentate, *both surfaces silky;* stipules small, deltoid, deciduous; peduncles elongate, *scapiform*, branched, bearing one or two leaves, the partial pluri-flowered; pedicels scarcely any, calyx densely villous, the tube much longer than the linear, obtuse segments. *DC. Prod.* 1. *p.* 665. *Bot. Mag. t.* 477. *P. primulinum, Sw. DC. l. c.*

HAB. North-West Coast of the Colony, *Hove* (1786). Opposite Ichaboe, Herb. Hook. (Herb. Hook., Holm.)

Remarkable for its thick, almost tuberous stem and leaves greatly tapering into the petiole. The nerves are 3, laterally branched. Flowers white, the petals sub-equal, each having a purple spot in the centre.

137. P. cortusæfolium (L'Her. Ger. t. 25); stem short, thick and fleshy, *rough with persistent stipules;* leaves from the crown, on long petioles, *cordate, inciso-lobate, undulate, unequally toothed,* pubescent; stipules subulate, their bases persistent; peduncles elongate, branched, with many-flowered umbels; calyx-tube four times as long as the ovate, reflexed segments; petals obovate, emarginate. *DC. l. c. p.* 665. *Sw. Ger. t.* 14. *Andr. Rep. t.* 121.

HAB. North-West of the Colony, lat. 23°. S., *Hove.* Introduced to England in 1786.

I have seen no specimens of this species.

138. P. reniforme (Bot. Mag. t. 493); stem shrubby, short, often scaly; branches succulent, *velvetty;* leaves on long petioles, *reniform or ovato-cordate*, obtuse, crenate or lobulate, velvetty above, albo-tomentose and prominently many-nerved underneath, the nerves digitate, forking; stipules from a broader base, subulate; peduncles elongate, often branched, the partial pluri-flowered; pedicels much shorter than the calyx-tube, which is lanoso-pubescent and thrice as long as the sepals; petals obovate. *DC. Prod.* 1. *p.* 666. *Sw. Ger.* 48. *Andr. Rep. t.* 108.

VAR. α. **reniforme**; leaves broadly reniform; flowers bright purple. *E. & Z.!* 599, 600.

VAR. β. **velutinum**; leaves ovato-cordate; flowers bright purple. *E. & Z.!* 598.

VAR. γ. **sidæfolium**; leaves ovato-cordate or reniform; flowers blackish purple or very dark. *Ger. sidæfolium, Thunb. Cap. p.* 518. *E. & Z.!* 597. *Drege!* 2257.

HAB. Among shrubs and in fields in the Eastern Districts. Uitenhage and Albany, common. All the varieties grow together. γ. also at Port Natal, *Sanderson!* (Herb. T.C.D., Hook., Sond.)

This well-marked species varies somewhat in the shape, and more in the size of the leaves and in the colour of the flower; but after examining large suits of specimens, I cannot consent to uphold as species any of the above-named varieties. E. & Z.'s "*C. rubro-purpurea*" is very unlike the figure to which they refer it, and is certainly nothing more than a starved condition of our var. α. Var. β. is exactly intermediate between α & γ. According to Dr. Atherstone this species is useful as an astringent in dysentery.

139. P. odoratissimum (Ait. Hort. Kew. 2. p. 419); stem very short, thick, scaly; *branches herbaceous, weak and straggling;* cauline leaves on very long petioles, *roundish-cordate, very obtuse*, entire or repando-lobulate, dentato-crenulate, *very soft and velvetty;* stipules *small, deltoid;* peduncles opposite the rameal leaves, filiform, 5–10-flowered; pedicels equal to the calyx-tube, which is longer than the lanceolate, villous segments; flowers small, petals not much exceeding the calyx. *DC. Prod.* 1. *p.* 659. *Dill. Elth. fig.* 138. *Cav. Diss. t.* 103. *f.* 1. *Sw. Ger. t.* 299. *E. & Z.!* 596. *Zey.!* 2079. *Drege*, 1298, 9516.

HAB. Cult. 1724. Common in the Eastern Districts; Uitenhage and Albany, *E. & Z.!* &c. (Herb. T.C.D., Hook., Sond.)

Caudex short, simple or branched; the flowering branches long and trailing, with distant nodes, from which spring the short peduncles. Petioles of the cauline leaves 6–10 inches long; the lamina 1–2 inches broad. Flowers very small, white. Leaves strongly and sweetly aromatic.

140. P. Sibthorpiæfolium (Harv.); stem short, fleshy; leaves on long petioles, *carnose, reniform, cordate at base*, crenato-lobulate, *minutely pubescent on both sides;* stipules . . . . . peduncles scape-like, branched, hairy; bracts ovate, tomentose; calyx-tube sub-sessile, tomentose, twice as long as the linear segments. *Drege*, 3232.

HAB. Hills, &c. near the mouth of the Orange River, under 600f. *Drege!* (Herb. Benth.)

I venture to describe from a single imperfect specimen. The stem is one inch long, 3 lines in diameter, and rough. Leaves less than an inch across, shaped like those of *Sibthorpia Europea;* the petiole twice or thrice the length of the lamina. The flowers have not been observed. Can this be *P. velutinum*, DC.?

141. P. alpinum (E. & Z.! 629); root-stock caulescent, woody and scaly; stems suffruticose, sub-simple, tomentose; leaves on long petioles, cordate-reniform, mucronato-crenate, *tomentose or densely pubescent* on both sides; stipules *ovate*, acuminate; peduncles filiform, elongate, branched, the partial 1-2-flowered, with broadly oval bracts; calyx-tube sub-sessile, curved, villous, thrice as long as the lanceolate, acuminate sepals; petals obovate. *Zeyher*, 190!

HAB. High mountains. Winterhoek, Worcester, *E. & Z.!* Witsenberg, *Zeyher!* (Herb. Sond., Hook.).

Woody portion of stem 3–4 inches; leafy portion 1–2 feet high, slender. Petioles 2–3 inches long, leaves about 1 inch, very obtuse or sub-acute. Bracts generally 4, remarkably broad in proportion to their length, membranous. Flowers pale, the upper petals with a purple spot. This seems to be a well marked species.

SECT. 15. **PELARGIUM.** Much-branched *shrubs.* Leaves simple or lobed, never pinnate. *Stipules* free. *Inflorescence* frequently panicled; the partial peduncles umbelliferous. Two upper petals longer and broader than the three lower, streaked or spotted. Fertile stamens 7. (Sp. 142–163).

* *Leaves oval, ovate or cordate, toothed or mult-angular.* (Sp. 142–145).

142. P. betulinum (Ait. Kew. 2. p. 429); shrubby, erect, the young twigs downy; leaves on short petioles, *oval or ovate, obtuse*, unequally toothed, *sub-glabrous* or scaberulous; stipules acuminate; peduncles opposite the leaves, deflexed, 3-4-flowered; pedicels and calyx silky; petals twice as long as the lanceolate sepals. *DC. Prod.* 1. *p.* 669.

Burm. Afr. t. 33. *f.* 2. *Bot. Mag. t.* 148. *E. & Z. No.* 611. *Zey. !* 197. *Drege,* 7451. *P. penicillatum, Willd. ?*

HAB. Mountain sides and hills in Cape and Stellenbosch districts, common. Klynriviersberg, Caledon, *E. &. Z.!* (Herb. T.C.D., Hook., Sond.)

Rather a slender, nearly glabrous shrub, downy or somewhat villous on the young twigs and flowerstalks, with small, neat foliage, the leaves seldom an inch long. Flowers purple, with dark streaks. Cultivated since 1786.

143. P. cordatum (Ait. Kew. 2. p. 427); shrubby, erect and much-branched, *either densely and softly villous or sub-glabrous;* leaves on long petioles, *cordate, acute,* denticulate and sometimes repando-lobulate; stipules subulate, from a broad base, deciduous; peduncles branched or panicled, the partial short and *many-flowered;* pedicels and calyx generally densely villous; petals twice as long as the sepals. *DC. Prod.* 1. *p.* 671. *L'Her. Ger. t.* 22. *P. cordifolium, Bot. Mag. t.* 165. *Cav. Diss. t.* 117. *f.* 3. *E. & Z. !* 618, 619, 620. *Drege,* 7450.

VAR. β. **lanatum**; leaves densely woolly on the lower surface. *Ger. lanatum, Thunb.! Fl. Cap. p.* 518.

VAR. γ. **rubrocinctum**; leaves nearly glabrous on both sides, or velvetty on the lower, often with a red margin. *P. rubro-cinctum, Link. DC. l. c. Drege,* 7449.

HAB. Mountain sides in the Langekloof and at Plettenberg Bay and the Knysna *E. & Z. !* near George and at Kayman's Gat, *Drege!* β. Van Staadensberg, *E. & Z.!* γ. Attaqua's Kloof, *Dr. Gill ! Drege!* Oliphant's River, *Thom, &c.* (Herb. T.C.D. Hook., Sond.)

A shrub, with the habit and flowers of *P. cucullatum,* but readily known by its cordato-ovate, acute, flat leaves. These are sometimes imperfectly multo-lobulate at the margin. Cultivated since 1774. β and γ. differ greatly in pubescence, but intermediate states may be found.

144. P. cucullatum, (Ait. Kew. 2. p. 426); shrubby, tall, and much branched, *densely and softly villous;* leaves on long petioles, *reniform-cupped,* denticutlate, *very soft;* stipules ovate-acute, withering; peduncles panicled, the partial densely many-flowered; pedicels and calyces densely villoso-sericeous; petals twice as long as the lanceolate-acuminate sepals. *DC. Prod.* 1. *p.* 671. *E. & Z. !* 621. *Cav. Diss. t.* 106. *f.* 1. *Drege!* 7269. *Zey. !* 2092.

HAB. Very common round Capetown and in the Western districts, where it is often used as an ornamental hedge-plant. Cult. in Europe since 1690.

This is a large and free growing shrub, from which many of the garden hybrids are derived. Flowers purple.

145. P. angulosum (Ait. Kew. 2. p. 426); shrubby, erect, much-branched, *densely hairy and harsh to the touch;* leaves on short petioles, truncate or broadly cuneate at base, *with* 3–5 *shallow, angular, acute toothed lobes, rigid,* and sometimes scabrous; flowers panicled, the partial peduncles 4–6-flowered; pedicels and calyces densely hairy and rough; petals twice as long as the acuminate sepals. *Dill. Elth. t.* 129. *f.* 156.

VAR. α. **angulosum**; leaves *truncate* or roundish-angular at base. *P angulosum, DC. Prod.* 1. *p.* 72. *E. & Z. !* 624, 625, 626. *Drege,* 7448.

G. acerifolium, Cav. Diss. t. 112. *f.* 2. *P. cochleatum, Willd.! En. p.* 48. *DC l. c.*

VAR. *β.* **acerifolium**; leaves *cuneate* at base, more deeply lobed, more scabrous and harsher than in *α*. *P. acerifolium, L'Her. Ger. t.* 21. *DC. Prod.* 1. *p.* 672. *E. & Z.!* 627, 622, 623.

HAB. Mountain sides, Cape, Stellenbosch districts. Table mountain, and Hott. Holl., *E. & Z.!* Simon's Bay, *Wright, W.H.H.* &c. (Herb. T.C.D., Hook., Sond.)

A large bush, resembling *P. cucullatum*, from which it is easily known by its sharply-angular, rigid and sharp leaves. Though very hairy, it is not villous. The flowers are in large panicles, purple, with dark streaks. This is one of the parents of many garden hybrids, and has been cult. since 1724.

** *Leaves obtusely 3–5-lobed, pubescent, not scabrous or viscid.* (Sp. 146–148.)

146. P. capitatum (Ait. Kew. 2. p. 425); suffruticose, *diffuse, or procumbent, densely and softly hairy and villous;* leaves on long petioles, cordate at base, 3–5-lobed, the lobes obtuse and rounded, toothed; stipules broadly cordate, membranous, pointed; peduncles longer than the leaves, simple, *densely many-flowered;* flowers sessile, the calyx-tube not half so long as the densely-villous and hairy, oblong, mucronate sepals; petals short. *DC. Prod.* 1. *p.* 674. *Cav. Diss. t.* 105. *f.* 1. *E. & Z.!* 630. *Thunb. Cap. p.* 521.

HAB. Common on the Cape Flats and around Table Mountain, &c. (Herb. T.C.D. Hook., Sond.)

Stems weak, trailing on the ground. Whole plant covered with long, soft, horizontally patent white hairs. Flowers rosy purple, in dense many-flowered heads. Very near the following. It has been cultivated in England since 1790.

147. P. vitifolium (Ait. Kew. 2. p. 425); suffruticose, *erect*, densely hairy and villous; leaves on long petioles, cordate at base, 3-lobed, *the lobes shallow, very obtuse and rounded,* toothed; stipules broadly cordate; peduncles longer than the leaf, simple, densely many-flowered; flowers sessile, the calyx-tube not half so long as the densely villous and hairy, oblong, aristate segments; petals short. *DC. Prod.* 1. *p.* 674. *E. & Z.!* 631. *Dill. Hort. Elth. fig.* 153.! *Cav. Diss. t.* 111. *f.* 2. *L'Her. Ger. t.* 19.

HAB. Among shrubs, near rivulets. Klapmuts and Hott. Holl. *E. & Z.!* Paarlberg, *Drege!* (Herb. T.C.D., Sond.).

Except in its erect growth and less deeply cut and harsher leaves this scarcely differs from *P. capitatum*. I retain it, as it has been acknowledged a species since the days of Dillenius, and was cultivated in England in 1724. Flowers small, purple, in densely many-flowered heads.

148. P. semitrilobum (Jacq.! Schoenb. t. 130); *shrubby*, much-branched, patently villous; leaves on longish petioles, concave, *bluntly 3-lobed,* the lobes short, toothed or sharply serrate, pubescent on one or both sides, the nerves radiating, branching, *prominent* and hairy; stipules cordate-acuminate; peduncles longer than the leaf, 2–5-*flowered,* with ovate, nerved bracts; calyx-tube *curved,* thickened at base, *equalling the pedicel* and the lanceolate, acuminate, villous segments; petals twice as long as the sepals. *DC. l. c. p.* 674. *E. & Z.* 612. *Zey.!* 2090, 2091.

VAR. *α.* **Jacquini**; leaves sparsely hispid above, villous below, cuneate at base. *P. trilobatum Schrad.? DC. l. c. p.* 677.? *P. adulterinum, E. & Z.!* 628. *Zey.!* 198.

VAR. β. **adulterinum**; leaves, softly hairy on both sides, cordate or truncate at base. *P. adulterinum, L'Her. t.* 34. *DC. Prod.* 1. *p.* 674. *Sw. Ger. t.* 22 *and t.* 25.

HAB. Rocky mountain sides. Klynriviersberg, *E. & Z.!* Zwarteberg, and near the Zonderende, *Zeyher!* Var. *a*, near Tulbagh by the Waterfall, *E. & Z.!* (Herb. Jacq.! Hook., Holm., Sond.).

A much-branched, hairy or silky shrub, well covered with leaves and flowers; variable in pubescence and in some minor characters. There are many cultivated varieties of this, and several handsome hybrids are partly traced to it.

*** *Leaves scabrous, acutely three-lobed, and sharply serrate.* (Sp. 149–150.).

149. P. crispum (Ait. Kew. 2. p. 430); shrubby, much branched, *very scabrous*, glandular; leaves *distichous*, on short petioles, *flabelliform, truncate* or cuneate at base, trilobulate or deeply 3-lobed, coarsely toothed, rigid and rough, the nerves prominent below, and bristly; stipules cordate-acuminate, rigidly ciliolate; peduncles short, 2-3-flowered; pedicels longer than the calyx-tube, which is glandular and scabrid; sepals oblong, acuminate; petals narrow. *DC. Prod.* 1. *p.* 677. *L'Her. t.* 32. *Sw. Ger. t.* 383. *Cav. Diss. t.* 109. *f.* 2. *E. & Z.!* 635. *Zey.!* 2088. *Drege!* 1304. *P. rigidum, Willd! Sp.* 3. *p.* 681.

VAR. β. **hermanniæfolium**; leaves on short petioles, mostly very small (but variable in size), *cuneate* at base, trifid or truncate. *P. hermanniæfolium, Jacq. Ic. t.* 545. *DC. Prod.* 1. *p.* 677. *E. & Z.!* 636. *Zey.!* 2087, 2085.

VAR. γ. **latifolium**; stem weak and straggling; leaves distant, and twice the usual size; peduncles long, filiform. *L'Her. t.* 33. *Drege!* 3259. *P. pustulosum, Sw. ? Ger. t.* 11.

HAB. Among shrubs. River Zonderende, *E. & Z.!* Both varieties also near Caledon and Gnadendahl, *E. & Z.!* Hassagaiskloof, *Zeyher!* (Herb. T.C.D., Sond., Hook.).

A very scabrous, slender shrub, with strongly-scented, curled, rough leaves, which vary much in size, being from ½ inch to 1½ inches in length and breadth. Their insertion is distichous, and the stipules are vertically, placed one over another, on each side of the branches. The scent is strong and like balm. Typical specimens of var. β. *with miuute, wedge-shaped, truncate leaves* look very different, but intermediate forms between these and the broad-leaved "*crispum*" of gardens, are common; and both grow intermixed, in a wild state. Drege's "*P. crispum*" belongs to our var. γ, which is well figured by L'Heretier.

150. P. scabrum (Ait. Kew. 2. p. 430); shrubby, much-branched, *harsh, hairy and glandular;* leaves on shortish petioles, *cuneate at base, deeply three-lobed, the terminal lobe often trifid and the lateral bifid, all coarsely toothed,* the nerves prominent underneath, and rough with short bristles; stipules ovate-acuminate, pubescent; peduncles lateral or panicled, *many-flowered;* the pedicels and calyx remarkably scabrous and bristly; petals small. *DC. Prod.* 1. *p.* 677. *Cav. Diss. t.* 108. *f.* 1. *Jacq.! Ic. t.* 542 *L'Her. Ger. t.* 31. *E. & Z.!* 634, 638. *Drege!* 7445 *Zey.!* 199, 2070.

VAR. β. **balsameum**; leaves deeply 3–5-fid or 3–5-parted, the segments lanceolate, acuminate, sharply serrate. *P. balsameum, Jacq.! Ic. t.* 543. *DC. l. c. p.* 679.

HAB. Mountain sides. Kochmanskloof, and in rocky places among shrubs, at

Berg Valley, Clanw., *E. & Z.!* Paarl, *Drege!* Stellenbosch, *W.H.H.* Slaay Kraal, Grahamstown, *Burke and Zeyher!* (Herb. Jacq., T.C.D., Hook., Sond.).

Remarkable for its great roughness, all parts feeling harsh to the touch. It is sometimes very glandular and viscid, sometimes nearly bare of glands. E. & Z.'s *P. cratægifolium* (638) has narrower leaves than common, and connects the normal state of the species with *P. balsameum Jacq.!* our var. β. Drege's 5460 is a feeble-growing variety. This species borders closely on *P. crispum*, but the leaves are larger, not distichous, more cuneate at base, and on longer petioles; and the stipules and inflorescence are different.

****** *Leaves cordate or hastate at base, more or less deeply 5–7-lobed, or sinuato-pinnatifid; often viscoso-pubescent.* (Sp. 151–155.)**

151. P. ribifolium (Jacq.! Ic. t. 538); shrubby, much-branched, *sparsely hairy* and glandular; leaves on long petioles, somewhat hastate at base, more or less deeply 5-7-lobed, the lobes shallow, obtuse, or acute, deeply 5-toothed, glandular, hispidulous or pubescent, the nerves prominent below; stipules cordate-acuminate; inflorescence panicled, cymose; the partial peduncles short, 4-5-flowered; pedicels and calyx setose; flowers small, *the 3 lower petals longer than the lanceolate sepals. DC. Prod.* 1. *p.* 671. *P. trilobatum, E. & Z.!* 633, *non Schrad. P. populifolium E & Z.!* 632. *Drege,* 7447, 7446.

HAB. Mountain sides, among shrubs and grass. Langekloof, George, and Zuureberg and Winterhoeksberg, Uit., *E. & Z.!* near Grahamstown, *Zeyher!* (Herb. T.C.D., Hook., Sond. Jacq.).

A much-branched, rather rigid shrub, 2–3 feet high, with inflorescence of *P. papilionaceum*, but much harsher and smaller leaves. Flowers white.

152. P. papilionaceum (Ait. Kew. 2. p. 423); suffruticose at base, herbaceous upwards, much-branched, hairy and glandular; leaves on long petioles, deeply cordate at base, *obtusely 5-7-lobed,* the lobes shallow or obsolete, crenate, *both surfaces densely hairy,* the nerves prominent below; stipules cordate, acute; inflorescence densely panicled, the partial peduncles short, glandular and hairy; pedicels and calyces patently hirsute; flowers small, 3 *lower petals linear, much shorter than the lanceolate sepals. DC. Prod.* 1. *p.* 671. *Dill. Hort. Elth. fig.* 155. *Cav. Diss. t.* 112. *f.* 1. *Sw. Ger. t.* 27. *E. & Z.!* 617, *also P. tomentosum, E. & Z.!* 613, *non Jacq. Zey.!* 2093.

HAB. In moist, shady, sub-alpine places. Districts of Stellenbosch, Caledon and Swellendam, *E. & Z.! Drege!* Grootvadersbosch, *Zey.!* Cult. since 1724. (Herb. T.C.D., Hook., Sond.).

3–4 feet high, with the habit of *P. hispidum*, but with blunter and obsoletely-lobed leaves, broadly cordate stipules and bracts, and very different flowers. The upper petals are twice as long as the sepals, strongly reflexed, purple, with a dark spot. Dillenius' figure, above quoted, is excellent. *E. & Z.* confounded this with *P. tomentosum*, and *Drege* with *P. hispidum*.

153. P. tomentosum (Jacq.! Ic. t. 537); half-herbaceous, *diffuse* much branched, *densely and softly tomentose;* leaves on long petioles deeply hastato-cordate at base, 5-7-lobed, the lobes broad and shallow obtuse, mucronulate, crenate, *velvetty on both sides;* stipules cordate cuspidate; inflorescence laxly panicled, the partial peduncles long, villous, many-flowered; pedicels many times longer than the calyx-tube; flowers small, the 3 lower petals longer than the sepals. *DC. Prod.* 1.

p. 671. *Bot. Mag. t.* 518. *Sw. Ger. t.* 168. *P. micranthum, E. & Z.!* 614. *Zey.!* 2095.

HAB. Mountain sides, among rocks and bushes. Voormansbosch, Swell., *E. & Z.!* Stellenbosch, *Drege!* (Herb. Hook., Sond.)

Cultivated since 1790, chiefly for its scent, which resembles that of *peppermint*. The soft, velvetty foliage is handsome, but the flowers are insignificant and white. The stems are several feet long, and spread widely. Leaves 2–3 inches long and broad.

154. P. glutinosum (Ait. Kew. 2. p. 426); shrubby, much branched, *hispidulous and viscoso-pubescent;* leaves on shortish petioles, cordato-hastate, 5-7-lobed, *the lobes deltoid, acute,* shallow, crenato-dentate, sparsely pubescent, the nerves netted and prominent below; stipules cordate acute, orten bifid; peduncles axillary, longer than the leaf, pluri-flowered, with ovate bracts; pedicels very short; sepals oblong, mucronate, hairy and glandular; petals twice as long as the calyx. *DC. Prod.* 1. *p.* 679. *Bot. Mag. t.* 143. *L'Her. t.* 20. *Jacq. Ic. t.* 131. *Ger. viscosum, Cav. Diss. t.* 108. *f.* 2.

HAB. Introduced to England 1777. C.B.S., *Dr. Pappe!* (v. v. c.; Herb. Hook.)

A leafy shrub, with a heavy, balsamic smell, and rosy purple or pale flowers, the upper petals with a dark spot or lines. E. & Z.'s No. 642, in Herb. Sond., seems to me to belong to a starved form of *P. quercifolium,* rather than to the present species.

155. P. quercifolium (Ait. Kew. 2. p. 422); shrubby, much branched, *hairy and glandular;* leaves on short petioles, *cordate at base, sinuato-pinnatifid, the lobes and sinuses rounded,* the margin wavy and crenate, both surfaces hairy, the nerves prominent underneath; stipules cordate, bifid; peduncles shorter than the leaf, deflexed, 3-5-flowered, with laciniate bracts; pedicels shorter than the calyx-tube; sepals elliptical, mucronate, half as long as the petals. *DC. Prod.* 1. *p.* 678. *L'Her. Ger. t.* 14. *Cav. Diss. t.* 119. *f.* 1. *P. panduriforme E. & Z.!* 640. *P. asperum, Willd.!* (a garden var.)

HAB. Cape, *Masson,* 1774. Langekloof, *E. & Z.!* (Herb. Sond.).

Of this well-known green-house shrub, "*the oak-leaf Geranium,*" I have seen no wild specimens, but those distributed by *E. & Z.!* In cultivation the leaves are marked with a dark, purplish spot, and are disagreeably scented. Flowers purple or pink. *P. panduriforme,* E. & Z., is more tomentose than usual; the leaves canescent on the underside, and the smaller ones fiddle-shaped.

***** *Leaves deeply palmatifid, fœtid, viscoso-pubescent or glabrous.* (Sp. 156–163).

156. P. graveolens (Ait. Kew. 2. p. 423); shrubby, much branched, hairy and glandular; leaves on long petioles, *palmately 5-7-lobed, or nearly partite, the lobes flat, deeply sinuato-pinnatifid, mostly obtuse,* crenate, both surfaces pubescent and hispid, the nerves prominent; stipules cordate acute; peduncles long or short, many-flowered; calyx subsessile, very hairy, its segments half as long as the petals. *DC. Prod.* 1. *p.* 678. *L'Her. Ger. t.* 17. *Ger. terebintaceum, Cav. Diss. t.* 114. *f.* 1. *Zey.!* 2094.

HAB. Introd. to England, 1774, *Masson.* Howison's Poort, *Zeyher!* (Herb. Hook., Sond.).

A densely-branched, leafy shrub, with balsamic odour. The wild specimens resemble *P. radula,* from which they chiefly differ in the broader and flatter leaf-segments. Many varieties or hybrids, partly derived from this species, are in English gardens.

157. P. viscosissimum (Sw. Ger. t. 118); "umbels capitate, many-flowered; leaves palmate, 5–7-lobed, viscid, the segments flat, sinuated and toothed, recurved at the apex; stem very viscid; petals oblong, obtuse; *calyx-segments very obtuse, the tube sub-sessile,* not much longer than the limb." *Sw. l. c. DC. Prod.* 1. *p.* 679.

HAB. Described from a garden plant, grown from Cape seeds, *Sweet.*

"Stem and branches covered with a viscid substance, which sticks to the fingers like bird-lime. Leaves clammy, 3–4 inches long. Stipules narrow lanceolate. Bracts ovato-cordate. Calyx segments unequal, orbicularly-obovate, concave, very blunt, with incurved points. Petals lilac or white, the upper marked with red lines." *Sw.*

158. P. hispidum (Willd. Sp. 3. p. 677); *suffruticose at base,* herbaceous upwards, much branched, *hairy and glandular;* leaves on long petioles, *palmately 5-7-lobed, the lobes acuminate, unequally sharply toothed and lobulate,* both sides hispid, the nerves prominent underneath; stipules cuspidate; inflorescence panicled, the partial peduncles longer than the leaves, glandular and hispid, 5-8-flowered; *pedicels longer than the calyx-tube,* which is shorter than *the lanceolate, acuminate,* hispid segments; flowers small. *DC. Prod.* 1. *p.* 679. *Cav. Diss. t.* 110. *f.* 1. *E. & Z.!* 616, *also* 615.

HAB. Moist, shady, alpine situations, in the Western Districts. Baviansberg, Gnadendahl and Winterhoek, Tulbagh, *E. & Z.!* Witzenberg, *Zeyher!* 200. Paarlberg, *Drege!* (Herb. T.C.D., Hook., Sond.).

A free-growing, much-branched, but imperfectly ligneous plant, 2–3 feet high. Leaves 4–5 inches broad, deeply divided, the divisions jagged and finely serrated. All parts of the plant are roughly hairy and viscid. Panicle terminal, much-branched, cymose. Flowers white, with streaks. Long cultivated in Europe.

159. P. Radula (Ait. Kew. 2. p. 423); *shrubby,* much branched, hispid and viscoso-pubescent; leaves on longish petioles, palmati-partite, roughly hispid on the upper, softly pubescent on the lower side, *the lobes narrow-linear, pinnatifid, obtusely lobulate, with revolute margins;* stipules ovate-acuminate; peduncles short, hispid, 4-5-flowered; flowers pedicellate, the short calyx-tube and lanceolate sepals densely setose and glandular. *DC. Prod.* 1. *p.* 679. *L'Her. Ger. t.* 16. *Cav. Diss. t.* 101. *f.* 1. *P. revolutum, Jacq.! Ic. t.* 133. *P. Radula, E. & Z.!* 644. *P. roseum, E. & Z.!* 645., *also* 646. *Zey.!* 2096, 2097. *Drege!* 7444.

HAB. Among shrubs on mountain sides, east and west. Tulbagh, *E. & Z.!* Common in Uitenhage and Albany. (Herb. T.C.D., Hook., Sond.)

A large, densely-branched, glandular, balsamic-scented bush, very rough and hairy with short, stiff bristles. Flowers small, pale purple, with dark streaks. E & Z.'s specimens distributed as "*P. denticulatum*" (No. 646) belong to this species, from which their "*P. roseum*" in no respect differs.

160. P. denticulatum (Jacq.! Hort. Schoenb. t. 135); *suffruticose,* much branched, erect, *slender, glutinous and scabrous;* leaves on long petioles, palmati-partite, *glabrous and viscid* above, hispidulous beneath, the lobes simple or pinnatifid, linear, flat, coarsely toothed; stipules ovato-lanceolate; peduncles short, hairy, 3-4-flowered; flowers sub-sessile, the short calyx-tube and the oblong, mucronate segments villous; upper petals emarginate or bifid. *DC. l. c. p.* 679. *Sw. Ger. t.* 109. *Drege!* 7443. *Zey.!* 196.

HAB. Introduced to England, 1789. *Masson!* Hex River, *Zeyher!* Rhinosterkopf, *Burke!* Dutoitskloof, *Drege!* (Herb. Hook., Sond.)

A tall, weak-stemmed, half-herbaceous species, with a rather succulent stem and much divided, clammy, balsam-scented, sharply toothed leaves. By these characters, and its smoothness, it is known from *P. Radula*, which is sometimes mistaken for it. Flowers lilac or purple-rosy, with dark streaks on the indented or 2-lobed upper petals.

161. P. jatropæfolium (DC.) "leaves palmati-partite, viscid, glabrous, the lobes lanceolate-linear, pinnatifid, the lobules toothed, distant, acuminate; umbels 4-flowered; calyx-tube very short; upper petals obtuse." *DC. Prod.* 1. *p.* 679. "Intermediate," according to De Candolle, "between *P. denticulatum* and *P. quercifolium.*"

162. P. delphinifolium (Willd. En. 708); "leaves scabrous, palmately 5-lobed, the lobes oblong, serrated, the medial 3-lobed; umbels few-flowered, compound; calyx-tube longer than the pedicel." *DC. l. c. p.* 679. *(quere, P. scabrum, var. ?)*

163. P. munitum (Burch Cat. 1240, Voy. Vol. 1. p. 225); "glabrous, leaves bi-pinnatifid; panicle dichotomous, hardening and becoming spinous after flowering [*spinoso-lignescente*]." *DC. l. c. p.* 682.

HAB. near the Juk River, *Burchell.* (Perhaps a "*Ligularia*"?)

[Little known and doubtful species, and all of the *ephemeral*, garden *Pelargonia* are here omitted. Nearly 200 of them are characterised in De Candolle's "*Prodromus*," Vol. 1. Many more may be found in Sweet's "*Geraniaceæ*" (5 vols. 500 coloured plates); and in other works on gardens and gardening. We see no necessity for burdening our pages with these truly "*trivial*" names:

"Names ignoble, born to be forgot."]

ORDER XXXII. LINEÆ, DC.

(By W. SONDER.)

(DC., Theorie, ed. 1. 217. Prod. 1. p. 423. Endl. Gen. cclv. Linaceæ, Lindl. Veg. Kingd. clxxxiii.)

Flowers perfect, regular and symmetrical. *Calyx* 4–5 parted, persistent, imbricated in æstivation. *Petals* 4–5, tapering at base, twisted in æstivation, caducous. *Stamens* as many as the petals, and alternate with them; filaments subulate, slightly connate at base, marcescent; anthers 2-celled, introrse. *Ovary* 4–5-celled, with two axile ovules in each cell: *styles* 4-5. *Capsule* globose, incompletely 8-10-celled. *Seeds* pendulous, compressed, exalbuminous; embryo, with flat cotyledons.

Annual or perennial herbs or suffrutices. Leaves alternate or opposite, rarely whorled, simple, sessile, very entire, veinless, exstipulate. Flowers panicled or cymose, blue, yellow, red or white, soon falling off.

A small Order, differing from Geraniaceæ chiefly in habit, and the absence of a beak-like torus to the fruit. It contains but four genera, and less than 100 species,

scattered over the globe, but much more numerous in Europe and North Africa than in other countries. The common *Flax (Linum usitatissimum)*, so valuable for its strong and finely divisible fibres and its oily seeds, is by much the most useful plant of the Order.

I. LINUM, L.

Sepals 5, entire. *Petals* 5. *Stamens* 5, perfect, alternating with as many tooth-like abortive filaments. *Styles* 5, rarely 3, separate or connate below; stigmata capitellate or linear. *Capsule* spuriously 10-celled. *DC. Prod.* 1. *p.* 423. *Endl. Gen.* 6056.

Herbs or suffruticose plants, scattered throughout the temperate zones, rare within the tropics. Leaves alternate, opposite or whorled, quite entire and most frequently glabrous, sessile. Flowers in cymes or panicles, yellow, blue, red, or white. Name from *Lin.*, thread, in Celtic and modern Gaelic.

** Styles connate to the middle or above the middle.*

1. **L. Africanum** (Linn. Mant. p. 360); shrubby, glabrous; leaves opposite, the upper alternate, either linear, lanceolate, oblong or ovate. acute; corymbs dichotomous, lax or contracted; sepals ovate, acute or lanceolate, acuminate, glanduloso-ciliate, more or less longer than the capsule.

VAR. α. **angustifolium**; leaves lanceolate-linear or narrow-linear, (5–12 lines long, ½-1 line wide); corymbs dichotomous, at length lax, and mostly elongate; sepals cuspidato-acuminate, but little longer than the capsule. *L. Africanum, Lin. Herb. Planch.! in Hook. Lond. Jour.* 7. *p.* 492. *Thunb. Fl Cap.* 277. *Jacq. Ic. Rar. t.* 353. *Bot. Mag. t.* 403. *E. & Z.!* 272, *ex parte. L. juniperifolium, E. & Z.!* 271 *L. africanum α, acuminatum, E. Mey.! Hb. Drege. L. pungens, Pl.! l. c. ex parte. L. adustum, E. Mey. a, b, c.*

VAR. β. **intermedium**; leaves broadly lanceolate, narrowed at base and apex (8-10 lines long, 1½-2 lines wide); corymbs dichotomous, lax, at length elongate; sepals lanceolate-acuminate, longer than the capsule. *L. Afric. β. latifolium, Bartl. Linn.* 7. *p.* 540. *excl. syn. Thunb. L. Bartlingii, E. & Z.!* 273.

VAR. γ. **litorale**; leaves crowded, more leathery, ovate, or ovate-oblong, acute (4–6 lines long, 1½-2 lines wide); corymbs contracted, more frequently many and densely flowered; sepals cuspidate, little longer than the capsule. *L. æthiopicum, Thunb. Fl. Cap. p.* 277. *E. & Z.!* 274, *and L. africanum, E & Z.!* 272, *ex parte. L. æthiopicum, Pl.! l.c. p.* 490. *E. Mey.! in Hb. Drege a, b, c.*

VAR. δ. **gracile**; leaves more leathery, obovate-oblong, or oblong, acute, the lower (about 6 lines long, 1½-2 lines wide) crowded, upper scattered, gradually smaller; corymbs lax, often few-flowered; sepals cuspidate, little longer than the calyx. *L. gracile, Pl.! l. c. p.* 496. *L. æthiopicum, β. tenue, E. Mey.! in pl. Drege.*

VAR. ε. **acuminatum**; lower leaves oblong or lanceolate (5–6 lines long, 1½-2 lines wide); upper scattered; corymbs lax and few-flowered; sepals much acuminate, twice as long as the capsule (3 lines long.) *L acuminatum, E. Mey.! L. pungens, Pl.! ex parte, in Hb. Hook.*

HAB. Among shrubs, on plains or mountains. *a*, on the Cape Flats, *Pappe*, *W.H.H.* Table Mountain and Devil's Mt. *Ecklon.* Hott. Holl. *Zey. !* 2099. Zwartkop's River, *Zeyher*, 2099, b. Kardow, *Zey. !* 201. Winterhoeksberg, *Preiss.* Onder Bokkeveld, Bergvalley and Langevalley; also in Albany, *Drege!* 6236. *β.* at Zwarteberg, Caledon, *E. & Z.!* *γ*, on the shore, Houtniquas, *Thunberg.* Cape Recief and Algoa Bay, *E. & Z. ! Zey!* 2100. Zwartkops, Kowie and Boschjesman's River, *Drege.* *δ*, Key River. *ε.*, Witsenberg, *Zey!* 202, c. Ezelbank, 3–4000f. *Drege!* Oct.–Jan. (Herb. Hook., E. Mey., T.C.D., Sond.)

A shrublet, 1–3 feet high. Primary stem mostly short, in *γ.* long, emitting many erect, simple or branched, virgate, striated or angular secondary stems. Leaves erecto-patent, rigidly membranous or coriaceous, green or glaucous, more numerous and mostly opposite in the lower, scattered in the upper portion of the branches; all one-nerved or in var. *γ.* sub-tri-nerved; the margin slightly inrolled, very entire. Stipulary glands 2, minute or thickened. Infl. cymose-corymbose, the lateral branches commonly racemose, most frequently lax, 2–3 inches long, sometimes 6 inches or more; in var. *γ.* contracted, 1–2 inches, the cymules short and leafy. Bracts leafy. Pedicels as long as the calyx or shorter. Sepals 3-nerved. Petals obovate, yellow, villous at base inside, twice or rarely thrice as long as the calyx. Styles equalling the petals, or shorter, connate for nearly half their length. Stigmas minute, capitellate. Capsule sub-globose, glabrous.

** *Styles separate to the base, or nearly so.* (Sp. 2–4.)

2. L. thesioides (Bartl. Linn. 7. p. 540); suffruticulose, glabrous; secondary stems numerous, ascending or erect, somewhat naked and paniculato-corymbose above; *leaves mostly alternate, crowded, erect, subimbricate, narrow-linear, acute;* stipulary glands none or rarely solitary or in pairs; sepals ovate, acute, pectinato-ciliate, about equalling the capsule. *E. & Z. !* 270. *L. spec. pl.! l. c. p.* 494. *L. africanum. β. minus, E. Mey. ! ex parte.*

HAB. Stony and sandy places at the foot of the Lion's Mt., Wynberg, and near Stellenbosch, *E. & Z. ! Mundt., Drege, W.H.H., Zey. !* 2026. Dec.–Feb. (Herb. T.C.D., Sond.)

Resembling the smaller specimens of *L. africanum;* this differs in the smaller size, the stems being commonly 3–6 inches, rarely 1 foot long, and more slender; in the smaller leaves, 3–4, rarely 5 lines long, ½ line wide; and smaller flowers. Root thick. Leaves evidently one-nerved below. Corymb short, few-flowered, an inch or rarely two inches long, sometimes 2–4 flowered. Pedicels as long as the calyx or shorter. Sepals one line long, remarkably ciliate. Petals yellow, twice as long as the calyx, equalling the styles.

3. L. Thunbergii (E. & Z. ! 275, excl. syn.); suffruticulose; secondary stems numerous, erect, simple, much-branched, glabrous, or pubescent below; lower leaves *opposite, rarely ternate or quaternate, oblong or oblongo-lanceolate, mucronulate,* intermedial sub-linear, opposite or alternate, shorter than the internode, uppermost and floral subulate; stipulary glands 2 or none; branches of the dichotomous cyme racemose; sepals ovate, acute, glandular-ciliate, about as long as the capsule. *L. africanum (virginianum), Reichb. Ic. Exot.* 1. *t.* 46. *opt. ! L. Reichenbachii, Pl. ! l. c. p.* 495. *L. racemosum, E. Mey. ! Zey. !* 203, 2101, 2102.

VAR. *β.* **paniculatum**; flowers disposed in an irregular, 6–12 inches long panicle. *L. quadrifol. β. paniculatum, E. Mey. !*

HAB. Constantia and Tokay, E. side of Table Mt.; hills at Adow, Uit.; and the Winterberg, Caffr. *E. & Z. !* Vanstaadensberg, Uit.; Salem, Albany, Hexrivier; and Caledon's Riv. *Zeyher !* Saureplaats, Sneeuberge and between Omsamwubo and Omsamcaba, &c. *Drege!* β. at the Key. Oct.–Feb. (Herb. Hook., T.C.D., E. Mey., Sond.)

Primary stem short; secondary 1–1¼ feet high. Branches, when present, long and slender. Leaves rigidly membranous, one-nerved, the lower 5–7 lines long, two lines wide; the upper gradually smaller and narrower. Corymb lax, its branches long and filiform. Pedicels as long as the calyx, in fruit rather longer. Sepals scarcely more than one line long. Petals yellow, according to Reichenbach 4 times as long as the calyx, but in the wild plant only 2–3 times as long. Styles twice as long as the stamens, with clavate stigmas. Capsule the size of a pepper-corn, green with white bands, 10-seeded, sub-acute?

4. L. quadrifolium (Linn. sp. 402); fruticulose, glabrous; secondary stems somewhat 4-angled, simple or virgately branched; *lower leaves 4 or 5 in a whorl, spreading, elliptical, or ovate-oblong, mucronulate,* or acute at each end, upper opposite or alternate, narrower; stip. glands two; corymb lax, compound, several-flowered; sepals ovato-lanceolate, acuminate, glandular-ciliate, equalling the capsule. *Thunb.! Cap. p.* 277. *Bot. Mag. t.* 431. *DC. Prod.* 1. *p.* 425. *E. & Z.!* 276. *Planch. l. c. p.* 496.

Hab. Among shrubs. Mountains round Capetown, *E. & Z.! W.H.H.* and Paarlberg, and at Keureboom's River, George, and Vanstaadensberg, Uit., *Drege!* Sep.–Nov. (Herb. T.C.D., Hook., Sond., E. Mey.).

1–2 feet high. Root slender. Secondary stems many, usually long, flexuous, simple or with a few opposite branches. Leaves patent or deflexed, 6–8 lines long, 3–4 lines wide, sometimes but half that size, rigidly membranous; the upper ovato-lanceolate, opposite, the very uppermost scattered and smaller. Corymb few-flowered or divided and elongate, with distant flowers. Sepals 1½ lines long. Petals yellow, 5–6 lines long. Fruiting pedicels shorter than the calyx. Capsule sub-acute. Easily known from *L. Thunbergii* by its more robust habit, broader and whorled leaves, and larger flowers.

Order XXXIII. BALSAMINEÆ, A. Rich.

(By W. Sonder.)

(A. Rich. Dict. Class. 2. 173. DC. Prod. 1. p. 685. Endl. Gen. cclvii. Balsaminaceæ, Lindl. Veg. Kingd. clxxxvi.

Flowers perfect, very irregular. *Calyx* 5-leaved, coloured, caducous; two *lateral* sepals outermost, opposite, small or minute, on one side imbricating the two *anterior*, which are also small, the *posterior* very large, saccate, produced at base into a spur or bag. *Petals* 5, more or less connate into 2 or 3, of which the anterior (opposite the spurred sepal) is much larger than the others. *Stamens* 5, hypogynous, imperfectly syngenesious. *Ovary* free, 5-celled; ovules indefinite, anatropous; stigma sessile. *Fruit* capsular (rarely *drupaceous* and fleshy), splitting elastically into 5 valves, and dispersing the exalbuminous seeds. *Embryo* straight, with flat, fleshy cotyledons, and a short superior radicle.

Herbaceous plants, with soft and juicy, fragile stems. Leaves alternate or opposite, sometimes all radical, simple, penninerved, exstipulate. Flowers axillary or in terminal racemes.

Two genera and about 120 species comprise this little group, which differs from *Geraniaceæ* chiefly in it highly irregular calyx and corolla, and its exstipulate leaves.

The mountains of Tropical Asia are the chief stations of these plants, a few of which are scattered in N. America, and Europe, and one in Madagascar. The common "*Balsam*" of gardeners is a familiar example of the Order. None are of much value to mankind.

I. IMPATIENS, L.

Ovary 5-angled or terete, 5-celled, with several ovules in each cell. *Capsule* 5-valved, splitting elastically. *Seeds* numerous or few. *Endl. Gen.* 6060.

Annual or perennial, succulent herbs, natives of the warmer, temperate and tropical zones, chiefly of the Northern Hemisphere. Leaves alternate, opposite or whorled, linear or lanceolate, serrate or toothed. Peduncles axillary, one or many-flowered. Flowers yellow, red, white or parti-coloured. Name, *impatient*, from the sudden opening of the capsule, when suddenly compressed. English name "*Touch-me-not.*"

1. I. capensis (Thunb. Prod. p. 41); leaves alternate, petiolate, ovate, acuminate at each end, with piliferous crenatures; peduncles axillary, solitary, capillary; spur longer than the flower. *Fl. Cap. p.* 187. *Balsamina capensis. DC. Prod.* 1. *p.* 686.

HAB. River banks in woods, in George, Uitenhage, at the Kat River, the Boschrivier, and on to Port Natal, *Thunberg, E. & Z.! Drege! Krauss, Gueinzius.* Oct.–Mar. (Herb. Sond., T.C.D., Hook.).

Stem herbaceous, weak, erect, fleshy, glabrous, simple, 1 foot high. Leaves 1½–3 inches long, 1–2 inches wide, narrowed into a sparingly piliferous petiole, with an obtuse acumination and piliferous crenatures, penni-nerved, glabrous or thinly hairy. Flowers from the axils of the upper leaves; the peduncles 1–2 inches long. Corolla ½ inch long, shorter than the slender spur, pale rosy.

2. I. bifida (Thunb. l. c.); leaves alternate, petiolate, oblong, acuminate at each end, serrated; spur very long, bifid. *Fl. Cap.* 187. *Balsamina bifida, DC. l. c.*

HAB. Cape of Good Hope, *Thunberg.*

"Stem herbaceous, weak, smooth, simple, erect, a span long. Leaves alternate, petiolate, tapering at each end, a finger long. Flowers axillary, pedunculate. Peduncles capillary, very lax, a finger long. Nectary horned, several times longer than the flower, uncial, curved," *Thunb.* Unknown to us.

ORDER XXXIV. OXALIDEÆ, DC.

(By W. SONDER.)

(Oxalideæ, DC. Prod. 1. p. 689, Endl. Gen. cclvi. Oxalidaceæ, Lindl. Veg. King. clxxxv.)

Flowers perfect and regular. *Calyx* 5-parted, imbricated in æstivation. *Petals* 5, hypogynous, clawed, free or connate at base, twisted in æstivation, deciduous. *Stamens* 10, 5 opposite the petals shorter than the others; filaments slightly connate at base; anthers 2-celled, introrse. *Ovary* deeply 5-lobed, 5-celled, few or many ovuled; styles 5, filiform, with capitate stigmas. *Fruit* either capsular or fleshy; seeds one or many, pendulous, with copious, fleshy albumen; embryo straight or curved.

Stemless or caulescent herbs, often with tuberous roots ; rarely suffrutices, shrubs, or even trees, with compound, exstipulate leaves ; natives of tropical and sub-tropical countries, a few straggling into the colder parts of both temperate zones. *Oxalis*, the largest genus and type of the Order has very many species in South Africa, and also abounds in extra-tropical S. America, where some of the species form rather tall shrubs. One is employed at Coquimbo, for house-building ! Its rodlike stems are very durable, and are made into a sort of wicker-work-skeleton of the house-walls, which are then strengthened and rendered water-tight by mortar and plastering. Oxalic-acid is the chief product of the Order, and gives the stems and foliage the sharp acidity which recommends them to the thirsty traveller. Some are used as salad. The tuberous roots of several of the American species are starchy and used for food. *Oxalideæ* is usually regarded as a member of the Geranial group of Orders, and by some botanists is included in *Geraniaceæ*. I am more disposed to consider them, with De Candolle, allied to *Zygophylleæ*, with which they agree in foliage and the albuminous seeds. By Planchon they are associated with *Leguminosæ*.

I. OXALIS. L.

Sepals 5, free or united at base. *Petals* 5, convolute, their claws conniving into a funnel-shaped tube. *Stamens* 10, connate at base, 5 alternate shorter. *Styles* 5, stigmas capitellate or pencilled. *Capsule* deeply 5-lobed, globose or oblong; seeds one or several. *DC. Prod.* 1. *p.* 690. *Endl. Gen.* 6058.

The S. African species of this large genus are, with one exception, bulbous-rooted perennials, stemless or caulescent, with alternate or fascicled leaves and scape-like peduncles. The leaves are mostly trifoliolate, rarely 1–2-foliolate, or digitately many-leafleted. Flowers red, purple, white, yellow, or streaked, appearing in the winter and early spring months. Name from *οξυς, acid* or *sharp ;* from the taste of the foliage.

KEY TO THE ARRANGEMENT OF THE SPECIES.

A. Peduncles one-flowered. (Sp. 1–98.)
I. Leaves simple. (Sp. 1–2.)
II. Leaves bi-foliolate, the petiole broadly winged. (Sp. 3–4.)
III. Leaves trifoliolate, the petiole not winged. (Sp. 5–89.)

* *Upper leaves crowded, mostly on long petioles. Peduncles mostly terminal.* (5–81.)

(1) Leaflets linear, lanceolate, oblong or obovate :—
(α.) Glabrous, or covered with *very short*, white hairs. (Sp. 5–32.)
(β.) Clothed with long, jointed, yellowish but *not glandular* hairs. (Sp. 33–37.)
(γ.) Clothed with *gland-tipped hairs*. (Sp. 38–43.)
(2.) Leaflets sub-rotund, obtuse or scarcely retuse. All stemless.
(α.) Flowers purple, rosy or white. (Sp. 44–56.)
(β.) Flowers yellow or yellowish. (Sp. 57–62.)
(3.) The medial, or all the leaflets obcordate.
(α.) Stemless or stipitate. (Sp. 63–69.)
(β.) Caulescent ; peduncles both terminal and axillary. (Sp. 70–75.)
(4.) Leaflets deeply two-lobed.
(α.) Caulescent. (Sp. 76–80.)
(β.) Stemless. (Sp. 81.)

** *All the leaves sessile or very shortly petiolate, the upper not crowded ; the petiole dilated at base. Peduncles axillary, never terminal. Caulescent.* (Sp. 82–89.)

IV. Leaves digitate; leaflets 5–19. (Sp. 90–98.)

B. Peduncles many-flowered. Leaves trifoliolate.

(1.) Root bulbous; flowers yellow. (Sp. 99–101.)
(2.) Root bulbous; flowers purple, rosy or white. (Sp. 102–107.)
(3.) Root fibrous; flowers yellow. Sp. 108.)

ANALYSIS OF THE SPECIES.

A. Peduncles one-flowered.

I. Leaves *simple* (unifoliolate)
- Leaflets obovate or elliptical, obtuse or sub-emarginate ... ... (1) **monophylla.**
- Leaflets obcordate, sub-bilobed ... ... (2) **Dregei.**

II. Leaves *bifoliolate*, the petiole broadly winged.
- Leaflets lanceolate ... ... (3) **asinina.**
- Leaflets obovate, emarginate ... ... (4) **fabæfolia.**

III. Leaves *trifoliolate*, the petiole not winged.

* Upper leaves crowded, mostly *on long petioles*. Peduncles terminal. (Sp. 5–81.)

(1.) *Leaflets linear, lanceolate, oblong, or obovate.* (Sp. 5–43.)

(a.) *Not glandular*; glabrous, or with *short*, whitish pubescence. (Sp. 5-32.)

Leaflets linear or sub-cuneate; *bulb* sharply *angular* ... ... (5) **goniorhiza.**

Leaflets linear, lanceolate, oblong or ovate; *bulb smooth.*

Sepals lanceolate, *setaceo-acuminate* ... ... (7) **leptocalyx.**

Sepals lanceolate or broadish:

Inner petioles *leafless.* (Stemless, glabrous) (18) **ligulata.**

All the petioles bearing three leaflets.

† *Leaflets* linear-terete:

Stem 6-12-uncial; leaflets *dilated at the apex* ... ... (13) **teretifolia.**

Stem 3-uncial; leaflets emarginate, not dilated ... ... (8) **Burkei.**

†† *Leaflets* linear, or linear-cuneate:

Stem or stipe *leafless*, scaly, rarely with a few leaves:

Bulb black;

Fl. purple; *leafl.* straight; *pedunc.* short ... ... (9) **linearis.**

Fl. purple; *leafl.* recurved; *pedunc.* longer ... ... (14) **falcata.**

Fl. pale, *with purple edges* ... ... (10) **versicolor.**

Bulb. brown;

Leafl. sub-glabrous, *not ciliate*, 6–12 lines long ... ... (11) **polyphylla.**

Leafl. glabrous or pilose, *ciliate* 3–4 lines long ... ... (6) **glabra,** *var.*

Leafl. hairy below, obtuse; petiole, peduncle and calyx *hairy* ... ... (16) **Mundtii.**

Stem elongate, leafy, *glabrous;* leaflets flattish ... ... (12) **gracilis.**

Stem long, leafy, *pubescent;* leafl. carinate (15) **tenuifolia.**

††† *Leaflets* linear-oblong, oblong, oval or obovate:—

Stemless; leaves *two or three* (puberulent) (25) **bifolia.**

Stemless; leaves *very numerous:*

Flowers white or red,

Pilose; leafl. oblong, *rigidly ciliate*... (26) **approximata.**

Pubescent; leafl. wavy, concave, *tomentose beneath* ... ... (24) **Algoensis.**
Glabrous: leafl. oblong or oval ... (27) **minuta.**
Very hairy; leafl. broadly ovate, wavy and curled ... ... (31) **hirsuta.**
Flowers yellow or straw-coloured:—
Pubescent; leafl. *narrow, undulate,* pilose on both sides ... ... (23) **crispula.**
Very thinly pubescent; leafl. lanceolate-oblong, *flat* ... ... (20) **laburnifolia.**
White-hairy; leaflets linear-oblong, punctate ... ... (21) **albida.**
Quite glabrous; leafl. linear-oblong (22) **Namaquana.**
Stemless or stipitate; leaflets obovate-cuneiform, flat, the lateral oblique ... (30) **mutabilis.**
Caulescent; the stem scaly or few-leaved below the crown:—
Leafl. linear-oblong, cuneate, *pubescent* on both sides ... ... (19) **cuneata.**
Leafl. linear-oblong, obtuse, *glabrous above,* villous below... ... ... (9) **linearis, β.**
Leafl. lin.-oblong, *white-hairy* ... ... (17) **polytricha.**
Leafl. exactly-oblong, obtuse, pubescent *underneath* ... ... (28) **ciliaris.**
Caulescent; leafy, leafl. cuneate-oblong or linear-oblong, carinate, emarginate; the plant glabrous, or pilose ... ... (6) **glabra.**
Caulescent; leafy, somewhat branched, *green,* puberulent; the leaflets obovate-oblong, flat ... ... (29) **virginea.**
Caulescent, much branched, *cano-pubescent;* leafl. carinate ... ... (32) **mutabilis.**

(β) *Not glandular;* clothed with *long, jointed,* yellowish hairs: (Sp. 33–37.)
Leaflets obovate, netted over with veins ... ... (33) **Meyeri.**
Leafl. oblong-cuneate, veinless above, not dotted (34) **florida.**
Leafl. linear-oblong, penni-nerved, veiny and dotted ... ... (35) **affinis.**
Leafl. linear, glabrous above, nerveless, hairy beneath ... ... (36) **angusta.**
Leafl. narrow-linear, with orange glands underneath ... ... (37) **adspersa.**

(γ) *Glandular,* with gland-tipped hairs: (Sp. 38–43)
Leaflets oblong, obtuse, black-dotted; stemless or stipitate ... ... (38) **glandulosa.**
Leafl. oblong-cuneate, emarginate; stem leafy... (39) **multicaulis.**
Leafl. linear-acute, callous at apex and margin; stem thickly glandular ... ... (40) **recticaulis.**
Leafl. linear-cuneate, broad topped, not callous; stem sparingly glandular ... ... (41) **clavifolia.**
Leafl. linear, glandular-ciliate; stem scaly, glandular ... ... (42) **droseroides.**
Leafl. linear, glabrous above, many dotted below; stem puberulous .. .. (43) **semi-glandulosa.**

(2) *Leaflets sub-rotund, obtuse, or scarcely retuse. All stemless.* (Sp. 44–62.)
(a) Flowers purple, violet, rosy or white. (Sp. 44–56.)
Petioles *compressed* (the inner stamens toothed) (46) **breviscapa.**
Petioles terete or furrowed above. *Pubescent or hairy.*

Leaflets wavy-crenate, glabrous above, or sub-pilose... ... ... (47) **collina.**
Leafl. anteriorly crenate, silky-villous above (50) **melanosticta.**
Leafl. flat, the medial not crenate, rigidly ciliate : corolla tubular . . .. (44) **purpurea.**
Leafl. flat, the medial cuneate,
Leafl. glabrous or pubescent beneath, minutely ciliate; corolla wide from the base ; inner stamens toothed ... ... (45) **variabilis.**
Leafl. tomentose, or glabrous above ; petioles and peduncles hairy ; inner stamens toothless ... ... (48) **pulchella.**
Leafl. tomentose on both sides, with black dots at margin and sides ; petioles and peduncles pubescent ... ... (49) **calligera.**
Petioles terete or furrowed above. *Glabrous.*
Bulb smooth,
Leafl. cellulose above, minutely dotted beneath, veiny, green ... ... (51) **commutata.**
Leafl. with impressed dots on both sides, pale beneath ; scape long ... ... (52) **convexula.**
Leafl. as in 52 ; scape shorter than the leaf ... ... (56) **grammopetala.**
Bulb angular ... ... (55) **punctata, β.**
Petioles terete or furrowed above. *Glandular.*
Scape and petioles pilose, not glandular ; leaflets glanduloso-ciliate ... ... (53) **setosa.**
Scape and petioles glanduliferous,
Leaflets roundish-obtuse, glandular at margin... ... ... (54) **adenodes.**
Leaflets cuneate-sub-rotund, emarginate, margin not glandular... ... ... (55) **punctata.**

(β.) Flowers yellow or yellowish. (Sp. 57–62.)
Bulb-scales *gummy,* not loose or separable ... (57) **balsamifera.**
Bulb-scales loose or separable :
Pubescent ; leaflets rounded ... ... (60) **Eckloniana.**
Pubescent; leaflets roundish-cuneate, hollow dotted beneath ... ... (58) **luteola.**
Pubescent ; leafl. roundish-cuneate, with black or golden dots and striæ ... ... (59) **stictophylla.**
Glabrous ... ... (61) **minima.**
Glandularly pilose ... ... (62) **glaucescens.**

(3) *The medial leaflet, or all the leaflets obcordate.* (Sp. 63–75.)
(α.) Stemless, or stipitate. (Sp. 63–69.)
Flowers white or rosy :
Leafl. glabrous above, beneath villous, not dotted ... ... (63) **Uitenhagensis.**
Leafl. pilose on both sides, dotted underneath ... ... (64) **imbricata.**
Leafl. woolly-tomentose on both sides ... (69) **lanata**
Flowers purple or violet-coloured :
Thinly pilose ; leafl. glabrous and dotted above, pilose beneath ... ... (65) **cruentata.**
Thickly pubescent ; leaflets pubescent and hollow-dotted ... ... (66) **obtusa.**
Very villous ; leafl. thick, triangular obcordate *silky-shining* ... . (67) **truncatula.**
Villous ; leaflets thin, roundish-obcordate, silky underneath ... ... (68) **holosericea.**

(β.) Caulescent. (Peduncles both terminal and *axillary.* (70–75.)
Petioles with stipuloid, basal dilatations:
Stem long, *floating*, few-leaved; with a terminal tuft ... ... (71) **natans.**
Stem *erect*, leafy, leaves distichous, crowded at top ... ... (72) **disticha.**
Petioles not dilated or winged at base:
Peduncles not bracteolated ... ... (74) **ebracteata.**
Peduncles bracteolated:
Stem and peduncles *puberulous* ... ... (73) **tenella.**
Stem and peduncles *glandular* ... ... (70) **aganophila.**
Stem and peduncles *glabrous* ... ... (75) **incarnata.**

(4) *Leaflets deeply two lobed.* (Sp. 76–81.)
(α) Caulescent:
Glabrous, sub-pruinose; stem leafy ... ... (76) **comosa.**
Glabrous, green; stem naked below ... ... (79) **bifida.**
Villous; leafl. reflexed, their lobes short, villous (78) **heterophylla.**
Whitish-tomentose; leafl. erect, forked, lobes silky beneath ... ... (80) **bifurca.**
Glandular, sub-viscid ... . (77) **caledonica.**

(β) Stemless; quite glabrous ... ... (81) **Smithii.**

** All the leaves sessile or very shortly petiolate, the upper not crowded; the petioles dilated at base. *Peduncles* axillary, never terminal. (Sp. 82–89.)

Flowers purple, or violet:
Pubescent or hairy:
Leafl. narrow; tube of petals narrow, 2–3ce as long as limb ... ... (82) **tubiflora.**
Leafl. narrow; tube of petals wide, not longer than limb ... ... (83) **hirta.**
Leafl. obovate, sub-rotund; stem and branches prostrate ... ... (84) **brevicaulis.**
Incano-pilose; hairs gland-tipped and simple; *leafl.* linear, pubescent beneath ... ... (88) **Meisneri.**
Glandular-viscid; *leafl.* linear or oblong cuneate, thinly glandular beneath ... ... (86) **viscosa.**
Glabrous, *slender*; leafl. linear-emarginate, with dots and lines ... ... (87) **pardalis.**
Glabrous, *stronger*; leafl. linear-cuneate, 1 line wide, obtuse, dotted beneath ... ... (85) **densifolia.**
Flowers yellow. Hairy tomentose ... ... (89) **cana.**

IV. Leaves *digitate*; leaflets five to nineteen. (Sp. 90–98.)
Flowers white, rosy, lilac, violet or purple:
Leafl. 12–19, oblong, tomentose (fl. white) ... ... (95) **tomentosa.**
Leafl. 9–11, minute, linear, canescent (fl. rosy) ... (94) **Zeyheri.**
Leafl. 5–7, lin.-setaceous, sharply emarginate, with black lines above ... ... (90) **capillacea.**
Leafl. 5, lin.-involute, not callous-dotted; bulb blackish ... ... (91) **pentaphylloides.**
Leafl. 5, lin.-involute, with two calli; bulb pale brown ... ... (92) **pentaphylla.**
Leafl. 5, obovato-cuneate, not callous-dotted ... (93) **quinata.**
Flowers yellow:
Leafl. 5–9, oblong-linear, channelled; sepals equal (96) **flava.**
Leafl. 5–9, lin.-lanceolate, or linear; sepals unequal (97) **flabellifolia.**
Leafl. 5–7, oblong or obovate oblong, margined; sep. unequal ... ... (98) **lupinifolia.**

B. Peduncles many-flowered. Leaves tri-foliolate. (Sp. 99–108.)

(1.) *Root* bulbous, *Flowers* yellow or yellowish. (Sp. 99–101.)
Leafl. silky *on both sides* ... ... (99) **sericea.**
Leafl. pubescent *beneath only;* petiole compressed, ciliate ... ... (100) **compressa.**
Leafl. pubescent *beneath only;* petiole terete, glabrous ... ... (101) **cernua.**

(2.) *Root* bulbous. *Flowers* blueish, purple, rosy or white. (Sp. 102–107.)
Leafl. obcordate ... . (104) **purpurata.**
Leafl. deeply obcordate-bilobed:
Stemless, pilose, glandular above; umbel 8–12-flowered ... ... (105) **semiloba.**
Stemless, glabrous; umbel 2–6-flowered ... ... (102) **caprina.**
Stipitate or caulescent; pedunc. terminal. . ... (107) **stellata.**
Caulescent; pedunc. axillary; leaf-lobes short, rounded .. .. (103) **livida.**
Caulescent; pedunc. axil.; leaf-lobes oblong or linear ... ... (106) **lateriflora.**

(3.) *Root* fibrous. *Stem* branched, diffuse or creeping (108) **corniculata.**

A. Peduncles one-flowered. (Sp. 1-98).
I. Leaves simple. (Sp. 1–2.)

1. O. monophylla (Linn. Syst. p. 432); stemless, puberulous; leaves *obovate or elliptical, obtuse* or subemarginate; scape one-flowered; sepals rather acute. *Thunb.! diss. No.* 1. *Tab.* 1. *Fl. Cap. p.* 532. *Jacq. Oxal. t.* 79. *f.* 3. *O. lepida, Jacq. l. c. t.* 21. *O. rostrata, Jac.! l. c. t.* 22. *DC. Prod. No.* 83, 84, 85. *E. & Z.!* 701.

VAR. *β.* **stenophylla**; leaves lanceolate or linear, obtuse. *O. stenophylla, Meisn.! in Hook. Lond. Jour.* 2. *p.* 54.

HAB. Sandy hills, round Cape Town and Cape Flats, common (*Thunberg, E. & Z.! &c.*) *β.* near Tulbagh, *Krauss!* Apr.-May. (Herb. Thunb. Holm., Vind., Meisn., T.C.D., Sond.)

Bulb globose, with a very delicate, torn, rusty-brown, soft, as if lanuginous coat. Leaves petiolate, the lamina ½–1 inch long. Peduncles 1–2 inches long or longer, equalling or exceeding the leaves. Corolla pale purplish, with a yellowish tube.

2. O. Dregei (Sond.); stemless, quite glabrous; leaves *obcordate-bilobed,* obtuse at base; scape one-flowered; sepals obtuse. *O. rostrata, E. Mey! non Jacq.*

HAB. Near rivulets, in the Kamiesberg Mts., Kl. Namaqualand, *Drege!* Aug. (Herb. E. Mey., Sond., Vind.).

Like the preceding, but smaller and entirely glabrous; and well distinguished by its cordate leaves broader than long. Leaves 4–5 lines wide, 3 lines long. Flowers as in *O. monophylla*, but rather smaller, and seemingly pale rosy. Sepals very obtuse.

II. Leaves bifoliolate; the petiole broadly winged. (Sp. 3–4).

3. O. asinina (Jacq.! l. c. No. 38. t. 24); stemless, substipitate, glabrous; leaflets two, *lanceolate*, cartilagineo-scabrous at the edge; scapes one-flowered. *DC. l. c. p.* 88. *O. lanceæfolia, Jacq.! l. c. No.* 40. *t.* 26. *DC. No.* 89.

VAR. β. **leporina**; leaflets elliptico-lanceolate; flowers white, yellow at base. *O. leporina, Jacq.! l. c. t.* 25. *DC. No.* 89.

HAB. Cape of Good Hope. (v. s. c. in Hb. Jacq.)

Bulb brownish. Petiole 2 inches long, with lanceolate wings, which are sometimes obsolete. Leaflets 2, rarely 3, 1½–2 inches long, ½–1 inch wide; in β. smaller and more obtuse. Scapes bibracteolate, equalling or exceeding the leaves. Corolla yellow, in β. white, with a reddish margin and yellow base.

4. O. fabæfolia (Jacq.! l. c. No. 41. t. 27); stemless, substipitate, glabrous; leaflets 2 or 3, *obovate*, emarginate, with a cartilaginous margin; scape one-flowered. *DC. No.* 90.

VAR. β. **crispa**; leaflets broadly obovate, with an undulate, more evidently cartilaginous margin. *O. crispa, Jacq. l. c. t.* 23. *DC. l. c. No.* 86.

HAB. Cape of Good Hope. (v. s. c. in Hb. Jacq.)

This species differs from the preceding only by its leaflets, 2–3 inches long, 1½–2 inches wide, and the petiole which is mostly ovate or subrotund. Flowers pale yellow, in β whitish, yellow at base, size of those of *O. asinina.*

III. Leaves trifoliolate, the petiole not winged. (Sp. 5–89).

* *Upper leaves, terminating the stem or branches, crowded, on long petioles. Scapes or peduncles terminal. Stemless or Caulescent species.* (Sp. 5–81).

(1). Leaflets linear, lanceolate, oblong, or obovate. (Sp. 5–43).

(*a*). *Not glandular;* glabrous, or clothed with short, whitish hairs. (Sp. 5–32).

5. O. goniorhiza (E. & Z.! 699); bulb *oblong, acutely* 5–10 *angled;* stem short, erect; leaflets narrow-linear, or subcuneate, keeled, with or without a callosity below the emarginate apex; peduncles longer than the leaves; sepals lanceolate, glabrous, 4–5 times shorter than the corolla. *O. pusilla, E. Mey! O. pallens, E. & Z!* 700, *ex parte.*

VAR. β. **semiglauca**; glabrous or downy; stem rather taller, subfoliate, leafless, with 2 or more callosities. *O. semiglauca, E. & Z!* 691. *Zey!* 245, 218.

HAB. Sands near Zwartland, *E. & Z.! Burke! Drege!* Berg River, Oliphant's R. & Krum Riv. *Zeyher!* β. Cape flats and at Caledon's Bath, *W.H.H., E. & Z.!* Jan. Aug. (Herb. Vind., E. Mey., Lehm., Sond., T.C.D.

2–4 inches high. Bulb acute at each end, 6 lines long, blackish. Leaflets 2–3 lines long. Peduncles with 2 glabrous bracteoles. Sepals mostly with 2 calli. Corolla ½ inch long, yellow at base, limb blueish, lilac or white.

6. O. glabra (Thunb. dissert. No. 17. f. 2); glabrous or pilose, with a small, *ovate, smooth bulb;* stem very short, leafless, or elongate, scaly or leafy; leaflets linear or oblong-linear, cuneate, with one or more callosities at the apex, or base; peduncles glabrous, longer than the leaves, bibracteolate at the apex; sepals *lanceolate,* 3–4 times shorter than *the externally glabrous corolla.*

VAR. α. **major**; stem tall (a span long), glabrous or sparingly hairy, simple or somewhat branched, leafy; leaflets linear-oblong or oblong, cuneate, glabrous, or ciliate, mostly bi-callous. *O. glabra, Thunb. l. c. O. venosa, Sav.! in Lam. Dic.* 4, *p.* 681. *DC. No.* 51.

VAR. β. **minor**; stem short (2–3 inches); leaflets oblong-linear, cuneate, with 2 or more calli, ciliated and pilose underneath. *O. glabra, Jacq.! l. c. t.* 76. *f.* 3. *DC. No.* 139. *E. & Z.!* 688. *O. cuneata, Thunb. Cap. p.* 540. *O. ciliaris, E. & Z.! O. linearis, E. & Z.!* 687. *O. minor,*

E. & Z. 689. *Herb. Un. No.* 589. *O. elongata, β. amoena and O. gracilis, α, E. Mey. O. arcuata, Meisn., in Pl. Krauss, non Jacq.*

VAR. γ. **pusilla**; stem very short, or scarcely any; leaflets linear or linear-cuneate, 1–2-callous, rarely nude? peduncles as long as the leaves or a little longer; flowers smaller, sepals mostly with callous dots. *O. pusilla, Jacq.! l. c. t.* 42. *E. & Z.! No.* 692. *O. pusilla, α, and O. miniata? E. Mey.! O minuta, E. & Z.!* 690. *Meisn.! in Pl. Krauss, non Thunb. Hb. Un. No.* 590.

VAR. δ. **albiflora**; somewhat hairy; stem short; flowers larger, white, with yellowish bases.

VAR. ε. **acuminata**; glabrous; stem short; leaflets linear, small, with several calli; sepals much acuminate, spreading, not callous tipped. *O. gracilis, β. E. Mey.*

HAB. Sandy places round Capetown, and many places in Cape, Stellenbosch, and Caledon Districts, common, *Thunberg, &c. &c.* δ. Berg river, *Zey!* 217. ε, Zwartland, *Drege!* Krumrivier, *Zey!* 219. Jul.–Aug. (Herb. Thunb., Lam., Jacq., E. Mey.; Meisn., Holm., T.C.D., Sond.)

Plant polymorphous. Bulb 3–4 lines long, brownish. Petioles dilated at base, glabrous or pilose, 2–4 lines long, rarely longer. Leaflets 3–4 lines long, ½–2 lines wide, emarginate, appearing finely dotted under a lens. Peduncles 1-2 inches long; in α, sometimes 3 inches. Corolla 8–10–12 lines long, with a yellow tube, and usually a violet-purple limb, rarely whitish. It varies with double flowers.

7. **O. leptocalyx** (Sond.); glabrous; bulb ovate, smooth; stem erect, somewhat branched, leafy; leaflets linear, subcuneate, emarginate; peduncles one-flowered, equalling the leaves; sepals lanceolate, *setaceo-acuminate,* not callous-tipped, ciliolate at the margin, as are the bracteoles; corolla *twice* as long as the calyx, *pilose externally.*

HAB. Moist places. Piquetberg, *Zey!* 216. June. (Herb. Sond.)

4 inches high, slender. Petioles 3–4 lines long. Leaflets 4–6 lines long. Sepals four lines long. Corolla yellowish at base, the limb violet colour. Allied to var. ε, of the preceding species.

8. **O. Burkei** (Sond.); glabrous or pilose; stem slender, nearly leafless; leaflets *linear, involute-subterete,* emarginate, with apical calli; peduncles glabrous, *longer than the leaf;* sepals ovate-acuminate, nude, ciliate (as well as the bracteoles) 6–8 *times shorter than the corolla.*

VAR. β. **multiglandulosa**; leaflets dotted and striated with blackish callosities.

HAB. Stony places in the Karroo, at Driekop, near Pinaarskloof, *Burke! Zey!* 257. β. same place, *Zey!* 256, and on hills by the Berg River, *Zey!* 252. May. (Herb. Hook., T.C.D., Sond.)

3 inches high, very slender. Bulb wanting. Petioles ½ inch long. Leaflets 3–4 lines long, ¼–½ line wide. Corolla 7–8 lines long, with a yellow tube and bluish-purple limb.

9. **O. linearis** (Jacq.! Ox. t. 32); bulb ovate, smooth; stem or stipe puberulous, procumbent, leafless below; leaflets *linear,* obtuse or subemarginate, *glabrous above, villoso-pubescent underneath;* peduncles *villous, equal to the leaves;* bracteoles near the calyx; sepals lanceolate, 5–6 times shorter than the corolla; claws of the petals longer than the spreading obovate lamina. *DC. l. c. No.* 134.

VAR. β. **latior**; leaflets oblongo-linear. *O. arcuata, Jacq. t.* 31. *DC. No.* 96.

VAR. γ. **minor**; leaflets shorter, linear-cuneate, obtuse or subemarginate, hairy underneath.

HAB. Cape, *Jacquin!* Acht valley, *Zeyher!* β Karroo, *Drege*, 3207. Between Grootriet and Eenkoker, *Zeyher* 241, *ex parte*. γ. Eenkoker, *Zeyher* 229. May. (Herb. Vind., E. Mey., Sond.)

Bulb blackish. Stem or stipe 3–4 inches long. Petioles numerous, 1–2 inches long. Leaflets 6–8 lines long, 1 line wide; in β. twice as wide; in γ. 3 lines long, 1 line wide. Corolla ¾ inch long, purple or red, yellowish at base. Very near *O. cuneata* (No. 27) which differs by its cuneate leaflets, pubescent on the upper side.

10. O. versicolor (Linn. Sp. 622); glabrous or somewhat hairy; bulb *ovate, with hard, black scales;* stem erect, with a few scales, or 1–2 leaves, many-leaved at the summit; leaflets *linear-cuneate*, channelled above, emarginate, bicallous; peduncles bibracteolate at the top, *longer than the leaves;* sepals lanceolate, acute; limb of the petals white or yellowish, *with a red border. Jacq. Ox. t. 36 & t. 77. f. 4. Thunb. Diss. No. 19, ex parte. DC. No. 140. E. & Z.! No. 693. O pallens, E. & Z! 700, ex parté. O. glabra, Meisn., non Thunb. O. tenuifolia, E. Mey! non Jacq.*

VAR. β. **flaviflora**; corolla yellow, limb with a deep red border; leaflets with two or more calli. *Zey!* 247.

VAR. γ. **elongata**; stem more slender, often branched and leafy; leaflets with or without calli; corolla with a yellow tube, and a white, pale-red border. *O. elongata, Jacq.! DC. No.* 141. *E. & Z.!* 694.

VAR. δ. **Meyeri**; leaflets shorter, with a minute callus; sepals more acuminate. *O. versicolor, E. Mey.! in Hb. Drege.*

HAB. Cape Flats, and round Capetown, very common. β. at Berg river, *Burke and Zeyher*. γ, α, and δ, in Zwartland, *Drege!* Ap.-June. (Herb. Thunb., Vind., T.C.D., Sond., &c.)

Stem 2–6 inches, simple or branched, solitary, or many from aggregated bulbs. Petioles 1–2 inches long, glabrous or pilose. Leaflets channelled, often subrecurved, 6 lines long, ½–1½ line wide, glabrous, paler and delicately punctulate above, acutely emarginate, pilose or glabrous beneath, below the apex marked with orange calli. Peduncles glabrous or minutely and sparsely pilose, erect or spreading, thicker than the petioles. Sepals 2 lines long, with or without orange calli. Corolla 4 times longer than the downy calyx, the claw equalling the red-bordered limb. Styles pilose. Capsule 5-angled, inclosed in the calyx. Var. γ. looks very like *O. gracilis*, Jacq., which has a very different corolla and bulb.

11. O. polyphylla (Jacq.! Ox. t. 39); bulb rather large, *subrotund, with brown, membranous scales;* stem or stipe erect, simple or branched, leafless and *scaly* below, many-leaved at the summit; leaflets *narrow-linear or filiform, channelled or involute*, emarginate, bicallous, glabrous at both sides or subpilose underneath; peduncles bibracteolate at the summit, longer than the leaves; sepals lanceolate, obtuse, at length recurved; *corolla with a purple limb. Thunb. Fl. Cap.* 541. *Burm. Afr. t. 27. f. 1. O. versicolor, Jacq.! Ic. Rar. t. 473. Hb. Lamk.! Hb. Un. It. 106. O. polyphylla, gracilis and filifolia, E. & Z.!* 695, 696, 697. *Zey!* 215. *Sieb. Cap. No. 120. O. polyphylla, α & β, Meisn.*

VAR. β. **filifolia**; leaflets very narrow, without calli, or with a very small callus. *O. filifolia. Jacq.! Schoenb. t. 273. DC. No. 144. O. revoluta and O. filifolia, E. Mey.*

VAR. γ. **pubescens**; densely pubescent in all parts, leaflets mostly without calli; sepals more acute. *O. polyph. β. glandulosa, E. & Z.!* 696, β.

HAB. Moist sandy places on the Cape Flats, Hott. Holland, &c., common. β, in the same places and at Drakenstein, *Dreye.* Vanstaadensberg, Uit., *Zey!* 2132. γ, on mountain sides near Brackfontein, Clanw., *E. & Z.!* Ap.-Jun. (Herb. Thunb., Holm., Vind., E. Mey., T.C.D., Sond. &c.)

Very like the preceding, with which, in many Herbaria, it is confounded, but readily known by the thin brown bulb-scales, narrower leaflets, and deep purple flowers. Var. β differs from *O. Burkei,* otherwise dissimilar, by its long petioles, twice as long as the leaflets, narrower corolla-tube, and blueish limb. *O, amoena* Jacq. ! Schoenb t. 206, seems to be a mere variety of *O. polyphylla,*, with a declinate, 1-2-leaved, downy stem, and rather larger flowers; it is wanting in Herb. Jacq.

12. O. gracilis (Jacq. ! Ox. t. 33); bulb roundish, brown; *stem long, glabrous, leafy,* mostly branched, many-leaved at summit; leaflets linear-obtuse or subemarginate, flattish, glabrous above, appressedly hairy underneath, *without calli;* peduncles equalling the leaves, or longer, bibracteolate at the top; sepals downy, *linear-lanceolate,* 4 times shorter than the *deep red or pale rosy corolla. DC. Prod. No.* 136. *P. versicolor.* γ, *gracilis Willd. Sp.* 2, *p.* 792.

VAR. β. **miniata**; the subacute limb of the corolla deep crimson inside. *O. miniata, Jacq. ! t.* 35. *DC. No.* 137.

VAR. γ. **reclinata**; stem very long, reclinate, glabrous or appressedly pilose; the rounded, obtuse limb of the corolla pale crimson. *O. reclinata, Jacq. ! t.* 34. *DC. No.* 135.

HAB. Moist sandy places. Between Oliphant's River and Knackisberg, *Zey. !* 211, (Herb. Vind., Sond.).

Stem 6–18 inches high, erect or declined, brown, mostly quite glabrous, with a few scattered leaves or short branchlets. Petioles glabrescent, 1½–2 inches long. Leaflets ½–1 inch long, 1–2 lines wide, green, flat or with the margin slightly rolled in; the lateral horizontal. Peduncles pilose. Corolla bell-shaped, ½–¾ inch long, glabrous. Filaments mostly glabrous. Styles very variable, even in Jacquin's specimens. The "pale yellow calli, scattered over the under surface of the leaflets," attributed by Jacquin to his *O. reclinata* are not visible on his own authenticated specimens. This species is known from *O. linearis,* Jacq. by its bulb, its long leafy stem, longer leaflets and shorter, campanulate corolla.

13. O. teretifolia (Sond.); bulb roundish, *yellow-brown*; stem simple, elongate, naked or scaly, *glabrous,* many-leaved at the summit; leaflets *narrow-linear-subterete, with involute margin, dilated at the points, with many calli underneath,* and appressedly pilose; peduncles *as long as the petioles,* bibracteate; sepals lanceolate, 6–8 *times shorter than* the corolla; tube of corolla narrow, limb obovate.

HAB. Muddy places at Eenkoker, *Zeyher,* 212. (Herb. Sond.)

6–18 inches high. Stem simple, rarely with a single branch. Petioles terete, glabrous, uncial. Leaflets 6 lines long, ¼ line wide, dotted with blackish calli. Sepals appressedly pilose, 1½ lines long, with a purple point. Petals glabrous, the tube yellowish, limb red. Known from the two preceding by its narrow leaves, *dilated at the point,* and elongate corolla; from *O. Burkei* by leaves twice as long, longer petioles, shorter peduncles and general aspect.

14. O. falcata (Sond.); bulb ovate, *with hard, black scales;* stem erect, *pubescent,* leafless, *rather scaly,* many-leaved at summit; leaflets *linear-cuneate,* complicate, shortly emarginate, *sub-recurved, ciliate,* glabrous above, appressedly pilose underneath, without calli; peduncles downy, much longer than the leaves, bibracteolate; sepals linear-lanceolate, pubescent, 4–5 times shorter than the somewhat hairy corolla, which has a yellowish tube and *an ample, violet-purple limb. O. elongata, E. & Z.!* 694, *non Jacq. O. elongata, α, E. Mey.! Zey!* 114. *Hb. Un. It.* 591.

VAR. β, **callosa**; leaflets a little broader, bicallous under the tip. *O. elongata, b, E. Mey.!*

HAB. Among shrubs, at the foot of the Lion's Mountain and other places round Capetown, *E. & Z.!, Drege! &c.* β. in Zwartland, *Drege!* Winter. (Herb. Vind., T.C.D., Sond.)

Bulbs often aggregate. Stems 2–3 inches long. Petioles uncial, or shorter. Leaflets 3–4 lines long, a line wide, dotted. Peduncles 2–3 inches. Cor. 9–10 lines long. Var. β has shorter and broader leaves, with calli. This differs from its allies by its pubescence, and from *O. versicolor* also by the colour of the flower.

15. O. tenuifolia (Jacq. ! t. 38); bulb ovate, blackish; stem erect, *leafy*, pubescent, many-leaved at the summit; leaflets *linear*, carinate, emarginate, subpilose, on the underside margined with many calli; peduncles longer than the leaves, pubescent, bibracteolate; sepals lanceolate, pubescent, thrice as short as the *campanulate* corolla; tube of corolla yellowish, *limb white with a red border. DC. No.* 142. *Lodd. Cab. t.* 712. *O. versicolor, δ, Thunb. O. Thunbergiana, patula, and Loddigesiana, E. & Z.!* 683, 684, 685. *O. limbata, E. Mey.!*

HAB. Grassy and stony places, near Constantia, and along the east sides of Table Mountain. Hott. Holland, the Paarlberg, Drakensteen, Palmiet River, and in Zwartland, *Thunberg, E. & Z.! Drege, W.H.H., &c.* Ap.-May. (Herb. Holm., Vind., E. Mey., T.C.D., Sond. &c.)

6–12 inches high. Stem rarely branched, leafy from the base or middle, with crowded leaves and undeveloped branchlets. Petioles of the lateral leaves rather shorter than the leaflets; of the terminal as long as the leaflets. Leaflets 4–6 lines long, scarcely ½ line wide. Peduncles 1–2 inch long, erect or spreading, deflexed after flowering. Sepals often callous-tipped. Petals 5–6 lines long, the limb rounded, white or flesh-coloured, with a red border. Style hairy. Capsule ½ as long as the calyx Very like the larger varieties of *O. versicolor*, but its tufted leaves have shorter petioles, the leaflets are narrower and the flowers smaller.

16. O. Mundtii (Sond.); bulb *ovate, with brown scales*; stem short, scaly, leafless or somewhat leafy above, many-leaved at the summit; *the petioles, under sides of leaflets, peduncles, and calyces, pubescent with very short hairs;* leaflets as long as their petiole, linear, complicate, obtuse, or minutely emarginate, glabrous above, without calli; peduncle bibracteolate, longer than the leaf; sepals ovate or sub-lanceolate, 4–6 times shorter than the corolla; tube of corolla yellowish, limb violet-purple.

VAR. α. **minor**; leaflets 2–3 lines long, ½–¾ lines wide; limb of the corolla violet purple.

VAR. β. **albiflora**; leaflets 2–3 lines long, ½–¾ lines wide; corolla white.

VAR. γ. **ferruginea**; leaflets 4 lines long, 1 line wide, rusty above; tube of the corolla shorter, limb pale, with red border.

HAB. Sandy and muddy places, α, at the Berg and Hex Rivers, *Zeyher!* 248, 255. β, at Berg R., *Zey.!* 250. γ. Swellendam, *Mundt.!* May. (Herb. Sond.)

Bulb nearly ½ inch. Stems 2–3 inches, with 2–3 very short branchlets below the crown. Petioles 4–6 lines long. Peduncles 1–1½ inches long. Cor. 8–9 lines long, with a wide limb. Known from *O. linearis* by the bulb, and hairy stem and leaves, and short petioles; from *O. Burkei* by the pubescence and want of calli.

17. O. polytricha (Sond.); bulb *subrotund*, with brown scales; stem short, leafly, pubescent, many-leaved at the summit; petioles equalling or exceeding the leaves, pilose; leaflets *linear-oblong*, keeled, emarginate, without calli, glabrous above, clothed below *with long white hairs, and ciliate at the margin;* peduncles equalling the leaves, bi-bracteolate, and the calyx hairy; sepals lanceolate, 6–7 times shorter than the corolla.

HAB. Achtvallei, near Oliphant's River, *Zeyher!* 226, *ex parte*. May. (Herb. Vind., Sond.)

Bulb ½ inch. Stem 2 inch, simple. Petiole 3–6 lines long. Leaflets 2–2½ lines long, ¾ lines wide, green above, very smooth. Corolla nearly uncial, with a narrow, yellowish tube, and a violet-coloured or red limb.

18. O. ligulata (E. Mey. !); stemless, glabrous; *outer leaves trifoliolate, with linear, emarginate, dotted leaflets;* inner leaves elongate, *without leaflets, with a lanceolate-ligulate* pellucid petiole; scape longer than the leaves, bi-bracteolate; sepals lanceolate, thrice as short as the large corolla.

HAB. Rocky places. Little Namaqualand, *Drege*, 3198. Aug. (Herb. Vind., E. Mey., Sond.).

A very remarkable species. Leaves numerous; the leafless petioles *(phyllodia)* a span long, 2 lines wide; the rest 3–4 inches. Leaflets inch long, 2 lines wide, some shorter and narrower, all either without calli, or below the tips and along the margins marked with a row of spots. Scape 1 foot high, with two sub-opposite bracteoles, three lines long. Flower nodding, at length sub-erect. Sepals 4–5 lines long, obtuse. Corolla campanulate, seemingly flesh-coloured or reddish, an inch or more in length, the tube four lines wide, the limbs of the petals rounded, obtuse.

19. O. cuneata (Jacq. ! tab. 40); stipitate or caulescent, pubescent; leaflets *flat, linear-oblong, cuneate,* emarginate, *thinly pubescent at each side,* the lateral equal; scapes as long as the petiole; sepals lanceolate, 5 times shorter than the corolla. *DC. No.* 131. *O. cuneifolia, Jacq.! t.* 41. *DC. No.* 132.

HAB. Cape. (v. s. Hb. Jacq.)

Bulb ovate, rather smaller than a hazel nut. Stipe or stem 2–3 inches. Petiole 2 inch long. Leaflets ½ inch long, two lines wide at the top. Sepals pubescent. Petals 8–10 lines long, white, yellow at base. Like *O. linearis*, but with cuneate leaflets, pubescent on both sides. It is known from *O. mutabilis*, Sond., by the narrower leaflets, the lateral equalling the medial, and all cuneate at base.

20. O. laburnifolia (Jacq. ! t. 28); stemless, *thinly pubescent;* leaflets oblong, obtuse, the medial oblong-lanceolate, narrowed at base, the lateral oblique at base; petiole semi-terete, compressed; scape as long as the leaves or shorter, bi-bracteolate; corolla yellow, 5–6 times longer than the acute sepals. *DC. No.* 91. *O. sanguinea, Jacq. ! t.* 29. *DC. No.* 92.

VAR. β. **angustata**; leaflets lanceolate, obtuse.

HAB. Cape, *Hb. Jacquin.* Sandy and stony places between Olifant's River and Knakisberg, *Zeyher.* β. same place, *Zey. !* 225. June. (Herb. Vind., Sond.)

Somewhat glaucous, drying to a dull green, at length glabrescent. Bulb ovate, black. Stipe subterranean, none or 1–2 inch long. Leaves numerous. Petioles 1–3 inch long, one line wide, hairy. Medial leaflets 1½–2 inch long, ½–¾ inch wide, longer than the lateral, all obtuse or very shortly emarginate, pubescent or subglabrous, ciliate, marked with reddish dots and striæ. Peduncles bi-bracteolate in the middle, deflexed after flowering, pubescent like the calyx. Flowers large, *yellow*. Var. *β*. is known by lanceolate leaflets twice as narrow, the medial 1½–2½ inches long, 3–4 lines wide, the lateral shorter but equally broad, and flowers a little smaller.

21. O. albida (Sond.); stemless, *densely hairy, with short, white hairs*; leaflets oblong-linear, *obtuse, with concave dots*, somewhat longer than the petiole; scape equalling or excelling the petiole; corolla six times longer than the sub-obtuse, gland-tipped sepals.

HAB. Sandy places between Hartebeeste R. and Eenkoker, *Zey.!* 228, May. (Herb. Vind., Sond.)

A minute, whitish plant. Bulb wanting. Stipe subterranean, very short. Leaves numerous. Leaflets 4–6 lines long, 1–1½ lines wide, thickly clothed with shining hairs, at length glabrate above, pellucid. Petioles sub-compressed, 2–4 lines long. Scapes hairy, becoming smooth, with setaceous bracteoles. Calyx tomentose, at length glabrate, with 2–4 reddish calli below the tips. Cor. *yellow*, large for the plant, inch long; petals with long claws and a wide limb.

22. O. Namaquana (Sond.); stemless or stipitate, *quite glabrous*; leaflets oblong-linear, acutely emarginate; scapes equalling the leaves; bracteoles ovate-acuminate, adhering to the calyx; sepals unequal, 3–4 times shorter than the corolla. *O. flava, β. trifoliata, E. Mey.!*

HAB. Kamiesberg, Kl. Namaqualand, *Drege!* Aug. (Herb. E. Mey., Vind., Sond.)

Stipes short or long. Petioles articulated above the dilated base, subterete, channelled, inch long. Leaflets petiolulate, six lines long, 1–1½ line wide. Bracteoles rather large. Sepals obtuse. Corolla yellow, nearly as in *O. flava*.

23. O. crispula (Sond.); stemless or stipitate, pubescent; leaflets linear-oblong, *undulate and curled at the margin, pilose on both sides*; scapes equalling the petiole; sepals lanceolate, five times shorter than the corolla.

HAB. Muddy places, between Eenkoker and Bitterfontein, *Zey.!* 230. (Herb. Vind, Sond.).

Minute, with an oblong bulb. Stipe scarcely any, or 2–4 inch long, with few scales. Petioles uncial. Leaflets six lines long, 1–1½ lines wide, obtuse, with a curled, sinuate margin. Sepals 1½ lines long. Petals yellow.

24. O. Algoensis (E. & Z.! 704); stemless, pubescent; leaflets with *a somewhat wavy and involute margin*, concave, dotted and *glabrous above, tomentose underneath*, the lateral oblong, unequal-sided, *the medial elliptic-oblong*; scapes excelling the petioles; sepals lanceolate-acuminate, thrice as short as the corolla.

HAB. Stony hills near Port Elizabeth, *E. & Z.!* Sep. (Herb. Vind., Lehm., Sond.).

Leaves numerous. Petioles two inch, pubescent. Leaflets six lines long, lateral one line, medial 2–2½ lines wide. Scapes densely pubescent. Bracts setaceous, under the calyx, or none. Sepals four lines long. Petals yellowish at base, the limb reddish, margined with lilac.

25. O. bifolia (E. & Z.! 725); stemless, downy; leaves *mostly two*, rarely 3 or 4; leaflets *oblong, emarginate, ciliate*, the lateral oblique at base, medial sub-cuneate; scapes ebracteolate, equalling or excelling the leaves; sepals lanceolate, 6–7 times shorter than the corolla.

HAB. Sandy places at Olifant's River, Clanw., *E. & Z.!* Langevallei, *Zey.!* 224. May.-June. (Herb. Vind., Lehm., Sond.).

Minute, with very few leaves. Bulb ovate, 3–4 lines long, brownish. Petioles mostly ½ inch, rarely 1–2 inches. Leaflets in the smaller specimens 3–4 lines long, two lines wide, in the largest nine lines long, three lines wide, cordate at base, acutely emarginate, dotted underneath, and minutely ciliolate. Corolla pale yellow, with a whitish or pale-rosy limb, inch long or longer.

26. O. approximata (Sond.); stemless or stipitate, hairy; leaves 3–6; leaflets oblong, obtuse, somewhat narrowed at base, with a slightly incurved margin, *ciliated with rigid hairs*, quite even above, *concave-dotted underneath*; scapes equalling the leaves, ebracteolate; sepals lanceolate, four times shorter than the corolla. *O. ciliaris, b. (non a.) E. Mey.!*

HAB. Piquetberg, Pietersfontein, *Drege!* 3209. Jul. (Herb. E. Mey., Sond.)

Nearly allied to the preceding, this differs in its brownish bulb twice as large, hairy petiole and scape, rigidly ciliate leaflets (eight lines long, 2½–3 lines wide) obtuse and rounded at each end, not emarginate; in the longer calyx, and corolla with shorter tube. Corolla yellow at base, with a blush-white limb.

27. O. minuta (Thunb.! diss. 2. t. 2); stemless, *slender, glabrous; leaves numerous;* leaflets *oblong or oval, obtuse or retuse,* quite even on both sides or minutely dotted underneath, the medial petiolulate; scapes longer than the leaves, bi-bracteolate; sepals lanceolate, 4–5 times shorter than the corolla. *Jacq. t.* 79. *f.* 2. *O. tenella, Lodd. Bot. Cab. t.* 1096, *non Jacq. O. litoralis, E. & Z.!* 703.

VAR. β. **major**; leaves larger, oval, or obovate, often ciliolate. *O. pratensis, E. & Z.!* 702.

HAB. Hills round Capetown and on Hott. Hollandsberg, *Thunberg*, &c. β. Paardevalley and Hott. Holl. *E. & Z.!* Ap.-Jul. (Herb. Thunb., Vind., T.C.D., Sond.)

1–2 inches high, with a small, ovate bulb. Petioles ½ inch. Leaflets 3–4 lines long, two lines wide, or less, without glands. Scape 2–3 times longer than the leaf, with setaceous bracteoles. Corolla uncial, the tube yellowish, limb white, sometimes edged with lilac externally. Var. β. has leaflets 6–7 lines long, 3–4 lines wide, with a naked or ciliolate margin.

28. O. ciliaris (Jacq.! t. 30); *caulescent*, pubescent; leaflets petiolulate, *all exactly oblong*, obtuse or shortly emarginate, glabrous above, below and at the margins pubescent; peduncles downy, longer than the leaves; sepals lanceolate, four times shorter than the corolla.

HAB. Cape. *Herb. Jacq.!* (v. s. cult.)

A span long, covered with short, soft hairs. Leaves along the stem few, simple and trifoliolate, numerous at the summit, with petioles 1–3 inches long. Leaflets 8-9 lines long, three lines wide, not cuneate. Scape 3–4 inch long. Bracteoles 2, below the calyx, narrow. Sepals about two lines long. Corolla purple with a yellow base.

29. O. virginea (Jacq. Schoenb. t. 275); *caulescent, downy in all parts;* bulb oblong, striate; stem erect, leafy, ramulose; leaves petiolate, the upper aggregate; lateral leaflets oblong, medial obovate-cuneate; peduncles shorter than the leaf; sepals oblongo-lanceolate, 3–4 times shorter than the corolla. *DC. No.* 47.

HAB. Cape, *Jacquin*. (Not preserved in Herb. Jacquin.)

Bulb inch long. Stem four inches to a span long. Petioles uncial or shorter. Medial leaflets 8–10 lines long, four lines wide, lateral half inch. Petals six lines long, sulphur-coloured at base, with a white limb. Possibly a caulescent form of the following.

30. O. mutabilis (Sond.); stemless or stipitate, pubescent, or appressedly pilose; leaflets *flat, obovate-cuneiform, obtuse* or emarginate, *the lateral oblique;* scapes equalling the petiole, or longer; sepals lanceolate, 5–6 times shorter than the corolla.

VAR. α.; somewhat whitish-pubescent; the medial leaflet cuneate obovate, sub-cordate, lateral elliptic-oblong or obovate. *O. fuscata, Jacq.! t.* 45. *DC. No.* 103. *O. strumosa, Jacq.! t.* 64. *DC. No.* 114.

VAR. β.; somewhat whitish-pubescent; middle leaflet cuneate-obovate, obtuse or sub-retuse, the lateral oblong. *O. undulata, Jacq.! t.* 44. *DC. No.* 100, (the leaflets often somewhat wavy). *O. ferruginata, Jacq.! Shoenb. t.* 274. *DC. No.* 98. *O. exaltata, Jacq.! Ox. t.* 49. *DC. No.* 101. *O. ambigua, Jacq.! t.* 43. *DC. No.* 99. *O. tricolor, Jacq.! t.* 47, 48. *DC. No.* 94. *O. rubro-flava, Jacq.! t.* 50. *DC. No.* 93. *O. flaccida, Jacq.! t.* 51. *DC. No.* 97.

VAR. γ.; leaflets cuneate-obovate, the lateral oblique, appressedly pilose underneath, the hairs projecting beyond the margin and somewhat pencilled at the apex.

HAB. Cape, *Jacquin!* α, in sandy places between Grootriet and Eenkoker, *Zeyher!* β. in muddy places at Bitterfontein, *Zeyher!* Liliefontein, *Drege* 3208. γ, at'Kamas, Bushmansland *Zey.!* 2128. (Herb. Vind., E. Mey., Sond.).

Bulb oval. Stipe very short, or several inches long. Petioles 1–3 inches long. Leaflets 4–7 lines long, 3 lines wide, sometimes spotted. Sepals 1½–2 lines long, sometimes unequal, one or two longer and more obtuse. Colour of the petals very variable, the tube in all yellowish, limb generally white, rarely yellowish or pale yellow, of one colour, or red-bordered externally. This species almost falls under the section with *roundish leaflets*, but the lateral leaflets are oblong.

31. O. hirsuta (Sond); stemless, *very hairy; leaflets broadly ovate, complicate, with curled and undulate margins* glabrous, and dotted on the upper, hairy on the under surface and margins; scapes equalling the petiole; sepals lanceolate, hairy, four times shorter than the corolla.

HAB. Karroo, on the road from Rhinosterkopf to Beaufort, *Zeyher.* Apl. (Herb. Sond.)

A very small plant, about an inch high, thickly clothed with white hairs. Stipes very short, with a few brown scales. Leaves numerous. Petioles scarcely ½ inch long. Leaflets 2–3 lines long, the lateral incurved. Sepals two lines long. Corolla seemingly violet at top, with a yellowish tube.—Very near *O. crispula,* Sond. from which its whitish pubescence, much smaller and proportionally broader leaves, and violet-coloured corolla, distinguish it.

32. O. ramigera (Sond.); *tall, caulescent, and branched from the base, clothed with very thin grey pubescence;* branches bearing a tuft of leaves and scapes at the summit; leaves *on long petioles*, the leaflets elliptical, obovate-cuneate, emarginate, complicate, glabrous above, *cano-pubescent underneath;* scapes equalling the leaves, bi-bracteolate; sepals lanceolate, four times shorter than the corolla.

VAR. β. **micromera**; leaflets and flowers but half the size.

HAB. On muddy soil between Bitterfontein and Mierenskasteel, *Zey.!* 207. β. Achvallei, Hartveld, Eenkoker and Bitterfontein, *Zey.!* 206. Grootriet and Lislap, *Zey.* 204. (Herb. Vind., Sond.)

A remarkable species. One or two feet high, with a few broad scales at the base, soon dividing into several longer or shorter branches. Petioles uncial. Leaflets about three lines long, 1½–2 lines wide. Bracteoles sub-opposite, setaceous. Calyx 2½ lines long. Corolla yellow at base, the limb pale rose-red. Var. β. differs in the smaller leaves, often not more than a line long, and flowers half the size, with blueish-red limb.—This nearly approaches the section whose species have roundish or obcordate leaves; but its *habit* is nearer that of this section.

(β). *Not glandular;* beset with long, *jointed*, yellowish hairs: flowers rosy, violet-purple or white. Xanthotrichæ. (Sp. 33–37).

33. O. Meyeri (Sond.); stipitate, scaly; leaves *on long petioles;* leaflets *obovate,* minutely emarginate, glabrous, *with impressed and netted veins,* but not dotted above, sparingly pilose underneath and ciliated; scapes equal to the petiole, bi-bracteolate; sepals lanceolate, four times as short as the corolla. *O. ciliaris, a., E. Mey.! non Jacq.*

VAR. β. **minor**; with shorter petioles, and smaller, rounded-obovate leaflets. *O. purpurea and O. speciosa, a., purpurea, E. Mey.!*

HAB. Hills near the Great Berg River. β. between Kanonenberg and Bergriver, Rhinostersbosch, *Drege!* (Herb. Vind., Lehm., E. Mey., T.C.D., Sond.).

Stipe hairy, several inches long, scaly in the upper half. Petioles slender, two inches long; in β. one inch, hairy. Leaflets petiolulate, sub-coriaceous, 9–12 lines long, 4–6 lines wide; in β. six lines, by five. Scapes hairy, with linear bracteoles. Sepals acuminate, hairy. Corolla white, inch long.

34. O. florida (E. Mey.!); stipitate or caulescent, rarely stemless; stem below with a few scales or leaves, many-leafed at summit; leaves *on short petioles;* leaflets *oblong-cuneate,* acutely emarginate, green above, glabrous, *veinless,* not dotted, hairy underneath and ciliated; scapes equalling or exceeding the petiole, bi-bracteolate; sepals lanceolate, 4 times shorter than the corolla.

VAR. α. flower red. *O. florida, E. Mey.!*

VAR. β. flower white. *O. virginea, E. Mey.! non Jacq.*

HAB. Hills in Rhinosterbosjes, Zwartland, *Drege!* Wolverivier, *Burke and Zeyher!* Krumm River, *Zey.!* 1890, 1891, June. (Herb. E Mey., Vind., Hook., T.C.D., Sond.).

Stipe or stem a span long, scaly from the base, the scales now and then leaf-bearing. Petioles of the upper leaves ½ inch long, hairy. Leaflets 4–6 lines long, two lines wide, the mid-rib impressed on the upper surface, but without visible veins. Scapes and sepals hairy. Corolla 10 lines long, with a yellowish tube; limb white, red or violet-purple.

35. O. affinis (Sond.); stipes hairy; leaves on *longish* petioles; leaflets *linear-oblong,* obtuse or sub-emarginate, glabrous above, *penninerved, veiny, and dotted,* hairy beneath and ciliolate; scapes equalling or excelling the leaves, bi-bracteolate; sepals ovate-*cuspidate-acuminate,* 4–5 times shorter than the corolla. *O. tricolor, E & Z.!* 742. *non Jacq.*

HAB. Clayey ground near Brackfontein, Clanw., *E. & Z.!* (Herb. Vind., Lehm. Sond.)

Stipe 1–2-uncial, with few scales. Petioles very hairy, 1–2 inch long. Leaflets 8–10 lines long, two lines wide. Scapes very hairy, with setaceous bracts. Calyx two lines long. Corolla with a yellow tube and pale blueish limb, bordered with deeper colour outside.—Known from the preceding by its thinner, veiny, longer-stalked leaflets, and by the mostly shorter, broader, and three-coloured corolla.

36. O. angusta (Sond.); stipitate or caulescent and often somewhat branched; leaves moderately petioled; leaflets *linear, minutely emarginate,* complicate, with recurved points, glabrous and *nerveless* above, pilose underneath; scapes rather hairy, equalling the leaves, bi-bracteolate; *sepals lanceolate acuminate,* four times shorter than the corolla. *O. cuneifolia, E. Mey.! non Jacq.*

HAB. Zwartland, on hills in Rhinosterbosje, *Drege!* Kalebaskraal *Zeyher!* Jun.–Jul. (Herb. E. Mey., Vind., Lehm., Sond.).

Stipes 1–2-uncial, rather hairy, with a few broad, brown, glabrous scales. Petioles ½–1-uncial, hairy. Leaflets 8–12 lines long, one line wide. Scapes several, reflexed after flowering. Calyx three lines long. Petals with yellow tube and white limb.

37. O. adspersa (E. & Z.! 743); stipes hairy; leaves on *long petioles;* leaflets *narrow-linear,* scarcely emarginate, *with involute margins,* glabrous and nerveless above, *sprinkled with orange-coloured glandular calli beneath;* scapes longer than the leaves, bi-bracteolate; sepals *ovate, cuspidate-acuminate,* hairy, six times shorter than the corolla.

HAB. In sandy and muddy soil, near Brackfontein, Clanw., *E. & Z.!* May. (Herb. Sond.).

Stipe 1–2-uncial, with few scales. Petioles glabrous, 1–2-uncial. Leaflets 12–15 lines long, scarcely one line wide, not hairy at either side. Scapes sub-pilose, with setaceous bracts. Corolla uncial, reddish-lilac or violet, with a yellow base.

(γ). More or less clothed with *gland-tipped* hairs. (Sp. 38–43).

38. O. glandulosa (Jacq.! Ox tab. 46); *stemless* or stipitate, *downy, with short gland-tipped hairs;* leaves on long petioles; leaflets *oblong, obtuse, black-dotted,* the medial cuneate, the lateral oblique at base; peduncles equalling the leaf, bi-glandular in the middle; sepals oblong-lanceolate, rather obtuse, without calli, four times shorter than the corolla. *DC. l. c. No.* 102.

HAB. Cape, Hb. *Jacquin!* (v. s. c. in Hb. Jacq.)

Stipes, if any, short. Petioles and scapes slender, 3–4 inches long. Leaflets 5–6 lines long, 3–4 lines wide, thin, green, sparsely puberulous on each side. Sepals 2 lines long. Petals white, yellow at base.

39. O. multicaulis (E. & Z.! 681); bulbs ovate-oblong, blackish; stems erect, *leafy, thickly glandular-pubescent;* terminal leaves on long petioles; leaflets *oblong, cuneate, emarginate,* complicate, glabrous above, appressedly pilose beneath; *at the apex and margins calloso-glandular;* peduncles equalling the leaves, bi-bracteolate, glandular; sepals lanceolate, glandular, 4 times shorter than the puberulous bell-shaped corolla. *O. tenuifolia, E. & Z.! 682. non Jacq.*

HAB. Stony and sandy places at the base of the Devil's Mt. and near Wynberg: also between Caledon and Klynriviersberg, *E. & Z.! Zeyher!* 2131. May–Aug. (Hb. Vind., Lehm., Sond.).

4–6 inches high, simple or somewhat branched, somewhat canescent. Bulb 6–8 lines long. Petioles of the terminal leaves ½–1 inch long, hairy; lateral or lower cauline leaves on very short petioles. Leaflets 2–2½ lines long, one line wide, green above. Flowers ½-uncial. Tube of cor. yellowish, limb white or flesh-coloured, bordered with red. Allied to *O. glabra,* Thunb. and *O. tenuifolia,* Jacq.; known from the former by a bulb twice as large, and flowers ½ the size; from the latter by leaflets 2–3 times broader and shorter; and from both by the glandular pubescence.

40. O. recticaulis (Sond.); bulbs aggregate, ovate, and black; stem straight, simple or somewhat branched, *leafy,* densely glandular-pubescent; terminal leaves numerous, *equalling the petiole;* leaflets *linear-cuneate,* keeled, emarginate, glabrous above, appressedly pilose beneath, *below the apex and along the margin bordered* with *glandular calli;* scapes *erect,* equalling the leaves, bi-bracteate, and the calyx glandular; sepals lanceolate obtuse, bicallous, five times as short as the corolla.

HAB. Piquetberg *Zeyher,* 209, June. (Herb. Sond.)

Bulb ½ uncial. Stem a span long, clothed from the middle to the apex with some leaves from scale-bases, ½ inch asunder. Petioles glandular, those of the lateral leaves very short. Leaflets 3–4 lines long, a line wide, thicker at the margin. Corolla 8–9 lines long, the tube yellowish, the lobes of the limb pale, with red border.

41. O. clavifolia (Sond.); bulb (unknown); stem erect, slender, *sparingly glandular, scaly, or with a few leaves;* terminal leaves crowded, *on long petioles;* leaflets linear-cuneate, carinate, *dilated at the point,* emarginate, *glabrous on both sides,* without calli, margined with very short, gland-tipped hairs; scapes *cernuous,* at length erect, bi-bracteolate, glandularly hairy; sepals acuminate, without calli, 5 times shorter than the corolla.

HAB. Knakisberg, *Zeyher!* (Herb. Sond.).

A weak plant, three inches high. Petioles uncial, glabrous, or with few glandular hairs. Leaflets 3–4 lines long, suddenly tapering from a cordate-emarginate apex 1½ lines wide. Younger peduncles short and cernuous, at length long and erect. Calyx one line long. Corolla six lines long, glabrous or few-glanded outside, the tube yellowish, with violet limb.—Like the preceding, but at first sight known by its complicate leaflets suddenly dilated, and thus seemingly club-shaped.

42. O. droseroides (E. Mey.!) *clothed in all parts with yellowish, gland-tipped hairs;* stem short, subsquamose; leaves equalling the petiole or shorter; leaflets *linear,* keeled, obtuse or subemarginate, *glandularly ciliate,* without calli; scapes equalling the leaves; sepals lanceolate, acuminate, 6 times shorter than the corolla.

HAB. Hills in Rhinosterbosjes, Zwartland, *Drege!* Jun. (Herb. E. Mey., Vind., Lehm., Sond.)

Bulbs unknown. Stem 2–3-uncial. Petiole ½–1-uncial. Leaflets 6 lines long, 1 line wide. Scapes in the specimens seen, without bracts; always so? Cor. 10–12 lines long, the tube yellow, limb purple-violet. Styles very long.

43. O. semiglandulosa (Sond.); stem erect, with few scales, or leafy in the upper part, *very thinly downy;* leaves longer than the petiole, leaflets linear, keeled, scarcely emarginate, *glabrous above, calloso-multipunctate and appressedly pilose beneath;* scapes equalling the leaves or longer, their upper portion and the calyx glandularly hairy; sepals lanceolate, 6 times shorter than the glandular corolla.

HAB. Muddy soil and in shady places at the Hex River, *Burke, Zeyher,* 249, 254, May. (Herb. Hook., Sond.)

3–4 inches high. Bulb wanting. Petioles downy, as long as the leaves or longer. Leaflets 4 lines long, 1 line wide, acute. Scapes glandular at the summit. Cor. ¾–1 inch long, with yellowish tube and violaceous red-margined limb. Very like the preceding, but with very different pubescence. The corolla in the former bell-shaped, in this tubular.

(2). Leaflets *sub-rotund,* obtuse or scarcely retuse. All *stemless.* (Sp. 44–62.)

(*a*). Flowers purple, rosy, violet or white. (Sp. 44–56.)

44. O. purpurea (Thunb! Cap. 535., excl. syn. Jacq.); pubescent or nearly glabrous; bulb rather small, ovate, blackish, often several in a cluster; leaflets *subequal,* roundish, emarginate, glabrous above, livid purple, dotted and more or less pilose beneath, ciliated, *the medial not cuneate;* petiole terete; scape *longer than the leaves,* bibracteolate; sepals lanceolate, acuminate, many times shorter than the corolla, the claws of the petals *united in a narrow tube* equalling the limb or longer. *E. & Z.!* 707. *O. speciosa, E. & Z.!* 706. *Herb. Un. It.* 585. *O. grandiflora, Sieb. Fl. Cap.* 123, *ex pte. O. laxula, E. & Z.* 708. *O. variabilis, a. & b., O. humilis and O. purpurea β. E. Mey.!*

VAR. *α*; petioles and scape pubescent; leaflets pilose underneath. *O. humilis, Thunb.! Fl. Cap.* 535.

VAR. *β*; pet. and scapes pubescent; leaflets glabrous, or hairy on the nerves.

VAR. *γ*; pet. and scapes nearly glabrous, leaflets glabrous, distantly ciliate, with long hairs; limb of the corolla purple or whitish. *O. nidulans, E. & Z.!* 724, *ex pte. O. fallax E. & Z.!* 716, *ex pte.*

HAB. Sandy places among shrubs, in the Cape Flats and round the town, common also in Cape and Stellenbosch districts. Zey. 234, 2113, 2112, 2115. (Herb. Thunb, T.C.D., Vind., E. Mey., Lehm., Sond.)

2–4 inches high. Bulb ½-uncial. Petioles 1-2-uncial. Leaflets 4–6 lines long and wide, all equal, or the medial a very little larger. Cor. slender, uncial, longer or shorter, with a yellowish tube and spreading, mostly deep purple limb: rarely pale or whitish.

45. O. variabilis (Lindl. Bot. Reg. t. 1505); downy; bulb rather large, ovate, blackish, often several together; leaflets roundish, glabrous or pubescent underneath, dotted, minutely ciliate, *the medial cuneate at base;* petiole terete; scape *equalling* or rarely excelling the leaves, bibracteolate below the middle; sepals lanceolate, erect, much shorter than *the wide-tubed* corolla, the claws of the petals shorter than the limb, rarely equalling it; inner stamens toothed.

VAR. *α.* **alba**; flowers white, sepals mostly sprinkled with glandular hairs. *O. variabilis, α, Jacq.! tab.* 52. *Willd. Sp. p.* 777. *excl. var. β. DC. No.* 108. *Thunb. Diss. No.* 12. *O. grandiflora, Jacq.! t.* 54. *Willd. Sp. p.* 778. *E. & Z.!* 712. *Bot. Mag. t.* 1683. *O. laxula, Jacq.! t.* 57. *O. rigidula. tab.* 59. *O. suggillata, Jacq.! t.* 61. *O. reptatrix, Jacq.! tab.* 20 (a garden monstrosity). *O. grandiflora, Un. It. No.* 584. *O. purpurea, L., ex Thunb. l. c.*

VAR. *β.* **rubra**; flowers red; sepals pilose, not glandular. *O. variabilis, β, Jacq.! t.* 53. *O. purpurea, Jacq.! t.* 56. *Hb. Un. It. No.* 583. *Sieb. Fl. Cap. No.* 123, *ex pte. O. speciosa, Jacq.! t.* 60. *Thunb. l. c. No.* 10. *O. humilis, E. & Z.!* 705, *non Thumb. O. speciosa. β, suggillata, E. Mey!*

VAR. *γ.* **nana**; bulb and leaves smaller; sepals rather obtuse; flowers white, turning rosy. *O. breviscapa, E. & Z.!* 714., *non Jacq. O. inscripta, E. Mey.!*

HAB. Cape Flats and Western districts, common. *Zey.!* 233, 235, 243, 2130, γ. Zwartland, and at the Zwartkop's River, Uit. *E. & Z.!* *Drege!* (Herb. T.C.D., Sond. &c.)

Very variable in size, 1–6 inches high. Bulb in α. & β. uncial. Petioles pilose or glabrous. Leaflets in the smaller specimens 3–4 lines, in the larger inch long, green or purple underneath, the medial larger and cuneate. Corolla uncial, in γ. half as large, with yellow claw and spreading obtuse limb. Styles very variable in length.

46. O. breviscapa (Jacq. ! tab. 58); downy ; bulb ovate, rather large, black ; leaflets roundish, glabrous on both sides, marked with black dots and striolæ, *the younger ciliate*, the medial cuneate, obtuse, or subretuse ; *petioles compressed ;* scape *twice as short as the petiole*, bibracteolate in the middle; sepals lanceolate, straight, 3–4 times shorter than the corolla ; claws of the petals equalling the lamina, *interior stamens toothed. DC. No.* 106. *E. & Z.* ! 714.

HAB. Cape, *Jacquin.* Among shrubs at Haazekraalrivier, *Drege!* Aug. (Herb. Vind., Lehm., Sond.)

From the preceding, which it much resembles, this differs in the glabrous leaflets, the flattened petiole, short scape, and white flowers half the size.

47. O. collina (E. & Z. ! 710); bulb ovate, blackish ; *petiole, scape and calyx densely pubescent ;* leaflets roundish-obovate, *somewhat wavy-crenulate,* thinly pilose above, at length glabrous, concave-dotted, *villoso-tomentose* underneath, the medial cuneate, the lateral smaller, often unequal; scape longer than the leaf, bibracteolate under the calyx ; sepals lanceolate, acuminate, thrice as short as the wide-tubed, funnel-shaped corolla ; claws of the petals equalling the limb. *O. depressa, pubescens, E. & Z.* 713.

HAB. Sand Hills at Port Elizabeth. Moist sandy places at Onaggaasvlakte, *E. & Z.!* Zwartkop's river, *Zey!* 2108. Apl.-Sep. (Herb. Lehm. Sond.).

Bulb 6-8 lines long. Petioles 1–2-uncial. Lateral leaflets rounded, oblique. Scapes 2–3-uncial. Calyx 4 lines long. Corolla uncial or nearly so, with yellowish tube, reddish white limb, often margined with purple externally and dotted here and there near the apex with black calli.

48. O. pulchella (Jacq. ! No. 69) ; pilose ; bulb rather large, oval, blackish ; leaflets *roundish, firm, hollow-dotted, veiny, hairy beneath, ciliated,* the medial petiolate, cuneate, subemarginate, the lateral smaller, rounded, oblique at base ; petiole *subcylindrical, hairy ;* scape equalling the leaf or shorter, bibracteolate ; sepals *ovate*, erect, *with reflexed points,* 6 times shorter than *the wide corolla ;* claws of the petals equalling the limb ; *stamens toothless. DC. No.* 116. *O. sulphurea, Jacq. ! t.* 63, *DC. No.* 105. (the flower white or pale yellowish). *O. marginata, Jacq.! t.* 68. *DC.* 115. (corolla white.)

VAR. β. **tomentosa** ; petioles very hairy ; leaflets tomentose at each side, above at length glabrate ; sepals somewhat hairy, ciliated, callous at tip. *O. pulchella, E. & Z. !* 711, *ex pte.*

VAR. γ. **glabrata**; leaves at length nearly glabrous.

HAB. Cape, Herb. *Jacquin!* β, sands at Grootepost, *E. & Z!* Achvallei ; and between Oliphant's Riv. and Knakisberg, *Zey!* 240, *ex pte.* Between Eenkoker and Hartebeeste Riv. *Zey!* 228, *ex pte.* γ. Langvalei below 1000f. *Drege!* Ap.-June. (Herb. Vind. Sond.)

Petioles 1–2-uncial. Leaflets 6–12 lines long, 5–9 wide, sometimes spotted with red underneath. Sepals glabrous or pilose, sometimes with terminal calli. Petals with a yellow tube, and rosy or white limb.—Allied to the preceding, and especially to the smaller forms of *O. variabilis*, but at once known by its firmer and thicker, conspicuously hollow-dotted leaflets, pale when dry, and by the white-hairy petiole.

49. O. calligera (Sond ;) bulb rather small, oblong, brownish-black; *petioles, scapes, and calyces pubescent*; *leaflets rounded-obovate*, obtuse or minutely emarginate, *tomentose on each side*, hollow-dotted, marked *at the apices and margins here and there with black calli*; the medial petiolate; scape shorter than the leaf, bibracteolate in the middle; sepals erect, obtuse, callous at the point, 6-times shorter than the corolla; claws of the petals united *in a tube twice as long as the calyx*, and equalling the lamina.

HAB. In Sandy places at Kamos, Bechuana Land, *Zey!* 238. (Herb. Sond.)

Bulb ½ inch long. Petioles 1-1½-uncial. Lateral leaflets oblique at base, the medial cuneate, 2–3 lines long, 2 lines wide, all (under a lens) pellucid-dotted, calligerous at the margin. Bracteoles rather long, setaceous. Calyx very thinly pubescent. Corolla nearly uncial, with a yellow tube and rosy limb.—Nearest to *O. pulchella*, but differing in the petioles and peduncles clothed with minute down, not with long hairs, and in the leaflets much smaller and more obovate. Perhaps, nevertheless, a mere variety?

50. O. melanosticta (Sond.); bulb small, ovate, brownish; petioles, scapes and calyces very hairy; leaflets roundish or subemarginate, *crenated in front, on the upper side silky with long, appressed, soft hairs, underneath somewhat pilose and sprinkled with black dots*; the lateral oblique, the medial equally cuneate; scapes half as long as the petiole, bibracteolate below the middle, deflexed after flowering; sepals lanceolate, four times shorter than the corolla, the claws of the petals equalling the limb.

VAR. β.; leaflets not conspicuously crenulate in front.

HAB. On rocky hills at Geelbeck, *Burke! Zey!* 258. β. Wolverivier, *Burke and Zey!* 265. May. (Herb. Hook., T.C.D., Sond.)

Bulb 3–4 lines long. Stipes, if present, ½–1 inch long, with a few brown scales. Petioles semi-uncial. Leaflets 4 lines long, 3 lines wide, the prominent nerves sometimes glabrescent above. Bracteoles setaceous. Corolla 7–10 lines long, with a short, yellowish tube, and a white or pale yellowish-white limb.

51. O. commutata (Sond.); *glabrous*; bulb ovate, smooth; leaflets sub-rotund, obtuse or emarginate, *cellular and whitish above, green, minutely punctate and veiny beneath*; scapes much longer than the leaves, bibracteolate above the middle; sepals lanceolate, acute, bicallous, 3–4 times shorter than the corolla; claws of the petals equalling the lamina or shorter. *O. tenella, E. & Z.!* 727. *E. Mey.! non Jacq. O. minima, E. & Z.!* 729 (incorrectly described). *O. erubescens, E. Mey.! O. livescens, E. Mey!* (leaves livid beneath).

VAR. β. **grandiflora**; flowers twice as large; purplish.

HAB. Sandy places round Capetown, and on the Cape Flats; Grootepost, Paarlberg, Zwartland, and Hex River, common. β. Wolverivier, Zeyher. (Herb. Holm., E. Mey., Vind., Hook., T.C.D., Sond.)

Varying from about an inch to nearly a span long. Leaflets 3–4 lines long. Corolla 8–10 lines, with a yellowish claw and violet or rosy limb.—Very like the following in all its parts, but differing in the thinner leaflets, evidently veiny, pale and cellular above, and not hollow dotted beneath.

52. O. convexula (Jacq.! Ox. t. 55); stemless or substipitate, *gla-*

brous; bulb oval, blackish, smooth; obtuse or subemarginate, *hollow-dotted on each side, veinless, paler underneath;* scapes longer than the leaves, bibracteolate above the middle; sepals lanceolate, obtuse, mostly with one or two apical calli, ciliolate, 4 times shorter than the corolla; claws of the petals equalling the lamina or shorter. *DC. No.* 110. *O. inops, E. & Z.!* 728 (leaflets round-topped). *O. depressa. E. & Z.!* 713. (leaflets retuse).

VAR. *β.* **dilatata**; leaflets broader than long, retuse; margin mostly dotted with black.

HAB. Sands near Port Elizabeth and fields by the Zwartkop's River, *E. & Z.!*, Zey! 2124. Sandy places at Hex R. and Wolverivier, *Burke, Zey!* 260. Gamke Riv. near Beaufort, *Burke, Zey.!* 28, 267, 269. *β.* Port Natal, *Miss Owen!* Zwartkop's R., *Zey.!* 2123. Caledon R. *Zey.!* 270. Feb.-Sep. (Herb. Vind., Lehm., Hook., T.C.D., Sond.)

Two inches to a span long, quite glabrous. Bulb scarcely ½ inch. Petioles 1–3 inches. Leaflets sub-convex or flat, 3–4 lines wide and long, in *β.* 5–6 lines wide, 3 lines long. Bracteoles alternate or opposite. Calyx 2–3 lines long. Corolla 8 lines to 1 inch long, with yellowish claw and rosy, white, or pale violet limb. Styles pilose or glandular.

53. O. setosa (E. Mey. !); stemless; *the long petiole and the scape setose with rigid, jointed hairs;* leaflets rounded, *wider than long,* rarely sub-retuse, thin, *veiny, dotted on each side,* ciliated with glands; scape equalling the leaf; sepals lanceolate, pilose and minutely callous, four times shorter than the corolla; lamina of the petals rounded.

HAB. Grassy places at Omsamwubo and Omsamcaba, 1000–2000f. *Drege!* 5248. Port Natal, *Sanderson!* Oct. (Herb. E. Mey., Hook., Sond.)

Petioles 3-uncial. Medial leaflet 5–6 lines wide, 4–5 long, the lateral equal or smaller. Corolla 6 lines long, with yellowish tube and pale limb.

54. O. adenodes (Sond.); stemless; *the terete, slender petioles, the scapes and calyces beset with very short, glandular hairs;* leaflets rounded or truncate-obtuse, the medial cuneate, petiolulate, the lateral oblique, *all above appressedly hairy, beneath punctate, glanduliferous at the margin;* scapes equalling or excelling the leaves, bibracteolate in the middle; sepals ovato-lanceolate, 6 times shorter than the corolla, the claws of the petals equalling the limb, or shorter.

HAB. At Lislap, *Zeyher.* (Herb. Sond.)

Bulb wanting. Stipe, if present, short. Petioles 1½–2 inches long. Leaflets about 2–2½ lines long, the medial shortly cuneate, larger than the lateral. Corolla 9–12 lines long, pallid.—Known from *O. luteola,* Jacq., by pubescence; from *O. glandulosa,* Jacq. by the roundish leaflets, 2–3 times shorter; from *O. punctata,* L. by the firmer leaves, calyx twice as long and much larger flowers.

55. O. punctata (Linn. f. p. 243); stemless; bulb ovate, *angular, furrowed;* leaflets *cuneate-rounded, subemarginate, hollow-dotted on each side, glabrous;* petiole, scape and calyx *glandularly downy;* scape longer than the leaves, bibracteolate above the middle; sepals ovate, 6 times shorter than the corolla, the claws of the petals equalling the lamina. *Thunb. Diss. No.* 3. *t.* 1. *Jacq.! tab.* 66. *Willd. Sp. No.* 33. *DC. No.* 113. *E. & Z.! No.* 730. *O. macrogonia, Sieb. Fl. Cap. exs. No.* 122. *O. calcarea, E. & Z.!* 731. *O. pygmæa, E. Mey!*

β. **glabrata**; petals and scape glabrous. *O. glabella, E. Mey!*

HAB. Hills near Capetown, Table Mountain, &c. Simon's Bay and Tulbagh, *Thunb., E. & Z.! &c.* *β.* Hills at Hollrivier, *Drege!* Mar.-Aug. (Herb. Thunb., Vind., E. Mey,, Lehm., T.C.D., Sond.)

2–3 inches high. Bulbs 5–6 lines long, blackish. Petioles very slender, 1–2 inches long. Leaflets green, sometimes purple underneath, the medial sub-retuse. Scape about twice as long as the leaves. Calyx 1 line long. Cor. 5–6 lines long, with yellowish claw, and white or pale flesh-coloured or violet limb.

56. O. grammopetala (Sond.); stemless, glabrous; bulb minute, ovate, smooth; leaflets rounded-cuneate, retuse, hollow-dotted on each side, veinless, paler underneath; scapes much shorter than the petiole, with two opposite bracteoles in the middle; sepals broadly ovate, 7–8 times shorter than the corolla; claws of the petals shorter than the striate lamina.

HAB. Sands near Gamke Riv., *Burke, Zey!* 266. (Herb. Hook., T.C.D., Sond.)

Bulb 3 lines long. Petioles inch long or shorter. Scape very short, 2–3 lines long, rarely longer. Calyx pale, about 1½ lines long. Cor. 10–12 lines long, yellowish on the claw, the limb violet-striate.

(*β*). Flowers yellow or yellowish. (Sp. 57–62.)

57. O. balsamifera (E. Mey. !); bulb *very large, covered with balsamiferous, black, adnate scales; petioles and scape hairy;* leaflets rounded, *obtuse,* pilose and ciliolate, minutely hollow-dotted underneath, the medial cuneate; scape shorter than the leaves, reflexed after flowering; sepals ovato-lanceolate, pilose, 6 times shorter than the corolla; claws of the petals suddenly widened, *with a short rounded lamina.*

HAB. Pietersfontein, Piquetberg, 2000f. *Drege!* Between Grooterivier and Eenkoker, *Zey!* 241. (Herb. Vind., Sond.).

Bulb inch long or more. Petioles uncial. Leaflets 2–3 lines long and wide, the nerves and veins not conspicuous, the medial mostly rather broader than long, the lateral oblique. Scapes ½ inch long. Corolla 7–8 lines long, tube yellowish, limb pale yellow.

58. O. luteola (Jacq.! tab. 65); stemless; bulb ovate, smallish, *with black, separable scales;* petiole *terete, pubescent;* leaflets roundish-cuneate, *shortly emarginate, at each side somewhat villous, the old ones somewhat glabrous above,* hollow-dotted underneath, the medial petiolulate, lateral oblique at base, sessile; scape pubescent, equalling the leaves, bibracteolate; sepals oblong, sub-obtuse, 6 times shorter than the corolla; claws of the petals equalling the limb; interior stamens toothed. *DC. No.* 119. *O. fallax, Jacq.! t.* 87. *E. & Z.!* 716, *ex pte.* *O. macrogonya, Jacq.! tab.* 70. *E & Z.!* 717. *O. laxula, Thunb.* *O. sulphurea, E. Mey.!*

VAR. *β.* **marginata**; petals rather large, externally bordered with violet. *O. rigidula, E. & Z.!* 709. *O. strumosa, E. Mey.!*

HAB. Sandy places on the Cape Flats, near Doornhoogte, Muysenberg, Simonstown, Hott. Holland, &c., both varieties in similar localities, *Thunb., E. & Z.! &c.* (Herb. Holm., Vind., E. Mey., Lehm., T.C.D., Sond.)

Bulb 4–6 lines long. Petioles in the smaller specimens an inch, in the larger 2 inches long, slender, at length glabrescent. Leaflets in the wild plant 2–4 lines long and wide, in cultivated mostly larger. Peduncles articulated above, with small bracts. Corolla entirely yellow, often drying green, 8–10 lines long, in *β.* uncial.

59. O. stictophylla (Sond.); stemless; bulb ovate, small, black; petioles *compressed and keeled,* pubescent or glabrous on both sides, *dotted with minute black or orange dots and striæ,* the lateral smaller, oblique at base; scape equalling the leaves, bibracteolate in the middle; sepals ovato-lanceolate, rather obtuse, ciliolate, 5 times shorter than the corolla; claws of the petals equalling the lamina.

VAR. β. **major**; petioles longer; leaflets 2–3 times as large, the medial more cuneate; flowers larger, yellow or sulphur coloured.

HAB. Muddy spots near Draag, Wolverivier, *Burke, Zey!* 261. β. at Eland's-fontein, Harteveld and Eenkoker, *Zey.!* 242. May. (Herb. Hook., Sond.)

Bulb blackish. Petioles ½–¾ inch long, in β. 2–3 inches. Leaflets 3–4 lines long and wide, very rarely sub-retuse; in β. 7–10 lines long, 5–8 lines wide, the medial remarkably cuneate. Sepals with orange striæ. Corolla 9–10 lines, in β inch long, yellow, the limb dotted with purple along the margin on the outside. Var. β. resembles *O. variabilis,* from which it is easily known by its yellow flowers.

60. O. Eckloniana (Presr.! Bot. Bem. p. 29); bulb ovate, *brownish;* petiole, scape, and calyx hairy; *leaflets rounded, sub-retuse,* glabrous above, *dotted underneath and pilose on the midrib, the margin ciliate with long hairs, all sub-equal;* scape equalling or excelling the leaf, bibracteolate; sepals *lanceolate acuminate, erect,* 3–4 times shorter than the corolla; claws of the petals equalling the ample limb. *O. sulphurea, Eckl. Herb. Un. It. No.* 516. *E. & Z.!* 715.

HAB. Sandy places at Greenpoint, *Ecklon.! W.H.H.* May.-June. (Herb. Vind, Lehm., T.C.D., Sond.)

Bulb about ½-uncial. Petiole uncial, shorter or longer. Sepals 4 lines long. Corolla inch long or more, entirely yellow, glabrous or pilose outside at the margin. Styles much longer than the inner stamens, or shorter than them. This resembles *O. purpurea,* Thunb., but the corolla is constantly bright yellow.

61. O. minima (Sond.); stemless, *quite glabrous; petiole keeled;* leaflets roundish, very smooth above, *minutely dotted underneath,* the medial truncate, scarcely retuse, lateral sub-equal, oblique at base; scape equalling or surpassing the leaves, bibracteolate at the apex; *sepals ovate, black-margined,* glabrous, 6–7 times longer than the corolla; claws of the petals equal to the limb.

HAB. Sandy and muddy places, Grootriet, *Zey!* 237. May. (Herb. Vind., Sond.)

An inch high. Petioles ½–¾ inch long. Leaflets green above, opaque underneath or livid purple, veiny, 1½–2 lines long and wide. Calyx scarcely more than 1 line long. Corolla yellow, bell-shaped. Known from *O. punctata* by its firmer, rounded, minutely dotted leaflets, much shorter scape, and larger, yellow corolla.

62. O. glauco-virens (Sond.); stemless; *the compressed petiole, scape, and calyx somewhat hairy, with acute and glandular hairs;* leaflets cuneate-rounded, or cuneate-obcordate, glaucous-green *sub-æruginous and glabrous* above, opaque beneath, *sparingly glanduloso-pilose at the margin;* scape equalling the leaves or shorter, bibracteolate above the middle; sepals lanceolate, *spreading,* 6–7 times shorter than the corolla; claws of the petals equalling the ample limb.

HAB. Sandy and muddy places between Eenkoker and Bitterfontein, *Zey.! No.* 239. May. (Herb. Sond.)

Bulb unknown. Petioles ½ inch long. Leaflets 4 lines long, 3 lines wide, the smaller 3 lines by 2, acutely emarginate, almost blueish-green above. Bracteoles setaceous, opposite. Corolla 9–12 lines long, yellow at base, the limb pale yellow, mostly violet-margined externally. Like *O. tenella,* Jacq., in the form of the leaves.

(3). The medial, or all the leaflets obcordate. (Sp. 63–75).

(a). Stemless or stipitate. (63–69.)

63. O. Uitenhagensis (Sond.); stemless; petiole and scape villous; leaflets obcordate, *quite glabrous above, villous underneath*, ciliated, *not dotted*; scape *much longer than the leaves*, with 2 setaceous, villous bracteoles above the middle; sepals lanceolate, acuminate, 3–4 times shorter than the corolla; capsule *glabrous*, twice as long as the calyx.

HAB. Sandy places near the Zwartkop's R., Uit., *Zey.!* 2119, 2121. Adow, *E. & Z.! Ox. No.* 4. Mar.-Aug. (Herb. T.C.D., Sond.)

A small, pale plant. Petioles ½–1 inch long. Leaflets 3–4 lines wide, 2–3 long. Scapes weak, 3–4 times as long as the leaves, sprinkled with long, soft villi and short capitate hairs. Calyx villous. Cor. 6–7 lines long, with yellowish tube and white limb. Caps. 4 lines long. Easily known from *O. imbricata*, E. & Z. by its smaller flowers and longer peduncles.

64. O. imbricata (E. & Z.! 736); shortly stipitate, villoso-pubescent; bulbs *numerous*, ovate, *smooth*, *imbricating each other*; leaflets obcordate, *pilose on both sides*, often purple underneath, *dotted*, ciliate; scape *equalling or excelling the leaves*, with 2 setaceous bracteoles above the middle; sepals lanceolate, 5 times as short as the corolla; capsule *pilose*.

VAR. β. limb of the corolla rose-coloured.

HAB. Fields near the Zwartkop's R., Uit., Koega Camma's kloof, Quaggasvlakte, and Oliphant's Riv., *E. & Z.! Mundt & Maire, Zey.!* 2110, 2116. β. near Caledon, *Zey.!* 2126. In the Bassutos country, and Somerset. May-June. (Herb. Vind., Lehm., Hook., T.C.D., Sond.)

Bulbs very numerous, ovate, with smooth, black scales aggregated in a bulb-cluster 2–3 inches or more in length. Stipe uncial, scaly. Petioles mostly 2 inch, with spreading soft hairs. Leaflets 3 lines long and wide, acutely emarginate, at length glabrate above, beneath with appressed white hairs. Scape rarely the length of the leaves, usually a little longer or nearly twice as long, villous. Calyx pubescent, ciliated. Cor. 10–12 lines long, yellowish at bottom, with a broad, white limb; in β. rosy or sub-violet.

65. O. cruentata (Jacq. f.! Eccl. t. 45); substipitate; bulb *ovate, minute*; petioles hairy; leaflets *obcordate, glabrous above, punctate, purplish underneath and pilose*; scape little longer than the leaves, villous, bibracteolate above the middle; sepals lanceolate, pilose, ciliate at the apex, thrice as short as the corolla. *DC. No.* 128.

HAB. Cape, Hb. *Jacq.!* Holls river, *Drege!* No. 7432, 9528. (Herb. Vind.)

Petioles 3–5 inches long, terete. All the leaves radical. Leaflets 5–6 lines long, 3–4 lines wide. Flowers 6–8 lines long, purple, with yellow tube. Sepals at point and margin sprinkled with bloody spots. Drege's plant differs from Jacquin's in having a glabrous scape, nearly twice as long as the leaves, and in the leaflets, otherwise identical, being pilose in the middle, underneath.

66. O. obtusa (Jacq.! tab. 79. f. 1.); stemless or stipitate, *densely pubescent, with an angular and pitted* bulb; leaflets obcordate, thickly puberulous above, *hollow-dotted on each side*; scape longer than the leaves, bibracteolate above the middle; sepals oblong, rather obtuse, 3–4 times shorter than the corolla. *Thunb.! Cap. No.* 8. *O. lanata, var.* 4. *Thunb. Diss. p.* 10. *Willd. Sp. p.* 29. *DC. No.* 137. *E. & Z.!* 732. *O. cuneata, Un. It. No.* 587. *O. thermarum, E. & Z.!* 733. *O. lacunosa, E. & Z.!* 734. *O. cuprea, Lodd. E. & Z.!* 735. *O. ciliaflora,*

E. & Z. 736. *O. pulchella, E & Z.* 711, *ex pte.* *O. plagiantha, E. Mey. Zey.!* 232, 2107, 2109, 2111, 2122.

VAR. α. **pubescens**; leaflets smaller, 2–3 lines long.

VAR. β. **sub-villosa**; leaflets larger, 3–5 lines long.

VAR. γ. **glabrata**; petioles, scape, and sepals, pilose or glabrous; leaflets small or large, glabrous above, pubescent beneath. *O. natans, E. & Z.!* 726, *non Thunb.* *O. stylosa, E. Mey.!*

HAB. Var. α. and β. common round Capetown and throughout the western districts: Uitenhage: and Klein-Namaqualand, &c., &c., *Thunberg, E. & Z.! Drege! &c.* (Herb. Holm., Vind., E. Mey., T.C.D., Sond.)

Bulb ovate, blackish. Stipe short, if any. Petioles in the smaller var. 1 inch, in the larger 2 inches long. Leaflets slightly or deeply obcordate, sometimes callous-dotted at margin. Scape 2–3 inches long. Bracteoles opposite, lanceolate, 2 lines long. Corolla 8–10 lines long, reddish-purple or brick-dust colour, with a yellow base, and ample, glabrous or pubescent and ciliolate limb. Readily known by its angular and deeply pitted bulbs.

67. O. truncatula (Jacq. tab. 62); stipitate; bulb broadly ovate-striate; petioles thickish, very villous; leaflets *rather thick, triangular, obcordate,* truncate, or emarginate, *on both sides silky with shining, appressed hairs,* green above, violet or rusty underneath, *rigidly ciliate*; scape hairy, longer than the leaves, bibracteolate near the middle; sepals lanceolate, villous, 5 times shorter than the corolla. *Willd. Sp. No.* 31. *DC. No.* 104. *O. cruentata, E. & Z.!* 719. *non Jacq.* *O. crassifolia, E & Z.!* 723. *O. holosericea, E. Mey.!*

HAB. Moist, sandy places on Mt. sides, Hott. Hollandsberg, near Grootepoost and Palmiet Riv. *E. & Z.!* Drakenstein, *Drege!* Mar.-Apl. (Herb. Vind., E. Mey., Lehm., Sond.)

Bulb rather large, blackish. Stipes 1–2 inches long, woolly. Petioles 1–2 inches. Leaflets 6 lines long and wide, the midrib impressed above, prominent beneath, without obvious veins, the lateral mostly oblique. Scapes 3–4 inches, with opposite, linear bracts. Corolla 10 lines long, downy externally, with yellowish tube, and obtuse violaceous or purplish limb. Styles hairy. Often entirely rust-coloured.

68. O. holosericea (Sond.); substipitate; petioles very villous; leaflets *roundish-obcordate, thin, veiny, above silky with very short hairs, beneath with appressed villi, the margin ciliated;* scape pubescent, much longer than the leaves, bibracteolate in the middle; sepals acute, villous, 5 times as short as the corolla.

HAB. Moist places at Knoblauch, *Zeyher!* (Herb. Sond.)

Whole plant shining and yellow. Stipes inch long, woolly. Petioles 2–3 inches. Leaflets petiolulate, ¾ inch long and wide, the lateral oblique at base, the medial emarginate, with reddish veins. Scape 5–6 inches. Corolla as in the preceding, from which this is known by its thinner leaves, twice the size, and weak scape.

69. O. lanata (Lin. f. Suppl. p. 244, excl. syn.); stemless or stipitate, *villous and woolly;* leaflets obcordate, *on each side woolly-tomentose* with long, appressed, soft hairs; scape longer than the leaves, bibracteolate in the middle; sepals lanceolate, four times shorter than the corolla. *Thunb.! Diss. No.* 6. *Fl. Cap. No.* 7. *Jacq.! Ox. t.* 77. *f.* 2. *Willd. Sp. No.* 30. *DC. No.* 118. *E. & Z.!* 721. *O. livida, Hb. Un. It.* 593. *O. luteola, E. Mey. !*

HAB. Clayey spots on mountain sides round Capetown, *Thunberg, E. & Z.! Drege!* July. (Herb. Thunb., Vind., E. Mey., Lehm., Sond.)

Bulb, according to Thunberg, "ovate, angular and deeply furrowed, hard and smooth." Stipe 1–3 inches long, woolly, with few scales. Leaflets crowded, the petiole ½–1 inch long, hairy. Leaflets 4 lines wide, 3 lines long. Scapes hairy. Bracts lanceolate, acute. Calyx hairy, the sepals acute. Petals 8 lines long, with a yellowish claw and obtuse, white limb. Styles hairy. Capsule cylindrical, obtuse, 5-angled, striate. Differs from *O. obtusa* in the bulb, the firmer leaves woolly on both sides and the white corolla; and from *O. truncatula* by the less thick leaflets, woolly not shining nor ciliate, and the white corolla.

(β). Caulescent; the peduncles *in some axillaxy* as well as terminal. (Sp. 70–75.)

70. O. aganophila (Sond.); caulescent, with few stem-leaves, top leaves crowded, on long petioles; the petioles *dilated at base, and the peduncles glandularly hairy;* leaflets obcordate, pale green and glabrous above, purple underneath and sparingly pilose, at length glabrate, *glandularly hairy at the margin;* scape longer than the leaves, bibracteolate above the middle; sepals lanceolate, acuminate, glandularly ciliate, 4–5 times shorter than the corolla, the claws of the petals twice shorter than the limb.

HAB. Grassy hills at Karrega River, *Zey.!* 2127. (Herb. Sond.)

A span long or less. Bulb wanting. Stipe and stem glandular. Petioles 2 inch. long. Leaflets 4–5 lines long and wide, equal, acutely emarginate. Peduncles 3–4 inches long, few or none from the axils of the lower leaves. Sepals mostly callous-pointed. Corolla 10–12 lines long, with a short, yellowish tube and violaceous limb. Like *O. incarnata* in habit and foliage.

71. O. natans (Lin. f. Suppl. p. 243); bulb small, ovate, with black-brown scales; stem elongate, flexuous, *submersed,* simple or branching, scaly, leafy at the summit, glabrous; leaves numerous, petiolate, *floating;* petioles *short,* wing stipuled, somewhat pilose; leaflets obcordate, glabrous, glaucous above; peduncles equalling the leaves, or slightly surpassing them, bibracteolate above the base; sepals obtuse, 3–4 times shorter than the corolla. *Thunb. Diss. No.* 4. *t.* 1. *Jacq.! tab.* 76, *f.* 2. *DC. No.* 126.

HAB. Stagnant waters on the Cape Flats and near Breede River, &c. *Thunberg, E. & Z.! &c.* Oct.-Dec. (Herb. Holm., Vind., T.C.D., Sond.)

Bulb 1 or 2, in a common tunic, 4 lines long. Stem filiform, sometimes 2–3 feet long or more, according to depth of water, sometimes leafy below the terminal tuft. Petioles equalling the leaves. Stipules wide, amplexicaul, whitish as in *O. disticha.* Leaflets 3–4 lines long, 2–3 lines wide. Sepals mostly with two golden lines at the point. Corolla ½ inch, white or whitish.

72. O. disticha (Jacq.! Ox. t. 18); bulb ovate, with black-brown scales; stem elongate, flexuous, somewhat branched, ascending, leafy, glabrous; *leaves alternate, distichous;* petioles *rather long,* wing-stipuled, glabrous; leaflets obcordate, glabrous, glaucous above; peduncles equalling the leaves, bibracteolate at the apex; sepals obtuse, 3–4 times longer than the corolla. *DC. No.* 50.

HAB. Cape, Herb. *Jacquin!*

Very similar to *O. natans,* but the stem is more leafy; the leaves distichous, not verticillate. Perhaps *O. natans,* altered by culture? Petioles 1–2 inch. long. Stipules ample, membranous, obtuse, connate. Leaflets quite as in *O. natans.* Cor. ½ inch long, pale yellowish at base, the limb paler.

73. O. tenella (Jacq.! tab. 19); bulb ovate, brown; stem weak, leafy above the middle, *very thinly puberulent;* lower leaves alternate, *uppermost crowded; petiole slender, not wing-stipulated at base;* leaflets obcordate, glabrous above, *puberulous underneath;* peduncles longer than the leaves, puberulous, *bibracteolate* above the middle; sepals lanceolate, hairy, 6 *times* shorter *than the corolla. DC. No.* 125. *O. divergens, E. & Z.!* 722. *O. cylindrica, E. Mey.*

HAB. Moist soil, near Brackfontein, Clanw., *E. & Z.!* Hills near Bergeriver, *Drege!* June. (Herb. Vind., E. Mey., Lehm., Sond.)

Bulb ½ inch. Stem 2–3 inches. Petioles slender, upper 1–2 inches, lower few and short. Leaves 3–4 lines long, 2–3 lines wide, in some specimens equally wide and long. Peduncles 2–3 inches, mostly terminal, a few axillary. Sepals acuminate, 1½–2 lines long, sometimes with minute, golden calli at the point. Corolla 8–10 lines long, pale violaceous.

74. O. ebracteata (Savig.! in Lam. Encycl. IV. p. 682. No. 15); bulb ovate; stem erect, weak, leafy from the base or the middle, *downy with glandular-hairs,* lower leaves alternate, uppermost crowded; petiole slender, not wing-stipuled at base; the leaflets obcordate or oblong-obcordate, *remotely glandular-ciliated;* peduncles axillary and terminal, *much longer than the leaves, ebracteolate, glandularly pubescent,* as well as the calyx; sepals lanceolate, acuminate, 6 times shorter than the corolla. *DC. No.* 52. *O. aganophylla, E. & Z.!* 672. *O. laxa, E. Mey.*

HAB. Heathy places near Brackfontein, *E & Z.!* Rocky hills at Langevallei and Piquetsberg, *Drege! Zey.!* 222. Jun.-July. (Herb. Lamarck, E. Mey., Vind., Lehm., Sond.)

Very like the preceding, but differs in the taller stem, the pubescence, the leaves remotely ciliate, glabrous at each side, and glandular on the midrib below; the peduncles mostly axillary, few terminal, weaker and 3–5 inch. long. Corolla white, with a yellowish tube, 8–10 lines long.

75. O. incarnata (Lin. Sp. 622); bulb ovate; stem erect, *branching, glabrous, leafy;* leaves petiolate, *here and there tufted or whorled;* leaflets widely obcordate, equal; peduncles *equalling the leaves,* nodding, with 2 oblong bracteoles above the middle; sepals oblong, subacute, callous, spreading, pilose, 3–4 times shorter than the corolla. *Thunb.! Diss. No.* 15. *Jacq.! tab.* 71. *Hort. Vind. t.* 71. *DC. No.* 49. *E. & Z.!* 650.

HAB. Rocky places round Table Mountain and Rondebosch. Woods at Puspas Valley and Grootevadersbosch, Swell., Thunberg! E & Z.! Zey. 121, 213 b. (Herb. Holm., Vind., Lam., Lehm., Sond.)

Green. Stem often bearing bulbs in the axils. Petioles 2 inch. long. Leaflets flat, ½ inch long, 5 lines wide, dotted underneath and red-purple. Bracteoles opposite. Flowers bell-shaped, 8 lines long, whitish or flesh-coloured. Inner filaments toothed.

(4.). Leaflets deeply two-lobed. *Scapes terminal.* (Sp. 76–81).

76. O. comosa (E. Mey.!); stem erect, branched, *glabrous, somewhat powdery,* leafy; leaves petiolate, *tufted or whorled at the ends of the branches, alternately tufted* along the side of the branches; leaflets equal, obcordate-two-lobed, the lobes obtuse, on each surface and at the margin glabrous, often callous-dotted; peduncles equalling the leaves, glabrous, with 2 setaceous bracts above the middle; sepals lanceolate, ciliolate at the point, 3–4 times shorter than the corolla.

HAB. Rocky places among shrubs. Uitkomst, Kl. Namaqualand, *Drege!* Aug. (Herb. E. Mey., Vind., Sond.)

More than a foot high, (the lower part of stem not seen). Stem as thick as a goose-quill, with alternate branches. Leaves very numerous in a tuft at the end of the branches. Petioles 1–2 inch long. Leaflets 6 lines wide, 4 lines long, the lobes 2–3 lines long. Peduncles thicker than the petioles, deflexed after flowering. Corolla inch long, pale rosy or white.

77. O. caledonica (Sond.); *glandularly-hairy, somewhat viscid;* stem erect, simple or branched from the base, leafy; stem-leaves on very short, terminal on long petioles, tufted; leaflets (minute) reflexed, obcordate, bi-lobed, hollow-dotted, underneath pilose on the nerve; peduncles longer than the leaf, bi-bracteolate above the middle; sepals lanceolate, *glandular*, 4 times shorter than the corolla

HAB. Hills between Caledon and Riv. Zonderende, *Zeyher!* 2133 b. Sept. (Herb. Sond.).

Stem 3–4 inches long, clothed with leaves from the base. Cauline leaves on shorter or longer petioles, the petioles of the terminal leaves $\frac{1}{2}$–1 inch long. Leaflets $1\frac{1}{2}$ lines long, bifid nearly to the middle, the lobes spreading, green, very minutely dotted above, hollow-dotted underneath. Peduncles mostly twice as long as the leaves. Flowers 6–7 lines long. Petals yellowish at base, the limb wide, pale violet, deeper coloured at the margin, glandular.

78. O. heterophylla (DC. No. 53); *villous;* stem erect, branched at base, leafy; cauline leaves sessile or very short-stalked (longer than the petiole), *obcordate, or sub-bilobed*, the upper crowded, on long petioles; leaflets *reflexed, 2-lobed*, dotted, lobes oblong, spreading, sub-acute; peduncles longer than the petiole, bi-bracteolate in the middle; sepals *oblong-lanceolate, hairy*, thrice as short as the corolla. *E. & Z.!* 667. *O. bifurca, E. & Z.!* 668, *non Lodd. Zey.!* 2133. *a.*

VAR. *β.* **procumbens**; stem filiform, procumbent, sparsely leaved, the cauline leaves mostly tufted. *O. anemonoides, E. & Z.!* 666.

HAB. Among shrubs. Hills near Caledon and Swellendam; between Zwarteberg and Klein Riviersberg and R. Zonderende. Aug.–Sep. *E. & Z.! β.* on the Zwarteberge. (Herb. Holm., Vind., Lehm., Sond.).

Bulb ovate, brownish. Stem a span long, in *β.* a foot or more, weak, and seemingly a woodland variety. Cauline leaves close or distant. Petioles terminal, 1–2 inch long, in *β.* longer. Leaflets 2–4 lines long, lobes villous on both sides $\frac{3}{4}$ line wide. Peduncles pubescent. Flowers 6–9 lines long, the tube yellowish, limb purple. It varies with stem and leaves villous and sub-tomentose, and stem hairy or glabrescent, with thinly villous leaves.

79. O. bifida (Thunb.! diss. No. 16. f. 1); bulb ovate; stem *filiform, naked below*, erect, *or decumbent*, somewhat branched, *the branches leafless, secund, glabrous;* leaflets crowded at the ends of the stem and branches; petiole slender; leaflets reflexed, obcordate-semi-bifid, *glabrous;* peduncles longer than the leaves, hairy at the apex, bi-bracteolate in the middle; sepals *oblong, obtuse, hairy, with black striæ*, thrice as short as the corolla. *Jacq.! tab.* 79. *f.* 4. *E & Z.!* 664. *O. filicaulis, Jacq.! Hort. Schoenb.* 2. *t.* 205. *DC. No.* 129. *E. & Z.!* 665.

HAB. In ditches, &c. Mountains round Capetown and at Grootepost, Hott. Holl., *Thunberg, E. & Z.! Drege!* &c. Ap.–Jun. (Herb. Holm., Vind., E. Mey., Lehm., T.C.D., Sond.).

Bulb yellow-brown, $\frac{1}{2}$–$\frac{3}{4}$ inch. Stem rarely simple, decumbent, a foot or more in length, scaly at base; branches striate. Leaves all terminal, the petioles an inch long. Leaflets 2–3 lines long, green, often gland-dotted at the margin, glabrous, or the petioles sprinkled with hairs, bifid to the middle, the lobes $\frac{1}{2}$ line wide. Peduncles twice as long as the leaves, with setaceous bracteoles near the middle. Flowers 8 lines long. Petals yellowish at base, with violaceous limb, the margin more deeply coloured.

80. O. bifurca (Lodd. bot. Cab. t. 1056); stem erect, simple or branched, *whitish-silky or tomentose*, leafy; cauline leaves shortly petiolate, *often tufted*, the terminal crowded, on long petioles; leaflets *erect, all deeply bifid or forked*, the segments spreading, *oblong or linear*, callous-dotted above, sub-silky, with a few scattered hairs below; peduncles equalling or surpassing the leaves, bi-bracteolate beyond the middle; sepals oblong-lanceolate, shortly hairy, four times shorter than the corolla.

VAR. α. **latiloba**; leaflets bifid to the middle, the lobes flattish, oblong, obtuse; peduncles equalling leaves; flowers purple. *O. bifurca, Lodd. l. c. O. heterophylla, E. Mey. ! non DC.*

VAR. β. **angustiloba**; stem very leafy; leaflets bifid to the middle or beyond it, the lobes involute, sub-linear; peduncles equalling or surpassing the leaves; flower purple. *O. argentea, E. & Z.!* 670. *O. Brehmiana, E. & Z.!* 671. *O. dissecta, E. Mey. !*

VAR. γ. **incana**; stem taller, leafy, whitish-tomentose; leaflets bifid to or beyond the middle, the lobes involute, linear-sub-filiform, silky-canescent; peduncles mostly surpassing the leaves; flower whitish, rarely purplish. *O. incana, E. & Z. !* 669. *Zey. !* 2134.

HAB. Var. α, in rocky places. Piquetberg, *Drege!* 5251. In grassy places between Gekau and Basche, and near Omsamculo, *Drege !* β. and γ Uitenhage and Albany, several places, *E. & Z. !* (Herb. E. Mey., Vind., Lehm., Sond., T.C.D.)

Stem short (three inches long), and densely leafy, or 1–1½ feet long, with subdistant leaves. Stem-leaves shorter than the petiole, some as long or longer; petioles of the terminal leaves 1–2 inch, appressedly silky. Leaflets in α, 5–6 lines long, the lobes one line wide or wider, in β and γ, 5–9 lines long, ½–¼ line wide. Bracts opposite, setaceous. Sepals 2–3 lines long, appressedly pilose or hairy or glabrous. Cor. 8–12 lines long, the tube yellowish, limb purple or whitish.

81. O. Smithii (Sond.) *stemless* or shortly stipitate, quite glabrous; bulb ovate, brownish; all the leaves on long petioles; leaflets forked or bi-lobed, *the lobes spreading, ovate, oblong or linear, hollow-dotted;* peduncles equalling or excelling the leaves, bi-bracteolate beyond the middle; sepals lanceolate, callous at the point, 3–4 times shorter than the corolla.

Var. α. **latiloba**; leaves obcordate-bilobed, the lobes ovate-oblong or oblong, the midrib obsolete underneath; flowers blueish or pale lilac. *O. tristis, E. & Z.! No.* 738. *O. Smithiana, E. & Z.!* 739. *O. gracilicaulis, E. & Z.!* 720. *O. bisulca, E. Mey.!*

VAR. β. **angustiloba**; leaflets deeply bi-lobed, the lobes linear, often elongate, with revolute margin, the midrib thicker. αα; flower blueish or lilac. *O. nemorosa, E. & Z. !* 741. *O. rectangularis, E. Mey. ! and Drege* 7436. ββ. flower white. *O. candida, E. & Z. !* 740. *and Drege* 7438, 5247.

HAB. Hill sides and mountains in Uitenhage, Albany and Caffraria. Near Port Elizabeth, and at Port Natal, various collectors. (Herb. Vind., E. Mey., Hook., Lehm., T.C.D., Sond.).

Two inches to a span long. Bulb inch long. Leaves mostly reflexed, in α, 6–8 lines long, the lobes 4–5 lines long, 2–3 wide; in *β*, as much as 1–1½ inches long, the lobes 1½–¾ line wide, divided sometimes nearly to the base, with divaricated lobes and a thick nerve vanishing below the point. Young peduncles nodding, the flowering erect, with opposite, linear bracts. Sepals very rarely without callus. Petals with a yellowish tube, and ample, rounded limb, 8–12 lines long.

** All the leaves *sessile* or very shortly petiolate, the upper not crowded; petioles *dilated at base*. All are caulescent, with *axillary* peduncles. Flowers *purple or violet*, except in *O. cana*, which has *yellow* petals. *Sessilifoliæ*, DC. (Sp. 82–89).

82. O. tubiflora (Jacq.! tab. 10); stem erect or declined, branching, leafy, pubescent; leaves sub-sessile, leaflets linear-cuneiform, obtuse or sub-emarginate, glabrous above, somewhat hairy beneath; peduncles axillary, longer than the leaf; sepals lanceolate, 5–6 times shorter than the corolla; tube of the petals narrow, 2–3 times as long as the limb.

VAR. *α*. **Jacquiniana**; limb of the corolla purple. *O. tubiflora, Jacq.! DC. No.* 39. *E. & Z.!* 673, *ex parte*. *O. macrostylis, Jacq.! tab.* 9. *DC. No.* 38.

VAR. *β*. **secunda**; leaflets broader; flowers secund, the violet limb thrice as short as the claws. *O. secunda, Jacq.! t.* 12.

VAR. *γ*. **canescens**; leaflets broader, 4–5 lines long, 1–1½ wide; flowers violet, the limb 1½–twice as short as the claws. *O. canescens, Jacq.! t.* 11. *E. & Z.!* 674. *O. secunda, E. & Z.!* 675.

VAR. *δ*. **minor**; stems very short (inch long); leaflets narrow, lanceolate, sub-acute, 2–3 lines long, ½ line wide; limb of the corolla purple, thrice shorter than the limb.

VAR. *γ*. **robusta**; stem taller (a foot high), somewhat branched, the larger leaflets (six lines long) oblong-cuneate, somewhat pilose underneath; limb of the corolla purple, thrice shorter than the claws. *O. macrostylis, E. Mey.!*

HAB. Western and N. Western districts. *α*, Piquetberg, *Drege*. *β*, Achtvalley and Heerelogement, *Zey.!* 205. *γ*, Brackfontein, Clanw. *δ*, Berg River, *Zey.!* 262. Hott. Holl., *Ecklon!* *ε*, Grootberge rivier, *Drege!* May-June. (Herb. Holm., Vind., E. Mey., Lehm., Hook., T.C.D., Sond.).

From a span to a foot long. Bulb rather large, roundish, brown. Leaflets 3–4–6 lines long. Peduncles mostly 1–2 inches long, in some specimens little longer than the leaves, toward the apex bi-bracteolate. Calyx 2–3 lines long. Corolla inch long, or more; in *γ* somewhat shorter; the claws yellowish, downy. It varies with denser or laxer pubescence, the leaflets sub-canescent or glabrescent.

83. O. hirta (Lin. spec. 623); stem erect or decumbent, branched, leafy, pubescent; leaves sub-sessile; leaflets oblong-cuneate, lanceolate or linear-cuneiform, obtuse or emarginate, glabrous above, hairy underneath; peduncles axillary, longer than the leaf; sepals lanceolate, 3–4 times shorter than the corolla; tube of the petals *widened from the base, equalling the limb or shorter*. *O. sessilifolia, Lin. Mant. p.* 241. *Thunb. Diss. No.* 18. *Burm. Afr. t.* 28. *f.* 1. *O. hirta, Lam. Herb.* *O. macromischos, Spr. cur. post. p.* 185. *O. hirta, Sieb. Fl. Cap. exs. No.* 121. *Herb. Un. It. No.* 594. *O. hirta, hirtella, multiflora, rubella, rosacea, E. & Z.! O. cuneata, Meisn. non Jacq.*

VAR. *α*. limb of the corolla pale violet. *O. hirta, Jacq.! tab.* 13.

DC. No. 42. *E. & Z. !* 676 *and* 677. *O. hirtella, Jacq. ! t.* 14. *DC. No.* 43. *O. multiflora, Jacq. ! tab.* 15. *DC. No.* 44. *E. & Z.* 678 (limb somewhat purple).

VAR. β. limb of the corolla purple. *O. rubella Jacq. ! t.* 16 (leaflets narrower). *DC. No.* 45. *E. & Z. !* 679. *Lindl. Bot. Reg. t.* 1031. *Burm. Afr. t.* 28. *f.* 2. *O. rosacea, Jacq. ! tab.* 17. *DC. No.* 46. *E. & Z.* 680 (leaflets wider). *O. tubiflora, E. & Z. !* 673, *ex parte.*

VAR. γ.; glabrescent; stem short, decumbent, downy; leaflets lanceolate. *O. fulgida, Bot. Reg. t.* 1073.

HAB. Paardeneiland, *Thunberg.* Lion's Mt. and Camp's Bay, and Hott. Holl. β. Muysenberg, and near Simonsbay. *Thunberg,* &c. (Herb. Holm., Vind., E. Mey., Lamarck., T.C.D., Sond.).

Near the preceding, but readily known by the form of corolla. It varies with stem and leaves hairy, hirsute or glabrescent, leaflets longer or shorter, wider or narrower, obtuse or sub-acute; peduncles longer or shorter, and filaments glabrous and glandular.

84. O. brevicaulis (Sond.); the *very short stem and prostrate branches ramulous,* leafy, hairy, at length glabrous; leaves petiolate; the petiole shorter than the leaflets, leaflets petiolulate, *broadly obovate or sub-rotund, obtuse, apiculate,* or emarginate, glabrous above, hairy underneath; peduncles axillary, much longer than the leaf, deflexed after flowering; the setaceous bracteoles and the calyces hairy; sepals lanceolate, 5–6 times shorter than the corolla, whose tube is 1½–twice as long as the limb. *O. reptatrix, E. Mey. !*

HAB. Among shrubs in Zwartland, *Drege !* June. (Herb. E. Mey., Vind., T.C.D., Sond.).

Bulb rather large, brownish. Stem branched nearly from the base, the branches two inches long. Petioles of the lower leaves equalling the lamina, of the upper gradually shorter. Leaflets two lines long and wide, obsoletely callous-dotted, as in the preceding. Peduncles ½–1 inch, longer after flowering. Corolla 10–12 lines long, with yellowish tube and purple or violet-coloured limb. Allied to *O. rosacea,* Jacq.

85. O. densifolia (Sond.); *quite glabrous;* stem simple or branched, leafy, the leaves short-stalked, *linear-cuneate, truncate-obtuse,* or shortly emarginate, *with involute margins, callous-dotted beneath;* peduncles axillary, much longer than the leaf, bi-bracteolate; sepals *lanceolate-acuminate,* striate, four times shorter than the *short-tubed* corolla.

HAB. South Africa, *Drege ! No.* 7426. (Herb. Vind., Sond.).

The specimens, deficient in the lower part, are four inches long, as thick as crow's quill. Leaves close together, from the axils of scales. Leaflets four lines long, a line wide at top. Peduncles slender, with setaceous, apical bracteoles, the lower two inches long, the upper gradually shorter. Calyx three lines long. Cor. 10 lines, the tube yellow, equalling the broad, purple limb. Dried leaves very fragile, and like the calyces turning blackish.

86. O. viscosa (E. Mey. !); *glandularly-viscid;* stem leafy, branched; leaves petiolate, petiole shorter than the leaf; leaflets *complicate,* oblong-cuneate, or linear-cuneate, acutely emarginate, glabrous above, sparingly glandular underneath, dotted on both sides, *glandularly-ciliate;* peduncles axillary, much longer than the leaf, bi-bracteolate, with the calyx glandularly pubescent; sepals lanceolate, 4–5 times shorter than the corolla; tube of the petals widened from the base, equal to thelimb, or shorter.

HAB. On hills, among shrubs, Zwartland and Pietersfontein, Piquetberg, *Drege!* Heerelogement, *Zey.!* 210. Jun.-Jul. (Herb. Vind., Lehm., Sond.)

Bulb rather large, brown. Stem ½–1–1½ foot high, with slender, erect branches, viscid with yellow hairs. Leaflets 3–4 lines long, one line wide, without calli, deeply emarginate. Peduncles two inch, with setaceous bracteoles. Cor. 6–8 lines long, with a pale limb, margined with violet. Like *O. droseroides*, E. Mey., but easily known by the very long, lateral peduncles.

87. O. pardalis (Sond.); *glabrous;* stem slender, simple, or branched, leafy, squamulose at base; leaves sub-sessile; leaflets *linear, emarginate, cuneate at base, marked underneath with callous, blackish lines and dots;* peduncles axillary, slender, much longer than the leaf, bi-bracteolate; sepals *ovate*, striolate, very shortly ciliolate, 5-6 times shorter than the corolla, whose tube equals or exceeds the limb

HAB. Sandy places at Hex River, *Burke & Zeyher*, 253. May. (Herb. Hook., T.C.D., Sond.)

2–3 inches high, slender. Leaflets 4–6 lines long, ½ line wide, much narrowed at base, paler above, with sub-involute margins. Peduncles very slender, 1–3 inches long. Calyx about one line long. Corolla 6–7 lines, whitish or pale rosy.

88. O. Meisneri (Sond.); *whitish-hairy;* stem erect, simple or slightly branched, leafy from the middle, pubescent, *as well as the peduncles and calyx, with simple and glandular hairs;* leaves on very short petioles; leaflets *linear-carinate*, shortly emarginate, glabrous above, appressedly pilose beneath, *bicallous at the apex;* peduncles much longer than the leaves, bi-bracteolate; sepals lanceolate-acuminate, bicallous, thrice as short as the corolla; tube of the petals *widened*, longer than the limb. *O. cuneifolia, Meisn.! in Pl. Krauss, non Jacq.*

HAB. Pastures near Tulbagh, *Krauss*, 1156. May. (Herb. Meisn.)

4 inches high, with a broadly ovate bulb, covered with rigid, blackish-brown scales. Leaflets 5–6 lines long, ¾–1 line wide. Peduncles 1–3 inch long. Corolla ten lines long, rosy violet. Near *O. polyphylla var. pubescens*, but easily known by the *lateral*, not terminal peduncles.

89. O. cana (Sond.); *hairy tomentose;* stem short, branching, leafy from the middle; leaves sub-sessile; leaflets linear, keeled, obtuse, or shortly emarginate, glabrous above, *hairy and canous* underneath, *neither furnished with calli, nor glandular;* peduncles much exceeding the leaves, bi-bracteolate; sepals lanceolate, acuminate, thrice shorter than the *yellow* corolla; tube of the petals widened from the base, equalling or surpassing the limb.

HAB. Sandy places at Hex Rivier, *Burke & Zeyher*. May. (Herb. Hook., Sond.)

Stem three inches long or more, with a few branches toward the apex. Petioles very short and broad. Leaflets three lines long, one-half line wide. Peduncles 1–2 inches or more, deflexed after flowering. Sepals 3½–4 lines long. Corolla entirely yellow, the limb wide, pubescent near the margin on the outside.—Known from *O. Meisneri* by the shorter and densely hairy, *not glandular*, leaves, and the yellow flowers.

IV. Leaves digitate, leaflets 5–12. Stemless or stipitate, the stipes naked, or with few scales. Scapes one-flowered. (Sp. 90–98).

90. O. capillacea (E. Mey.!); stipitate; leaves crowded; the petiole dilated at base and setoso-ciliate; leaflets 5-7, *linear-setaceous*, involute, *acutely emarginate*, without calli, *marked above with blackish lines*; scapes longer than the leaves, with two setaceous bracteoles; sepals lanceolate, striate, 3–4 times shorter than the corolla.

VAR. *α*, **Meyeri**; stipe pubescent; the petiole, the sub-capillaceous leaflets, and the bracteoles thinly pilose; sepals glabrous, ciliate.

VAR. *β*. **major**; petiole, scape and sepals pubescent; leaflets linear-sub-setaceous.

VAR. *γ*. **glabra**; base of the petioles and sepals ciliate, otherwise glabrous; leaflets linear, emarginate; scapes slender; fl. smaller.

HAB. Hills at Great Berg Riv. *Drege!* Hex R. *Burke, Zey.* 251. *β*. Brede R., *Zey.!* 259. *γ*, Krumm Riv., *Zey.!* 213. May-June. (Herb. E. Mey., Vind., Hook., Sond.).

Bulb ovate, brown-black, smaller than a hazel nut, the scales much acuminate, uncial. Stipe 1–4 inch, with few scales. Petioles 1–2 inches. Leaflets 4–6 lines long. Sepals 1½ lines, in *β*, three lines, in *γ*, 1–2 lines long. Petals with a whitish limb, externally margined with violet; in *β*., judging from dried specimens, entirely yellowish. Capsule oblong, hairy, rather shorter than the calyx.

91. O. pentaphylloides (Sond.); stipitate, hairy; bulb *blackish*; leaflets 5, *linear*, channelled-involute, *obtuse*, green on both sides, *without calli*; scapes longer than the leaves, bi-bracteolate; sepals *lanceolate*, 5–6 times longer than the corolla. *O. heterophylla, E. & Z.!* 744, *non DC.*

VAR. *β*. **glabriuscula**; leaflets at length sub-glabrous. *O. pentaphylla, E. Mey.! non Sims.*

HAB. On muddy and sandy soil near Brackfontein, Clanw., *E. & Z.!* *β*. hills among Rhinosterbosches (*Elytropappus*) in Zwartland, *Drege!* May-June. (Herb. E. Mey., Vind., Lehm., T.C.D., Sond.).

Bulb ovate, blackish, with glabrous, acuminate scales ½ inch long. Flowers pale lilac. Leaflets rarely only four.—Like the preceding, but with a different bulb, obtuse leaflets, not lineate, non-striate sepals and longer petals.

92. O. pentaphylla (Sims. Bot. Mag. t. 1549); stipitate, glabrous or downy; bulb *pale-brown*; leaflets 5–8, linear, obtuse, scarcely emarginate, *with two calli*; scapes longer than the leaves, bi-bracteolate; sepals *oblong, obtuse*, bi-callous, four times shorter than the corolla. *DC. No.* 145. *O. pentaphylla, and falcata, E. & Z.!* 698, 749. *O. digitata, Poir. Suppl.* 4. *p.* 254, *excl. syn.*

HAB. Sandy and stony mountain places, Muysenberg and near Simonsbay, *E. & Z.!* May-June. (Herb. Vind., Lehm., Sond.).

Bulb oblong; scales slender, acuminate. Stipe short or long. Petioles a finger long or shorter. Leaflets 1 inch long, ½–1 line wide, sub-falcate. Sepals downy. Corolla lilac, or rosy-fleshcoloured, with yellowish base.

93. O. quinata (Savig! in Lam. Dic. 4. p. 688); *hairy, canescent*; stipes long and scaly; leaflets 5, *obovate-cuneate*, emarginate, hairy on both sides, without calli; scapes longer than the leaves, bi-bracteolate near the apex; sepals lanceolate, four times shorter than the corolla. *DC. No.* 54.

HAB. Kamos, Bushmansland, *Zeyher!* May. (Herb. Lam. (Roep.); Sond.).

Stipe decumbent, 4–6 inches long. Petioles ½–1 inch long. Leaflets sessile, 3–4 lines long, 1–1½ line wide. Bracteoles opposite, sub-setaceous, sometimes in the middle of scape, a line long. Corolla uncial, externally minutely pubescent, with a narrow yellow tube and purplish or lilac limb.

94. O. Zeyheri (Sond.); in all parts *canescent with appressed hairs;* stipes long and scaly; leaflets 9–11 (minute), linear-cuneate, complicate, obtuse, callous at the apex; scapes downy, surpassing the leaves, bi-bracteolate; sepals lanceolate, four times shorter than the corolla.

HAB. Clayey soil, at Bergrivier, *Zey. ! No.* 248. ex pte. May. (Herb. Sond.).

This resembles the preceding in habit and colour, but differs in the more numerous and generally much smaller leaflets, 1½–2 lines long, ¼–½ line wide, callous at point. Petioles ½ inch long. Peduncles an inch long, with alternate, setaceous bracts. Flowers 4½ inch long, whiteish, or lilac? thinly downy outside, the tube yellowish, with five red streaks.

95. O. tomentosa (Linn. Syst. p. 434); almost stemless, *tomentose;* leaflets peltate, 12–19, *oblong-cuneate,* convolute, obtuse, or sub-emarginate, without calli; petioles and scapes hairy, of nearly equal length; sepals lanceolate, six times shorter than the petals. *Pluck. Almath. t.* 350; *f.* 3. *Thunb. Diss. No.* 24. *Jacq. ! Ox. t.* 81. *DC. No.* 152. *E. & Z. !* 750.

HAB. Stony places round Capetown and Stellenbosch, Common. Clanwilliam, *E. & Z. !* Zey. ! 227. Wolveriver, *Burke & Zeyher !* 244. Ap.-May. (Herb. Holm. E. Mey., Vind., Lehm., T.C.D., Hook., Sond.).

Bulb ovate, smooth. Stipe scarcely any, or 1–3 inches long, with numerous terminal leaves and peduncles. Leaflets six lines long, a line wide. Scapes equalling or excelling the leaves, with setaceous bracts. Corolla large for the size of the plant, white, yellowish at base. Calyx silky.

96. O. flava (Linn. Sp. 621); stipitate, glabrous; leaflets 5–9, oblong-linear, or linear, *channelled-connivent,* obtuse or acute, with or without calli; scapes bi-bracteolate, equalling or excelling the petioles; sepals *equal,* oblong, acute, 4–5 times shorter than the *yellow* corolla.

VAR. *α.* **Thunbergiana**; leaflets oblong-linear, channelled, recurved at point. *Burm. Afr. t.* 27. *f.* 4. *Thunb. Diss. No.* 24. *Jacq. ! t.* 73 & 78. *t.* 2. *DC. No.* 149. *E. & Z !* 747.

VAR. *β.* **pectinata**; leaflets linear, with involute margins. *O. pectinata, Jacq. ! t.* 75. *E. & Z. !* 748. *Burm. Afr. t.* 30. *f.* 1.

HAB. Sandy places round Capetown and at Greenpoint, common. *β.* Oliphant's R., *E. & Z. !* Zwartland, *Drege !* Ap.-June. (Herb. Sond., T.C.D., &c.)

Bulb ovate, smooth. Stipe 1–3 inches long, with rather large scales. Petioles terete, mostly reddish, in *α,* 1–2 inch., channelled, in *β,* 2–4 inch. long. Leaflets ½–¾ inch long, 1–2 lines wide, in *β,* 1–2 inch. long, 1–3 lines wide, Scapes slender. Sepals acute or obtuse, often callous-tipped, glabrous or ciliate. Corolla yellow, in *β,* mostly paler.

97. O. flabellifolia (Jacq. ! Ox. t. 74); stemless, or shortly stipitate, glabrous; leaflets 5–9, linear or linear-lanceolate, obtuse or subemarginate; scapes equalling the petiole, bibracteolate; sepals recurved at point, *unequal,* 1–2 *larger and spathulate,* 3–4-times shorter than the sulphur-coloured corolla. *DC. No.* 151. *E. & Z. !* 746.

VAR. *β.* **latior**; leaflets oblong, cuneate.

HAB. Cape, *Jacquin !* Sandy places near Oliphant's R., *E. & Z. ! Zey. !* 231, *ex pte.* *β.* Lislap and Kamapus. *Zeyher !* and between Grootriv. and Eenkoker, *Zey.* 231. May. (Herb. Vind., Lehm., Sond.)

Very near *O. flava*, from which it is chiefly to be known by its sepals recurved at the point, one or two conspicuously larger than the rest. Petioles larger than the leaflets. Flowers sulphur-coloured, with a deeper limb. β. is smaller, with broader leaves, showing a transition to the following species.

98. O. lupinifolia (Jacq. ! tab. 72); stemless, glabrous; leaflets 5–7, *spreading in a circle, oblong or obovate, obtuse or subacute, with a cartilaginous margin*, rigid, *discoloured underneath;* scapes equalling the petiole, bibracteolate; sepals sub-equal, oblong, 3–4 times longer than the yellow corolla. *E & Z.!* 745, *DC. No.* 148.

HAB. Grootepost, *E. & Z.!* Between Knakisberg and Oliphant's Riv. *Zey.* 231 *ex pte.* Wolverivier, *Burke & Zeyher!* 263, 264. Ap.-June. (Herb. Vind., Hook., Sond.).

Quite like the preceding, and chiefly distinguishable by the broader, more rigid, and cartilage-edged leaflets. Petioles broad, as long as the leaflets. Leaflets flattish, 6–9 lines long, 4–6 lines wide, mostly blotched at base. Flowers yellow.

B. Peduncles umbellately many-flowered. Leaves trifoliolate. (Sp. 99–108).

(1). Root bulbous. Flowers *yellow*. (Sp. 99–101).

99. O. sericea (Lin. f. Suppl. p. 243); stemless, or shortly stipitate; leaves on long petioles; petioles and scapes *hairy;* leaflets obcordate, *silky;* scape longer than the leaves, umbellate; flowers nodding; sepals lanceolate, hairy, 2–3 times shorter than the corolla. *Thunb. Diss. No.* 113. *Jacq.! tab.* 77. *f.* 1. *DC. No.* 59. *E. & Z.!* 658. *Herb. Un. It.* 592. *Zey.!* 120.

HAB. In ditches, by road sides, and on the hills round Capetown and Wynberg, &c. Common among shrubs at Koopman's River, *E. & Z.!* Ap.-Jun. (Herb. Sond., T.C.D., &c.)

Stipe, if present, short and scaly. Petioles numerous, 1–3 inch. long. Leaflets wider than long, hairy above, at length glabrous, silky-tomentose underneath, the larger 8 lines broad, 6 lines long. Scapes twice as long as the leaves, or longer, umbelliferous; pedicels 4–12, unequal, the longer uncial, with two ovate bracts at base. Corolla ½ inch long, yellow.

100. O. compressa (Thunb. ! diss. No. 7); stemless; all the leaves on longish, *compressed, ciliated*, petioles; leaflets *obcordate*, ciliated, *glabrous above, hairy or pubescent and paler below;* scape *hairy*, equalling the leaves, umbellately 2–4 flowered; sepals lanceolate, 4 times shorter than the corolla. *Jacq. Ox. t.* 78. *f.* 3. *DC. No.* 73. *E. & Z.! No.* 539. *O. cernua, d. Drege!*

HAB. Sandy fields beyond Capetown, in wet places, *Thunberg!* Doornhoogde, *E. & Z.!* May.-Sep. (Herb. Holm., Vind., T.C.D., Sond.)

Petioles 1–2 inch long. Leaflets 5 lines long and wide. Scapes equalling the leaves or rather longer: pedicels bibracteate at base. Cor. yellow, ½ inch long. Near the following, from which it differs in the compressed petioles, hairy leaflets, not bilobed, and shorter, few-flowered scape. Perhaps a mere variety.

101. O. cernua (Thunb. ! diss. No. 12. t. 2); stemless, or shortly stipitate; leaves on long, *terete, glabrous* petioles; leaflets *obcordate-bilobed*, glabrous at each side, or villous underneath; scapes *elongate, glabrous;* pedicels downy, the younger cernuous, erect in flower; sepals lanceolate, bicallous at the apex, 5 times shorter than the corolla. *Jacq.! t.* 6. *Willd. Sp. No.* 46. *DC. No.* 72. *E. & Z.!* 657. *O. Pes Capræ, Savig. O. Burmanni, Jacq.! No.* 20. *Burm. Afr. t.* 74, *f.* 29. *O. lybica, Viv.*

VAR. β. **Namaquana**; smaller, stemless; leaflets obcordate. sub-bilobed, below hollow-dotted, sub-pilose; scapes 2–3 flowered; pedicels glabrescent, ½ inch long; flowers smaller. *O. caprina, E. Mey.! non Lin.*

HAB. In cultivated ground and by road sides, very common round Capetown, &c. Kamps bay and Hott. Holland. β. Sand hills in Kl. Namaqualand, *Drege!* June–Aug. (Herb. Holm., Vind., E. Mey., Lehm., Sond., T.C.D.)

Frequently bulbiferous at base. Petioles 3–6 inches. Leaf-lobes divaricate, obtuse, the larger leaflets inch wide, 8 lines long, dotted and paler beneath. Scape twice as long as the petiole, with a many-flowered umbel; pedicels ½–1 inch long. Flowers yellow, 10–12 lines long, the tube equalling the calyx. It varies with double flowers.

(2.) Flowers purple, flesh-coloured, blueish or white. (Sp. 102–107.)

102. O. caprina (Lin. Syst. p. 433); *stemless*, glabrous; bulb ovate-triangular; leaves on longish petioles; leaflets obcordate, sub-bilobed, keeled and purple beneath; scapes twice as long as the leaves, umbellately 2–6-flowered; pedicels erect; sepals oblong-lanceolate, callous-pointed, 2–3 times shorter than the corolla. *Thunb.! diss. No.* 11. *Burm. Afr. t.* 28, *f.* 3. *Willd. Sp. No.* 45. *DC. No.* 71. *O. macrophylla, Horn. Hort. Hafn.* 1. *p.* 428. *E. & Z.!* 656. *O. erecta, Savig.! in Lam. Dic.* 4. *p.* 685. *O. dentata, E. & Z.!* 655.

HAB. In ditches, waste places and on the hills round Capetown, &c. Gardens and moist places near Zwartkop's R., Uit. *E. & Z.! Zey.!* 2105. Graaf Reynet and Somerset, *Mrs. F. W. Barber.* (Herb. Holm., Vind., Lehm., Hook., T.C.D., Sond.)

Bulb small. Petioles 2–3 inches, glabrous or thinly hairy. Leaflets 4 lines long, 6–7 lines wide, the lobes roundish. Scapes slender, ½–1 foot long, terminal; pedicels glabrous or pilose, inch long or longer. Bracts 2, minute, at the base of the pedicels. Corolla ½ inch long, with yellowish tube, and oblong, obtuse, blueish limb.

103. O. livida (Jacq. Ox. t. 8); *caulescent*, glabrous; bulb ovate; stem scaly at base, *leafy at the crown and lengthening beyond the tuft*; petioles crowded, elongate; leaflets obcordate, sub-bilobed, green above, livid-purple and glabrous or sparsely pilose underneath; peduncles *lateral*, longer than the petiole, umbellately 2–5-flowered; pedicels and calyx villous; sepals oblong-lanceolate, callous, 3–4 times shorter than the corolla. *DC. No.* 75. *E. & Z.!* 654. *O. caprina, E & Z.!* 653.

VAR. β.; one or two sepals sometimes toothed. *O dentata, Jacq.! t.* 7. *DC. No.* 74.

HAB. Mountains round Capetown and Simon's bay and Muysenberg, *Thunberg, E. & Z.! &c.* (Herb. Holm., Vind., T.C.D., Sond.).

Bulb inch long, brown, with striate scales. Stipe 1–3 inches, stem ½–1 inch long. Petioles 1–1½ inch, nearly glabrous. Smaller leaflets 3 lines wide, 2½ long; larger ½ inch long and wide, or wider, with 2 short, rounded-obtuse lobes. Peduncles from the axils of the upper scales, or lower leaves, *never terminal*, thicker than the petiole, glabrous or pilose; pedicels inch long. Corolla flesh-colour, with a yellowish tube.

104. O. purpurata (Jacq.; Schoenb. t. 356); *stemless*; bulb ovate, brownish; leaves on long petioles; petioles pilose, or very shortly pubescent; leaflets *roundish-obcordate, the medial larger*, glabrous above, appressedly pilose or downy underneath, ciliated; scape longer than the leaves, pilose or puberulent, umbellately 3–12-flowered; pedicels and calyx downy, often gland-bearing; sepals oblong, acute, or lanceolate, thrice as short as the corolla.

VAR. α. **Jacquini**; pilose, leaflets ½ inch long, livid or reddish, the younger silky; scape 2–3 times as long as the leaves, 3–6-flowered; flowers white, rosy, or purple. *Jacq.! l. c. DC. No.* 76. *E. & Z.!* 652.

VAR. β. **petiolaris**; leaflets larger, ½–¾ inch long; petioles elongate, equalling the foot-long scape; sepals calloso-striate; flowers rosy. *O. petiolaris, E. Mey.! in Hb. Drege.*

VAR. γ. **Bowiei**; whole plant minutely pubescent; leaflets very large, the medial 2 inch long, or longer; petioles glabrescent, much shorter than the umbellately 8–12-flowered scape; flowers larger, deep rose-red. *O. Bowieana, Lodd. Cab. t.* 1782. *O. Bowiei, Lindl. Bot. Reg. t.* 1585.

HAB. Cape, Hb. *Jacquin.* α. mountains near Klipplaat's R., Caffr., *E. & Z.!* Salem, Alb., *Zey.!* 2125. β. between Omtendo and Omsamculo, *Drege!* γ. Shrubberies and woods at Adow, Uit., *E. & Z.!* July. (Herb. Vind., E. Mey., Sond., T.C.D.)

6 inches to a foot or more in height. Bulb moderate. Leaflets commonly semiuncial, the larger inch long and wide; the medial sometimes 2 inch long, the lateral oblique at base. Pedicels of the umbel bibracteate at base, ½–1 inch long or longer, pilose or glabrate, or glandularly pubescent. Corolla 6–8 lines long, varying from white to deep rosy-purple. Like *O. holosericea*, Sond., in the form of the leaves, but with very different inflorescence. Var. γ. is a favourite border flower in England.

105. O. semiloba (Sond.); stemless; leaves on long glabrous or pilose petioles; leaflets *obcordate-bilobed, the lobes divaricate, rounded, glabrous*, ciliated; scape twice as long as the leaves, umbellately 8–12-flowered; pedicels and calyx glandularly downy; sepals oblong, obtuse, callous, thrice as short as the corolla.

HAB. Shady and rocky places near Caledon River, *Burke, Zey.!* 271. Somerset, *Atherstone!* (Herb. Hook., Sond.)

Near the preceding. 6–8 inches long. Leaflets 6–8 lines wide, 5–6 lines long; bifid nearly to the middle and much more wedge-shaped than the leaflets of *O. purpurata*, which are rounded. Flowers half-inch long. Sepals with yellow calli, thickish. Limb of the corolla purple.

106. O. lateriflora (Jacq.! Schoenb. t. 204); *caulescent*, glabrous; bulb ovate, striate; stem somewhat branched, *naked at base, ascending*; petioles somewhat crowded at the summit; leaflets *cuneate, bilobed*, lobes oblong or sub-linear; peduncles *very long, lateral*, umbellate; sepals lanceolate, callous, 3–4 times shorter than the corolla. *Willd. Sp. No.* 78. *DC. No.* 37. *E. & Z.!* 662.

VAR. β. **liniflora**; lobes of the leaves linear; corolla whitish, margined with violet. *O. liniflora, E. & Z.!* 663. *O. phellandroides, E. Mey.! Meisn. in Hook. Journ. Bot.* 2. *p.* 54.

HAB. Heathy places among shrubs on mountain sides, near Brackfontein, Clanw. *E. & Z.!* Drakenstein, *Drege!* β. near the Oliphant's R., Clanw., *E & Z.!* (Herb. Vind., E. Mey., Lehm., Sond.)

Bulb inch long, brown. Stem 6 inches long or longer, leafy at the summit. Petioles slender, 2–5 inches long. Leaflets 4–6 lines long, bifid to the middle or beyond it, the lobes in the typical form 2 lines long, 1 line wide, in β. 4 lines long, ¾–1 line wide. Peduncles from the lower, naked portion of the stem, 6 inches or more than a foot in length, glabrous, or thinly hairy. Pedicels with short bracts. Corolla 8 lines long, the tube yellowish, the limb violet; in β. pale, with a deeper margin. In var. *β*, occasionally the peduncle is short, 1-2-flowered, from the axil of the upper leaves.

107. O. stellata (E & Z.! 661); stipitate or caulescent, glabrous; bulb ovate, blackish, with smooth scales; stem somewhat branched at base, scaly; terminal leaves crowded, with slender petioles; leaflets *deeply bilobed*, lobes *linear, divaricate;* peduncles terminal, longer than the leaves, umbellately 2–6-flowered; sepals lanceolate, callous at the point, 3–4 times shorter than the corolla.

VAR. *β*. **violacea**; nearly stemless or shortly stipitate, with violet-coloured flowers. *O. polyscapa, E. & Z.!* 660.

HAB. In fields and clayey places, of Uitenhage, Albany, and Somerset, *E. & Z.! Zey.!* 2135, 2137, 2138. (Herb. Vind., E. Mey., Lehm., Hook., TC.D., Sond.).

Bulb inch long. Stipe or stem sometimes ½ foot long. Petioles mostly 1–2 inch long, in some specimens more than 4 inches. Leaflets 4–6 lines long, divided to ¾ their length, the lobes 1 line wide. Scapes mostly twice as long as the leaves. Pedicels of the umbel uncial, shorter or longer. Cor. 8–12 lines long, white, rarely pale-rosy, with yellowish tube and ample; in *β*. violaceous and less wide.

(3.) Root *fibrous*, not bulbiferous. Stem branching. Leaves trifoliolate. Flowers yellow. (Sp. 108).

108. O. corniculata (Linn. Sp. 624); root fibrous, branching; stems diffuse or creeping, filiform, pubescent, rooting at base; leaves on long petioles; leaflets obcordate, ciliate, villous or glabrous underneath; stipules oblong, adnate; peduncles axillary, 2–4-flowered, shorter than the leaf or longer; pedicels refracted in fruit; capsules pubescent. *Thunb. diss. No.* 20. *Jacq.! t.* 5. *E & Z.!* 648. *O. pusilla, Salisb. Linn. Soc. Trans.* 2. *p.* 243. *t.* 23. *f.* 2. *O. repens, Thunb. Diss. No.* 14. *t.* 1. *Fl. Cap. No.* 22. *Jacq.! t.* 58. *f.* 1. *DC. No.* 33. *E & Z.!* 649. *O. ceratilis, E. Mey.! Un. It. No.* 595.

HAB. Everywhere in cultivated ground, throughout the colony and in Caffraria. (Herb. Sond., T.C.D., &c.)

Dull green or whitish. Stems 6–12 inches long, herbaceous, with short branches. Petioles inch long or more, villous. Leaflets 3–4 lines long and wide. Peduncles bracteolate at the forking. Corolla small, yellow, about twice as long as the calyx.

ORDER XXXV. ZYGOPHYLLEÆ. R. Br.

By W. SONDER.

(R. Br. in Flinder's Voy. DC. Prod. 1. p. 703. Endl. Gen. No. ccliii. Zygophyllacea, Lindl. Veg. King. No. clxxx.)

Flowers perfect and regular. *Calyx* free, persistent or deciduous, 4–5-parted, imbricate or valvate in æstivation. *Petals* 4–5, clawed, twisted in æstivation, rarely wanting. *Stamens* twice as many as the petals, hypogynous; filaments broader at base, naked or furnished with scales; anthers introrse, bilocular. *Ovary* on a glandular or annular disc, 4–5 angled and celled; ovules axile, either in pairs, or several in a double row, very rarely solitary; styles (except in *Seetzenia*) connate, sometimes very short. *Fruit* capsular or fleshy, rarely indehiscent, often winged, tubercled, spiny or hispid, loculicidal. *Seeds* with or without albumen, the embryo straight and green. Radicle superior.

Herbs, under-shrubs, shrubs or trees, often with oppositely branched, nodose or distinctly jointed, rigid stems. Leaves opposite, compound, or rarely simple, abruptly or impari-pinnate, often bifoliolate, the common petiole produced at the top into a short point. Leaflets sessile, opposite, impunctate, generally unequal-sided, flat or fleshy, and sometimes terete; the margin often cartilaginous. *Stipules* twin, at the base of the petioles, persistent, often spinous, rarely deciduous. Peduncles one-flowered, axillary, or terminal, ebracteate.

Natives of dry, desert-places and sea-shores in the hotter parts of the temperate zones both north and south. *Tribulus terrestris* is a common sub-tropical weed of cultivation. The only valuable plants of this Order belong to *Guaiacum*, a shrubby or arborescent genus exclusively American, and which furnishes the *Lignum-vitæ*-wood and the *Gum-Guaiacum* of commerce. The *Zygophylla* generally indicate saltness in the soil: many of them are detestably scented.

TABLE OF THE SOUTH AFRICAN GENERA.

Tribe I. Tribuleæ. *Seeds* without albumen.

I. **Tribulus.**—*Petals* 5. *Fruit* thorny, indehiscent. *Leaves pinnate.*

II. **Sisyndite.**—*Petals* 5. *Fruit* capsular, 5-angled, clothed with golden hairs, dehiscent. *Leaves simple.*

III. **Augea.**—*Petals* none. *Fruit* capsular, 10-angled, glabrous. *Leaves simple.*

Tribe II. Zygophylleæ. *Seeds* albuminous.

IV. **Zygophyllum.**—*Petals* 4–5. *Stamens* 8–10. *Style* 1. *Leaves simple or bifoliolate.*

V. **Seetzenia.**—Petals none. *Stamens* 5. *Styles* 5. *Leaves trifoliolate.*

I. TRIBULUS, Tourn.

Calyx 5-parted, deciduous. *Petals* 5, spreading, longer than the calyx. *Stamens* 10, hypogynous; filaments subulate, the 5 opposite the calyx-lobes with a gland, externally at base; anthers cordate, introrse. *Ovary* sessile, on a short 10-lobed urceolus, hairy, 5-celled; ovules 3–4 in each cell. *Style* short or none; stigma large, 5-angled-pyramidal. *Fruit* depressed, 5-angled, of five indehiscent, dorsally tuberculated, spinous or winged, spuriously plurilocular carpels; each locellus one-seeded. *Seed* exalbuminous. *Endl. Gen.* 6030.

Weeds and weed-like, diffuse, procumbent, or prostrate herbs, dispersed through the warmer temperate and tropical zones. Leaves opposite, one usually much larger than the other, bistipulate, abruptly pinnate; leaflets opposite, quite entire, in several pairs. Peduncles axillary, one-flowered; petals pale yellow or white. Name from τρεις, *three*, and βολος, *a point*; each carpel is often armed with 3–4 larger spines.

* *Carpels not winged.* (Sp. 1–2).

1. T. terrestris (Linn. Sp. p. 554); leaflets in 5-8 pairs, sub-equal, rather acute; peduncle *from the axil of the shorter leaf;* petals *rather longer* than the calyx; stigma sub-sessile, *hemispheric;* carpels with 2–4 larger spines, and prickly on the back. *Lam. Ill. t.* 346. *f.* 1 *Schk. Handb. t.* 115. *DC. Prod.* 1. *p.* 703. *Juss. Mem. Mus. xii. tab.* 14. *f.* 1. *Thunb. Fl. Cap. p.* 543. *E. & Z.!* 751. *T. albus, Poir. T. murex, Presl. Drege,* 7159.

VAR. *β*. **hispidissimus**; stem hairy with spreading hairs; leaves and calyx hirsute; stipules and sepals narrower and rather longer. *T. hispidus, Presl.*

VAR. *γ*. **desertorum**; fruits mostly twice as small; the spines equal or unequal, sometimes very short. *E. & Z.!* 751. *β*. *T. terrestris, β. E. Mey. T. parvispinus, Presl.*

HAB. A weed of cultivation in Cape and Uitenhage. *β*, at Basche, at the river mouth and in gardens, *Drege!* *γ*, Sandy Karroo land near Gauritz R., Swell., *E. & Z.!* Nieuweveld, *Drege,* 9532. Springbokkeel, Boschemansland, *Zey.!* 273. Dec.-Mar. (Herb. T.C.D., Sond., Lehm.)

A decumbent or prostrate annual, with stems 6–12 inches long or longer, villous or glabrescent, in *β*. hispid or hirsute. Leaves opposite, unequal, the longest 1–1½ inch; leaflets oblong, acute, green above, sparingly appressedly villous, below appressedly pubescent, often whitish or silky. Stipules ovato-lanceolate, 2–3 lines long. Peduncles axillary, solitary, shorter than the leaf, the fruit-stalk nearly as long, villous or, in *β*., hairy. Flowers yellow or whitish. Sepals 1½–2 lines long. Petals ¼ as long again as the calyx, obtuse. Stigma roundish-conical, 5-rayed, with inconspicuous style. Carpels crested and tubercled on the back, or prickly and hairy, in *γ*. more coarsely hairy, armed with 4, rarely 3–2 marginal spines, the two upper spines much divaricate, 3 lines long; 2 lower turned downwards and shorter. In var. *γ*. the upper spines are not more than 1 line long, often much shorter, sometimes very short and confounded with the dorsal prickles. Var. *β*. is not unlike *T. cistoides*, &c., which, however, is readily known by its much larger flowers and long styles.

2. T. Zeyheri (Sond.); leaflets in 6–8 pairs, sub-equal, obtuse; petiole *in the axil of the 2–3 times shorter leaf;* petals nearly *thrice as long as the calyx;* stigma *pyramidal,* twice as long as the short style; carpels with 4 spines and prickly at back.

HAB. Springbokkeel, *Zey.!* 272. Feb.-Mar. (Herb. Sond.)

Nearly allied to the preceding, especially to var. *γ*., with which it agrees in the silky-canescent leaves and smaller fruits, but differs in the peduncles twice as long, larger petals and stigma. Stems weak, 1–2 feet long, villous or glabrescent. Stipules ovato-lanceolate. Leaves unequal, one uncial, the other (from whose axil the peduncle springs) thrice as short. Leaflets 1½–2 lines long, green above. Peduncles uncial. Sepals ovato-lanceolate, appressedly silky. Petals cuneate, 5–6 lines long. Style twice or thrice as short as the stigma, stigma 1 line long. Fruit smaller than a pea; the spines mostly unequal, the larger nearly ½ line long; tubercles numerous, hard, acute or sub-spinulous, mostly hairy. This new species entirely agrees in habit with *T. lanuginosa,* Miq. in Metz. Pl. Ind. Or. No. 779, which nevertheless is known by its constantly 2-spined carpels, and long style. *T. lanuginosus,* L. Sp. p. 553, according to Burman's figure, and the specimen collected by Dr. Wright, is only *T. terrestris. T. cistoides,* L. is much more robust, and has larger leaves, flowers and fruits, besides a style 2 lines long, and a short, terminal stigma, by which character, as already stated by Schlechtendal in Bot. Zeit. 1851, p. 844, it is known from other *Tribuli. T. cistoides* occurs in S. America, Cape Verd, and in the East Indies.

*** Carpels two-winged.* (Sp. 3–4.)

3. T. pterophorus (Presl, Bot. Bem. p. 30); leaflets in about 6 pairs, sub-equal, acute; peduncle in the axil of the 2–3 times shorter leaf; petals broadly cuneate, thrice as long as the calyx; style short, stigma pyramidal; carpels *pubescent, two-winged at the margin, the wings very large, somewhat toothed;* disc of the carpels prickly and setose. *T. alatus? Drege! non Del.*

HAB. On the Gareep, near Verleptram, *Drege!* Sep. (Herb. Sond., T.C.D., Lehm.)

Stems flexuous, 1–2 feet long, with the branches, stipules, peduncles and calyx *hirsute*. Internodes uncial. Stipules lanceolate, acuminate, 2 lines long. Leaves unequal, the larger inch-long. Leaflets silky, green above. Peduncles erect, 8–12 lines long. Sepals acuminate, 4 lines long. Petals 6–7 lines long. Stamens as long as the calyx. Stigma conical-oblong, 1 line long. Ripe carpels 3 lines long, including the wings 6 lines wide; the wings rounded, transversely striate, toothed, the teeth at length more evident and mostly unequal, one 3 lines, the other 2–2½ lines long.—Nearly related to *T. alata*, Del., from North Africa, but in Sieber's specimens of that species the stems and branches are not hairy, the stipules are shorter and broader, the peduncles very short, and the disc of the carpels is wider, and the wings narrower.

4. T. cristatus (Presl, l. c. p. 30); leaflets in about 5 pairs, sub-equal, acute; peduncles in the axil of a 3–4 times shorter leaf; petals broad, cuneate, twice as long as the calyx; style short; stigma pyramidal; *carpels with two marginal-crested, unequally spinulose wings;* disc of the carpels prickly.

HAB. At the Gareep, on stony hills at Verleptram, below 1000f. *Drege.!* 7160. Sep. (Herb. Sond., T.C.D., Lehm.)

Very near the preceding, and also hairy, but the hairs are stiffer, less dense and the colour is greener. Stipules acuminate, 2–3 lines long. Leaf an inch or more in length. Peduncles ½–1 inch long. Flowers large, the petals in some 6 lines, in others nearly an inch long. Carpels 3-seeded, emarginate at summit, the ripe ½ inch wide and high; the wings membranous, twice as wide as the disc, armed at the margin with spines, ½–2 lines long. Seeds oblong.

II. SISYNDITE, E. Mey.

Calyx 5-parted, the lobes somewhat imbricate. *Petals* 5, at first short and truncate, at length oblong, longer than the calyx. *Stamens* 10, hypogynous; filaments subulate, glabrous, equalling the calyx; anthers linear, versatile, two-celled, longitudinally slitting. *Hypogynous-scales* 5, circling the ovary, opposite the sepals. *Ovary* sessile, very hairy, 5-angled, 5-celled; ovules solitary, erect. *Style* filiform, hairy, thickened into a clavate 5-furrowed stigma. *Fruit* capsular, 5-lobed, the carpels compressed, ovate, acute, transversely furrowed, in all parts clothed with long, golden-yellow hairs, at length separating and opening by the ventral sutures. *Seed* compressed, erect; testa membranous; embryo without albumen; cotyledons thick; radicle short, superior.

A suffrutex, with the habit of *Spartium junceum*, di-tri-chotomous and quite glabrous and glaucous, with very few leaves. Leaves on short petioles, remote, mostly approaching in pairs, simple, obovate, retuse or shortly apiculate, leathery. Stipules 2, short. Peduncles solitary, in the forks of the branches, one-flowered. Name, perhaps from συνδέω, to *link together?*

1. S. spartea (E. Mey.! in Herb. Drege).

HAB. Between Natvoet and the Gariep, 1000–1500f. *Drege!* Sep. (Herb. Sond.)

Several feet high, erect, with terete, virgate branches and ramuli. Adult leaves about 8 lines long, 3 lines wide, very entire; petiole 1 line long. Stipules ovate, obtuse, 1 line long, membranous or somewhat coloured, albo-tomentose within. Peduncles erect, 4–6 lines long. Calyx segments unequal, ovato-lanceolate and lanceolate, 5–6 lines long, the wider tomentose within. Petals broader than the calyx and nearly twice as long, yellow, striate. Hyp.-scales glabrous, 1 line long, bifid, the lobes denticulate on the inner side. Style thick, 3–4 lines long. Carpels 6 lines long, 3 lines wide. Seed 4 lines long.

III. AUGEA, Thunb.

Calyx 5-cleft, persistent, the segments ovato-lanceolate, with valvate æstivation. *Petals* none. *Hypogynous-disc* cup-shaped, membranous, 10-toothed, with subulate-setaceous teeth. *Stamens* 10, inserted between the teeth of the disc; filaments very short, broad, trifid, the medial segment antheriferous, the lateral subulate, longer than the anthers; anther fixed below the middle, oblong, 2-celled. *Petaloid scales* linear (white) bifid to the middle, outside the stamens and opposite them, rather longer than the calyx. *Ovary* free, somewhat angular, glabrous, 10-celled; ovules pendulous, fixed to the central angle. *Style* short, filiform; stigma simple. *Capsule* oval-oblong, 10-angled, 10-valved, the valves dehiscing. *Seeds* solitary, oblong, with a thickish testa; embryo exalbuminous; cotyledons flat, thickish; radicle superior. *Thunb. Fl. Cap. p.* 389.

An annual, glabrous, fleshy herb, with the aspect of a *Mesembryanthemum*. Root fusiform, fibrous. Stem simple or divided at base, jointed, with terete, erect, alternate branches. Leaves opposite, simple, connate, terete, obtuse, flattish above. Stipules short. Flowers axillary, solitary, or 2–3 together. Peduncles 1-flowered, bibracteolate at base. Name unexplained by Thunberg.

1. A. capensis (Thunb. l. c.)

HAB. Karroo between Oliphant's River and Bocklandsberg, *Thunberg.* Salt soil at Bitterfontein, Bechuana's land, *Zey.!* 281. Apl. (Herb. Sond.)

Stem $\frac{1}{2}$–1 foot long, as thick as a goosequill. Lower internodes 1–1$\frac{1}{2}$ inch long, Leaves 1–1$\frac{1}{2}$ inch long, 2 lines wide. Upper stipules ovato-lanceolate. Peduncles 3–4,—in fruit 6 lines long. Bracteoles very slender. Calyx 3–4 lines long. Scales (or petals) bifid, with erect, obtuse segments. Capsules inclosed in the bottom of the calyx, 7–8 lines long, 4–5 wide.

IV. ZYGOPHYLLUM. L.

Calyx 4–5-parted, persistent or deciduous, imbricate. *Petals* 4–5, clawed, twisted-imbricate. *Stamens* 8–10; filaments subulate, with an entire or bifid, or bipartite scale at base. *Disc* fleshy, 8–10-angled. *Ovary* 4–5-angled or lobed, 4–5-celled; ovules 2 or more, axile. *Style* furrowed, continuous with the ovary; stigma minute. *Capsule* 4–5-angled or winged, 6–5-celled, few or several seeded. *Embryo* straight, in thin albumen. *DC. Prod.* 1. *p.* 705. *Endl. Gen.* 6036.

Small shrubs or undershrubs, with rigid, sometimes spiny branches and fleshy or thinly membranous leaves. Stipules membranous or spinous. Leaves simple or bifoliolate, sessile or petiolate. Peduncles one-flowered. Petals white or yellowish, rarely red, mostly with a deep-coloured basal spot, and radiating-coloured nerves. Name from ζυγος, a *yoke*, and φυλλον, a *leaf*; the leaves are opposite.

ANALYSIS OF THE SPECIES.

* Leaves simple.

Scales at the base of the filaments *undivided* ... (1) cordifolium.

Scales at the base of the filaments *bipartite* :
shrubby ; capsule prismatic ... ... (2) **prismatocarpum.**
annual ; capsule obcordate ... ... (3) **simplex.**

** Leaves bifoliolate, *sessile*.
Leaflets *linear* : capsule oval, 5-lined ... ... (4) **pygmæum.**
,, ,, : capsule globose, 5-angled ... ... (5) **spinosum.**
,, *linear-cuneiform* ; caps. oblong, 5-winged (8) **cuneifolium.**
,, *oblong-cuneate* ; caps. globose, 5-angled... (9) **flexuosum.**
,, *lanceolate-oval*, with *scabrous* margins ... (6) **sessilifolium.**
,, *lanceolate-oval*, with *smooth* margins ... (7) **fulvum.**
,, *obovate-roundish* ... ... (10) **divaricatum.**

*** Leaves bifoliolate, *petiolate*.
Scales at the base of the filaments *undivided* :
Capsule orbicular, very large, *widely-winged* ... (11) **Morgsana.**
Capsule with narrow wings, or merely angular :
Arborescent ... ... (16) **dichotomum.**
Shrubby ; with tufted leaves ... ... (18) **incrustatum.**
Shrubby ; with opposite leaves :
leaflets linear-lanceolate ... ... (12) **maculatum.**
leaflets narrow-elliptical ... ... (19) **microcarpum.**
leaflets oblong or elliptical :
filaments $\frac{1}{3}$ longer than the *scales*... ... (17) **glaucum.**
filaments 2–4 times longer than the scales :
leaflets inch long ; caps. sub-globose, with obtuse angles ... ... (13) **Uitenhagense.**
leafl. $\frac{1}{2}$–1 inch long ; caps. oblong, 4–5 winged ... ... (14) **debile.**
leafl. $\frac{1}{2}$–1 inch long ; caps. sub-globose, 4–5 winged ... ... (20) **Lichtensteinianum**
leafl. 1–2 lines long ; caps. winged, emarginate ... ... (15) **microphyllum.**
leaflets obovate, on longish petioles :
branches 4-angled ; *scales* of stamens *ciliate* ... ... (22) **Meyeri.**
branches terete, scarcely striate ; scales *serrate* ... ... (23) **foetidum.**
branches terete, furrow-striate ; scales *ciliate* ... ... (21) **leptopetalum.**

Scales at the base of the filaments *bipartite* :
Peduncles axillary ; leaflets 2–3 lines long ... (24) **retrofractum.**
Flowers cymose ; leaflets 6 lines long... ... (25) **Dregeanum.**

---

* *Leaves simple.* (Sp. 1–3.)

1. Z. cordifolium (Lin. f. Suppl. p. 232) ; shrubby ; leaves simple, sessile, cordate-orbicular ; flowers axillary, solitary ; filaments *twice* as long as *the truncate-lacerate* scales ; capsule oblong, *four-winged*. *Thunb. Fl. Cap. p.* 543. *DC. Prod.* 1. *p.* 705. *E. & Z. !* 752.

HAB. Sandy places at Saldanha Bay, *Thunberg*, *E. & Z. !* Olifant's R. and Ebenezer, *Drege !* Jul.-Nov. (Herb. Sond., T.C.D., E. Mey.)

Stem a foot or more high, ash-coloured. Branches and branchlets opposite or alternate, spreading, terete or somewhat angular. Internodes of the branchlets $\frac{1}{2}$–1 inch long or longer. Leaves mostly sub-cordate at base, some oblique at base, or half-cordate, these 1–3-nerved, the mid-nerve subdivided, those mostly 5-nerved from the base, all very entire, somewhat fleshy, very delicately netted with veins, about an inch long and wide, the upper smaller. Stipules lanceolate, 1 line long. Flowers 4–5-cleft. Peduncles about as long as the leaves. Sepals ovate-oblong, 3 lines long. Petals yellowish, broad, thrice as long as the calyx. Stamens ½ the length of the petals. Style subulate. Capsule 8 lines long, 4 lines wide, netted, obtuse at each end, the wings 1 line long. Seeds not seen.

2. Z. prismatocarpum (E. Mey. !) *shrubby;* leaves simple, sessile, roundish-obovate, narrowed at base; flowers panicled at the ends of the branches; filaments 8–10 *times* longer than the *deeply bifid, obtuse scales;* capsule prismatic, sharply 5-angled.

VAR. β. **diffusum**; stem shorter, diffuse; leaves scarcely narrowed at base; panicle more leafy.

HAB. Between Kaus, Natvoet and Doornpoort; β. between Natvoet and the Gariep, 1000–2000f. *Drege!* Sep.-Oct. (Herb. Sond., T.C.D., E. Mey.)

Stem erect, terete, with grey bark; branches erecto-patent, angularly-compressed, greenish, ramulose. Internodes ½–1 inch long. Leaves spreading, sub-amplexicaul, with the midrib conspicuous at the base only, carnoso-coriaceous, glaucous-green, fragile when dry, 6–8 lines long, 4–6 lines wide at the apex, 1 line at base, some smaller; in β. with the base sub-cordate, often not narrowed, 5–6 lines long and wide. Stipules minute, often wanting. Upper flowering branchlets crowded in a pyramidal panicle, which in our specimens is 2–3 inches long. Pedicels solitary in the forks, nodding, 1–2 lines long. Sepals oval, 1½ lines long. Petals cream-coloured, twice as long as the calyx. Stamens as long as the petals. Scales very minute, scarcely visible to the naked eye, bifid to the base, the segments truncate, obtuse, very entire, hyaline. Capsule brownish-purple, oblong, cuneate, 4–5 lines long, 1½ lines wide above, obtuse, tipped with a short style, deeply 5-furrowed and somewhat winged.

3. Z. simplex (Linn. Mant. p. 68); *annual;* leaves simple, fleshy, cylindrical. *Willd. Sp. Pl.* 2. *P.* 1 *p.* 560. *DC. Prod.* 1. *p.* 705. *Z. portulacoides, Forsk. descr. p.* 88. *Ic. t.* 12. *B.*

VAR. **Capense** (Sond); stems stronger, ascending-erect, 6–12 inches high, much-branched, leaves obovate or cuneate-oblong, obtuse. *Z. microphyllum, E. & Z.!* 771, *non Thunb. Z. microcarpum, E. Mey.!, non Lichtenstein. Z. Dregeanum, Presl, non Sond.*

HAB. Clayey soil in Bosjesmansland, at the Orange R., *E. & Z.!* near Verleptpram, *Drege!* Armaedsfontein and Kamos, *Zey.!* 280. Sep.-Jan. (Herb. E. Mey., Sond., Lehm.)

A succulent annual, fragile when dry. Root perpendicular, as thick as a pigeon's quill. Stems as thick, many from the same root, decumbent at base, then erect, terete, pale, with alternate, divaricating branches, and alternate, rarely opposite, short, slender branchlets. Internodes in the stem ½–1 inch long, shorter in the branches. Leaves very widely spreading, sometimes 1½–2 lines, sometimes 4–5 lines long, a line wide, narrowed at base, sub-sessile, very fleshy. Stipules hyaline, small, triangular. Flowers minute, 5-cleft. Peduncles axillary, solitary, ½–1 line long. Petals yellowish, clawed, obovate, twice as long as the obtuse sepals. Stamens as long as the petals. Scales bipartite to the base, hyaline, the segments oblong, obtuse, thrice as short as the filaments. Capsule mostly pendulous, 1 line long and wide, emarginate at the apex, obcordate, sharply 5-angled; style very slender, half as long as the capsule or longer. The flowers and fruits in the Arabian and Egyptian specimens entirely agree with those in the Cape plant.

** *Leaves bifoliolate, sessile.* (Sp. 4–10.)

4. Z pygmæum (E. & Z. ! 769); fruticulose, glabrous; leaves sessile, bifoliolate; leaflets subincurved, *oblong, with revolute-margins, subterete, obtuse,* fleshy; stipules very short, recurved, spiny; filaments 3 times as long as the serrato-ciliate scale; capsule *oval, marked with 5 lines.*

HAB. Among shrubs at Gauritz Riv., Swell, *E & Z.!* Nov.-Dec. (Herb Sond).

A small bush, 6–12 inches high, rigid, much-branched, greyish; with alternate, terete primary branches, as thick as a goose-quill, and alternate or opposite, slender, curved, somewhat angular branchlets. Internodes 4–6 lines long, the lower longer. Leaflets erect, green, 3–4 lines long. Peduncles solitary, rather longer than the leaves. Sepals ovate, 2 lines long. Petals obovate, thrice as long as the sepals, yellowish, with yellow bases. Filaments shorter than the petals. Capsule oval, shortly apiculate, surrounded by the persistent calyx, 4 lines long, somewhat angular, with 5 raised lines. *Zygophyllum*, Drege, 7172 (in flower) seems to belong to the present, rather than to the following species.

5. Z. spinosum (Linn. Sp. p. 552); fruticulose, glabrous; leaves sessile, bifoliolate; leaflets linear, *flat or with the margin sub-revolute, acute,* fleshy; stipules short, at length spiny; filaments thrice as long as the obovate, recurved, serrate scale; capsule *globose, somewhat winged or sharply 5-angled. Burm. Afr. p. 5. t. 2. f. 2. Thunb. Fl. Cap. p. 544. E. & Z. !* 770.

HAB. In sandy places and between shrubs in the Cape Flats, near Zekoevallei and Rietvallei, and at Saldanha Bay, *Thunberg, E. & Z.!* Groenrivier and Bergvallei, *Gueinzius! Drege,* 7171. Muysenberg, *Zey.!* 2149. Langvallei, *Zey.!* 274. Sep.-Oct. (Herb. Sond., E. Mey., Lehm.)

A bush 1–2 feet high, much-branched from the base; the branches alternate, grey; branchlets alternate or opposite, terete or angular, filiform, flexuous. Internodes 4 lines, sometimes 1–2 inches long. Leaflets 4–10 lines long, ½–1 line wide, drying blackish. Stipules triangular, very acute, reflexed, about ½ line long. Flowers 5-cleft, nodding. Peduncles equalling or surpassing the leaves, sepals ovate, 2 lines long. Petals obovate, 2–3 times longer, yellowish or cream-colour, and from base to middle red-streaked, or with a purple spot at base. Capsule 4–5 lines long.

6. Z. sessilifolium (Linn. Sp. p. 552); fruticulose; stem weak, prostrate, kneebent, branched; branches and branchlets herbaceous, somewhat angled, or semi-terete; leaves sessile, bifoliolate; leaflets *lanceolate-oval, acute,* or mucronulate, narrowed at base, with cartilaginous *denticulate-scabrid* margin; stipules ovato-lanceolate; filaments twice as long as the ciliato-serrate scales; capsules *sub-globose, 5-angled, obtuse, apiculate. Burm. Afr. p. 4. t. 2. f. 1. Thunb. Fl. Cap. p. 544. var. 1, E. & Z.!* 763. *Z. Capense, Lam. Z. limosum, E. & Z.!* 661. *Z. Commelini, E. & Z. !* 765, *ex pte.*

HAB. Sandy places on the Cape Flats and shore below Lion's Mt., Table Mt. and Kamp's Bay and in Hott Holl.berge, *E. & Z.!* &c. Ap.-Aug. (Hb. Sond., E. Mey., T.C.D.,, Lehm.)

Root white. Stems prostrate, as if creeping, whitish, leafless, 6–12 inches long, breaking up into ascending or erect, weak branches, 4–6 inches long or longer, of which the shorter are mostly undivided, the longer alternately or oppositely branched. Internodes 4–12 lines long. Leaflets fleshy, flat, 6–8 lines long, about 3 lines wide, the upper smaller. Stipules small, reflexed. Peduncles longer than the leaves, ½–1 inch long. Sepals ovate, acute, fleshy, mostly purplish, 3 lines long, at length reflexed. Petals cuneate-obovate, crenulate, twice as long as the calyx, white, with purple streaks, to the middle or coloured nearly to the point. Stem shorter than the petals. Capsule 6 lines long, 5 wide.

7. Z. fulvum (Linn. Sp. 1. p. 386); shrubby, erect; branches angular, the younger semi-terete; leaves sessile, bifoliolate; leaflets *lanceolate-oval, acute,* somewhat narrowed at base, *with smooth margins;* stipules lanceolate; filaments thrice as long as the ciliato-lacerate, dorsally papilla-margined scales; capsules *ovate-oblong, acute,* furrowed and 5-angled. *Z. sessilifolium β., Lin. Sp. ed. 2. p.* 552. *Burm. Afr. p.* 6. *t.* 3. *f.* 1. *Z. fulvum, E. & Z.!* 764. *Z. maritimum, E. & Z.!* 757, *ex pte. Z. semiteres, E. & Z.!* 762. *Z. sessilifolium, Bot. Mag. t.* 2184. *Z. alatum, E. Mey! Zey.!* 275, 2143, 2144.

HAB. Sandy and stony places, near Capetown, *Gueinzius!* Witsenberg; Klein Rivier Kloof and Caledon's Bay, *Zeyher!* Howhoek, Hott. Holl., and Brackfontein, *E. & Z.!* Also at the Zwartkop's R. and Port Elizabeth. Langvallei, Wupperthal and Gnadendahl, *Drege!* Jul.-Dec. (Herb. Sond., E. Mey., Lehm., T.C.D.)

Allied to the preceding, but known by the more shrubby, taller stem, thicker branches, and particularly by the semi-terete, almost winged branchlets; larger, fulvous flowers, and longer, not globose capsules. Stem and primary branches as thick as a goose-quill, terete, greyish, somewhat woody; branches and ramuli alternate or opposite, erect, patulous, 3–4 inch. to a foot long, herbaceous or purplish, striate, 1 line thick, one side flat, the other convex; sometimes with the opposite angles prominent, two-winged or ancipital. Internodes ½–2 inch. long. Larger leaflets 10–12 lines long, 4 lines wide, the rest 5–8 lines long, 2 lines wide, very fleshy, one-nerved, the cartilaginous edge not denticulate, acute or mucronate. Stipules reflexed. Peduncles ½–1 inch long, reflexed after flowering. Sepals 4 lines long, widely ovate, acute, sub-carnose, at length reflexed, persistent. Petals twice as long as the calyx, broadly obovate, fulvous or yellow, with a red blotch at base. Filaments half as long as the petals. Style subulate. Capsule 10 lines long, 8 lines wide, 5-furrowed when dry, the angles narrow-winged. Flowers in a few specimens whitish; changed in drying?

8. Z. cuneifolium (E. & Z.! 767); shrubby; branches woody; leaves sessile, bifoliolate; leaflets *linear-cuneiform, obtuse,* oblique at base, fleshy; stipules small, acute; sepals ovato-lanceolate; filaments *twice* as long as the oblong, lacero-ciliate scales; capsules *oblong, 5-winged. Z. rigescens, E. Mey.!*

HAB. Stony mountains, near the mouth of the Orange River, *E. & Z.!* Karroo hills at Mierenkasteel, below 1000f. *Drege!* Aug.-Nov. (Herb. Sond., E. Mey.)

Like a broad-leaved state of *Z. spinosa,* but distinguishable by its cuneate leaves and non-globose capsules. A rigid, much-branched, greyish, glabrous, bush, with short terete branches, and sub-angular ultimate branchlets. Internodes 3–6 lines long, or longer. Leaflets 5–6 lines long, about 1 line wide, pale green, one-nerved. Peduncles 5–6 lines long, recurved in fruit. Sepals 2 lines long, at length reflexed, persistent. Petals twice as long as the calyx, yellowish. Ovary oblong. Style subulate. Capsule 4 lines long, narrowly 5-winged, apiculate.

9. Z. flexuosum (E. & Z.! 768); shrubby; branches herbaceous, angular; leaves sessile, bifoliolate; *leaflets obovate-oblong or linear-oblong, obtuse,* or apiculate, narrowed at base, smooth-margined, flat, fleshy; stipules small, subulate; sepals oval, obtuse, apiculate; filaments thrice as long as the oblong, serrato-ciliate scales; *capsules globose, furrowed and 5-angled.*

VAR. α.; leaves linear-oblong, cuneate at base. *Z. Commelini, E. & Z.!* 765, *ex pte. Z. flexuosum, E. & Z.!*

VAR. β.; leaves cuneate-oblong or obovate-oblong. *Dill. Hort. Elth. t.* 116. *f.* 142. *Z. Commelini, E. & Z.! ex pte. Z. retrofractum, Jacq. Schoenb. t.* 354. *non Thunb.*

VAR. γ. **pruinosum**; leaves obovate, cuneate, thick, powdery.

HAB. Hills near Caledon; β. at Constantia and about Capetown, and at Saldanha Bay, *E. & Z.! Pappe!* River Zonder Ende and Hassagais Kloof, *Zey.!* 2148 γ. in sandy places, between Oliphant's R. and Knakisberg, *Zey.!* 276. Jan.-Sep. (Herb. Sond.)

A shrub, 1–4 feet high, almost intermediate between *Z. spinosum* and *Z. sessilifolium*; and in var. γ. approaching the following. Var. α. nearly approaches *Z. cuneifolium*, but that is a smaller and more rigid plant, and specially differs in the fruit; var. γ., of which the fruit is unknown, may possibly be a distinct species. Stem terete; primary branches terete or angular, often flexuous; branchlets angular, short or virgate. Internodes ½–1 inch long. Leaflets 6 lines long, 1½ line wide, in γ. twice shorter, fleshy, the nerve often not conspicuous. Stipules recurved. Peduncles equalling or excelling the leaves, nodding. Sepals 2 lines long. Petals obovate, cuneate, thrice as long as the calyx, yellowish, with an orange or purple spot at base. Filaments shorter than the petioles. Capsule 4 lines long and wide, or a little narrower, tipped with the base of the style. Drege's No. 7166 differs little from this, but the specimens are imperfect.

10. Z. divaricatum (E. & Z.! 766); shrubby; branches divaricate; leaves sessile, bifoliolate; leaflets *obovate-rounded or elliptical, not cuneate at base, unequal-sided*, fleshy; stipules minute, at length spinous; sepals oval or oblong-ovate; filaments thrice as long as the oblong, sub-recurved, lacero-ciliate scales; capsules *oval-globose, apiculate, bluntly 5-angled.*

HAB. Calcareous hills between the Zondag and Coega Rivers, Uit., *E. & Z.!* Coega-Kamma's Kloof, *Zeyher!* 2147. May.-June. (Herb. T.C.D., Sond.)

A greyish, sub-erect shrub, a foot or more in height; branches mostly alternate, curved, terete, with angular branchlets. Internodes 4–8 lines long. Leaflets mostly 3 lines long, 2–2½ lines wide; some 4 lines, others not more than 2 lines long, rounded, rarely minutely pointed, quite entire. Peduncles twice, in fruit thrice as long as the leaf. Sepals 1½ lines long, in fruit a little longer, reflexed. Petals obovate, 4–5 lines long, yellow, with a purple spot at base. Stamens half as long as the petals. Style subulate. Capsule 4–5 lines long, nearly as broad, obtuse, tipped with the base of the style, pentagonal with raised lines and scarcely impressed furrows.

*** *Leaves bifoliolate, petiolate.* (Sp. 11–25).

11. Z. Morgsana (L. Sp. p. 551); shrubby; leaves bifoliolate, shortly petiolate; leaflets *obovate, obtuse*, sub-oblique at base; calyx glabrous, sepals oval; capsules orbicular, deeply emarginate at each end, 4–5-winged, *the wings very wide, membranous, netted. Dill. Elth. t.* 116. *f.* 141. *Burm. Afr. t.* 3. *f.* 2. *Willd. Sp.* 9. *p.* 562. *E. & Z.!* 753. *Drege!* 391, 7161, 7162, 7163, *Zey.!* 2145. *Z. Lichtensteinianum, E. & Z.!* 754, *ex pte.*

HAB. Sandhills, Zwartland, at Saldanha Bay; and at the Zwartkops, Uit., *E. & Z.!* Winterhoek, and Kampsbay, *Krauss!* Blauwberg, Bergvalley, Olifant's River and between Kenda and Zwartebergen, *Drege!* Jul.-Nov. (Herb. Sond., E. Mey., Meisn., Lehm.)

A shrub, 4 feet high. Stem terete, with sub-angular, greyish-green, alternate or opposite branches. Internodes in the primary branches 1–2½ inch long, in the branchlets ½–1 inch. Petioles ½–1 inch long. Larger leaves 1–1½ inch long, 8–9 lines wide, the rameal mostly one-half inch long or less. Stipules lanceolate, reflexed. Flowers nodding, yellow, 4–5-cleft. Peduncles 6–9 lines long, axillary. Sepals 2 lines long. Petals obovate, thrice as long as the calyx. Staminal scales oblong, lacerate. Capsule very large, mostly 1–1½ inch long and wide, sometimes 2 inches wide, 1½ long, the basal and apical excision nearly one-half inch deep; the wings delicately membranous, with prominent, netted veinlets. Seeds blackish, subpyriform; embryo with little albumen.

12. Z. maculatum (Ait. Hort. Kew. 2. p. 60); shrubby; branches terete, furrow-striate; leaves bifoliolate, shortly petiolate; leaflets *linear-lanceolate or linear, obtuse,* somewhat narrowed at base; stipules short, acute; filaments twice as long as the fimbriato-ciliate scales; ovary 5-angled.

HAB. Pinaarskloof, *Zeyher!* May. (Herb. Sond.).

Branches herbaceous, alternate, erecto-patent, six inches or longer, with shorter branchlets. Internodes in the larger branches 1–1½ inch long, in the branchlets 6–8 lines. Petioles ½–1 line long, with a sharp, recurved point. Leaflets (of the branchlets) 4–6 lines long, ¾–1 line wide. Fl. axillary, solitary, 4–5 cleft, the peduncles equalling or excelling the leaves. Sepals ovate, two lines long, at length reflexed, persistent. Petals obovate, five lines long, yellow, with a heart-shaped, red spot at base, and (according to Aiton) a transverse red line above the spot in the three upper petals. Stamens half as long as the petals. Style subulate. Capsule wanting.—Although I have not seen an authentic specimen of Aiton's plant, I doubt not ours is the same. It much resembles *Z. spinosum*, L., but differs in several points.

13. Z. Uitenhagense (Sond.); suffruticose; branches herbaceous, angular, striolate, the upper semi-cylindrical; leaves short-stalked, leaflets *elliptical* or *elliptic-oblong, mucronulate,* sub-oblique at base, thickish; calyx glabrous; sepals elliptical or ovate; filaments thrice as long as the obovate-oblong, apically serrulate scales; capsules sub-globose, 5-angled, *with prominent, thickened, obtuse* angles. *Z. maritimum, E. & Z.!* 757, *ex pte. Z. insuave, E. & Z.!* 756, *non Sims. Drege,* 1265. *Zey.!* 2146.

HAB. Woods at the Zwartkop's R. and rocks near the sea-side at Cape Recief, and Port Elizabeth, *E. & Z.! Drege!* Oct.–Feb. (Herb. Sond., T.C.D., Lehm.).

A suffrutex or perennial, somewhat decumbent herb, often with 1–2 foot long, lax, ramulose branches; the branchlets opposite or alternate. Petioles ½–1 line long, with a recurved point. Leaflets 10–12 lines long, 4–5 lines wide, the upper smaller, 5–6 lines long, 2–3 lines wide. Stipules lanceolate, one line long, at length hardened and recurved. Peduncles axillary, 6–12 lines long, sometimes shorter. Flowers 5-cleft. Calyx two lines long, at length reflexed, persistent. Petals 5–6 lines long, cuneate-obovate, emarginate, or denticulate, yellowish. Seed oval, one line long. Embryo with little albumen. Nearly allied to the following, but with different fruit.

14. Z. debile (Cham. and Schl. Linn. V. p. 45); shrubby; branches herbaceous, angular, striate; leaves short-stalked; leaflets elliptical, or elliptic-oblong, mucronate, sub-oblique at base, thin; calyx glabrous, sepals elliptical; filaments thrice as long as the oblong, toothed scales; capsules *oblong, obtuse at each end,* 4–5-*winged, the wings narrow, thin. E. & Z.!* 759. *Z. foetidum, E. & Z.!* 758., *non Schrad. and Wendl.*

HAB. Among shrubs and in stony places. Groeneberg, *Mundt.* Greenpoint, *Zey.!* 2147. Karrega Riv., Albany, Coega, and Zondag Rivers, Uit., *E & Z.!* Zwarteberge, *Drege* 7169. June–Oct. (Herb. Sond., r. Berol., Lehm.).

A shrublet or perennial herb. Stems many, short, lignescent, terete, as thick as a fowl's quill, soon dividing into alternate or opposite, quadrangular, ramulose branches 1–2 feet long; branchlets somewhat angular. Internodes 1–2½ inch long. Petioles 1–2 lines long, with a sharp, reflexed point. Leaflets obtuse, mucronulate, or acute, or acuminate, one-nerved, veiny, more pellucid at the margin, the larger 10–12 lines long, 4–5 lines wide, usually much smaller. Stipules small, triangular, acuminate, reflexed, at length hardened. Flowers 4–5-cleft. Peduncles ½–1 inch long. Sepals two lines long, at length reflexed, persistent, the outer acute. Petals thrice as long as the calyx, cuneate-obovate, yellowish or yellow, with a purple-violet basal spot. Stamens ½ as long as the petals. Capsule 7–8 lines long, 4–5 lines wide, with thin wings, one line in width.

15. Z. microphyllum (Linn. f. Suppl. p. 232); fruticulose, much-branched; branches lignescent, with sub-spinescent branchlets; leaves shortly petiolate; *leaflets minute, ovate or obovate,* sub-oblique, coriaceo-carnose; peduncles longer than the leaves; sepals ovate, obtuse; petals thrice as long as the calyx, obovate, clawed; filaments more than twice as long as the obovate, lacerate scales; capsules *sphæroidal,* 4–5 *winged, truncate at base and apex,* sub-emarginate. *Thunb. Cap. p.* 545. *Z. horridum, Cham. and Schl. l. c. p.* 46.

HAB. Hantam and Onderste Roggeveld, *Thunberg, Drege,* 7173. Bitterfontein, *Zey.* 277. May. (Herb. Thunb., r. Berol., Sond.).

A shrub, two feet high or less, the primary branches as thick as a fowl's quill, greyish; secondary spreading, rigid; branchlets erecto-patent, opposite or alternate, terete, yellowish, with internodes 4–8 lines long. Leaflets mostly 1½–2 lines long, a line wide, some longer and broader. Petioles as long as the leaflets or shorter, with a triangular point. Stipules minute, deciduous. Peduncles axillary, 4–6 lines long. Flowers 4–5-fid. Sepals 1½ lines long, reflexed, persistent. Lamina of the petals elliptical, yellowish, spotted at base. Stamens shorter than the petals. Capsule 2½ lines long and wide, the wings veiny and glabrous.

16. Z. dichotomum (Lichst.; Cham. and Schl. l. c. p. 48); *arborescent;* branches dichotomous; branchlets sometimes opposite, *terete,* short; leaves short-stalked; leaflets minute, widely obovate or obliquely obcordate, coriaceous.

HAB. On the Gareep River, *Lichtenstein.* June. (Herb. Reg.-Berol.)

A tall tree, called "*Witgat*" by the colonists, *Licht.*—The single branch, in Herb. Berol., is about eight inches long, as thick as a goose-quill, eight times forked, sub-trichotomous toward the extremity, with spreading branchlets. The few leaves that remain are carnoso-coriaceous, the leaflets 1½–2 lines long. Flowers and fruits wanting.

17. Z. glaucum (E. Mey.); *fruticulose;* branches lignescent; branchlets angular; leaves shortly petiolate; leaflets oblong or elliptic-oblong, sub-oblique at base, coriaceous, *glaucous;* sepals obtuse, glabrous; petals obovate, acute, clawed; filaments one-third longer than the oblong-cuneate, *fimbriato-ciliate* scales; (unripe) capsule sub-globose.

HAB. Karroo-land, in Klein Namaqualand, 3000f. Sep. *Drege!* 3189. (Herb. E. Mey. Sond.)

A low, much-branched, rigid, greyish undershrub, with alternate branches. Petioles mostly but one-half as long as the leaflets, with a short reflexed point. Leaflets 4–5 lines long, 2–3 lines wide, some smaller, one-nerved. Stipules small, triangular, acute. Peduncles 3–4 lines long. Sepals one line long, at length reflexed, persistent. Petals 3–4 times longer than the calyx, yellowish. Stamens shorter than the petals, the larger part covered with the fimbriate scales. Unripe capsule 2½ lines long, smooth or furrowed at base, brownish, tipped with a rigid, subulate style one line long.

18. Z. incrustatum (E. Mey.); fruticulose, with alternate, rigid, lignescent, densely albo-papillose branches, the ultimate branchlets short and spiny; leaves *faciculate or geminate,* petiolate; leaflets elliptic-oblong, obtuse, leathery; sepals obtuse, glabrous; petals obovate, narrowed at base; filaments 3–4 times longer than the *oblong, inciso-serrate* scales; capsules *ovate, acute at each end, 5-winged.*

HAB. Stony places, at Zeekoe Riv., 5000f.; between Beaufort and Rhinosterkopf. 2500–3000f. *Drege!* Cradock, Tarka River 3000–4000f. and Fish River, *Zey.!* 278. Bassuto-land, *Meisner.* Oct.-Dec. (Herb. Sond., E. Mey., T.C.D.)

A small, rigid, much-branched bush, remarkable for its white-warted bark, at length cleft, and its spiny branchlets. Spines 2–6 lines long, simple or bifid. Primary branches as thick as a goose-quill. Leaves rarely solitary, almost always clustered at the base of the spinous branchlets. Leaflets 2–3 lines long, 1–1½ line wide. Petioles equalling the leaf. Sepals nearly two lines long, reflexed in fruit, persistent. Petals more than twice as long as the calyx, yellow. Stamens as long as the petals. Capsule on a pedicel 1½–2 lines long, about 4–5 lines long, two lines wide, tipped with a subulate style; the wings nearly one line wide.

19. Z. microcarpum (Lichtenst. herb., Cham. & Schl. l. c.); shrubby; the branches and branchlets opposite or alternate, *terete, articulate;* leaves petiolate; leaflets *narrow elliptical, sub-acute,* fleshy, nearly equal to the petiole; stipules hyaline; sepals ovate, acute; petals 1½-twice as long, narrow-obovate, clawed; filaments twice as long as the broadly cuneate scales; ovary tomentose; *capsule pilose, broadly 5-winged, deeply excised at the apex,* the wings sub-orbicular. *Z. gariepense, E. Mey.! in Hb. Drege.*

VAR. β. **prostratum**; stem prostrate, leaves mostly smaller. *Z. prostratum, E. Mey.! non Thunb.*

HAB. Olifant's River, *Lichtenstein.* Gariep, *Drege!* Rhinosterkopf, near Beaufort, *Zey.!* 279. β. Limoenfeld, Winterveld, 3000–4000f. *Drege!* Oct.-Jan. (Herb. reg.-Berol., T.C.D., Sond.).

Stems and branches woody, as thick as a fowl's quill, yellowish, grey below, thickened at the joints. Internodes 3–4 lines to one inch long, Branchlets spreading, greenish. Petiole with a small, weak point. Leaflets 2–4 lines long, 1–1½ lines wide. Stipules short, whitish, subulate-acuminate from a broad base. Flowers axillary, short-stalked, 5-cleft. Sepals two lines long. Petals whitish. Stamens as long as the calyx, the scale much dilated and toothed at top. Capsule shortly pedicellate, 3–4 lines wide, 2 lines long; wings veiny; style very slender, one line long, in the indention of the capsule.

20. Z. Lichtensteinianum (Cham. and Schl. l. c. p. 47); shrubby; branches lignescent, *terete, without striæ;* leaves petiolate; leaflets *elliptical* or *elliptic-oblong, rounded-obtuse,* sub-retuse, oblique at base, leathery; sepals obtuse, downy; petals obovate, cuneate; filaments thrice as long as the obovate-oblong, fimbriato-lacerate scales; capsule (unripe) sub-spheroid, 4–5-winged. *Z. Thunbergianum, E. & Z.!* 763, *excl. syn.*

HAB. Cape *Lichtenstein.* Among shrubs in Karroo, at Gauritz River, Swell. *E. & Z.!* Dec. (Herb. reg. Berol., T.C.D., Sond.)

An erect, greyish shrub, with alternate or opposite, erecto-patent, curved branches; the branchlets angular. Petioles 2–3 lines long, dilated, with a short point. Leaflets green (not glaucous), the larger 6–7 lines long, three lines wide, but mostly 4–5 lines long, three lines wide or less. Stipules minute, caducous. Flowers 4–5-fid. Peduncles axillary, solitary, nodding, 5–6 lines long. Calyx 1½ lines long, reflexed in fruit. Petals thrice as long as the calyx, pale yellow. Stamens as long as the petals or shorter. Capsule (unripe) obtuse at both ends, two lines long, 1½ line wide, broadly winged, tipped with a subulate style. Known from *A. Morgsana* by the longer petioles; flowers twice as small, and much smaller capsule.

21. Z. leptopetalum (E. Mey. !); suffruticose; branches herbaceous, *terete, furrow-striate, somewhat jointed;* leaves on long petioles; leaflets *obovate, obtuse* or *mucronulate, oblique at base;* calyx at length glabrous; petals oblong-cuneate; filaments twice as long as the oblong, fimbriate scales; ovary glabrous, furrowed; capsule sub-rotund, 5-angled.

HAB. Rocky places. Silverfontein, 2000–3000f. *Drege!* Sept. Oct. (Herb. E. Mey., Sond., T.C.D., Lehm.).

Stems erect, as thick as a goose-quill; branches opposite or alternate, branchlets gradually more slender, with internodes ½–2 inch long. Petioles 4–6 lines long, with a short, recurved point. Leaflets pale green, 8–10 lines long, six lines wide, some smaller. Stipules lanceolate, a line long, deciduous. Pedunc. axillary, simple or bifid, 4–6 lines long. Sepals ovate, one line long, at length reflexed, persistent. Petals four lines long, one line wide, yellow. Stam. shorter than the corolla. Capsule brownish, three lines long and wide.

22. Z. Meyeri (Sond.); suffruticose; branches herbaceous, *primary 4-sided;* leaves on long petioles; leaflets obovate, obtuse, oblique at base; calyx *glabrous,* the younger downy; sepals roundish, obtuse; petals obovate; filaments twice as long as *the oblong, ciliato-fimbriate* scales; ovary globose, glabrous, *smooth;* capsule? *Z. foetidum, E. Mey. non Schrad.*

HAB. Modderfontein and Mierenkasteel, 1000-2000f. *Drege!* Aug. (Herb. E. Mey., Sond.).

Allied to the preceding and following species; known from the first by its 4-sided, not furrowed branches; from the latter by its remarkably quadrangular branches, glabrous calyx and the long cilia of the staminal-scales. Branches and branchlets opposite, curved. Petioles in the lower or larger leaves half inch long, in the upper shorter. Larger leaflets 12–14 lines long, ten lines wide, very oblique at base; upper 4–6 lines long, 3–4 wide. Stipules ovato-lanceolate, one line long. Flowers 5-fid, yellow, little smaller than in the following. Petals narrower and not lacerate at the point. Stamens shorter than the petals. Style subulate.

23. Z. foetidum (Schrad. and Wendl. Sert. Han. p. 17. t. 9); suffruticose; branches herbaceous, *the primary terete, scarcely striolate;* leaves on long petioles, leaflets obovate, obtuse, oblique at base; *calyx pubescent,* sepals rounded, obtuse; petals broadly obovate, *inciso-dentate;* filaments 3–4 times longer than the *oblong, ciliato-serrate* scales; ovary sub-globose, *sub-sulcate;* capsules oval-sub-rotund, 5-angled. *DC. Prod.* 1. *p.* 705. *Z. Fabago, Thunb. Cap. p.* 543, *non L. Z. insuave Sims, Bot. Mag. t.* 372 (a form with narrower petals). *Z. retrofractum E. & Z.!* 755 *non Thunb.* Z. *Lichtensteinianum, E. & Z.!* 754, *ex pte.*

HAB. Interior regions, *Thunberg.* Among shrubs at the Zwartkop's River, Uit. and Gauritz R., Swell., *E. & Z.!* Stony places between Zwarteberg and Kendo, 2000-3000f. *Drege.* Aug.-Dec. (Herb. Thunb., Sond., T.C.D., Lehm.)

A small shrub, 2–4 feet high, mostly with elongated branches and branchlets. Petioles ½ inch long, keeled, with a lanceolate, finally reflexed point. Larger leaflets 1–1½ inch long, 6–10 lines wide, obliquely narrowed at base, penninerved, the midrib thickened, obtuse or mucronulate, upper shorter, 4–6 lines long, 3 lines wide. Stipules ovate, acuminate, about 1 line long. Peduncles nodding, 6 lines long. Sepals 2 lines long, pubescent at each side, at length reflexed, persistent. Petals four times longer than the calyx, orange yellow, with a purple spot at base; in *Z. insuave,* with a violet cordate spot. Filaments rather shorter than the petals. Squamules reflexed at margin, of one colour, unequally ciliato-serrate. Capsules 4–5 lines long and wide, obtuse, shortly apiculate, brownish, obtusely angled.

24. Z. retrofractum (Thunb. Fl. Cap. p. 545); suffruticose; branches alternate, twigs articulate, sub-secund, flexuous, terete; leaves shortly petiolate; leaflets (minute) obovate, carnose; *peduncles axillary, shorter than the leaf;* sepals oblong; petals *little longer than the calyx;* filaments *thrice as long as the hyaline, deeply bifid scales;* capsule spheroidal, 4–5-winged. *Z. retrofractum, E. Mey. Hb. Drege. Z. horridum, E. & Z.!* 772. *Z. microcarpum, E. & Z.!* 773.

HAB. Karroo, below Bockland, *Thunberg.* Uitflugt, between Limoenfontein, Brackvalei and Buffelrivier, 3000-4000f., and in the Nieuweveld, *Drege!* Between Kochmanskloof and Gauritz River, Kannaland, and at Olifant's R., Clanw. *E. & Z.!* Nov. Dec. (Herb. Thunb., E. Mey., T.C.D., Sond.).

A low-growing (decumbent, *Thumb.*), rigid, squarrose, greyish or yellowish undershrub; the ultimate ramuli 1–2 inches long. Internodes 3–4 lines long. Petioles nearly as long as the leaflets. Leaflets 2–3 lines long. Peduncles shorter than the calyx. Calyx one line long. Petals clawed, equalling the filaments. Segments of the minute staminal scales obovate, obtuse. Ripe capsule 1½ line long and broad, tipped with a short style; wings glabrous, one-half line wide. A remarkable species, having sometimes *simple*, obovate, narrow-based leaves, intermixed with the usual bifoliolate ones.

25. Z. Dregeanum (Sond.); suffruticose; branchlets terete; leaves petiolate, bifoliolate, the floral mostly simple; leaflets (rather large) obovate, obtuse, somewhat fleshy; *flowers cymose;* petals narrow, clawed, *twice as long as the oblong sepals;* filaments *six times as long* as the minute, hyaline, bifid scales; ovary 4–5-winged.

HAB. On a mountain at Trado, 2000f. *Drege!* 7164. (Herb. Sond.).

Branch 6–8 inches long, whitish, with sub-secund branchlets, compressed toward the upper end, the lower 2–3 inches long, the upper gradually shorter. Leaves at the base of the branchlets petiolate, the petiole equalling the leaf. The larger leaves in our specimens are ½ inch long, 4 lines wide; some smaller. Stipules linear subulate, hyaline. Branchlets alternately divided, with 2–4 ramelli, opposite the leaves, converted into 3–8-flowered cymules. Bract-leaves of the cymules smaller than the rest, simple, obovate-oblong. Pedicels 1½–2 lines long. Sepals 1 line long. Petals whitish, oblong. Filaments equalling the petals. Segments of the staminal scales obovate, entire. Ovary spheroidal, tipped with a filiform style.

V. SEETZENIA, R. Br.

Calyx 5-parted, with valvate æstivation. *Corolla* none. *Stamens* 5, hypogynous, opposite the calyx-segments; filaments subulate, naked; anthers introrse, 2-celled, sub-globose-didymous, longitudinally dehiscing. Ovary sessile, oblong, 5-celled. Ovules solitary, pendulous, axile, anatropous. *Styles* 5, terete, reflexed, with capitate stigmas. *Capsule* ovoid, penta-coccous; the cocci separating from a central 5-angled axis, each with a narrow, linear, dorsal sarcocarp, and brittle endocarp, otherwise nude, one-seeded. Seeds inverse, oval, compressed, with a crustaceous testa, inside a mucous investing layer. Embryo lying in thin albumen, orthotropous; with thickish cotyledons and a minute, superior radicle. *R. Br. in Denh., Oud. and Clapp. Narr. p.* 231. *Endl. Gen.* 6042.

A woolly or glabrous suffrutex, native of Southern Africa. Branches jointed. Leaves opposite trifoliolate, the leaflets flat, apiculate, the terminal larger, obovate. Stipules intrapetiolar. Peduncles axillary, one-flowered, pendulous in fruit. Name in honour of *Seetzen*, a meritorious African traveller.

1. **S. africana** (R. Brown, l. c.); *Zygophyllum prostratum, Thunb. Cap. p.* 543, *S. prostrata, E. & Z.!* 774.

HAB. Interior provinces, *Masson*. Sandy ground at Heerelogement, Clanw., *E. & Z.!* Sept. (Herb. Thunb., T.C.D., Sond.).

Stem 1–2 feet long, knee-bent, as stick as a pigeon's quill, terete, papilloso-scabrid, as well as the leaves. Branches and branchlets divaricate. Internodes ½ inch long in the stem, shorter in the branches and ramuli. Leaves petiolate, the petiole 1 line long, equalling the petiolulate leaflets. Peduncles longer than the leaves, ½ inch long, nodding. Calyx 3 lines long, the segments 1 line wide. Stamens as long as the calyx or longer. Anthers versatile. Styles 5, rarely 6–7. Capsule broadly ovoid, 3 lines long, of 5, rarely 6–7 carpels: carpels obtuse and green at back. Seeds 2 lines long.

ORDER XXXVI. MELIANTHEÆ, Planch.

(By W. SONDER.)

(Planch. Trans. Linn. Soc. Vol. xx. p. 414. Melianthaceæ, Endl. Gen. Suppl. iv. p. 80. Zygophylleæ spuriæ, alternifoliæ, DC. Prod. 1. p. 707. Zygoph. Tribe 3, Melianthеæ, Harv. Gen. p. 48.)

Flowers perfect (or sometimes polygamous?), more or less irregular. *Calyx* 5 parted, the odd segment posterior, all quincuncially imbricate in æstivation. *Petals* 4–5, alternate with the calyx-lobes, clawed, naked, or having on the claw minute fleshy tubercles. *Stamens* 4–5, alternate with the petals; filaments thick; anthers fixed at the back above the base, 2-celled, cells adnate to a dorsal connective, introrse, slitting. *Disc* between the petals and stamens, either horse-shoe-shaped or imperfectly annular. *Ovary* 4–5-celled; ovules 2–4 in each cell, in two rows, affixed above the middle of the inner angle, horizontal or ascending, or solitary and ascending from the base of the inner angle of the loculus. *Style* subulate, incrassated, with 4–5 stigmatic lobes or teeth. *Capsule* loculicidally 4–5-valved (the dehiscence, however, not constantly along the dorsal suture of the carpel.) *Seeds* mostly solitary, with a crustaceous testa, and copious horny albumen. *Radicle* of the straight, axile embryo near the hilum, linear-clavate, longer than the linear-elliptical, plano-convex, face-to-face cotyledons. *Pl. l. c.*

Evergreen shrubs, with naked buds. Leaves alternate or sub-opposite, unequally pinnate; the petiole often winged or margined between the leaflets. Stipules two, often concrete in one intra-axillary stipule, sometimes lateral and free. Racemes terminal and axillary.

Tribe I. EU-MELIANTHEÆ, Planch.

Flowers resupinate. *Calyx* large, conspicuously irregular, the segments separate. *Petals* shorter than the calyx, evidently perigynous, the claws of the four uppermost (i.e., the two posterior and two lateral) connivent, and cohæring by means of their woolly covering, but not truly concreted. *Filaments* of the two (strictly *lateral*, but seemingly *posterior*) stamens, together with the rudiment of the abortive or obsolete posterior stamen concrete at base. *Ovary* 4-celled, the cells 2–4-ovuled; ovules attached above the middle of the cell. Stigmatic teeth minute. *Capsule* papery or leathery, mostly 4-winged; the carpels 2–1 seeded. Seeds exarillate. *Racemes bracteate. Flowers nodding. Odour of the bruised foliage fœtid.*

I. **MELIANTHUS**. Tournef.

Calyx laterally compressed, with or without a saccate gibbosity at base, and furnished within with a nectariferous gland. *Petals* 5, the anterior one abortive. *Stamens* 4, didynamous. *Endl. Gen.* 6043. *(Melianthus and Diplerisma, Pl.).*

Large, shrubby or suffruticose, disagreeably scented plants, natives of South Africa and Nepaul. Leaves alternate, stipulate, imparipinnate; leaflets unequal sided, toothed. Flowers in axillary or terminal racemes, secreting honey in abundance, whence the generic name, from *μελι, honey,* **and** *ανθος,* **a** *flower.*

***** ***Stipules*** **concrete into one large, intra-axillary piece, attached to the lower part of the petiole.** ***Calyx*** **gibbous at base.** ***Capsule*** **papery, four-lobed at the apex.** *(Melianthus, Pl. l. c.)*

1. M. major (Linn. Sp. p. 892); glaucous, glabrous; the raceme sometimes downy; leaves coarsely serrate. *Thunb. Fl. Cap. p.* 489. *Lam. Ill. t.* 552. *DC. Prod.* 1. *p.* 708. *Bot. Reg. t.* 45. *Juss. Mem. Mus. xii. p.* 28. *n.* 48. *icon. E. & Z.!* 775. *Planch. l. c. p.* 416. *t. xx. fig.* 4-13. *Pappe, Fl. Cap. Med. Prod. p.* 6.

HAB. Generally, throughout the Colony. Aug.-Dec. (Herb. Sond., T.C.D.).

Stem shrubby, flexuous, glabrous, several feet long, with a widely-creeping root. Leaves a foot long or longer, the upper cauline ones smaller. Stipules acuminate from a cordate base, several inches long. Petioles naked at base, with cuneate wings between the leaflets. Leaflets 9–11, paler underneath, many veined, 3–4 inches long, two inches wide, acutely and often doubly serrate, the teeth pointed. Racemes simple, densely flowered, a foot or more in length, glabrous or pubescent. Bracts ovate, acuminate, equalling the flower-stalks, shorter than the fruit-stalks. Flowers brown-red, inch long. Upper calyx-segments produced at base into a very short, obtuse calcar, rarely ecalcarate, and having instead a very wide, obtuse, and little-conspicuous gibbosity. Claws of the petals pubescent. Capsule 1–1¼ inch long, membranous, netted-veined, glabrous, the valves compressed, four-winged, girt at base with the persistent calyx, deeply emarginate and four-lobed at the apex, the lobes opening by their inner sutures. Seeds two in each cell, black and shining.

****** ***Stipules*** **2, subulate, lateral, free.** ***Calyx*** **not conspicuously gibbous at base.** ***Capsule*** **obtuse at each end, scarcely four-lobed.** *(Diplerisma Pl. l. c.* 416.)

2. M. comosus (Vahl, Symb. iii. p. 86); leaflets lanceolate, serrate, villoso-pubescent above, with stellate hairs, at length glabrate, albo-tomentose underneath; stipules lanceolate-subulate; *racemes nodding, the flowers alternate;* bracts *cordate-acuminate;* upper calyx-segment lanceolate, acuminate; petals oblong or spathulate, acute; capsule membranous, canescent, oval, four-winged. *Commel. rar. t.* 4. *M. minor, Houtt. Linn. Pfl. Syst. iv. p.* 108. *Bot. Mag. t.* 301. *Thunb. Cap. p.* 489. *E. & Z.!* 777. *Diplerisma comosum, Pl. l. c. p.* 416. *Tab. xx. fig.* 16-19. *D. minor, Pl. fig.* 20. *(fruit).*

HAB. Near Langvallei and the Karroo below Bokkeveld and elsewhere. ***Thunberg.*** **Banks of the Camtour's River, Uit.; in the Karroo, at Gauritz R., George, and mts. near Philipstown, Cafr.,** ***E. & Z.!*** **mountain near Graaf Reynet, 3–4000f.**

Drege! 7176., and on Los Tafelberg, 6–7000f. *Drege!* 7177! Zuureberg, *Burke.* Oct.-Dec. (Herb. Sond., T.C.D.).

A shrub with round, alternate, greyish-white branches; branchlets canescent or villous. Leaves 4–6 inches long. Stipules six lines long, semi-cordate at base. Petiole winged between the leaflets. Leaflets 1½–2½ inches long, 6–8 lines wide. Racemes frequent between the leaves on the branches, the uppermost sometimes axillary, three inches long. Flowers green at base, the sepals and petals orange-yellow within, the larger marked externally with a red spot. Bracts wide, canescent or villous, 6–8 lines long, longer than the flower-stalks, equalling the fruit-stalks. Calyx canescent, the two lower segments ovate-acuminate, eight lines long, the upper segment lanceolate, ½ inch long, lateral linear. Petals 4, equalling the calyx, connate at the claws, the lamina oblong or spathulate-lanceolate. Longer stamens equalling the calyx. Capsule one inch long, downy-canescent, netted, sub-emarginate at base and apex, with compressed valves, four-winged. Style subulate, in the notch. Seeds mostly 2, sub-globose, black, shining. Embryo in fleshy or horny albumen. In a few specimens I have seen a fifth petal placed between the lower lobe of the calyx, very slender, clawed, with a lamina one line long, tapering into a three-line-long claw.

3. M. minor (L. sp. p. 892, excl. syn. Com.); leaflets lanceolate, serrate, glabrous above, albo-tomentose underneath; stipules lanceolate-subulate; *racemes erect; flowers approaching in whorls;* bracts *lanceolate-subulate-attenuate;* upper calyx segment cuspidate-acuminate; petals lanceolate-linear; capsule toughish-membranous (like parchment), whitish-tomentose, *sub-globose,* 4-*angled. Vahl, symb. iii. p.* 85. *Thunb. Fl. Cap. p.* 489. *E. & Z.!* 776. *Diplerisma minus, Pl. l. c. p.* 416. *tab. xx. fig.* 14–15 *(excl. fr.).*

HAB. Sands at Saldanha Bay, at Kange River, *Thunberg, E. & Z.!* Stony hills at Ebenezer and in Langvalei, below 1000f. *Drege!* Nov.-Jul. (Herb. Lehm., Sond.).

A shrub, very similar to the preceding. Branchlets and inflorescence rather canescent. Leaves 5–6 inches long. Stipules narrow, four lines long. Petiole winged between the leaflets. Leaflets 1½–2 inches long, 6–10 lines wide. Racemes 6–12 inches long, sub-terminal. Flowers dull red. Peduncles ⅓ inch long, bracts 8 lines. Lower calyx-segments 8 lines long, 2 lines wide at base; upper 6 lines long, linear, abruptly taper-pointed from the middle (3 toothed, the middle tooth longest); lateral segments lanceolate-subulate. Petals 4, longer than the calyx, 8 lines long, 1 line wide, clawed, the claws downy, connate, lamina glabrous; the outer above the base, on each side furnished with a linear tooth. Disc horse-shoe-shaped. Longer stamens equalling the calyx. Capsule 8 lines long, loculicidally four-valved, the valves not compressed, with a dorsal longitudinal furrow. Confounded by all the older authors and by Linnæus with the preceding, but well distinguished by Vahl, on account of the inflorescence and bracts. The capsule, hitherto undescribed, is more rigid, acutely angular, but in no respect winged.

4. M. Dregeana (Sond.); leaflets lanceolate, serrated, villous-downy above, with stellate hairs, albo-tomentose underneath; stipules lanceolate, subulate-acuminate, half-cordate at base; racemes *nodding, flowers alternate;* bracts *cordate*-acuminate; upper calyx-segment lanceolate, acuminate; petals . . . . . .; capsule *coriaceous-sub-lignescent,* tomentose, *depressed-globose, obtusely four-angled.*

HAB. Grassy places between Kachu, Geelhoutrivier and Zandplaat, 1000-2000f. *Drege!* 4473. Jan. (Herb. Sond.).

A shrub, like the preceding, with woody branches, branchlets leafy at the summit, bearing racemes between the upper leaves and in the axils of the leaves. Leaves with 3–4 pair of leaflets, the petiole winged between the leaflets, which are quite like those of *M. comosa.* Stipule ½ inch long. Racemes flexuous, 1½–2 inches long. Peduncles 6–8 lines long, the fruit-bearing a little thickened at top. Calyx as in *M. comosa*, the upper segment 6 lines long, the lower wide, lateral narrow. Rudiments only of the petals have been observed. Ovary tomentose, depressed at each end, obtusely quadrangular. Style subulate. Fruit remarkable in the genus. Capsule 4 lines long, loculicidal beyond the middle and to the very base 4-valved, the valves inflexed at the point, when dry hard, sub-ligneous, septiferous in the middle, turgid, but longitudinally impressed at back. Seeds two in each cell, rarely solitary, exarillate, sub-globose, black, smooth and shining. Embryo straight, in the midst of horny albumen.—A remarkable species, approaching *Bersama* and *Natalia* in the nature of its capsule.

Tribe II. BERSAMEÆ. Pl. l. c. p. 416.

Flowers not resupinate. *Calyx* neither large nor conspicuously irregular, the two anterior sepals more or less concreted together. *Petals* 5, longer than the calyx, the anterior narrower than the rest, all free and equidistant. *Stamens* 4–5. *Ovary* 4–5-celled; cells 1-ovuled; ovules axile, ascending from the base of the cell. Stigmatic-lobes 4–5, thick, approaching in a cone. Capsule coriaceous, loculicidally splitting into 4–5 septiferous valves. Seeds with a cup-shaped, fleshy adnate arillus. *Bracts minute. Flowers spreading or nodding.*

II. NATALIA, Hochst.

Calyx with its two anterior segments connate in one bidentate lobe, four-parted, unequal. *Petals* 5, oblong, clawed. *Gland* fleshy, half-moon-shaped. *Stamens* four, hypogynous, two anterior dilated at base and connate, the posterior free. *Ovary* bluntly four-angled, four-celled. *Style* ascending; stigma pyramidal. *Capsule* coriaceous, sub-globose, loculicidally 4-valved. *Hochst. in Flora*, 1841, *p.* 663. *Pl. in Hook. Ic. t.* 780. *Endl. Gen. suppl.* 4. *p.* 80.

Shrubs, or small trees, with alternate, imparipinnate leaves; the leaflets in two or many pair, very entire or serrulate. Stipules intra-axillary, connate in one. Flowers racemose, the pedicels subtended at base by a minute bract.

1. **N. lucens** (Hochst. l. c.); *Rhaganus lucidus, E. Mey.! in Hb. Drege.*

HAB. Woods at Port Natal, *Drege, Krauss, T. Williamson, Gueinzius.* Mar.-Apr. (Herb. Sond., Hook., T.C.D.).

A shrub, 8–10 feet high, with glabrous branches. Leaves petiolate. Leaflets in two pairs, with an odd one, minutely petiolulate, obovate, obtuse, mucronulate or retuse, quite entire, shining, sub-coriaceous, penninerved, the smaller 1½–2 inches long, an inch wide, the larger 3 inches long, 1½–1¾ wide. Racemes 3–5 inches long, minutely downy. Pedicels 3–6 lines long. Calyx 1½ lines long. Petals white, downy at both sides, 3 lines long. Capsule ½ inch long; valves thick, woody, septiferous in the middle, corrugated externally. Seeds 3 lines long, blackish, with a white arillus covering one-third of the seed.

ORDER XXXVII. RUTACEÆ, DC.

(By W. SONDER.)

(Rutæ, Juss. Gen. 296, ex pte. Rutaceæ, DC. Prod. 1. p. 709. Endl. Gen. cclii. Lindl. Veg. King. No. clxxvi. Diosmeæ, Br. in Flinder's Voy., Endl. Gen. No. ccli.

Sub-Order, DIOSMEÆ.

Flowers perfect and regular (very rarely unisexual and irregular). *Calyx* 4–5-cleft or parted, its segments imbricate in æstivation. *Petals* 4–5, (rarely wanting), separate, or united by the edges below, twisted-convolute or rarely valvate in æstivation, inserted under a hypogynous disc. *Disc* saucer-shaped or urceolate, free, or attached to the calyx, sometimes obsolete. *Stamens* hypogynous, as many or twice as many as the petals; in the latter case those opposite the petals are *sterile*, or at least feebler than the rest; anthers introrse, 2-celled, often tipped with a gland. *Ovary* of 3–5 carpels, sessile or stipitate, syncarpous or apocarpous; ovules 2, rarely 4 in each carpel, collateral or obliquely superposed; *styles* united, at least above; stigma simple or capitate. *Fruit* of 1–5 capsules or cocci, distinct or united at base, often horned below the apex, mostly one-seeded; the dry walls of the pericarp separating at maturity into an outer and inner shell. *Seeds* with hard coats, usually exalbuminous; radicle superior.

Shrubs or small trees, very rarely herbaceous. Leaves opposite or alternate, coriaceous, simple or compound, almost always pellucid-gland-dotted beneath, entire or serrulate, balsam-scented. Stipules none. Inflorescence various. A large tribe very abundant in S. Africa, but having its centre in Australia, where the generic types are much more diversified. A few outlying genera occur in S. America, and one (*Dictamnus*) in S. Europe. All are remarkable for a strong, often offensive odour and for their bitterness. These properties reside in an essential oil, contained in the transparent, glandular dots, which, in greater or less abundance, are sprinkled over the leaves, calyx, petals and seed-vessels. Several of the Cape species pass currently under the name "*Buku*"; but *Barosma crenulata* is considered to possess the medical virtues of the tribe in a stronger degree than others. *See Pappe, Fl. Med. p.* 7, 8.

TABLE OF THE SOUTH AFRICAN GENERA.

I. *Fruit syncarpous.*

I. **Calodendron.**—*Calyx* deciduous. *Flowers* perfect, with 5 petals and sepals.

II. *Fruit apocarpous.*

A. *Flowers complete, (rarely polygamous), 5-parted.*

§ *Style short. Stigma capitate.*

II. **Euchaetis.**—*Disc* adnate with the calyx. *Petals* oblong, attenuated at base and transversely bearded. *Sterile* filaments none.

III. **Diosma.**—*Disc* free above, 5-lobed. *Petals* naked, sessile. *Sterile* filaments none.

IV. **Coleonema.**—*Petals* longitudinally *channelled. Filaments* 10; 5 fertile; 5 sterile naked, infolded in the channel of the petals.

V. **Acmadenia.**—*Petals* clawed, the claws bearded within (rarely naked). *Filaments* 5 or 10; 5 sterile on the margin of the disc, filiform, or short. *Anthers* with a sessile, erect gland.

VI. **Adenandra.**—*Petals* naked, with very short claws. *Filaments* 10, shorter than the calyx; the 5 sterile filiform, tipped with a gland. *Anthers* with a stalked, finally reflexed gland.

§§ *Style as long as the petals. Stigma simple (not capitate).*

VII. **Barosma.**—*Calyx* segments equal. *Petals* sub-sessile. *Inflorescence* axillary *Leaves* almost always opposite.

VIII. **Agathosma.**—*Calyx* segments unequal. *Petals* with long claws, rarely sub-sessile. *Infl.* terminal. *Leaves* alternate.

§§§ *Style lengthening greatly after flowering, slender at the base. Stigma not capitate.*

IX. **Macrostylis.**—*Disc* closing over the ovary, perforated by the style.

B. *Flowers incomplete, monoecious, 4-parted.*

X. **Empleorum.**—*Petals* none.

I. CALODENDRON, Thunb.

Calyx short, 5-parted, deciduous. *Petals* 5, much longer than the calyx, oblongo-lanceolate. *Stamens* 10, inserted under a short, tubular disc, 5 fertile; 5 alternate sterile and petaloid. *Capsule* shortly stipitate, rough, 5-angled, 5-celled, septicidally 5-valved; seeds two in each cell. *Endl. Gen. No.* 6014. *DC. Prod.* 1. *p.* 712.

Only one species known. The name is derived from καλος, *beautiful*, and δενδρον, *a tree.*

1. C. capense (Thunb Nov. Gen. P. 2. p. 41, 42, 43); *Prod. p.* 44. *Fl. Cap. p.* 197. *Jus. Mem. Mus. xii.* 469. *t.* 19. *f.* 15. *Pallasia capensis, Houtt. Dictamnus capensis, Lin. f. Suppl. p.* 232.

HAB. Woods in the Eastern districts and Caffirland, extending to Port Natal, *Thunberg, E & Z. Mr. H. Hutton, &c.* Nov.-Jan. (Herb. T.C.D., Hook, Sond. &c.)

A large and handsome tree, the "Wilde Kastanien" of the colonists. Branches and branchlets opposite or in threes, divaricate, terete. Leaves decussate, petiolate, ovate, obtuse, retuse or acute, with parallel nerves, pellucid-dotted, evergreen, green above, paler underneath, 4–5 inches long. Petioles thick and short, a line long. Flowers in terminal panicles; the peduncles mostly trichotomous. Petals white, linear-oblong, reflexed, 1½ inch long, 2 lines wide, stellato-pubescent externally, sprinkled with purplish glands. Ovary pedunculate. Valves of the capsule thick and hard, 1½ inch long; seeds larger than a hazel nut, black and shining, 8 lines long.

II. EUCHAETIS, Bartl. and Wendl.

Calyx 5-parted. *Disc* adnate. *Petals* oblong at base, oblongo-lanceolate toward the point, with a transverse beard. *Filaments* 5, shorter than the calyx; sterile none; *anthers* roundish, with an adnate gland. *Style* short. *Stigma* capitate. *Fruit* of 5 cocci, shortly horned at the summit. *Endl. Gen.* 6018. *B. & W. p.* 15.

Small shrubs with *scattered*, very rarely opposite, lanceolate, keeled leaves, and *terminal*, capitate, or glomerate flowers. Name from ευ, *well*, and χαιτη, *a hair or bristle;* in allusion to the bearded petals.

1. E. glomerata (B. & W.! l. c. p. 16); stem and branches erect, rather patent, glabrous; leaves erect-appressed, *lanceolate*, sharply keeled, tipped with an incurved mucro, with a pellucid, *rigidly-ciliate* margin; bracts *ovate, sub-lanceolate,* shorter than the calyx; calyx-lobes lanceolate, petals longer than the calyx, bearded, ciliate to the base, the exserted half oblongo-lanceolate, naked. *Diosma glomerata, G. F. W., Mey. mss. Euchaetis abietina, E. & Z!* 821.

HAB. Cape, *Hesse.* Cederberg Mts., Clanw., *E. & Z.!* (Herb. Bartl., T.C.D., Sond.)

1–2 feet high, leafy, with reddish branches and yellowish twigs. Leaves 2–3 times longer than the internodes, sub-imbricate, sub-sessile, obtuse at base, lanceolate-attenuate upwards, sub-cartilaginous and acute at the point, concave, below with a double row of black dots along the keel, green at both sides, 4 lines long, 1 line wide. Heads terminal, or by innovation, spuriously lateral, smaller than a hazel nut. Calyx glabrous, pale, its lobes keeled, 1 line long. Disc almost entirely attached to the calyx, the margin 5-crenate. Petals white, rather longer than the calyx, white-bearded on the throat. Capsule glabrous, $3\frac{1}{2}$ lines long; the round-backed carpels rugulose, tricarinate at point, their horns divergent, half as long as the carpel.

2. E. elata (E. and Z.! 819); stem and branches erect, glabrous; leaves erecto-patent, *lanceolate-linear*, convexo-carinate, inflexed, *obtusely-mucronate*, the hyaline margin *ciliolate;* bracts *lanceolate*, half the length of the calyx; calyx-lobes lanceolate; petals *oblongo-lanceolate*, longer than the calyx, bearded on the face, ciliate to the base, the sub-exserted portion acuminate, naked.

VAR. β. **hirsuta**; branches, leaves, bracts and calyces hairy.

HAB. Mountain sides, Hott. Holland, Stellenb., *E. & Z.!* June. (Herb. Lehm., Hook., Sond.)

Habit of the preceding. Lower leaves sub-remote, upper closer, sub-imbricate, 4 lines long, $\frac{1}{2}$ line wide. Calyx 2 lines long. Petals scarcely narrowed at base. Style very short. *Capsule* shining, 2 lines long, the carpels scarcely beaked. This differs from the preceding by the twice narrower leaves, scarcely ciliate at margin, obsoletely dotted at the keel underneath; the heads rather laxer; bracts twice narrower, and smaller, more shortly rostrate capsules.

3. E. linearis (Sond.); stem and branches straightly erect, glabrous; leaves appressed, *narrow-linear*, acute, convexo-carinate, the hyaline margin *serrulate;* bracts lanceolate, longer than the calyx; calyx-lobes lanceolate, *subulate-attenuate;* petals *lanceolate*, narrowed at base, transversely bearded in the throat, bordered with long cilia below, the exserted half acuminate, naked. *E. glomerata. E. & Z.!* 820, *non B. & W.*

HAB. In grassy places at the Zwarteberg, near Caledon Bath, *E. & Z.! Drege*, 7116. (Herb. Lehm., Sond.)

A slender shrub, 1 foot high. Leaves crowded, but not imbricate, quite appressed, the lower 5–6 lines long, upper 3–4 lines, $\frac{1}{3}$–$\frac{1}{2}$ line broad, concave above, with 2 rows of dots beneath. Heads terminal, 8–12-flowered, the outer shortly pedicellate, equalling the leaves. Bracts green. Calyx pale rosy, 2 lines long, the lobes ciliate, naked at the tip. Throat of the petals densely bearded, the exserted half involute. Fertile stamens half as long as the petals. This differs from *E. glomerata* in the longer and narrower leaves, and the subulate-acuminate calyx.

4. E. flexilis (E. & Z.! 822); stem and branches erecto-patulous, glabrous; leaves quadriferiously imbricate, erect, *lanceolate-oblong*, convexo-carinate, *sub-obtuse*, with a minute incurved mucro, the hyaline margin ciliolate; bracts *oblong*, half as long as the *ovate, obtuse* calyx-lobes; petals rather longer than the calyx, *obovate*, bearded, ciliate to the base, the exserted half obtusely acuminate, naked.

HAB. Baviansberg, near Genadendal, Caledon, *E. & Z.!* Jul.-Aug. (Herb. Sond.)

1–2 feet high, rather more slender than the preceding, the branches sometimes bent. Filaments half as long as the calyx. Capsule shining, brown, $1\frac{1}{2}$ lines long, with a very short beak. Easily known by its smaller leaves ($1\frac{1}{2}$–2 lines long) with a wider hyaline margin, laxer inflorescence and more obtuse calyces and bracts. Leaves with a double row of glands at back.

5. E. dubia (Sond.); glabrous, 2–3-chotomous, the branches virgate, erecto-patulous; leaves close, erect, strongly appressed, oblong or oblongo-lanceolate, sub-acute, concave above, convex beneath; flowers sessile, capitato-glomerate; bracts obtuse, half as long as the ovate, obtuse calyx-lobes; petals longer than the calyx, obovate, apiculate, narrowed at base, bearded, the exserted part obtuse, apiculate.

VAR. β. **Dregeana**; lower leaves opposite, those of the upper branchlets alternate, evidently dotted at back; petals rather narrower.

VAR. γ. **pauciflora**; flowers solitary or in pairs.

HAB. Lang valley, *Zey.!* 291. β. in the Cape Flats, *Drege!* γ. Ratelklip, alt. 2. June. *Zey.!* 292. (Herb. Sond., Holm.)

2 feet high or more, branches and branchlets slender, terete, purplish. Leaves dull green, longer than the internodes, not conspicuously keeled, the lower 2–2½ lines long, ¾ line wide, upper half as long, oblong. Flowers 6–8 in a cluster, in β. few, terminal or pseudo-lateral at the ends of short ramuli. Calyx 1 line long, its segments with a widely membranous margin, ciliolate. Petals reddish, oblong-obovate, the claw not ciliate at the margin, but the disc from the middle to the throat furnished with a long, white beard. Filaments equalling the calyx; gland of the anther sessile. Disc with a free margin, 5-lobed, equalling the ovary. Style glabrous, at length twice as long as the calyx; stigma capitate. Capsule glabrous, shining, 3 lines long, the 3–5 carpels round-backed, with very short beaks. Var. β. agrees with α., except in the points above indicated. In γ. all the leaves are alternate and much shorter: otherwise, it agrees with α.

A doubtful member of the present genus, approaching *Macrostylis*, from which it differs in the stamens, the style not elongating, nor more slender at base, and in the capitate stigma. From *Diosma* it is known by the petals narrowed to the base and bearded.

III. DIOSMA, L.

Calyx 5-parted. *Hypogynous-disc* free at the margin, sub-campanulate, 5-sinuate, 5-plaited. *Petals* 5, naked, sessile. *Filaments* 5, shorter than the corolla; sterile none; anthers roundish, with an adnate gland (in *D. cupressina* sub-pedicellate). *Style* short; stigma capitate. *Capsule* longer than the calyx, 5-coccous; the cocci horned at the outer point. *Juss. l. c. p.* 472. *Endl. Gen.* 6017.

Small shrubs, with alternate or opposite, linear-acute, channelled, serrulate or sometimes ciliate, glandularly dotted leaves; white or reddish, terminal, sub-solitary or corymbose, pedicellate flowers. Pedicels bracteate. Name from διος, *divine*, and οσμη, a *smell*.

I. *All the leaves constantly opposite.* (Sp. 1–2.)

1. D. succulenta (Berg.! pl. cap. p. 63, excl. syn. pl.); branchlets opposite, pubescent; leaves thickish, opposite, quadrifariously imbricated at base, the lower lanceolate or linear, upper oblong, *complicate, keeled, somewhat 3-sided, papilloso-punctate,* ciliate; flowers 3 or more, sub-corymbose, the calyx-lobes *ovate,* acute, ciliate.

VAR. α. **Bergiana**; all the leaves oblong, triquetrous, sub-obtuse or the lower longer, lanceolate, acute. *D. succulenta, Berg. l. c. Thunb.! Cap. p.* 224. *B. & W.! p.* 25. *E. & Z.!* 839. *D. decussata, Lam.! encycl.* 2. *p.* 284. *D. rigidulum, Willd.! D. oppositifolia, R. & Sch. V. p.* 457. *excl. syn. pl.*

VAR. β. **Lamarckiana**; leaves linear, complicate, mucronate, the upper shorter, oblong. *D. scabra, Lam.! l. c.* 283. *B. & W.! l. c. p.* 29. *E. & Z.!* 838. *D. succulenta, Wendl. coll.* 1. *p.* 2. *t.* 1. *Willd. En. p.* 258. *R. & Sch. p.* 456. *D. oppositifolia, E. & Z.!* 837, *ex. pte.*

HAB. Common in the Cape Flats and throughout the Western Districts. Nov. Apl. (Hb. Berg., Thunb., Lam., Sond., T.C.D., &c.)

1–2 feet high, erect, more or less branching. Leaves very rarely and in the uppermost branches only, alternate, erecto-patent, minutely petiolate, dull green, scabrid, sometimes smoothish, deeply channelled above, 2–4 lines long in α., in β. the lower 8–12 lines, the upper 2–4 lines long. Peduncles shorter than the calyx, glabrous, bibracteolate. Calyx 1 line long. Petals about twice as long as the calyx, elliptical, white or reddish. Ripe capsule 4–5 lines long; the 4–5 carpels subcompressed, glabrous, obtusely and shortly horned. D. succulenta β. and γ. of E. & Z.! are starved specimens, with small leaves.

2. D. cupressina (L. Mant. p. 50 & 343); much-branched, the branchlets glabrous or minutely downy; leaves sessile, opposite, erect, *quadrifariously imbricate, oblongo-lanceolate,* or the upper ovate, *convexo-carinate, acute, glabrous,* minutely ciliate; flowers terminal, sub-solitary, sessile; calyx-lobes *lanceolate;* petals oblong, acuminate. *Thunb. Fl. Cap. p.* 225. *Wendl.! coll.* 11. *p.* 59. *t.* 61. *B. & W.! p.* 50. *E. & Z.!* 853. *Brunia uniflora, Lin. Hort. Clif. p.* 71. *D. dichotoma, Berg.! Cap. p.* 63.

HAB. Cape Flats and Hott. Holland, *Thunberg, E. & Z.!* Klipfontein, *Zey.!* 290. Nov.-Apl. (Herb. Wendl., Lehm., T.C.D., Mart., Sond.)

Erect, 1 foot or more high. Branches ternate or quaternate, slender, di-tri-chotomous, the ultimate short. Younger leaves close, adult sub-remote, with the apex spreading or sub-recurved, underneath bifariously multiglandular, 1–2 lines long. Flowers solitary or 2–3–4. Calyx $1\frac{1}{2}$ line long, its lobes ciliolate. Petals at length sub-involute, twice as long as the calyx. Fil. glabrous. *D. cupressina* of Herb. Lam. is a *Berzelia!*

2. *Leaves alternate; occasionally some are opposite.* (Sp. 3–11).

(α). *Leaves narrow, very acute or mucronate.* (Sp. 3–6.)

3. D. vulgaris (Schl. Linn. 6. p. 201); branchlets minutely pubescent; leaves scattered, rarely opposite, linear, *convexo-carinate, subulate-acuminate,* serrulato-scabrous or ciliate at the margin; flowers subcorymbose; calyx-lobes sub-obtuse, acute or acuminate, with membranous, *ciliate* margins; petals obtuse or sub-acute.

VAR. α.; leaves opposite or alternate; calyx acuminate; petals subacute. Leaves mostly opposite, *D. oppositifolia, Lin. Sp. p.* 286. *B. & W.! l. c. p.* 31, *excl. syn. Thunb. D. subulata, Wendl. Coll.* 1. *p.* 31. *t.* 8. *D. pinifolia, Fisch.* Leaves mostly alternate, *D. sp. Zey.!* 2162.

VAR. β.; leaves mostly opposite; calyx acute; petals rather obtuse. *D. oppositifolia, E. & Z.!* 837.

VAR. γ. **rubra**; leaves scattered, more rigid, erect, often sub-appressed, branches fastigiate, corymbose; calyx sub-obtuse. Peduncles nearly glabrous; flowers reddish, *D. rubra, Lin. Sp.* 287. *Thunb.! Fl. Cap. p.* 221, *Willd.! Sp.* 1. *p.* 1134, *excl. syn. Berg. Bot. Reg. t.* 563. *B. & W.! p.* 40. *D. ericifolia, Andr. Bot. Rep. t.* 541. *D. ericoides, E. & Z.!* 842. *Coleonema album, E. Mey! in Hb. Drege. and Sieb. Fl. Cap.* 61. *Fl. Mixt.* 86. Peduncles pubescent; flowers white. *D. ambigua, B. & W.! p.* 42. *E. Mey.! D. rubra, E. & Z.! ex pte.*

VAR. δ. **longifolia**; branches virgate, often flaccid; leaves mostly longer, narrow, erect or patulous, glabrous; the upper mostly ciliate; flowers white; capsules mostly shortly horned. Leaves erect, rigid, *D. ericoides, Sims, Bot. Mag.* 2332. *D. Simsii and D. longifolia, E. & Z.! ex pte.*, 847, 846. *D. hirsuta, E. & Z.!* 843. *D. glabrata. E. Mey!* Leaves spreading, flaccid, *D. pectinata, Thunb.! Cap.* 222. *D. longifolia, Wendl.! Coll.* 1. *p.* 61, *t.* 19. *B. & W. l. c. p.* 43. *D. rubra, Lam. excl. syn. D. tenuifolia, Willd.! . R. & Sch.!* 454. *D. rubra and ericoides, E. & Z! ex pte.*

VAR. ε. **hirsuta**; villous, at length sub-glabrous. *D. hirsuta, Linn. Sp.* 286. *Thunb.! Cap. p.* 222. *Willd.! Sp.* 1. *p.* 1134. *B. & W. p.* 38.

HAB. In stony ground and on hills throughout the Colony, common, flowering throughout the year. (Herb. Thunb., Lam., Sond., T.C.D., &c.)

1–2 or more feet high, erect, much-branched. Leaves ½–1 inch long, paler and bifariously punctate underneath, with a very narrow, diaphanous margin. Cal.-lobes ½–1½ line long. Petals about twice as long as the calyx, elliptical, white, or red on the outside. Capsule 3–4 lines long, the carpels compressed, punctate at back, with shorter or longer, straight or hooked, obtuse horns.

4. **D. aspalathoides** (Lam.! l. c. p. 286); branchlets minutely downy; leaves alternate, erect, sub-appressed, linear or oblong-linear, *carinato-trigonous*, glabrous, *with a recurved mucro*, serrulato-scabrous at the margin; flowers sub-corymbose, the peduncles and calyx *quite glabrous*; petals obtuse. *R. & Sch. l. c. p.* 455. *D. ericoides, Lam.! l. c. p.* 285. *D. glabrata, Meyer. B. & W. l. c. p.* 34. *E. & Z.!* 850. *E. Mey.! Hb. Drege. D. ambigua, E. & Z.!* 844. *D. acmæophylla, E. & Z.!* 849. *D. depressa, E. & Z.!* 851.

VAR. β.; leaves longer. *D. linearis, E. & Z.!* 845, *non Thunb. and D. longifolia, E & Z.!* 846, *ex pte.*

HAB. Cape Flats, and many places in the Western Districts. β. near Tulbagh. (Herb. Lam., Wendl., Lehm., T.C.D., Sond.)

Six inches to 1 foot high, rigid. Leaves 3–4 lines, in β. 4–6 lines long, the mucro evident in the upper, often deficient in the lower leaves. Flowers mostly larger than in *D. vulgaris*. Petals white. Capsule glabrous; the carpels sub-compressed, gland-dotted on the back, transversely striate at the sides, 4 lines long, tipped with patent, obtuse horns. Known from the last by its dwarfer habit, shorter and erect leaves with a recurved point, and calyx not ciliate at edge. Perhaps a mere variety?

5. **D. Eckloniana** (Sond.); much-branched, branches and branchlets *quite glabrous*; leaves alternate, *spreading*, linear, concave above, *convex but not keeled* below, with very entire margins, *glabrous*, tipped *with a straight, pungent mucro*; flowers few, crowded, terminal and axillary; peduncles and the acute calyx quite glabrous; petals obtuse.

HAB. Dry hills near Zonderende R., Swell., *Ecklon*. Sep. (Herb. Sond.)

Some feet high, with many alternate branches. Leaves rigid, often horizontally patent, 4–8 lines long, ½ line wide, with one or two rows of glandular dots, and a slender mucro. Flowers 2–3, on minute pedicels. Calyx-lobes broadly ovate, acute, ½ line long. Petals twice as long, elliptic-oblong, purplish. Disc 5-lobed. Unripe capsule 3 lines long, glabrous, with obtuse horns 1½ lines long.

6. **D. virgata** (G. F. W. Mey. ap. B. & W.! p. 46); branches and branchlets erect, slender, quite glabrous; leaves alternate, *straight, appressed*, linear, mucronate, convex beneath, with a narrow *diaphanous*,

serrulato-scabrid or ciliolate margin; flowers sub-glomerate or subsessile; calyx acute, ciliolate; petals *acute*. *E. & Z.!* 848.

HAB. Cape, *Hesse!* Mountains near Brackfontein, Clanw., *E. & Z.* Langevalei, *Pappe.* Jul. (Hb. Wendl., Lehm., T.C.D., Sond.)

Very slender, a foot or more high, with filiform, pale or reddish branches. Leaves crowded, the upper sub-imbricate, narrow, glabrous, obtusely channelled above, bifariously dotted below, 3–6 lines long, minutely petiolate. Flowers 2–3 or more, shortly pedicellate at the end of the branchlets: the bracts keeled, ciliate, scarcely a line long. Calyx a line long, with keeled segments. Petals twice as long, obovate-elliptical, narrowed at base, sub-carinate, acute, white. Stamens about equalling the calyx. Unripe capsule 3 lines long, glabrous, with filiform, obtuse, incurved horns. Habit nearly that of a *Passerina.*

(b.) *Leaves shorter, obtuse.* (Sp. 7–11).

7. D. ramosissima (B. & W.! l. c. p. 48, excl. syn. Lam.); branchlets downy; leaves alternate, sessile, *straight,* imbricate, *linear, 3-angled, very obtuse,* convexo-carinate, glabrous; flowers terminal, sub-solitary, sessile; calyx-lobes ovate, obtuse; petals obtuse. *Pluk. Almag.* 136. *t.* 279. *f.* 2.

HAB. Cape, *Hesse!* (Herb. Wendl.)

A foot or more in height, much-branched; branches mostly ternate or quaternate, filiform, glabrous, the ultimate branchlets shortened, sub-umbellate, fastigiate. Leaves rather remote on the branches, crowded on the twigs, erect-appressed, flat and furrowed above, obtusely keeled, bifariously dotted, glabrous or (under a magnifying power) minutely scaberulous, 1–2 lines long. Flowers solitary or twin, minute. Cal. lobes ½ line long. Petals white or reddish, elliptical, a line long. Fil. glabrous. Capsule 4–5 lines long, the carpels with thick, obtuse horns. *D. ramosissima* does not exist in Herb. E. & Z.; the habitat given in their Enumeratio must therefore be erased.

8. D. teretifolia (Link. enum. p. 237); much-branched, with downy twigs; leaves opposite or alternate, petiolate, crowded, *spreading, terete, obtuse,* pubescent, hispidulous, narrow-furrowed above; flowers terminal, solitary, or few together, sessile; calyx-lobes ovate, obtuse; petals obovate, obtuse, narrowed at base. *B. & W. l. c. p.* 206. *D. ferulacea, Hort. Kew. Agathosma? teretifolia, Don.*

VAR. β. **glabrata**; twigs and leaves glabrous or nearly so. *Acmadenia obtusata, E. & Z.!* 827, *excl. syn.*

HAB. In the Interior, *Niven.* Dutoit's Kloof, 2–3000f. *Drege!* β. Winterhoeksberg near Tulbagh, Worcester, *E. & Z.!* Sep.–Jan. (Herb. Sond.)

A foot or more high, with leafless branches, and di-tri-chotomous, terete twigs, the ultimate short and leafy. Leaves densely set, 4–6-farious, shortly petiolate, with involute margins, callous-tipped and very obtuse, sparingly glandulous below, 2–3 lines long, ½ line wide. Flowers solitary, or 2–4, clustered. Disc free, lobed. Calyx covered by the upper leaves, downy, nearly 2 lines long white-edged. Petals white, 3 lines long, narrowed into a very short claw. Stamens included, 5 fertile filaments subulate, glabrous; anthers with a minute sessile gland; the place of the sterile filaments occupied by small sessile glands. Style glabrous; stigma capitellate. Fruit unknown.

9. D. ericoides (L. Sp. p. 287); much-branched, the branches and twigs *quite glabrous;* leaves alternate, crowded, *recurvo-patent, oblong, obtuse,* keeled, pointless, *glabrous;* flowers terminal, 2–3 together, very shortly pedicellate; calyx-lobes ovate, obtuse, ciliolate; petals elliptic-oblong, obtuse. *Berg. ! Cap. p.* 65. *Thunb. ! prodr. p.* 43. *Fl. Cap. p.* 223. *B. & W. ! l. c. p.* 36. *D. recurva, Cham. & Schl. ! Linn.* 5. *p.* 51. *D. thyrsophora, E & Z.!* 840.

HAB. Cape, *Thunberg*. Cango, *Mundt & Maire*. Heathy mountain-sides near Riv. Zonderende, Swell., *E. & Z.!* Feb.-Mar. (Herb. Thunb., Berl., Lehm., Sond.)

1–2 feet high, sub-dichotomous below, branches erect, leafy, twigs short, mostly crowded. Leaves erect, appressed at base, recurved at point, the upper sub-imbricate, keeled, somewhat 3-angled, concave and nerveless above, bifariously dotted below, 2–2½ lines long, ½–1 line wide, thickish, with entire margins. Peduncles 2–3, rarely 4–9, in a crowded cyme. Calyx turbinate, 1½ lines long, calyx-lobes broadly ovate, sub-acute. Petals sub-concave, elliptic-oblong, or obovate, narrowed into a short claw, twice as long as the calyx, reddish, often with a few hairs on the disc, and very short marginal cilia. Disc free, 5-lobed.

10. D. obtusifolia (Sond.); much-branched, branches glabrous, twigs erecto-patent, *downy ;* leaves sessile, either opposite, ternate or alternate, imbricate, appressed at base, recurved at apex, the lower *oblong-lanceolate*, upper *ovate-obtuse*, furrowed above, keeled below, *the margin and keel rigidly ciliolate ;* flowers terminal, *solitary,* sessile ; calyx-lobes broadly ovate, obtuse, keeled, ciliate ; petals ovate, obtuse.

HAB. S. Africa, *Drege,* 7114. (Herb. Sond.)

A shrub, with the habit of the preceding. Branches terete ; ultimate twigs ½–1 inch long. Leaves smooth above, with 2–4 rows of glands beneath, the lower more distant, 2–3 lines long, a line wide, upper a line long. Cal. 1 line long. Petals 2 lines, with a minute claw, reddish. Unripe capsule 2 lines long, with obtuse horns equally long.—Allied to *D. ericoides* and *D. cupressina ;* from the first it differs by the downy twigs, more appressed leaves, the upper shorter, ovate, sub-trigonous, and ciliate at back and margin ; from the latter by the thicker, blunt-backed, ciliate leaves, the calyx and petals. From *D. ramosissima* it differs by the differently shaped leaves and capsule.

11. D. passerinoides (Steud. ! Flora, 1830, p. 549) ; much-branched, the branches patent ; *twigs, leaves, calyx, and capsule minutely velvetty ;* leaves alternate, sessile, erect, quadrifarious, sub-imbricate, ovate ; petals elliptical, obtuse, or apiculate.

HAB. Mountains near Basson, *Baron v. Ludwig.* Zwarteberg, at Vrolyk, 3–4000f. *Drege!* Aug. (Herb. Sond., reg. Stuttgardt.)

A robust, greyish, erect shrub, seemingly several feet high, with sub-ternate, terete branches, the ultimate twigs uncial or shorter. Leaves 1 line long, dull green, few-glanded underneath, the lower ones sub-distant, but always as long as the internodes. Flowers always terminal. Calyx 1 line long, minutely bracteate at base, 5-cleft to the middle, the segments broadly ovate, acute, spreading. Disc lobed, the lobes equalling the ovary. Petals reddish, twice as long as the calyx or longer. Ovary hairy. Style about as long as the calyx. Capsule 4 lines long, the beaks of the carpels compressed, erecto-patent, 1 line long.

Diosma squalida, E. Mey ! in Hb. Drege is a *Bruniacea.*

IV. COLEONEMA, B. & W.

Calyx 5-parted. *Hypogynous-disc* slightly free at the edge. *Petals* 5, tapering and longitudinally channelled below. *Filaments* ten ; the 5 fertile equalling the sepals, the sterile filiform, naked, *hid in the channels of the petals.* *Anthers* roundish, with an adnate gland. *Style* short ; stigma capitate. *Capsule* 5-coccous, the cocci shortly horned at the outer apical angle, compressed, rough with dots. *B. & W. l. c. p.* 55. *Juss. l. c. p.* 471. *Endl. Gen.* 6016.

Shrubs, with scattered, linear leaves. Flowers axillary, solitary, the peduncles caliculate with several appressed bracts. Name from *κολεος*, a *sheath*, and *νημα*, a *filament;* the barren filaments are sheathed by the channelled petals.

A. *Flowers white.* (Sp. 1–3.)

1. C. album (B. & W. ! l. c. p. 56); much-branched, the twigs very thinly pubescent; leaves crowded, sub-erect, linear-lanceolate, or linear, channelled above, convex-carinate beneath, *with a straight, pungent mucro,* the diaphanous margin serrulate; bracts and calyx-lobes *ovate, obtuse or acute,* pubescent-ciliate at the edges; petals spathulate, roundish-obtuse, shortly apiculate. *E. & Z. !* 831. *D. rubra, Berg. Cap. p.* 62. *excl. syn. D. alba, Thunb. Prod. p.* 84. *Fl. Cap. p.* 221. *Willd. ! Sp. Pl.* 1134. *Hb. Un. It.* 239. *Adenandra alba, R. & Sch. Drege, No.* 7138, 7139.

HAB. Common on hills and mountain sides round Capetown, and through the Cape and Stellenbosch districts. (Herb. Thunb., Sond., T.C.D., &c.)

One or more feet high, erect, nearly glabrous, the branches erect or ascending, twigs leafy. Leaves about 6 lines long, ½ line wide, bi-quadri-fariously pellucid-dotted beneath. Flowers white, axillary, solitary on very short branchlets, sub-racemose, rarely crowded at the tips. Bracts 6–8, about a line long, ovato-lanceolate, acute, half as long as the calyx or longer. Petals more than twice as long as the calyx, the claw equalling the calyx, the limb spreading. Sterile filaments sub-acute, not gland-tipped. Capsule thrice as long as the calyx, the carpels acutely carinate, rough-dotted, with short, straight or slightly patent horns, ½ line long.

2. C. aspalathoides (Juss. l. c. p. 471); much-branched; twigs very thinly *pubescent;* leaves rather crowded, sub-erect, linear, keeled and sub-triangular, *with a recurved mucro,* the diaphanous margin serrulate; bracts and calyx-lobes *acuminate;* petals oblongo-spathulate, *acuminate.*

HAB. In the Karroo, at Rietkail, Swell., *Zey. !* 2159. Oct. Interior regions, *Niven.* (Herb. Sond.)

6 inches to 3 feet high. Leaves 3–3½ lines long, ⅓ line wide. Flowers axillary or terminal, as large as in *C. album,* from which this species is known by its more slender branches, smaller and narrower leaves, with recurved points, rather narrower and more acuminate petals, and smaller capsules, not twice as long as the calyx.

3. C. juniperinum (Sond.); much-branched, branches and twigs slender, virgate, *glabrous;* leaves sub-erect, narrow-linear, with *a short, straight mucro,* concave above, *convex* underneath, the diaphanous margin scarcely scabrid; bracts minute, oblong; calyx-lobes *broadly ovate, apiculate;* petals *obovate, obtuse,* or sub-apiculate, with very short claws. *Diosma juniperina, Spreg. Herb. Zey.* 397. *D. virgata, E. Mey.! D. Meyeriana, Steud. Coleonema juniperifolium and C. virginianum, E. & Z.!* 834, 835.

VAR. β.; calyx rather longer.

HAB. Hott. Hollandsberg and Howhoek, Stell.; and Zwarteberg and Klynriviersberg, Caledon, and near Tulbagh, Worcester, *E. & Z.! Zey. !* 2150, *ex pte.* β. Klyn-Howhoek, *Zey !* 2150, *ex pte.* Aug.-Dec. (Herb. Lehm., Sond., T.C.D.).

A small shrub, 1–2 feet high, shining, branched from the base. Flowers very shortly pedunculate, half as long as the leaves. Sterile filaments thickish, and shorter than in the other species.—Known by its long, filiform branches and twigs, narrow and short leaves (3–4 lines long, scarcely ⅟₄ wide) and chiefly by the smaller flowers, about 1 line long.

B. *Flowers red.*

4. C. pulchrum (Hook. Bot. Mag. t. 3340); much-branched, branches virgate, very thinly pubescent; leaves erecto-patent, patent, or recurved, linear or linear-subulate, with a straight pungent mucro, flattish above, convexo-carinate beneath, the diaphanous margin serrulate; bracts and calyx-lobes acuminate, ciliate; petals spathulate, apiculate, or sub-acuminate. *E. & Z.!* 836. *C. album, β. virgatum, and γ. gracile, Schl. Lin.* 6. *p.* 199. *C. virgatum, E. & Z.!* 832. *C. gracile, E. & Z.!* 833. *D. calycina, Steud. D. oppositifolia, E. Mey.! Hb. Drege. D. tenuifolia, Presl. Col. Dregeanum, Presl. l. c. Drege,* 2251, 7143.

HAB. Hills near Bethelsdorp, and in the flats, under the Vanstaadensberg; at Cape Recief, Algoa Bay; and in Swellendam District, *E. & Z.! Drege! Zey.!* 2151. Jul.-Nov. (Herb. Holm. (*Sparmann*), Lehm., Sond., T.C.D.)

Leaves in some specimens ½ inch, in others 1½ inch long, ¼ line wide. Bracts mostly as long as the calyx, subulate, narrowed. Petals clawed, the claw equalling the stamens. Capsule twice as long as the calyx, the carpels rough-cast at back, horns short, straight, obtuse.—Mostly taller than *C. album*, with leaves often longer and narrower, flowers mostly larger, more numerous, racemose, and beautifully red, (drying whitish).

V. ACMADENIA, B. & W.

Calyx 5-parted. *Hypogynous-disc* slightly free at the edge. *Petals* clawed, the claws bearded within (in *A. psilopetala* naked). *Filaments* either 10 or 5; the 5 fertile equalling the claws of the petals, the sterile inserted on the margin of the disc, filiform, short or none. *Anthers* ovate or oblong, with an erect, sessile, conical gland. *Style* shorter than the filaments; stigma capitate, faintly 5-furrowed. *Capsule* 5-coccus, the cocci compressed, horned externally at the apex. *B. & W. l. c. p.* 59. *Juss. p.* 473. *Endl. Gen.* 6019.

Leaves mostly imbricate, very rarely sub-remote. Bracts or floral leaves submembranous, ciliate, imbricating the calyx. Limb of the petals sub-oblique, alternately imbricated. Name, ακμη, a *point*, and αδην, a *gland*; the anthers are tipped with pointed glands.

(1.) Flowers terminal, solitary, or rarely 2–3 together. (Sp. 1–13.)

A. Sterile filaments alternating with the fertile. (Sp. 1–10.)

* Leaves linear, three-angled-keeled, imbricating. (Sp. 1–6.)

1. A. juniperina (B. & W.! l. c. p. 61); lesser branches and twigs downy, leafy; leaves *opposite*, erect, sub-incurved, linear, sub-trigonous, acute, sub-pungent, glabrous, bifariously punctate beneath, *the pellucid margin scabrid;* flowers solitary, the two *ovate, sub-acute bracts* half the size of the calyx-lobes; limbs of the petals elliptical, acute, claws bearded; fertile filaments 5–6 times longer than the gland-like sterile ones. *Diosma obtusata, Thunb. Prod. p.* 84. *Fl. Cap.* 222. *Acmadenia juniperina, et muraltioides, E. & Z.!* 829, 830.

HAB. Cape, *Thunberg!* Sea-side dunes between Cape L'Agulhas and Potberg,

Swell.; Zoutendaals Valley, Caledon, and hills between Sondag and Coega Rivers, Uit.; and in Caffraria, *E. & Z.! Drege*, 7146. June-Oct. (Herb. Thunb., Wendl., Hook., Lehm., T.C.D., Sond.)

A shrub, 1–2 feet high. Stem glabrous, as thick as a fowl's quill, with many erect or ascending, opposite or alternate branches, 6–12 inches long; twigs very short. Leaves sometimes alternate on young shoots, densely imbricated, or sub-remote, the sides convex or flat, with a hard cartilaginous point, acutely channelled on the upper side, 3–6 lines long. Flowers sessile, among the uppermost leaves. Bracts 1½ lines long, membranous, with a green dorsal nerve. Cal.-lobes 2½ lines long. Petals purple, white when dry. Capsule rather longer than the calyx; carpels smooth, gland-dotted, transversely striate, with short blunt horns. It varies with the upper leaves downy.

2. A. densifolia (Sond.); branches and twigs downy; leaves *mostly opposite, densely imbricated*, appressed, sub-incurved, linear, trigonous, acute, *somewhat hairy, the sub-pellucid margin and keel ciliated*, with 2-4 rows of dots underneath; leaves of the lower branches glabrous; flowers solitary, the two *lanceolate-acuminate bracts and the calyx hairy at the apex;* calyx-lobes oblong, keeled, obtusely mucronate, membranous-ciliate; petals elliptical, mucronate-acute, ciliate, tapering into a long bearded claw; sterile filaments very short.

HAB. Woods at Kromriver, *Drege!* 7145. May. (Hb. Sond.)

A shrub, allied to *A. juniperina*, but more robust, with more closely placed, hairy leaves, longer bracts, and larger, evidently ciliate petals, with a longer and thicker beard on the claw. Leaves 4–7 lines long. Bracts 3 lines. Cal.-lobes about 3 lines, petals 5 lines long. Capsule twice as long as in *A. juniperina.*

3. A. Niveni (Sond.); branches and twigs opposite, *glabrous*, leafy; leaves opposite, erect, appressed, linear, 3-angled, acute, glabrous, ciliate at the sub-pellucid margin, bifariously dotted beneath; flowers solitary; *bracts* 4–6, *rather large*, ovate, *obtusely acuminate*, membrane-edged, ciliate; petals elliptical, acute, glabrous, narrowed into a shortish bearded claw; *sterile filaments subulate, twice as long as the fertile.*

HAB. Interior regions, *Niven!* (Herb. Mart., Sond.)

Near *A. juniperina*, from which its numerous bracts and sterile filaments distinguish it. More than a foot high, with rod-like branches. Leaves 3–5 lines long, the uppermost strongly ciliate, otherwise as in *A. juniperina*. Bracts 3 lines long. Calyx 4 lines long, its lobes with an excurrent green nerve. Petals twice as long as the calyx, purple.

4. A. psilopetala (Sond.); straight, branched from the base, the branches alternate or opposite; leaves sub-imbricate, linear-trigonous, acute, sub-pungent, glabrous, glandularly multipunctate beneath, ciliate, afterwards naked; flowers solitary; bracts 4-6, ovate, *acuminate*, the lower ¼, the upper ½ as long as the calyx; calyx-lobes *oblong-lanceolate, keeled, sub-trigonous at the point*, ciliate; petals elliptical or roundish, apiculate, tapering into a *beardless* claw of the length of the lamina; *sterile filaments very short. Adenandra trigona, E. & Z.!* 792.

HAB. Mountain sides near Gauritz R., Swell., *E. & Z.!* Oct.-May. (Herb. Lehm., Sond.)

Slender, 2 feet high; the lateral branches ascending-erect, yellow-purplish. Leaves 6–8 lines long, acutely channelled above. Flowers terminal or rarely axillary. Peduncles 2 lines long, with 2 lanceolate bracts below the apex. Calyx 3 lines long. Capsule short-stalked; the carpels 3 lines long.

5. A. alternifolia (Cham. & Schl.! Linn. 5. p. 52); much-branched, the branches and twigs leafy, downy; leaves *alternate,* erect, sub-incurved, linear, *three-cornered,* acute, sub-pungent, bifariously punctate beneath, *scabrid;* flowers terminal, *about 3 together;* bracts leafy, ciliate; calyx-lobes villoso-ciliate, *ovate-acuminate, the leaf-like apex recurved;* petals obovate, narrowed into a short-bearded, ciliate claw; fertile filaments thrice as long as the sterile.

HAB. Hanglip, *Mundt* & *Maire!* Cape District, *Ecklon!* (Herb. reg. Berol., Sond.)

Very like *A. juniperina* in habit and foliage, but readily known by the constantly alternate leaves, with concave sides, smaller flowers, narrower hook-pointed calyx-lobes, and petals with a *scarcely bearded* claw.

6. A. obtusata (B. & W.! l. c. p. 63. excl. syn. Thunb.); twigs downy; leaves *opposite,* quadrifariously imbricate, *linear-lanceolate, or linear,* with a narrow channel, sub-trigonous, *obtuse,* ciliate; flowers solitary; the broadly-ovate, sub-acute, bracts and the calyx downy; calyx-lobes oblong, keeled, ciliate, membrane-edged; petals sub-orbicular, downy within, the hairy claw equalling the calyx; sterile filaments very short. *Diosma obtusata, Willd. Sp.* 1. *p.* 1134. *R. & Sch. Syst.* 5. *p.* 452. *Acm. pungens, E. & Z.!* 828, *non B. & W.*

HAB. Clayey hills at Zwellendam, *Zeyher!* 2160. Dr. Pappe! Hassagaiskloof & Breederiver, *E. & Z.!* Speerbosch, *Drege!* 7144. Aug.-Sep. (Herb. reg. Berol., T.C.D., Sond.)

Stems many from a thick root, 6–8 inches long, much-branched above. Leaves on very short petioles, 2–3 lines long, thinly dotted at the back. Flowers nearly as in *A. juniperina,* but the bracts and calyces are downy. Filaments glabrous, the fertile thrice as long as the sterile. Capsule equalling the calyx. It varies with downy or glabrous leaves.

** Leaves roundish, or elliptic-oblong, complicato-carinate, imbricate. (Sp. 7–9.)

7. A. tetragona (B. & W.! l. c. p. 65); twigs pubescent; leaves *opposite,* decussate, *rhomboid-sub-rotund,* complicate, *acutely keeled,* sub-acute, ciliate, or roughened at margin, the uppermost densely imbricate; flowers solitary; calyx-lobes and floral leaves cuneate, ciliate, dilated and ear-shaped at top, *sub-recurved, apiculate;* petals thrice as long as the calyx, obovate, apiculate, ciliolate, narrowed into a villous-bearded claw; sterile filaments filiform, equalling the fertile. *Diosma tetragona, L. Suppl. p.* 155, *Thunb. Cap. p.* 224. *Bucco tetragona, R. & Sch. V. p.* 444, *Acm. strobilina, E. Mey.! in Hb. Drege.*

HAB. Cape, *Hesse!* Between Capetown and the Paarl, *Thunberg.* Attaqua's Kloof, 1000–2000f. *Drege!* Jan. (Herb. Wendl., Lehm., Mart., T.C.D., Sond.)

A foot or more high, with glabrous or downy branches, and short, crowded branchlets. Leaves minutely petioled, rather broader than long, spreading, dotted underneath, minutely scabrid on the keel, 2 lines long, 2½ wide. Flowers sessile, rather large. Calyx 2½ lines long, covered by the very similar floral leaves. Petals 6 lines long, 3 lines wide. Fil. 1½ line long. It varies with leaves downy or glabrous.

8. A. cucullata (E. Mey.! in Hb. Drege); sub-dichotomously branched, the twigs thinly downy; leaves *alternate,* densely imbricate, *elliptical-obtuse,* complicate, with flat sides, very smooth, the hyaline margin not ciliate; flowers solitary; bracts ovate, calyx-lobes oblong, *obtuse, not* *keeled*, ciliolate; lamina of the petals roundish, pointed, with a bearded claw; sterile filaments very short.

HAB. Limestone hills between Cape L'Agulhas and Potberg, *Drege!* Aug. (Herb. Lehm., T.C.D., Sond.)

A foot or more high. Twigs very short. Leaves 4–5-ranked, glabrous, 2 lines long, acute at back. Flowers rather large, pale rosy or white. Bracts ½ as long as the calyx, not keeled, coloured and membranous, as well as the calyx. Cal. 3 lines long; petals 5–6 lines. Fertile filaments 1 line long; sterile gland-like. Capsule as long as the calyx. This has the habit of *A. lævigata*, from which it differs in the leaves being alternate, membrane-edged and not ciliate, the round-backed sepals, and the presence of sterile filaments.

9. **A. Mundtiana** (E. & Z.! 825); di-tri-chotomous, the twigs minutely pubescent; leaves alternate, densely imbricate, *oblong*, obtuse, complicate, sub-incurved, with flat sides, *pubescent or glabrous, the hyaline margin and the keel ciliate*; flowers solitary; bracts ovate; calyx-lobes obtuse, without keel, ciliolate; the limb of the petals roundish, pointed, bearded on the claw; sterile filaments very short.

HAB. Mountain sides near Swellendam, *Mundt*. Oct. (Hb. Sond.)

Next the preceding, but with narrower leaves, 4 lines long and 2 lines wide, pubescent and ciliate, the adult glabrous with a rough margin; the stem also is taller and stronger. Flowers nearly as in *A. cucullata*, but the petals downy and ciliolate.

*** Leaves flat, sub-distant, not imbricate. (Sp. 10).

10. **A. flaccida** (E. & Z.! 823); quite glabrous, di-tri-chotomous; branches terete, twigs compressed; leaves sub-remote, opposite, oblongo-lanceolate, or lanceolate, very acute, flat, one-nerved, obsoletely glanduloso-denticulate; flowers in the forks of the upper branches, pedunculate, the peduncle once or twice three-forked, bracteate at the divisions; pedicels compressed, longer than the calyx, bibracteolate at base; calyx-lobes oblong, acute, keeled; claws of the petals thinly bearded, limb obtuse; sterile filaments very short.

HAB. Heathy spots on the Mts. near Brackfontein, Clanw., *E. & Z.!* Oct. Herb. Lehm., Sond.).

A foot or more high, with erecto-patent branches, and slender yellowish twigs. Leaves equalling the internodes, or shorter or longer, patent, green, glaucous and with the nerve prominent beneath, 6–7 lines long, 1½–2 lines wide. Infl. cymoid. Bracts like the leaves, but smaller; pedicels 2–3 lines long. Cal.-lobes 1–1¼ line long, sub-unequal. Disc erect, obsoletely sinuate. Petals 1½ lines long, the claw equalling the calyx, the limb obtuse. Fertile filaments ½ line long; sterile gland-like. Capsule 6 lines long, with horns as long as the carpels.

B. Sterile filaments none. Leaves imbricate. (Sp. 11–13.)

11. **A. lævigata** (B. &. W.! l. c. p. 64); twigs minutely downy; leaves mostly opposite, densely imbricated, elliptical, *obtuse*, complicate, with flat sides, very smooth, ciliated; flowers terminal, solitary, or sub-aggregate; bracts oblong, obtuse; calyx-lobes broadly ovate, acut.e keeled, glabrous, ciliolate; petals *sub-rotund*, with bearded claws. *D, tetragona, Willd. Sp.* 1. *p.* 1139. *Acm. tetragona, and lævigata, E & Z.* 824, 826.

HAB. Rhinosterfontein, Sebastian Bay. Near the Paarl, Stell., and Zoutendal, Swell., *E. & Z.!* June-Oct. (Herb. reg. Ber., Lehm., Sond.)

A foot or more high, erect, much-branched, with crowded branchlets. Leaves 4–5-ranked, straight, acutely keeled, the thin margin slightly recurved, with gland-dots along the keel, 2–3 lines long, 1½ line wide. Flowers as in *A. juniperina*, much smaller than in *A. tetragona.* Calyx 2 lines long, longer than the bracts. Limb of petals 1 line long, the claw as long as the calyx. Disc free, campanulate, somewhat lobed.

12. A. pungens (B. & W.! l. c. p. 64); twigs minutely downy; leaves opposite, densely imbricate, appressed, *patent-recurved from the middle, broadly elliptical, very acute,* complicate, ciliolate; flowers solitary or sub-aggregate; bracts and calyx-lobes broadly ovate, *cuspidate,* sub-carinate, *ciliate;* petals *oblong,* longer than the calyx, with bearded claws.

HAB. Cape, Hb. *Willd.!* Bosjesveld, near the Doornriver, under 1000f. *Drege!* Oct. (Herb. reg. Ber., Sond.)

Known from the last by the leaves being appressed from the base to the middle and then suddenly curving back, by their very acute points and the ciliate calyx. Erect, greyish, di-tri-chotomous, the lower branches spreading; ultimate branchlets 1–1½ inch long. Leaves underneath impresso-punctate along the margin and keel, 2–2½ lines long, 2 lines wide, with a short, hard and sharp mucro, the uppermost bract-like. Flowers sessile, 1–3 together. Calyx-lobes 2 lines long, glabrous or minutely downy externally. Claws of the petals chiefly bearded at the throat; the limb oblong, spreading, 1 line long.

13. A. assimilis (Sond.); twigs minutely downy; leaves 4-ranked, imbricate, *patent, cordate-sub-rotund,* sub-acute, *obtusely mucronate,* complicate, punctate underneath at the margin and keel, denticulato-ciliolate; flowers solitary, or few together; calyx-lobes obovate-cuneate, *truncate-obtuse, keeled,* patent at the point, ciliolate; limb of the petals glabrous, roundish, with a bearded claw. *A. lævigata, E. Mey.! in Hb. Drege, non B. & W.*

HAB. Between Cape L'Agulhas and Potberg, *Drege!* Aug. (Herb. Sond.)

6 inches or more high, the branchlets 3–4 inches long or shorter. Leaves decussate, acutely keeled, one line long and wide, the uppermost closer, covering the calyx. Calyx 1½ lines long, the segments having the dorsal nerve prolonged into a mucro. Petals 2 lines long, the claw below hairy and ciliate, densely bearded at the summit; limb rounded and pointed, glabrous. Fert. filaments 1 line long. Disc free, somewhat lobed.

II. Flowers aggregated in many-flowered heads.

14. A. rosmarinifolia (Bartl.! Lin. 17. p. 355); twigs sub-fastigiate, pubescent; leaves alternate, crowded, patent, linear-lanceolate, obtuse, with revolute margins, glabrous, the younger downy; heads terminal; the linear calyx-lobes and the petals ciliate; fertile filaments hairy, the sterile pubescent, half as long, *Barosma foetidissima, E. & Z.!* 813.

HAB. Between shrubs and stones, on Babylon's Toorensberg, Caledon, *E. & Z.!* Aug. (Herb. Lehm., Sond.).

1–2 feet high, simple below, dichotomously branched above the middle, as thick as a goose-quill. Leaves crowded, the lower spreading, upper sub-erect, sometimes linear-oblong, dull green above, paler, one-nerved, and impunctate beneath, glandularly-punctate at the margin, 4–6 lines long, 1–1½ wide; petiole ½ line long. Heads 5–9-flowered, ½ inch in diameter. Bracts membranous, linear, ciliate, 1–2 lines long. Calyx 2½ lines long, glabrescent at base, the segments channelled, ciliate and hairy at the apex. Petals oblong-obovate, acute, 3½ lines wide, downy at the base inside, and thinly pilose. Fert. filaments 2 lines long. Capsule glabrous, shorter than the calyx. This, which has the habit of a *Barosma*, differs, by its capitate flowers and beardless petals, from others of this genus.

VI. ADENANDRA, Willd.

Calyx 5-parted. *Hypogynous-disc* adnate. *Petals* naked, with very short claws. *Filaments* 10, shorter than the calyx; 5, opposite the petals, sterile, tipped with a concave or globose gland, 5 alternate, shorter, fertile; anthers large, oblong, erect, bearing a pedicellate, spoon-shaped or rarely globose, erect, at length refracted gland. *Style* short, cylindrical. *Stigma* capitate, 5-crenate. *Capsule* equalling the calyx (rarely longer), 5-coccous, the cocci obtuse or horned, glandularly muricated above. *Willd. En. Ber. p.* 256. *B. & W. l. c. p.* 69. *Juss. Mem. Muss. vol. xii. p.* 470. *t.* 19. *n.* 16. *Endl. Gen.* 6015.

Small, virgate or much-branched shrubs, with scattered, rarely opposite, pellucid-dotted leaves. Flowers larger than in other Cape genera of this sub-order, handsome. The name is derived from αδην, a *gland*, and ανηρ, here meaning a *stamen*.

ANALYSIS OF THE SPECIES.

I. Flowers *sessile*, crowded in a dense head.

Heads clammy:

leaves oblong, glabrous, 2 lines long ... ... (3) **obtusata.**
leaves oblong, or ovate-oblong, glabrous, 4–6 lines long ... ... (4) **viscida.**
leaves oval-oblong, pubescent underneath ... (2) **pubescens.**

Heads not clammy:

leaves obovate, concave and keeled ... ... (5) **gracilis.**
leaves rounded, flat ... ... (1) **rotundifolia.**

II. Flowers *on short stalks*: the peduncles equal to the upper leaves, or scarcely longer.

Leaves opposite (very rarely in a variety, alternate). (13) **coriacea.**
Leaves alternate, (very rarely sub-opposite).
Leaves underneath and at the margin, with *prominent* glands ... ... (14) **macradenia.**
Leaves underneath, with *immersed* glands:
Glands in *two* rows. (Leaves oblong, obtuse, 3–4 lines long) ... ... (12) **biseriata.**
Glands *scattered*, more or less numerous:
leaves lanceolate, *with revolute margins* (11) **uniflora.**
lvs. lin. oblong, flat, 6–12 inches long; *petals* obovate-oval ... ... (7) **umbellata.**
lvs. oval, tapering to both ends, crenulate, 3–5 lines long; *pet.* roundish apiculate ... ... (8) **amoena.**
lvs. ovate, or ov.-oblong, acuminate, entire, flat; *calyx* ciliate ... ... (6) **cuspidata.**
lvs. ovate or ov.-oblong, sub-obtuse, concave, *calyx* glabrous ... ... (9) **Kraussii.**
lvs. obovate-oblong, crenulate, ciliate; *calyx* hairy ... ... (10) **ciliata.**

III. Flowers *on long stalks*, umbellate or crowded in a corymb.

Leaves, at least the upper, roundish or oval, 1–2 lines long ... ... (15) **brachyphylla.**

Lvs. oval or oblong, hard-pointed, 3-5 lines long; *pedunc.* and *cal.* glabrous ... ... (16) **mundtiæfolia.**
Lvs. ovate-oblong or obl., hard-pointed, 4-6 l. long; *pedunc.* and *cal.* villous ... ... (17) **lasiantha.**
Lvs. linear-oblong, uncial.. ... ... (18) **fragrans.**
Lvs. ovate, or cordate-lanceolate, acuminate, *hook-pointed* ... ... (19) **serpyllacea.**
Lvs. ovate, or cordato-ovate; the upper lanceolate, or lin. lanceolate, sub-obtuse.
calyx-segments obtuse, reflexed ... ... (20) **marginata.**
calyx-segments acute, erect ... ... (21) **humilis.**

I. *Flowers sessile, capitate.* (*Sp.* 1–5).

1. A. rotundifolia (E. & Z. ! 798); the erect stem and the branches *glabrous;* twigs minutely pubescent; leaves closely imbricate, appressed at base, with recurved points, *rounded, calloso-mucronate, flat, glabrous,* glandularly crenulate at the thickened margin, one-nerved beneath, few-dotted or impunctate; flowers capitate, close-set; bracteoles linear-carinate, dilated upwards, coloured, ciliate; calyx-segments glabrous, obovate, apiculate, *with naked, coloured edges;* petals twice as long as the calyx, oval, on longish claws. *Drege,* 7148.

HAB. Mountains near Swellendam, *E. & Z!* (Herb. Lehm., Sond.).

1–2 feet high. Stem as thick as a goose-quill, scarred; twigs short, ending in heads of flowers. Leaves 2 lines long and wide, rarely petiolate, pale underneath, the uppermost, surrounding the flowers, on longer petioles, ciliolate. Flowers six or more in a head. Bracts adnate to the pedicel below the calyx, 3 lines long. Calyx 3 lines long, five cleft scarcely beyond the middle, glandular. Claws of the petals 1½ lines long, gradually dilating into a three-line-long limb. Filaments hairy; anthers small. Style short.

2. A. pubescens (Sond.); *branches pubescent;* leaves imbricate, appressed, *elliptical or oval-oblong, obtuse, concave,* glabrous above, *pubescent beneath,* nerveless, obsoletely impresso-glandular; flowers capitate; bracteoles lanceolate; calyx glabrous at base, 5-fid, the leaves ovate, sub-obtuse, downy towards the point, *ciliate;* petals . . . ? *A. villosa, E. & Z. !* 797., *non Lichtenst.*

HAB. Sandy places near Grietjesgat, Palmiet River, Stell., *E. & Z.!* Sept. (Herb. Sond.).

Branches 3–4-uncial, densely clothed with closely appressed leaves. Leaves two lines long, 1½ lines wide, nearly sessile. Flowers four or more, involucrated by the uppermost leaves. Bracts keeled, ciliate, wider upwards. Calyx purplish at the tip, 4½ lines long. Petals and the remaining parts of the flower are wanting in our specimens. It varies with leaves externally glabrous, and obsoletely nerved and green-streaked in the upper part of the branches.

3. A. obtusata (Sond.); the erect stem and the branches glabrous; twigs thinly downy; leaves very dense, erecto-imbricate, *oblong, obtuse at each end,* thickened at the margin, *glabrous,* one-nerved beneath, impunctate; flowers capitate, *glutinous;* bracteoles lanceolate, *obtuse;* calyx glabrous, its lobes ovate-oblong, obtuse, sub-carinate, erect, ciliolate; petals sub-orbicular, sub-retuse, thrice as long as the calyx; capsule twice as long as the calyx. *A. cuspidata, E. & Z. !* 796, *non E. Mey.*

HAB. West side of Table Mountain, *E. & Z.! Drege!* 7147. Oct. (Herb. Lehm., Sond.).

A foot or more high. Stem thick, scarred. Twigs leafy. Leaves two lines long, one line wide, shortly petiolate, underneath with a green, percurrent nerve, but not apiculate. Flowers 3–4 in a head, involucrate by the uppermost leaves. Bracts 2 lines long, leafy, keeled. Calyx 2½ lines long. Limb of the petals 3 lines long, white above. Anthers oval, with a minute recurved gland. Capsule 3 lines long, obtuse.—Resembling *A. biseriata* in habit, but easily distinguished by its capitate, gummy flowers.

4. **A. viscida** (E. & Z.! 795); the straight, sub-simple stem and the *angular* branches *glabrous;* leaves crowded, erecto-patent, *oblong or ovate-oblong*, the upper often ovate, tipped with an obtuse, callus, glabrous, with revolute margins, few-dotted or impunctate beneath; flowers capitate, *glutinous;* bracts cuneate, dilated upwards, keeled, *the keel produced into a thickened, obtuse point;* calyx glabrous, its lobes ovate-elliptical, obtuse, with recurved points, not glandular; petals roundish, thrice as long as the calyx. *Bartl. Linn.* 17 *p.* 358. *A. Bartlingiana, E. & Z.!* 794, *excl. syn. A. uniflora, var.? Schl.*

HAB. In stony sub-alpine places. Near Hout Bay, *E. & Z.!* Hemel en Aarde, Klynriviersberge and Zoutendals Valei, Swell., *Miss Joubert.* Aug.-Sep. (Herb. Lehm. Sond.)

1½ feet high. Leaves on very short petioles, 4–6 lines long, 1½–2 lines wide, the upper mostly smaller and less crowded, pale green underneath, shining, one-nerved, with impressed glands at the margin. Heads 4–8-flowered, involucred by the broadly-ovate or sub-rotund uppermost leaves. Bracts appressed to the calyx, 2½ lines long, complicato-carinate, membrane-edged, prolonged into an obtuse, three-angled cusp. Calyx 2½ lines long, its lobes thickened and purple upwards. Petals white within, flesh-coloured or reddish externally, glabrous.

5. **A. gracilis** (E. & Z.! 799); stem straight, simple or branched, glabrous; leaves crowded, erecto-patent, *oval or obovate*, obtuse, glandular-crenulate, *concave, somewhat keeled,* glabrous, impunctate; flowers terminal, sessile, capitato-aggregate; bracteoles *obtuse;* calyx glabrous, its lobes ovate-elliptical, obtuse, keeled, *diaphanous* at the margin, glandular at back; petals sub-orbicular, thrice as long as the calyx.

HAB. Stony places on mountain sides at the Zonderende River, near Appelskraal, Swell., *E. & Z.! Zey.!* 2155. Sep. (Herb. Lehm., T.C.D,. Sond.).

1–2 feet high, with slender, yellow-purplish branches. Leaves on very short petioles, 1½–2 lines long, 1 line wide, thickened at the margin. Flowers 4–6 together, not glutinous. Bracts appressed to the calyx, and half its length. Calyx 2 lines long. Petals obtuse or apiculate, white, reddish underneath. Capsule equalling the calyx, obtuse, glandular.

II. *Flowers on short peduncles; peduncle shorter, or scarcely longer than the uppermost leaves.* (*Sp.* 6-14).

6. **A. cuspidata** (Meyer.—B. & W. l. c. p. 87); erect, branching, glabrous or pubescent; leaves alternate or opposite, *ovate* or *ovate-oblong, acuminate,* flat, *thickish at the edge,* punctate beneath, the lower erecto-patent, the uppermost *densely imbricated;* flowers terminal, sub-capitate, crowded; calyx pubescent, its oblong, acute lobes and the floral-leaves ciliated; petals obovate-elliptical, apiculate, downy on both sides, once and half as long as the calyx.

VAR. *α.* **glabra**; branches and leaves glabrous. *A. cuspidata, Mey. l. c., Herb. Un. It. No.* 4. *ex pte.*

VAR. *β.* **villosa**; branches and leaves pubescent. *A. villosa, Lichtenst! B. & W. l. c.* 85, *excl. syn. Thunb., A. umbellata, α, speciosa, Eckl.! Linn.* 1831. *p.* 198. *Hartogia villosa, Berg.! Cap. p.* 70 *et Herb! spec.* 2 *(sed tertium differt et ad A. biseriatam pertinet.)*

HAB. On the Cape Flats; round Table Mt., and at Campsbay, common. Mountains near Swellendam. Aug.-Oct. (Herb. Holm., Wendl., r. Ber., Lehm., T.C.D., Sond., &c.).

1–2 feet high, dichotomous, with slender branches. Leaves 3–6 lines long, 1½–3 lines wide, with a percurrent nerve beneath. Petioles ½ line long, mostly glandular at base. Peduncles hairy, 3–7 in a cluster, 3 lines long. Calyx 5–6 lines long, its lobes 2 lines wide, pellucid-dotted. Petals on short claws, the limb 6 lines long, white above, reddish underneath. Fil. hairy, the sterile and fertile of equal length.—Very near *A. umbellata* and *A. uniflora*, and perhaps a variety of the former, from which, however, it differs, at first sight, by the smaller, acuminate, and evidently margined leaves, densely imbricated at the ends of the branches.

7. A. umbellata (Willd. Enum. Hort. Bert. p. 257); the erect stem and the branches glabrous, twigs downy; leaves *linear-oblong*, and *oblong*, sub-acute, *sub-ciliate, many-dotted beneath*, the lower spreading, the upper erect, *sub-imbricate;* flowers terminal, crowded; calyx glabrous or pubescent, its lobes linear-oblong, glandular-denticulate, ciliate; petals *obovate-elliptical*, ciliate, twice as long as the calyx.

VAR. *α.* **speciosa** (B. & W. p. 81); leaves linear-oblong, ciliate. *Glandulifolia umbellata, Wendl.! Coll.* 1. *p.* 37. *t.* 10. *Diosma speciosa, Bot. Mag. t.* 1271. *A. speciosa, Link. A. Wendlandiana, E. & Z.! Diosma rugosa, Don! Hort. Cant.* (with sub-solitary flowers.)

VAR. *β.* **glandulosa** (B. & W.); leaves oblong, glabrous. *D. uniflora, Lin. f.! non Linn. pater. Hartogia uniflora, Berg. Cap. p.* 71. *Gland. uniflora, ovata, Wendl. Coll. t.* 33. *A. glandulosa, Licht. E. & Z.!* 786.

VAR. *γ.* **sub-pubescens**; calyx pubescent.

HAB. Common round Capetown. At Hott. Holland, *Ludwig.* Wupperthal, *Drege!* 7072. Jul.-Oct. (Herb. Wendl., Sond., T.C.D., &c.)

1–2 feet high or more; the branches mostly crowded, erecto-patent, leafy above. Leaves alternate, glandular-denticulate at the margin, in var. *α.* 9–12 lines long, in *β.* more oblong, 6–8 lines long, 1½–2 lines wide. Petioles one line long. Flowers solitary, or two together, the largest and handsomest in the genus. Bracts several, narrow, leafy. Calyx 5–6 lines long, its lobes acute, rarely sub-obtuse. Petals on short claws, the limb apiculate, white, with a purple streak within, reddened beneath. Sterile filaments as long as the fertile, hairy; fertile shorter than the anthers.—In starved specimens the leaves are shorter and wider, and the flowers smaller.

8. A. amoena (B. & W.! l. c. p. 80); stem erect, branching; twigs leafy, downy; leaves *elliptical, tapering at both ends*, sub-obtuse, obsoletely *glandular-crenulate* at the margin, quite glabrous, punctate and obsoletely ribbed beneath; flowers terminal, solitary, sub-sessile; bracts ovate, sub-acute, ciliate; calyx glabrous, its lobes ovate, ciliate, impresso-punctate beneath; petals orbicular, pointed, twice as long as the calyx. *Bot. Reg.! t.* 553, *Lodd.! Cab. Bot. t.* 161. *A. ovato Link. A. amoena and acuminata E. & Z.!* 791, 790.

HAB. Tulbaghskloof and Witsenberg, Worcester, *E. & Z.!* Oct. Foot of Devil's Mt., *Drege!* (Herb. Wendl., Lehm., Sond.).

Allied to *A. umbellata* and *A. cuspidata,* from both which it differs in the dwarfer and weaker stem, more slender branches, and smaller leaves and flowers. Twigs leafy, even to the calyx. Leaves alternate, 3–5 lines long, 1½ wide, ovate-oblong, with a short, obtuse point. Petiole very short. Calyx four lines long, green-red. Petals clawed, twice as long as the calyx, white, with a purple streak above, deep red underneath. Filaments hairy, the sterile twice as long.

9. **A. Kraussii** (Meisn. ! pl. Krauss.); the erect stem and branches glabrous, the twigs downy and leafy; leaves spreading, *ovate-oblong* or *ovate, sub-obtuse,* pointless, nearly nerveless, concave above, glabrous, beneath flat, punctate and *downy;* flowers terminal, sub-sessile, solitary or in pairs; bracts *keeled,* cuspidate; calyx *glabrous,* its lobes ovate, sub-acuminate, *glabrous at the edge, glandular at apex,* purplish; petals glabrous, twice as long as the calyx, the sterile filaments longer than the fertile.

HAB. Outeniquasberg, George, *Dr. Krauss.* June. (Herb. Sond.).

Branches sub-dichotomous, the twigs crowded. Leaves alternate 2–3 lines long, 1½ wide, obsoletely nerved, and scarcely paler beneath, many-dotted, with a slightly thicker margin. Calyx 2¼ lines long, surrounded by the uppermost bract-like leaves. Petals oval-orbicular, white above, reddish beneath. Sterile fil. hairy.

10. **A. ciliata** (Sond.); branches and twigs very thinly pubescent, leafy; leaves *horizontally patent, obovate-oblong, obtuse,* flat, *crenulate near the point,* quite glabrous above, paler beneath, impunctate or thinly sprinkled with immersed glands, downy, *ciliate;* flowers 2–4, subsessile at the ends of the branches; calyx *hairy,* its lobes oblong, *obtuse,* ciliate; petals elliptic-orbicular, downy externally, ciliated, twice as long as the calyx.

HAB. Muysenberg, Cape, *v. Ludwig.* May. (Herb. Sond.).

2 feet high or more, clothed from the base with very minute pubescence. Branches and twigs few, dichotomous, sometimes elongate. Leaves 4–5-ranked, alternate, on very short, ciliate petioles, 4–5 lines long, 2 lines wide, the uppermost shorter, evidently crenulate from the middle to the apex, nerveless. Flowers rather large. Calyx 4 lines long, with a few glandular dots. Petals apiculate. Fertile filaments very short; sterile, with long hairs, taller than the anthers.

11. **A. uniflora** (Willd. enum. p. 256); erect; branches glabrous, twigs downy, leafy; leaves horizontally patent, *lanceolate, mucronulate, with revolute margins,* quite glabrous, paler beneath, punctate; flowers solitary, sub-sessile; calyx sub-pubescent, its lobes ovato-lanceolate, acuminate, ciliate from the base to the middle; petals obovate-subrotund, ciliolate at the apex, twice as long as the calyx. *B. & W.! p.* 77. *E. & Z. !* 793. *Diosma uniflora, Linn.! Sp.* 287. *Thunb. Cap. p.* 228. *Bot. Mag. t.* 273. *D. cistoides, Lam. D. acuminata, Lodd. Bot. Cat. t.* 493. *Eriostemon capense, Pers. Syn.* 1. *p.* 465.

VAR. β. **pubescens**; altogether covered with soft hairs, or the leaves glabrous above. *A. villosa, Hort.*

VAR. γ. **linearis**; leaves lanceolate, or linear-lanceolate, sparingly ciliate; calyx lobes mostly ciliate from base to apex. *Diosma linearis, Thunb. Fl. cap. p.* 226. *A. linearis, Juss. l. c. p.* 471.

HAB. Stony and rocky places round Capetown, common. β., a cultivated state.

γ., near the Pot River, *Zey.!* 2153. Cape Flats, *Drege,* 7075, *Gueinzius, &c.* Jul.-Oct. (Herb. Thunb., Lam., Wendl., Lehm., T.C.D., Sond.).

A foot high; branches di-tri-chotomous, twigs crowded, sub-umbellate. Lower leaves often deflexed, upper more distant and wider, with a straight or recurved point. Calyx 4 lines long, its lobes ovate-acuminate, or lanceolate, glandular denticulate, or sub-entire near the point. Petals short-clawed, 6 lines long, white within, purple without. It varies with tetramerous flowers. Var. γ., with ciliate leaves passes into var. β.

12. A. biseriata (Meyer, in B. & W. l. c. p. 75); erect; branches glabrous, twigs leafy, minutely downy; leaves alternate, erect, sub-imbricate, *oblong, obtusely mucronate,* with revolute margins, quite glabrous, paler beneath, *bifariously punctate;* flowers terminating the twigs, solitary; the short peduncles and calyces *villous;* cal. lobes ovate, acute, ciliated; petals elliptic-sub-rotund, retuse, twice as long as the calyx. *A. coriacea,* B. & W.! l. c. p. 3, non Lichtenst. *A. uniflora β., Schlecht. A. cistoides, E. & Z.!* 787, *non Lam.*

HAB. Sandy places on the Cape Flats, and mountain sides, Hott. Holland, *Hesse, E. & Z.! Zey.!* 283. Drakenstein, *Drege,* 7076. Oct.-Nov. (Herb. Wendl., Bartl., Lehm., T.C.D., Sond.).

1–2 feet high, slender. Leaves very short, petiolate, closely set, erect, 3–4 lines long, 1–1½ line wide; the upper shorter and broader. Flowers on the ultimate, crowded branchlets. Bracts not distant from the calyx, keeled, cuneate. Calyx 3½ lines long, the lobes villous at back, naked at the tip. Petals white, with a purple streak, reddish underneath. Capsule as long as the calyx, muricated with stipitate glands. Seeds black, oval.—It varies with a smooth calyx.

13. A. coriacea (Lichst.! in R. & Sch. Syst. v. p. 452); erect; branches glabrous, twigs slightly downy; leaves *opposite, ovate, sub-acute,* flat, with revolute margins, one-nerved beneath, pale, multipunctate, *glabrous;* flowers terminal, sub-ternate, pedunculate; calyx glabrous, its lobes *oblong, obtuse,* crenulate; petals roundish, twice as long as the calyx.

VAR. β. **oblongifolia**; leaves oblong, or linear-oblong.

HAB. On the Skurfdeberg, Bokkeveld, 27 Oct., 1816, *Dr. Lichtenstein.* β. in Dutoitskloof, *Drege,* 7073, 7074, 7085, *ex pte.* (Herb. r. Berol., T.C.D., Sond.).

A foot or more high. Leaves spreading, minutely petiolate, 7 lines long, 3½–4 wide, green above, densely dotted with minute glands beneath, the uppermost smaller. Peduncles 4 lines long, glabrous, with 2 opp., oblong, glabrous bracts above the middle. Calyx 3 lines long, pellucid-dotted. Petals white, externally purple. Stamens pilose. Ovary papillose. Stigma capitate, 5-lobed. Var. β. differs in having leaves either opposite or alternate, 6–8 lines long, 2 lines wide, and obtuse calyx-lobes; in other respects it resembles the normal form.

14. A. macradenia (Sond.); branches spreading, twigs *villous;* leaves alternate, short-stalked, erect, imbricate, concave, *elliptic-oblong,* or oblong, obtuse, glabrous, very smooth above, beneath paler, one-nerved, *irregularly sprinkled with raised glands, toward the apex glandular-denticulate;* flowers terminal, on short stalks, the peduncles and calyces *villous;* calyx-lobes ovate, acute, keeled, ciliate, *glanduliferous* at the point; capsules muricate, longer than the calyx, with slender, spreading horns. *A. coriacea, E. & Z.!* 789, *non Licht., nec B. & W.*

HAB. Stony places near the Tulbagh waterfall, Worcester, *E. & Z.!* Dec. (Herb. Sond.).

A shrub, several feet high, with the aspect of a *Barosma.* Branches terete, brown-black, twigs leafy, the ultimate 1–2 inches long. Leaves 4 lines long, 2 wide, erect-appressed, from the middle to the apex, or at the apex alone denticulate with globose glands. Peduncles 2 lines long, bracteolate at base, villous, as well as the calyx. Petals and stamens unknown. Capsules 4 lines long, glandularly muricate. Style, 1 line long. Stigma capitate, crenulate.

III. *Flowers on long peduncles, corymbose.* (*Sp.* 15–21).

15. A. brachyphylla (Schl. ! Linn. vi. p. 199); erect, sub-dichotomous, twigs downy; leaves alternate, on very short petioles, *sub-orbicular or oval*, mucronate, with revolute, thickened margins, paler beneath and *punctate*, all equal, or the lower ones oblong; flowers terminal; peduncles bibracteate; calyx glabrous or pubescent, its lobes *erect, broadly ovate, obtuse*, sub-carinate, *thinner* at the margin; petals roundish, thrice as long as the calyx. *E. & Z. !* 780.

VAR. *α.* **isophylla**; all the leaves oval or roundish, few-glanded beneath, the upper appressed at base, spreading at the apex; peduncles and calyces glabrous. *A. brachyphylla, Schl. l. c.*

VAR. *β.* **glandulosa**; all the leaves rounded, many-glanded beneath, the upper with spreading points; peduncles and calyces pubescent.

VAR. *γ.* **heterophylla**; lower leaves patent, oblong, upper oval or rounded, glabrous or pubescent; ped. and cal. mostly pubescent.

HAB. *α* & *γ*, on mountain tops. Howhoek and Klynriviersberg; Babylon's Toorensberg, Caledon; Potrivier. *E. & Z. ! Wahlberg, Zey. !* 2152. *β*, Muysenberg, *Niven, W.H.H.* Jul.-Nov. (Herb. Holm., Lehm., T.C.D., Sond.).

A shrub, 1–2 feet high. Leaves 1-nerved, the lower in *γ*., 4–5 lines long, $1\frac{1}{2}$ wide, the upper in *α* & *γ*, either scattered or imbricate, sub-bifariously glandular, 1–2 lines long, $1-1\frac{1}{2}$ wide; in *β*. mostly a little larger, all roundish, many-glanded. Peduncles 4–6 lines long, towards the middle furnished with cuneate, obtuse bracts. Calyx about 3 lines long, glandular at back, leafy, the margin, especially in *α* and *γ*, much thinner, naked and coloured, in *β* ciliated. Petals on longish claws, obtuse, sub-retuse, white within, partly purple externally; the claw pubescent on the inside. Sterile filaments, and the much shorter fertile pilose.

16. A. mundtiæfolia (E. & Z. ! 779); erect, branching, the twigs downy; leaves alternate, spreading, on very short petioles, *oval-oblong, obtuse, callous-pointed*, with revolute margins, pallid beneath, shining, glabrous, *mostly impunctate;* flowers terminal, few, pedunculate; peduncles bi-bracteolate, glutinous, and the *calyces glabrous;* cal. lobes *ovate acute, with recurved points, glandular-dotted* at the margin; limb of the petals twice as long as the calyx, obovate-rounded, the claw downy above. *Bartl. Linn. l. c.* 357.

HAB. Mountains between Hassagaiskloof and the Breede River, Swellendam, and at Knobluch, *E. & Z. !* Sep. (Herb. Lehm., T.C.D., Sond.).

A foot or more high, the twigs mostly fastigiate. Leaves scattered, rarely sub-opposite, oblong or oval, 3–5 lines long, $1\frac{1}{2}-2\frac{1}{2}$ lines wide, impunctate, rarely bifariously few-dotted. Peduncles 4–5 lines long, purplish. Bracts spathulate, a little below the middle of the peduncle. Calyx 3 lines long, glutinous at base. Petals white, reddish beneath, glabrous, sub-glutinous. Sterile filaments villous. Capsule glandularly-muriculate, equalling the calyx.

17. A. lasiantha (Sond); branches and twigs downy; leaves crowded, spreading, *ovate-oblong, or oblong, hard-pointed*, with revolute margins, quite entire, glabrous, glandular beneath; peduncles terminal, few, corymbose, much longer than the leaves, and the *calyx villous*, bracteate; calyx-lobes *oblong, obtuse;* petals sub-orbicular, downy externally, more than twice as long as the calyx.

HAB. Among stones, Klynriviersberge, Caledon, *E. & Z.!* Aug. (Herb. Sond.)

A foot or more high. Leaves alternate, 4–6 lines long, 2 wide, the uppermost shorter. Peduncles 2–4, about 6–8 lines long. Calyx 4 lines long, the lobes pellucid-dotted, the margin paler, quite entire, villous up to the apex. Petals white within, reddish without, the claw nearly equalling the calyx. Sterile filaments very hairy at the point, longer than the pubescent anthers.

18. A. fragrans (R. & Sch. l. c. p. 451); erect, glabrous; branches and twigs leafy, sub-umbellate-aggregate; leaves alternate, horizontally patent, *linear-oblong, obtuse,* mucronulate, *quite glabrous,* flat, one-nerved and punctate beneath; petioles glutinous; peduncles terminal, umbellate, naked or bracteolate, and the calyces glabrous; cal. lobes *roundish, at length reflexed;* petals broadly-elliptical, thrice as long as the calyx. *B. & W.! l. c. p.* 89. *Diosma fragrans, Bot. Mag. t.* 1519.

VAR. β. **amoena;** corolla red. *A. amoena, Link, enum. p.* 239 excl. syn.

HAB. In the interior, *E. & Z.!* (Herb. Sond.).

Sparingly branched; branches erect, the young twigs pale, glutinous toward the top. Leaves uncial, 2 lines wide, dull green above, pale beneath, one-nerved, glandularly-serrulate. Petioles 1 line long. Peduncles mixed with sterile, leafy twigs, 1–2 inches long (shorter and clammy in the wild specimens), naked, or with a pair of subulate, appressed bracts below the middle. Calyx semiquinquefid, 2 lines long, with rounded, membrane-edged, jagged lobes. Petals clawed, emarginate, slightly crenulate, white above, reddish, or red underneath. Filaments villous. Ovary glabrous. Style very short.—There are cultivated specimens in Herb. Wendl., Martius, and Lehm.

19. A. serpyllacea (Bartl.! Linn. l. c. p. 359); erect, branched from the base, twigs downy, leafy; leaves opposite, very rarely alternate, with very short petioles, *ovate-acuminate* or *lanceolate, mucronate,* obtuse or obcordate at base, flattish, or with revolute margins, often sub-emarginate, glabrous, the upper mostly *ciliolate or downy;* peduncles corymbose, much longer than the leaves, bracteate above the middle; bracts linear-subulate; calyx pubescent, its lobes *ovato-lanceolate,* half as long as the *elliptic-obovate* petals. *A. fragrans, E. & Z.!* 781, *non R. & Sch.*

VAR. β.; branches and leaves pubescent. *A. uniflora, E. Mey. & Drege,* 7078.

HAB. Mountain sides near Klapmuts, Stell., *E. & Z.!* Paarlberg & Drackensteen, *Drege,* 7078, b., *W.H.H.,* Sep.-Oct. (Herb. T.C.D., Sond.)

An undershrub, 6–12 inches high. Lower leaves often reflexed and larger, 6 lines long, 2–3 wide; the rest patulous, 2–4 lines long, a line wide, all smooth above, paler and punctate beneath, the glandular margin more or less revolute and the mucro reflexed. Peduncles pubescent or nearly glabrous, uncial or shorter. Bracts 1½–2 lines long. Calyx 2 lines long, ciliolate. Petals obtuse or mucronulate, white above, reddish beneath, downy towards the point. The leaves, as in *A. marginata* & *A. humilis,* are sometimes pellucid-margined.

20. A. marginata (R. & Sch. l. c. p. 452); erect, glabrous; twigs downy, leafy; leaves opposite or alternate, *the lower ovate or sub-cordate,* upper lanceolate, *sub-obtuse,* quite glabrous, *with revolute membranaceous margins,* sub-punctate beneath; peduncles corymbose, many times longer than the leaves, bibracteate, and the calyces glabrous; cal. lobes ovate, *obtuse, reflexed;* petals *roundish,* thrice as long as the calyx. *B. & W.! l. c. p.* 92. *Diosma marginata, Linn. f. Suppl. p.* 155. *Thunb. Cap. p.* 229. *D. rosmarinifolia, Lam.! Ency.* 11, *p.* 286 (*pars superior.*)

VAR. β. **angusta**; leaves mostly or all lanceolate or linear-lanceolate. *A. intermedia, E. & Z.!* 782.

HAB. Between the Cape and Drakenstein, near French Hoek, and at the Paarl, *Thunberg!* β. at Winterhoeksberg, Tulbagh, *E. & Z.!* Nov. (Herb. Thunb. Sond.)

6–12 inches high, the branches and twigs scattered or subumbellate-aggregate, short, filiform, erecto-patent. Leaves sub-remote, spreading, reflexed at the sides, pellucid at the margin, gland-dotted, dull green above, paler beneath, 4–6 lines long; in β. the lower longer and linear. Flowers the smallest in the genus. Upper peduncles 6 lines, lower 12 lines long, with two small bracts above the middle. Calyx 1 line long, the lobes very obtuse, membrane-edged. Petals white, red outside. Capsule twice as long as the reflexed calyx.

21. A. humilis (E. & Z.! 784); erect, branching; twigs downy, leafy; leaves alternate, rarely opposite, on very short petioles, spreading or erect, *lanceolate or linear,* sub-obtuse, glabrous, with revolute, membranous margins, one-nerved beneath; peduncles corymbose, much longer than the leaves, bibracteolate; calyx glabrous, its segments *ovate or sub-lanceolate, erect in flower and fruit,* one-third as long as the ovate-sub-cordate petals.

VAR. α. **glabra**; *A. humilis, E. & Z.! A. linifolia, Bartl. Linn. l. c.* 360. *A. marginata, E. & Z.!* 783. *Schl. l. c. p.* 199.

VAR. β. **imbricata**; glabrous; leaves erect, 5–6-fariously imbricate, obtuse; calyx-lobes subcordate-ovate, obtuse.

VAR. γ. **pubescent**; branches and leaves thinly pubescent; pedunc. and calyces glabrous.

HAB. Stony places above the baths at Zwarteberg and Klynriviersberg, Caledon; Palmiet R., Howhoek, and on to Potrivier, Swell., *E. & Z.! Niven.* Paarl, *Mad. Pallas.* Babylonstoorensberg, *Zey.!* 2154. Jul.-Oct. (Herb. Lehm.; Sond.)

6–8 inches to a foot high, with erect or ascending branches. Lower leaves 4–6 lines long, 1 line wide at base, upper 3–4 lines long. Peduncles 3–8 at the end of the twigs, uncial or shorter. Bracts ½ line long, subulate, obtuse. Calyx 1½ lines long, glandular. Petals obtuse or mucronulate, white, reddish beneath. Capsule 4 times as long as the calyx, muricate, obtuse.—Very close to *A. marginata,* R. & Sch., if not a mere variety.—*A. marginata,* E. Mey.! by its acute, *erect* calyx would belong to this species; by its shorter, ovate or oblong, minutely apiculate leaves, to the preceding. Is it a species distinct from both? [or, rather, an intermediate form, *connecting* both? H.]. I have seen but few specimens, gathered at Tradow, near Gnadendahl.

VII. BAROSMA, Willd.

Calyx 5-cleft or 5-parted. *Hypogynous-disc* very short, with a nearly entire, scarcely free margin. *Petals* 5, oblong, subsessile. *Filaments* 10; the alternate sterile, petaloid, shorter than the fertile. *Style* equalling the petals; stigma punctiform, obtuse. *Capsule* 5-coccous, the cocci eared at the apex, on the outer angle, gland-dotted. *B. & W. l. c. p.* 97. *Juss. l. c. p.* 474. *Endl. Gen. No.* 6020. *Baryosma, R. & Sch. Parapetalifera, Wendl.*

Leaves mostly opposite, rarely in threes or scattered, coriaceous, glandularly serrulate. Flowers on axillary twigs, resembling peduncles, solitary or tufted, very rarely sub-umbellate at the end of the branchlets. Possibly, it would be better to unite this genus with *Agathosma*. Name, from *βαρυς*, *heavy*, and *οσμη*, a *smell*.

Sect. I. **Eubarosma.** *Style* villous below. *Peduncles* short, terminating the floral twigs. *Barosmæ veræ, B. & W. l. c. p.* 97. (Sp. 1–3)

1. B. serratifolia (Willd. Enum. p. 257); branches and twigs glabrous; leaves opposite, *linear-lanceolate*, *equally narrowed at both ends*, truncate-obtuse, *sharply* serrulate, glabrous, paler beneath, and pellucid-dotted in the incisures; peduncles axillary, 1–3 flowered, mostly leafy; style villous below. *B. & W.! excl. var. β. E. & Z.!* 802. *O. Berg. Bot. Zeit.* 1853, *p.* 911, *t.* xii., *fig. R.–T.* (*folia*); *Darstellg. offic. Gew.* ii. *a. opt. Diosma serratifolia, Curt. Bot. Mag. t.* 456. *DC. Prod.* i. *p.* 714. *Parapetalifera serrata, Wendl. coll.* 1, *p.* 92, *t.* 34. *Baryosma serratifolia, R. & Sch. Adenandra serratifolia, Link. En. p.* 239.

HAB. Mountain sides above Duyvelsbosch and Grootvadersbosch, *E. & Z.! Drege!* Voormansbosch, Swell., *Zey.!* 2173. Sep.-Oct. (Herb. Wendl., Lehm., Sond.)

An erect shrub, several feet high, with rod-like branches and angular twigs. Leaves very rarely sub-alternate, shortly petiolate, twice as long as the internodes, spreading, flat, a little wider in the middle, 1–1½ uncial, 2–3 lines wide, 3-nerved, with sub-incurved serratures. Peduncles (flowering twigs) shorter than the leaves, 4–6 lines long, glabrous, leafy at the apex, and bibracteolate under the calyx, or leafless and 4–6 bracted; bracts 1 line long; pedicels very short. Calyx-lobes ovato-lanceolate, obtuse, keeled, 1 line long. Petals 3–4 times longer than the calyx, 1½ lines wide, oblong, obtuse, ciliolate at base, punctate at back, white. Stamens equalling the corolla, at length recurved; the anther with a minute gland. Sterile filaments linear-oblong, pubescent at the margin, gland-tipped. Style as long as the stamens. Capsule 4 lines long, shortly 5 horned.

2. B. crenulata (Hook. Bot. Mag. t. 3413); branches and twigs glabrous; leaves opposite, *oblong*, *oval*, or *obovate*, crenulate or serrulate, glabrous, paler beneath, the incisures pellucid-dotted; peduncles axillary, 1–3 flowered; style villous below. *Diosma crenulata, Lin. Amoen, IV. p.* 308. *D. crenata, Lin. Sp. p.* 287. *Thunb. Fl. Cap.* 227. *D. serratifolia, Burch. Trav.* 1, *p.* 476, *icon. D. latifolia, Lodd. Bot. Cat. t.* 290. *D. odorata, DC.* 1, *p.* 714. *Barosma odorata, Willd. Baryosma odorata, R. & Sch. Bucco crenata, R. & Sch. l. c.* 444. *Adenandra cordata, Link. Bar. serratifolia, β. B. & W. p.* 99. *B. crenata, Kunze., E. & Z.!* 800. *B. crenulata, crenata & Eckloniana, O. Berg, Bot. Zeit.* 9, *p.* 910–911, *tab. xii. B. crenulata, O. Berg. Darst. Off. Gew.* 1, *c. Icon. Opt.*

HAB. Fissures of mountains, Tablemountain, Zwartland and Hott. Holland; also near Tulbagh, *Thunberg, Hesse, E. & Z., Pappe, Drege*, 7079, &c. Oct. (Herb. Sond. &c.)

Very nearly allied to the preceding, and seemingly only differing in the constantly broader leaves. Branches and twigs quite glabrous; the sterile branches sometimes velvetty and furnished with broader leaves (*B. Eckloniana*, Berg.) Leaves mostly 5-nerved, uncial, 3–6 lines wide, crenate or crenulate-serrulate, but with less sharp and less prominent serratures than in the preceding. Pedunc. 1–3 or 5-flowered, leafy at the summit; pedicels very short. Calyx, corolla and capsule as in *B. serratifolia*. This is the true *Buku* bush.

3. B. betulina (B. & W.! l. c. p. 102); branches and twigs glabrous; leaves opposite, coriaceous, *cuneate-obovate*, *with obtuse*, *sub-recurved points*, sharply and closely denticulate, glabrous, scarcely paler beneath, pellucid-dotted in the incisures; peduncles axillary, the upper 1–3 flowered, equalling the leaves or shorter, lower longer, 3–5 flowered, leafy; style villous below. *E. & Z.!* 801. *O. Berg, Darst. off. Gew.* 1, *opt.!* *D. betulina, Thunb. Cap. p.* 227. *D. crenata, Lodd. Cab. t.* 404. *Hartogia betulina, Berg.! Fl. Cap. p.* 67. *Bucco betulina, R. & Sch. l. c. p.* 433.

HAB. On the Roodesand mountains, and elsewhere, *Thunberg.* Kardouw, *Hesse.* Mts. near Brackfontein, Clanw., *E. & Z.! Drege,* 7079, c. & 7080. July. (Herb. Thunb., Wendl., Lam., Lehm., Sond.)

A much branched shrub, with rod-like branches. Leaves very shortly petiolate, opposite, rarely sub-alternate, erecto-patent, rhomboid-obovate, mostly 5–nerved, the midrib very prominent, cauline 8–10 lines long, 5–7 wide, rameal smaller, sub-acute. Flowering-twigs axillary, 2–4 bracted, the lower bracts leafy. Flowers and fruit as in *B. crenulata,* from which this is easily known by the cuneate-obovate leaves, closely cartilaginous-toothed.

SECT. 2. **Trichopodes,** *B. & W.* *Style* glabrous (in Sp. 13 & 14 minutely downy). Peduncles 1-4, one flowered, bractless (very rarely bibracteate at base), from axillary buds. Calyx-segments erect. (Sp. 4-14.)

4. B. latifolia (R. & Sch. l. c. p. 449, excl. syn. Andr.); *pubescent;* leaves scattered, sub-opposite on the branches, *ovate, crenate, with glandular incisures,* dotless beneath; peduncles axillary, one-flowered, solitary or twin, quite glabrous, longer than the leaves, racemose at the ends of the branches; style glabrous. *B. & W.! l. c. p.* 105. *E. & Z.!* 803. *D. latifolia, Lin. f. Suppl. p.* 154. *Thunb.! Cap. p.* 229. *DC. Prod.* 1. *p.* 714. *D. odoratissima, Montin.! Act.-Lund.*

HAB. Rocky, mountain situations near Olifants river and Tulbagh, *Thunberg, E. & Z.!* Nieuwekloof, *Drege.!* Oct.-Nov. (Heb. Thunb. Lehm. Wendl., Sond.)

More than a foot high, entirely pubescent, the branches and twigs alternate, erecto-patent. Leaves short-stalked (petiole 1 line long) concave, veiny, ovate or subcordato-ovate, slightly recurved at margin, 6 lines long, 3–4 wide. Pedunc. 6 lines long. Calyx glabrous, its lobes obtuse, ciliolate, 1 line long. Petals twice as long as the calyx, oblong, white. Fertile fil. rather longer than the petals, ciliate at base; the anthers with a minute gland; sterile half as long, subulate, with a globose gland. Capsule 3 lines long, glabrous; its horns 1 line long, spreading.

5. B. pulchella (B. & W.! l. c. p. 107); twigs pubescent; leaves scattered, *ovate, with a thickened recurved margin,* glandularly crenate, *quite glabrous,* punctate beneath; peduncles capillary, axillary, one-flowered, solitary or in pairs, longer than the leaf, quite glabrous, racemose at the ends of the branches; style glabrous. *D. pulchella, L. sp. p.* 288. *Thunb.! Cap.* 229. *Bot. Mag. t.* 1357. *D. graveolens, Lichtenst.! R. & Sch. l. c. p.* 461. *Hartogia pulchella, Berg. Pl. Cap. p.* 69. *Bucco pulchella, R. & Sch. p.* 442.

VAR. *β.* **major**; leaves and flowers larger; fertile fil. ciliate at the base, sterile to the middle or nearly to the apex. *B. pulchra, Cham. & Schl.! Linn. V. p.* 53. *B. alpina, E. & Z.* 806.

HAB. Between the Cape and Drakenstein, near French Hoek, and the Paardeberg, *Thunberg.!* Table mountain, *E. & Z.! Hesse, W.H.H.* Dutoitskloof, *Drege.!* Witsenberg, *Zey.!* 284, Nov. (Herb. Thunb., Wendl., r. Ber., Lehm., T.C.D., Sond.)

A shrub, 3 or more feet high, much branched, the branches and twigs slender, crowded, straightish, leafy. Leaves alternate, in the younger twigs often opposite, erecto-patent, shining; the lower ovate-oblong, sub-obtuse, 5–6 lines long, 2–2½ wide, upper ovate, sub-acute, 2–3 lines long, 1½–2 wide. Flowers numerous toward the ends of the branches, purplish-white. Calyx glandular, its segments oblong, obtuse, ciliolate, ½ line long. Petals thrice as long as the calyx, oblong, a little narrowed at the very base, sparingly pellucid dotted. Filaments glabrous, in β. sometimes piloso-ciliolate at base. Capsule 2 lines long; the horns ½ line long, erect.—This differs from the preceding by its very slender branches, small, shining, glabrous, thick-edged leaves, and smaller flowers.

6. **B. venusta** (E. & Z.! 807); twigs downy; leaves scattered, *obovate-rounded*, cuneate at base, crenato-serrate, *punctate and glabrous beneath;* flowers axillary, tufted, racemose-crowded towards the end of the branches; pedicels downy, *bracteolate above the base;* sterile filaments and style glabrous; capsule very shortly horned.

HAB. Among shrubs on the Vanstaadensberg, Uit. *E. & Z.! Zey.!* 2156. Albany, *Williamson!* May-Jul. (Herb. T.C.D., Sond.)

A shrub, more than a foot high, with crowded rod-like branches, and slender, leafy twigs. Leaves erecto-patent, mostly alternate, rarely in threes, or opposite on the young shoots; in the lower part of the stem obovate or roundish, 4–5 lines long, 3 wide, the rest 2–3 lines long and wide, paler beneath, with impressed marginal glands; petioles ½ line long. Flowers white or reddish, in the axils of the upper leaves, 2–5 or more together; peduncles 2 lines long. Calyx obtuse, ½ line long, the lobes erect, glabrous. Petals 3–4 times longer than the calyx, oblong, much attenuate at base, glandular near the summit, outside. Filaments twice as long as the petals, glabrous. Capsule 2 lines long, with very short, obtuse, spreading horns.—Very like *B. pulchella,* but at once known by its rounded, gland-dotted leaves.

7. **B. acutata** (Sond); twigs very minutely downy; leaves sub-sessile, scattered, or approaching by threes, ovate, *acute, with a hard, acute, mostly recurved mucro,* the margin slightly recurved, pale underneath, and dotted with black glands, glabrous; peduncles axillary, one-flowered, 2-3 together, equalling the leaf, or shorter; calyx lobes broadly ovate, obtuse; sterile filaments ciliolate; style glabrous.

HAB. Interior regions, *Niven!* (Herb. Mart., Sond.)

A shrub, several feet high, with erect, glabrous branches. Larger leaves 6 lines long, 2–3 wide, smaller 3 lines long, 1½ wide, broad based, often sub-cordate. Flowers produced along the whole branch; like those of *B. ovata.*—Near the following, differing in the shortly petioled mostly alternate, leaves, with wide bases, and tapering into a sharp, rarely a blunt point, much paler underneath.

8. **B. ovata** (B. & W.! l. c. p. 109); branches glabrous, twigs minutely downy; leaves petioled, *ovate, obovate or roundish,* with a recurved obsoletely-crenulate margin, glandular beneath, *rather large,* punctate, glabrous or the upper downy; peduncles axillary, one-flowered, solitary or in pairs, equalling the leaf or shorter; calyx-lobes *ovate,* obtuse; sterile filaments ciliolate; style glabrous; capsule rough with glands, the sub-acute horns half as long as the capsule. *Diosma ovata, Thunb.! Cap. p.* 227. *Willd.! Sp. p.* 1139. *Bucco ovata, R. & Sch. Drege,* 7081, 7082, 7083.

VAR. α; leaves ovate, sub-acute, with sub-recurved margins; pedunc. glabrous. *Diosma pulchella, Houtt. L. Pl. Syst.* 3. *p.* 288. *t.* 21. *f.* 2. *B. ovata, E. & Z.!* 805.

VAR. β.; leaves obovate-oblong, obtuse, flat; pedunc. glabrous.

VAR. γ.; leaves obovate, obtuse; peduncles minutely pubescent. *D. ovata, Andr. Bot. Rep. t.* 464. *Bot. Mag. t.* 1616. *DC. l. c. p.* 714. *Bucco ovata, Wendl. Coll. t.* 20.

VAR. δ.; leaves cuneate, roundish or elliptical, 5-nerved; peduncles glabrous. *D. punctata, Lichtenst.!—R. & Sch. l. c. p.* 461. *B. graveolens, E. & Z.!* 804.

HAB. var α, among stones on the Zwarteberg, near Caledon Baths; and on the Vanstaadenberg, Uit.; β, in Duyvelsbosch; γ, near Gnadendahl; δ, Duyvelsbosch and Zwarteberg, *Thunberg! E. & Z.! Dreye.!* (Herb. Thunb., Wendl., r. Ber. Lehm., Sond.)

One or two feet high, erect, much branched, and ramulose. Leaves erecto-patent, opposite, very rarely alternate, coriaceous, 3–6 lines long, flat on both sides, dull green above, shining, pale green beneath, and multi-punctate, the broader 5 nerved; petiole 1–2 lines long. Cal. lobes ½ line long. Petals white, 2 lines long. Fert. filaments glabrous. Capsule 4 lines long.—Thunberg's specimens nearly all belong to our var δ.

9. B. oblonga (E. & Z.! 810); branches glabrous, twigs minutely downy; leaves opposite, *oblong or elliptical-oblong, sub-acute,* obtusely mucronulate, with sub-recurved, obsoletely glandular-crenulate margins, glabrous or the uppermost downy, *minutely* punctate underneath; peduncles axillary, one-flowered, solitary or in pairs, mostly shorter than the leaf; calyx-lobes *lanceolate,* obtuse; sterile fil. *ciliate;* style glabrous; capsule rough-glanded, with sub-acute horns half its length. *Diosma oblonga, Thunb.! Cap. p.* 227. *R. and Sch. p.* 460. *B. Ecklониana, Barth. Lin.* 17. *p.* 363.

HAB. Cape, *Thunberg!* Among stones and shrubs on the Vanstaadensberg mts., *E. & Z.! Zey.!* 2157. Aug-Sep. (Herb. Thunb., Lehm., Sond.)

A shrub having the habit of the preceding, with angular twigs. Leaves spreading, shining above, pale green beneath, 4–6 lines long, 2-lines wide, one-nerved; the nerve gland-bearing below the point. Petiole 1 line long. Petals reddish underneath. Capsule as in *B. ovata,* but the horns longer.—Very near *B. ovata,* from which it differs in the narrower, more oblong leaves, dotted below with smaller glands, calyx lobes twice as long, and petals scarcely twice as long as the calyx.

10. B. scoparia (E. & Z.! 809); branches glabrous, twigs very thinly downy; leaves opposite, *elliptical and oblong, obtuse,* sub-recurved at the margin, obsoletely glandular-crenulate, minutely downy or glabrous, minutely punctate beneath; peduncles axillary, one-flowered, solitary or in pairs, shorter or longer than the leaf, densely racemose towards the ends of the branches; calyx segments minute, *ovate, sub-rotund, obtuse;* sterile fil. *pubescent* beyond the middle; style glabrous. *B. oblonga, B. & W.! l. c. p.* 112 *excl. syn. B. dioica, β. Schl. in Linn. B. ovata, E. Mey.! D. microphylla, Spr.*

VAR. β, **ternata**; branches virgate; leaves mostly ternate, glabrous. *B. ternata, E & Z.!* 811. *Diosma spartiifolia, Steud.*

VAR. γ. **pauciflora**; branches very slender; leaves oblong, narrowed at base; flowers axillary, few. *B. pauciflora, E & Z.!* 812.

HAB. Cape, *Hesse!* Flats near Port Elizabeth, and Vanstaadensberg, Uit., *E & Z.!* Albany, *Mrs. T. W. Barber!* β. Uitenhage, *E. & Z.!* γ, Chumiesberg, Caffr. *E. & Z.!* June.-Sep. (Herb. Wendl., γ. Stuttg., Hook., T.C.D., Sond.)

A shrub 2 or more feet high, with opposite or whorled branches and twigs. Leaves opp., or 3–4 in a whorl, sometimes alternate on young shoots, spreading, on the stem and branches 4–6 lines long, 2 wide, on the twigs, or on dwarf specimens, 2–3 lines long, 1–1½ wide, all narrower towards the obtuse point, pale and one-nerved beneath. Petioles very short. Twigs leafy, floriferous towards the apex. Peduncles glabrous, mostly longer than the leaf. Calyx ⅓ line long, glabrous or downy. Petals oblong, 2 lines long, white, reddish underneath. Stam. longer than the petals, the filaments glabrous, 4 times longer than the sterile. Capsule 2 lines long, with erecto-patent horns 1 line long.—The specimens with leaves mostly oblong constitute B. & W.'s var. *α*, *macrophylla;* and those with smaller and broader leaves their var *β.*, *microphylla.*

11. B. lanceolata (Sond.); branches mostly opposite or ternate; twigs very thinly pubescent; leaves opposite; *lanceolate or linear, acute, with recurved or revolute margins,* glabrous or downy, sparsely punctate beneath; peduncles 1-4, one-flowered, axillary, equalling the leaves or shorter; calyx-lobes ovate or ovate-oblong, obtuse; ster. filaments hairy; style glabrous; capsule much longer than the calyx, glabrous, impressed-glanded, its horns sub-acute and nearly half as long. *Diosma lanceolata, Thunb.! Fl. Cap. p.* 226.

VAR. *α.* **linifolia** (B. & W.); leaves lanceolate, mucronate, with recurved margins and an obtuse or acute, straight point. *D. dioica, Ker. Bot. Mag. t.* 582. *B. dioica, B. & W.! p.* 114. *E. Mey.! in Hb. Drege.*

VAR. *β.* **hamata** (B. & W.); leaves mucronate, with revolute margins and a hooked point. *Bucco hamata, Wendl.! pat. in Hb.*

VAR. *γ.* **Natalensis**; leaves linear, sub-obtuse, with recurved glandular-crenulate margins. *B. Kraussiana, Buch.! in Krauss, Beytr. p.* 40.

VAR. *δ.* **obtusifolia**; leaves linear, with revolute margins, obtusely truncate or emarginate, often with a pellucid gland at the point. *B. angustifolia, B. & W.! p.* 116. *B. dioica, E. & Z.!* 816.

HAB. Mountains of the Langekloof, near Wolwekraal, *Thunberg!* Kromme R., *Krauss! γ.* on the Tafelberge, Natal, *Krauss! δ*, woods at Adow, and Olifantshoek, Uit.; and near Grahamstown, &c. *E. & Z.! Atherstone! Williamson! Zey.!* 2155. Aug.-Nov. (Herb. Thunb., Wendl., Meisn., Hook., Lehm., T.C.D., Sond.)

A grayish-brown shrub, 1–3 feet high, with rodlike, glabrous branches. Leaves sometimes ternate, rarely on the young shoots sub-alternate, dull green above, paler beneath, the lower spreading or reflexed, 6–8 lines long, 1 line wide, upper more erect, incurved, mostly oblong-lanceolate, with short petioles. Flowers small, on capillary peduncles. Calyx minute, semiquinquefid; petals 3–4 times as long, white, reddish at back, few glanded. St. longer than the corolla, glabrous; the sterile fil. ⅓ of fertile. Ovary glabrous, sometimes abortive. Capsule 2 lines long. It varies in the same specimen with wider or narrower, obtuse or acute, downy or glabrous leaves. Ecklon gathered specimens with double-flowers, near Grahamstown.

12. B. pungens (E. Mey! in Hb. Drege); branches glabrous, twigs pubescent, leafy; leaves opposite, close-set, erecto-patent, *lanceolate, acuminate-pungent, mucronate, carinate-concave,* very smooth above, gland-dotted beneath, with thickened, glabrous or basally *ciliate* margins; peduncles axillary, 1-3, one-flowered, shorter than the leaves; calyx-lobes lanceolate, obtuse; sterile filaments pubescent; style glabrous.

HAB. Zwarteberg, at the Klaarstrom, 3000f. 4000f. *Drege.!* July. (Herb., Lehm., Sond.)

More than a foot high, with terete branches and twigs. Leaves from a broad base, gradually narrowed to a very acute point, the lower 8–10 lines long, 1 line wide at base, upper mostly 6 lines long, and incurved, one-nerved beneath, with several rows of dots. Petioles very short. Pedunc. slender, ebracted, hairy or glabrous, 3 lines long. Calyx 5-parted, the segments glandular at back, and tipped with a gland, ciliate. Petals oblong, 1½ lines long. Stamens scarcely longer than the petals, the filaments glabrous; sterile ½ as long, gland-tipped. Style equalling the stamens; stigma obtuse.

13. **B. microcarpa** (Sond.); *entirely clothed with a very short down;* branches rod-like; leaves opposite or ternate, on the young shoots subalternate, *linear-oblong, obtuse,* with revolute margins, sparsely glanded beneath; peduncles 1-4, axillary, one-flowered, shorter than the leaves; calyx-lobes lanceolate, obtuse; sterile filaments pubescent; style *downy;* capsule scarcely longer than the calyx, *very short-horned,* glandular-muricate at back.

HAB. Gauritz river, and between Vanstaadensberg and Bethelsdorp, Uit., under 1000f. *Drege.!* 7084. Dec. (Herb. Sond.)

More than a foot high, the primary stem as thick as a goose quill, grayish-pubescent. Branches erect, with spreading twigs. Leaves 4–5 lines long, ¾-1 line wide, obtuse, neither mucronate nor gland-tipped, erect or spreading, convex above from the strongly revolute margin. Pedunc. ebracteate, capillary, thickened upwards in fruit. Calyx 5-parted, the lobes 1 line long, downy within and without, and ciliate. Petals oblong, twice as long as the calyx. Stamens scarcely longer than the petals; the fertile filaments glabrous; sterile ⅓ of petal, oblong, tipped with a minute gland. Style equalling the stamens, to the middle or the apex downy; stigma obtuse. Capsule scarcely more than a line long, the 5 carpels compressed, prominently-glanded; horns erect, acute, very short.—Very similar to *B. lanceolata.* From *B. foetidissima,* which it resembles in foliage, it differs by inflorescence and calyx.

14. **B. Niveni** (Sond.); branches glabrous, twigs slender, downy near the ends; leaves opposite, *elliptical or oblong,* acute at each end, *flat,* or somewhat channelled toward the point, *acutely-mucronulate,* with thickened glandular-crenulate margin, punctate beneath, softly downy on both sides, nerveless; peduncles *axillary and terminal,* 1-4, equalling or excelling the leaves, without bracts, downy; calyx downy, its lobes oblong-lanceolate; sterile filaments linear, downy, tapering into a glabrous point; *style pubescent.*

HAB. Interior regions, *Niven!* Rocky places on the Modderfonteinsberg, Roodeberg and Ezelskop, 4000-5000f. *Drege.!* Nov. (Herb. Mart., Sond.)

A slender shrub, several feet high, with erecto-patent, opposite, ternate or whorled branches and twigs. Leaves equalling the internodes, 3–4 lines long, 1½ wide, nerved when young, the nerve gradually vanishing, bright green above, paler and many-dotted beneath, thickened, but neither recurved nor revolute at the margin. Peduncles 3–4 lines long. Calyx 5-parted, its lobes downy on both sides, at length spreading, obtuse, 1½ lines long. Petals oblong, not narrowed at base, white, reddish at back, not much longer than the calyx. Stam. about as long as the petals, or little longer; the sterile filaments equalling the calyx. Style as long as the stamens.—In aspect not unlike an *Adenandra.* Allied to *Agathosma* by its pseudo-axillary and terminal peduncles; but the leaves are always opposite.

SECT. 3. **Agathosmoides,** *B. & W. l. c. p.* 118.—*Style* glabrous. *Peduncles* terminal, one-flowered, without bracts, sub-umbellate-crowded. *Calyx-segments* reflexed during flowering.

15. **B. fœtidissima** (B. & W.! l. c.) branches glabrescent, twigs very thinly pubescent; leaves sub-ternate, linear, with revolute margins, minutely downy; flowers umbellate at the ends of the branches; calyx-segments ovate, obtuse. *Ag. fœtidissima,* Hort.

HAB. Cape, *cultivated* in Europe (v. v. cult. and in Herb. Wendl.)

A slightly branched shrub, 2 feet high. Branches scattered or ternate, ascending, sub-flexuous, terete, as thick as a pidgeon-quill, grayish, leafy; the twigs 1–2-uncial, filiform. Leaves ternate, rarely scattered, little longer than the internodes, mostly horizontally patent, shortly petiolate, oblong-linear, or linear, impresso-punctate at the margin, sparsely punctate beneath, 4–6 lines long, the lower 6–8 lines, ¾–1 line wide. Peduncles 8–10, equal, 4–6 times long, 2–3 times as long as the obtuse floral leaves, minutely downy, as well as the calyx. Calyx gland-dotted, the segments one line long, reflexed. Petals ⅓–½ longer than the calyx, oblong, obtuse, marked with a pellucid gland under the apex, white. Stam. longer than the petals, with glabrous filaments; sterile fil. ⅓ as long as petals, linear-oblong, ciliate, narrowed to the point and tipped with a small gland. Ovary glabrous. Style rather shorter than the stamens.—Allied to *B. microcarpa.*

VIII. AGATHOSMA, Willd.

Calyx 5-parted, with sub-unequal segments. *Petals* clawed. *Filaments* 10, the 5 alternate petal-like. *Style* as long as the petals, or very rarely longer; stigma obtuse. *Capsule* mostly 3-coccous, rarely 2–4-coccous; the cocci horned at the external angle. *B. & W. l. c. p.* 121. *Diosmæ, Sp. Linn., Thunb. &c.*

Leaves alternate, very rarely opposite intermixed. Flowers at the ends of the branches capitate or more commonly densely umbellate, in one species axillary, towards the ends of the branches, as in *Barosma.* Peduncles short or long, one-flowered, mostly 2-bracted in the middle. Sterile filaments sometimes much expanded and like the petals, and as long, mostly inserted on the disc at the base of the petals, very rarely connate with the claw. The name is derived from αγαθος, *good*, and οσμη, *a smell.*

ANALYSIS OF THE SPECIES.

I. **Barosmoides.**—Inflorescence of *Barosma.* *Petals* with long claws. (1) **tabularis**

II. **Capitato-racemosæ.**—*Flowers* crowded in a dense, capitate-raceme. *Calyx* angular. *Leaves* hairy. *Style* elongate ... (2) **hirta**

III. **Alares.**—*Umbels* sessile or pedunculate in the forking of the branches. *Petals* clawed. (Sp. 3–6.)

(*a.*) *Leaves* of two forms; the lower lanceolate:

upper leaves ovate-subcordate, with *glabrous* edges (3) **alaris.**
upper leaves oblong, with *villous* edges ... ... (4) **affinis.**

(*b.*) *Leaves* uniform:

leaves ovate, keeled, gibbous at the point; style *hairy* ... ... (5) **leptospermoides.**
lvs. ovate. acuminate, flat, not gibbous; style *glabrous* ... ... (6) **pentachotoma.**

IV. **Involucratæ.**—*Heads* of flowers involucred. Inner involucral leaves coloured. (Sp. 7–10.)

leaves subcordate-ovate, acute, not thick-edged, *ciliate* (10) **cephalotes.**
lvs. ovate-oblong or ovate-lanceolate, gibbous at point, 3-cornered, the margin *thickened* ... ... (7) **involucrata.**
lvs. lanceolate, or linear-lanceolate, with *incurved* points (8) **sabulosa.**
lvs. linear, with involute margins, round-backed, with *straight* points ... ... (9) **Hookeri.**

V. **Pseudostemon.**—*Sterile-filaments* villous, *attached* to the claws of the petals. (Sp. 11–13.)

Flowers *capitate:*
- leaves lanceolate, keeled, 3-angled, obtuse, scabrous; *ster.-fil.* filiform ... ... (11) **collina.**
- lvs. ovate-oblong, channelled and gibbous at point, ciliato-pilose; *ster.-fil.* dilated, petaloid ... (12) **Schlechtendalii.**

Flowers in a few-flowered, dense *umbel* ... ... (13) **humilis.**

VI. **Diplopetalum.**—*Sterile-filaments* like the petals, as long and *free.* (Sp. 14–27.)

(*a.*) Leaves linear-oblong:
- *obtuse*, sharply keeled, *ciliate* ... ... (14) **umbellata.**
- *sub-obtuse*, convex-keeled; pedunc. and calyx *ciliated* (16) **anomala.**
- *sub-acute*, round-backed, *glabrous;* pedunc. and calyx *glabrous* ... ... (15) **lediformis.**

(*b.*) Leaves linear:
- flowers capitate ... ... (21) **Dregeana.**
- flowers umbellate:
  - leaves flat:
    - peduncles *hirsute;* petals narrow; ster.-fil. not glanded ... ... (24) **filipetala.**
    - peduncles *glabrous;* pet. oblong; ster.-fil. externally glandular ... ... (25) **linifolia.**
  - leaves channelled, more or less three-angled:
    - twigs with prominent glands ... ... (17) **monticola.**
    - twigs not glanded:
      - petals not much longer than the calyx ... (19) **parviflora.**
      - petals twice or thrice as long as the calyx:
        - leaves thickly gibbous at the apex ... (23) **gonaquensis.**
        - leaves not gibbous at the apex; *quite glabrous:*
          - claws of petals *glabrous;* leaves *obtuse* (26) **virgata.**
          - claws of petals *hairy;* leaves *acute* (27) **commutata.**
        - leaves not gibbous; the margin and keel *ciliate*, at length sub-glabrate:
          - claws of the sterile filaments *glabrous* (18) **gracilicaulis.**
          - claws of the sterile filaments *pilose:*
            - calyx lobes *glabrous*, longer than the claws of petals .. .. (20) **montana.**
            - cal. lobes *ciliate*, as long as the petal-claws ... ... (22) **nigromontana.**

VII. **Barosmopetalæ.**—*Flowers* in dense umbels. *Petals* cuneate at base, with a very short claw. *Sterile-filaments* short, thickish, subfiliform, or half cylindrical, gland-tipped. (Sp. 28–45.)

Leaves neither mucronate nor piliferous:
- *lvs.* obovate-cuneate, convex above, *velvetty* ... (36) **fraudulenta.**
- *lvs.* obovate-oblong, flat, *glabrous* ... ... (37) **craspedota.**
- *lvs.* elliptical, convex above, shortly *hairy* ... (30) **blaerioides.**
- *lvs.* ellipt. or ovate, concave above, glabrous (young downy) ... ... (38) **recurvifolia.**
- *lvs.* ellipt.-oblong, concavo-canaliculate, *pubescent* (33) **pubigera.**
- *lvs.* ovate-oblong, obtuse, *flat, glabrous* ... ... (32) **planifolia.**
- *lvs.* oblong, sub-acute, convex above, *glabrous* (29) **Niveni.**
- *lvs.* lanceolate, sub-obtuse, sub-cordate at base, *hairy* ... ... (31) **pubescens.**
- *lvs.* lanceolate, *very acute*, glabrous ... ... (28) **Mundtii.**

Leaves mucronate :

lvs. lin-lanceolate or lanceolate, downy ... ... (39) **gnidioides.**
lvs. oblong-acute, downy beneath, straight-pointed (40) **barosmoides.**
lvs. ovate-oblong, acute, glabrous, hook-pointed (41) **acutifolia.**
lvs. ovate, cuspidate, spine-pointed, glabrous, *imbricated* ... ... (42) **Martiana.**
lvs. cordate-ovate, spine-pointed, downy beneath, *spreading* ... ... (45) **spinosa.**
lvs. ellip.-rounded, with a short, obtuse, *straight* mucro ... ... (43) **punctata.**
lvs. rounded, with a short, acute, *recurved* mucro (44) **mucronata.**

Leaves tipped with a long hair :

lvs. ovate-subcordate, with *reflexed* edges ; *pedunc.* equalling calyx ... ... (34) **apiculata.**
lvs. oblong-elliptical, *flat ; pedunc.* 3–4 times longer than the calyx ... ... (35) **pilifera.**

VIII. **Imbricatæ.**—*Leaves,* at least the upper ones, densely imbricate, ovate, or roundish-ovate, acuminate. *Flowers* sub-umbellate-capitate. *Petals* with a capillary claw, and a roundish limb. *Sterile filaments* short. (Sp. 46–48.)

leaves roundish-ovate, obtuse, underneath gland-tubercled (46) **squamosa.**
lvs. roundish-ovate, cuspidate, tomentose underneath ... (47) **lycopodioides.**
lvs. ovate, acuminate, ciliate or pubescent ... ... (48) **imbricata.**

IX. **Eu-agathosmæ.**—*Flowers* terminal, umbellate or sub-capitate. *Petals* with long claws, the claw equalling or exceeding the calyx; lamina oblong. *Sterile filaments* free, equalling or exceeding the calyx, mostly filiform. *Leaves,* in the broad-leaved species, not imbricate. (Sp. 49–100.)

A. *Leaves roundish, ovate, oblong, or lanceolate.* (*Sp.* 49–70.)

* Flowers umbellate :

limb of the petals sub-cordate at base ... ... (56) **latipetala.**
limb of the petals oblong, narrowed at base :

Leaves, at least the upper, *imbricate or appressed :*

lvs. ovate or ovate-oblong, not ciliate ; *pedunc.* hairy ... ... (59) **tenuis.**
lvs. oblong, or ov.-oblong, ciliate ; *pedunc.* glandular ... ... (58) **propinqua.**
lvs. ovate-lanceolate, or lanceol., hairy beneath ; straight-pointed ; *pedunc.* equal (66) **Ventenatiana.**
lvs. lanceolate, hairy beneath, incurved at point ; *pedunc.* unequal ... ... (67) **villosa.**

Leaves spreading or reflexed :

leaves lanceolate, at the apex beneath *gibbous,* 3-angled ... ... (69) **serpyllacea.**

Leaves not gibbous at the apex :

leaves roundish, flat ... ... (49) **orbicularis.**
lvs. elliptic, sub-concave ; peduncles *glabrous* (50) **minuta.**
lvs. oval or obovate, flat ; pedunc. *pubescent* (51) **thymifolia.**
lvs. cordate-ovate, flat, pubescent ; pedunc. glandular ... ... (52) **glandulosa.**
lvs. ovate-subcordate, concave above, glabrous ; *pedunc.* pubescent ... ... (53) **foliosa.**
lvs. ovate-oblong, or lanceolate, ciliate, *rugulose* above ; the claw of the petals *ciliate* ... ... (54) **rugosa.**

lvs. ovate-oblong, or obl. lanceolate, *velvetty*, not rugulose; claw of the petals *glabrous* ... ... (55) **marifolia.**
lvs. oblong, sub-acute, *pubescent* on both sides, 3 lines long; *peduncles* pubescent (57) **elegans.**
lvs. oblong, obtuse, *glabrous*, 1–1½ lines long; *pedunc.* glabrous ... ... (60) **microphylla.**
lvs. obl.-lanceolate, *flat*, 2–3 lines long; claws of petals as long as the calyx ... (61) **florida.**
lvs. ovato-lanceolate, 5–10 lines long, with *recurved* margins, ciliate; claws of petals equalling the calyx ... ... (64) **ciliata.**
lvs. oblong-lanceolate, or lanceolate, flattish; the claw of the petals thrice as long as the calyx ... ... (68) **barosmæfolia.**
lvs. oblong-lanceolate, acute, flattish above, keeled beneath, 2–3 lines long; peduncles villous ... ... (62) **cerefolium.**
lvs. ovato-lanceolate, or lanceolate, acuminate, concave and keeled, 3–4 lines long, pedunc. hairy ... ... (65) **ambigua.**
lvs. lanceolate, acute, flat above, keeled, 2–3 lines long; peduncles villous ... (63) **Thunbergiana.**

** Flowers capitate:

leaves lanceolate, pungent, channelled, acutely keeled ... ... (70) **lancifolia.**

B. *Leaves narrow, linear lanceolate or linear*, 4–10 *lines long*, (*Sp.* 71–85.)

Leaves long, linear, quasi-*filiform* (*revolute margins*) (72) **acerosa.**
Leaves linear or lin. lanceolate, *flat*:
lvs. glabrous; peduncles glandular ... ... (71) **juniperifolia.**
lvs. hairy-pilose; peduncles glabrous ... ... (73) **Eckloniana.**
Leaves linear nerve-keeled, and bisulcate below (with thick edges):
minutely downy; cal. lobes ovate, obtuse; petals roundish ... ... (74) **bisulca.**
pubescent-hispid; cal. lobes linear lanceolate; petals oblong ... ... (75) **hispida.**
Leaves linear, or lanceolate-subulate, keeled, *not bisulcate*:
twigs and the recurved-patent leaves glabrous (81) **melaleucoides.**
twigs pubescent or hairy;
sterile-fil. linear-oblong, with a gland *below* the apex ... ... (77) **robusta.**
sterile-fil. with a *terminal* gland;
pedunc. shorter than the calyx or nearly as long;
lvs. glabrous; sterile-fil. glabrous-pointed (83) **rubra.**
lvs. ciliate; sterile-fil. ciliate ... ... (85) **juniperina.**
pedunc. *at least* twice as long as the calyx;
lvs. linear, incurved, *hairy at both sides* (76) **hirtella.**
lvs. linear, incurved-pointed, *glabrous above*, the margin and keel ciliate ... ... (78) **Joubertiana.**
Leaves linear, or lin.-subulate, straight-pointed:
sterile-fil. *equalling* the calyx:
leaves very acute, flattish above ... ... (84) **cuspidata.**
lvs. rather obtuse, channelled above ... (82) **ericoides.**
sterile-fil. *twice* as long as the calyx:
cal. lobes lanceolate; petals oblong-linear (80) **prolifera.**
cal. lobes ovate, sharply keeled; pet. obovate .. .. (79) **adenocaulis.**

C. *Leaves small (1½–3 lines long) oblong or lanceolate or sub-linear, 3-cornered, flat above or slightly channelled. Umbels often few flowered; flowers rarely capitate.* (*Sp.* 86–96.)

Flowers capitate:
- twigs downy; leaves glabrous, ciliate ... ... (88) **capitata.**
- branches, leaves and calyx roughly tubercled ... (94) **asperifolia.**

Flowers umbelled, 2 or 3 or more times longer than the calyx:
- twigs pubescent or hairy;
  - erect;
    - *lvs.* narrow, obtuse. bluntly keeled, *pubescent-scabrid* ... ... (86) **patula.**
    - *lvs.* narrow, incurved-pointed, margin and keel *ciliate* ... ... (87) **variabilis.**
    - *lvs.* oblong-linear, obtuse, *glabrous* ... ... (89) **erecta.**
    - *lvs.* ovate or ovate-oblong, velvetty ... ... (93) **fastigiata.**
  - decumbent or prostrate;
    - *lvs.* many-dotted below; twig-pubescence spreading ... ... (91) **decumbens.**
    - *lvs.* few-dotted below; twig-pubescence reversed ... ... (92) **nigra.**
- twigs and leaves glabrous; peduncles *downy* ... (90) **chortophila.**
- twigs, leaves and peduncles *quite glabrous:*
  - *lvs.* spreading, opp. and alternate, keeled; stem 2–3 uncial ... ... (95) **Gillivrayi.**
  - *lvs.* imbricate, alternate, convex beneath; stem 1 foot high ... ... (96) **glabrata.**

D. *Leaves small* (2–3 *lines long*), *thick or thickish, oblong or linear-oblong, very obtuse, convex beneath, round-backed, obtuse or with a short, terminal three-cornered gibbosity, slightly furrowed or concave above.* (Sp. 97–100.),

Leaves quite glabrous:
- peduncles 2–3 times as long as the calyx; claws of petals hairy ... ... (100) **crassifolia.**
- pedunc. 4–6 times as long as the calyx; claws of petals glabrous ... ... (99) **elata.**

Leaves ciliate:
- calyx ovate-acute, villoso-ciliate... ... ... (98) **sedifolia.**
- calyx lanceolate, glabrous ... ... (97) **florulenta.**

SECTION. I. **Barosmoides.** Inflorescence as in *Barosma. Petals* with longish claws. (Sp. 1.)

1. A. tabularis (Sond.); twigs pubescent; leaves ovate, glandular-crenate, quite glabrous, not dotted beneath; peduncles one-flowered, 2 or 3 together in the axils of the upper leaves, simulating a long, leafy raceme; claws of the petals filiform, glabrous, equalling the calyx. *Barosma pulchella, E. & Z.! ex pte.*

HAB. Between stones on the top of the Table-mountain, *Ecklon!* Nov. (Herb. Sond.)

A much branched shrub, very like *Barosma pulchella,* and scarcely to be distinguished except by the flowers. Leaves scattered, rarely sub-opposite, ovate, subcordate, sub-acute, margined with large glands, 2–4 lines long. Peduncles equalling or slightly exceeding the leaves. Calyx deeply 5-fid, ½ line long, the lobes ovate, obtuse, with large dorsal glands, and naked margins. Petals thrice as long as the calyx, glabrous, with an oblong, obtuse lamina. Fertile filaments as long as the petals, glabrous or sparsely hairy; sterile linear-filiform, equalling the claws of the petals, ciliate, not gland-tipped. Carpels mostly 3. Style glabrous, equalling the stamens; stigma obtuse.

SECT. II. **Capitato-racemosæ.** *Flowers* crowded at the end of the branches, simulating an ovate, leafy raceme. *Calyx* angular. *Leaves* hairy. *Style* elongate. (Sp. 2.)

2. A. hirta (B. & W. l. c. p. 188); twigs pubescent; leaves imbricate, with spreading or recurved points, lanceolate or ovato-lanceolate, acuminate, concave and glabrous above, convex beneath, impresso-punctate, and pilose; flowers racemoso-subcapitate; pedicels angular upwards, glabrous; calyx-lobes keeled, villous-bearded at the apex; petals thrice as long as the calyx, the claw villous within, limb obovate; sterile filaments twice as long as the calyx, narrow linear, villous, glabrous-pointed; ovary pilose, style glabrous. *Diosma hirta, Lam.! Encyc.* 2, *p.* 286. *Bucco hirta, R. & Sch. Syst. V. p.* 446. *A. biophylla, E. & Z.!* 861. *Bartl. Linn.* 17, *p.* 373.

HAB. C.B.S., Hb *Lam.! Krebs!* Flats under the Vanstaaden Mts. *E. & Z. Zey.* 2166. Stony places near Port Elizabeth, *Zey!* 2164. *Drege,* 7096. Dec.-Feb. (Herb. Lam., r. Berol., T.C.D., Sond.)

A shrub, 2 feet high or more, with terete branches, and crowded, sub-angular and pale, sub-umbellate, short twigs. Leaves 3–5 lines long, 1 line wide, close-set on the branches and twigs, short-stalked, appressed at base, then spreading or recurved, lanceolate, triangular-keeled at the point, softly hairy beneath, the hairs springing from glands. Flowers numerous, in the axils of the uppermost leaves; pedunc. 2-bracted, equalling the floral leaves. Calyx 1 line long, turbinate, acutely 5-angled, the lobes oblong, obtuse, ciliate. Petals white or reddish, the slender claw rather longer than the calyx, the limb narrowed at base, not glanded. Ster-filaments wider upwards, gland-tipped; fertile longer than the petals, pilose below. Style nearly *twice* as long as the petals! Carpels 3 or 4, about 2½ lines long, impresso-punctate at back, with short blunt horns.—It varies with shorter, wider and glabrescent leaves. *A. villosa,* Willd. is readily distinguished by its narrower, keeled leaves, with incurved points, inflorescence, &c.

SECT. III. **Alares.** Flowers umbellate-aggregate; umbels sessile or pedunculate in the forks of the branches. *Petals* clawed. *Ster.-filaments* villous (in one sp. unknown). (Sp. 3–6.)

3. A. alaris (Cham. & Schl. Linn. 5, p. 56); twigs scarcely downy; lower leaves lanceolate from an ovate base, *the rameal broadly ovate or sub-cordate, glabrous-margined;* umbels axillary; pedicels bibracteolate in the middle; calyx-lobes ovate, obtuse; petals obovate-oblong; ster.-filaments *very villous;* style glabrous.

HAB. Cape, *Mundt & Maire!* (Herb. r. Berol.)

More than a foot high, erect; the branches or twigs in threes or fours, leafy. Leaves sub-imbricate, petiolate, thickish, with a thick nerve beneath, glandularly downy and resinous, the lower 4 lines long, 1 line wide, the upper 2 lines long, 1½ wide. Flowers numerous, in the forks of the branches, densely umbellate. Pedicels resinous, scabrid, with linear bracts. Calyx 1 line long, downy, the lobes keeled, obtuse, ciliate. Petals white, twice as long as calyx; the claw villous, limb oblong, obtuse. Fertile stamens downy at base, glabrous upwards, longer than the petals; sterile slightly longer than the calyx, sub-linear, gland-tipped. Style shorter than the stamens.

4. A. affinis (Sond.); twigs downy; lower leaves lanceolate, rameal *oblong*, *villoso-ciliate*, all sub-carinate, with incurved points; umbels axillary, the pedicels 2-bracted in the middle; calyx-lobes ovate, obtuse; petals *oblong;* the sterile filaments villous; style glabrous.

HAB. Zwarteberg, at Vrolyk, in wet, mountain places, *Drege!* Aug. (Herb. Sond.)

Very like the preceding, from which it may be known by the much narrower (2 lines long, ¾ line wide) rameal leaves, not cordate at base, villous, especially toward the point, the shorter pedicels, the fertile filaments altogether glabrous, the rather longer sterile, and the style longer than the stamens.

5. A. leptospermoides (Sond.); twigs downy; leaves *all equal, ovate keeled, gibbous* under the point, the younger villoso-ciliate, the umbels axillary and lateral; pedicels 2-bracted *at base;* calyx lobes ovate, obtuse; petals elliptic-oblong; sterile filaments villous; style *pilose. A. apiculata, E. Mey.! non G.F. Mey. in B. & W.!*

HAB. Clayey hills, at Port Elizabeth and by the Vanstaaden River, *Drege!* Dec.-Jan. (Herb. Lehm., Sond.)

A shrub, several feet high, with 2–3-chotomous, reddish, leafy branches. Leaves sub-imbricate, petiolate, ovate, acute at each end, 3 lines long, 2 wide, glabrous, with a prominent nerve, gibbous under the tip, with a single row of glands at nerve and margin beneath, the younger villoso-ciliate near the point. Umbels many flowered, pedunculate on a very short branchlet; pedicels minutely pubescent, 2-bracted at the very base, 2 lines long. Calyx 1 line long, the segments keeled, obtuse, ciliate. Petals 2 lines long, the hairy claw as long as the limb or longer. Fert. filaments very hairy, longer than the petals; sterile linear-filiform, equalling the petals, glabrous and gland-bearing at the point. Style equalling the stamens. Capsules 2–3, sub-compressed, glabrous, not conspicuously glandular, pubescent at top, 2 lines long, with a short horn.

6. A. pentachotoma (E. Mey.! in Hb. Drege); twigs scarcely downy; all the leaves equal, *ovate, acuminate, flat,* glabrous, *glandular-denticulate;* umbels axillary; pedicels very short; calyx segments ovate, *acute;* petals . . . . . ; style glabrous.

HAB. Dutoit's Kloof, 2000–3000 f., *Drege!* Oct.-Jan. (Herb. Sond.)

A slender shrub, 2 feet or more high; branches in threes, fours or fives, reddish, glabrous. Lower leaves remote, upper closer; the lower 3 lines, the rameal 2 lines long, shortly acuminate, the midrib scarcely prominent beneath, the margins sparingly glandular. Flowers less numerous than in the preceding, and on shorter pedicels. Calyx deeply divided, the segments 1 line long, ciliated. Style glabrous, 2 lines long. Capsule 2 lines long; cocci 3–4, round-backed, glabrous, with a marginal row of prominent glands; horn short.

SECT. IV. **Involucratæ.** Flowers crowded at the end of the branchlets; sub-capitate, on short peduncles or sub-sessile; heads involucrate. Inner leaves of the involucre somewhat coloured. (Sp. 7–10.)

7. A. involucrata (E. & Z.! 858); twigs downy; leaves *ovate-oblong* or *ovate-lanceolate,* concave-channelled above, convex beneath, scabro-puberulous, gibbous under the tip; flowers *sub-sessile,* capitate, the inner leaves of the involucre *broadly obovate,* apiculate, coloured, with membranous margins; calyx lobes *linear,* the claw of the petals *equalling* the calyx; sterile filaments narrow-linear, wide in the middle, pilose, tapering into a glabrous apex. *Bartl. Linn.* 17, *p.* 380. *A. cephalotes, E. Mey.! in Hb. Drege.*

HAB. In sandy places, Berg-valley, Clanw., *Niven, E. & Z.!* Between Bergevallei and Langevallei, at Zwartbastkraal, *Drege!* Sep.-Nov. (Herb. Lehm., Sond.)

A shrub, 2–3 f. high, erect, branching, with crowded, sub-fastigiate twigs. Leaves on very short petioles, sub-imbricate, thick-edged, gland-dotted beneath, 3-angled at the point, 2½–3 lines long. Heads 4–6 lines in diameter. Outer invol. leaflets like the leaves, but villoso-ciliate, inner (about 10) in two rows, 2½ lines long, broadly obovate, membranous and purplish, with a lanceolate, acute, green middle, minutely pointed, villoso-ciliate; the inmost ones smaller, narrower and more evidently ciliate. Calyx 5-partite, the segments linear, 1½ lines long, sub-obtuse, villous at point. Petals twice as long as the calyx, the linear claw sub-ciliate, the linear-oblong limb obtuse. Fert. fil. equalling the petals, glabrous; sterile narrowed at base and apex. Ovary and style glabrous.

8. **A. sabulosa** (Sond.); twigs downy; leaves *lanceolate or linear-lanceolate, incurved* at point, concave-channelled above, convex-*keeled* beneath, minutely and thickly pubescent; flowers *shortly pedunculate*, capitate; inner involucral leaves somewhat coloured, *ovate-oblong*, or oblong-spathulate, ciliate; calyx-lobes *oblong*, shorter than the claw of the petals; sterile filaments narrow-linear, hairy from base to the middle.

HAB. Interior regions, *Niven!* (Herb. Sond.)

Very like the preceding. It differs in the longer leaves (4-5 lines long, 1 line wide), incurved at point and with an inconspicuous gibbus; but chiefly in the twice as narrow, not broadly margined, and less coloured inner involucral leaves, more evidently pedicellate flowers and long clawed petals.

9. **A. Hookeri** (Sond.); twigs pubescent; leaves imbricate, *linear, with straight, blunt points* and involute margins, *round-backed, hairy;* flowers short-stalked, capitate; inner involucral leaves somewhat coloured, clawed, *elliptical, acute,* ciliate; calyx-lobes spathulate, *acute,* ciliate; claw of the petals equalling the calyx or longer; sterile filaments narrow-linear, glabrous at base and apex, hairy in the middle.

HAB. Cape. (Hb. Hooker!)

Known from the preceding by its linear, nearly terete leaves, with strongly involute margins. About a foot high, 2-3-chotomous, with naked branches and very leafy twigs. Leaves 4-5 lines long, ½ line wide, on petioles 1 line long, ciliate, bluntly keeled, with two rows of glands, covered with short, spreading hairs, the uppermost dilated at base, membranous, and forming an outer involucre. Heads ½-inch diameter. Outer leaves of the inner involucre broadly obovate, membranous, scarious, mucronulate under the apex, 3 lines long, inner as long, clawed, the glabrous claw longer than the limb. Peduncles glabrous, 1 line long, 2 bracted below the calyx. Cal. 2 lines long, the lobes spathulate, narrowed at base. Petals on long claws, the limb short, oblong or obovate. Fert. filaments glabrous, longer than the petals; sterile shorter.

10. **A. cephalotes** (E. Mey! b., in Hb. Drege); twigs pubescent; leaves *ovate-subcordate*, acute, concave above, nerve-keeled beneath, *gibbous at tip,* pubescent and ciliate; flowers sub-sessile, capitate; inner leaves of the involucre coloured, obovato-cuneate, cuspidate, the inmost apiculate and ciliate; calyx-lobes *linear*, obtuse; claws of the petals *long;* sterile filaments glabrous at the point.

HAB. Between Bergevalei and Langevalei, at Zwartbastkraal, *Drege!* Nov. (Herb. Lehm., Sond.)

6-12 inches high, with crowded branches. Leaves sub-imbricate, 3-4 lines long, 2-3 lines wide, with a very short, broadish petiole, glabrous above, many dotted beneath, with a very entire, not thickened margin; the dorsal nerve thickening under the tip. Outer invol. leaflets very like the leaves, but longer and more acuminate, inner in two rows, silky-canescent outside, as long as the flowers. Calyx-lobes ciliolate, bearded at point. Petals 2 lines long. The rest of the flower as in *A. involucrata.*

Sect. 5. Pseudostemon. Sterile filaments villous, connate with the claw of the petals at base or in the middle. Flowers terminal-capitate or umbellate aggregate. (Sp. 11-13).

11. A. collina (E. & Z.! 860); twigs hairy; leaves imbricate, thickish, *keeled-three-angled, obtuse,* little incurved at the point, scabrous *beneath;* flowers capitate, the peduncles equalling the calyx; calyx-lobes *ovate-oblong, obtuse, ciliate;* petals twice as long as the calyx, with a villous claw; sterile filaments *linear-filiform,* villous, glabrous at point and gland-tipped, adnate with the claws of the petals to the middle. *A. graveolens, Meisn.! in Fl.* 27. 1. *p.* 302.

Hab. Hills near Swellendam, *Mundt.!* Mountain sides near Genadenal, *Krauss, Thom.* Oct.-Dec. (Hab. Meisn., Hook., Lehm., Sond.)

A foot or more high, with erect, crowded branches and twigs, the ultimate short, yellowish. Leaves 5-6 farious, imbricate, erect, lanceolate or linear-oblong, 3-angled, channelled above, scabrous beneath and at the margin, 1½-2 lines long. Petiole very short. Heads densely flowered, 3-4 lines in diameter; the peduncles glabrous, with linear, ciliate bracts at the apex. Calyx 1 line long, its lobes keeled, with membranous, ciliate margins. Petals white, 1½ lines long, with oval-oblong limb and a capillary, villous claw slightly longer than the calyx. Fertile filaments as long as the petals, pilose; sterile shorter; style equalling the stamens.—In drying this turns dull green or blackish. It varies with leaves at length glabrate, and limb of the petals obovate or elliptical.

12. A. Schlechtendalii (Sond.); twigs glabrous, hairy at the base of the petiole; leaves *ovate-oblong, or oblong, acute,* glabrous above, beneath at the keel and margin *ciliato-pilose, gibbous at tip;* flowers capitate; peduncles shorter than the calyx; calyx-lobes *oblong-lanceolate,* with *naked* margins; petals twice as long as the calyx, with villous claws; sterile filaments *dilated upwards, petaloid,* with villous, filiform bases, and connate with the claws of the petals. *Diosma eriantha, Steud.! Fl.* 1830. *p.* 550. *A. serpyllacea, latifolia, Schl. Linn.* 6. *p.* 205. *A. Schlechtendaliana, E. & Z., ex pte.*

Hab. Calcareous hills between the Breede and Duyvenhoeks rivers, Swell., *E. & Z.!* Oct. (Herb. r. Stuttg., Lehm., Sond.)

A robust, much branched shrub, with branches and twigs aggregate-umbellate, short, pale. Leaves 2-3 lines long, 1 line wide, densely sub-imbricate, erect, channelled above towards the point, beneath keeled with a raised nerve and furnished with a thick gibbosity at the apex, flat at the margin and ciliate with long white hairs; petiole very short. Heads densely flowered, 4-6 lines in diameter. Floral leaves equal to the others, shorter than the flowers. Peduncles glabrous, with ciliate, linear bracts at point. Calyx 5-parted, the lobe nearly 1½ lines long. Limb of the petals oblong, obtuse, 1 line long, white; the claw twice as long, carrying the sterile filaments in the middle. Fert. filaments longer than the petals, pilose; sterile linear-oblong, gland-bearing below the point, shorter. Style equalling the petals, glabrous. Capsule 2 lines long; carpels 2-3, glandular at back, with a straight horn.

13. A. humilis (Sond.); twigs pubescent; leaves imbricate, erecto-patent, *lanceolate, obtuse,* concave above, glabrous, *convex and pubescent* beneath; flowers in few-flowered, dense umbels; peduncles equalling the leaves, pubescent as well as the calyx; cal. lobes lanceolate-oblong, obtuse; petals with pilose claws, bearing the sterile filaments in their middle; the limb ovate-oblong; ovary 2-coccous, pubescent; style pilose.

HAB. Rocky places on the Blaawberg, 4000 to 5000 f., *Drege.!* Dec.-Jan. (Herb. Sond.)

A very dwarf shrub, 3-4 inches high, much branched, with crowded bi-uncial, twiggy branches. Leaves 2¼ lines long, ¾ line wide, close on the branches and twigs, short-petioled, narrowed from a wider base, glabrous or thinly pilose above, clothed with patent hairs beneath; the younger with longer hairs. Flowers 4-10, at the ends of the branchlets; the peduncles slender, villous, equalling the leaves. Calyx-lobes unequal, 1 line long. Petals white, thrice as long as the calyx. Capsule minutely pubescent, 2 lines long.

SECT. 6. **Diplopetalum.** *Sterile-filaments* resembling the petals, and equally long or longer. *Flowers* terminal, pedunculate, closely umbellate or capitate. (Sp. 14-27).

14. A. umbellata (Sond.); twigs glabrous; leaves imbricate, linear-oblong or linear-lanceolate, *obtuse, acutely-keeled,* sub-incurved at the point, ciliate; umbel many flowered, the peduncles glabrous, longer than the leaves; calyx-lobes not ciliate; sterile filaments equalling the petals, pilose at base, mostly gland-bearing below the apex. *D. umbellata, Thunb.! Cap. p.* 224. *D. bifida, Jacq.! Coll.* 3. *t.* 20. *f.* 1. *Ag. bifida, B. & W.! l. c. p.* 152. *E. Mey.! in Hb. Drege, b.* (*non a*). *A. bifida & serruriæfolia, E. & Z.* 855. 856.

HAB. Cape, *Thunberg!* Hills between Caledon and Genadendal, and on sand dunes by the shore, near the mouth of the Klynrivier and Cape L'Agulhas, *E. & Z.!* Babylon's Toorenberg and Hott. Holl. in steep places, *Zey!* 2172. Franschehoeks-kloof, *Drege,!* Nov.-Feb. (Herb. Thunb., Willd., Lehm., Sond.)

An erect shrub, 1-2 feet high, the branches rough with leaf-scars; twigs crowded, 1-3 uncial. Leaves close-set, erect, concave above, prominently nerved beneath, punctate at margin, 5-6 lines long, ¾-1 line wide, the younger and upper piloso-ciliate, older glabrate. Petiole ½-¾ line long. Umbel larger than in other species. Peduncles purple, 4-6 lines long, 2-bracted above the base. Calyx 1¼ line long; its segments very unequal, linear-oblong, glabrous, gland-dotted at back. Petals white, glabrous, twice as long as the calyx, linear-oblong, with narrow, pilose claws. Fert. fil. shorter than the petioles, glabrous. Carpels 3, pilose at top. Style equalling the stamens, glabrous.—It varies in the length of the claws of the petals and staminodia, whose limbs are either glanded or glandless.

15. A. lediformis (E. & Z. 854); twigs glabrous, leaves imbricate, linear-oblong, or lanceolate-attenuate, *sub-acute,* concave above, *convex* beneath, glabrous-margined; umbel many flowered; peduncles *glabrous,* longer than the leaves, calyx-lobes *glabrous-margined;* sterile filaments equalling the glabrous petals, pilose at base, *glandless* at the point. *Bartl. Linn.* 17. *p.* 372. *A. bifida, E. Mey.! c.*

HAB. In heathy ground near Brackfontein, Clan., *E. & Z.!* Oct. (Herb. Lehm., Sond.)

Very like the preceding. It differs by its round-backed, not keeled leaves, more equably acute, and glandless petals and sterile filaments. Perhaps a mere variety?

16. A. anomala (E. Mey.! in Hb. Drege); twigs pubescent or sub-glabrous; leaves imbricate, *linear-oblong, sub-obtuse,* concave above, *convex-keeled* beneath, glabrous-margined; umbel many flowered; peduncles *villous,* equalling the leaves or rather longer; calyx-lobes *ciliate;* petals *pilose;* filaments pilose, the sterile as long and thrice as broad as the petals, and like them with a gland below the apex; ovary glabrous; style pilose. *A. anomala & A. bifida? d., E. Mey. in Hb. Drege.*

HAB. Mountain at Stellenbosch, and at Dutoitskloof, 3000-4000 f. *Drege!* Oct.-Jan. (Herb. Sond.)

A shrublet with glabrous branches and reddish, crowded twigs. Leaves dense, glabrous, with 2-3 rows of glands beneath, 4-5 lines long, 1 line wide or wider; petiole 1 line long. Umbel smaller than in the preceding, the peduncles 2 lines long, with 2 linear, ciliated bracts. Calyx 1½ lines long, the segments ovate-lanceolate, ciliate, glandular at back. Petals on long glabrous or pilose claws; limb oblong, hairy. Sterile filaments petaloid. Carpels 3.

17. A. monticola (Sond.); twigs glabrous, *with raised glands;* leaves imbricate, *linear, sub-acute, keeled, 3-angled,* the younger ciliate, the adult minutely toothed; umbel many flowered; peduncles *pilose,* longer than the leaves; calyx-segments oblong, ciliate; petals on long claws, glabrous, glandless; fertile filaments glabrous, sterile as wide and as long as the petals, gland-bearing below the apex, the claw hairy; ovary setose; style glabrous.

HAB. Hills near Caledon, and at Zwartberg, *Zey.!* 2170. Nov. (Herb. Sond.)

A slender shrublet, 2 feet high, with long branches and short, crowded twigs. Leaves crowded on the twigs, erect, short-stalked, thickish, quite glabrous, 3-4 lines long, ½ line wide; upper or floral ones ½ as long. Peduncles 2-3 lines long. Calyx-lobes unequal, 1 line long. Petals more than twice as long, the capillary claw with a gland at its summit, and a few hairs at the base, longer than the obovate, obtuse limb. Sterile filaments quite like the petals.

18. A. gracilicaulis (Sond.); twigs rather hairy, *glandless;* leaves erect, alternate or sub-opposite, linear, sub-acute, *complicate-carinate, villoso-ciliate,* at length glabrate, punctate beneath; umbel many flowered; peduncles *glabrous,* rather longer than the leaves; calyx-lobes ovate at base, *ciliate,* acuminate-subulate, keeled, *and tipped with an obtuse gland;* the petals twice as long as the calyx, and the fertile filaments glabrous; sterile as long, and half as broad as the petals, with a *glabrous* claw; ovary pilose; style glabrous.

HAB. Stony mountain places near Riv. Zonder Ende, *Zey.!* 2167. (Hb. Sond.)

A very slender shrublet, 1-2 feet high; branches virgate, glabrous; twigs short, sparingly pilose. Lower leaves sometimes opposite, the rest alternate, longer than the internodes, less crowded than in the preceding, 4-5 lines long, ½ line wide. Peduncles 2 lines long. Calyx segments membrane-edged at base. Petals with a filiform claw, rather shorter than the calyx, and an oblong limb. Carpels 3, the younger pilose at top, adult glabrous. Style equalling the stamens. Capsule little longer than the calyx; the carpels punctate, with short straight beak.

19. A. parviflora (B. & W. l. c. p. 181); branches virgate, glabrous; twigs filiform, *downy;* leaves erecto-patent, *narrow-linear, acuminate,* mucronulate, *glabrous,* concave above, convex-keeled beneath; flowers 10-15 in a dense umbel; peduncles equalling or surpassing the leaves; calyx-segments *glabrous,* lanceolate, keeled, *petals a little longer than the calyx,* with glabrous claws, and obovate-oblong, spreading, recurved limb; ster. filaments quite like the petals, glabrous; carpels 2; style glabrous. *Diosma parviflora, Willd.! reliq. in R. & Sch. Syst. v. p.* 462.

VAR. *β.* **glabrata**; twigs and peduncles glabrous; leaves mostly longer. *A. parviflora, E. & Z.!* 896.

HAB. Mountain sides near Swellendam, *Mundt!* Oct. (Herb. Willd., Lehm., Sond.)

Slender, a foot or more high, with long branches and short twigs. Leaves rather close, with 2 rows of glands beneath, 3-6 lines long, in β 4-8 lines, ½ line wide. Peduncles 3-4 lines long. Calyx ½ line long. Petals scarcely half longer than the calyx, glabrous, pure white. Ster. fil. mostly equalling the petals, spreading or recurved, glandless. Fert. fil. glabrous. Style as long as calyx.—Readily known from neighbouring species by its small flowers.

20. A. montana (Schl.! Linn. 6. p. 207); twigs sub-glabrous, glandless; leaves imbricate, *linear, keeled 3-angled,* acute, *the margin and keel long-ciliate,* at length glabrate; umbel many flowered, peduncles glabrous, longer than the leaves; calyx-segments *lanceolate-linear,* glabrous-edged; petals twice as long as the calyx, *short-clawed,* glandless at point; fertile filaments glabrous, sterile half as wide as the petals, gland-tipped, the claw pilose; ovary with long hairiness; style glabrous. *E. & Z.!* 866.

HAB. Mountain sides, Baviansberg, near Genadendal, Caledon, *E. & Z.!* Nov. (Herb. Lehm., Sond.)

Very like *A. monticola,* but the branches are not elongate nor the twigs glandular, the leaves are more sharply keeled, the calyx segments narrow, 1½ line long and the petals oblong, narrowed into a filiform, pilose claw. The ster. fil. are also narrower, with lin. oblong limb. Leaves 3–4, sometimes 6–8 lines long, sub-incurved. Old peduncles pale, 4 lines long. Style at length twice as long as the calyx.

21. A. Dregeana (Sond.); virgate branches and twigs *glabrous;* leaves erect, *linear,* channelled above, convex beneath, *not keeled,* punctate, *quite glabrous,* the younger ciliate; flowers *capitate;* peduncles half as long as the calyx, glabrous, 2-bracted; calyx-lobes obtuse, ciliate; petals twice as long as the calyx, with shortish, glabrous claws; ster. fil. linear-sub-spathulate, equalling the petals, the claw filiform, pilose.

HAB. Giftberg, 2000–2500 f. *Drege!* Nov. (Herb. r. Berol., T.C.D., Sond.)

A much branched shrub, 2 feet or more high. Branches mostly crowded, 3–5 inches long. Leaves short-stalked, longer than the internodes, sub-appressed or spreading, impresso-punctate and obsoletely one-nerved beneath, 4-4½ lines long. ½ line wide, the rameal shorter than the uppermost. Heads of 12–16 white flowers. Calyx-segments 1 line long, unequal, ovate-oblong, pale, green at point, the margin hyaline, ciliate. Petals unequal, obovate-cuneate, or oblong, sub-acute, with glabrous, shortish claw. Fert. fil. longer than petals, white, glabrous; sterile like the petals but narrower, glandless. Ovary glabrous. Style equalling the stamens, glabrous; stigma obtuse. Carpels 2, shortly horned.

22. A. nigromontana (E. & Z.! 867); twigs glabrescent; leaves spreading, linear, 3-angled, *acute,* shortly and obtusely mucronate, the margin and keel *ciliato-pilose,* at length glabrate; flowers *umbellate,* peduncles glabrous scarcely longer than the floral leaves; calyx segments lanceolate, keeled, *long-ciliated;* petals *twice* as long as the calyx, the claw glabrous, filiform, equalling the calyx, limb oblong; sterile filaments equalling the petals, oblong-linear, with a gland, and a filiform, hairy claw; ovary hairy; style glabrous.

HAB. Mountain sides, Howhoek, Stell. and Zwarteberg, Caledon, *E. & Z.!* Jul. (Herb. Sond.)

A span long, branched from the base; the branches erecto-patent, sparingly divided, leafy, glabrous or thinly hairy. Leaves patent or recurved, shortly petiolate, acute at each end, impresso-punctate beneath, 5–6 lines long, ½ line wide. Flowers numerous, dense, the peduncles scarcely exceeding the calyx, at length twice as long, with 2 linear ciliate bracts above the base. Calyx segments unequal, incurved, obtuse, the shorter 1 line, the longer 1½ long. Petals white, the capillary claw glabrous or few haired, the limb about equalling the claw. Fert. fil. the length of the petals, glabrous; sterile with claws shorter than their limb. Carpels 3, setoso-pilose. Style glabrous, equalling the stamens.

23. A. gonaquensis (E. & Z.! 864); twigs very thinly downy; leaves erect, appressed or incurvo-patent, linear, 3-angled, *obtuse, glabrous, gibbous under the tip;* flowers umbellate; peduncles rather shorter than the floral leaves, glabrous; calyx glabrous, its segments linear, keeled, long-ciliate, petals *thrice* as long as the calyx, the pilose claw *twice* as long as the calyx, limb oblong; sterile filaments equalling the petals, linear-spathulate, with a gland below the apex, the capillary claw hairy; ovary and style glabrous. *Bartl. Linn.* 17. p. 375.

HAB. Sandy plains between Krakakamma and Vanstaadensriversberg, *E. & Z.!* Rocky places in Koegaskoppe, Algoa Bay, *Zey!* 2165. Glenfilling and Kovisriver, *Drege*, 7097, 7110. Albany, *T.W.* July. (Herb. Lehm., T.C.D., Sond.)

A foot high, with sub-erect, crowded branches, and short, rigid, leafy twigs. Leaves dense, tapering into the petiole, channelled above, nerve-keeled and impresso-punctate beneath, 4-6 lines long, ½ line wide. Flowers numerous, involucrated by the upper leaves; peduncles 1½-2 lines long, with deciduous bracts Calyx 1½ lines long, the segments sub-unequal, obtuse. Petals white. Fert. filaments equalling the petals, hairy at base; sterile narrow. Style equalling the stamens. Capsule glabrous, 3 lines long. Carpels 3-5, with recurved horns.

24. A. filipetala (E. & Z. 881); twigs downy; leaves close, erecto-patent, *linear, flat, sub-obtuse,* calloso-incrassate, glabrous, with 3-4 rows of glands beneath; umbel many flowered; peduncles longer than the leaves, *somewhat hairy;* calyx-lobes glabrous, lanceolate-linear, sub-obtuse; petals and sterile filaments equal, with long, hairy claws, limb linear-oblong, *glandless;* ovary and style glabrous. *Diosma stenopetala, Steud.! Flora,* 1830, *p.* 549.

HAB. Limestone hills at Grasrugg, between the Coega and Zondag-river, *E. & Z.!* Algoa Bay, *Forbes!* Aug. (Herb. r. Stuttg., Hook., Lehm., Sond.)

Erect, much branched, the branches and twigs 2-3 inch long, aggregate, sub-ternate. Leaves glabrous on both sides, with marginal, pellucid dots, the rameal mostly 4 lines long, ¾ line wide, upper shorter. Peduncles 2 lines long. Calyx-lobes sub-linear, obtuse, ciliolate, 1½ line long. Pet. and ster. fil. identical, twice as long as calyx. Fertile fil. equalling petals, hairy at base. Carpels 3.—Very like *A. linifolia*, Licht., and *A. juniperifolia*, Bartl.; differing from both in the blunter leaves, calyx and petals.

25. A. linifolia (Lichtenst.! B. & W. l. c. p. 187); twigs minutely downy; leaves patent, linear-lanceolate or linear, *acute, flat,* glabrous, pellucid dotted at the margin, the younger ciliate with soft hairs; flowers umbellate; peduncles and calyces *quite glabrous;* cal. lobes linear-lanceolate, sub-obtuse; petals twice as long as the calyx, the sub-ciliate claw a little longer than the calyx; limb *oblong, obtuse,* shorter; sterile filaments equalling the petals; the claw capillary, pilose; the spathulate limb with a *gland;* ovary and style glabrous. *E. & Z.!* 877. *Bucco linifolia, R. & Sch. l. c. p.* 448.

HAB. In Duyvelsbosch, *Lichtenstein!* Mts. near Swellendam, above Voormansbosch and Duyvelsbosch, *E. & Z.!* Oct. (Herb. r. Berol., Wendl., Sond.)

More than a foot high, erect, with scattered, erecto-patent branches. Leaves sub-remote, sometimes scaberulous under a lens, bright green above, paler and minutely punctate beneath. Pedunc. 12-20, slender, 2-3 lines long. Calyx 1 line long, the segments tipped with an obtuse gland. Petals white, spreading. Fert. filaments rather longer than the petals, glabrous; sterile half as wide, otherwise like the petals, with a small brown gland under the tip. Style glabrous. Capsule glabrous, of 2 compressed carpels, 1½ lines long, with spreading horns. This differs from *A. juniperifolia* by the longer and broader leaves, 6–8 lines long, 1-1½ wide, irregularly minutely punctate beneath; by the glabrous peduncles and form of the sterile filaments.

26. A. virgata (B. & W.! l. c. p. 139); quite glabrous; twigs filiform; leaves sub-imbricate, very narrow-linear, trigonous, *obtuse;* flowers 4-10, sub-umbellate or aggregate; peduncles 2-3 times as long as the leaves; calyx segments ovato-lanceolate, glabrous; petals *glabrous*, the claw equalling the calyx; sterile filaments equalling the petals, rather narrower, gland-tipped, the claw pilose, longer than the calyx; ovary and style glabrous. *Diosma virgata, Thunb. Prod. p.* 84. *Fl. Cap.* 222 *and Hb. fol. α, non β. Bucco Lamarkiana, R. & Sch. l c.* 447. *A. berzeliæfolia, E. & Z.* 906 *and A. aulonophila, E. & Z.!* 903, *ex pte.*

Hab. Cape, *Thunberg!* Heathy ground on the sides of Tulbagh Mt., near Winterhoek, Worcester, *E. & Z.!* Nov.

A foot or more high, erect, much branched, with sub-dichotomous branches, and crowded ramuli, the uppermost fastigiate, pale, glandular. Leaves erect, crowded on the twigs, shortly petiolate, with two rows of glands beneath, the lower 2–3, the upper 1½-2 lines long. Flowers umbellate, sometimes racemoso-corymbose. Peduncles very slender, 2-bracted at base. Calyx 1 line long, the segments sub-acuminate, punctate. Petals thrice as long as the calyx, the limb linear-oblong, obtuse, narrowed at base. Style equalling the stamens; stigma obtuse.—"Schaapen-boeku" of the colonists.

27. A. commutata (Sond.); quite glabrous; twigs slender; leaves imbricate, linear-subulate, 3-angled, *acute*, mostly spreading, or recurve-pointed; flowers 10-20, umbellate, the peduncles twice as long as the leaves; cal. segments lanceolate, glabrous edged; petals glabrous, with a *hairy* claw as long as the calyx; ovary and style glabrous. *A. virgata, E. Mey.! E. & Z.* 904, *non B. & W. A. aulonophila, E. & Z.!* 903, *ex pte.*

β. ovary hairy at top; *A. virgata, E. & Z.!* 904, *non B. & W.*

Hab. Among shrubs in mountain valleys above the Waterfall, Tulbagh, *E. & Z.!* Rocky places at Picquetberg, 1500-3000 f., *Drege!* Jul.-Nov. (Hb. Sond.)

Allied to the preceding by characters, but with a different habit. It is taller and more robust, with leaves twice as long, and tapering to an acute point, with 4 rows of glands beneath, a many flowered umbel, and hairy claws to the petals. Calyx 1 line long. Petals twice as long as the sepals, the limb linear-oblong, obtuse. Fert. filaments little longer than the petals, glabrous, or hairy at base. Carpels 3, rarely 2. Style as long as the stamens. Stigma obtuse.

Sect. 7. **Barosmæpetalæ.** Flowers terminal, umbellate-aggregate. *Petals* obovate-oblong, or oblong, cuneate below, *with a very short or scarcely any claw.* Sterile-filaments short, thickish, cylindrical-filiform or semicylindrical, gland-tipped. (Sp. 28–45).

28. A. Mundtii (Cham. & Schl.! Linn. 5. p. 56); twigs hairy; leaves spreading, *lanceolate, very acute, glabrous,* with revolute margins; peduncles *glabrous;* calyx segments glabrous, obtuse, keeled; petals oblong, narrowed into a very short *downy* claw; sterile filaments *twice* as long as the calyx, semicylindrical, glabrous, ciliolate.

HAB. Cape, *Mundt!* Swellendam, *Ecklon!* (Hb. r. Berol., Sond.)

Erect, much branched; branches 2–3–chotomous, densely leafy. Leaves 3 lines long, ½ line wide, shortly petioled, one-nerved beneath, impresso-punctate at the margin. Bracts minute, at base of peduncles. Pedunc. numerous, 3 lines long, longer than the floral leaves. Cal. ½ line long. Petals thrice as long as the calyx. Stamens at length longer than the petals, glabrous. Ster. filaments tipped with an acute gland. Style glabrous, as long as the stamens. Capsule blackish, of 2 carpels, 2 lines long, with a short straight beak.

29. A. Niveni (Sond.); twigs hairy; leaves spreading, *sub-incurved, oblong, sub-acute,* with revolute margins, impresso-glandular, convex above, *glabrous;* peduncles glabrous; calyx-lobes obtuse, sub-carinate, glabrous, ciliolate; petals obovate-oblong, with a *glabrous* claw; sterile filaments *scarcely* longer than the calyx, semicylindrical, ciliate towards the apex.

HAB. On rocks, Devilskop, *Niven!* (Herb. Sond.)

Known from the preceding by the denser and thicker leaves, tipped with an obtuse mucro, by the shorter peduncles and broader petals; from the following by the glabrous surface of all its parts. It is procumbent, much branched, with erect branches and twigs. Leaves 2½-3 lines long, ¾-1 line wide, thickish, with a minute blunt mucro; the revolute margin not furnished with stalked glands, but crenulate with impressed points. Peduncles 1½-2 lines long, equalling the floral leaves. Flowers minute, a line long. Petals twice as long as the calyx, obovate-oblong, with a very short claw. Fert. filaments as long as the petals; style at length longer. Ovary glabrous.

30. A. blaerioides (Cham. & Schl.! l. c. p. 55); twigs hairy; leaves spreading, *sub-incurved, elliptical, obtuse,* the revolute margins *set with stalked-glands,* convex above and *hairy;* peduncles hairy; calyx-segments hairy, obtuse; petals downy at base; sterile filaments *little* longer than the the calyx, semicylindrical, ciliolate.

HAB. Langekluft Mountain, Mordkuilshoogdte, *Mundt and Maire!* Olifants, R., George, *Dr. Gill.* (Herb. r. Berol., Hook.)

Much branched, a foot high, allied to *A. Mundtii,* but easily known by its smaller leaves, 1½ line long, slightly incurved, hairy above, with revolute gland-bearing margins. Flowers as in *A. Mundtii,* but the peduncles shorter, 1½ line long. Petals thrice as long as the calyx. Fert. fil. glabrous, longer than the petals; sterile half-cylindrical, concave on the inside, tipped with a minute gland. Capsule 1½ line long.

31. A. pubescens (Sond.); twigs. leaves, peduncles and calyx hairy; leaves spreading, *lanceolate, sub-obtuse, sub-cordate at base,* with revolute *impresso-glanded* margins; calyx-lobes keeled, obtuse; petals downy at base; sterile filaments little longer than the calyx, cylindrical-filiform, glabrous, pilose towards the base. *Gymnonychium pubescens, Bartl. Linn.* 17. p. 354, *cum icone.*

HAB. Interior regions, *Niven!* (Herb. Bartling).

Very like *A. blaerioides,* and perhaps merely a variety. It differs in the longer, lanceolate-attenuate lower leaves, sub-cordate at base, and destitute of *stalked* marginal glands, or having them very rarely on the uppermost leaves. It agrees in all other respects. Leaves 2½-4 lines long, rarely sub-opposite.

32. A. planifolia (Sond.); twigs pubescent; leaves spreading, *ovate-oblong, obtuse, flat,* with very entire margins, sparingly punctate beneath, *glabrous;* flowers umbellate; peduncles pubescent; calyx downy at base, the segments oblong, obtuse; petals twice as long as the calyx, with a very short, downy claw; ster. filaments shorter than the calyx, oblong-linear, downy; ovary and style *glabrous*.

HAB. Interior regions, *Niven!* (Herb. Mart., Sond.)

A much branched, erect shrub, with crowded, slender branches and twigs. Leaves very shortly petioled, coriaceous, 2-2½ lines long, 1-1½ line wide, quite flat, paler beneath, 1 nerved. Peduncles 6-10, 2 lines long. Calyx-segments keeled, green, ciliate. Petals obovate-oblong, 2 lines long, 1 line wide. Fert. filaments as long as the petals; sterile minute, gland-tipped. Style lengthening.

33. A. pubigera (Sond.); *branches, twigs, leaves, peduncles and calyx very minutely pubescent;* leaves spreading, elliptic-oblong, sub-acute, *concave and at length glabrate above, keeled beneath,* and multi-punctate; flowers umbellate; calyx-segments keeled, obtuse; petals twice as long as the calyx, obovate-oblong, narrowed at base, almost sessile; sterile filaments . . . . ; ovary *pubescent;* style *pilose. Drege,* 7120.

HAB. Rocky places on mountain tops, Ezelsbank, 4000-5000 f. *Drege!* (Herb. Sond.)

1½ foot high, yellow-green. Branches and twigs erecto-patent. Leaves 2 lines long, 1 line wide, keeled, especially towards the point. Peduncles 6-10, 2-3 lines long. Calyx ¾ line long. Petals 1 line long. Stamens wanting in our specimens. Style lengthening to 2 lines long.—Allied to *A. obtusa* by habit and leaves; but sufficiently distinct in character.

34. A. apiculata (G. F. W. Meyer, ap. B. & W.! l. c. p. 176); twigs minutely pubescent; leaves very patent or reflexed, *ovate, sub-cordate,* with reflexed, impresso-punctate margins, *tipped with a setiform, inflexed mucro, impunctate* beneath, glabrous; flowers umbellate; peduncles *equalling* the calyx or shorter, pubescent; calyx-segments ovate, acuminate, mucronate; petals rather longer than the calyx, oblong, with very short claws; sterile filaments equalling the calyx, very narrow, filiform at apex, glabrous; ovary glandular; style glabrous. *Barosma apiculata, E. & Z.!* 815.

HAB. Cape, *Hesse!* Among shrubs and sand dunes by Algoa Bay, not far from Cape Recief and Port Elizabeth, *E. & Z.! Drege!* Comtours River, *Dr. Gill!* Sep.-Oct. (Herb. Wendl., Lehm., Hook., Sond.)

Erect, much branched, the secondary branches and sub-umbellate twigs crowded, leafy. Leaves mostly reflexed, quite glabrous, rugulose above, 2 lines long; petioles appressed. Flowers umbellate-sub-capitate, peduncles 1 line long. Calyx 1¼ line long, pubescent at base, the segments membrane-edged. Petals pubescent-ciliate below, the obtuse-lamina ⅓ longer than the calyx. Fert. fil. longer than the petals, glabrous. Style equalling or exceeding the stamens. Capsule 2 lines long. glandular-dotted at back, set at margin and top with long-stalked glands; horns short and straight.

35. A. pilifera (Schl. Linn. 6, p. 206); twigs pubescent; leaves patent, *elliptical* or oblong-elliptical, acute, *with a long, hair-like, inflexed, mucro,* glandular-crenate, impunctate, glabrous; flowers umbellate; *peduncles 3–4 times longer than the calyx;* calyx tube velvetty, segments oblong, acute; petals obovate-oblong, narrowed at base; sterile filaments shorter than the calyx, downy, with an incurved, glabrous point; ovary and style glabrous. *E. & Z.!* 884. *Diosma pilifera, Steud.! Flora,* 1830, *p.* 549. *D. apiculata, Spreng.*

HAB. Heathy soil on mountain sides, Winterhoeksberg, Uit., *E. & Z.!* Sep.-Oct. (Herb. r. Berol., Stuttg., Lehm., Sond.)

Slender, 1 foot high, with slender, sub-virgate branches and filiform twigs. Leaves 2–2½ lines long, 1 line wide, sub-horizontal, flat, the mucro ½ as long. Peduncles 8–12, 3 lines long. Calyx 1 line long, afterwards longer, the segments downy at the base, some obtuse, others acute. Petals 1½ line long. Fert. fil. rather longer than the petals, glabrous; sterile oblong-linear. Style at length longer than the stamens. Capsule 2 lines long; carpels 3, punctate at back, glandless at top, with a short, obtuse beak.

36. A. fraudulenta (Sond.); branches, twigs, leaves, peduncles, and calyx *velvetty;* leaves cuneate-obovate, sub-acute, *rather convex* above, gland-dotted beneath, thickened at margin, *obsoletely crenulate;* flowers umbellate, peduncles twice as long as the leaves; calyx segments lanceolate-linear, obtuse with a gland, spreading or reflexed; petals elliptic-oblong, scarcely longer than the calyx, downy above the base within, with very short claw; sterile filaments linear-oblong, ciliolate, downy, *half the length* of the calyx; ovary *downy;* style glabrous.

HAB. Cape District; exact locality unknown. (Hb. Sond.)

A shrub with terete branches, and erect or ascending, sub-angular purplish twigs. Leaves close together, longer than the internodes, erect or spreading, coriaceous, 4–5 lines long, 1½ line wide; petiole 1 line long. Peduncles 4–6 lines long. Calyx segments unequal, erect in fruit. Petals 2 lines long, 6–8 glanded. Sterile filaments tipped with a black gland. Style as long as the stamens. Capsule of 3–5, compressed, downy, glandular carpels, 3 lines long; with erect, obtuse horns.—Quite like a *Barosma* in habit.

37. A. craspedata (E. Mey.! Hb. Drege); branches and twigs very minutely downy; leaves *obovate-oblong, sub-acute,* narrowed at base, *glabrous at both sides,* very smooth above, gland-dotted beneath, *flat,* the thickened margin sub-recurved, *glandular-denticulate;* flowers umbellate; peduncles twice as long as the leaves, and the calyx *glandular;* calyx-lobes broadly ovate, obtuse, ciliolate, at length reflexed; petals twice as long as the calyx, elliptical, downy at base, sub-sessile; sterile filaments equalling the calyx, oblong, ciliate, downy in front; ovary and style glabrous.

HAB. Rocky places on the Blaauwberg, 3000–5000 f. *Drege.!* Nov.-Jan. (Herb. Sond.)

Next the preceding in habit and foliage; but the leaves are more glabrous, as long as wide, flat, with thick, gland-toothed margin, less numerous dots, peduncles not pubescent, calyx tube wide, scarcely cleft beyond the middle, petals not glandular and sterile filaments twice as narrow, tipped with a black gland. Hyp. disc completely adnate. Style equalling the stamens.

38. A. recurvifolia (Sond.); twigs downy; leaves quadrifarious, *ovate* or *elliptical,* with a short, obtuse, acumination, *recurvo-patent* or *reflexed,* minutely downy, becoming glabrous, *concave* above, dotted with large glands beneath, with a *raised, pale margin,* quite entire; flowers umbellate; peduncles and calyces velvetty; calyx segments *ovate, obtuse;* petals *thrice* as long as the calyx, narrowed into a very short, downy claw; sterile filaments linear-semiterete, but little longer than the calyx, downy, tipped with an oblong, acute gland; ovary velvetty, 2-carpelled; style glabrous.

HAB. Stony mountains. Zwarteberg at Klaarstroom, 3000–4000 f. *Drege.!* 7124. Jul. (Herb. Sond.)

A robust, much branched shrub, with spreading, 2–3-chotomous branches, and short twigs Leaves very smooth above, dark green, 1½–2 lines long. Peduncles 4–6, 1–1½ lines long. Calyx ½ line long. Petals obovate-oblong.

39. A. gnidioides (Schl.! Linn. 6, p. 206); clothed with a very short pubescence; branches sub-ternate, erect; leaves *linear-lanceolate or lanceolate*, acute, subpungent-mucronate, flat, with the margin thickened or sub-recurved, with 2 rows of gland-dots beneath; flowers umbellate; peduncles downy; calyx segments *lanceolate*, reflexed (in anthesis); petals a *little longer* than the calyx, oblong, with a very short, downy claw; sterile filaments half as long as the petals, linear, pubescent, tipped with a large gland. *Diosma dubia, Spreng. D. puberula, Steud.! Fl.* 1830, *p.* 548. *Bar. gnidioides, E. & Z.!* 814.

VAR. β. **coriacea**; leaves broader, more coriaceous, convex above; calyx lobes with an obtuse gland. *Bar. mucronata, Meisn. in Krauss. Beytr. p.* 40.

HAB. In heathy soil, between Beaufort and Graafreynet, and between shrubs over Elandsriver, *Zeyher!* 2161. Mountain sides, Winterhoeksberg, Uit., *E. & Z.! Drege,* 7089. β. north east side of Winterhoeksberg, *Krauss.! E. & Z.!* Ap.-Aug. (Hb. r. Stuttg., Meisn., Lehm., Sond.)

Over a foot high. Leaves erect, dense, 7–8 lines long, 1 line wide, much longer than the internodes, flat or sub-convex, tipped with a very acute mucro nearly 1 line long, marked above with a furrow, and beneath with a raised midrib, having a row of glands at each side. Peduncles 4–6 lines long. Cal. lobes unequal, 1½ line long. Petals few-glanded. Capsule of 3 carpels, 2½ lines long, with a straight beak.

40. A. barosmoides (Sond.); twigs velvetty; leaves erecto-patent, *oblong, acute, mucronate, flat,* very entire, with a *straight* point, glabrous above, velvetty, one nerved and dotted beneath; flowers umbellate; peduncles longer than the leaves, and the calyx velvetty; calyx-lobes ovate, keeled, acuminate; petals *twice* as long as the calyx, obovate-oblong, cuneate at base, with a very short, downy claw; sterile filaments *shorter than the calyx,* downy; ovary downy; style glabrous.

HAB. Sandfontein, *Zey!* 285. Nov. (Herb. Sond.)

Over a foot high, with the aspect of a *Barosma.* Branches erecto-patent, glabrous; twigs short. Rameal leaves crowded, 3–3½ lines long, 1½ wide, quite flat, paler beneath. Peduncles 4–8, 3–4 lines long. Calyx semi-5 fid, the lobes with membranous edges. Petals 2 lines long, punctate. Sterile filaments with a gland sunk in the emarginate apex. Carpels 3. Style equalling the stamens. Capsules 2 lines long, gland-dotted, with straight beaks, ½ line long.

41. A. acutifolia (Sond.); twigs downy; leaves erecto-patent, *ovate-oblong,* acute, mucronate, flat, *recurved at the points, margined,* glabrous, dotted with large glands beneath; flowers umbellate; peduncles velvetty; calyx *glabrous,* its segments ovate, obtuse; petals *thrice* as long as the calyx, downy at base, gradually narrowed, with scarcely any claw; sterile filaments *longer than the calyx,* oblong-linear, downy, tipped with an oblong reversed gland, the 3 carpels and the style *glabrous.*

HAB. Stony places, Aasvogelberg. 2000–3000 f. *Drege.!* Aug. (Hb. Sond.)

A foot high, much branched, resembling *A. recurvifolia.* Leaves close-set, 3 lines long, 1 line wide, pungent-mucronate. Pedunc. 4–8, about 1½–2 lines long. Calyx-lobes ½ line long, membrane-edged, ciliolate. Petals obovate-oblong, few glanded. Fert. fil. longer than the petals, glabrous.

42. A. Martiana (Sond.); twigs downy; leaves imbricate, *ovate, sud-*

denly acuminate, spinous-pointed, very entire, glabrous dotted beneath; flowers umbellate; peduncles longer than the leaves, and with the calyx-tube *villoso-pubescent;* cal. lobes lanceolate, *keeled, acuminate,* ciliate; petals *equalling* the calyx, oblong, downy at base within, sub-sessile; sterile filaments *more than twice as short* as the calyx, oblong-linear-semiterete, ciliolate, tipped with a large gland; ovary and style glabrous.

HAB. Interior regions, *Niven!* (Herb. Mart.)

Erect, with spreading, greyish, crowded, glabrous branches; twigs 2–4 inches long, leafy. Leaves erect, 2 lines long, 1 line wide, very smooth above, one nerved beneath, paler, with 16–24 blackish dots, the marginal pellucid. Pedunc. 3–4 lines long, 2-bracted at base. Calyx 2 lines long, its lobes glabrous, reflexed in anthesis. Petals dotted. Fert. fil. glabrous, equalling the petals. Like *A. imbricata* in aspect, but readily known by the sharply mucronate leaves and sub-sessile petals.

43. **A. punctata** (Sond.); twigs downy; leaves spreading, *elliptic or suborbicular, with a short obtuse mucro,* flat, the apex sub-recurved, underneath minutely downy and dotted with *large* glands; flowers umbellate; peduncles and calyx velvetty; calyx-lobes oblong, obtuse; petals *thrice as long as the calyx,* obovate-oblong, narrowed into a very short claw; sterile filaments linear-filiform, *rather longer* than the calyx, throughout ciliato-puberulous, glabrous externally; ovary velvetty; style glabrous. *A. acuminata, E. Mey.! Hb. Drege, non Willd.*

HAB. Rocky places between Zwarteberg and Kendo, and in Witpoortsberg, 2000–4000 f. *Drege!* 7122. Jun.-Aug. (Herb. T.C.D., Sond.)

Two feet or more high, erect, slender, 2–3-chotomous, the ultimate twigs scarcely uncial, leafy. Leaves 1½–2 lines long, ½ line wide, horizontal, quite smooth and glabrous above, obsoletely one nerved beneath, with many raised glands. Pedunc. 3–6, 2–2½ lines long. Calyx semi-fid, its segments 1 line long, ciliolate. Petals 2½ lines long, scarcely clawed, downy above and at the margins. Fertile fil. scarcely longer than the petals, glabrous; sterile ½ as long, tipped with a large gland. Style as long as the stamens.

44. **A. mucronulata** (Sond.); twigs scarcely downy; leaves erecto-patent, orbicular, *with a short, acute, recurved mucro,* flat, underneath minutely downy, and sprinkled with large glands; flowers umbellate; peduncles and calyx velvetty, the calyx lobes ovate, keeled at point setaceo-acuminate; petals little longer than the calyx, obovate, narrowed into a downy base, with scarcely any claw; sterile filaments linear, thickish, *shorter than the calyx,* downy within, glabrous externally; ovary sub-velvetty; style glabrous.

HAB. Rocky places in Zwaanepoelspoortberg, 3000–4000 f. *Drege!* Aug. (Herb. T.C.D., Sond.)

Very like the preceding.—Branches trichotomous, erect. Leaves 1½ lines long and wide or smaller, at length glabrate, more evidently petiolate than in *A. punctata.* Pedunc. 3–5, about 1½ line long. Cal. lobes 1 line; petals 1½ line long. Sterile filaments tipped with a minute gland.

45. **A. spinosa** (Sond.); twigs pubescent; leaves quadrifarious, patent, sub-recurved, *cordate-ovate, spinous-mucronate,* concave and very smooth above, convex, minutely downy and roughly many-glanded beneath, with a thickened *pale* margin; flowers umbellate; peduncles pubescent, as long as the calyx; cal. lobes mucronate; petals oblong,

obtuse, sub-sessile; carpels 2, truncate, pubescent, with a short, straight horn.

HAB. South Africa (no station given), *Drege!* 7153. (Herb. Sond.)

A span long, grey, with robust branches, and slender, uncial, leafy twigs. Leaves sub-imbricate, the cucullate base appressed to the branch, mostly curved back, not keeled beneath, 1½–2 lines long, and tipped with a mucro ½–1 line. Peduncles few. Calyx pubescent at base, the lobes ovate, pungent-mucronate, ciliated, few-glanded. Petals downy, tapering into a very short claw. Capsule densely pubescent, 3 lines long; the beaks 1 line. Style glabrous, filiform in the ripe fruit and as long as the carpels.—In habit this resembles *Acmadenia pungens.*

SECT. 8. **Imbricatæ.** All the leaves, or at least those of the branchlets, *densely* imbricated, roundish-ovate, or ovate, obtuse or acuminate, concave above, ciliate. Flowers sub-umbellate-capitate. Petals with a long, capillary claw, and a roundish or obovate limb. Sterile filaments not longer than the claws of the petals. (Sp. 46–48.)

46. A. squamosa (B. & W.! l. c. p. 141); twigs sub-pubescent; leaves closely imbricate, roundish-ovate, *obtuse*, concave, glabrous, *gland-tubercled* beneath, villoso-ciliate; flowers capitate; calyx lobes ovate-oblong, ciliate; petals twice as long, the capillary claw *equalling* the calyx, limb oblong, obtuse; sterile filaments linear, villous. *Diosma squamosa, Willd.! in R. & Sch. Sqat. V. p.* 462.

HAB. Cape of Good Hope. (Herb. Willdenow.)

A dwarf shrublet, with 2–3-chotomous branches and short twigs. Leaves ½–¾ line long, appressed, sub-sessile, dark green, convex externally, with large glands. Head as large as a *pea;* with sessile flowers. Calyx ¾ line long; petals with a capillary, glabrous claw. Fert. fil. quite glabrous; sterile equalling the claws of the petals, gland-tipped. Style glabrous.

47. A. lycopodioides (B. & W.! l. c. p. 148); twigs pubescent; leaves closely imbricate, *roundish-ovate, cuspidate, tomentose beneath;* flowers sub-capitate; peduncles downy or sub-glabrous; calyx lobes oblong, obtuse, pilose, and ciliate; claw of the petals *twice as long* as the calyx, limb *roundish;* sterile filaments linear, *villous*, with glabrous points, gland-tipped. *Diosma lycopodioides, Willd.! ap. R. & Sch. l. c. p.* 461.

HAB. Cape, Hb. *Willdenow!* Cape flats, near Klipfontein, *Zeyher*, Nov. (Herb. r. Berol. Sond.)

A foot high, erect, much branched. Leaves minutely petiolate, 1–1½ line long and wide, glabrous above, the nerve beneath prominent near the point, impresso-punctate. Heads 4–6 lines diameter. Pedunc. 1 line long. Calyx 1 line long. Petals with very long claws and short limb, hairy at base. Ovary hairy. Style glabrous, as long as the stamens.

48. A. imbricata (Willd.! enum. Berol. p. 259); twigs pubescent or glabrate; leaves erect, imbricate, the lowest sometimes loose, *ovate-acuminate*, ciliate or pubescent; flowers sub-capitate; peduncles pubescent or glabrous; calyx lobes obtuse or sub-acute, keeled at the point, ciliate; the claw of the petals 2–3 *times as long as the calyx*, limb roundish or obovate; sterile filaments linear-spathulate, *ciliate below. B. & W. l. c. p.* 144.

VAR. α. **reflexa**; leaves ciliate, the lower erecto-patent; calyx lobes ovate. *B. & W. l. c. Hartogia ciliata, Berg. cap.* 68. *H. imbricata, Linn.*

Mant. p. 124. *Diosma reflexa, Soland. in Herb. Lamk.! D. imbricata, Thunb.! prod. p.* 43. *Fl. Cap. p.* 230. *Bucco imbricata, Wendl.! Coll. t.* 9. *Ag. imbricata and acuminata, E. & Z.!* 870, 869.

VAR. β. **virgata**; leaves ciliate, imbricate; peduncles longer than the calyx; calyx-segments oblong; twigs virgate. *A. virgata, Spreng.*

VAR. γ. **obtusata**; leaves all over pubescent, imbricate; peduncles longer than the calyx; cal. lobes linear-oblong. *Bucco obtusata, Wendl. Coll. t.* 76.

VAR. δ. **acuminata**; larger leaves ovate, or sub-cordate, with long acumination, ciliate, the lower very patent; calyx lobes ovate; limbs of the petals obovate or obovate-oblong. *Bucco acuminata, Wendl.! Coll. t.* 28. *Ag. acuminata, Willd.! Enum. p.* 260. *B. & W.! l. c. p.* 147. *Diosma cordata, Mart. Hort. Erl. p.* 67.

VAR. ε. **vestita**; leaves imbricate, evidently keeled beneath, glabrous, ciliate; peduncles glabrous or sparsely pilose, equalling the calyx; cal. lobes sub-acute; limb of the petals broadly obovate. *Ag. vestita, Willd.! B. & W.! l. c. p.* 142. *E. & Z.!* 871.

HAB. Cape flats and mountains round Capetown. In Zwartland, at Saldanha Bay and Hott. Holland; Tulbagh; Winterhoeksberg, &c. *Thunberg.! E. & Z., Pappe, &c. Zey.!* 298, 299. *Drege.* 1882, 7095, 7127. (Herb. Sond., T.C.D., &c.)

Erect, much branched, 3 feet or more high; twigs fastigiate, leafy. Leaves sparingly dotted beneath, flattish above, prominently nerved beneath, variable in pubescence, 1–3 lines; in δ., 4–5 lines long. Flowers numerous; pedunc. longer or shorter than the calyx, 2-bracted. Calyx 1 line long, glabrous or roughly pubescent. Petals with long claws pubescent at base, equalling the sterile filaments. Ovary pilose. Style glabrous. Capsule 2 lines long, the carpels hairy at the summit.

B. obtusata is only known to me from Hb. Wendland, but Sieber's plant is very near it, and only differs by the peduncles equalling the calyx. *A. vestita,* Willd., is certainly not specifically different, for the twigs, in the original specimen, are not glabrous, but glabrescent, and the peduncles are not always glabrous, but mostly downy or pilose at base. Many specimens of *A. imbricata,* collected in the Cape flats, belong to this variety.

SECT. 9. **Eu-Agathosma**. *Flowers* at the ends of the twigs umbellate, the peduncles mostly many times longer than the calyx; rarely shorter, and the flowers sub-capitate. *Petals* on long claws, with the limb oblong. *Sterile filaments* free, equalling or surpassing the calyx, not as long as the petals.—*Leaves,* in the narrow-leaved sometimes im bricate, in the wide-leaved not imbricate. (Sp. 49–100.)

A. *Leaves roundish, ovate, oblong, or lanceolate.* (Sp. 49–70.)

49. A. orbicularis (B. & W.! l. c. p. 175); branches and twigs minutely pubescent; leaves (minute) patent and reflexed, thickish, *roundish,* or broadly ovate, obtuse, *flat,* glabrous, *impunctate* beneath; flowers umbellate; peduncles *pubescent;* calyx segments roundish, claws of the petals longer than the calyx, limb elliptic, obtuse, narrowed at base; sterile filaments twice as long as the calyx, oblong-linear, pubescent in the middle; ovary and style glabrous. *E. & Z.!* 885 & 886, *ex pte. Diosma orbicularis, Thunb. Cap. p.* 230. Willd. Sp. Pl. I. p. 1140.

HAB. Among shrubs near the hot baths, Zwarteberg, Caledon, *Thunberg! E. & Z.!* Aug. (Herb. Thunb., Lehm., Sond.)

2 feet or more high, much branched, white-barked, with filiform flexuous, patent branches, and short erect twigs. Leaves either orbicular, cordate, elliptical or broadly ovate, one nerved beneath, 1 line long or shorter. Flowers 6–10, very small. Peduncles capillary, 2 lines long. Calyx downy, $\frac{1}{2}$ line long. Petal $1\frac{1}{2}$ lines. Sterile filaments narrowed at point, gland-tipped, fertile longer than the petals, glabrous. Carpels 3. Capsule $1\frac{1}{2}$ line long, glandular, short-beaked.

50. A. minuta (Schlect! Linn. 6, p. 206); twigs *downy;* leaves (minute) patent or reflexed, thickish, *oblong-elliptical,* or *oval,* obtuse, *concave* above, somewhat keeled and *sparsely gland-sprinkled* beneath; flowers umbellate; peduncles and calyx *glabrous;* cal.-lobes *roundish;* claws of the petals *equalling* the calyx, limb obovate-elliptical, narrowed at base; sterile filaments linear, wider and pubescent in the middle; ovary and style glabrous. *E. & Z.!* 886 *exclus. sp. from Zwarteberg.*

HAB. Kars-river and Potberg, *E. & Z.!* Between Cape L'Agulhas and Potberg, *Drege!* 7121. Aug-Oct. (Herb. Lehm., T.C.D., Sond.)

Scarcely 6 inches high, in ramification and whitish colour of the branches resembling the last, differing in the leaves, never orbicular, but oval, thicker, more strongly nerved beneath, and prominently keeled, scarcely more than $\frac{1}{2}$ line long. Umbels 4–8 flowered; peduncles glabrous, $1–1\frac{1}{2}$ line long. Petals thrice as long. Stamens longer than the petals; the sterile $\frac{1}{3}$ shorter. Carpels 3, cylindrical. Style lengthening.

51. A. thymifolia (Schlecht! l. c. p. 205); twigs glabrous; leaves patent, *oval* or *obovate,* obtuse, *flat,* glandular beneath; flowers umbellate; peduncles *downy at base;* calyx glabrous, its lobes *oblong,* obtuse, keeled; claws of the petals *shorter* than the calyx, glabrous; limb obovate-oblong, narrowed at base; sterile filaments linear-filiform, twice as long as the calyx, pilose at base; ovary and style glabrous. *E. & Z.!* 883.

HAB. Sand dunes near Saldanha Bay, *E. & Z.!* Groenekloof and Saldanha Bay, *Drege!* 7119, Aug.-Oct. (Herb. Sond.)

A span-high, erect, dichotomous, or with ternate or clustered branches and short twigs. Leaves $1–1\frac{1}{2}$ line long, 1 line wide, rather longer than the internodes, very smooth above, one nerved and with 4 rows of glands beneath, the marginal glands pellucid. Umbels 6–10 flowered, pedunc. $2–2\frac{1}{2}$ lines long, downy below or to the middle. Calyx lobes $\frac{3}{4}$ line long. Petals 2 lines long, the claw half longer than the calyx. Fert. fil. longer than the petals, glabrous; sterile half as long. Style as long as the stamens.—Easily known from the preceding by the larger flowers and leaves.

52. A. glandulosa (Sond.); twigs *pubescent;* leaves *spreading, cordate-ovate, acute, flat,* with *reflexed* margins, *minutely pubescent on both* sides, impunctate, glandular at margin; flowers umbellate; peduncles and calyces *glandular-pubescent;* calyx lobes ovate, keeled; petals more than twice as long as the calyx, glabrous, with a linear claw equalling the calyx, limb elliptic-oblong; sterile filaments *twice* as long as the calyx, lanceolate, subulate-acuminate, ciliate; ovary and style *pilose. Diosma glandulosa, Thunb.! Fl. Cap.* 229.

HAB. Cape, *Thunberg!* (Herb. Thunb.)

Several feet high, with terete, glabrous, flexuous-erect branches and sub-umbellate twigs, the ultimate 3–4 inches long, reddish, with partially glandular pubescence. Leaves sub-remote on the branches, shortly petioled, coriaceous, shining above, rugulose, and with a medial, impressed line, opaque beneath with a raised nerve, flat, $2\frac{1}{2}–3$ lines long, $1\frac{1}{2}$ wide. Umbel 10–12 flowered, pedunc. 3 lines long. Calyx $1\frac{1}{2}$

line long, the lobes keeled, sub-acute, unequal. Petals rosy, limb 2 lines long. Fert. filaments as long, glabrous; sterile shorter. Style pilose with a glabrous point.

53. A. foliosa (Sond.); twigs downy; leaves *reflexed, ovate-acuminate, sub-cordate* at base, *concave and glabrous above, minutely downy beneath,* ciliate; flowers umbellate; peduncles and calyces *pubescent;* calyx lobes ovate-oblong, obtuse; petals twice as long as the calyx, *glabrous,* the filiform claw equalling the calyx, the limb obovate-oblong; sterile filaments *scarcely longer* than the calyx, from a downy, lanceolate base, tapering into a *filiform glabrous point;* ovary and style *glabrous.*

HAB. Sandy hills between Langevalley and Bergvalley, *Zeyher!* 297. (Herb. Sond.)

Erect, with crowded, glabrous branches, and sub-fastigiate twigs, the ultimate very short. Leaves all reflexed, 1½–2 lines long, 1 line wide at base, shining above, opaque and sparsely punctate beneath, nerve-keeled. Umbel many flowered, peduncles 1½ line long. Calyx 1 line long, ciliolate. Petals 2 lines long. Fertile filaments equalling the corolla; sterile ciliolate, gland-tipped. Carpels 3, glabrous or few-haired. Style equalling the petals.

54. A. rugosa (Link. enum. p. 238); twigs pubescent and glandular; leaves spreading, *ovate-oblong, oblong, or oblongo-lanceolate, sub-obtuse, flattish* above, somewhat channelled, *rugulose,* underneath slightly keeled, *piloso-ciliate* at the margin and keel; flowers umbellate; peduncles glandular, villous or glabrous; calyx-lobes *obtuse;* petals *thrice* as long as the calyx, the *ciliate claws* scarcely longer than the calyx, limb oblong, attenuate at base, glabrous; sterile filaments *twice* as long as the calyx, attenuate to each end, villous in the middle; ovary and style glabrous.

VAR. α.; leaves ovate-oblong; peduncles villous. *Hartogia ciliaris, Linn. Syst. p.* 223. *Diosma rugosa, Thunb.! Fl. Cap.* 226. *D. ciliata, Lam.! Encycl.* 2, *p.* 287 (*excl. syn. Pluk.*) *Bucco obtusa, Wendl. Coll. t.* 13. *Ag. pubescens, Willd.! enum. p.* 259. *D. thymifolia, Willd.! in R. & Sch. V. p.* 462. *Ag. obtusa,* α., *B. & W. l. c. p.* 169, *E. & Z.* 882. *A. cerefolium, E. Mey., non. B. & W.*

VAR. β.; leaves oblongo-lanceolate, or sub-lanceolate; peduncles villous. *Hartogia lanceolata, Lin. Diosma ciliata,* β. *Lam. l. c. Bucco obtusa, oblonga, Wendl. Coll. t.* 14. *Ag. pubescens,* β., *Willd. Ag. obtusa,* β. *B. & W.!*

VAR. γ.; leaves oblong, clothed on both sides and at the margin with long, soft hairs, at length glabrate; peduncles villous. *A. mollis, B. & W. l. c. p.* 168, *excl. syn. D. tomentosa, Lee, which belongs to Adenandra uniflora,* β.

VAR. δ.; leaves oblong or oblong-lanceolate; peduncles sparingly pilose or glabrous. *A. hybrida, B. & W.! l. c. p.* 167.

VAR. ε.; leaves lanceolate; sub-acute, pubescent on both sides; peduncles glabrous. *A. hybrida, E. & Z.!* 880.

HAB. Hills and mountains near Capetown, and at Onrust River, *Thunberg, &c. Zey.!* 2171, *Drege.* 7126. δ. East side of Table Mt., *Pappe.* ε. At mouth of Klynriver, Caledon, *E. & Z.!* Apl.-Aug. (Herb. Sond., T.C.D. &c.)

Erect, a foot or more high, the branches and twigs whorled, crowded, punctate, rough and pubescent. Leaves 2–4 lines long, 1 line wide, patent or reflexed, shining above, pale, with the nerve prominent beneath, gibbous at the tip, the recurved margin impresso-punctate. Umbels 10–20 flowered. Pedunc. 2–2½ lines long. Calyx-lobes unequal, lanceolate, ciliate, 1 line long. Fert. fil. glabrous. Carpels 4–5, glabrous. Style equalling the stamens. Capsule 2 lines long.—The pubescence is very variable, the adult leaves often glabrate. *A. mollis*, B. & W., is a form with very villous leaves, the upper nevertheless glabrate. The peduncles which are mostly villous, with or without glands, in var. δ. are sometimes quite glabrous, as in a wild specimen from Dr. Pappe; and sometimes sparingly covered with short hairs, as in the authentic sp. of *A. hybrida*, B. & W.

55. A. marifolia (E. & Z.! 887); twigs *downy;* leaves reflexed, *ovate-oblong* or *oblong, obtuse, flat, with revolute, glandular-subcrenulate margins,* thickish, *velvetty on both sides;* flowers umbellate; peduncles and calyces pubescent; calyx-lobes ovate-oblong, obtuse; petals twice as long as the calyx, the claws *glabrous,* equalling the calyx, limbs *obovate-oblong, narrowed at base;* sterile filaments *but little* longer than the calyx, ovato-lanceolate, downy, with a spreading, narrow, glabrous point; ovary and style glabrous.

VAR. β., **lanceolata**; leaves oblongo-lanceolate, or lanceolate, acute; flowers rather longer.

HAB. Sandy places near Klipfontein, Clanw., *E. & Z.!* Knakisberg, *Zey.!* 296. β. Knackerberg, 1000–1500 f. *Drege!* 7088. June–Aug. (Herb. Lehm., T.C.D., Sond.)

A foot high, with spreading or erect scattered, grey branches and short twigs. Leaves close set, 1½–2 lines long, ½ line wide, prominently nerved beneath. Peduncles 6–12, 1½ line long, in front 2–3 lines. Calyx ½ line; petals 1½ line long. Fert. filaments as long, glabrous. Style short. Capsule 1½ line long, glabrous. Var. β. is 18 inches high, with leaves 3–5 lines long, ¾–1 line wide, glabrous, the upper smaller, pubescent; peduncles 3–4 lines long, and petals 2 lines long, with filiform claws; in other characters it coincides.

56. A. latipetala (Sond.); twigs minutely pubescent; leaves reflexed, ovate-oblong or oblong, obtuse, *flat* above, prominently nerved beneath, gland-dotted and *remotely ciliate* at margin, *thickish, glabrous on both sides;* flowers umbellate, the peduncles minutely pubescent; calyx glabrous, its lobes keeled, sub-acute; petals twice as long as the calyx, glabrous, with sub-filiform claws *shorter* than the calyx, limb elliptical, *obtuse or sub-cordate at base;* sterile filaments longer than the calyx, downy at base, glabrous at the apex; ovary and style glabrous.

VAR. β. **glabrata**; leaves sub-ciliate; peduncles downy at the base only.

HAB. Interior regions, *Niven!* (Herb. Mart., Sond.)

A small shrub with alternate, greyish white branches, and short crowded ramuli. Leaves close, 1 line long, ½–¾ line wide, with about 6 marginal glands and longish cilia, in β. often wanting. Peduncles 6–8, about 2–2½ lines long. Calyx partite, the segments 1 line long, ovate, with membranous ciliolate edges. Petals with a limb 1½ line long, 1 line wide, broad at base, and a very narrow claw, ½ line long. Fert. filaments as long as the petals, glabrous; sterile lanceolate at base, filiform at the apex. Style equal to the petals.—Near the preceding but differs in the mostly shorter, glabrous leaves, margined with pellucid glands and ciliate, and the broad-based petals.

57. A. elegans (Cham. & Schl.! Linn. 5, p. 55); twigs hairy; leaves *spreading, oblong, sub-acute,* pointless, with reflexed margins, *convex above,* paler underneath, *almost impunctate, minutely pubescent* on both sides; flowers umbellate; peduncles and calyx pubescent; cal.-lobes ovate, keeled, sub-obtuse; petals twice as long as the calyx, *with a downy claw* equalling the calyx; sterile filaments rather longer than the calyx, linear, ciliate in the middle; style glabrous.

HAB. Cape of Good Hope, *Mundt & Maire!* Swellendam, *Ecklon!* (Herb. r. Berol. Sond.)

Erect, much branched, with virgate branches and short, crowded, leafy ramuli. Leaves coriaceous, green above, 3 lines long, 1 line wide, the marginal glands often immersed. Flowering branches often fastigiate, forming a sort of corymb, sometimes a raceme. Peduncles 8–12, 2½–3 lines long. Calyx ⅗ cleft, 1 line long, with ciliolate segments. Petals obovate-oblong, the limb tapering into the claw. Fert. filaments glabrous. Style equalling the stamens.

58. A. propinqua (Sond.); twigs downy; leaves *appressed*, oblong or ovate-oblong, *acute*, flattish above, *keeled beneath, piloso-ciliate at the margin and keel;* flowers umbellate; peduncles and calyx *glandular;* cal.-lobes oblong-ovate, obtuse; petals thrice as long as the calyx, with glabrous claw as long as the calyx, limb oblong, narrowed at base; sterile filaments nearly twice as long as the calyx, lanceolate-linear, downy as far as the middle, the narrow apex glabrous; ovary *pilose;* style glabrous.

HAB. Sands between Tigerberg and Zandhoogte, *Drege*, 7092. Oct.-Dec. (Herb. Sond.)

Erect, 1 foot high, with scattered or crowded branches, and purplish 1–2 uncial twigs. Leaves 1½–2 lines long, ¾ line wide, sub-imbricate, glabrous above, with a very prominent nerve beneath, and quickly deciduous cilia. Pedunc. 8–12, slender, 2 lines long, 2-bracted. Calyx purplish, deeply cleft, 1 line long, the lobes keeled, pilose at back and apex, the hairs deciduous. Petals 2½ lines long. Fert. fil. as long. Style as long as the petals.

59. A. tenuis (Sond.); twigs downy; leaves crowded, erect, *appressed, spreading or recurved at the point, ovate or ovate-oblong, callous-pointed,* concave above, convex beneath, *keeled* with a slender prominent nerve, *glabrous;* flowers umbellate; peduncles *slightly longer than the calyx, pilose;* calyx glabrous, its lobes ovate-oblong, keeled, long-ciliate; petals thrice as long as the calyx, the claw *pilose, longer than the calyx,* dilating into an oblong, *basally ciliate* limb; sterile filaments twice as long as the calyx, linear, hairy to the middle, with a filiform, glabrous point; ovary hairy at the summit; style glabrous.

HAB. Sands between Paardeneiland, Blauwberg, and Tigerberg, *Drege*, 7111 ex pte. May-Aug. (Herb. Sond.)

A foot high, erect, slender, with ternate branches and twigs, the ultimate a finger long. Leaves conspicuously petioled, few glanded, 1–1½ line long, ½ line wide. Umbel 6–10 flowered; pedunc. clothed with very short, patent hairs. Calyx 1 line long; petals 3 lines. Fert. filaments as long, glabrous; sterile very hairy below.—Near *A. glabrata*, but the leaves and flowers are different. It differs from *A. fastigiata* by the leaves not being three-angled.

60. A. microphylla (G. F. W. Meyer, ad B. & W.! l. c. p. 173); twigs scarcely downy; leaves spreading, *oblong, obtuse,* flat above, keeled beneath, *thickish, gibbous-pointed, glabrous,* lower often longer; flowers umbellate; peduncles *glabrous;* calyx lobes oblong, obtuse; petals *twice* as long as the calyx, glabrous; claw as long or longer than the calyx limb obovate-oblong; sterile filaments twice as long as the calyx, linear-lanceolate, *pubescent-ciliate;* ovary and style glabrous. *E. & Z.!* 890.

VAR. β. **stadensis**; calyx-lobes keeled, with an incurved point; limb of the petals oblong, ster. filaments gland-tipped. *A. stadensis, E. & Z.!* 889.

HAB. Cape, *Hesse!* Sand-hills near Cape L'Agulhas, Swell. *E. & Z.!* β. in the plains under the Vanstaaden Mts., Uit., *E. & Z.!* Oct.-Nov. (Herb. Bartl., r. Stuttg., Lehm., Sond.)

2 feet high or more, quite glabrous, erect, much branched, the branches and twigs filiform, crowded, reddish-yellow. Lower leaves sub-remote, upper close, shining above, pale beneath, keeled with a thick nerve, the margin flat, with pellucid glands; lower 1½ line, upper 1–1½ line long, ½ line wide. Flowers 6–15, pedunc. 2–3 lines long, ebracteate, the younger nodding. Calyx glabrous, gland-dotted, scarcely more than ½ line long. Petals 2 lines long, punctate towards the apex. Fert. filaments glabrous; sterile 1 line long, attenuate to a glabrous point.—Allied to *A. glabrata*, but differs in the smaller, spreading leaves, and in the sterile filaments.

61. A. florida (Sond.); twigs scarcely downy; leaves spreading, *oblong-lanceolate* or *lanceolate, quite flat,* paler beneath, glabrous, *the younger fringed with long, soft hairs;* flowers umbellate; peduncles and calyces glabrous; calyx segments oblong, keeled, sub-acute; petals *thrice* as long as the calyx, the claws *pilose,* equalling the calyx, limb narrow-oblong, narrowed at base; sterile filaments *nearly equalling* the petals, lanceolate-linear, villous from base to middle, filiform and glabrous at the point; ovary crowned with hairs, style glabrous. *A. mollis, E. & Z.!* 879, *non B. & W.*

HAB. Stony hills near Rietkuyl, Swell., *Mundt.!* Sep. (Hb. Sond.)

Slender, 1 foot high, with scattered, erect branches, and short, crowded twigs. Leaves horizontal, 2½–3 lines long, ½–¾ line wide. Flowers numerous, peduncles 1½–2 lines long. Calyx lobes ½ line; petals 2 lines long. Fertile filaments exserted, glabrous; sterile shorter. Capsule glabrous; the 3 carpels with filiform horns, 1 line long.

62. A. cerefolium (B. & W.! l. c. p. 159); twigs pubescent; leaves spreading or recurved, oblong-lanceolate, acute, flattish above, *keeled* beneath, *piloso-ciliate at keel and margin;* flowers umbellate; peduncles *pubescent-villous;* calyx segments lanceolate, acute; petals 2–3 times longer than the calyx, the claw rather exceeding the calyx, limb narrow-oblong; sterile filaments scarcely longer than the calyx, linear, pubescent-ciliate below; ovary crowned with hairs; style glabrous. *Diosma cerefolia, Vent. Malm. t.* 93. *Bucco cerefolium, R. & Sch. l. c.* 439. *Ag. suaveolens, E. & Z.!* 888. *A. Bartlingiana, E. & Z.!* 898, *excl. syn.*

VAR. β. **glandulosa**; pubescence of the branches and peduncles glandular; calyx-lobes more obtuse, glandularly ciliate at keel and margin, often hair-pointed.

VAR. γ. **glabrata**; leaves glabrescent; peduncles with a thick, nearly glandless pubescence, calyx lobes lanceolate-oblong. *A. obtusa β., E. Mey.!*

HAB. Mountain tops near Puspas valley, Swell.; β. in mountains at Klynrivier, Caledon, *E. & Z.!* γ. rocky places near Gnadenthal, 2000-3000 f. *Drege! J. D. Hooker.* Aug.-Oct. (Herb. Wendl., r. Stuttg., Lehm., Hook., Sond.)

6-12 inches high, or more, much branched, with spreading branches and filiform twigs. Leaves close, prominently nerved beneath and thicker at the point, margined with glands and very slender hairs, or the uppermost sometimes with minute glands, 2–3, rarely 4 lines long, ½ line wide. Peduncles 1½–2 lines long, 2 bracted. Calyx ½–1 line long, ciliolate. Petals oblong, or linear-oblong, narrowed at base, 2 lines long. Fert. filaments exserted, glabrous; sterile downy at base, tapering to a filiform, glabrous, recurved point. Ovaries 3. Capsule 1½–2 lines long; the carpels obtuse, glabrous, with a very short beak. Var. *β*. is chiefly distinguished by its smaller leaves, 1–1½ line long, at margin and back furnished with stalked glands; some, however, quite glabrous.

63. A. Thunbergiana (Sond.); twigs *minutely pubescent;* leaves (small), erect or spreading, lanceolate, acute, pointless, flat above, keeled beneath, piloso-ciliate; peduncles *pubescent;* calyx pubescent at base, the lobes lanceolate, keeled, ciliate; petals about twice as long as the calyx, the glabrous claws shorter than the calyx, limb oval-oblong, obtuse, narrowed at base; sterile filaments a little longer than the calyx, lanceolate, villous, with a subulate, glabrous, recurved point; ovary and style glabrous. *Diosma ciliata, Thunb. Fl. Cap.* 225. *A. platypetala, E. & Z.!* 915. *Bartl. Linn.* 17. *p.* 371.

VAR. *β*. **patula**; leaves more patent, naked at margin or ciliate and piligerous towards the apex; peduncles glandular-downy; calyx rather smaller.

VAR. *γ*. **hirtula**; leaves appressed or patent, lanceolate, shortly ciliate at keel and margin; calyx downy; ovary pilose at top.

HAB. Cape, *Thunberg!* Winterhoeksberg, near Tulbagh, Worcester; var. *β*. on sand dunes near Cape L'Agulhas; var. *γ*. on mountains near Tulbagh, *Zeyher!* Nov. (Herb., Thunb., Holm., Stuttg., Sond.)

Erect, 2 feet high, much branched, the branches and twigs crowded, yellowish, rather hairy, at length glabrescent. Lower leaves sub-distant, upper close, punctate beneath, with prominent nerve, 1½–3 lines long, ⅓–½ wide. Peduncles 4–12, filiform, 3–4 lines long. Calyx lobes 1 line long. Capsule 2 lines long, of 3 carpels.—*A. platyphylla*, E. & Z.! only differs from Thunberg's specimen of *A. ciliata*, by the leaves in that often, in this rarely, concave. Our var. *β*. is very like *A. cerefolium β*., but differs by the acuminate leaves. Var. *γ*. is known by denser pubescence, and ciliated keels to the leaves, etc.

64. A. ciliata (Link! enum. 1. p. 238); twigs *hairy;* leaves patent, *ovato-lanceolate* or lanceolate, *acute, flat*, glabrous on both sides, *with recurved, denticulate, ciliate margins;* flowers umbellate; peduncles *glabrous;* calyx-lobes ovato-lanceolate, obtuse; petals about twice as long as the calyx, the claws glabrous, about equalling the calyx, limb elliptic-oblong; sterile filaments *twice* as long as the calyx, narrow and pilose below, above exserted, wider, glabrous; ovary pilose at top; style glabrous. *B. & W.! l. c. p.* 155. *D. ciliata, Linn. Sp. p.* 287. *Bot. Reg.* 366. *D. myrsinites, Lam. D. rugosa, Willd.*

HAB. Among shrubs on Table Mt., *Hesse, E. & Z., Drege, Pappe.* Dutoitskloof, *Drege*, 7086. Boschman's River, Albany, *Ecklon*, Ap.-June. (Hb. Holm., Lam., Wendl., Lehm., Sond.)

1–2 feet high, with reddish or yellowish, scattered or crowded, often long branches, and short twigs. Leaves close, broad-based, acuminate, paler beneath, rugulose, one-nerved, pellucid-dotted at margin, rigidly ciliate, 5–10 lines long, 1–2 lines wide. Peduncles numerous, capillary, 3 lines long. Calyx 1½ line long, ciliate. Capsule 2½ lines long, carpels mostly 2, with spreading horns, 1 line long.

65. A. ambigua (Sond.); twigs hairy; leaves erect or spreading, sub-imbricate, ovato-lanceolate, or lanceolate, *acuminate*, *patent-mucronulate*, *concave* above, keeled beneath, *piloso-ciliate* at *keel* and margin; flowers umbellate; peduncles *rather longer* than the floral leaves, *very hairy;* calyx-lobes ovato-lanceolate, keeled; petals twice as long, with glabrous claws equalling the elliptic-oblong limb; sterile filaments *longer* than the calyx, lanceolate-linear, pubescent to the middle, spreading; ovary piloso-setose at top; style glabrous. *A. Thunbergiana, Schl.! Lin. 6. p. 204, excl. syn. Thunb., non. B. & W.*

HAB. Stony places on mountain sides, Lion's and Table Mt.; and Hott. Holland and Elandskloof, *E. & Z.!* Oct.-Nov. (Herb. Sond.)

More than a foot high, much branched, with crowded branches and short pubescent or hairy twigs. Leaves 3 lines long, ¾ line broad at base, the upper mostly smaller, broad based, nerve-keeled; glabrate when old. Pedunc. 12–20, 1½–2 lines long. Calyx lobes unequal, acuminate, often ciliate, 1½ line long. Petals sometimes downy above the base. Fert. fil. exserted. Ovaries 3.—Known from *A. Thunbergiana* by the hairy branches and peduncles and the sub-concave, acuminate and mucronate leaves; from *A. variabilis* by the leaves not 3-angled; and from *A. cuspidata* by the wider and shorter, not narrowed leaves.

66. A. Ventenatiana (B. & W.! l. c. p. 161); twigs pubescent; leaves erecto-patent, the uppermost sub-imbricate, *ovato-lanceolate*, straight-pointed, flattish, or slightly channelled above, *with a medial row of hairs, keeled and hairy* beneath; flowers umbellate; peduncles equal, *villous, thrice as long as* the floral leaves; calyx-lobes lanceolate, sub-acute; petals, more than twice as long as the calyx, with a *subciliate* claw equalling the calyx, and an oblong limb; sterile filaments *twice* as long as the calyx, pubescent to the middle; style glabrous. *D. hirta, Vent. Malm. t. 72, excl. syn. Lam. Bucco Ventenatiana, R. & Sch. l. c.* 442.

HAB. Cape, *v. v. cult.* (Herb., Wendl., Sond.)

A much branched bush, 1 foot or more high, the branches and twigs rather crowded. Leaves 3–4 lines long, 1 line wide, the upper smaller, closer and more evidently channelled. Peduncles several, slender, 2–2½ lines long. Calyx 1 line long. Petals lilac, purple, or white. Fert. fil. equalling the petals, glabrous. Carpels 2.—Very near *A. villosa*, and perhaps merely a cultivated variety.

67. A. villosa (Willd.! enum. p. 259); twigs pubescent; leaves imbricate, erect, *sub-incurved* and *complicate-triquetrous*, lanceolate, glabrous above, *channelled*, keeled and *hairy* beneath; flowers sub-umbellate; peduncles mostly *unequal*, villous, longer than the floral leaves; calyx-lobes ovate-oblong, sub-acute; petals more than twice as long as the calyx, with a *glabrous* claw equalling the calyx, and an elliptic-oblong limb; sterile filaments longer than the calyx, linear, pubescent to the middle, narrowed and glabrous at point; ovary and style glabrous. *B. & W.! l. c. p. 163, excl. syn. D. villosa, Th. E. & Z.!* 907. *D. corymbosa, Montin! D. pubescens, Thunb.! Cap. p. 225. Bucco villosa, Wendl. coll. p. 14. t. 2. D. Wendlandiana, DC. Prod 1. p. 715. Ag. hirta, Bot. Reg. No. 369. Ait. Kew. 2. p. 30, excl. syn.*

VAR. *α.* **vera**; leaves hairy beneath; pedunc. unequal, scarcely longer than the floral leaves, petals twice as long as the sub-acute calyx.

VAR. *β.* **Niveni**; leaves glabrescent, green; peduncles very villous, sub-equal; calyx lobes ovato-lanceolate; petals thrice as long as the calyx, the claw twice as long as the limb.

VAR. γ. **laxa**; twigs lax; leaves villous-pubescent beneath; peduncles equal, 2-3 times longer than the floral leaves; calyx lobes ovato-lanceolate; petals twice as long as the calyx, the claw equalling the limb. *A. laxa, B. & W.! l. c. p.* 162.

HAB. Very common on the mountains round Capetown, &c., and on the Cape Flats, *Thunberg, &c. Drege,* 7090, 7093. *Zey.* 298, *ex pte.* (Herb., Thunb., Willd., Sond., T.C.D. &c.)

A foot or more high, with scattered or sub-ternate branches, and filiform twigs. Leaves 3–4 lines long, the upper shorter. Peduncles numerous, the inner much longer than the rest. Calyx 1½ line long, pubescent, the segments unequal, keeled at back. Petals flesh coloured, lilac or white, 2–2½ lines long. Ovaries 2–3, glabrous.

68. A. barosmæfolia (E. & Z.! 874); twigs and leaves *minutely pubescent;* leaves spreading, *oblongo-lanceolate* or lanceolate, *sub-mucronate,* narrowed at base, flat, *leathery, thickish at the margin,* sparsely gland-dotted beneath; flowers sub-umbellate; peduncles and calyces *pubescent;* calyx segments oblong-lanceolate, obtuse; petals *four times,* with a capillary glabrous claw thrice, as long as the calyx and an elliptic-obovate limb; sterile filaments thrice as long as the calyx, linear, villous, glabrous-pointed; ovary and style glabrous. *Bartl. Linn.* 17. *p.* 378.

HAB. Sandy ground near Brackfontein, Clanw., *E. & Z.!* July. (Herb. Hook., Sond.)

A foot high, 2–3–chotomous, with crowded, even-topped twigs. Leaves narrowed at base into a line-long petiole, shining above, pale beneath, 6–10 lines long, 1½ line wide. Flowers numerous; pedunc. 3–4 lines long, 2 bracted. Cal. 1 line long. Petals white, the limb roundish.

69. A. serpyllacea (Lichtenst! B. & W.! l. c. p. 153); twigs hairy, pubescent and ciliate; leaves erecto-patent, *lanceolate, acute,* with recurved margins, *trigonous, the nerve beneath gibbous at point,* flat above or obsoletely channelled; flowers umbellate; peduncles and calyces *hairy;* calyx-lobes ovato-lanceolate, sub-obtuse; petals *twice* as long as the calyx, the claws *much shorter* and limb oblong-elliptical; sterile filaments equalling the calyx, linear, broader at base, pubescent. *Bucco serpyllacea, R. & Sch. l. c. p.* 447. *Ag. blaerioides, E. & Z.!* 876, *non Cham. & Schl.*

VAR. α.; leaves pubescent; pedunc. hairy, twice as long as the calyx. *B. serpyll. Licht. and D. stricta, Willd. Hb.*

VAR. β.; leaves long-ciliate at keel and margin, incurved-erect; peduncles scarcely longer than the calyx.

VAR. γ.; leaves glabrous; peduncles pubescent. *A. serpyllacea, E. & Z.!* 909, *ex pte. A. Bartlingiana, E. & Z.!* 898, *ex pte.*

VAR δ.; leaves and peduncles glabrous. *A. serpyll. var. glabra, Schlecht. A. hyponeura, E. & Z.!* 894. *A. glabra, E. & Z.* 908.

VAR. ε.; leaves oblongo-lanceolate, gibbous-thickened at apex, the nerve and margins with long hairs; peduncles glabrous, twice as long as the calyx.

HAB. Hott. Hollandsberg, *Lichtenstein.* Zondagsflakte, *E. & Z.!* var. β. in sandy places near Potriver, *Zey.!* 2169. γ. flats near Tigerberg, and Zoutendalsvalley, *E. & Z.!* Between Groenekloof and Saldanha Bay, *Drege* 7087. Var. δ. near Cape L'Agulhas, Swellendam and Hott. Holland, *E. & Z.!* ε. at Gauritz River and Vishbay, *Drege.* Aug.-Oct. (Herb. r. Berol., Wendl., Lehm., T.C.D., Sond.)

A much branched bush, 1 foot high, greyish below, with reddish branches, and crowded, short twigs. Leaves close, dark green, paler beneath, 3–5 lines long, 1 line wide. Peduncles 6–12 or more, slender, twice as long as the calyx or shorter in β. Calyx 1½ line long. Petals white or lilac, glabrous or subciliate. Var. ε is possibly a distinct species, differing from the others in the leaves quite glabrous and cucullate-contracted at the apex above and hairy, having a thick short gibbus beneath: peduncles 3 lines long, with two long-ciliate minute bracts; the flowers do not differ.

70. A. lancifolia (E. & Z.! 857); twigs *scarcely downy;* leaves erect, sub-imbricate, lanceolate, acuminate, pungent, channelled above and glabrous, *acutely keeled* beneath, the margin and keel *rough* and *hispidulous;* flowers *capitate;* peduncles half as long as the calyx, *glabrous;* calyx-lobes ovate, acute, ciliate; petals twice as long as the calyx; sterile filaments longer than the calyx, shorter than the claw of the petals, linear-filiform, pilose to the middle, glabrous at apex, style glabrous.

HAB. Stony mountain-sides near Olifant's river, Clanw., *E. & Z.!* Oct. (Herb. Lehm., Sond.)

A foot high, with grey branches, and short, yellowish, sub-umbellate twigs. Leaves close-set on the twigs, thrice as long as the internodes, tapering from a broad base into a hard, triangular point, 4–6 lines long, 1–1¼ line wide, with a petiole nearly 1 line long. Floral leaves shorter than the rest, spreading, with a broad, membranous base. Calyx 1 line long, widely membrane-edged and ciliate.

B. *Leaves narrow, linear-lanceolate, or linear, 4–10 lines long.* (Sp. 71–85.)

71. A. juniperifolia (Bartl.! Lin. 17. p. 376); twigs downy; leaves patent, *linear-lanceolate,* acute, *flat, very glabrous, pellucid-dotted at the margin,* the younger ciliate with long, soft hairs; flowers umbellate; peduncles and calyx *glandular;* calyx-lobes lanceolate, sub-acute; petals twice as long as the calyx, with *sub-ciliate* claws *equalling* the calyx, and oblong limb; sterile filaments rather longer than the claws of the petals, linear, villous below, glabrous above, ovary and style glabrous.

HAB. Cape, *Dr. Gueinzius!* (Herb. Sond., a beato *Kunze* com.)

More than a foot high, with scattered or crowded, brownish grey branches and twigs. Leaves horizontal or recurved, dull green above, pale beneath, 4–6 lines long, 1 line wide. Pedunc. 12–20, 3 lines long. Calyx 1 line long, glandular at base.

72. A. acerosa (E. & Z.! 892); twigs and leaves minutely downy; leaves erect, sub-patent, *elongate-linear, with revolute margins, sub-filiform,* one-furrowed *beneath, mucronate;* flowers sub-umbellate; peduncles and calyces downy; calyx-lobes ovate-lanceolate, acute; petals 4 times as long as the calyx, with a capillary claw twice as long as the calyx, and an obovate limb; sterile filaments twice as long as the calyx, linear-spathulate, pubescent-ciliate below; ovary pubescent; style glabrous. *Bartl. Linn.* 17. p. 364.

HAB. Sandy places near Herculesfontein, Clanw. *E. & Z.!* Sep. (Herb. Lehm., Sond.)

An erect small bush, with ascending-erect, terete, glabrous branches, and even-topped twigs. Leaves erect, but not imbricate, straight or incurved, linear-terete, impresso-punctate at the margin, 8-14 lines long, ½ line wide: petiole 1 line long. Flowers tufted; pedicels capillary, 5-6 lines long. Calyx 1 line long, its lobes membrane-edged. Disc. cup-shaped, glabrous. Style elongate.

73. A. Eckloniana (Schlecht! l. c. 207); twigs downy; leaves spreading, *linear, acute, flat, hairy on both sides, with revolute margins,* the midrib prominent beneath; flowers umbellate; peduncles twice as long as the calyx, *glabrous;* calyx-lobes linear-lanceolate, sub-obtuse, ciliate, petals twice as long as the calyx, their claws *equalling* the calyx, glabrous, limb oblong, obtuse; sterile filaments *equalling* the calyx, linear, broader at base, ciliate, glabrous above; ovary and style glabrous.

HAB. Among stones on mountain sides, Howhoek, Stell., *E. & Z.!* July. (Herb. Lehm., Sond.)

8-12 inches high, with greyish, sub-erect branches and scattered or whorled slender twigs, with patent pubescence. Leaves sub-horizontal but closely set, 4-5 lines long, nearly ½ line wide, with strongly revolute, thickened and ciliate margins. Pedicels 10-16, slender, 2 lines long. Calyx 1 line long. Petals white, the capillary claw equalling the limb, which has a gland below the apex. Ovaries 3, few-haired at top. It dries dark green: and is known from allied species by its narrow-linear, hairy, revolute-margined leaves, glabrous peduncles and white petals.

74. A. bisulca (B. & W.! l. c. p. 129); twigs *and leaves clothed with very short pubescence;* leaves erect, incurved, sub-imbricate, *with recurved points,* thickish, linear, acute, *concave above, convex-keeled,* and two-furrowed beneath, *with thickened margins;* flowers sub-corymbose; peduncles and calyx downy, the calyx-lobes *ovate, obtuse;* petals 4 times longer than the calyx, with a capillary claw *more than twice as long as the calyx,* and a *roundish* limb; sterile filaments *more than twice as long as the* calyx, spathulate, ciliate, glabrous-pointed, recurved; ovary downy at the point; style glabrous. *Diosma bisulca, Thunb.! Cap. p.* 22, *Willd. sp.* 1. *p.* 1136. *Ag. prolifera, E. & Z.!* 893, *non B. & W.*

VAR. *β.*; leaves broader. *A. glauca, E. & Z.!* 875.

VAR. *γ.* **glabra**; twigs, leaves, peduncles, and calyces glabrous or nearly so.

HAB. Cape, *Thunberg!* Sandy places at Bergriver, Worcester, *E. & Z.!* *β.* at Bergvalley, Clanw., *E. & Z.!* *γ.* in the interior, *Niven!* Between Kromrivier, Piquetberg and Bergvalley, under 1000 f. *Drege* 7100. Aug.-Sep. (Herb. Thunb., Lehm., Mart., Sond.)

Erect, a foot or more high. Branches flexuous-erect, leafy; twigs scattered or aggregate, leafy, 1-2 uncial. Leaves with 4 rows of glands beneath, 6-10 lines long, ½ line wide; in *β.* 1-1½ line wide. Peduncles numerous, racemoso-corymbose, the lower axillary, 3-5 lines long. Calyx-lobes 1 line long, convex. Petals white, the claw slightly hairy at base, the limb sometimes downy. Carpels 3, 2 lines long, with very short beak.

75. A. hispida (B. & W.! l. c. p. 132); twigs and leaves *pubescent-hispid;* leaves recurvo-patent, sub-imbricate, linear, channelled above, keeled and bisulcate beneath, with thickened margins; flowers umbellate; peduncles and calyces *velvetty;* calyx-lobes *lanceolate-linear,* obtuse; petals thrice as long as the calyx, the capillary claw glabrous, *twice as long* as the calyx, limb *oblong,* obtuse; sterile filaments longer than the calyx, linear, downy, with a glabrous recurved apex; ovary velvetty; style glabrous. *Hartogia capensis, Lin. Sp.* 288. *Diosma capensis, Lin.*

Mur. Syst. ed. xiv. p. 239. *D. hispida, Thunb.! Cap. p.* 222. *Willd. Sp.* 1. *p.* 1135, *Bucco hispida, R. & Sch. l. c. p.* 446. *Ag. trachyphylla, E. & Z.!* 865.

HAB. Sands between Piquetberg and Verlooren Valley, and elsewhere, common, *Thunberg!* Between Zwartland and Paardeberg, Stell., *E. & E.! Pappe, &c.* Paarl, *Rev. W. Elliott, Drege* 7099. Jul.-Sep. (Herb. Thunb., Wendl., T.C.D., Sond.)

A foot or more high, erect, much-branched, brownish-grey, with scattered and sub-umbellate, erecto-patent, purplish, leafy branches. Leaves 3–6 lines long, with 4 rows of glands beneath. Peduncles numerous, 2–2½ lines long, purplish. Calyx purplish, 1 line long. Petals glabrous. Carpels 3.—It varies with the upper leaves incurved-erect and shorter peduncles.

76. A. hirtella (Sond.); twigs *hairy-pilose*; leaves spreading, *incurved*, linear, acute, channelled above, keeled beneath, *hairy on both sides, piligerous at summit;* flowers umbellate; peduncles 4 times as long as the calyx, minutely pubescent, ebracteate; calyx pubescent, its segments oblong, *obtuse;* petals twice as long as the calyx, the claw *equalling the calyx,* glabrous, limb oblong; sterile filaments longer than the calyx, lanceolate-linear, downy; ovary somewhat hairy; style glabrous.

HAB. At Kardow, *Zeyher*. (Herb. Sond.)

A foot high, erect, much branched, the branches scattered or crowded, downy; twigs slender. Leaves mostly horizontal, sub-complicate, gland-dotted, 3–4 lines long. Pedicels 4–6, downy, 2½–3 lines long. Calyx ⅓ line long. Petals glabrous. Ovary few-haired, at length glabrate. Carpels 3-2, glabrous, 2 lines long, shortly and bluntly horned.—Differs from *A. juniperina* by its sub-complicate, not acuminate leaves.

77. A. robusta (E. & Z.! 863); twigs pubescent; leaves erect, sub-imbricate, linear, channelled and *glabrous* above, convex-keeled and *villous* beneath, *with an incurved point furnished beneath with a small gibbosity;* flowers umbellate; peduncles and calyces *villous;* calyx lobes ovato-lanceolate, obtuse, keeled; petals twice as long as the calyx, the *pilose* claw equalling it, the limb elliptical, narrowed at base; sterile filaments slightly longer than the calyx, linear-oblong, with a *villous* claw; style glabrous.

HAB. Hills near Swellendam, *Mundt!* Oct. (Hb. Sond.)

A span high, with a thick root, and erect grey stems as thick as a goose quill, divided above into branches 2–3 inches long. Leaves 3–4 lines long, the lower somewhat lanceolate. Peduncles numerous, 4–6 lines long. Calyx-segments 1 line long, membrane-edged. Petals with a capillary claw equalling the limb. Style twice as long as the calyx.—At first sight this looks like *A. villosa* and *A. hirtella,* but has very distinct characters.

78. A. Joubertiana (Schl.! Linn. 6. p. 207); twigs hairy; leaves erect, sub-imbricate, lanceolate or linear, with an *acute* incurved point, channelled above, *sharply keeled* beneath, *the margin* and *keel ciliate,* afterwards *scabrous;* flowers umbellate; peduncles *glabrescent;* calyx *glabrous,* its lobes *lanceolate;* petals twice as long as the calyx, claw *glabrous* equalling the calyx, limb oblong; sterile filaments *equalling* the calyx, narrow, linear as far as the middle and downy, filiform and glabrous above, straight; ovary piloso-setose at summit; style glabrous. *E. & Z.!* 868.

HAB. Fields near the sea, Zoutendalsvalley, Swell., *Miss H. Joubert.* Oct. (Herb. Sond.)

Allied to the preceding. Six inches high or less, with naked branches and leafy twigs. Leaves 3-4 lines long, acutely keeled, the younger rigidly-ciliate, the old scabrous-edged and keeled. Umbel rarely few-flowered; pedunc. 2-3 lines long. Calyx 1 line long.

79. A. adenocaulis (E. & Z.! 902); twigs downy; leaves spreading, linear, acute, channelled and quite glabrous above, *convex-sub-carinate* beneath, piloso-ciliate, the short petiole *springing from a large gland;* flowers umbellate; peduncles twice as long as the calyx, *downy*, bracteolate; calyx glabrous, its lobes *ovate, angularly-keeled,* ciliate; petals twice as long as the calyx, the claw a little longer than the calyx, rather hairy, limb *obovate;* sterile filaments *longer than the calyx,* linear, downy to the middle, glabrous tipped; ovary and style *glabrous.*

HAB. Hills near Krakakamma, Uit, *E. & Z.!* July. (Herb. Sond.)

A slender bush, the branches a span long, with short twigs. Leaves erecto-patent or patent, 4-6 lines long, the upper smaller, sub-imbricate. Peduncles numerous, 1½-2 lines long. Calyx 1 line long.

80. A. prolifera (B. & W.! l. c. p. 185); twigs *pubescent;* leaves spreading, *lanceolate-subulate, somewhat 3-angled,* channelled and glabrous above, keeled beneath, and ciliate at keel and margin; flowers umbellate; peduncles *pubescent-hispid,* thrice as long as the calyx; calyx glabrescent, its lobes *lanceolate,* ciliolate; petals twice as long as the calyx, the claw equalling the calyx, limb *oblong-linear;* sterile filaments twice as long as the calyx, linear-lanceolate, *ciliate* to the middle, style glabrous. *Bucco prolifera, Wendl.! coll. iii. 9. t. 77. R. & Sch. l. c. p.* 445.

HAB. Cape. (Herb. Wendl., sp. cult.)

Branches leafy; twigs crowded, short, the upper one lengthening beyond the flowers. Leaves close, sub-erect or spreading, pellucid-dotted at edge, the lower 3-5 lines long, ½ line wide, upper half that size. Peduncles 10-15, 2-3 times longer than the floral leaves. Calyx 1 line long. Petals glabrous.—In the shape of its leaves it resembles *A. Thunbergiana* and *A. cerefolium.*

81. A. melaleucoides (Sond.); twigs glabrous; leaves *recurvo-patent,* lanceolate-subulate, *sub-obtuse,* channelled above, keeled beneath, *glabrous;* flowers umbellate; peduncles *glabrous,* 3-4 times as long as the calyx; calyx glabrous, its segments *ovate-oblong, obtuse;* petals thrice as long as the calyx, the claw *sparsely-ciliate, twice as long* as the oblong limb; sterile filaments twice as long as the calyx, linear-ciliate, style glabrous.

HAB. Cape. (Herb. Hook.)

Branches and twigs scattered or ternate, the ultimate crowded, ¾-1 inch long. Petiole short, on a persistent gland. Leaves crowded, 2-3 lines long, ½ line wide, gland-dotted at margin and keel; the floral smaller. Umbel 4-12 flowered; pedunc. 2 lines long, slender. Calyx ½ line long. Petals with a gland.—Allied to *A. prolifera;* differing in the glabrous, recurved leaves and long-clawed petals.

82. A. ericoides (Schl.! in Linn. 6. p. 206); twigs pubescent; leaves erect-subpatent, *linear, keeled, and trigonous,* channelled above, subobtuse, *shortly pilose all over;* flowers umbellate; peduncles, *downy,* longer than the glabrous calyx; cal. lobes *linear-lanceolate,* ciliate; petals twice as long as the calyx, the *glabrous* claw equalling the oblong limb; sterile filaments *equalling* or slightly exceeding the calyx, linear villoso-ciliate to the middle, *attenuated and glabrous* towards the point, ovary pilose to the summit; style glabrous. *E. & Z.!* 906. *D. pubescens, Thunb.! in Hb. Holm.*

VAR. β. **glabrata**; leaves glabrous, the upper mostly pilose; peduncles glabrous or pubescent at base.

VAR. γ.; cauline and lower branch leaves flattish; peduncles glabrous. *A. geminifolia, E. & Z.!* 901.

HAB. Sandy places between Krakakamma and Vanstaadensriversberg, *E. & Z.!* β. in the same place, *Zey.!* 2161 *ex pte. Drege!* 7109. Mts. near Swellendam, *Zey.!* 6168. Driefontein, Mosselbey, *Drege!* 7094. γ. Vanstaadensberg, *E. & Z.!* Sep.-Oct. (Herb. Holm., Lehm., Sond.)

A bush 1–1½ foot high, with erect, greyish branches and level-topped, rufescent twigs. Lower leaves sub-distant, 6–8 lines long, the rest closer, 3–6 lines long, ½ line wide, broader or narrower. Peduncles 2-4 lines long, with linear bracts. Calyx-lobes 1 line long, broad at base, membranous and ciliate. Claw of the petals either a little shorter or longer than the calyx.

83. A. rubra (Willd. et Licht.! ap. B. & W.! l. c. p. 178); twigs pubescent; leaves erecto-patent, *linear-subulate*, channelled above, sub-trigonous, *mucronate, glabrous*, with a thickened, incurved margin, flowers umbellate-*sub-capitate; peduncles shorter than the calyx or nearly as long, pubescent;* calyx-lobes lanceolate-linear, ciliate; petals longer than the calyx, the claw nearly equalling it, limb oblong; sterile filaments equalling the calyx, linear, obtuse, pubescent, glabrous pointed; ovary pilose at the crown; style glabrous. *A. cuspidata, E. & Z.!* 897, *non B. & W.*

HAB. River Zonderende, *Lichtenstein!* Hott. Holland, *Mundt and Maire!* Hills near Groenekloof and in Zwartland, *E. & Z.!* Oct. (Herb. r. Berol., Wendl., Sond.)

An erect bush, more than a foot high, much branched, the branches and twigs crowded, densely leafy. Leaves very narrow, punctate at margin, 4–6 lines long, the uppermost 3–4 lines, ⅓ line wide. Peduncles shorter than the floral leaves, rarely as long, ebracteate. Calyx 1½ line long, pubescent at base, the segments narrow from a broader base. Petals glabrous-clawed, or pilose at base. Carpels 3, glandular, with long horns.

84. A. cuspidata (B. & W.! l. c. p. 182); twigs pubescent, leaves erecto-patent, the lower often very patent, linear-subulate, *very acute, flat or slightly channelled above, keeled* beneath; flowers *umbellate;* peduncles *more than twice as long as the calyx;* calyx-lobes lanceolate-linear, ciliate; petals twice as long as the calyx, the claw equalling it, limb oblong; sterile filaments *longer* than the calyx, pubescent *recurvo*-attenuate beyond the calyx, glabrous; ovary and style glabrous.

VAR. α. **vera**; most of the leaves erect, glabrous; peduncles pubescent. *Bucco cuspidata, Wendl.! Coll. t.* 81. *D. mixta, Hort. Kew. D. acuta, Lec. Ag. cuspidata*, α, *B. & W. l. c.*

VAR. β. **glabra**; branches, leaves, and peduncles glabrous. *A. patentissima, E. & Z.!* 900.

VAR. γ. **bruniades**; leaves piloso-ciliate, mostly spreading; peduncles pilose. *D. hispida, Hort. D. bruniades, Link.! Enum. p.* 237.

HAB. Var. α, Cape, *Drege!* 7103. Interior regions, *Niven!* β. near the Waterfall, Tulbagh, *E. & Z.!* γ. Stellenbosch, *W.H.H.* (Herb. r. Berol., Wendl.. Lehm., Mart., T.C.D., Sond.)

Several feet high, with the aspect and leaves of the preceding to which it is nearly allied, but from which it differs by its very acute leaves, and longer peduncles. Leaves very narrow, punctate-glandular beneath, 4–8 lines long. Peduncles even-topped, 3–4 lines long. Calyx 1½ line long. Carpels 3, punctate, long-horned, Easily known from *A. ericoides* by its very sharp leaves, flattish above.

85. **A. juniperina** (Sond.); twigs *hairy;* leaves erect, or erecto-patent, linear, subulate-acuminate, sub-acute, *flattish* and glabrous above, *convex* beneath, nerve-keeled, the margin and keel shortly ciliate; flowers *umbellate;* peduncles *minutely downy, equalling or a little exceeding the calyx,* scarcely exceeding the floral leaves; calyx glabrescent, the lobes lanceolate, keeled, ciliate; petals twice as long as the calyx, the claws glabrous, shorter than the calyx, limb oblong; sterile filaments lanceolate-linear, little longer than the calyx, ciliate; ovary and style glabrous.

HAB. Winterhoeksberg, 2–3000 f. *Drege!* 7102. Jan. (Herb. Sond.)

A small, slender, rather hairy erect bush, with spreading branches and short twigs. Leaves 4–6 lines long, ½ line wide, the upper shorter and more erect. Peduncles few, 1–1½ line long. Calyx 1 line long, with very unequal, keeled segments. Sterile filaments very narrow.—Very like *A. cuspidata β.*, but known by its flatter leaves, convex-keeled beneath, not very acute, and ciliate or hirtulate with shorter hairs, and the shorter peduncles. From *A. prolifera* it differs by its longer leaves. shorter peduncles and wider petals. It is also very similar to *A. Eckloniana.*

C. *Leaves small (1–3 lines long), oblong or lanceolate, or sub-linear-trigonous, flat or slightly channelled above.* (Sp. 86–96.)

86. **A. patula** (G. F. W. Mey. ap. B. & W. l. c. p. 134); twigs *pubescent-scabrid;* leaves spreading, sub-trigonous, channelled above, *lanceolate-linear,* or *linear, obtuse, convex, sub-carinate, and pubescent-scabrid* beneath; flowers umbellate; the *ebracteate* peduncles and the calyx pubescent; calyx lobes lanceolate, obtuse; sterile filaments twice as long as the calyx, *oblong,* taper-pointed, everywhere hairy; ovary scabrid; style glabrous.

HAB. Cape, *Hesse.* (Unknown to us; not in Herb. Bartl. and Wendl.)

A much branched, greyish bush, with scattered or crowded branches and twigs, flexuous-erect, leafy upwards. Leaves close, the lowest very patent, thickish, shining above, obsoletely impresso-punctate beneath, with a thickened margin, often ciliate, 2 lines long. Peduncles 4 lines long. Calyx 1 line long, its lobes unequal, keeled, glandular. Petals unknown.—Seemingly allied to the following, but not quite the same.

87. **A. variabilis** (Sond.); twigs *pubescent, or hirsute;* leaves imbricate, linear-lanceolate, or subulate-trigonous, *the apex incurved, apiculate,* at length sub-obtuse, *slightly channelled* and glabrous above, convex-keeled beneath, *piloso-ciliate at keel* and margin; flowers umbellate; peduncles *little longer than* the floral leaves, pubescent, or hirsute; calyx lobes lanceolate, ciliate; petals twice as long as the calyx, the claw equalling the calyx, the limb oblong; sterile filaments equalling the calyx, *lanceolate,* villous; style glabrous. *A. Thunbergiana, B. & W.! l. c. p.* 150, *excl. syn.*

VAR. *α.*; leaves erect, lanceolate-subulate, ciliate; peduncles minutely pubescent.

VAR. β. leaves erect (short), oblong-linear or lanceolate, at length glabrescent; peduncles pubescent.

VAR. γ.; leaves longer, with spreading points, pilose at keel and margin; peduncles hirsute.

VAR. δ.; lower leaves spreading, lin. lanceolate, upper appressed, imbricate, lanceolate, ciliate at keel and margin, afterwards scabrous; peduncles somewhat hirsute. *A. patula, E. & Z.!* 911, *non Mey.*

HAB. α. in Dutoitskloof, 1000-2000 f. *Drege,* 7108. β. same place and at Paarlberg, *Drege,* 7106. γ. (no station given) *Drege,* 7105. δ. mountain sides near Tulbagh Waterfall, *E. & Z.!* Oct.-Jan. (Herb. Wendl., Sond.)

An erect shrub, a foot or more high; γ. smaller, much branched, with crowded branches and twigs. Lower leaves more distant and longer, 2–2½ lines, upper 1–1½ line long, flattish or channelled, sparsely punctate beneath. Flowers mostly surmounted with new branches, 8–16, sub-umbellate; peduncles 2–3 lines long. Calyx 1–1½ line long, hairy or glabrous-ciliate.—Var. α. is near *A. prolifera* from which it differs in the erect, imbricate, trigonous-pointed leaves, not dorsally-ciliate, twice shorter peduncles and broader petals. Var. β. is more like *A. erecta;* but differs in the acuminate, acuter leaves, more hirsute peduncles and twigs, and innovations exceeding the leaves. Var γ. approaches *A. rubra,* but the leaves are much shorter and pubescence different. From *A. ericoides* it differs in the acute and shorter leaves, always glabrous above.

88. A. capitata (Sond.); twigs scarcely downy; leaves imbricate, *lanceolate-trigonous, sub-obtuse,* acutely keeled, *long-ciliate* at keel and margin, otherwise glabrous; flowers *capitate;* peduncles equalling the calyx, glabrous, bracteolate below the apex; calyx-lobes lanceolate, *keeled,* glabrous at back and margin; petals longer than the calyx; sterile filaments equalling the claw of the petals, linear filiform, glabrous at base and apex, pilose in the middle; style glabrous. *A. erecta, E. Mey. in Hb. Drege, non B. & W.*

HAB. By water-courses, Piquetberg, 2000–3000 f. *Drege!* Nov. (Herb. Sond.)

A small, much branched bush, a foot high, with greyish glabrous branches, and short, about uncial twigs. Leaves appressed, channelled, 1½–2 lines long. Heads sessile, 10–20 flowered: pedunc. 1 line long, bracts lanceolate, ciliate, separate, with 2 ciliolate bracteoles at the apex. Calyx-lobes concave-keeled. Petals with a filiform, glabrous claw, and a shorter lamina. Ovary pilose at top. Carpels 2, glabrous, with ciliate horns.

89. A. erecta (B. & W.! l. c. p. 135); erect, much branched; twigs clothed with *very short hairs;* leaves imbricate, *oblong-linear, obtuse, trigonous, minutely gland-dotted,* the younger ciliolate at keel and margin, adult glabrous; flowers umbellate; peduncles *villous;* the segments of the glabrous calyx ovato-lanceolate, obtuse; petals twice as long as the calyx, the claw equalling the calyx, the limb oblong, or elliptical; sterile filaments equalling the calyx or rather longer, linear, pubescent, glabrous-pointed; ovary and style glabrous.

VAR. α.; peduncles short, equalling the floral leaves. *Bucco erecta, Wendl. Coll.* 1. p. 77. t. 3.

VAR. β.; peduncles equalling, or a little exceeding the floral leaves; flowers twice as long as in α. *D. brevifolia, Lam. Encycl.* 2. *p.* 285.

VAR. γ.; peduncles 2-4 times longer than the floral leaves. *D. thuy-*

oides, Willd.!—R. & Sch. p. 462. *A. erecta, β. Wendl. excl. syn. Lam. D. tenuissima, Lodd.—Willd. Sup. p.* 12. *Ag. tenuissima, Otto! D. Mey.!*

VAR. δ. **pilosa**; leaves long-ciliate at margin and keel. *Schl. Lin.* 6. *p.* 204.

HAB. Among shrubs in the Cape flats, and Zwartland. γ. at the Paarl. Nov.-Dec. (Herb. Willd., Wendl., T.C.D., Sond.)

A much branched bush, a foot high, with scattered or crowded twigs. Lower leaves sub-distant, upper closer, flat or channelled, keeled, 1½-2 lines long, umbels many flowered, in var. α, often few flowered, the peduncles variably pubescent, with or without bracts. Calyx 1 line long or longer, keeled.—*D. brevifolia*, Lam., according to an authentic specimen sent by Prof. Roeper, is more robust and has larger flowers than the other varieties, but does not otherwise differ. Var δ, pilosa, Schl. is unknown to me, and may belong to some other species. *A. tenuissima*, Otto! cannot be distinguished from specimens of *A. thuyoides*, in Hb. Willd.

90. A. chortophila (E. & Z.! 914); *glabrous;* leaves erect, sub-appressed, sub-incurved at point, *oblong-linear*, sub-acute, trigonous, with 2-4 *rows of tuberculoid-glands* beneath; flowers umbellate; peduncles downy or sub-glabrous; calyx *gland-tubercled,* its lobes sub-obtuse; petals twice as long as the calyx, the claw equalling the calyx or longer; sterile filaments longer than the calyx, lanceolate-linear, pubescent-ciliate, with a recurved point; ovaries and style glabrous. *Bartl. Linn.* 17. p. 367.

VAR. α.; leaves sub-distant, 2-3 lines long; peduncles downy or half-glabrous; petals oval. *A. chortophila, E. & Z.!*

VAR. β.; leaves sub-distant, 1 line long; pedunc. sub-glabrous; petals oval-oblong or oblong.

VAR. γ.; leaves sub-distant; the younger sub-imbricate, 1½–2 lines long; peduncles minutely downy; petals oblong. *A. cyminodes, E. & Z.!* 916. *Bartl.! l. c. p.* 368.

VAR. δ. **ciliolata**; leaves, especially the lower, minutely ciliate.

HAB. Among grass and *Restiaceæ* at the base of the mountains near the Tulbagh Waterfall, *E. & Z.!* Var β. same place, *Drege* 7107. Var. γ. near Swellendam and Koschmanskloof; δ. with var. α. *E. & Z.!* Nov.-Dec. (Herb. Lehm., Sond.)

Erect, many stemmed, from a thick root, a foot or more high, glabrous, except the youngest twigs. Branches and twigs slender, fastigiate. Leaves tuberculated with long convex-glands, umbels 5-8 flowered; pedicels 3-4 lines long. Calyx ¾-1 line long, glabrous, with oblong or lanceolate segments. Sterile filaments in vars. α. and β. but little exceeding the calyx, in γ. twice as long, narrowed at point.

91. A. decumbens (E. & Z.! 912); stem *decumbent-prostrate,* branches *divaricate,* twigs and leaves *minutely-pubescent;* leaves *sub-imbricate, linear-trigonous,* obtuse, with thickened margins, 2-*furrowed* and *impresso-multiglandular* beneath; flowers umbellate; peduncles and calyx *glabrous;* calyx-segments lanceolate, sub-obtuse; petals twice as long as the calyx, the *claw shorter than the calyx,* limb obovate; sterile filaments *linear,* pubescent-ciliate, tapering to a recurved, glabrous point, longer than the calyx; ovary pilose at the summit; style glabrous.

HAB. Calcareous and argillaceous hills at Grasrugg, near Coega River Uit., *E. & Z.!* Aug. (Herb. Sond.)

Stems numerous, from a thick root, prostrate or rooting, 1-1½ foot long, much branched, with greyish branches and short, sub-umbellate twigs. Leaves straight or sub-recurved, with 4 rows of glands beneath, the lowest sub-distant, 2–3 lines, upper

1½–2 lines long, ½ line wide. Peduncles few, 2–3 lines long. Calyx ½ line long.—Readily known from allied species by its pubescence and prostrate growth.

92. A. nigra (E. & Z.! 917), branches *weak* prostrate; twigs downy, *with reversed hairs;* leaves sub-imbricate, erect-appressed, *linear, obtuse,* 3-angled, *sparingly impresso-punctate* beneath, *glabrous,* rarely ciliolate at keel and margin; flowers umbellate; peduncles longer than the floral leaves, minutely pubescent or glabrate; calyx glabrous, its segments ovato-lanceolate, keeled; petals twice as long as the calyx, the claw *half* its length, limb obovate-oblong; sterile filaments scarcely exceeding the calyx, *oblong-linear,* pubescent-ciliate to the very apex; ovary pilose at apex; style glabrous. *Bartl. Linn.* 17. p. 366.

HAB. Hills between Zwartberg and Klynriviersberg, Caledon, *E. & Z.!* Aug. (Herb. Lehm., Sond.)

A dwarf bush, with prostrate or ascending, greyish branches, 6–12 inches long, and erect, fastigiate reddish twigs. Leaves sub-distant on the branches, crowded on the twigs, channelled, 1½–2 lines long. Umbel 3–6 flowered, pedunc. 2–2½ lines long, never quite glabrous. Calyx segments ½ line long, not ciliate. Petals quite glabrous, with an apical gland. Carpels 2–3, 1 line long, with short, straight horns.

93. A. fastigiata (E. & Z.! 918); branches *fascicled;* twigs clothed with very short, *patent* hairs; leaves close, not imbricate, sub-appressed, *ovate or ovate-oblong, inflexed and acute* at the point, rather concave above, acutely nerve-keeled beneath, sub-trigonous, sparingly impresso-punctate, glabrous; flowers umbellate; peduncles velvetty; calyx-lobes glabrous, ovato-lanceolate, *ciliolate;* petals twice as long as the calyx, the glabrous claw *equalling it,* limb obovate-oblong; sterile filaments twice as long as the calyx, lanceolate, villous, *with a glabrous filiform apex;* ovary and style glabrous.

HAB. Mountain sides of the Langekloof, George, *E. & Z.!* Dec. (Herb. Sond.)

Many stemmed, erect, 6–8 inches high, with slender, virgate, yellowish, fastigiate twigs, 2–4 inches long. Leaves 1½ line long, ½ line wide. Umbels few flowered, sometimes aggregate; pedunc. longer than the floral leaves, 2 lines long. Calyx gland-dotted without, the lobes keeled, sub-obtuse, ½ line long. Petals glabrous; the limb 1 line long.—Like *A. erecta,* but differs in the sub-simple branchlets, leaves not imbricate, more ovate and acute, velvetty peduncles and smaller calyx.

94. A. asperifolia (E. & Z.! 872); twigs *hairy;* leaves sub-imbricate, *oblong-lanceolate* or lanceolate, carinate-trigonous, obtuse, *gland-tubercled, rough;* flowers *densely capitate,* sub-sessile; segments of the *rough* calyx oblong, obtuse, ciliate; petals thrice as long as the calyx; sterile filaments not equalling the claw of the petals, villoso-ciliate, glabrous pointed; style elongate, pilose. *Bartl. Linn.* 17, *p.* 379. *A. salina, E. & Z.!* 873, *ex pte.*

HAB. Stony mountain-sides, Tulbagh's berg, near the Waterfall, *E. & Z.!* Interior regions, *Niven!* Between Nieuwekloof and Ylandskloof, 1000–2000 f. *Drege!* Nov.-Jan. (Herb. Mart., Lehm., Sond.)

2 or more feet high, much branched, with erect branches, and crowded, short leafy twigs. Lower leaves spreading, upper imbricate, rough with tubercles and bristles, 1½–2 lines long. Heads as large as a pea. Calyx 1 line long, the keeled segments membrane-edged below. Petals glabrous, with a capillary claw twice as long as the calyx, limb obovate, with a gland under the apex. Ster.-fil. narrow-linear, attenuate at base, gland-tipped. Style at length elongating.

95. A. Gillivrayi (Sond.); *dwarf, erect, quite glabrous;* leaves erecto-patent, scattered or opposite, *oblong or linear-oblong, obtuse,* channelled above, keeled and sub-trigonous beneath, with a *sub-hyaline, pellucid-dotted* margin; flowers *umbellate;* peduncles 3–4 times longer than the leaves; calyx-segments ovate, *obtuse;* petals thrice as long as the calyx, the glabrous claw equalling it; sterile filaments twice as long as the calyx, linear, downy in the middle; ovary and style glabrous.

HAB. Simon's bay, *M'Gillivray! Baron v. Ludwig!* (Herb. Hook,, Sond.)

2–3 inches high, many-stemmed, with a long, white root. Stems erect, filiform, rufescent. Leaves decussate, alternate on the young twigs, bright green, with a few glands along the keel and margin, 1½ line long. Umbels 4–6 flowered; pedunc. 4 lines long. Calyx 1 line long, the lobes membrane-edged, naked. Petals glabrous. limb obovate, obtuse, narrowed at base, gland dotted.—Allied to *A. erecta,* and *A. decumbens,* but sufficiently distinct from both.

96. A. glabrata (B & W.! l. c. p. 165); quite glabrous; leaves imbricate, thickish, *oblongo-lanceolate* or *oblong,* obtuse, *convex* beneath, *gibbous and trigonous* at the apex; flowers umbelled; peduncles 2–4 times as long as the calyx; calyx-lobes *ovato-lanceolate, sub-carinate;* petals twice as long as the calyx, the claw equalling it or a little shorter, glabrous; limb elliptical, narrowed at base; sterile filaments *a little longer than the calyx,* linear-spathulate, ciliate at base; ovary and style glabrous. *A. thuyoides, E. & Z.! 910, non. Willd. A. erecta, δ., glaberrima, Steud.!*

VAR. β. **Eckloniana**; upper leaves shorter, sub-acute, densely imbricate; calyx lobes ovate, acute; petals wider; ster. fil. linear, not wider upwards.

HAB. Cape flats near Duykervalley and Doornhoogde, and in Zwartland, *E. & Z.!* Bergvalley, *Zeyher! β. Cape flats, W. H. H.! Zey.!* 298. (Herb. Wendl., Lehm., T.C.D., Sond.)

A foot high, erect, with straightish, scattered or crowded branches, and short, leafy twigs. Leaves dense, erect, quite glabrous, concave above, convex-keeled beneath, sparsely punctate, glaucous-green, the lower 3–4 lines, the upper 3–2 lines long; in var. β. the upper 1½–1 line long. Umbel many flowered; pedunc. 3 lines long. Calyx 1½ line long; in β. 1 line, ciliolate or glabrate. Petals 1½ line long, white or lilac. Carpels 2, glabrous, with short, straight horns.

What is probably a variety of this species with minutely pubescent peduncles, is preserved in Herb. Thunb. under the name *D. virgata, b.;* the habit is the same, and neither leaves nor flowers differ materially. I have not seen this plant in other collections.

D. *Leaves small (2–3 lines long), thick or thickish, oblong or linear-oblong, very obtuse, convex-round-backed or with a short trigonous swelling under the point; concave above, or rather furrowed.* (*Sp.* 97–100.)

97. A. florulenta (Sond.); twigs *minutely downy;* leaves erect, subpatent, *thick,* oblong-linear, or oblong, obtuse, *slightly channelled* above, *convex-keeled, obtusely trigonous* beneath, remotely *ciliate,* at length glabrate; flowers umbellate; peduncles *downy;* calyx glabrous, its segments *lanceolate,* keeled, *not ciliate;* petals more than twice as long as the calyx, the claw equalling the calyx, glabrous; limb elliptical, obtuse, narrowed at base; sterile filaments *twice* as long as the calyx, oblong-linear, downy below, linear and glabrous above; ovary and style glabrous.

HAB. Hills near Zoutendaals valley, Zwell. *Miss Joubert.* Oct. (Herb. Sond.)

Dwarf, 4–5 inches high. Stems several, erect, with scattered branches and many short twigs. Leaves sub-distant, 2–3 lines long, the upper close, but not imbricate, 2–2½ lines long, rather more strongly keeled, and with a minute apical swelling. Umbel 8–16 flowered; pedunc. twice as long as the outer floral leaves. Calyx 1 line long, with unequal segments. Petals glabrous, white.—Like *A. sedifolia*, Schl., from which it differs in the conspicuously keeled and sub-trigonous leaves, with remote, very short and deciduous cilia; and especially in the calyx.

98. **A. sedifolia** (Schl.! l. c. p. 206); twigs glabrous; leaves erecto-patent, *thick, linear-oblong,* channelled, *sub-cucullate* at point, obtuse, obtusely keeled, *the margin and keel piloso-ciliate,* then scabrous, otherwise glabrous; flowers umbellate; peduncles and calyces *glabrous; calyx* segments *ovate, acute, keeled, villoso-ciliate;* petals twice as long as the calyx, the claw short, downy, and limb obovate; sterile filaments rather longer than the calyx, linear, pilose below, narrowed and glabrous above, curving back at the point. *E. & Z.!* 862.

HAB. Hills near Zoutendaals valley, Zwell., *Miss Joubert!* Oct. (Herb. Sond.)

A dwarf, glabrous bush, with leaves of a *Sedum* (stonecrop). Leaves very obtuse and thick, rugulose when dry, 3 lines long. Umbel 8–12 flowered; pedunc. not longer than the outer floral leaves, 2½–3 lines long. Calyx 1½ line long. Petals white, obovate, narrowed into a short claw, downy above.

99. **A. elata** (Sond.); quite glabrous; leaves close, not imbricate, erect, sub-appressed, *oblong, obtuse, little attenuated at base,* concave above, convex beneath, *not gibbous at point;* flowers umbellate; peduncles 4–6 times longer than the calyx; cal. lobes ovate, *obtuse;* petals twice as long as the calyx, the claw equalling the calyx, *glabrous,* lamina elliptic-oblong; sterile filaments *twice* as long as the calyx, linear, a little wider at base, *ciliate quite to the apex;* ovaries and style glabrous.

VAR. β. **Niveni**; twigs at the end, and peduncles minutely pubescent; petals rather wider.

HAB. Giftberg, 1500–2500 f. *Drege!* β. in the Interior, *Niven!* Nov. (Herb. Mart., Sond.)

A slender shrub, several feet high, much branched; branches spreading, twigs crowded, yellow green, 1–3 inch long. Leaves 1–1½ line long, pale green, or glaucous, oblong, wider upwards, the nerve obsolete beneath. Peduncles capillary, pale, glabrous, 3–4 lines long. Calyx ½ line long. Known from *A. glabrata,* to which it is allied, by the smaller leaves, not swollen at tip, little attenuate at base, the more slender peduncles and smaller flowers.

100. **A. crassifolia** (Sond.); quite glabrous; leaves *erecto-patent,* oblong, obtuse, *semi-cylindrical,* or *slightly furrowed* above, with a few impressed glands beneath; flowers umbelled; peduncles 2–3 times longer than the calyx; calyx segments ovate, *very obtuse;* petals thrice as long as the calyx, the claw *pilose,* longer than the calyx, limb elliptic-oblong, glabrous; sterile filaments twice as long as the calyx, linear, *pilose beyond the middle, glabrous at the point;* ovary and style glabrous. *A. glabrata, E. Mey.! in Hb. Drege, non B. & W.*

VAR. β.; leaves and flowers smaller; sterile filaments equalling the petals.

HAB. Mountains at Hexrivier-kloof, 1000–2000 f. *Drege!* Sep. β. *Drege* 7117. (Herb. Sond.)

A small shrub, 3–5 inches long, branched from the base, the branches crowded, erecto-patent; in β. ascending, twigs ½–1 inch long. Leaves on the branches close, oblong or oval-oblong, very obtuse, with 3–10 gland-dots, 1½–2 lines long; some oval, smaller, scarcely 1 line long. Peduncles glabrous, 1½–2 lines long, twice as long as the floral leaves. Calyx 5-fid, its lobes ciliolate, ½ line long. Petals 2 lines long, in β. 1½ line; the claw linear, limb, in β. shorter.—Known from the rest by its half-cylindrical leaves.

(Doubtful species.)

A. reflexa (Link, Enum. Hort. Berol. p. 238). *Diosma reflexa, Lodd. Cat. B. & W. p.* 207.

"A glabrous shrub. Leaves 6–8 lines long, densely set, sparse, recurvo-patent, flat, narrow, linear, tapering to each end, acute, muticous, slightly crenulate with marginal pellucid dots, the younger sometimes hair-tipped, dull green above, much paler beneath, with two rows of minute dots. Petioles a line long, pale, appressed, not glandular at base. Flowers unknown."—No doubt this is either a variety of *A. cerefolia*, or of some allied species. It is altogether a dubious species; for in Herb. Berol., the specimen preserved under the name "*reflexa*," belongs to *A. apiculata.*

IX. MACROSTYLIS, B. & W.

Calyx 5-parted. *Hypogynous-disc* closing over the ovaries, perforated by the style. *Petals* 5, tapering at base, bearded in the middle. *Filaments* 5, exserted; sterile none; anthers roundish, with a minute, adnate gland. *Style* lengthening after *anthesis*, slender at base; stigma obtuse. *Capsule* longer than the calyx, of 3–5 horned carpels. *B. & W. l. c. p.* 69. *Juss. l. c. p.* 476. *Endl. Gen. No.* 6022.

Small bushes, with alternate or opposite, short, nerve-keeled leaves, pellucid dotted along the margin and nerve. Flowers small, white or rosy, sub-umbellate at the ends of the twigs; peduncles short, bracteate at base. Name from μακρος, *large* or long, and στυλος, a *style.*

1. M. villosa (Sond.); twigs short, pubescent; leaves scattered, erect, the uppermost imbricate, *lanceolate-acute, flattish above,* glabrous, keeled beneath, *triquetrous at the point,* with a short pubescence; flowers umbellate-capitate; petals longer than the calyx, bearded in the middle. *Diosma villosa, Thunb.! Cap. p.* 225, *excl. syn. Berg. D. barbata, Berg.! Herb. Ag. barbata, Spreng.! pugill.* 1, *p.* 20. *Bucco barbata, R. & Sch. V. p.* 445. *Macrostylis lanceolata, E. & Z.!* 817, *ex pte.*

VAR. β. **glabrata**; twigs downy; leaves glabrous or sub-glabrous. *M. lanceolata, B. & W.! l. c. p.* 194, *E. & Z.!* 817, *ex pte.*

HAB. In sandy places among shrubs on the Cape Flats, near Doornhoogde, Wynberg and Muysenberg, *Thunb.! Bergius, Hesse, Niven, E. & Z., Drege,* Apl.-May. (Herb. Thunb., Sond.)

6–12 inches high, sub-decumbent, grey. Branches mostly aggregate, twigs leafy, 1–2 uncial. Leaves close, lanceolate, rarely linear-lanceolate, very entire, nerve-keeled, punctate, sub-glaucous, 4–6 lines long, shorter on the twigs. Flowers terminal, 8–10 together, on very short pedicels. Calyx glabrous, its lobes obtuse, ciliolate. Petals 2 lines long. Capsule glabrous, 5 lines long, of 3–5 round-backed carpels; the beaks 1 line long, recurved at point.

2. M. decipiens (E. Mey.! in Hb. Drege); twigs short, glabrous or minutely downy; leaves scattered, erect, imbricate, *linear, obtuse, concave above, convex-keeled beneath,* glabrous; flowers capitate; petals twice as long as the calyx, bearded in the middle.

HAB. Rocky places at Piquetberg, 1500–2000 f. *Drege!* At 24-rivers, *Zey.!* 293. Nov.-Dec. (Herb. E. Mey., Lehm., Sond.)

6–12 inches high and more, many stemmed, much branched, the twigs uncial or shorter, erecto-patent, terete. Leaves sessile, erecto-appressed, 1½–2 lines long, subcucullate at apex, margin nude or roughish. Flowers 6–8 sessile. Calyx glabrous, 1 line long, its lobes obtuse, ciliolate.—Readily known from the preceding by its narrow and smaller leaves.

3. M. squarrosa (B. & W. p. 198); twigs short, minutely pubescent; leaves scattered, sessile, spreading, recurved or squarrose, basally appressed, *ovate, or ovate-oblong, obtuse,* flat above, the younger pubescent-hispid, the adult mostly glabrate; flowers capitate; petals longer than the calyx, obovate, obtuse, bearded in the middle. *Diosma squarrosa, Wendl. in Hb. D. obtusa, G. F. W. Mey. M. obtusa, B. & W.! l. c. p.* 197.

HAB. Oliphants River, *Hesse!* Sandy ground on mountain sides near Brackfontein, *E. & Z.!* Rocky places at Uienvalei, 2000–3000 f. *Drege!* Aug.-Dec. (Herb. Bartl., Wendl., Mart., E. Mey., Sond.)

6–12 inches high or more, much branched; branches erect or flexuous, mostly aggregate, terete, the younger leafy. Leaves mostly alternate, close, or sub-distant, with the margin flat or reflexed, flat on each side, smooth and green above, paler beneath, with a strong nerve, 1–2 lines long. Flowers 6–10, sessile or pedicellate. Calyx glabrous, 1 line long, segments obtuse. Petals reddish, twice as long as the calyx. Style 4 times as long as the calyx.—*M. obtusa,* Hb. Bartl. and *M. squarrosa,* Hb. Wendl. are the same; leaves patent, or squarrose-recurved, are found in the same specimen.

4. M. crassifolia (Sond.); twigs scarcely downy; leaves scattered, *horizontally patent from the base, sessile,* ovate, obtuse, *sub-trigonous,* flat above, *convex* beneath, *fleshy, glaucous,* glabrous; flowers sessile, capitate-aggregate; petals twice as long as the calyx, obovate, obtuse, bearded in the middle.

VAR. β. **affinis**; leaves erecto-patent, concave above; twigs downy or glabrous, *Drege!* 7154.

HAB. Bergvalley, *Zey.!* 294. β. (station not given) *Drege!* 7157. Nov.-Dec. (Herb. Sond.)

A greyish shrub, taller and stronger than the preceding, with more rigid branches, mostly trichotomous; it differs also in having smaller leaves (½–1 line long), all equal, horizontal, laxly set on the branches, thick, rigid, sub-globose beneath, and three-angled by the prominent nerve, with several rows of glands. Flowers nearly as in *M. squarrosa,* but a little smaller. Calyx sub-angulate, the lobes obtuse, ciliolate. Petals reddish, long-bearded. Style thrice as long as the calyx. Capsule glabrous, 5–6 lines long, with 3–5 oblong carpels, with reflexed horns.—Var. β., besides the characters given above, has shortly pedunculate flowers, and petals not much longer than the calyx.

5. M. tenuis (E. Mey.! in Hb. Drege); twigs erecto-patent, minutely downy; leaves scattered, sessile, *appressed,* ovate or ovate-oblong, with a straight point, sub-obtuse, *the hyaline margin ciliate-scabrid,* convex beneath, and *somewhat nerve-keeled,* glabrous; flowers capitate-aggregate; petals longer than the calyx, bearded in the middle.

HAB. Rocky places on the Cederberge, 3000–4000 f. *Drege!* Dec. (Herb. E. Mey., Sond.)

More than 1 foot high, erect, with slender, mostly ternate branches and uncial or shorter twigs. Leaves 1 line long, uppermost half that size, closely appressed, sometimes with spreading tips. Flowers sessile, 6–10 together. Calyx 1 line long, the lobes obtuse, ciliolate. Limb of the petals patent, roundish, pointed, glabrous. Style 4 times as long as the calyx.—This differs from the others of the genus by its altogether appressed leaves.

6. M. ovata (Sond.); twigs patent, scarcely downy; leaves scattered, close, *sub-imbricate*, sessile, *ovate*, sub-acute, sub-recurvo-patent, *concave above*, glabrous, nerve-keeled beneath, and *hairy at the margin;* flowers sub-sessile, capitate-aggregate; calyx lobes lanceolate, obtuse, hairy, ciliate at the hyaline margin; petals longer than the calyx, obovate-cuneate, sub-acute, bearded in the middle.

HAB. Interior regions, *Niven!* (Herb. Mart., Sond.)

A foot or more in height, with patent, glabrous branches, the ultimate twigs uncial. Leaves 2–2½ lines long, 1½ line wide, with flat margins, the uppermost somewhat imbricate. Flowers 4–6 together. Calyx 2 lines long, the lobes concave. Petals ½ longer than the calyx.—Style eventually elongate, a little thickened upwards. Habit nearly that of an *Acmadenia.* It differs from *A. squarrosa,* B. & W. by the more robust growth, 2–3 times larger leaves, &c.

7. M. hirta (E. Mey.! in Hb. Drege); the spreading twigs and leaves *hairy;* leaves *scattered,* sub-imbricate, sessile, patent, *cordate-acute* flattish above, glabrous, keeled; flowers very shortly pedicellate, *capitate-aggregate;* calyx lobes oblong, obtuse; petals longer than the calyx, obovate-oblong, sub-acute, bearded in the middle.

VAR. β. **glabrata**; leaves and calyx-segments glabrous.

HAB. Between Bervalley and Langevalley, at Zwaartbastkraal, *Drege!* Nov. (Herb. E. Mey., Sond.)

Scarcely a foot high, differing from the preceding by pubescence and the cordate, horizontal slightly keeled leaves. Leaves 1 line long and wide, or a little longer or shorter. Flowers 4–6 together. Calyx ¾ line long, the segments obtuse, hairy beneath, ciliate. Petals and disc, &c. as in others. Capsule in β. of 2–3 round backed carpels, 3 lines long, tipped with 1 line-long straight horns.

8. M. barbigera (B. & W.! l. c. p. 195); *quite glabrous;* branches and twigs spreading; leaves sessile, *opposite,* cordate, acute, glaucous; flowers *umbellate. Diosma barbigera, Linn. Sup. p.* 155. *Thunb.! Cap. p.* 230. *Lam.! Encycl.* 2, *p.* 287.

HAB. Zwartland, *Thunb.!* Kardow Mt., *Hesse, Zeyher* 295. Dec. (Herb. Thunb., Lam., Wendl., Sond.)

Erect, 1–2 feet high, flexuous, 2–3-chotomous; branches mostly opposite, glaucous, twigs terete, yellowish, leafy. Leaves cordate at base and very patent. equalling the internodes or longer, pale green, and striate above, one nerved and glaucous beneath, the margin minutely gland-dotted, 3–5 lines long, 2–4 lines wide. Flowers 6–10, in terminal, shortly peduncled umbels. Calyx 2 lines long, its lobes lanceolate, acute, keeled. Petals tapering to each end, red, white-bearded in the middle. Style thrice as long as the calyx.

X. EMPLEURUM, Soland.

Flowers monœcious. *Calyx* 4-cleft. *Hgpogynous-disc* obsolete. *Corolla* none. *Stamens* 4; anthers 4 sided, emarginate, with a sessile, globose gland. *Ovary* of 1, rarely 2 carpels: *style* short; *stigma* elongate, cylindrical. *Carpels* mostly solitary.—*Lam. Ill. t.* 86. *B. & W. l. c. p.* 200. *Juss. l. c. p.* 476. *T.* 19. *n.* 21. *Endl. Gen. No.* 6023.

Only one species known. The generic name is composed of εν, *in*, and πλευρον, the pleura or membrane surrounding the lungs, because the seeds are (as in others of this tribe) contained within the membranous inner *hull* of the ripe capsule.

1. E. serrulatum (Ait. Kew. ed. 1, vol. 3, p. 340); leaves scattered, linear-lanceolate, flat, serrate; flowers axillary. *B. & W.! l. c. p.* 200. *Diosma ensata, Thunb. Cap. p.* 226. *Emp. ensatum et serrulatum, E. & Z.!* 919, 920.

HAB, In mt. vallies, Tulbagh, near the Waterfall and in the Langekloof, *Thunb.* Hott. Holland, by the Palmiet River and near Hanglipp, Stell. *E. & Z.!* Vanstaadensberg; Dutoitskloof; and in the Zwarteberg, *Drege!* Jul.-Dec, (Herb. Thunb., Wendl., Hook., T.C.D., Sond., &c.)

A shrub, 2–3 feet high or more, entirely glabrous, with angular, reddish brown branches and erect, rod-like, filiform, yellowish twigs. Leaves close, acute, 1–2 uncial, 1½–2 lines wide, glabrous, shining, crenate-serrulate, pellucid-glanded in the crenatures; petioles 1 line long. Flowers ♂ and ♀ on different branches, axillary, solitary or in pairs. Peduncles erect, shorter than the leaves, 2–5 lines long, thickened upwards, one flowered. Calyx 1 line long, punctate, the lobes obtuse. Filaments very short, afterwards twice as long as the calyx; anthers erect, linear, 4-angled. Fruit of a single, obliquely oblong, compressed carpel, nearly uncial, punctate, glabrous, ending in a compressed, sword-shaped, obtuse horn. Seed solitary, shining. Embryo straight, inverse. Radicle short, superior. Cotyledons oblong, convex, fleshy.

APPENDIX. *(Genus requiring further examination.)*

EMPLEURIDIUM, Sond.

Flowers diœcious. *Calyx* 4-partite, persistent, the sepals acute, imbricating in æstivation. *Petals* 4, deciduous, sessile, ovato-sub-rotund, inserted underneath the edges of a fleshy 4-lobed disc. ♂ *stamens* 4, on the margin of the disc and alternating with its lobes; filaments subulate, shorter than the petals; anthers didymous, without terminal gland or appendage. An abortive style-like ovary in the centre of the disc. ♀: *Ovary ?*—*Capsule* oblong, follicular, dehiscing at the side, and tipped with a short, persistent style. *Seed* solitary, ascending-erect ?

A small suffrutex, with rigid, triquetrous, scattered, impunctate leaves, resembling those of a *juniper* or of a *Muraltia*, and minute, axillary, solitary flowers.—Name, an alteration of *Empleurum*, to which genus the present seems allied.

1. E. juniperinum (Sond. & Harv.) *Harv. Thes. Cap. t.* 77.

HAB. Near Caledon, *Ecklon!* (Herb. Sond.)

Root simple, filiform. Stems several from the crown, 6–8 inches high, simple or branched from near the base, erect, or sub-erect; branches virgate, rugulose. Leaves scattered, patent, acicular-triquetrous, 6–12 lines long, ⅓ line in diameter, scabrous at the margin and keel, acute, the younger hair-pointed, glabrous, dull dark green, without pellucid dots. *Flowers* axillary, very minute, solitary; *pedicels* 1–1½ line long, bibracteate at base, ending in the conical base of the calyx. *Sepals* 4, ovate, cuspidate, somewhat keeled. Petals scarcely longer than the sepals, concave, very obtuse, broadly ovate-sub-rotund. Stamens (in the male flowers) opposite the sepals and nearly equalling them. Ovary not seen. Ripe capsule brown, 2½ lines long, opening laterally: seed not examined. Whole plant glabrous.

ORDER XXXVIII. PITTOSPOREÆ, R. Br.

(By W. SONDER.)

(Pittosporeæ, R. Br. in Flinder's Voy. 2. 542. DC. Prod. 1. p. 345. Endl. Gen. No. ccxxxiv. Pittosporaceæ, Lindl. Veg. Kingd. No. clxi.)

Flowers perfect, mostly regular. *Calyx* 5 parted or cleft, its segments imbricate in æstivation. *Petals* 5, hypogynous, free or with connate claws, imbricate, deciduous. *Stamens* 5, hypogynous, free; anthers 2-celled, introrse. *Ovary* syncarpous, free, 2–5-celled; ovules numerous, axile; *style* single, terminal; *stigma* obtuse or capitate. *Fruit* capsular or fleshy and pulpy. *Seeds* numerous (often lying in pulp), with copious albumen and a minute embryo.

Trees or shrubs, erect or climbing, very numerous in Australia, with outlying species in the Islands of the Pacific, Japan, Tropical Asia, the Mauritius, South Africa, and the Canary Islands. The typical genus, *Pittosporum*, has as wide a range as the Order; the 10 or 12 others associated with it are all confined to Australia. Leaves alternate, petiolate, simple, entire or pinnatifid, sub-coriaceous, exstipulate. Flowers axillary or terminal, solitary or in racemes, corymbs or cymes; often sweetly scented, white, red, blue, yellow, or *greenish*.

I. PITTOSPORUM, Soland.

Calyx 5-cleft or parted. *Petals* 5, their claws erect, connivent; limbs spreading. *Filaments* subulate. *Ovary* sessile, imperfectly 2–5-celled; ovules many; style short, stigma sub-capitate. *Capsule* sub-globose or obovate, with leathery, thick valves, carrying the septa on the face. *Seeds* lying in viscid resin and often massed together in a lump. *Endl. Gen.* 5661. *DC. Prod.* 1. *p.* 346.

Shrubs or trees, widely dispersed. Leaves entire or toothed. Flowers terminal or axillary, solitary or corymboso-cymose.—Name, from *πιττη*, *resin*, and *σπορος*, *a seed;* because the seeds lie in resinous pulp.

1. P. viridiflorum (Sims, Bot. Mag. t. 1684); leaves obovate, tapering into the petiole, acute or rounded, or sub-emarginate at the apex, very entire, leathery, shining, reticulated beneath; panicle terminal, sub-umbellate, dense, glabrous or downy; capsules sub-globose. *DC. Prod.* 1. *p.* 346. *E. & Z.!* 236. *P. sinense, Desf. Cat.* 231. *P. commutatum and viridiflorum, Putt. Syn. Pitt. p.* 10. *Celastri sp. Drege! No.* 6181. *Psychotriæ sp. Drege! No.* 3470.

HAB. Woods near the Knysna, George; Chumieberg, and at the Kat River, *E. & Z.!* Galgebosch, Vanstaadensberg, Bethelsdorp and Omtata, Caffr., *Drege!* Albany, *Mrs. F. W. Barber!* (Herb. Sond., T.C.D., &c.)

An erect shrub. Branches crowded, glabrous; young twigs often downy. Young leaves downy, adult glabrous, 1½–3 inches long, 1–1¼ inch wide; flat or with revolute margins. Peduncles uncial, densely corymbose, alternately branched, often surmounted by fresh shoots, downy or glabrous. Sepals acute. Petals yellow-green, 5–6 lines long. Ovary downy. Style 1 line long; stigma capitate. Ripe capsule glabrous, as large as a pea, 2–6 seeded. It varies with smaller flowers and panicles.

ORDER XXXIX. AURANTIACEÆ, Corr.

(By W. H. HARVEY.)

(Aurantiaceæ, Corr. Ann. Mus. 1805.—DC. Prod. 1. p. 536. Endl. Gen. No. ccxxiv. Lindl. Veg. Kingd. No. 457.)

Flowers mostly perfect, regular. *Calyx* free, very short, 4–5 toothed or sub-entire. *Petals* 4–5 (very rarely but 3), broad at base, free or slightly cohering below, erect or spreading, deciduous, slightly or strongly imbricating in the bud. *Stamens* hypogynous, as many or twice as many as the petals, or a higher multiple of them; filaments free or connate, flat, tapering; anthers introrse, terminal or fixed at back. *Ovary* on a short, fleshy stipe, often surrounded by a crenulate disc, 5 or many celled; ovules solitary or in pairs, or numerous, pendulous, anatropous. *Style* terminal, simple, thick; stigma obsoletely lobed. *Fruit* pulpy, indehiscent, with a leathery rind, 1 or several-celled. *Seeds* with a membranous coat, and very distinct raphe and chalaza, exalbuminous; embryo straight, cotyledons thick and fleshy, radicle short.

Trees or shrubs, chiefly of the Eastern Continents, everywhere sprinkled with pellucid dots, filled with a strong-scented, essential oil. Leaves alternate, impari-pinnate, sometimes reduced to a single terminal leaflet. Stipules none. Branches often thorny. Flowers white, and sweetly scented. The genus *Citrus*, the type of this Order, includes the various species and sub-species of orange, lemon, citron, lime, shaddock, &c. now commonly cultivated in all warm countries of both hemispheres. Lime-juice, containing a large per centage of citric acid, is an invaluable preventive of scurvy, in long voyages.

I. MYARIS, Presl.

Calyx minute, 4-parted. *Petals* 4, concave, free, spreading, imbricate in the bud. *Stamens* 8, hypogynous, equal; filaments free, from a thickened base, subulate; anthers sagittate, incumbent. *Ovary* raised on a cylindrical, fleshy torus, obtusely 3-lobed, 3-celled; ovules two in each cell, axile, collateral; style short, thick, deciduous; stigma 3-lobed. *Berry* thinly fleshy, abortively 1–2 seeded. *Presl. Bot. Bem. p.* 40.

This genus is proposed by Presl. for *Elaphrium inæquale.* DC., *Amyris inæqualis*, Spr. and E. & Z., a South African shrub. It is nearly allied, in many respects, to *Clausena*, but the ovules are collateral, not superimposed. The name is an anagram of *Amyris.* The genus was first indicated and characterized, but not *named*, by Prof. Walker Arnott in *Hook. Journ. Bot. Vol.* 3. p. 152.

1. Myaris inæqualis (Presl. Bot. Bem. p. 40); *Amyris inæqualis, Spr. Syst.* 2. *p.* 218. *E. & Z.!* 1139. *Elaphrium inæquale, DC. Prod.* 1. *p.* 724. *Rhus obliqua, E. Mey. in Herb. Drege, litt. d.*

HAB. Common in woods, &c. from the districts of George, through Uitenhage and Albany, to Caffraria, and extending to Port Natal. (Herb. T.C.D. &c.)

A small tree, 10–20 feet high, with strongly scented foliage, fœtid if rubbed in the hand. The very young twigs, petioles and leaflets are minutely downy, the full grown, glabrous. Leaflets often approaching in pairs, but not opposite, ½–1 inch long, 2–5 lines wide, unequal sided, the upper side rounded at base; nervation obvious but not prominent. Panicles scarcely as long as the leaves, or much shorter, few flowered. Fl. small, white; the petals very concave or hoodshaped. Stamens equalling the petals and style. Berry the size of a pea, with a thin, fleshy sarcocarp. Testa of seed highly reticulated, thinly membranous, easily separating. Embryo lenticular, with very thick and large cotyledons, eared at base; a well developed plumule; and a minute radicle.

ORDER XL. XANTHOXYLEÆ, Nees & Mart.

(By W. H. HARVEY.)

(Xanthoxyleæ, Nees and Mart. 1823. A. de Juss. 1825.—Endl. Gen. No. ccl. Xanthoxylaceæ, Lindl. Veg. Kingd. p. 472.)

Flowers mostly unisexual, regular, small. *Calyx* free, 4-5 parted. *Petals* as many as the calyx-lobes, deciduous, twisted in æstivation. *Stamens* as many or twice as many as the petals, on the torus; filaments free, subulate; anthers 2-celled. A rudiment of an ovary in the ♂ flowers. In the ♀: stamens none, or rudimentary, hypogynous. *Carpels* 3-5, on a *gynophore* or stipe, separate or more or less cohering into a plurilocular ovary; ovules 2–4 in each carpel; styles united or more or less distinct, sometimes very short or obsolete. *Fruit* fleshy or membranous, 2–5-celled, or of 1–5 drupes or follicles. *Seeds* mostly solitary, albuminous, with a shining (black) coat, rarely winged; embryo the length of the albumen, with flat cotyledons and a short, superior radicle.

Trees or shrubs, chiefly tropical, common in America and Asia, much less numerous in S. Africa and Australia. Branches, twigs, petioles, and leaflets often armed with strong and sharp prickles. Leaves alternate or opposite, abruptly or imparipinnate or pinnately trifoliolate; leaflets mostly pellucid dotted, entire or serrulate. The properties are aromatic and pungent. The fruit of *X. capense*, called "*Wild Cardamon*" by the colonists, is sometimes prescribed for flatulency and paralysis, *Pappe, Fl. Med. p.* 8. Some of the American species have similar, but much more powerful qualities.

TABLE OF THE S. AFRICAN GENERA.

I. **Xanthoxylon.**—*Stamens* as many as the petals. *Ovary* of *one* or several one-seeded carpels; styles distinct; stigmas capitate.

II. **Toddalia.**—*Stamens* as many as the *petals*. *Ovary* ovoid, 4–5-celled; stigma sub-sessile, 4–5 lobed.

III. **Vepris.**—*Stamens* twice as many as the petals. *Ovary* sub-globose, 4-celled; stigma sessile, broad and peltate.

I. XANTHOXYLON, L.

Flowers polygamous. *Calyx* 4 (3 or 5) parted, small. *Petals* hypogynous, as many as the calyx-lobes, and alternate with them, imbricate in æstivation. ♂: *stamens* as many as the petals and alternate; filaments free, subulate. A rudimentary ovary. ♀: *stamens* none or abortive. *Carpels* 1–5, on a fleshy torus, separate or sub-coherent; ovules 2 in each carpel, collateral. *Styles* terminal, cylindrical, long or short; stigma capitate. *Capsules* leathery, 1 5, sessile or stipitate, 2 valved, 1–2 seeded; seeds black and shining.—*W. & A. Prod.* 1. *p.* 148. *Endl. Gen. No.* 5972.

Trees or shrubs, of both hemispheres, armed with sharp prickles on the branches, and often on the petioles and nerves of the leaflets. Leaves alternate or opposite, rarely simple or trifoliolate, mostly abruptly pinnate, often pellucid dotted. Flowers small, in axillary or terminal panicles. The name is compounded of ξανθος, *yellow*, and ξυλον, *wood;* the roots of some contain a yellow dye. The Cape species belong to the sub-genus RHETZA, W. & A. (*Fagarastrum*, Don.), characterized by a 4-parted, tetrandrous flower, and a single carpel.

1. X. (Rhetza) capense (Harv.); branches armed; petioles *unarmed*, channelled; leaves paripinnate, leaflets elliptical, obovate, ovate, or ovato-lanceolate, obtuse or acute, crenulate, sessile, sub-unequal at base; panicles axillary and terminal, puberulous; flowers 4 parted; ovary single, ovoid; style filiform, equalling the petals; stigmas capitate. *Fagara capensis, Thunb. Fl. Cap. p.* 141. *E. & Z.! No.* 921. *Fagarastrum capense, Don.—Pappe, Fl. Med. p.* 8. *Elaphrium capense, DC. Prod.* 1. *p.* 724. *Rhus obliqua, litt. c. E. Mey.! (non Thunb.)*

HAB. In woods, in the districts of George, Uitenhage, and Albany, common; also in Caffirland, and at Port Natal. (Herb. T.C.D., Hook., Sond.)

A large, much branched shrub or small tree, the branches and twigs armed with sharp, strong, horizontal, solitary prickles, usually placed under the insertion of the petiole. *Leaflets* in 4–5 pairs, opposite or alternate, ½–1½ inch long, 2–6 lines wide, the lowest smallest, the rest gradually larger, very variable in shape, with a translucent gland in each serrature, and sometimes sparsely pellucid dotted. Panicles many flowered, shorter than the petioles. Petals of the ♂ flowers oval, concave, very obtuse; of the ♀, ovato-lanceolate, sub-acute, keeled.

2. X. (Rhetza?) Thunbergii (DC. Prod. 1. p. 726); branches, *petioles, and nerves* armed; leaves paripinnate, leaflets elliptical, oblong, or ovato-lanceolate, obtuse, or acute, or acuminate, crenulate, sessile, subunequal at base; flowers and fruit unknown. *Fagara armata, Thunb. Fl. Cap. p.* 141. *E. & Z.!* 922.

VAR. α. **grandifolia**; lower leaflets elliptical or oblong, upper lanceolate, acuminate, 2–3 inches long.

VAR. β. **obtusifolia**; leaflets elliptical or obovate, obtuse, ¼–¾ inch long.

VAR. γ. **multijuga**; leaflets in 10–12 pairs, linear-oblong, 3–5 lines long, 1½–2 lines wide.

HAB. Var. α. In woods near the Knysna, *Bowie!* Winterhoeksberg, above Uitenhage, *E. & Z.!* β. Macallisberg, *Burke and Zeyher.* γ. Zuureveld, *Dr. Gill!* (Herb. Hook., Sond.)

A very imperfectly known species, and whether the three varieties here enumerated, which agree in having prickly petioles, belong to one or more species, it is impossible to say. The flowers and fruit of none of them are known to me.

3. X.?? alatum (Steud.); twigs *unarmed*, virgate, glabrous and glossy; petioles *winged between the leaflets;* leaves paripinnate, leaflets opposite, in 4–6 pairs, obovate, mucronulate, undulate, very entire; flowers and fruit unknown. *Fagara alata, E. & Z.!* 923.

HAB. In Karroo-regions near Louisfontein, Clanwilliam, *E. & Z.!* (Herb. Sond.)

The genus of this plant is wholly uncertain. The petioles resemble those of *Hippobromus alatus;* but the twigs are glossy, and dark brown, the leaflets quite glabrous, with prominent mid-rib, but not conspicuously veiny: they seem not to be fully developed on E. & Z.'s specimens.

II. TODDALIA, Juss.

Flowers unisexual. *Calyx* short, cup-like, 4–5 toothed or crenate. *Petals* 4–5, much longer than the calyx, spreading, sub-imbricate in æstivation. ♂: *stamens* 4–5, equalling or exceeding the petals, inserted round the base of the gynophore, which bears a rudimentary 4-5 angled pistil. ♀: abortive *stamens* very short. *Ovary* shortly stipitate, ovoid, 4–5 celled; stigma sub-sessile, petals 4–5, lobed. *Fruit* fleshy, dotted, 2–5 celled; cells one-seeded, some often abortive. *W. & A. Fl. Penins. p.* 149. *Endl. Gen. No.*

Shrubs or small trees, natives of Bourbon, Mauritius, and South Africa. Leaves alternate, exstipulate, trifoliolate, pellucid-dotted; leaflets entire, glabrous, netted-veined and glossy. Flowers small, in axillary or terminal racemes, spikes or panicles. *Kaki-Toddali* is the Malabar name of *T. aculeata.*

T. natalensis (Sond.); unarmed; leaves digitately trifoliolate (sometimes uni- or bi-foliolate) on sub-terete petioles; leaflets petiolate, lanceolate or oblongo-lanceolate, tapering to base and apex, glossy, netted-veined; spikes (of male flowers) axillary, about equalling the petiole, few flowered, simple or slightly branched; flowers sessile, quadrifid.

Hab. Port Natal, Gueinzius! No. 67 & 650. Frontier of Albany, *Mrs. F. W. Barber!* (Herb. Sond., T.C.D.)

A glabrous shrub much resembling *Vepris lanceolata*, but with a different inflorescence and flowers of diverse structure, petiolulate leaflets, &c. Bark cinereous. Petioles 1–1½ inch long, nearly terete. Leaflets usually 3, sometimes but 2 or 1, 3–5 inches long, 1–2½ inches broad, tapering at base into distinct petiolules 1–3 lines long, more or less acuminate, closely netted and multi-punctate. Peduncles ½–1 inch long, simple or slightly branched, bearing 4–8 or more sessile flowers. Calyx sub-urceolate, 4-crenate. Petals 4, thrice as long as the calyx, oblong, spreading, round-backed, concave above, obtuse. Stamens about as long as the petals. Abortive ovary 4-angled. Female flowers unknown.

III. VEPRIS, Commers.

Flowers unisexual. *Calyx* short, 4–parted. *Petals* 4, much longer than the calyx, twisted-imbricate in æstivation. ♂: *stamens* 8; 4, opposite the petals shorter, inserted round the base of the gynophore, which bears a rudimentary 3–4 angled pistil. ♀: *ovary* shortly stipitate, sub-globose, fleshy, 4–celled; stigma sessile, peltate. *Fruit* fleshy, 4 (or 3) celled and furrowed; cells one-seeded, some often abortive. *Walk-Arn. in Hook. Journ. Bot.* 3, p. 154. *Endl. Gen.* 5976.

Shrubs or small trees, natives of the Mauritius, Bourbon, and South Africa. Leaves alternate, trifoliolate; leaflets entire, glabrous, with netted veins, pellucid-dotted. Flowers in terminal panicles. The name is an alteration of *vepres*, a bramble.

1. **V. lanceolata** (A. Juss.) unarmed, glabrous; leaves digitately trifoliolate, the leaflets lanceolate, acute or obtuse, undulate, sessile, cuneate at base; panicle terminal, much branched. *Boscia undulata, Thunb. Cap. p.* 159, *E. & Z.! No.* 1140. *Asaphes undulata, DC. Prod.* 2. *p.* 90. *Toddalia lanceolata, Lamk. Ill. n.* 2760. *DC. Prod.* 2. *p.* 83.

Hab. In primitive woods. Districts of George and Uitenhage, common. Macalisberg, *Burke.* Jan.-Feb. (Herb. T.C.D., Hook., Sond.)

A much branched, small tree or large shrub, glabrous or nearly so. Twigs with grey, wrinkled bark, densely leafy. Leaves spreading; petioles 1–1½ inches long, angular; leaflets lanceolate, oblongo-lanceolate, or oblong, obtuse or acute, undulate, closely reticulate, and multi-punctate, 2–2½ inches long, 8–10 lines wide, very entire. Panicles terminal, many flowered, with alternate, sub-trichotomous and divided branches, the male and female on different branches (or trees?). Fruits the size of a pea, 4-lobed, gland-dotted. If correctly referred to *V. lanceolata*, Juss., this shrub is also a native of the Mauritius. I follow *Planchon*, in Herb. Hook., in so referring it. Probably *V. paniculata* should also be enumerated among the synonyms.

ORDER XLI. OCHNACEÆ, DC.

(By W. H. HARVEY.)

(Ochnaceæ, DC. An. Mus. 17. p. 398. Prod. 1. p. 735. Endl. Gen. No. ccxlviii. Lindl. Veg. King. clxxviii. Pl. in Lond. Journ. Bot. vol. v. p. 584, 644, and vol. 6, p. 1.)

Flowers perfect, regular. *Calyx* free, 4–6 parted, mostly persistent, frequently coloured, imbricate in æstivation. *Petals* 4–6 (rarely twice as many), deciduous, inserted at the base of a fleshy torus (*gynophore*). *Stamens* definite or indefinite, hypogynous, free; anthers 2–celled, elongate, erect, hard and dry, articulated with the end of the filament, opening by pores or splitting. *Ovary* of 4–6 separate or connate, one or many ovuled carpels; style single or none. *Fruit* drupaceous, or berry-like; rarely capsular. *Seeds* exalbuminous, or rarely with a thin fleshy albumen.

Tropical and sub-tropical trees or shrubs of both hemispheres, mostly glabrous, with glossy foliage and bright yellow or orange flowers. Leaves alternate, simple, shortly petioled, coriaceous or membranous, entire or serrulate, finely veined. *Stipules* separate and deciduous, or united in one intra-axillary, persistent stipule. Bitter tonics, and stomachics; but only in local use. About 120 species, of which ¾ are American, ⅛ African, and ⅛ Asiatic, are known.

I. OCHNA, Schreb.

Calyx 5-6 leaved, coloured, persistent. *Petals* 5-10, deciduous. *Stamens* numerous; anthers linear, erect, basifixed, opening by short or long pores. *Ovary* deeply 3-6 lobed, on a hemispheric torus; *style* central, between the lobes, sub-entire or deeply 3-6-cleft. *Drupes* 3–6 or fewer, separate, on an enlarged, fleshy torus. *Seed* exalbuminous, *DC. Prod.* 1. *p.* 735. *Endl. Gen.* 5959. *Diporidium, Wendl.*

African and Asiatic trees and shrubs, chiefly tropical. Flowers yellow; the calyx frequently red or vinous purple, brightening as the fruit advances. Name, οχνη, the *wild-pear*, fancifully applied to this genus.

1. **O. atropurpurea** (DC. An. Mus. 17. p. 398); leaves elliptical, oblong or ovate, obtuse, coriaceous, *sharply serrulate;* pedicels solitary or racemulose, on depauperated, naked branchlets, shorter than the leaves; sepals elliptical, obtuse, *enlarged and deep red in fruit;* anthers longer than the filaments, opening by short, terminal pores. *DC. Prod.* 1. *p.* 736. *Diporidium atropurpureum, Wendl.—E. & Z.! No.* 925. *D. serrulatum, Hochst! in Pl. Krauss, No.* 473.

β. **natalitia**; leaves larger, lanceolate-oblong; flowers corymboso-racemulose. *Diporidium Natalitium, Meisn! in Hook. Lond. Journ.* 2. *p.* 58.

HAB. Woods in the eastern districts of Caffraria, frequent; extending to Port Natal. β., at Port Natal, *Krauss! Gueinzius!* (Herb. T.C.D., Hook., Sond.)

A shrub, 2–10 feet high. Twigs with greyish bark, often rough with minute pustules. Petioles ½–1½ line long. Leaves ½ inch to 1½–2½ inches long, 4–6 lines wide, rigid, glossy, but with slender, closely set, raised veinlets, either rounded at each end, or tapering to a more or less acute point. Pedicels ⅓–¾ inch long, from the ends of minute, denuded branchlets, ¼–1 inch long, solitary; or in β. 4–6 or more in a short, imperfect racemule; bracts scale-like, caducous. Calyx enlarging after the petals fall, spreading. Drupes 1–2 or more, dark, as large as peas, widely separated on the hemispherical torus. Style splitting in fruit.—Var. β. appears to be merely a luxuriant form, with larger leaves and more abundant flowers; through Hochstetter's *O. serrulata* it is connected with the ordinary state. Its characters probably depend on local influences.

2. O. arborea (Burch. Cat. No. 4012); leaves elliptical or ovate-oblong, obtuse or sub-acute, *sub-entire or obscurely denticulate;* pedicels subternate or *in short, 3-4 flowered racemules;* sepals ovate-elliptical, concave, enlarged and red-brown in fruit; anthers longer than the filaments, opening by small, terminal pores; *style sub-entire. DC. Prod.* 1. *p.* 736. *Diporidium arboreum, Wendl. E. & Z.!* 925. *D delagoense, E. & Z.!* 926.

HAB. Woods in Uitenhage and Albany, Caffraria, and Port Natal, frequent. Delagoa Bay, *Forbes!* (Herb. Hook., T.C.D., Sond.)

A tree, 20–40 feet high. Twigs with greyish bark, often cracked or pustulated. Petioles 1 line long or less. Leaves 1–2½ inches long, ½–1 inch wide, rigid, reticulated with more or less prominent veinlets, either quite entire or with very shallow or obsolete serratures, variable in shape. Pedicels either axillary or ending short branchlets.—Known from the last by its arborescent habit, and larger and *nearly* entire and paler leaves. Shoots from the root occasionally produce flowers, and on one of these Ecklon and Zeyher found their *O. Delagoense.*

3. O. pulchra (Hook! Ic. Pl. t. 588); leaves elliptic-oblong, sub-acute at each end, *minutely spinuloso-ciliate, at length quite entire; racemes many-flowered, pendulous, longer than the leaves;* sepals 5-6, elliptical-obovate, enlarged and red-orange in fruit; anthers *shorter* than the filaments, opening by terminal pores; style *deeply 5-6 cleft. Pl. in Lond. Journ.* 5. *p.* 655.

HAB. Macallisberg, *Burke!* Kalighari Desert, Lake Ngami, *J. M'Cabe.* (Herb. Hook., T.C.D.)

A shrub, 10–11 feet high. Twigs fulvous, somewhat angular, densely leafy. Petioles 2–3 lines long. Leaves 2½–3 inches long, 1–1¼ inch wide, pale when dry, finely reticulated with slender veins, obtuse or acute, tapering at base into the petiole; the margin, in the young leaf, set with very slender, appressed, minute, subulate teeth, afterwards deciduous, leaving a perfectly entire margin. Racemes 3–4 inches long, 12–14 flowered. A very beautiful shrub. According to Mr. M'Cabe, the Bushmen grease their heads with an oil expressed from the seeds.

SUB-CLASS II.

CALYCIFLORÆ.

Calyx and corolla generally present. *Calyx* gamosepalous. *Petals* separate, or united in a monopetalous *corolla,* either perigynous or epigynous. *Stamens* inserted on the calyx (*perigynous*), or on a perigynous or epigynous corolla. *Ovary* either free, or more or less adnate to the calyx-tube.

ORDER XLII. CHAILLETIACEÆ, DC.

(By W. H. HARVEY.)

(Chailleteæ, R. Br. Cong. p. 442. Chailletiaceæ, DC. Prod. 2. p. 57. Endl. Gen. No. ccxl. Lindl. Veg. Kingd. No. ccxxiii.)

Flowers (small) regular or irregular, perfect or polygamous. *Calyx* 5-cleft or parted, coloured within, with imbricate æstivation. *Petals* (or barren stamens?) 5, inserted in the base of the calyx, and alternating with its divisions, equal or unequal, simple or bifid. *Stamens* 5, opposite the calyx segments, inserted with the petals, and combined with them at base; anthers 2-celled, introrse. *Ovary* free, sessile, 2-3-celled; ovules in pairs, collateral, axile, pendulous. *Styles* 2-3, filiform, separate or connate. *Fruit* a capsule or drupe, 2 or 1 celled; seed solitary, exalbuminous: cotyledons fleshy.

Trees or shrubs, natives of the tropics of Africa, Asia, and America. Leaves alternate, penninerved, entire, coriaceous, with small deciduous stipules. Flowers in axillary di-trichotomous, much branched and many flowered cymes or fascicles. Of this small Order but three genera, including 15–20 species are known: only one species is South African.

I. CHAILLETIA, DC.

Calyx deeply 5-parted. *Petals* sub-equal, emarginate or bifid. *Stamens* 5, alternating with the petals. *Perigynous-glands* 5, opposite the petals. *Ovary* 3-celled; styles 3, filiform, distinct, or connate at base. *Drupe* coriaceous, dry, abortively one-seeded. *DC. Prod.* 2. *p.* 57. *Endl. Gen. No.* 5758.

Shrubs or trees, natives of the tropics of both hemispheres. Named in honour of M. Chaillet, a Swiss botanist.

1. C. cymosa (Hook. Ic. Pl. t. 591); twigs hairy; leaves narrow-oblong, obtuse, somewhat narrowed at base, glabrous, coriaceous, netted with veins; cymes pedunculate, panicled, shorter than the leaves; calyx lobes linear, obtuse, erect; petals costate, deeply bifid; stamens as long as the calyx-lobes; styles three, connate for half their length, albo-tomentose below.

HAB. Aapges River, *Burke and Zeyher!* Oct. (Herb. Hook., T.C.D.)

A very dwarf shrub or suffrutex, under a foot in height, sub-simple or slightly branched. Branches and twigs rough, villoso-pubescent, densely leafy towards the extremity. Leaves 3–3½ inches long, 6–10 lines wide, erect, closely set, with prominent ribs and veinlets. Peduncles 1–1½ inches long, below the oblong panicle. Calyx-segments nearly 4 lines long, ½ line wide, slightly imbricate in æstivation, villoso-tomentose externally. Petals as long as the calyx-lobes, cloven to the middle, the lobes linear. Fruit unknown.—A remarkable plant, very different in habit from others of the genus, but apparently not differing generically, unless the fruit afford a character.

ORDER XLIII. CELASTRINEÆ, R. Br.

(By W. SONDER.)

(Celastrineæ, R. Br. in Flinders.—DC. Prod. vol. 2. p. 2. Endl. Gen. No. ccxxxvi. Celastraceæ, Lindl. Veg. Kingd. No. ccxxv.)

Flowers mostly complete, regular. *Calyx* 4-5-lobed or parted, persistent, with imbricate æstivation. *Petals* 4-5, inserted under the margin of a fleshy, 4-5 crenate disc that clothes the bottom of the calyx, sessile, flat, patent, deciduous, imbricate in æstivation. *Stamens* 4-5, alternate with the petals and inserted into or under the margin of the disc; filaments subulate; anthers introrse, 2-celled, short. *Ovary* more or less sunk in the disc, free, or slightly adnate at base, 2-3-5-celled; ovules solitary or in pairs, or several, collateral, erect or pendulous; style short, thick; stigma lobed. *Fruit* 2-5 celled, either a dehiscent *capsule* or an indehiscent *drupe*, or a *samara*. *Seeds* fewer than the ovules, mostly with a fleshy or membranous arillus, albuminous (rarely exalbuminous); embryo straight, with flat cotyledons and a radicle next the hilum.

Trees or shrubs, erect or climbing, often spiniferous. Leaves opposite or alternate, simple, entire or toothed, coriaceous, penninerved, often glaucous and mostly glabrous. Flowers small or minute, white or greenish, in axillary cymes or panicles. Aril of the seeds often bright red or orange. Natives of the warmer parts of Europe, N. America, and Asia, with outliers in temperate S. America and in Australia; numerous in Southern Africa. None are remarkably useful. The fruit of the drupaceous genera is sometimes edible.

TABLE OF THE SOUTH AFRICAN GENERA.

Tribe I. EUONYMEÆ. *Fruit* capsular, loculicidally dehiscent.

I. **Celastrus.** *Leaves* alternate. *Valves* of the capsule *simple.*
II. **Pterocelastrus.** *Leaves* alternate. *Valves* of the capsule *dorsally winged.*
III. **Methyscophyllum.** *Leaves* opposite. *Disc* hypogynous.

Tribe II. ELÆODENDREÆ. *Fruit* fleshy, indehiscent.

* *Seeds* exalbuminous.

IV. **Hartogia.** *Leaves* opposite. *Disc* hypogynous. *Ovary* 2-celled. *Seeds* exarillate.

** *Seeds* albuminous. *Ovules* pendulous.

V. **Maurocenia.** *Leaves* opposite. *Disc* hypogynous. *Ovary* 3-celled. *Seeds* exarillate.

*** *Seeds* albuminous. *Ovules* erect.

† Seeds without an arillus.

VI. **Cassine.** *Leaves* opposite. *Flowers* 4-5 parted, corymbulose. *Drupe* juicy with a thin, crust-like putamen.
VII. **Elæodendron.** *Leaves* opposite, or the upper alternate. *Fl.* 4-5 parted, corymboso-paniculate. *Drupe* rather dry, with a very hard ligneous putamen.
VIII. **Lauridia.** *Leaves* opposite. *Flowers* 4-parted, racemose. *Drupe* rather dry, with a crust-like putamen.

IX. **Mystroxylon.** *Leaves* alternate. *Flowers* 5-parted, umbellate. *Drupe* dry, with a crust-like putamen.

†† Seeds with an arillus.

X. **Scytophyllum.** *Leaves* alternate. *Flowers* pedunculate; peduncle 1-flowered, or dichotomously panicled. *Ovary* 2-celled.

I. CELASTRUS, L.

Calyx 5, rarely 4-parted. *Petals* 5, rarely 4. *Stamens* 5, rarely 4, inserted under an orbicular, perigynous disc, alternate with the petals; filaments subulate; anthers introrse, 2-celled, longitudinally slitting. *Ovary* half-sunk in the disc, 2-3-celled. *Ovules* 2, rarely 5-6 in each cell, fixed to the central angle, ascending. *Style* short; stigma 2-3 lobed. *Capsule* coriaceous, sub-globose, obovate or obcordate, sharply or bluntly 3-angled, 3-celled, loculicidally 3-valved: valves placentiferous. *Seeds* in pairs or solitary, erect, with a complete or incomplete arillus. *W. & A. Prod. Ind. Or. p.* 158. *Putterlickia, Catha & Celastrus, Endl. Gen.* 5674, 5678, 5679.

Unarmed or spiny shrubs, natives of the warmer parts of both hemispheres. Leaves alternate. Flowers axillary, sub-umbellate, cymose or panicled, small, white or greenish. The name is from κηλας, the *latter season;* because the species are late in ripening their fruit.

TABLE OF THE SOUTH AFRICAN SPECIES.

A. Cells of the ovary 6-*ovuled:*

Branches smooth ... ... (1) **pyracanthus.**
Branches densely *warted* ... ... (2) **verrucosus.**

B. Cells of the ovary with 2-*ovules:*

* Unarmed (not spiny):

Capsule obovate or obcordate, *blunt-edged.*
Umbels on *long* peduncles ... ... (3) **peduncularis.**
Umbels on *short* peduncles; leaves obtuse ... ... (4) **cordatus.**
Umbels *sub-sessile;* leaves acuminate ... ... (5) **acuminatus.**

Capsule 3-angled, or sharply 3-edged.
Leaves oval or sub-orbicular, quite entire-edged ... (10) **lucidus.**
Leaves elliptic-oblong, cuneate, toothed, 8–10 lines long ... ... (11) **Zeyheri.**
Leaves ovate or elliptical, acute, *spinous-toothed,* 1½–2 inches long ... ... (12) **procumbens.**
Leaves obovate-cuneiform or rhomboid, wavy-toothed, *obtusely serrate,* 1½–2 inches long... ... (13) **undatus.**

** Armed with *spines:*

Leaves minutely *downy* ... ... (9) **tenuispinus.**

Leaves quite glabrous: *narrow* or *small,*
lvs. linear, 1-nerved, leathery, 2–3 inches long, 2 lines wide ... ... (6) **linearis.**
lvs. lanceol., sub-coriaceous, penninerved, 2½–3 inches long, 4–6 lines wide; panicle ¼ as long as the leaf ... ... (8) **lanceolatus.**
lvs. linear-cuneate, membranous, penninerved, 1–1½ inch long, 3 lines wide; panicle about as long as leaf ... ... (7) **polyacanthus.**

lvs. linear-oblong, cuneate, coriaceous, very smooth above, 1½ inch long, 3 lines wide (17) **ellipticus.**
lvs. oblong, thick, veinless, entire, ½–1 inch long, 2–3 lines wide... .. .. (14) **integrifolius.**
lvs. obovate-cuneate, truncate-obtuse, or obcordate, *spinuloso-serrate* ... ... (21) **angularis.**
lvs. obovate-cuneate, coriaceous, very smooth above, entire, ½–1 inch long, 3–4 lines wide; fl. *capitate* ... ... (15) **capitatus.**

Leaves quite glabrous; *broader and larger.*
lvs. obovate-oblong, cuneate at base, membranous, crenulate; ½–1 inch wide, ... (18) **buxifolius.**
lvs. obovate-cuneate or ellip. lanceolate, spinuloso-serrate, 5–9 lines wide ... ... (16) **heterophyllus.**
lvs. elliptical, obtuse, acute at base, crenulate, 1–1½ inch wide... ... ... (19) **nemorosus.**
lvs. (the upper) rhomboid, obtuse, serrate, 8–10 lines wide ... ... (20) **rhombifolius.**

SECT. 1. **Putterlickia.** *Disc* somewhat raised. *Cells* of the ovary with several (about 6) ovules. *Capsule* 3-angled. (Sp. 1-2.)

1. C. pyracanthus (L. Sp. 285); spinous; branches terete, glabrous, *even;* leaves obovate, serrate or entire, *coriaceous,* tufted; flowers axillary, panicled; pedicels divaricate; capsule 3-angled, 3-valved, 6-seeded. *Thunb.! Fl. Cap. p.* 220. *DC. Prod.* 2. *p.* 8. *Bot. Mag. t.* 1167. *E. & Z.! No.* 935. *C. integer, Thunb.! l. c. C. obtusus, Thunb.! l. c. p.* 217. *C. campestris, E. & Z.!* 937. *Catha campestris, Presl. Bot. Bem. p.* 34.

HAB. Cape Flats and about Table Mt., beyond Salt River, Hout Bay, Zwartkops R., Adow and Quaggasflakte, *Thunberg, E. & Z.! &c. Zey.,* 2176. Mosselbay and Zuureberge, *Drege!* (Herb. Thunb., Holm., Lehm., T.C.D., Sond.)

A shrub, 2 or more feet high, with curved branches. Spines frequent on the stem and branches, horizontal, short or 1–2 inches long, naked, rarely leafy or bearing flowers, sometimes wanting on the upper branches. Leaves tufted, rarely solitary, elliptical, more often obovate-cuneate, emarginate, rarely acute, with revolute margin, penninerved and netted-veined, 1–1½ inch long or less. Peduncles solitary or 2–4 together, ½–1 inch long, panicled at the apex, the panicle 3–12 flowered; pedicels 4–6 lines long, minutely bracted at base. Calyx minute, obtuse. Petals oblong, 2 lines long. Stamens as long. Ovary conical, 3, rarely 4 celled, ovules biseriate. Capsule red, 6–9 lines long, 3-valved; the valves thick, septiferous. Seeds ovate, concealed within a coloured arillus. *C. integer,* Thunb! Herb. is a very spiny branch, and the same as *C. campestris,* E. & Z! *C. obtusus,* Thunb! Herb. is a spineless branch.

2. C. verrucosus (E. Mey! in Hb. Drege); spinous; branches terete, *warted, the younger angular;* leaves *sub-coriaceous,* obovate-spatulate, gradually narrowed into a short petiole, obtuse, emarginate, the margin revolute, spinuloso-denticulate or serrulate; flowers axillary, panicled; pedicels divaricate; capsule 3-angled, 3-valved, 6-seeded.

HAB. Woods between the Keiskamma and Buffel's River, 1–2000 f., and at Port Natal, *Drege, Gueinzius, T. Williamson.* (Herb. E. Mey., T.C.D., Sond.)

This resembles *C. buxifolius* in its leaves, and *C. pyracanthus* in its fruit. Stem curved, very warty. Spines strong, horizontal, 2 inches long on the branches, ½ inch on the twigs, and slender. Leaves 1–½ inch long, 6–9 lines wide, subundulate, narrower and less coriaceous than in the preceding. Panicle pedunculate, few-flowered, equalling the leaf. Flowers 4 or 5-fid. Petals 1 line long. Capsule and seeds as in *C. pyracanthus.*

SECT. 2. **Eucelastrus.** *Disc.* orbicular. *Cells* of the ovary 2-ovuled. *Capsule* obovate, sub-globose, three-cornered or 3-edged. (Sp. 3–21.)

* *Capsule obovate, obcordate or sub-globose (not angular.)* (Sp. 3–9.)

(*a.*) *Unarmed, broad leaved.* (Se. 3–5.)

3. C. peduncularis (Sond.); twigs sub-angular, thinly *pubescent;* leaves petiolate, *ovate-oblong or ovate, obtusely acuminate,* obtuse or narrowed at base, crenato-serrate, glabrous, the younger downy on the nerves underneath; umbel axillary, *on a long peduncle,* few flowered, bibracteate, pubescent; pedicels equalling the leafy bracts; capsule obcordate-sub-globose. *Ilex flexuosa, E. Mey! in Hb. Drege.*

HAB. Mountain woods at Howison's Poort, near Grahamstown, *H. Hutton! Zey!* 2187. Between the Key and Gekau, *Drege!* (Herb, T.C.D., Sond.)

A tree, 15–20 feet high, with terete ashen branches, and curved twigs. Leaves 2–3 inches long, 1–1¼ inch wide, ovate, obtuse at base or more oblong and narrowed to each end, obtuse or emarginate, the upper sometimes sub-lanceolate, paler beneath; petioles furrowed, 2–3 lines long. Common peduncle 6–9 lines long. Umbel or cyme bracted with 2 oblong leaflets 1–1½ line long; pedicels simple. Calyx ovate, obtuse, pubescent, ciliolate. Petals rounded, twice as long as the calyx. Disc somewhat lobed. Stamens 5, short. Style very short. Capsule 4 lines long, 2-celled, or abortively 1-celled, 2 seeded. Seed erect, elliptical, compressed, with a thin arillus.

4. C. cordatus (E. Mey! in Hb. Drege); *quite glabrous;* twigs curved, angular; leaves very shortly petiolate, *cordate-ovate, obtuse,* sub-emarginate, or very shortly mucronulate, serrulate or sub-entire, paler underneath; umbel axillary, *shortly pedunculate,* few flowered, glabrous, flower-pedicels equalling the peduncle, fruit-pedicels exceeding it; capsule *sub-globose or didymous.*

HAB. Among shrubs near the mouth of the Omsamcaba, *Drege!* (Herb. Sond. E. Mey.)

A shrub or tree, with terete, brownish branches, and acutely angular twigs. Leaves 1½–2 inches long, 1–1¼ inch wide, veiny at both sides, with sub-revolute margins. Petiole 1 line long. Common peduncle 1 line long; pedicels bracteolate at base, as long, or in fruit twice as long. Calyx very obtuse. Petals roundish, narrowed at the very base. Stamens 3–4 times as short as the corolla. Capsule glabrous, as large as a *pea,* 3–2–1-celled, cells 1-seeded. Arillus thin. Seed compressed.

5. C. acuminatus (Linn. Suppl. 154); quite glabrous; twigs angular, striate; leaves petiolate, *elliptical-acute, or elliptic-oblong-acuminate, cuneate at base,* serrated, paler beneath; umbel sessile or nearly so, about 3-flowered; capsule *obovate or obcordate,* 1-2 seeded. *Thunb. Fl. Cap. p.* 218. *E. & Z.!* 927. *Pappe, Sylv. Cap. p.* 8. *C. populifolius, Lam! Illustr. No.* 2698. *C. rupestris and mucronatus, E. & Z.!* 920, 930. *C? Plectronia, DC. Prod.* 2. *p.* 9. *Catha acuminata and rupestris, Presl. l. c.*

VAR. *β.* **microphyllus;** leaves smaller, ½ inch long, 3 lines wide, acuminate, narrowed at base. *C. microphyllus, E. & Z.!* 928, *non Thunb.*

HAB. Among shrubs, East side of Table Mt., *Pappe, Krauss.* Zwarteberg, Genadendal, Grahamstown, Winterberg and Kat River, *E. & E.!* Drackenstein and the Paarl, *Drege, W.H.H.* Winterhoeksberg, *Krauss.* Krakakamma and Howison's Poort, *Zeyher!* 2186, 2188, 2189, and several other localities. *β.* Zitzekamma, Stormberg and Omtata, *Ecklon, Drege,* 6745, 6746. (Herb. Thunb., Lam., E. Mey., T.C.D., Sond.)

A shrub 5–6 feet, or a tree 12–15 feet high, with a trunk 7–12 inches diameter, the branches and twigs spreading. Leaves unequal, sub-coriaceous, veiny on both sides, the margin flat or recurved, the serratures acute or obtuse; petiole 2 lines long. Peduncles 2–4, capillary, one flowered, 2–3 lines long, or cymose or umbellate on a common peduncle. ½–1 line long, rarely proliferously many flowered. Calyx 4–5 fid, obtuse. Petals 4–5, elliptic-orbicular, 4 times longer than the calyx. Stamens 4–5, very short. Ovary 3–2 celled; style scarcely any, stigma thick, obsoletely lobed. Capsule 3 lines long, 2–3 lobed, bivalve, abortively one-seeded. Seed ovate-oblong, with a thin aril.—Two sub-varieties may be distinguished by the form of leaf:—*α.* with ovate or elliptic rhomboid, sub-acute or acuminate leaves, sub-obtuse at base and minutely or coarsely serrate, the larger 2–3 inches long, 1–1½ inch wide: *β.* with elliptic-oblong, long-acuminate, sometimes obliquely mucronate leaves, cuneate at base and quite entire, the larger 2–3 inches long, ¾–1 inch wide.

(*b.*) *Armed with spines. Leaves narrow.* (Sp. 6–9.)

6. C. linearis (Linn. f. Suppl. 153); spiny, quite glabrous, glaucous; leaves short-stalked, *linear,* remotely toothed or entire, *one-nerved, veinless, coriaceous;* panicles axillary, cymose, many flowered, 2-3 times shorter than the leaf; capsules ovato-sub-globose, bivalve, 2-celled, 2-1 seeded. *Thunb. Fl. Cap. p.* 219. *C. stenophyllus, E. & Z.!* 955. *Eucentrus linearis, Presl. Bot. Bem. p.* 33. *Polyacanthus stenophyllus, Presl. l. c.*

HAB. Woods at Zondagriver, Graaf Reynet and Uitenhage. *Thunberg, E. & Z.!* Camdebosberg, *Drege!* Dec.-Jan. (Herb. Thunb., E. Mey., Lehm., T.C.D. Hook., Sond.)

A greyish white shrub, with curved branches. Spines horizontal, 1–2 inches long, sometimes opposite, leafy. Leaves spreading, acute or obtuse, flat, 2–3 inches long, 2 lines wide. Panicle 2 inches long, the branches spreading, level-topped. Flowers minute, white, 4–5 fid on the same branch. Calyx persistent. Petals oblong, as long, or longer than the stamens. Ovary 3, rarely 2 celled. Style trifid. Capsule 2 lines long, mucronulate or obtuse, 2–3 valved, the cells 1, rarely 2 seeded. Seed oval-oblong, apiculate, raphe linear, testa shining chesnut, aril incomplete, cotyledons rounded, radicle short, cylindrical.—Does *C. linearis, Burch. Trav.* 2. *p.* 133, from the Snowy mountains, belong here?

7. C. polyacanthus (Sond.); quite glabrous, very spiny; spines strong; branches warted; leaves tufted or solitary, minutely petioled, *linear-cuneate, or linear-oblong, obtuse or emarginate,* crenulate or quite entire, *membranaceous, penni-nerved at each side,* panicles axillary, cymose, many-flowered, equalling the leaves or half as short; stamens equalling the petals, ovary sub-globose, 2-3 celled, cells one-ovuled. *Celastrus linearis, E. & Z.!* 946, *not Thunb. Polyacanthus angustifolius, Presl. l. c.*

VAR. *β.* **inflexus**; branches incurved, sub-unilateral. *C. inflexus, Zeyher.*

HAB. Woods by the Zwartkops and Zondag Rivers, *E. & Z.!* Port Natal, *Gueinzius!* (Herb. Lehm., T.C.D., Sond.)

A grey, rigid shrub, with curved, straight or inflexed branches. Spines horizontal, 1–2 inch long, often bearing leaves and flowers. Leaves green, mostly tufted, but at the ends of branches and on the spines often solitary, 1–1½ inch long, 3 lines wide at the points, rarely wider, but often smaller. Panicle semi-uncial, and flowers minute, like those of the preceding. Fruit unknown.—*C. Willemetiæ*, E. Mey! in Hb. Drege, seems the same as this. The few branches collected by Drege are *unarmed;* the leaves are tufted, uncial, 2–3 lines wide, the panicle equalling the leaf, and the capsule obovate or sub-globose, 2–3 valved, 2-seeded.

8. C. lanceolatus (E. Mey! in Hb. Drege); glabrous, spiny; leaves solitary, *lanceolate*, narrowed at base, blunt, mucronulate, closely denticulate at the margin, penninerved on both sides, *glaucous;* panicles axillary, cymose, many flowered, 4-5 times shorter than the leaf; petals oblong, longer than the stamens; ovary 2-celled; cells 2-ovuled. *Catha patens, Presl. l. c.*

HAB. Mouth of the Gariep, *Drege!* Sep. (Herb. Sond. T.C.D.)

Branches greyish-white; twigs lax, virgate. Spines 1–2 uncial. Leaves patent, sub-coriaceous, 2½–3 uncial, 4–6 lines wide, some smaller; petiole 1–2 lines long. Panicle ½ inch long. Fl. minute, 5-parted. Ovary 2, rarely 3-celled. Style very short, stigma 2-lobed. Fruit unknown.—Differs from *C. polyacanthus* in the long-lanceolate leaves, much longer than the panicle; from *C. linearis* by the much broader and penninerved leaves.

9. C. tenuispinus (Sond.); *glabrous, except the leaves*, spiny; spines slender, short; leaves tufted, *linear-oblong or oblong, narrowed into a short petiole*, mucronulate, or sub-emarginate, minutely denticulate, one-nerved, nearly veinless, *minutely downy beneath and on the petiole;* panicle axillary, few flowered, much shorter than the leaf.

HAB. Macallisberg, *Burke* 20. *Zey!* 305. (Herb. Hook., T.C.D., Sond.)

A greyish shrub, with erecto-patent branches, and sub-terete flexuous twigs. Spines 3–6 lines long, leaves often not tufted on the young branches, membranous, scarcely coriaceous, uncial, 2–3 lines wide. Petiole 1 line long. Panicle or cymule on a peduncle 2–3 lines long. Flowers minute, unopened. Readily known by its *downy* leaves. The capsule being unknown, its exact position in the genus is uncertain.

** *Capsule sub-globose, 3-angled or sharply 3-edged. Catha, Endl.* (Sp. 10–21.)

10. C. lucidus (L. mant. p. 49); erect, glabrous; leaves minutely petiolate, erecto-patent, *oval or sub-orbicular*, coriaceous, shining, netted, *quite entire;* peduncles axillary, crowded, very short; capsule trigonous, 3-4 valved, 2-3 seeded. *Thunb. Fl. Cap. p.* 218. *Meerburg, Pl. t.* 12. *L'Her. Stirp. t.* 25. *DC. Prod.* 2. *p.* 6. *C. concavus, Lam. Illustr. n.* 2695. *Cassine concava, Lam. Dict.* 1. *p.* 633.

HAB. About Capetown, *Thunberg, Ecklon*, &c. Kamp's bay, *Pappe, Krauss.* (Herb. Thunb., Vind., Lehm., T.C.D., Sond.)

A much branched shrub, one or more feet high, with greyish-brown, terete branches and angular twigs. Leaves elliptical, orbicular or obovate, obtuse or with a short, obtuse acumination, rather concave, with a revolute margin, the larger uncial, smaller ½-uncial, petiole 1 line long. Pedicels 2–4, rarely more, 1–2 lines long, jointed above the base. Flowers small. Petals 5, ovate-oblong, twice as long as the obtuse calyx. Stamens shorter than the petals. Ovary 2–4, mostly 3-celled. Capsule as large as a *pea*, brown. Seed angular, purplish, with a thin aril.

11. C. Zeyheri (Sond.); glabrous; leaves very shortly petioled, *squarrose, spreading or recurved*, elliptic-oblong, or elliptical, obtuse or subacute, *cuneate at base*, rather concave, coriaceous, shining, veiny, *dentato-serrulate*, peduncles 1–4, axillary, very short, capsule 2–3 valved, 1–3 seeded.

HAB. Grootrivier and Trompeterspoort, Bervalley, Uit. *Zey! Celast. No.* 5. Nieuwveldsbergen at Beaufort and Camdebosberg, *Drege!* 6728. (Herb. T.C.D., Sond.)

A dwarf, much branched shrub, with short, pale, angular twigs. Leaves unequal, the larger 8–10 lines long, 3–4 wide, smaller 4–6 lines long, 2–3 wide, one-coloured, keeled-convex, the margin not revolute, somewhat toothed towards the extremity. Petiole 1 line long. Pedunc. 1 line long, jointed below the middle. Fl. as in the preceding. Capsule the size of a *pea*, globose-trigonous, rarely 4-valved. Seed thinly arillate.

12. C. procumbens (Linn. f. Suppl. 153); glabrous; leaves shortly petioled, spreading, ovate or elliptical, acute or obtusely recurve-pointed, with *the margin revolute and spinous-toothed*, coriaceous, shining, netted; peduncles axillary, 1-flowered, crowded; capsule sub-trigonous, 2–3 valved, 1–3 seeded. *E. & Z.!* 934. *C. ilicinus, pl. Krauss. Cassine articulata, E. Mey.! in Hb. Drege.*

HAB. Sand Hills at Mosselbay, *Thunberg!* Port Elizabeth, *E. & Z.!* Zitzekamma, *Krauss! Zey.!* 2205. Between Omtendo and Port Natal, *Drege! T. Williamson! Sanderson!* Ap.-Jun. (Herb. Thunb., Hook., Vind., T.C.D., Sond.)

A span or more in height, procumbent, with erect branches. Leaves unequal, paler underneath, mostly 1½–2 inches long, 9–14 lines wide, on other branches ¾ inch long, obovate, sharply dentate. Pedunc. few or many, 3–4 lines long. Flowers small, white. Capsule as big as a *pea*, rarely 4-celled, 1–2 cells abortive. Aril thin. *C. ilicinus, Burch. Trav.* 1. *p.* 340, from the character given, seems to belong rather to this species than to the next.

13. C. undatus (Thunb! Prod. p. 42); glabrous; leaves petiolate, spreading, *obovate-cuneiform* or sub-rhomboid, obtuse or sub-acute, *wavy, toothed*, or *coarsely and obtusely serrate*, coriaceous, shining, netted; peduncles axillary, one-flowered, crowded; capsule trigonous, 2–3 valved, 1–3 seeded. *C. collinus, E. & Z.!* 931. *C. dumetorum, E. & Z.!* 932. *C. ilicinus, E. & Z.!* 933. *C. cymatodes, Spr. Syst. p.* 775.

HAB. Woods at Zwartkops River, *Zey!* 2190. Botasberg, *E. & Z!* Boschesmans River, *Zey!* 2185. *Drege,* 5612, ex pte. Feb.-Nov. (Herb. Thunb., Lehm., T.C.D., Sond.)

An erect shrub, "*Koko*" of the Hottentots. Branches terete, twigs angular or compressed. Leaves unequal, of various shapes, but mostly obovate, obtuse, or shortly acuminate, the point obtuse or sub-acute, some wider, rhomboid-ovate or elliptical, equal or unequal sided at base, but all cuneate at base or tapering into a 2 line long petiole, wavy above, paler beneath, the larger 1½–2 uncial, 1 inch wide, the smaller 8–12 lines long, 3–6 lines wide. Pedicels 3–4 lines long, slender. Petals oblong. Capsule trigonous or triquetrous.

14. C. integrifolius (L. f. suppl. 153); glabrous, very spiny; branches warted; leaves tufted or solitary, minutely petioled, ovate-oblong or oblong, obtuse or emarginate, quite entire, *coriaceous*, veinless or somewhat veiny beneath, the upper reflexed; panicles axillary, lax, longer than the leaves; stamens not equalling the petals; capsules triquetrous, 3-celled, 3-seeded. *Thunb. Fl. Cap. p.* 219. *E. & Z.* 936.

HAB. In Carroid scrubs beyond Hex Rivier, *Thunberg!* Between Kochman's Kloof and Gauritz R., Kannaland, *Mundt!* Namaqualand, *v. Schlicht.* Olifant's R., *Dr. Pappe!* Somerset, *Atherstone!* Oct.-Nov. (Herb. Thunb., Hook., T.C.D., Sond.)

A rigid shrub, bristling with strong, horizontal spines 1–2 inches long, and often leaf or flower bearing; branches unarmed at the end, flexuous Leaves sub-opposite on the spines, flat, thick, very smooth above, minutely punctate under a lens, sometimes obsoletely veiny beneath, 6–12 lines long, 2–3 lines wide. Panicle 1–1½ inch long, cymose, with spreading branches. Flowers white. Petals ovate-oblong. Style scarcely any, stigma 3-lobed. Ovules in pairs. Capsule 3 lines long, scarcely attenuate at base. Seed oval-oblong, imperfectly arillate.—This resembles *C. polyacanthus*, but has thicker, not penninerved leaves, larger panicles, and larger flowers, stamens, and ovary.

15. **C. capitatus** (E. Mey! in Hb. Drege); glabrous, very spiny, the spines strong; leaves tufted, *obovate-cuneate*, emarginate, *very entire*, coriaceous; panicles axillary, *capitate-glomerate*, thrice as short as the leaf; capsules *triquetrous*, 3-celled, 3-seeded. *C. rigidus, E. & Z.!* 947. *E. Mey. in Hb. Drege.*

HAB. Woods at the Zwartkops River, *E. & Z.!* Karroo at Koega Kammaskloof, *Zey.!* 2177, 2178. Graaf Reynet, *Mrs. F.W. Barber!* Fish River and near Basche, *Drege!* (Herb. E. Mey., Hook., T.C.D., Sond.)

A rigid shrub, 2 or more feet high; the branches secund, twigs very short and leafy. Spines horizontal, 1–2 uncial, terete, sometimes leafy. Leaves ½–1 inch long, 3–4 lines wide, flat, very smooth above, sometimes obsoletely veiny beneath, the smaller obcordate; petioles very short. Fl. 5-fid, small, white, numerous, crowded: peduncles branched or simple. Petals 1 line long, equalling the stamens. Style trifid. Capsule apiculate, with hard valves, 3 lines long. Seeds oval, incompletely arillate.

16. **C. heterophyllus** (E. & Z.! 943); glabrous, spiny; the branches angularly striate; spines slender, sometimes floriferous; leaves tufted, sub-solitary on the younger twigs, elliptical, obovate-cuneate or elliptico-lanceolate, obtuse, acute or sub-acuminate, *spinuloso-serrate* or subentire, netted beneath; panicles axillary, *cymose, short*, often glomerate; capsules rather large, *trigonous*, 3-celled, 3–1 seeded. *C. cymosus and parvifolius, E. & Z.!* 948 & 949. *Catha heterophylla and cymosa, Presl. l. c.*

VAR β. **glomeratus**; leaves obovate or sub-orbicular apiculate, dentato-serrate; panicles glomerate, equalling the leaf or shorter. *C. glomeratus, E. Mey.! in Hb. Drege.*

HAB. Woods at Olifant's Hoek and Bosjesmansriver, in Adow, Zwarthoogdens, Hassagaisbosch, and Grahamstown, *E. & Z.! T. Williamson, Zey!* 2181. *H. Hutton! Drege,* 6730. β. between Gekau and Basche, *Drege!* (Herb. Hook., T.C.D., Sond.)

A much branched shrub, 2–4 ft. high, with spreading, spiny, or sub-unarmed branches, the spines patent, 4–6 lines, rarely 1 inch long, sometimes leaf and flower bearing. Leaves very unequal, polymorphous, always larger than in the preceding, some 1–1½ inches long and 6–9 lines wide, others 8–9 lines long and 5 wide; in var. β. the lower ¾–1 inch long and wide, obtuse at base, with a recurved or acute point, upper 6–4 lines long. Petiole 1 line long. Panicle 3–5 flowered or 7–12 fl. subsessile, or pedunculate, always shorter than the leaf. Fl. small and white, 4–5 fid. Stamens short. Capsule as large as a *pea*, reddish; seed obovate, with a thin, yellow aril.—This may possibly be *C. saxatilis, Burch. Trav.* 2. *p.* 264, from Kor Rock Fountain.

17. **C. ellipticus** (Thunb. Fl. Cap. p. 218); glabrous, with small spines; leaves tufted, the upper solitary, *linear-oblong*, obtuse, *greatly narrowed towards the base*, crenulate or quite entire, coriaceous, *thick*, very smooth above, veiny and wrinkled beneath; panicles axillary, cymoid; capsules globose, trigonous, about 3-seeded. *C. integer, E. & Z.!* 945, *non Th.*

HAB. Woods at Amsterdamvlakte, at Zwartkops R. and Vanstaadensberge, *E. &*

Z.! Thunberg, Zey.! 2180 *and Cel. No.* 6. Enon and Zuureberg, *Drege!* 6729, 6734. (Herb. Thunb., Sond.)

An erect, greyish shrub, called, "*Kammassie-hout*" by the colonists. Spines 3–4 lines long. Leaves from an obtuse or mucronulate or emarginate apex, tapering gradually into the petiole, 1½ inch long, 3 lines wide or less glaucous. Panicle pedunculate, longer or shorter than the leaf, branches spreading. Flowers small, 5 fid, rarely 4–fid. Petals oblong, 1 line long. Capsule as large as a pea, mucronulate. It differs from *C. polyacanthus* by its short, less frequent spines and thick leaves, without veins above, rugulose beneath; from *C. integrifolius* by the short spines, long-tapering leaves, often crenulate, and trigonous, not *triquetrous* capsules.

18. C. buxifolius (Linn. Sp. 285); glabrous, spiny; branches angular; leaves tufted, ovate or *obovate-oblong, tapering at base,* obtuse or emarginate, serrulato-dentate or crenulate, veiny on both sides; panicles axillary, corymbose, pedunculate, shorter or longer than the leaf; capsules sub-globose, trigonous, 3-celled, 3-2 seeded. *Thunb. prod. p.* 42. *Fl. Cap.* 220. *Houtt. Pfl. Syst.* 3. *t.* 21. *f.* 1. *Catha buxifolia, Presl. l. c.*

VAR. α. **genuinus;** spines mediocre, small or obsolete; leaves obovate-oblong, emarginate, cuneate, 1–2½ inch long, ¾–1 inch wide or wider, panicle mostly many-flowered: either (αα) equalling the leaf or longer. *C. buxifolius, humilis, and goniecaulis,* E. & Z.! 942, 944, 940. *Drege!* 6735. *Zey!* 304: or (ββ) panicle solitary or several together, many-flowered, half as long as the leaf, or equalling it. *C. cymosus, Sol. Bot. Mag.* 2070, *spinis nudis. C. multiflorus, E. & Z.!* 951, *spinis floriferis.*

VAR. β. **laxiflorus;** spines mediocre, small or obsolete; leaves obovate or obov -oblong, obtuse or sub-acute, sub-membranaceous, 2½ inch long, 1 inch wide; panicle lax and mostly few-flowered. *C. patens and spathæphyllus, E. & Z.!* 939, 950. *C. leptopus, Bernh. in Hb. Krauss. Catha patens and spathophylla, Presl.*

VAR. γ. **venenatus;** spines strong, very long, the lower 6 inches; leaves obovate-oblong, 2–2½ inches long, 6–9 lines wide; panicles aggregate, many flowered, shorter than the leaf. *C. venenatus, E. & Z.!* 952, *Zey!* 2182. *Drege,* 6733. *Catha venenata, Presl.*

VAR. δ. **empleurifolius;** spines small or obsolete; leaves obversely lanceolate, obtuse or sub-acute, 2 inches long, 4–6 lines wide, panicle equalling the leaf or twice shorter. *C. empleurifolius, E. & Z.* 953, *fol. membranaceis. C. polyanthemos, E. & Z.!* 954, *fol. coriaceis. Zey!* 2179.

VAR ε. **glomeruliflorus;** almost unarmed; leaves obovate-oblong, obtuse, crenulate, much attenuate at base, 2–3 inches long, 6–8 lines wide; panicle much branched, sub-sessile, dense, much shorter than the leaf.

HAB. Common from Capetown to Port Natal, in one or other of its multifarious shapes. (Herb. Thunb., Sond., T.C.D., &c.)

A greyish shrub, several feet high. Spines naked or leafy, rarely floriferous, patent. Twigs often unarmed. Leaves unequal, obtuse or excised, rarely acute, on short petioles. Flowers small, white, 5–fid. Petals oblong, much longer than the calyx, equalling the stamens. Capsules the size of a small *pea,* rarely larger. Var. ε. (from Camtous river) may be a species; but we have seen but a single, unarmed twig of it. *C. multiflorus,* Lam. Enc. Meth. 1. 661. seems to be a var. of *C. buxifolius;* it does not exist in Herb. Lamk.

19. C. nemorosus (E. & Z.! 938); glabrous, spiny; spines strong; leaves tufted, solitary on the twigs, *elliptical, rounded, obtuse*, or shortly emarginate, *margined, dentato-serrate, shortly cuneate* at base, sub-coriaceous, veiny; panicles axillary, cymose, shorter than the leaf, or equalling it; capsules trigonous, 3–2-seeded.

VAR. β. panicles lax, few flowered. *C. laxus, E. Mey.!*

HAB. Krakakamma, Adow and Olifantshoek, Hassagaisbosch, Howisons Poort, and Port Natal, *E. & Z.! Drege*, 9533. *T. Williamson, Zey.!* 2183, 2184. (Herb. Lehm., T.C.D., Sond.)

Very like the preceding, but differing in the broader and rounder leaves, not much attenuate at base, evidently serrate or toothed. Also allied to *C. verrucosus* which is distinguished by its foliage, warty branches, and different fruit. Branches terete, sometimes sub-verrucose; twigs somewhat angular. Leaves glaucous above, livid below, with raised nerves, the larger 2–2½ inches long, 1–1½ inch wide; the smaller uncial. Panicle many flowered, in β. looser and longer. Flowers of *C. buxifolius*. Capsule 3 lines long.

20. C. rhombifolius (E. & Z.! 941); glabrous, spiny; twigs subquadrangular; leaves tufted, solitary on the twigs, those of the tufts obovate-oblong, gradually tapering into a petiole, crenato-dentate; those *of the twigs rhomboid, obtuse, shortly cuneate, serrated*, coriaceous, netted; panicles axillary, cymose, equalling the leaf; capsules small, sub-orbicular, trigonous, 3–2-seeded. *Herb. Un. It. No.* 161, *ex pte.*

HAB. North and east side of Devil's Mt., Capetown, *E. & Z!* (Herb. Sond.)

A shrub, with very short spines. The tufted leaves are quite like those of *C. buxifolius*, 1½–2 inches long, 8–10 lines wide, emarginate, with larger crenatures or teeth than in that species; the upper 1½ inch long, inch wide, coarsely serrate, obtuse and apiculate. Panicle on a peduncle ½ inch long, with divaricate branches, the pedicels 2 lines long. Flowers 5 fid. Capsule the size of a pepper-corn, tipped with a very short style.

21. C. angularis (Sond.); glabrous, spiny; stem and branches *angular-striate;* spines slender; leaves tufted, cuneate-obovate, *truncate-obtuse*, or *obcordate, spinuloso-serrulate*, or denticulate, very smooth above, with raised nerves beneath, coriaceous, glaucescent; panicles axillary, cymose, longer than the leaf; capsules small, globose-trigonous, apiculate.

HAB. Grassy hills at Vanstaadensberg, *Zey.!* 2182. Grahamstown, *Dr. Atherstone.* Transvaal, *Dr. Sutherland!* Feb. (Herb. Hook., Sond.)

A much branched shrub, with terete, ribbed and furrowed branches. Spines ½ inch long. Leaves rarely solitary, ½ inch long, 3 lines wide, obtuse, emarginate or tipped with a recurved point, entire at base; petiole very short. Panicles solitary or several together, ½–1 inch long. Fl. small, 5 fid. Petals oblong, equalling the stamens. Capsule as large as a *pepper-corn*, furrowed, 3 valved, 3–2 seeded.

Doubtful Species.

22. C. rhamnoides (Poir. Enc. Meth. Suppl. 2. 145); leaves ovate or lanceolate, acute, sharply serrate; flowers minute, axillary, tufted; peduncles simple, unequal, 1 flowered. *R. & Sch. Veg. V. p.* 422.

HAB. Cape. (Herb, Juss.)

Branches diffuse, grey, numerous. Leaves alternate sub-petioled, glabrous, paler beneath, teeth very short and acute; capsule 3 celled, as large as a pea; seeds blackish.

Excluded from the genus Celastrus.

Celastrus cernuus, *Thunb. prod.* 42 ; omitted in Fl. Cap. ; wanting in Hb. Thunb.		
C. tetragonus, *Thunb. prod.* 42	=	**Cassine scandens**, *E. & Z.*
C. filifolius, *Thunb. l. c.*	=	**Astephanus frutescens**, *E. Mey.* (*Asclepiadeæ*).
C. crispus, *Thunb. Fl. Cap.* 217	=	**Eucleæ** sp. (*Ebenaceæ.*)
C. microphyllus, *Thunb. prod.* 42	=	**Ehretia Zeyheriana** *Buek* (*Boragineæ*).
C. excisus, *Thunb. Fl. Cap.* 219	=	**Scutia capensis**, *Brg.* (*Rhamneæ*).
C. flexuosus, *Thunb. l. c.* 220	=	**Dovyalis rhamnoides**, *H.* (*Bixaceæ*).
C. rotundifolius, *Thunb. prod.* 42	=	**Dovyalis rotundifolia**, *H.* (*Bixaceæ*).
C. rigidus, *Thunb. Fl. Cap.* 220	=	**Rhigozum brachiatum**, *E. M.* (*Bignoniaceæ.*)
C. ovata, *E. Mey.!*	=	**Cassinopsis capensis**, *Sond.* (*Ilicineæ*).
C. *No.* 6747, *Herb. Drege*	=	*Rubiacea.*

II. PTEROCELASTRUS, Meisn.

Calyx minutely 5-parted. *Petals* 5, elliptical-orbicular. *Stamens* 5, inserted between the lobes of a fleshy disc, shorter than the petals. *Ovary* half sunk in the disc, 3 celled ; ovules in pairs, erect. *Style* 1 ; stigma 3-lobed. *Capsule* 2-3 valved, 3-6 winged, the valves septiferous in the middle. *Seeds* 1-3, erect, with a membranous aril, testa coriaceous, shining, brown ; raphe linear. *Embryo* in horny albumen ; the radicle inferior. *Meisn. Gen.* 1. *p.* 68. *Endl. No.* 5682. *Asterocarpus, E. & Z.! Enum. p.* 122.

Trees or shrubs, natives of S. Africa, erect, glabrous, with erecto-patent branches and alternate, coriaceous leaves. Flowers axillary, cymose or panicled, rarely subsessile. Name from πτερον, *a wing,* and *Celastrus;* this genus differs from *Celastrus* by its winged capsule.

1. *Cyme sessile or shortly pedunculate, few flowered.* (Sp. 1-2.)

1. P. tricuspidatus (Sond.) ; leaves ovate-elliptical, or obovate, obtuse, narrowed at base, veinless or obsoletely veiny ; cyme about 3-flowered ; capsule 3-winged, the *wings erect or inflexed,* lanceolate, subulate at the point, *sub*-trigonous, with horizontal margins. *Burm! Afr. t.* 97, *f.* 1.

VAR. α. **Lamarckiana** ; wings of the capsule trigono-lanceolate, mostly entire. A shrub or tree, 8-10 f. high. *Celastrus tricuspidatus, Lam.! DC. Prod.* 2. *p.* 5. *Asterocarpus typicus, E. & Z.!* 956, *excl. syn. DC. Pterocel. typicus, Meisn.—Pappe, Sylv. cap. p.* 9.

VAR. β. **litoralis** ; wings of the capsule trigono-lanceolate, all or one or two bifid, or subtrifid. A shrub, 1-3 feet high. *C. peterocarpus, DC. l. c. Asteroc. Burmanni, E. & Z.!* 958.

HAB. Among shrubs, round Table Mts. and Rondebosch ; and between Breede and Duivenhocks River. Var. β., in Hott. Holland, near Saldanha Bay, Zwartland, and in Zwellendam, *E. & Z.! Sieber, Mund, Pappe, &c.* (Herb. Lam., Lehm., T.C.D., Sond.)

Branches terete, twigs angular. Leaves uncial or biuncial, pale green when dry, obtuse or emarginate, rarely sub-acute. Petiole thick, 1-2 lines long. Cyme sessile, or on a peduncle 1-2 lines long. Fl. small, white. Calyx obtuse. Petals oval-oblong, patent. Capsule tricuspidate, 3 lines long. Wings 1-1½ lines wide at base, sometimes in var. β., by a twist, sub-vertical.

2. P. litoralis (Walp. Rep. 1. p. 535); leaves ovate or obovate-oblong, obtuse, narrow at base, veinless; cyme sub-sessile, few flowered; capsule 3 winged, the wings *spreading, flat, bipartite* or bifid, with horizontal margins. *Asterocarpus litoralis, E. & Z.!* 957.

HAB. Sand hills near Port Elizabeth, *E. & Z.!* July. (Herb. Sond.)

Very like the preceding, and only differing in the capsule, which with the wings is 4 lines long; the wings 2 lines wide at base, flat, and not trigonous with a raised keel. Leaves, flowers, and seeds as in the preceding.

2. *Panicle cymose, shortly pedunculate, many flowered.* (Sp. 3–5.)

3. P. variabilis (Sond.); leaves elliptical, obovate, or obovate-oblong, obtuse, sub-emarginate, narrowed at base, veinless or somewhat veiny; panicle 4 times as short as the leaf; capsule 3 winged, wings *ovate, mostly bifid,* the lobes *acute or obtuse,* with vertical margins.

VAR. α. **acutilobus**; wings or their lobes ovate or acuminate. *Aster. nervosus, E. & Z.!* 959. *Pter. nervosus, Walp. l. c.*

VAR. β. **obtusilobus**; lobes of the wings broad, obtuse. *A. arboreus, E. & Z.!* 960. *Pt. arboreus, Walp. l. c.*

VAR. γ. **armatus**; lobes of the wings wide, toothed or lacerate, as well as the capsules here and there furnished with narrow appendages. *Ast. tetrapterus, E. & Z.!* 962. *Pt. tetrapterus, Wapl. l. c.*

HAB. In woods. Var. α. in Grootvadersbosch and Wagenmakers Bosch, *E. & Z.! Mundt! Pappe! Drege,* 6727. *Zey.* 2194. β. Zwartkops River, Bothasberg and Howison's Poort, *E. & Z.! Pappe, Hutton, Zey.* 2197. *Drege,* 6726. γ. Philipstown and Kat River, *E. & Z.!* Oct.-Nov. (Herb. Lehm., T.C.D., Sond.)

A tree, 15–20 f. high. Leaves 1½–4 inches long, either quite veinless, or strongly veined, the veins diverging, arching within the flat or revolute margin. Petiole 2–4 lines long. Panicle ½ inch long, the common peduncle 2 lines long, branches forked, spreading, bracteolate; ultimate pedicels 1 line long. Flowers white. *Caps.* 4 lines long, in var. α. not unlike that of *P. tricuspidatus,* but all the wings or 2 of them are bifid, the third sometimes entire. The lobes in var. β. are very divergent, as wide as the cells of the capsule, mostly obtuse, occasionally acutely toothed, and at the base, as on the capsule itself, furnished with short wing-like appendages; in γ. wide lobes are intermixed with narrower, the margin sharply toothed or lacerate, and as well as the capsule itself armed with numerous appendages or horns.

4. P. stenopterus (Walp. l. c.); leaves elliptical, *obovate,* or *ovate-oblong,* obtuse or emarginate, narrowed at base, mostly somewhat veiny; panicle axillary, cymose, many flowered, ¼ as long as the leaf; capsule 3 winged, wings *bipartite,* with vertical margins, lobes *linear-lanceolate,* very patent. *Asteroc. stenopterus, E. & Z.!* 961.

HAB. Primitive woods. Krakakamma and Adow, Uit., *E. & Z.!* Nov. (Herb. T.C.D., Sond.)

A tree, 15–20 feet high. Leaves as in the preceding (of which this is perhaps a *variety*), with revolute margin, nearly always marked beneath with simple veins arching within the margin. Common peduncles as long as the petiole, rarely longer, divaricate, forked; pedicels 1 line long. Body of the capsule smaller than in the other species, 1 line long; wings bipartite to the very base, so that the capsule is apparently 6-winged, or one pair being abortive, 4-winged; lobes of the wings divaricate, 3–4 lines long, 1 line wide, flat.

5. P. Dregeanus (Sond.); leaves (*small*) ovate or oblong-ovate, flat, nerveless, *striato-rugulose and papilloso-multi-punctate beneath;* panicles cymose, nodding, shortly pedunculate, 2-3 times shorter than the leaf; ovary 3 celled. *Elæodendron Dregeanum, Presl, l. c.*

HAB. Rocky places at Wupperthal, 2-3000 f. *Drege!* 6725. Dec. (Herb. Lehm., T.C.D., Sond.)

A small, much branched shrub, with terete, greyish branches, and angular, spreading twigs. Leaves erecto-patent, 1 inch long, 4-6 lines wide, some small and narrow, all thick, very smooth and shining above, concave, narrowed at base and apex, obtusely or slightly emarginate, nerveless and veinless; petiole 1 line long. Common peduncle 3 lines long, angular, the pedicels simple or once or twice forked, the ultimate thick, 1 line long. Flowers as in the preceding. Petals thrice as long as the stamens. Ovary 3 celled; style very short. Ovules in pairs, Fruit unknown.—Readily known by its small leaves, minutely white-dotted over the whole under surface. At first sight it is more like a species of *Scytophyllum*, but the ovary is trilocular.

3. *Panicle on a long peduncle; its branches fastigiate.*

6. P. rostratus (Walp. l. c.); leaves oblong or oblong-lanceolate, obtusely acuminate, narrowed at base, with revolute margin, veiny; panicle axillary *on a long peduncle*, half as long as the leaf or longer, many flowered; its branches divaricate; capsule armed with numerous, compressed, wing-like, patent or recurved horns.

VAR. *α.* **Thunbergianus**; capsule ovato-globose, horned toward the summit. *Celastr. rostratus, Thunb.! Fl. Cap.* 218. *Asteroc. rostratus, E. & Z.!* 963. *Pter. rostratus. Presl,*

VAR. *β.* **polyceras**; capsule sub-globose, from base to apex horn-bearing. *Asteroc. tricuspidatus, E. & Z.!* 964, *excl. syn.*

HAB. In woods. River Zonderende, Puspas Valley, Grootvadersbosch, and Voormansbosch. *β.* in Duyvelsbosch, E. & Z.! Thunberg! Pappe! Zeyher, 2193. Aug.-Sep. (Herb. Thunb., Lehm., T.C.D., Sond.)

A tree, 20 feet high, with a trunk 1-2 feet diameter, and whitish grey bark. Twigs yellow-purplish. Leaves spreading, produced into an obtuse often emarginate acumination, narrowed at base into a 3-4 line long petiole, shining above, dull green and paler beneath, 2½-4 inches long, 1-1¼ inch wide, on some branches 2½ long and 8 lines wide. Common peduncle uncial, panicle much branched. Fl. white. Capsule 4 lines long, 3 valved, the valves horned at back; the horns unequal, 1-2 lines long, some subulate from a wider, compressed base, incurved or recurved; some broader, wing-like, acute. In var. *α.* the capsule has 4-8 apical horns; in *β.* almost the whole capsule is covered over with numerous, spine-like, unequal horns.

III. METHYSCOPHYLLUM, E. & Z.

Calyx 5-lobed, persistent. *Petals* 5, oblong, rounded, imbricate in æstivation. *Disc* 5-lobed, obtuse. *Stamens* 5, inserted between the lobes of the disc, alternate with the petals. *Ovary* sessile, surrounded by the disc, 3 celled; ovules in pairs, erect, oblong, netted-dotted. *Style* short, thick; stigma 3 lobed. *Capsule* oblong, obtuse, loculicidally 3-valved; 3-2-1 seeded. *E. & Z.! Enum. p.* 152. *Endl. Gen.* 5942.

A resinous shrub, with opposite leaves and axillary, dichotomous panicles of small white flowers. This is the "*Bosjesmans-thee*" of the colonists. The leaves, chewed to excess by the Bosjesmen, have intoxicating effects: a moderate infusion is said to be good as *tea*, and also as a remedy for asthma. The properties perhaps are similar to those of the *Paraguay tea*, a shrub not distantly related to the present. The name is from *μεθυσκω, to intoxicate*, and *φυλλον, a leaf.*

1. M. glaucum (E. & Z.! l. c.); *Hartogia Thea, E. Mey.! in Hb. Drege.*

HAB. Mountains on the right bank of the Zwartkey River and Windvogelsberg Mt., Tambukiland, *E. & Z.!* Klipplaat R. and Zwartkey, *Drege!* Nov. (Herb. Lehm., T.C.D., Sond.)

A glabrous shrub, 8–12 feet high, with the aspect of *Hartogia*, with terete branches, and compressed, opposite, fastigiate, greyish glaucous twigs. Leaves opposite, lanceolate, acuminate at each end, sub-emarginate, with revolute margin, repando-serrate, coriaceous, glaucous, netted with veins, 2–4 inches long, 6–9 lines wide. Panicles axillary, 8–12 flowered, much shorter than the leaf, dichotomous, fastigiate; the pedicels divaricate, bracteolate at base. Calyx obtuse. Petals 1 line long. Capsule (unripe) 2 lines long. Seeds unknown.

IV. HARTOGIA, Thunb.

Calyx 4-5 fid. *Petals* 4-5, obovate-oblong, patent. *Hypogynous-disc* fleshy, 4-5 lobed. *Stamens* 4-5, inserted between the lobes of the disc. *Ovary* 2-celled; cells 1-ovuled. *Style* short, thick; stigma 2-lobed, the lobes emarginate. *Drupe* dry, ovate, 2-celled, 2-seeded, often abortively one-seeded. *Seed* erect, exarillate, with a coriaceous coat; albumen none; cotyledons leafy, elliptical; radicle short, cylindrical, inferior. *Endl. Gen.* 5687.

Shrubs with opposite, coriaceous, glaucous leaves, with revolute, serrulate margins. Peduncles axillary, cymose or panicled; flowers minute, white, pedicellate. Named in memory of *John Hartog*, an early Dutch traveller in South Africa and Ceylon.

1. H. capensis (Thunb. Diss. nov. pl. Gen. 5. p. 35, cum icone); leaves petiolate, ovato-lanceolate or lanceolate, acuminate, obtusely emarginate, serrate; panicles axillary, pedunculate. *Thunb. Fl. Cap. p.* 142. *E. & Z.!* 980. *Schrebera schinoides, Th. nov. act. Ups.* 1. *p.* 91. 5. *f.* 1.

VAR. α. **lanceolata**; leaves lanceolate, narrowed at each end, panicle lax, cernuous, much shorter than the leaf. *H. capensis, E. & Z.!* 980.

VAR. β. **latifolia**; leaves larger, ovate-oblong, obtuse, more or less narrowed at base; panicle lax, much shorter than the leaf. *H. riparia, E. & Z.!* 982.

VAR. γ. **multiflora**; leaves smaller, obovate or ovate-oblong, obtuse, rather narrowed at base; panicle cernuous or sub-erect, many flowered, half as long as the leaf. *H. multiflora, E. & Z.!* 981. *H. capensis* β., *E. Mey.!*

HAB. Grootvadersbosch, and woods near the Zonderende, *Thunb.!* Breede River, Klynriviersberge, and 24 rivers, *E. & Z.! Zey.!* 2206, 307. Dutoitskloof, *Drege!* β. near Saldanha Bay, *E. & Z.!* γ. at Paarlberg, *Drege!* Tulbagh, *E. & Z.!* Aug.-Dec. (Herb. Thunb., Lehm., T.C.D., Sond.)

A much branched, glabrous, small tree, with glaucous grey bark, alternate, terete branches, and virgate, opposite, angular twigs. Leaves very smooth, shining, little paler beneath, with the midrib prominent at each side, reticulated with veins, serrate, the serratures incurved or obtuse, close or remote, very entire at base, in α, 2–3 inches long, 6–10 lines wide; in β, 2½–3 inches long, 1–1½ inch wide; in γ, 1½–2 inch long, 8–9 lines wide. Petiole 2 lines long. Common peduncle 3–4, in γ, 6–8 lines long, trichotomous or panicled; pedicels bracteolate, the ultimate very short, so that the unopened flowers look glomerated. Petals 1½ line long. Calyx ⅓ line. Drupe 4–5 lines long, with a terminal or sub-lateral point. Seeds ovate; testa black and shining. Var. γ, has some look of *Cassine capensis*.

V. MAUROCENIA, L. (*ref.*)

Calyx minute, 5–parted. *Petals* 5, longer than the calyx. *Hypogynous-disc* annular, circling the ovary. *Stamens* 5, alternate with the petals, and longer than them; anthers roundish. *Ovary* sessile, 3-celled; ovules in pairs (or solitary), pendulous from the inner angle of the cell. *Style* very short; *stigma* 3–lobed. *Drupe* somewhat juicy, 2, 3, or 1 seeded, with a hard stone (*putamen*). *Seed* inverted; testa coriaceous; aril none. *Embryo* in fleshy albumen; cotyledons flat; radicle superior. *Maurocenia and Cassine, sp. Mill. Dict. Cassine sp. Linn. Gen., Lam., DC.*

A glabrous shrub, with quadrangular twigs, opposite, coriaceous, quite entire leaves, and axillary cymoid inflorescence of small, white flowers. Name, in honour of a noble Venetian horticulturist, *F. Mauroceni.*

M. capensis (Sond.) *Cassine Maurocenia, Linn. Sp. p.* 385. *Willd. Sp.* 1. *p.* 1493. *Thunb. Fl. Cap. p.* 268. *E. & Z.!* 984. *Hook. Ic. Pl. t.* 552. *Frangula sempervirens, &c., Dill. Elth.* 146. *t.* 121. *f.* 147.

HAB. East side of Table Mountain, and about Camps Bay, *E. & Z.! &c.* common. Dec. (Herb. Sond., T.C.D., &c.)

An erect shrub, 4–5 feet high. Branches terete; twigs purplish, angular. Leaves very short-petioled, sub-orbicular, elliptical, ovate or obovate, obtuse, slightly emarginate, with the margin revolute, thick, shining, often livid beneath, 2–2½ inches long, 1½–2 inches wide. Peduncles crowded, 3–4 lines long, 1 flowered or branched, panicled. Calyx obtuse, fimbriate. Petals toothed. Drupe oval, as large as a *cherry.* Engl. name, "*Hottentot Cherry.*"

VI. CASSINE, L. (ex pte.)

Calyx minute, 4–5 parted. *Petals* 4-5, longer than the calyx. *Hypogynous-disc* annular, thickish, sub-sinuate, circling the ovary. *Stamens* 4-5, inserted under the disc, shorter than the petals or nearly as long; anthers roundish. *Ovary* sessile, 2–3 celled; ovules in pairs, erect. *Style* very short; *stigma* 2–3 lobed. *Drupe* somewhat juicy, 2–1 seeded, with a hard stone (*putamen*). *Seed* erect; aril none; embryo in fleshy albumen; cotyledons flat; radicle inferior.

Glabrous, erect or climbing shrubs, with 4-angled twigs, opposite, leathery, very entire or serrated leaves, and axillary cymoid inflorescences of small white flowers. The generic name is unexplained.

1. *Ovary trilocular; stigma 3-lobed.* (Sp. 1.)

1. C. affinis (Sond.); leaves petioled, ovate, acuminate, obtuse, remotely crenato-serrate, narrowed and quite entire at base, flat, coriaceous, paler beneath, netted-veined; panicles half as long as the leaf; ovary 3-celled; stigmas 3. *C. Colpoon, E. & Z.! ex pte.*

HAB. Mountains near Brackfontein and Olifants R., Clanw., *E. & Z.!* April (Herb. Sond.)

Branches alternate or ternate, terete, grey; ultimate twigs somewhat angular. Leaves 2–2½ inches long, 1 inch wide, narrowed into a 3-line long petiole, produced at the apex into a blunt point. Panicle sub-trichotomous; pedicels 2–bracted, 1 line long. Calyx obtuse. Petals oblong. Stamens equalling the petals. Ovary roundish; stigmata sub-sessile. Young drupe 2–seeded.—Allied to *C. capensis.*

2. *Ovary bilocular; stigma 2-lobed.* (Sp. 2–7.)

* *Leaves quite entire.*

2. C. parvifolia (Sond.); twigs angular; leaves short-petioled, ovate-oblong or oblong, acute at each end, sub-inflexed at the apex, emarginate, with revolute margin, concolorous, veiny beneath; veins confluent within the margin; pedicels axillary, one flowered, 2-3 times longer than the petiole.

HAB. Witsenberg, *Zeyher!* Dec. (Herb. Sond.)

A dwarf shrub, with terete, brownish purple branches. Leaves equalling the internodes or longer, very smooth above, beneath furnished with a thick midrib and 6–8 divergent veins, not unlike those of a *Scytophyllum*, 1–1½ inch long, 6–8 inches wide. Petiole 2 lines long. Peduncles slender, 3–4 lines long. Flowers small. Calyx segments ciliolate. Petals 3–4-times as long as the calyx. Hyp. disc fleshy. Stam. shorter than the petals. Fruit unknown.

** *Leaves serrate: erect shrubs.* (Sp. 3–4.)

3. C. Capensis (L. Mant. 220); twigs patent, *sub-angular;* leaves *petioled*, ovate or obovate, very obtuse, emarginate, *narrowed at base*, crenato-serrate, flat or with sub-recurved margins, netted, livid beneath; panicles axillary, ½ as long as the leaf; style bifid. *Burm. Afr. t.* 85. *Dill. Elth. t.* 236. *Thunb.! Fl. Cap. p.* 269. *E. & Z.!* 986. *C. Æthiopica, E. & Z.!* 988, *non Thunb.* *C. Kraussiana, Bernh.*

VAR. β. **Colpoon** (DC. Prod. p. 12); panicle looser, nearly as long as the leaf. *Euonymus Colpoon, Lin. Mant.* 200. *Burm. Afr. t.* 86. *C. Colpoon, Thunb.!—E. & Z.* 987, *ex pte.*

HAB. Woods, &c. in the Cape, Swellendam, and Uitenhage districts, and at Kat River, *Thunberg, E. & Z.! &c. Zey.!* 2204, *and* 2199, *ex pte.* (Herb. Thunb., Sond., T.C.D., &c.)

A glabrous, greyish shrub with opposite, alternate or clustered terete branches. Leaves entire at base, upwards closely or distantly bluntly serrated, 1½–2 inches long, ¾–1 inch wide, or smaller; petiole 2–3 lines long. Panicle fastigiate, corymbose, mostly semi-uncial; in β, an inch or more in diameter, pedicels spreading, 1 line long. Flowers 4–5 fid, white. Disc somewhat lobed. St. shorter than the petals. Drupe obovate-roundish, purple, mostly 1 seeded.

4. C. barbara (L. sp. 385); twigs *exactly* 4-*angled*, the opposite sides channelled; leaves *sub-sessile*, thick, broadly ovate or sub-orbicular, *sub-cordate at base*, obtuse or emarginate, crenato-serrate, *smooth above*, livid and netted-veined beneath; panicles axillary, dichotomous, *many flowered*, shorter than the leaf; drupe 1-2 seeded. *Thunb.! Fl. Cap. p.* 269. *E. & Z.!* 985.

HAB. Hills. About Muysenberg and in Zwartland, also on hills near Saldanha Bay, *Thunberg, E. & Z.! Zey.* 2202. (Herb, Sond.)

A shrub, 2 or more feet high, with terete or line-angled, erecto-patent, rigid, short branches. Leaves obtuse or rounded, flat or with reflexed margins, mostly uncial, and 10 lines wide; some smaller, some 1½–2 inches long; petiole thick, 1 line long. Fl. small. Calyx 4 parted. Petals 4. Disc fleshy. St. 4, equalling the petals. Drupe roundish, as large as a *pea*.

*** *Leaves serrate; climbing shrubs.* (Sp. 5–7.)

5. C. latifolia (E. & Z. 990); glaucous; stem flexuous, sub-scandent, *terete;* twigs divaricate, quadrangular; leaves sub-sessile, sub-coriaceous, *cordate-ovate,* obtuse, sub-emarginate, undulato-serrate, *netted at both sides,* livid beneath; cymes on long peduncles, *few flowered;* drupe 1-2 seeded.

VAR. β. **heterophylla**; lower leaves ovate, upper ovate-oblong, cordate at base. *Celastrus heterophyllus, E. Mey.!*

HAB. Woods at Plettenberg Bay, *E. & Z.!* β. Rocky places on the Witberg, 5000 f. *Drege!* (Herb. E. Mey., Sond.)

A tall shrub with alternate, brownish grey branches, and lax, mostly trifid twigs. Leaves equalling the internodes, 1¼–1¾ inch long, 8–15 lines wide, some smaller, flat or with sub-revolute margins, serrulate, the teeth often spinigerous on the young leaves, mostly obtuse on the older; nerves prominent on both surfaces. Petiole ½ line long. Common peduncles 6–9 lines long; cyme 6–12 flowered. Flowers minute, 5 fid. Drupe as large as a *pea.* In var. β. the upper leaves are 2–2½ inches long, inch wide, with short obtuse acumination; the petiole ½–1 line long. This differs from *C. barbara* by the long, slender, flexuous, subscandent branches, much thinner leaves, veiny at both sides, and the lax, few flowered panicle.

6. C. Albanensis (Sond.); stem *terete,* flexuous, sub-scandent; branches divaricate; twigs angular; leaves sub-coriaceous, petiolate, *ovate,* obtuse or obtusely acuminate, emarginate, undulato-serrate, netted-veined at both sides, *obtuse or a little narrowed at base;* panicle trichotomous, *half as long as the leaf;* ovary bilocular.

HAB. Albany, *Mrs. F. W. Barber!* (Herb. T.C.D.)

Near the preceding, but apparently distinct. Petiole 2 lines long. Leaves 1½–2 inches long, 1 inch wide, not cordate but mostly narrowed at base, green above, mostly livid beneath. Panicles lax; common peduncle 3–4 lines long, twice or thrice trichotomous, its branches bracteate at base, the ultimate pedicel 3-flowered. Flowers sessile, minute, 5 fid. Petals oblong. Fruit unknown.

7. C. scandens (E. & Z.! 989); glaucous; stem flexuous, scandent, 4-6-*winged;* branches divaricate, 4-angled; leaves short-petioled, *oblong,* obtuse or sub-acute, sub-cordate at base, undulato-serrate, sub-coriaceous, netted-veined on both sides; cymes axillary, pedunculate, 2–3 times shorter than the leaf; drupe sub-globose, mostly one-seeded. *Celastrus tetragonus, Thunb.! prod. p.* 42. *Rhamnus tetragonus, Thunb.! Fl. Cap. p.* 196, *excl. sp. fructif. Cel. refractus, E. Mey.!*

VAR. β. **latifolia**; leaves ovate-oblong, or ovate, obtuse, sub-cordate at base. *Cel. tetragonus, E. Mey.!*

HAB. Woods at the Zwartkops River and Adow, also near Galgebosch, *E. & Z.! Drege!* β. Vanstaadensberg, *Drege! Zey.!* 2203. Sep.–Dec. (Herb. Thunb., Lehm., T.C.D., Sond.)

A shrub, 18–20 feet high. Stems evidently winged; twigs horizontal, ascending, leafy. Leaves very unequal, mostly 1½–2 inches long, 6–9 lines wide, but the lower are wider and larger, the upper smaller and narrower. In var. β. they are 2–2½ inches long, 14–18 lines wide; all with sub-revolute margin, and short, incurved-obtuse sometimes spinuligerous serratures. Common peduncles 4–6 lines long, equalling the 12–20 flowered cyme or longer; ultimate pedicels with clustered flowers at their tips. Flowers minute, 5 fid. Calyx fimbriate, ⅓ as long as petals. St. short. Drupe as large as a pea, drying black, mostly one seeded. Seed rugulose.

VII. ELÆODENDRON, *Jacq.*

Calyx minute, 4–5-lobed. *Petals* 4–5, patent, broad at base. *Stamens* 4–5, alternate with the petals. *Hypogynous-disc* fleshy, circling the ovary. *Ovary* 2–4-celled; ovules in pairs, erect. *Style* short: stigma obsoletely lobed. *Drupe* dry or nearly so, with a hard, woody, 5-celled, sometimes 2–1 celled stone. *Seeds* solitary, oblong, exarillate. *Embryo* in fleshy albumen; cotyledons leafy; radicle inferior. *DC. Prod.* 2. *p.* 10. *Lam. Ill. t.* 132. *Elæodendron and Chloroxylon, E. & Z.!*

Trees or shrubs, with glabrous, mostly opposite leaves, and axillary, sub-corymbose peduncles; flowers small, white.—Name from ελαια, an *olive*, and δενδρον, a *tree*; from some resemblance in foliage to the olive. Eng. name, "*Olive-wood.*"

1. E. Capense (E. & Z.! 979); glabrous; branches rough with points; leaves sub-opposite, petioled, coriaceous, *elliptic-ovate or ovate-oblong*, sub-acute or acuminate, quite glabrous at both sides, netted-veined, remotely serrate, with revolute margin; panicles simple, dichotomous; drupe *ovate or ovate-oblong, 2–3-celled. Hook. Bot. Mag. t.* 3835. *E. papillosum, Hochst.! in Krauss, Beytr.* 42. *Ilex crocea, E. Mey.! in Hb. Drege. Cassine crocea, Presl.!*

HAB. Primitive woods of Zitzekamma and Krakakamma, and at Gauritz River, *E. & Z.!* Swellendam, *Mundt!* Vanstaadensberg, *Zey.!* 2191. Port Natal, *Krauss!* Jul. (Herb. Vind., Lehm., T.C.D., Sond.)

An erect shrub. Leaves 2–3 inches long, 1–1½ inch wide, opposite, or sub-alternate; petiole 3–4 lines long, furrowed. Panicle few or many flowered, inch long. Pedicels bracteolate. Flowers 4-parted. Ovary sunk in a flat disc. Drupe ½–1 inch long, red, rugged, with a hard woody nucleus. Seeds mostly 2, sub-compressed, red-brown, shining.

2. E. croceum (DC. Prod. 2. p. 11); glabrous; branches verrucoso-scabrid; leaves mostly opposite, petioled, *oblong*, obtuse at each end, or *obovate, narrowed at both ends*, obtuse or retuse, quite glabrous at both sides, netted-veined, serrato-crenate; panicles shorter than the leaf; *drupe globose, 4–5-celled. Pappe, Sylv. Cap. p.* 10. *Ilex crocea, Thunb.! Fl. Cap. p.* 159. *Crocoxylon excelsum, E. & Z.!* 983. *Rhamnus capensis, Spreng! Cassine parvifolia, E. Mey.! and Drege No.* 6740. *Elæodendron verrucosum, Kunth! Linn.* 19. 393. *C. Zeyheri, Turcz. Bull. Mosc.* 1858. 2. *p.* 452.

HAB. Primitive woods of Krakakamma, Olifantshoek, Adow, Vischriver and Kat River, *Thunberg, E. & Z.! Drege! &c.* Fl. Jul. fr. Apl.-May. (Herb. Lehm., Sond.)

A tree, 20–40 f. high, 2–4 f. in diameter. Wood hard, yellow, "*Saffranhout.*" Branches terete, grey, alternate or crowded; twigs sub-angular. Leaves 2 inch long, ¾ inch wide, less coriaceous than in *E. capense*, but more netted, the sub-recurved margin with small serratures, spinulous on the young leaves. Peduncles in a corymbose panicle ½ or ⅓ of the leaf; or several undivided one flowered pedicels, jointed in the middle. Calyx 4 parted, obtuse. Petals 4, 1½ line long. St. 4, shorter than the petals. Ovary sunk in a 4 lobed fleshy disc, 4 angled, 2 celled; cells 2 ovuled. Style short; stigma obsoletely lobed. Drupe the size of a small cherry. Seeds 1–2.—*Salacia Zeyheri, Planch.*—*Harv. supra*, page 230, proves to be merely an abnormal state of this tree, with *triandrous* flowers, *hypogynous* stamens, &c.!

VIII. LAURIDIA, E. & Z.

Flowers sessile, 3-bracteolate. *Calyx* 4-fid. *Petals* 4, ovate, reflexed, scarcely narrowed at base. *Stamens* 4, inserted at the margin of a 4 lobed disc, shorter than the petals. *Ovary* sub-immersed in the disc. *Style* very short; stigma 2 lobed. *Drupe* rather dry, 2 celled, often abortively one seeded, with a crustaceous stone. *Embryo* albuminous. *E. & Z.! Enum. p.* 124. *Elæodendron, sp. Endl.*

A shrub with terete branches, opposite, ovate, very entire or sparingly toothed, netted veined leaves, and panicled, axillary racemes of small white flowers. Remarkable by its general habit and inflorescence; possibly hereafter to be united with *Cassine*, unless better characters should be found in its imperfectly known fruits and seeds. The name is an alteration of *Laurus;* because the foliage resembles that of a laurel.

1. L. reticulata (E. & Z.! 968.) *L. rupicola, E. & Z.! 969.*

HAB. Among shrubs at Zoutpanshoogde, at Zwartkops R., and in Winterhoek, Uit.; also at Bothasberg, Grahamstown. *E. & Z.! T. Williamson!* Hills at Elands River, *Zey.!* 2198. (Herb. Lehm., T.C.D., Sond.)

An erect, glabrous shrub, 8–10 feet high, trichotomously branched, with grey bark; the twigs angular. Leaves 1–2 inches long, very variable in form, erect, ovate, obovate, sub-orbicular, ovate-oblong or oblong, often acute, rarely obtuse or emarginate, rounded or somewhat narrowed at base, very smooth above, netted-veined beneath, and traversed by a thick midrib, coriaceous, rigid, with sub-revolute margin. Petiole thickish, 2 lines long. Racemes panicled, ½ inch long. Flowers white, about 1 line long, the upper ones on the same rachis sessile, the lower sub-ternate at the tips of short pedicels. Rachis and pedicels short, angular, furnished with minute, acute, persistent bracts. Calyx obtuse. Petals twice as long. Drupe, of which very few have been examined, as large as a small pea, roundish and reddish.

IX. MYSTROXYLON, E. & Z.

Calyx minute, 5-lobed. *Petals* 5, spreading, roundish-ovate. *Stamens* 5, short. *Ovary* sitting on a furrowed, fleshy disc, 2 celled; ovules in pairs, erect. *Style* very short; *stigma* obsoletely bi-lobed. *Drupe* dry or fleshy, globose or ovato-globose, with a hard, 2-celled or one-celled stone, 1–2 seeded, seed erect, exarillate; raphe linear; embryo in sub-horny albumen. *E. & Z.! p.* 125. *Elæodendron, sp. Endl.*

Trees or shrubs with alternate leaves, and small, umbellate, mostly densely crowded flowers. The name is a translation into Greek of the colonial "*Lepelhout;*" *spoon-wood.*

1. *Fruit quite juiceless, small.* (Sp. 1–5.)

* *Leaves serrate or crenate; umbels pedunculate.* (Sp. 1–4.)

1. M. confertiflorum (Tulasn. Ann. Sc. Nat. 1857, Bot. p. 106); branches terete, the young glabrous or downy; leaves petioled, ovate, obovate or elliptical, obtuse, shortly emarginate, *cuneate at base*, crenate or callous-serrulate, netted-veined at both sides, shining, coriaceous, glabrous; umbels short-peduncled, axillary, dense, glabrous or downy; drupe *broadly ovate*, apiculate; seed ovate. *Cassine æthiopica, Thunb.! Fl.Cap.* 269. *Mystr. athranthum, spilocarpum and sessiliflorum, E. & Z.!* 973, 974, 975. *M. sphærophyllum, Pl. Krauss!*

VAR. β. **leptocarpum**; drupe oblong, apiculate; seed oblong.

HAB. Woods at the Zwartkops River, in Adow, Olifantshoek, and at the Zeekoe R., *E. & Z.! Krauss! Zey.!* 2199, *ex pte.* Zuureberge, *Drege,* 6736. Port Natal, *Gueinzius!* β, Zwartkops River, *Zey.!* 2201. Nov.–Dec. fr. July. (Herb. Thunb., Vind., Lehm., Hook., T.C.D., Sond.)

A tree 15–20 feet high, or a much branched shrub; branches ashen; twigs leafy. Leaves 1–1½ inch long or less, varying from orbicular to oblong, some very thick, others thinner, sub-coriaceous, with a flat or undulato-crenate margin, or serrulate or sub-entire. Petiole 1–2 lines long. Pedunc. solitary, rarely 2, 1–3 lines long (in *M. sessiliflorum*, E. & Z.! not shorter than in the others). Umbels 12–20 flowered, pedicels ½–1 line long. Flowers 2 line long. Stamens ⅓ of petals. Drupe as large as a *pea*, one seeded. Var. β., in other respects like α., differs in having a drupe 4–6 lines long, 2–2½ wide.

2. M. sphærophyllum (E. & Z.! 976); *glabrous;* leaves petioled, elliptical or sub-orbicular, emarginate, *sub-cordate at base*, the recurved margin crenulate or callous serrulate, veiny at both sides, coriaceous; umbels axillary, pedunculate, glabrous; drupe *globose.*

VAR. β. **litorale**; leaves elliptical or oblong-elliptical, and the twigs glaucous. *M. kubu, E. & Z.!* 978. *M. pubescens, Pl. Krauss!*

HAB. Kat River, near Philipstown, *E. & Z.!* Swellendam, *Mundt!* Zwartkops-river. *Zey.!* 2200. Vanstaadens River, *Drege*, 6732, 6742? β. Sea-shore woods at Cape Recief and Zitzekamma, *E. & Z.!* Feb.–Jul. (Herb. Vind., Hook., Lehm., T.C.D., Sond.)

A tree, 30–40 f. high. It differs from the preceding in the shorter-petioled leaves, not cuneate at base, thicker, more evidently serrulate (the larger 1½–2 inches long), in the peduncles 3–5 lines long, often shortly bifid at the summit, in the flowers not glomerated, on pedicels 2 lines long, and in the globose fruit. The flowers themselves are like those of the preceding, but the petals a little narrower at base, and the stamens twice as long. Drupe as large as a pea; seed similar in shape.

3. M. pubescens (E. & Z. 977); twigs *pubescent;* leaves short-petioled, elliptical or orbicular, emarginate, glabrous above, *softly pubescent beneath*, coriaceous, veiny at both sides, the recurved margin crenate or dentato-serrate; umbels axillary, pedunculate, glabrous; drupe obovate, pointed.

HAB. By the Zwartkops River, *E. & Z.! Zey.!* 2200. Oct. (Herb. Lehm., Sond.)

This differs from the last by its velvetty-pubescent twigs and leaves, even the older ones. Larger leaves 1½ inch long and wide, in some specimens scarcely uncial, but with a few larger intermixed. Common peduncles 3–4 lines long; the umbel always few-flowered, the pedicels 2 lines long. Fl. of the last. Drupe oval or obovate; seed similar.

4. M. Burkeanum (Sond.); *branches, leaves, peduncles, and calyx densely clothed with short pubescence;* leaves very short-petioled, elliptical, or orbicular, emarginate, sub-entire or minutely serrulate, coriaceous, veiny at both sides; umbels axillary, shortly peduncled; petals glabrous.

HAB. Banks of the Crocodile River, *Burke!* 378, *Zey.!* 309. Nov. (Herb. Hook., Sond.)

The terete branches and horizontal twigs grey with short pubescence. Larger leaves uncial, 8–12 lines wide, mostly smaller, 6–9 lines long, 5–8 wide, acutely emarginate, the marginal teeth scarcely discernible by the naked eye. Petiole ½–1 line long. Common peduncle equalling the petiole, or longer. Calyx lobes obtuse. Petals twice as large, elliptic-orbicular, longer than the stamens. Ovary immersed in a sinuous disc. Style short. Fruit wanting.

** *Leaves sub-entire; umbels sessile.* (Sp. 5.)

5. M. eucleæforme (Sond.); glabrous; leaves petioled, ovate, or elliptical, obtusely acuminate, sub-emarginate, narrowed at base, coriaceous, shining above, paler beneath, veiny, the revolute margin quite entire or repand-toothed; peduncles axillary, umbellate, aggregate, 1 flowered; drupe sub-globose, apiculate. *M. filiforme, E. & Z.!* 970, *excl. syn. M. oligocarpum, E. & Z.!* 971, *and M. eucleæforme, E. & Z.!* 972.

HAB. Zitzekamma, George, *Mundt.!* Vanstaadensberge, Winterhoeksberge, and Krakakamma, Uit., *E. & Z.!* Jan.–Jul. (Herb. Lehm., Sond.)

A shrub or tree, with terete, erecto-patent branches and angular, leafy twigs.

Leaves longer than the internodes, the larger 2–2½ inches long, 1–1¼ inch wide; the smaller 1–1½ inch long, 6–8 lines wide. Petiole 2 lines long. Pedunc. of fruit 2–4 together (the flowering probably more numerous), 1 line long. Drupe the size of a small pea.

2. *Fruit of large size, almost juiceless.* (Sp. 6.)

6. M. macrocarpum (Sond.); branches and twigs *minutely pubescent;* leaves short-petioled, elliptic or elliptic-oblong, obtuse, or sub-excised, the recurved margin serrate, glabrous above, *pubescent beneath;* umbels axillary, pedunculate; peduncles pubescent; drupe *very large,* globose, one-seeded.

HAB. Between the Omtata and Omsamwubo, below 1000 f. *Drege!* 5612, b. (*non a. Celastr. undatus*). (Herb. Sond.)

A shrub or tree with terete, grey branches. Larger leaves 2½ inches long, 1½–2 inches wide, coriaceous, netted at both sides, the margin acutely toothed. Petiole 2 lines long. Common peduncle 2–4 lines long. Calyx persistent, 5 parted, with obtuse lobes. Petals wanting. Drupe blueish, as large as a *Bullace-plum,* or 8–9 lines in diameter. Seed erect, with a shining chestnut skin.

X. SCYTOPHYLLUM, E. & Z.

Calyx minute, 5-parted. *Petals* 5, sub-orbicular, narrowed at base. *Disc* fleshy, 5-lobed. *Stamens* 5, inserted between the lobes of the disc. *Style* short; stigma 2-lobed. *Ovary* sub-immersed in the disc, 2-celled; the cells 2-ovuled. *Drupe* rather dry, oblique, 1–2 celled, one-seeded, rarely 2-seeded. *Seed* erect, covered with a thin, membranous *aril;* testa coriaceous. *Embryo* in fleshy albumen; radicle inferior. *E. & Z.! Enum. p.* 124.

Glabrous shrubs, with alternate, short-petioled, very entire, thick and leathery veinless leaves, and axillary, crowded, one-flowered, forked or panicled peduncles. Flowers sub-sessile, deciduous. Name, from σχυτος, *leather,* and φυλλον, a *leaf:*= "*Leather-leaf.*"

1. S. laurinum (E. & Z.! 966); twigs angular; leaves *ovate-oblong,* or oblong, obtuse, sub-emarginate, with sub-revolute margin; peduncles panicled, the panicle much shorter than the leaf. *Celastrus laurinus, Thunb.! Fl. Cap.* 217. *Hb. Un. it. No.* 162. *Scyt. obtusum, E. & Z.!* 965, *excl. syn. Cel. laurinus and C. obtusus, Drege. Herb. Cel. oleoides, Lam. Ill. No.* 2695.

VAR. β. **minus**; stem shorter; leaves narrower, often sub-acute; cymes few flowered. *S. oleoides, E. & Z.!* 967.

HAB. Common on hills round Capetown; also at Witsenberg, Muysenberg, Kardow, Paardeberg, Heerelogement, and near Tulbagh, &c., *Thunberg, E. & Z.! Drege,* &c. *Zey.!* 2192, 306 *a.* β. Bergvallei and Langvallei, *Zey.!* 306 β., *Drege!* 6724, and 1259 β. Jan.–Nov. (Herb. Thunb., Lam., Lehm., T.C.D., Sond.)

A shrub, 8–10 feet; in β. 1–2 feet high. Stem and terete branches ashen-brown; twigs erecto-patent. Leaves erect, ever-green, obtuse or emarginate, sometimes oblique at base, one-nerved, paler beneath and obtusely nerved after being dried, unequal in size, some 2 inches long, inch wide, others half the size; in β. uncial, 4 lines wide, narrowed at each end. Panicle 6–10 lines long, 8–20 flowered on a peduncle 2–3 lines long; at first dense, then open. Flowers minute, white. Drupe obovate, roundish, 3 lines long; in β. rather smaller. "*C. oleoides*" Krauss! from Natal, does not belong to the *Celastrineæ;* its stamens are opposite the petals.

2. S. angustifolium (Sond.); twigs sub-angular; leaves *lanceolate*, obtuse, narrowed into a petiole; peduncles panicled, much shorter than the leaf. *Elæodendron angustifolium, Presl, Bot. Bem. p.* 34.

HAB. Mountains near Capetown, *Ecklon!* Albany, *Mrs. F. W. Barber!* (Herb. T.C.D., Sond.)

A shrub, with the aspect of *S. laurinum*, with erecto-patent branches, and mostly powdery twigs. Leaves very short-petioled, about 1¼–1½ inches long, ½–¾ inch broad, paler underneath, obsoletely veined, tipped with a short, often inflexed point. Peduncles 4–6 together, 1–2 lines long. Flowers white, like those of *S. laurinum*. Fruit wanting.

ORDER XLIV. AQUIFOLIACEÆ, DC.

(By W. SONDER.)

(Aquifoliaceæ, DC. Theorie, 1813. Lindl. Veg. Kingd. No. ccxxx. Ilicineæ, A. Brongniart, 1826. Endl Gen. No. ccxxxviii.)

Flowers mostly perfect, regular. *Calyx* small, 4–6 cleft or parted, persistent, with imbricate æstivation. *Corolla* mostly monopetalous (in *Monetia* pleio-petalous), regular, 4–6-cleft, with a very short tube, the segments imbricate or sub-valvate in æstivation. *Stamens* alternate with the lobes of the corolla and inserted on its tube (in *Monetia* inserted on the margin of a perigynous disc); filaments subulate, short; anthers introrse, 2-celled, the cells adnate. *Ovary* sessile, fleshy, subglobose, 2–6–8 or many celled; ovules solitary, pendulous or erect, anatropous; stigma sub-sessile. *Drupe* fleshy and juicy, with 2 or more indehiscent, one-seeded stones. Seed (except in *Monetia*) with copious, fleshy albumen, and a small or minute embryo.

Evergreen trees or shrubs, natives of North and South America, and of the temperate regions of Europe and Asia, rare within the tropics. Leaves alternate or opposite, petioled, simple, coriaceous, glabrous and shining, penninerved, entire or spinous-toothed, exstipulate. Flowers small, white or greenish, axillary, either solitary, clustered, or in cymes or fascicles. The common Englsh *Holly* is the type of this Order, which is nearly allied to *Celastrineæ*, but known by its usually monopetalous corolla, small embryo and copious albumen; characters which do not apply to *Monetia*, a genus placed here *provisionally* until some better place may be found for it. The Paraguay Tea (*Ilex paraguayensis*), so largely used in South America, is the most famous plant of the Order. The wood of most, as of the English holly, is very compact and fine grained, well suited for turning and for ornamental cabinet work. When stained black it is a cheap substitute for Ebony.

TABLE OF THE SOUTH AFRICAN GENERA.

Sub-Order 1. ILICINEÆ.—*Corolla* monopetalous. *Ovules* pendulous. *Seeds* with copious, fleshy albumen and a minute embryo.

I. **Ilex.**—Ovary 4–6 celled, with sessile stigmata.
II. **Cassinopsis.**—Ovary one-celled, with a style.

Sub-Order 2? MONETIEÆ.—*Corolla* polypetalous. *Ovules* erect. *Seeds* exalbuminous.

III. **Monetia.**—Flowers unisexual. Spines axillary, spreading.

I. ILEX, L.

Flowers perfect. *Calyx* small, 4–6 toothed, persistent. *Corolla* hypogynous, rotate, 4–6-parted, the segments obtuse, imbricated in æstivation. *Stamens* as many as the lobes of the corolla and alternate with them; filaments subulate; anthers introrse, erect, longitudinally dehiscent. *Ovary* sessile, 4–6-celled; ovules solitary or in collateral pairs, pendulous from the central angle of the cell. *Stigmata* 4–6, sessile. *Drupe* baccate, sub-globose, crowned with the stigmata. *Seed* inverted; embryo minute, in the apex of fleshy albumen. *Endl. Gen.* 5705.

Evergreen shrubs or trees, natives of the temperate and sub-tropical regions of both hemispheres. Leaves alternate, coriaceous, entire or toothed. Peduncles axillary, numerous, often tufted, one or many flowered, bracted; flowers white. There is but one Cape species. The English "*Holly*" is a well known example of this almost cosmopolitan genus.

1. I. Capensis (Sond. & Harv.); glabrous; leaves on channelled petioles, oblongo-lanceolate or lanceolate, or obovate-oblong, acute at base, either acute, acuminate, or sub-obtuse and mucronate at the apex, shining above, paler beneath; peduncles fascicled, shorter than the petiole, few flowered; corolla 5-cleft; berry 5, or by abortion, fewer-seeded. *Leucoxylon foliis laurinis, &c., Burm. Afr. t.* 92. *f.* 2. *Sideroxylon mite, Jacq. Coll. Bot. Mag. t.* 1858. *Manglilla Milleriana, Pers.* 1. *p.* 237, *Scleroxylon mite, Willd. Enum. p.* 249. *Myrsine? mitis, Spreng. Pappe, Sylv. Cap. p.* 22. *Leucoxylon laurinum, E. Mey. in Hb. Drege. Celastrus Sieberi, Bernh. in Hb. Krauss.*

HAB. Swellendam, *Mundt!* Tulbaghskloof and Witsenberg, *E. & Z.!* Winterhoeksberg and Nieuwekloof, *Drege!* Chumiberg, Kat River, &c., *E. & Z.!* Near Uitenhage, *Zey!* 3365, 3366. Macalisberg, *Zey!* 1129. Constantia, *Krauss! Pappe!* Drakenstein Waterfall, *W.H.H.* (Herb. T.C.D., Sond., &c.)

A large shrub or small tree, 6–12 feet high or more, densely leafy, with whitish or pale bark and angular twigs. Petioles ⅛–⅜ inch long, slender, channelled. Leaves 2–3–3½ inches long, ½–1½ broad, very variable in form, passing from narrow-lanceolate to broadly-oblong, sub-obovate. Flowers crowded, sub-umbellate; peduncles jointed at base, or racemose, 3–4 flowered. Calyx urceolate 5–6 fid. the lobes ciliolate. Corolla monopetalous; segments ovate-oblong, obtuse, spreading or recurved, imbricate in bud. Stamens on the throat of the corolla, exserted, filaments subulate, scarcely as long as the corolla-lobes. Drupe globose, tipped with a thick 4–5 lobed stigma. Seeds trigonous. Embryo minute.

II. CASSINOPSIS, Sond.

Calyx 5-fid. *Corolla* 5-fid, the segments oblong, sub-acute, sub-imbricate in æstivation. *Stamens* 5, inserted on the base of the corolla, alternate with its lobes; filaments subulate; anthers ovate, longitudinally dehiscent. *Disc* none. *Ovary* sessile, one-celled, 2–1 ovuled; ovules unequal, pendulous from the central angle of the cell. *Style* simple; stigma obsoletely lobed. *Drupe* nearly dry, ovato-globose. compressed, apiculate, with a one-celled, one-seeded crustaceous stone. *Seed* inverted, compressed; embryo minute, in the apex of copious, fleshy albumen; radicle superior.

A glabrous shrub, with alternate branches and twigs, opposite leaves and axillary, dichotomously panicled, minute flowers. The name signifies "resembling a *Cassine.*"

1. C. Capensis (Sond.) *Cassine ilicifolia, Hochst. Pl. Krauss. Hartogia ilicifolia, Hochst. in Krauss, Bertr. p.* 42. *Cassine mucronata, Turcz. Bullet. Mosc.* 1858. *No.* 11. *p.* 455.

HAB. Primitive Woods of Outeniqua, George, *Krauss.* Grootvadersbosch, *Zey!* 2209. (Herb. Vind., Sond., T.C.D.)

A shrub. Branches terete, shining, flexuous, glabrescent. Spines terete, very acute, ½ inch long, in the forks of the branches or opposite the leaves. Leaves shortly petioled, those of the lower branches ovate, sub-acuminate, tipped with an obtuse mucro, ½ line long, obtuse at base, at length shining above, paler beneath, opaque, with alternate, prominent, primary veins, remotely spinuloso-serrulate or sub-entire, glabrous, 1½–¾ inch long, 12–14 lines wide. Leaves of the flowering branches smaller and narrower, 1–1½ inch long, 6–9 lines wide, narrowed at base. Petiole keeled, 2–3 lines long. Panicle or corymb scarcely half inch long, with a common peduncle 2–3 lines long. Flowers small. Corolla 1 line long, thrice as long as the pubescent, ovate, sub-acute calyx-lobes, its segments oblong, or oblong-lanceolate, downy externally. Stamens shorter than the petals. Style short. Drupe tipped with the style, compressed, as big as a large *pea.* Seed compressed.

III.? MONETIA, L'Her.

Flowers dioecious. *Calyx* campanulate, shortly 4–5-cleft, the segments separate or cohæring, valvate in æstivation. *Petals* 4–5, hypogynous, linear-lanceolate, spreading, longer than the calyx, with valvate æstivation. *Stamens* as many as the petals and alternate with them, inserted into the margin of a fleshy, hypogynous disc; filaments erect, subulate; anthers erect, introrse, 2 celled, versatile.—♀ *Calyx* and *corolla* as in the ♂. *Stamens* abortive. *Ovary* free, 2-celled, the septum thin and membranous, ovules solitary, anatropous, erect or ascending from the base of the cell; stigma sessile, globose, sub-bilobed. *Berry* 1–2-seeded. *Seed* exalbuminous, with a thin testa; embryo lenticular, *green;* cotyledons fleshy, cordate and eared at base, the ears over-lapping and concealing the long, straight radicle. Endl. Gen. 5711.

An erect, glabrous shrub, with opposite branches and twigs. Leaves opposite, petioled, with axillary, solitary or binate spines. Flowers small and greenish, in dense axillary fascicles. Name in honour of *M. Monet,* better known as the *Chevalier de Lamarck,* a very distinguished botanist and zoologist of the last century.

1. M. barlerioides (L'Her. Stirp. Nov. fasc. 1. p. 1). *Spreng. Syst. Veg.* 1. *p.* 442. *Azima tetracantha, Lam. Encycl. Bot.* 1. *p.* 343. *Illustr. t.* 807. *Zey.! No.* 4653. *Drege,* 6749, *a.*

HAB. Zwartkopsriver, E. & Z.! Drege! Nov.–Jan. (Herb. T.C.D., Sond. &c.)

An erect, glabrous shrub, 3 or 4 feet high; branches opposite, 4-angled, divaricated, aculeate; leaves opposite, on short petioles, ovate, pungent, entire, glabrous, coriaceous, about an inch long. Spines solitary or geminate, linear, spreading, an inch long. Flowers small, white, sub-pedunculate; males in axillary glomerated spikes; females sub-solitary or in 3-flowered cymes, axillary or terminal. Berry roundish, white.

ORDER XLV. **RHAMNEÆ**, R. Br.

(By W. SONDER.)

(Rhamni, Juss. Gen. 376. Rhamneæ, DC. Prod. 2. p. 19. Endl. Gen. No. ccxxxix. Rhamnaceæ, Lind. Veg. Kingd. No. ccxxii.)

Flowers perfect, regular. *Calyx* tubular, free, or more or less adnate with the ovary, 4–5 fid, with valvate æstivation. *Disc* fleshy, lining the calyx tube. *Petals* 4–5, in the throat of the calyx, small, hood-shaped or flat, sometimes wanting. *Stamens* as many as the petals, and opposite to them. *Ovary* free or adnate, 2–4-celled; ovules solitary, erect, anatropous; styles more or less connate. *Fruit* fleshy or capsular; seeds erect; embryo lying in fleshy albumen, rarely exalbuminous; radicle inferior.

Trees or shrubs, often spiny, common in the warmer parts of the temperate zones and especially so in the southern hemisphere; rare within the tropics and the colder parts of the temperate zone. Flowers small or minute, cymose, umbelled, spiked or capitate. Leaves simple, alternate, rarely opposite, 3-nerved or penni-nerved, entire or serrated; stipules if present minute. The fleshy fruits of several are violent purgatives; those of others, like the Jujube, are wholesome and agreeable to the taste. The wood of the arborescent species is compact and applied to various purposes. About 250 species are known.

TABLE OF THE SOUTH AFRICAN GENERA.

* Limb of the calyx deciduous after flowering. *Fruit* superior or half superior.

I. **Zizyphus.**—*Disk* flat, 5-angled, expanded. *Ovary* sunk in the disk. *Styles* 2–3. *Fruit* fleshy.

II. **Rhamnus.**—*Disk* thin, lining the calyx-tube. *Ovary* free. *Styles* 3–4. *Fruit* fleshy.

III. **Scutia.**—*Disk* fleshy, lining the calyx-tube. *Ovary* not sunk. *Style* short, simple, obconical. *Fruit* juicy.

IV. **Noltea.**—*Disk* very thin, lining the calyx-tube. *Ovary* half inferior. *Style* simple, 3-angled. *Fruit* dry, 3-valved, the valves with a dorsal ridge.

** Limb of the calyx persistent; or ovary crowned with a fleshy disc. *Fruit* inferior.

V. **Helinus.**—Climbing shrubs, with broad leaves. Flowers cymose, glabrous.

VI. **Phylica.**—Erect shrubs, with small, mostly linear leaves. Flowers pubescent, spicate or capitate.

I. **ZIZYPHUS**, Tournef.

Calyx spreading, 5-cleft. *Petals* 5, obovate, unguiculate, convolute. *Stamens* 5, exserted; anthers ovate, 2-celled. *Disc* flat, pentagonal, expanded, adhering to the tube of the calyx. *Ovary* 2–3-celled, immersed in the disc and adnate to it. *Styles* 2–3, diverging or combined. *Fruit* fleshy, containing a 1–2-celled nut. *Seeds* sessile, sub-orbicular, compressed, very smooth. *Tourn. inst. t.* 403. *DC. l. c.* 2. *p.* 19. *Brogn. Mem. Rhamn. p.* 47. *t. I.* 2. *Endl. Gen.* 5717.

Shrubs, with alternate, 3-nerved leaves and spiny stipules. Flowers axillary, cymose. Fruit mucilaginous, eatable, more or less grateful.—Zizouf, in Arabic, is the name of the Lotos, Zizyphus Lotus, Lam.

1. **Z. mucronata** (Willd.! En. Berl. p. 251); branchlets, petioles, and

panicles *thinly pubescent;* prickles in pairs, one of them recurved; leaves petiolate, ovate, or *cordate-ovate, obtusely-acuminate,* mucronulate, *crenate-serrate,* hairy beneath at the nerves or quite smooth; cymes. axillary, about the length of the petioles; stigmas 2, recurved; drupe spherical, smooth, shining. *DC. l. c. p.* 19. *E. & Z.* 991. *Z. bubalina Licht. in Roem. et Schult. Syst. Veg. V. p.* 334.

VAR. *β.* **pubescens**; branchlets and cymes covered with a dense, short tomentum; leaves obtuse, slightly pubescent beneath.

VAR. *γ.* **glabrata**; branchlets glabrous; leaves smooth on both sides, obtuse or obtusely acuminate-mucronate.

HAB. Near Gariep river, *Lichtenstein.* Vish and Katriver, near Philipstown and Fort Beaufort. *E. & Z.!* Keiskamma and near Port Natal, *Drege, Plant.* Bavians-river and Sanddriftsport, near Garieprіver, *Zeyh.* 312. Var. *β.*, Port Natal, *Gueinzius, T. Williamson.* Var. *γ.*, near Gariep, Drege, 6748. Macalisberg, *Burke and Zeyher,* 311 *and* 313. Dec. (Herb., reg. Berol., T.C.D., Sond.)

A tree, 25 feet high. Branches greyish, flexuose. Leaves 1½–2 inches, petioles furrowed above, 4–6 lines long. Cymes many-flowered, sometimes few-flowered. Calyx 1 line long, segments acute. Petals obovate, yellowish. Drupe scarcely the size of a cherry, red. Seeds 2.

2. Z. Zeyheriana (Sond.); branches and branchlets flexuose, as well as the petioles, hairy pubescent; prickles in pairs, recurved or one of them straight; leaves short, petiolate, *ovate* or *ovate-oblong, obtuse,* mucronate, *callous-serrulated,* coriaceous, reticulate, *glabrous or hairy on the nerves;* cymes axillary, about the length of the petioles, hairy, at length glabrous; stigmas 2, recurved.

HAB. Macalisberg, *Burke, Zeyh.!* 310 partim. Oct. (Herb,, Sond., T.C.D.)

A bush, 1–2 feet high; branches greyish; branchlets usually yellowish, old ones glabrate. Leaves 1–2 inches long, pale green; petiole 2 lines long. Prickles 2–3 lines long. Cymes mostly few-flowered. Calyx and petals as in the preceding, from which it differs by the dwarfer habit, and the sharply and callous-serrate leaves. *Z. trinervia,* Roxb., is unarmed, and its leaves are crenulate-serrated. *Z. spina Christi,* Willd., is a tall shrub, with whitish branches.

3. Z. helvola (Sond.); branches and branchlets flexuose, tomentose with brown-yellowish hairs, prickles in pairs, one of them recurved; leaves short-petioled, *elliptic* or *sub-orbicular,* acute, serrulate-dentate on both surfaces, adpressedly *villous with yellowish hairs,* coriaceous; cymes axillary, villous, about the length of the petioles.

HAB. Grassy-stony places near Aapgesriver. *Zeyh.* 310 *partim.* Oct. (Herb., Sond.)

A shrub, similar to the preceding, but very different by the yellowish-brown or somewhat orange pubescence. Prickles yellowish, 3–6 lines long. Leaves 1 inch long, 8–10 lines wide. Petioles 2 lines long. Calyx 1 line long; segments acute. Petals and stigmas as in the preceding.

II. RHAMNUS, Lam.

Calyx urceolate, 4–5 cleft. *Petals* wanting, or 4–5, emarginate. *Stamens* 4–5, with ovate 2-celled anthers. *Disc* thin, covering the tube of the calyx. *Ovary* free, 2–4 celled. *Styles* 3–4, connected or free. *Fruit* fleshy, containing 2–4 indehiscent nuts, one of them occasionally abortive. *Seeds* oblong. *Lam. ill. t.* 128. *DC. l. c.* 2. *p.* 23. *Brogn. l. c. p.* 53. *Endl. Gen.* 5722.

Shrubs or small trees, with opposite or alternate, bistipulate, short-stalked, entire or toothed, usually smooth leaves, which are permanent and coriaceous, or caducous, with feather nerves. Plants possessing strong purgative qualities. Name, *ῥαμνος*, signifying a tuft of branches.

1. R. prinoides (l'Herit. sert. angl. 6, t. 9); unarmed, branchlets *minutely pubescent;* leaves *alternate, ovate-lanceolate,* or *ovate-oblong, acuminate,* mucronulate, serrate, coriaceous, glabrous, shining; pedicels 2–4 axillary, about the length of the petioles; calyx 5–cleft; petals and stamens 5. *Willd. Spec. 1. p. 1099, excl. syn. DC. l. c. 2. p. 24. E. & Z.! 992.*

HAB. Katriver near Philipstown and Fort Beaufort, *E. & Z.!* Grootvadersbosch, Krakakamma, Wolvekop, *Zey.!* 315. Vanstadensrivier and in Stormberg, 5–6000 f. *Drege!* 9123. Outeniqua, George, *Krauss.* Oct.–Dec. (Herb, Sond., T.C.D.)

A moderate tree. Branches and leaves, when dry, blackish. Wood white. Leaves 1½–3 inches long, ¾–1 inch wide, paler and with prominent nerves beneath; petioles carinate, 3–6 lines long. Pedicels 4–6 lines long. Calyx 1 line long, segments triangular acute, as long as the sharply emarginate petals and the stamens. Styles 3, connected to the middle, then diverging. Berry scarcely the size of a pea. sub-spherical. Seeds 3, not furrowed. Embryo flat.

2. R. Zeyheri (Sond.); unarmed, branches and leaves *opposite, smooth;* leaves short, petiolate, *elliptic* or *sub-orbiculate,* obtuse or slightly emarginate, quite entire, paler beneath; pedicels axillary, aggregate, glabrous, twice as long as the petiole; calyx 5–cleft; petals and stamens 5.

HAB. Macalisberg, in woods, *Zeyher!* 316. Nov. (Herb. Sond., T.C.D.)

A moderate tree. Branches greyish. Leaves opposite, rarely sub-alternate, about inch long, 8–12 lines wide, or smaller, elliptic or oblong, with about 10 reddish nerves beneath. Petiole about 1 line long. Peduncles smooth. Calyx 1 line long, segments acute. Petals slightly emarginate, as long as the calyx and the stamens. Styles 3, connected to the middle, then diverging. Allied to *R. Frangula L.*, but the leaves are much smaller. *Rhamnus celtifolius,* Thunb. is a Celtis.

III. SCUTIA, Commers.

Calyx urceolate, 5–cleft; segments erect. *Petals* nearly flat, emarginate. *Stamens* short; anthers ovate, 2–celled. *Disc* fleshy, covering the tube of the calyx, closely surrounding the ovary, but not adnate to it. *Ovary 2–3 celled. Style* short, simple. *Stigmas* 2–4. *Fruit* globose, serrate, 2–3-coccous, girt at the base by the persistent entire tube of the calyx. *Brogn. l. c. p. 55. Endl. Gen. 5724.*

Smooth shrubs, with alternate leaves approximating by pairs, and nearly opposite, quite entire, or slightly serrulated, coriaceous, feather-nerved, bistipulate; stipules minute, deciduous. Prickles wanting or arched, about equal in length to the petioles, rising from the axillæ of the lower leaves. Flowers axillary, disposed in few-flowered, simple umbels, scarcely longer than the petioles. Name from *scutum,* a shield.

1. S. Commersoni (Brogn.! l. c. p. 56); branchlets decussate, spreading, slender, armed with somewhat opposite, recurved prickles; leaves petiolate, sub-distichous, elliptic, ovate or obovate, obtuse, mucronate or retuse, quite entire, shining above, paler beneath; umbels pedunculate; peduncle of the length of the petiole. *Rhamnus Capensis, Thunb.! fl. cap. p. 197. Zizyphus Capensis, Thunb.! apud Poiret, Encyc. meth.*

Suppl. 3. *p.* 193. *Ceanothus Capensis et C. circumscissus.* β. *DC. l. c. p.* 30. *Olinia cymosa Herb. un. itin. p.* 305. *Scutia Capensis, E. & Z.!* 994. *Pappe. Sylv. cap. p.* 11. *S. natalensis Hochst. in Krauss. Beyt. p.* 44.

HAB. Common in the woods of Swellendam, Uitenhage, and Albany, *E. & Z.! Drege! Zeyh.!* 2210. Port Natal, *Krauss!* Dec.–Jan. (Herb. Thunb. Sond., T.C.D.)

A shrub, 4–5 feet high, Katdorn; branches sub-angulate. Leaves 1–2 inches long. Petiole 2–3 lines long. Umbels 4–10 flowered. Calyx 1 line long. Segments ovate, acute. Fruit of the size of a pea, apiculate by the short style, 2-celled. Seeds elliptic-ovate, compressed; testa coriaceous, very smooth. Nearly allied to *S. indica*, from which it differs by the mostly retuse, not acute and serrulate leaves.

IV. NOLTEA, Reichenb.

Calyx urceolate, tube adnate to the ovary at the base, but free above; limb 5–cleft, erect. *Petals* 5, cucullate, sessile. *Stamens* 5, enclosed within the petals; anthers ovate, 2–celled. *Disc* very thin, lining the tube of the calyx. *Ovary* half-inferior, 3–celled. *Style* simple, triangled. *Stigma* 3–lobed. *Fruit* inferior, spherical, dry, 3–coccous, 3–valved; valves with a small rib-like dorsal wing in their upper part; dehiscence septicidal; central axis 3–partite. *Seeds* solitary, erect; testa hard and thick; embryo orthotropus, straight; cotyledons green, leafy, sub-orbicular; radicle short, but well defined; albumen fleshy. *Harvey, gen. p.* 61. *Endl. gen.* 5725. *Willemetia, Brogn. l. c. p.* 63.

A perfectly glabrous shrub, with erect branches, alternate, oblong-lanceolate, more or less obtuse, serrated, feather-nerved leaves, and small white flowers, arranged in few-flowered, terminal or axillary panicles. Name in honor of Professor Nolte of Kiel.

1. N. Africana (Reichb. consp. n. 3800); *Ceanothus Africana, L. spec. p.* 284. *Pluk. L.* 126. *f.* 1. *Seba, thes.* 1. *t.* 22. *f.* 6. *Thunb.! fl. cap. p.* 196. *DC. l. c. p.* 32. *Willemetia Africana, Brogn. l. c. p.* 63. *t.* 5. *f.* 1. *E. & Z.!* 995. *Vitmannia Africana, Wight & Arn. Prod. p.* 166.

HAB. Table mountain, woods near Stoningklipp, Goudsriver; Zwartkopsriver, Katriver, Buffaljadsriver, and Stellenbosch; not uncommon in gardens near Capetown. Aug.–Sept. (Herb. Thunb. Sond, T.C.D., &c.)

10–12 feet high. Branches brownish purple. Leaves 2–2½ inches long, 6–8 lines wide, or smaller, paler beneath. Petiole 3–4 lines long. Cymes or panicles about 1 inch long, the pedicels 2 lines, the calyx 1 line long. Fruit the size of a large pea, often apiculate by the short style.

V. HELINUS, E. Meyer.

Calyx obconical, adnate to the ovary, limb spreading, 5–cleft. *Petals* 5, obovate-oblong, convolute, inserted in the margin of a fleshy, epigynous, obsoletely 5–angled disc. *Stamens* 5, anthers 2–celled. *Ovary* inferior, 3–celled. *Style* 1, with 3 spreading-recurved stigmas. *Fruit* obovate-globose, areolate at the top, 3–coccous; cocci crustaceous, at length dehiscent, with a 3–partite central axis. *Seeds* solitary, erect *Funiculus* short. *Embryo* in a fleshy albumen. *Cotyledons* flat. *Radicle* short, inferior. *Endl. Gen.* 5745.

A climbing shrub, with alternate, petiolate, sub-orbicular, mucronate, feather-nerved, entire leaves, and axillary, cymose peduncles. Name from ελινος, a *branch*.

1. H. ovata (E. Mey.! in Herb. Drege); branches somewhat angular; branchlets spreading, pubescent, at length smooth; tendrils spiral; leaves ovate or elliptic, obtuse, mucronate, sub-cordate at base; cymes about the length of the peduncles; flowers smooth. *Rhamnus mystacinus Ait. Kew.* 1. *p.* 266. *Willemetia scandens, E. & Z.!* 996.

VAR. *β.* **rotundifolia**; leaves orbicular (1 inch long and wide).

HAB. Keiskamma, Keyriver and Gekau, *Drege!* Zuurberge (Uitenh.), Katriver, near Philipstown and in Caffraria, *E. & Z.!* Dornkop, Bechuana country, *Zeyh.!* 317. Port Natal, *Gueinzius*, 400, *Krauss.*; var *β.* Port Natal, *Gueinzius*, 399. Nov.–April. (Herb. Sond., T.C.D.)

Flowering branches divaricate, 3–4 inches long. Petioles in sterile branches 6–8 lines, in flowering branchlets 2–3 lines long. Upper leaves 1½–2 inches long, 12–15 lines wide, ciliate or pubescent; lower ones 1 inch long, or shorter, pubescent or glabrous. Peduncles 6–9 lines long; pedicels 2–3 lines long, in fruit longer. Fruit the size of a pea, quite smooth, with a large area. Seeds convex-trigonous, shining.

VI. PHYLICA, L.

Calyx obconical, urceolate or cylindrical, 5-cleft, its limb persistent. *Petals* wanting or bristle-like, or cucullate. *Stamens* 5, filaments very short, anthers ovate or kidney-shaped, 2-celled or 1-celled. *Disc* more or less distinct. *Ovary* inferior or adnate to the tube of the calyx, 3-celled. *Style* simple, stigma 3-lobed or 3-toothed, in some entire. *Fruit* inferior, crowned by the permanent calyx, and marked above by a small or large areola, 3-coccous. *Seeds* solitary, erect, with a thick, fleshy umbilicus; albumen fleshy; embryo flat. *L. Gen. n.* 266. *Lam. ill. t.* 127. *Trichocephalus, Phylica, and Soulangia Brogn. l. c. t.* 6. *i. ii. iii. Walpersia, Reiss. Petalopogon, Reiss. Pylanthus, Reiss. Endl. Gen.* 5636–5740.

Small, much branched shrubs, with alternate, crowded, simple, quite entire, ovate, lanceolate or linear leaves and small flowers.—Name from φυλλικος, *leafy.*

ANALYSIS OF THE SPECIES.

I. Leaves stipulate ... ... (1) **stipularis.**

II. Leaves without stipules:

A. Flowers mostly pedicellate, in the axilla of leafy bracteæ, solitary, panicled, spiked or in heads. Bracteolæ wanting. Calyx short, campanulate, obconical or subtubular. Petals cucullate. Anthers 1-celled. Area mostly as large as the fruit.

leaves ovate or cordate-ovate, heads short-stalked ... (2) **buxifolia.**
leaves ovate or ovate-lanceolate; racemes panicled ... (3) **paniculata.**
leaves ovate or oblong-lanceolate; axillary loose racemes ... ... (4) **oleoides.**
leaves linear lanceolate; flowers solitary or in loose racemes ... ... (6) **rosmarinifolia.**
leaves linear sub-imbricate; spikes elongated with very long foliaceous bracts ... ... (13) **ambigua.**
leaves lanceolate spreading; flowers longer than the bracteæ, in lax terminal racemes ... ... (9) **lutescens.**
leaves lanceolate spreading; flowers shorter than the bracteæ, in short axillary racemes; calyx hirsute... (10) **axillaris.**
leaves linear spreading; flowers shorter than the bracteæ, in oblong racemes; calyx appressed pubescent (12) **villosa.**
leaves linear sub-imbricate; racemes capitate ... ... (14) **Willdenowiana.**

leaves linear-lanceolate, spreading; small heads (6–10 flowered) ... ... (5) **lasiocarpa.**
leaves linear-lanceolate, spreading; heads many-flowered ... ... (11) **purpurea.**
leaves linear secund; glomerate spikes or corymbose peduncles ... ... (7) **cryptandroides.**
leaves linear erect; terminal, short, ovate, or pyramidal panicle ... ... (8) **rigidifolia.**

B. Flowers sessile or sub-sessile, spiked or in heads, each of which is sub-tended by 2 or 3 lateral bracteolæ.

a. Calyx obconical or cylindrical, segments about as long as the tube, or not much shorter.

a. Plumosæ. Flowers involucrated with very long, villous, or feathery bracts, forming mostly a beautiful plume at the summit of the branches. Anthers 2-celled.

* Petals cucullate.
hairs of the calyx tube reversed ... ... (15) **plumosa.**
hairs of the calyx tube not reversed ...
flowers in oblong or cylindrical spikes:
flower leaves about 4 times as long as the cylindrical calyx; leaves silky, downy ... ... (16) **velutina.**
flower leaves about 4 times as long as the campanulate calyx; leaves villous pubescent ... ... (17) **recurvifolia.**
flower leaves about twice as long as the calyx; leaves villous or glabrous ... ... (18) **excelsa.**
flowers in globose heads:
leaves linear; flowerheads glabrous (19) **Meyeri.**
leaves linear; flowerheads rigid ciliate (20) **affinis.**
flowers in hemispherical heads:
leaves linear, acerosē ... ... (22) **tortuosa.**
leaves ovate-cordate ... ... (21) **reflexa.**

** Petals setaceous or wanting.
heads sub-globose:
twigs tomentose; leaves lanceolate, mucronate, cordate at base ... ... (24) **rigida.**
twigs pubescent; leaves lanceolate, acute, obtuse at base ... ... (23) **capitata.**
heads hemispherical:
leaves lanceolate, scabrous; calyx segments bearded on the throat ... (28) **trachyphylla.**
leaves lanceolate, smooth; calyx segments naked on the throat ... ... (27) **reclinata.**
spikes ovate or cylindrical:
leaves ovate-cordate ... ... (26) **spicata.**
leaves lanceolate ... ... (25) **cylindrica.**

β. Ericoides. Flowers in spherical or hemispherical heads; involucre leafy, flat or hemispherical, shorter or not longer than the flowers.

* Petals setaceous or wanting.
leaves cordate-ovate, obtuse ... ... (29) **retrorsa.**
leaves cordate-lanceolate, apiculate ... (31) **debilis.**
leaves linear-lanceolate ... ... (30) **montana.**

** Petals roundish, dilated, ciliate, barbate.
leaves cordate-lanceolate, hamate-inflexed, calyx glabrous .. .. (32) **brevifolia.**
leaves cordate-lanceolate or cuspidate, mucronate; calyx woolly, heads globose ... ... (33) **cuspidata.**

leaves cordate-lanceolate, mucronate; calyx appressed-silky, heads ovate ... (34) **Thunbergiana.**

*** Petals roundish or cucullate, quite glabrous.

leaves cordate, ovate, or oblong ... ... (35) **callosa.**

leaves cordate, ovate, smooth above (1 line long); calyx woolly ... ... (38) **humilis.**

leaves cordate, lanceolate, villous; calyx glabrous ... ... (39) **acmæphylla.**

leaves cordate, lanceolate (2–3 lines long), scabrous above, calyx glabrous ... ... (37) **alpina.**

leaves cordate, lanceolate (4–5 lines long), smooth; calyx glabrous ... ... (36) **atrata.**

leaves cordate, linear-lanceolate, villous, ciliate; flowers solitary, heads large ... (40) **gracilis.**

leaves cordate, linear lanceolate, villous, ciliate; flowers corymbose or sub-paniculated, heads small ... ... (41) **virgata.**

leaves not cordate at base.

calyx clothed with reversed hairs:

leaves linear-lanceolate, hirsute; heads dense many-flowered ... ... (47) **bicolor.**

leaves linear-lanceolate, appressed-pubescent; heads lax, 5–8 flowered (49) **strigulosa.**

calyx pubescent, villous, or woolly:

leaves lanceolate, apiculate; scales of the involucre lanceolate, hirsute; calyx woolly ... ... (42) **diosmoides.**

leaves lanceolate, acute, scabrous, scales of the involucre ovate, acute; calyx downy (2 lines long) ... (44) **comosa.**

leaves lanceolate, smooth, involucre foliaceous; calyx tomentose (½ line long) ... ... (45) **rubra**

leaves linear-terete, obtuse, with inflexed points, glabrous ... ... (52) **selaginoides.**

leaves linear or lanceolate, terete, obtuse, glabrous ... ... (54) **brachycephala.**

leaves linear-terete, obtuse, scabrous; calyx tomentose; capsule ovate ... (56) **cephalantha.**

leaves linear-acute, sub-villous; calyx pubescent; capsule roundish ... (55) **eriophoros.**

calyx glabrous:

leaves lanceolate, acute, glabrous; involucre hemispherical ... ... (43) **propinqua.**

leaves lanceolate-attenuate, very acute, margins ciliate ... ... (46) **apiculata.**

leaves linear-lanceolate, glabrous; involucre flat ... ... (48) **nigrita.**

leaves linear or lanceolate-subulate, smooth; heads on sub-umbellate twigs; petals concave, cucullate ... (50) **ericoides.**

leaves lanceolate-subulate, pilose, scabrous; heads on panicled twigs; petals concave, cucullate ... ... (51) **parviflora.**

leaves lanceolate-subulate; very smooth; heads solitary, petals mitriform ... ... (53) **disticha.**

b. Calyx cylindrical, segments about 4 times as short as the tube.

leaves compressed, sub-ancipital, heads mostly aggregate ... ... (57) **gnidioides.**

leaves linear, sub-terete, obtuse, with inflexed point; heads solitary ... ... (58) **abietina.**

SEC. 1. STIPULARES.—Leaves stipulate.

1. **P. stipularis** (L. mant. 208); twigs hoary-velvetty; leaves spreading, short-stalked, lanceolate-linear, acutish, smooth above, hoary-velvetty beneath, with revolute edges; stipules awl-shaped; heads of flowers terminal, sessile, very downy. *Burm. afr.* 117. *t.* 43. *f.* 2. *Thunb. fl. cap. p.* 200. *Wendl. coll. t.* 32. *Spreng. Berl. Mag. viii. p.* 104. *t.* 8. *f.* 3. *Herb. Un. itin. n.* 114, *et* 632. *Trichocephalus stipularis, Brogn. l. c. t.* 6. *i. E. & Z.! n.* 1008. *Walpersia Dregeana, Presl. Bot. Bem. p.* 38.

VAR. β. **robusta**; leaves broader, obtuse or acutish, somewhat cordate at the base; heads somewhat larger. *Trichocephalus ripophorus, E. & Z.! n.* 1007. *Walpersia rip. Presl. Tylanthus diosmoides, Presl.*

HAB. Common on the Cape Flats and on mountain sides through the Cape and Stellenbosch districts. *Zeyh.* 2213, 2214. *Drege,* 6752 *a.* Var. β. on mountain sides near Brackfontein, Clanw. *Drege,* 6752 *b.* Mar.–Jul. (Herb., Thunb., Holm., Wendl., T.C.D., Sond.)

Shrub, 2–3 feet, greyish; branches erecto-patent. Young leaves sub-tomentose, old ones very smooth, convex above, 6–9 lines long, 1–1½ line wide; in var. β. 2–3 lines wide. Petioles tomentose. Stipules brown, 1–1½ line long. Heads of flowers the size of a hazel nut, involucre hemispherical, its scales brown, shorter than the flowers. Calyx with long, subulate, spreading segments, woolly on the outside. Petals bristle-like.

SEC. 2. EXSTIPULATÆ.—Leaves without stipules.

A. Flowers mostly pedicellate, in the axils of leafy bracteas solitary, panicled, spiked or in heads. Bracteolæ none. Calyx short, campanulate, obconical or sub-cylindrical. Petals cucullate. Anthers 1-locular. Areola mostly as large as the fruit. *Soulangia, Brogn.* (Sp. 1–14.)

2. **P. buxifolia** (L. spec. 283); twigs minutely pubescent; leaves spreading, *ovate or cordate-ovate, flat* and smooth above, tomentose beneath, with sub-recurved margins; *heads of flowers* small, axillary, stalked. *Burm. Afr.* 119. *t.* 44. *f.* 1. *Thunb. fl. cap. p.* 204. *Lodd. bot. cab. t.* 848. *Wendl. coll. t.* 26. *Sieb. fl. cap. exs. n.* 69. *Herb. Un. itin n.* 108. *P. cordata, L. spec.* 283. *Commel. pracl. p.* 62. *t.* 12. *Soulangia buxifolia and cordata, Brogn. l. c. p.* 71. *E. & Z.!* 1047.

HAB. Mountains near Capetown and in Caledon and Stellenbosch districts. *Thunb. E. & Z.! Drege,* 9535. Mar.–Aug. (Herb., Thunb., Holm., T.C.D., Sond.)

Shrub, 2–10 feet high, greyish-brown. Twigs often verticillate. Leaves yellowish or whitish tomentose beneath, 6–9 lines long, 4–6 lines wide. Petioles 1 line long. Flower heads, the size of a pea, involucrated by some small leaves, on stalks of ½–1 inch long. Flowers minute, tomentose on the outside; segments ovate, acute. Fruit woolly, at length sub-glabrous, the size of a pea, with a large area.

3. **P. paniculata** (Willd.! spec. 1. p. 1112); branches virgate, twigs puberulous; leaves spreading, *ovate or ovate-lanceolate,* obtuse or subcordate at the base, shining above, downy white beneath, with subrecurved margins; flowers disposed in leafy terminal, *panicled racemes. Thunb. fl. cap. p.* 203. *DC. l. c. p.* 36. *P. myrtifolia, Poir. dict.* 5. *p.* 293. *Soulangia paniculata, Brogn. l. c.* 5. *S. arborescens et paniculata, E. & Z.!* 1045, 1046.

VAR. β. **condensata**; upper leaves crowded, sub-linear; panicle short and dense. *Soulangia marifolia, Bernh. in pl. Krauss.*

HAB. Stony places of Mt. Bothasberg, Grahamstown, Albany, Boschmansrivier, *Zeyh.* 2224 *a.*; Winterhoek and Vanstadensriviersberge, Uitenhage, *E. & Z.! T. Williamson.* Kromrivier, *Drege.* Elandsrivier, *Niven.* Port Natal, *Gueinzius.* Var. β. Winterhoek, *Krauss.* Elandsrivier, *Ecklon.* Ap.–Jun. (Herb., reg. Ber., Holm., Sond., T.C.D.)

Shrub, 2–10 feet high. Leaves short, petiolate, spreading or reflexed, 5–6 lines long, 2 lines wide, or the larger ones 8–12 lines long, 3 lines wide, upper shorter. Flowers about ½ line long, in the axil of a subulate bractea, tomentose, sub-turbinate; segments ovate, acute. Fruit 2½–3 lines long, obovate, smooth, blackish when dry; area as large as the capsule.

4. P. oleoides (DC. l. c. p. 36); twigs velvetty, pubescent, at length smooth; leaves spreading, coriaceous, *ovate-oblong, mucronate, acute,* pubescent or smooth and papillose above, downy beneath, with sub-recurved margins; flowers in *loose racemes at the top of the branches; racemes axillary,* about the length of the leaf; calyx tomentose, pedicellate; capsule *globose,* with a large area. *P. oleæfolia, Vent. Malm. n.* 57, *obs.* 2. *P. spicata, Lodd. Bot. Cab. t.* 323. *non L. Soulangia oleæfolia, Brogn. l. c. p.* 71. *E. & Z.!* 1048.

VAR. β. **angustifolia**; leaves lanceolate, lower ones oblong-lanceolate; racemes shorter than the leaves.

HAB. Among shrubs on mountain sides of the valley Tulbagh, Worr. *E. & Z.!* Heerelogement, *Zey.* 327. Nieuwe Kloof, Wupperthal, 1200, *Drege.* Var. β. Stoffkraal, *Zeyh.* 326. Mar.–Oct. (Herb., Holm., Sond., Lehm., T.C.D.)

Over 2 feet high, greyish-brown, twigs often crowded, sub-fastigiate. Leaves often veined beneath, 8–10 lines long, 4 lines wide, others 12–15 lines long, 3–6 lines wide, upper mostly half an inch. Petiole 1 line long. Racemes 4–6 lines long. Calyx obconical or campanulate, of the length of the pedicel or somewhat shorter; segments acute. Capsule blackish when dry, nearly the size of a small cherry; area as large as the capsule. The var. β. has sub-virgate twigs and smaller leaves.

5. P. lasiocarpa (Sond.); branches and twigs minutely pubescent; leaves spreading, short-petiolate, *linear-lanceolate or linear, glabrous,* shining above, tomentose beneath, with revolute margins; *flower heads terminal* (small), tomentose, involucrated by the uppermost leaves; calyx campanulate, tomentose; segments short, acute; capsule ovate, *tomentose.*

VAR. α. **lediformis**; leaves 6–8 lines long, 1 line wide, flattish above. *Soulangia ledifolia, E. & Z.!* 1044. *excl. syn.*

VAR. β. **parvifolia**; leaves 3–4 lines long, ½–¾ line wide, obtuse or apiculate, papillose-punctate above. *P. lanceolata, E. & Z.!* 1012. *non Thunb. P. parviflora, E. Mey. c. ex pte. P. nitida, E. Mey. ex pte. non Lam.*

HAB. Var. α., Hott. Holl. berge, on the highest mountains, *E. & Z.!* Var. β., Hott. Holl. kloof, *E. & Z.!* Klynriviersberge, *Zeyh.* 2221. Between Langevalei and Heerelogement, *Drege.* Jun.–Aug. (Herb., Sond., Lehm., T.C.D.)

1–1½ feet high; branches red or greyish. Leaves horizontal, the upper sub-erect Petiole ½ line long. Heads 6–10 flowered, the size of a small pea. Calyx 1 line long. Capsule greyish, short but densely tomentose, about 4 lines long; area small.

6. P. rosmarinifolia (Thunb.! fl. cap. p. 203); much branched, branches virgate; twigs pubescent; leaves spreading, *linear-lanceolate,* acute, the lower sub-linear, smooth and shining above, whitish tomentose beneath, with revolute margins; flowers in the axil of the uppermost leaves sessile, *solitary or spiked, spike lax,* shorter than the leaves; calyx minute, tomentose; capsule obovate-turbinate, *smooth,* with a *small* area. *Dissert. de Phylica, p.* 8. *P. parviflora e. E. Mey. in Herb. Drege, not of Bergius.*

HAB. Mount. Zwartebergen, near Vrolyk, 3-4000 ft., *Drege.* Aug. (Herb., Thunb., Sond., T.C.D.)

Over 2 feet high. Branches and twigs filiform, abbreviate, greyish. Lower leaves about 1 inch long, 1½ line wide, the upper 5-3 lines long, ½-¾ line wide. Flowering branches paniculate. Flowers in the lower axils spiked; spike few-flowered, bracteate, 4-6 lines long; in the upper axils solitary, whitish downy. Calyx about 1 line long; segments spreading, acute. Capsule blackish when dry, smooth, the size of a pea.—Very near *P. paniculata,* but the leaves and the area are much smaller.

7. **P. cryptandroides** (Sond.); branches rigid; twigs simple, *subglabrous;* leaves *spreading,* secundate, *linear,* acute, with revolute margins, *convex,* glabrous and *papillose* above, whitish downy beneath, young ones adpressed-pubescent; flowers sessile, *glomerate or spiked* at the summit of *corymbose* peduncles; calyx whitish tomentose; capsule ovate-globose, woolly, at length glabrate; *area as large as the capsule. P. rosmarinifolia, E. Meyer in Herb. Dreg. c.*

HAB. Roodeberge, 2-3500 ft. Nov. *Drege.* Kardow, *Zeyh.* 323. (Herb., Sond., T.C.D.)

Shrub, 2-3 feet high; branches erect, virgate. Leaves circ. 6 lines long, 1 line wide. Corymb. ½-1 inch long and wide; peduncles at the upper part with 6-10 sessile flowers, formed in a head-like spike. Calyx campanulate, 1 line long, segments acute. Capsule of the size of a pea.

8. **P. rigidifolia** (Sond.); branches rigid; twigs simple, *minutely downy;* leaves erect, *sub-appressed, linear, acute,* convex, papillose above, whitish downy beneath; flowers pedicellate, disposed in a *short, ovate or pyramidal, coloured panicle;* pedicels as long as the tomentose bracteas; calyx yellowish, tomentose; capsule truncate ovate, woolly, at length glabrate; area as large as the capsule. *P. rosmarinifolia, E. Meyer, b.*

VAR. β. **thymifolia**; leaves shorter, spike ovate, calyx whitish. *Soulangia thymifolia, Presl.*

HAB. Vanstaadensberg, Bloedrivier, *Drege,* 6771. Uitkomst, sandy places, *Zeyh.* 324. Var. β. stony places near Kendo, 3-4000 ft. *Drege,* 6772. Jun.-Oct. (Herb. Wendl., T.C.D., Sond.)

1-1½ ft. high; branches erect. Leaves rigid, 6 lines long, 1 line wide, sulcate beneath. Panicle dense, 6-9 lines long. Bracteas lanceolate, acute, the lower, foliaceous. Calyx obconical or campanulate, 1 line long, as long as the pedicels or a little shorter; segments acute. Capsule about 3 lines long, apiculate by the short, permanent style.

9. **P. lutescens** (Sond.); much branched, branches fastigiate, corymbose; twigs pubescent; leaves *spreading, lanceolate* or linear-lanceolate, *obtuse,* callous, puberulous, at length smooth above, tomentose beneath, with revolute margins; flowers pedicellate, axillary, solitary, *longer than the foliaceous bractea,* disposed in a *lax, terminal, leafy raceme;* calyx yellowish-tomentose. *Soulangia lutescens, E. & Z.!* 1043.

HAB. Mountains between Grahamstown and Hassagaybosch, Albany, *E. & Z.!* Zuureberge, Howisonspoort, *Zeyh.* 2223. *Drege,* 6770. Jul.-Sept. (Herb., Sond., T.C.D.)

Shrub, 1–2 feet high, known by the very numerous, filiform, fastigiate branches. Leaves about 3 lines long, 1 line wide, the uppermost 1 line long. Racemes ½–1 inch long, in the upper twigs often paniculate. Pedicels spreading, 1–1½ lines, calyx 1 line long, campanulate, segments acute.

10. P. axillaris (Lam.! ill. n. 2615); twigs pubescent; leaves spreading, *lanceolate* or linear-lanceolate, *acute*, young ones hairy, old ones smooth, shining above, flattish, tomentose beneath, with revolute margins; flowers pedicellate, *axillary, solitary, racemose, shorter* than the leafy bractea; calyx campanulate, *hirsute*. *Spreng. Berl. Mag. viii. p.* 104. *t.* 8. *f.* 4. *Roem. & Schult. Syst. Veg. v. p.* 485. *DC. l. c. p.* 36. *Soulangia axillaris, Brogn. l. c. p.* 70. *P. rosmarinifolia, Willd. Herb. fol.* 4604. *fol.* 2. *E. & Z.!* 1036. *ex pte.*

VAR. β. **hirsuta**; twigs sub-hirsute; leaves appressed, pubescent, the lower at length glabrate; flowers shorter than the foliaceous bracteæ. *P. hirsuta, Thunb.! fl. cap. p.* 202. *Soulangia hirsuta, E. & Z.!* 1041.

VAR. γ. **pedicellaris**; flowers sometimes larger, on longer pedicels, of the length of the foliaceous bractea.

VAR. δ. **parvifolia**; leaves and flowers twice smaller. *Drege,* 6757.

VAR. ε. **glabrata**; leaves linear-lanceolate, smooth. *P. pinea, Thunb.! fl. cap. p.* 201. *ex pte.*

HAB. Hills near Port Elizabeth, Krakakamma and Van Staadensrivier, Uit., *E. & Z.! Niven, Drege,* 6758 *and* 6768. *Zeyh.* 2218 *and* 2222. Hassaquaskloof, *Drege.* Feb.–Jul. (Herb., Thunb., Holm., Lam., Willd., Wendl., T.C.D., Sond.)

Shrub, 1 or more feet high. Leaves alternate, rarely sub-opposite, 4–6 lines long, 1 line wide, rarely longer, in var. δ. shorter, obtuse or sub-cordate at base, smooth or hairy, or sub-villous above. Raceme ovate, about 6 lines long, at length lax, inch long or longer. Flowers white. Pedicels 1–4 lines long. Calyx campanulate, segments acute, ferruginous on the inside. Capsule ovate-globose, the size of a pea; area as large as the capsule or smaller.

11. P. purpurea (Sond.); twigs pubescent; leaves linear-lanceolate, callous-acute, sub-villous, at length smooth and shining above, tomentose beneath, with revolute margins; flowers on short pedicels, collected in a *terminal, many-flowered head;* calyx campanulate, tomentose. *P. rosmarinifolia, Lodd. Bot. Cab. t.* 849. *Herb. Willd. n.* 4606. *fol.* 1. *Soulangia rubra, Bot. Reg. t.* 1498. *exc. synon.*

VAR. β. **reclinata**; leaves reclinate, silky above, bearded at the apex; flower stalk equalling the calyx. *P. reclinata, Wendl. collect.* 2 *p.* 49. *t.* 56. *R. & Sch. l. c. p.* 485.

HAB. Cape, *Niven;* var. β. cultivated in Herrenhausen Bot. Gard. (Herb., Willd., Wendl., Sond.)

Very similar to the preceding, but the flowers are collected in a head and longer than the bracteæ; perhaps a variety. Leaves 6 lines long. Calyx whitish tomentose, limb ferruginous or purple on the inside.

12. P. villosa (Thunb.! fl. cap. p. 202); twigs minutely pubescent; leaves *spreading, mostly secundate, linear,* acute, convex above, sulcate and tomentose beneath, with recurved margins, lower ones glabrous, *papillate-punctate,* upper ones villous; flowers axillary, solitary, shorter than the foliaceous bracteæ, disposed in a *lax or dense, oblong-raceme;*

calyx campanulate, *appressed-pubescent;* capsule ovate-globose, at length glabrate; area as large as the capsule.

VAR. β. **glabrata**; leaves glabrous, the lower ones somewhat larger. *P. pinea, Thunb.! l. c. ex parte. E. & Z.* 1042. *P. ericoides, E. Meyer, d.*

VAR. γ. **squarrosa**; leaves glabrous, recurvo-patent; raceme lax, few-flowered. *Soulangia microphylla, E. & Z.* 1040.

VAR. δ. **pedicellata**; more robust; leaves 1 inch long, villous, at length glabrate, papillose-punctate; flowers mostly larger, longer, pedicellate; racemes often corymbose. *P. pedicellata, DC. l. c. p.* 36. *E. Mey. in Hb. Drege. Soulangia pedic. Presl.*

HAB. Piquetberg, Thunberg. Boshrivier, *Drege!* 6767. Hills near Port Elizabeth and Vanstaadensriviersberge, *E. & Z.!* Kromrivier, *Zeyher!* Var β. Brackfontein, *E. & Z.!* Var. γ. between Coega and Zondagsrivier, *E. & Z.!* Var. δ. between Langvalley and Olifantsrivier, *Drege!* interior reg. *Niven.* Feb.–Nov. (Herb. Thunb., Hook., Sond., T.C.D.)

Very near *P. axillaris*, but the leaves are mostly secundate, very small, and not wider at the base, the upper leaves and the calyx with appressed hairs, the flowers usually smaller. The var. δ. is covered with greyish, appressed hairs, the lower ones glabrous, the flowers larger and more woolly. Leaves about 6 lines long, ½ line wide, in var. δ. somewhat longer and wider, in var. β. about 3–4 lines long.

13. P. ambigua (Sond.); much branched; twigs minutely pubescent; leaves erect or secundate, sub-imbricated, *linear*, acute, with revolute margins, convex, glabrous and papillose-punctate above, tomentose beneath, *floral ones spreading, silky-pubescent, much longer than the flowers;* spike racemose, cylindrical; calyx pedicellate, *campanulate*, downy; segments ovate-acute, hirsute on the outside; capsule roundish, smooth; area twice smaller than the capsule. *P. plumosa, E. Mey. in Hb. Drege, non L. or Thunb. Soulangia plumosa. Presl.*

HAB. Blauwberg, 3–4000 feet, *Drege!* Nov. (Herb., Sond., T.C.D.)

Branches crowded, scabrous; twigs short. Leaves 6–3 lines long, ½ line wide, the younger appressed pubescent. Raceme about 1 inch long; floral leaves or bracteæ about 4–5 lines long, recurved at top. Pedicel ½ line, calyx 1 line long, segments short. Capsule blackish, smooth, the size of a pea. Seed brown, about ½ line long. In habit and by the long, silky bracts it approaches *P. plumosa, L.*, but the flowers are very different and the bracteolæ wanting.

14. P. Willdenowiana (E. & Z.! 1035); branches virgate; twigs fastigiate, velvetty-pubescent; leaves erect, sub-imbricate, rigid, linear, mucronate, with revolute margins, glabrous and papillose-scabrous above, the younger appressed, villous; *raceme capitate;* bracteæ lanceolate, *equalling the flowers* or a little longer; calyx *sub-cylindrical*, tomentose; segments acute; capsule ovate-globose, smooth. *P. rosmarinifolia, Schlecht. Linn.* 6. *p.* 195. *non Thunb. nec aliorum, Tylanthus Pres.*

HAB. Mountain sides, Vanstadensriviersberge (Uit), *E. & Z.! Zey.!* 2220. Elandsrivier, *Zey.!* 2219. Jul.–Sept. (Herb. Lehm., T.C.D., Sond.)

1–2 feet high, branches erect, greyish. Lower leaves 4–6, upper about 3 lines long, ½–¾ line wide, sulcate and tomentose beneath. Petiole ½–1 line long. Flowers about 12–16 in a head-like raceme. Pedicels 1 line long. Calyx cylindrical, sub-campanulate, 2 lines long; segments erecto-patent, 3 times shorter than the tube. Style half as long as the calyx. Capsule the size of a small pea.

B. Flowers sessile or sub-sessile, spiked or in heads, each sub-tended by two or three lateral bracteolæ. (Sp. 15–58.)

a. Calyx obconical or cylindrical, segments about the length of the tube, or not much shorter. (Sp. 15–56.

a. Plumosæ: flowers involucrated with very long, villous or feathery bracteæ, forming mostly a beautiful plume at the summit of the branches. Anthers 2-celled. (Sp. 15–28.)

* *Petals cucullate.* (Sp. 15–22.)

15. P. plumosa (Thunb.! fl. cap. p. 203); branches and twigs *pubescent;* leaves linear-lanceolate or lanceolate, awl-shaped, *smooth above,* tomentose, hairy beneath, with revolute margins; floral leaves villous or hirsute, spreading, much longer than the flowers; spike oblong or roundish; bracteolæ villous, twice as short as the tubular, minutely *pilose* calyx; hairs of the tube *reversed, appressed;* segments ovate-lanceolate, erect, twice shorter than the tube.

VAR. *α.* **horizontalis**; spike small, roundish; leaves sub-linear, villous, spreading or erect, floral ones silky-plumose. *P. horizontalis Vent. Malm. n.* 57. *obs.* 3. *DC. l. c. p.* 35, *E. & Z.!* 1028. *P. plumosa, Spreng. l. c. p.* 105. *f.* 7. *Lam. ill. t.* 127, *f.* 4. *Drege!* 6777 *b.* *Zey.!* 2212.

VAR. *β.* **squarrosa**; spike roundish, leaves spreading, sub-lanceolate, villous or glabrate; floral ones silky-pubescent, often gold-coloured; spike small, leaves villous, linear-lanceolate. *P. squarrosa, Vent. l. c. Bot. Cab. t.* 36. *P. Commelini, Spreng. l. c. t.* 8, *f.* 6. *P. pumila, Wendl.! in Willd. enum I. p.* 252. *P. pubescens, Lodd. Bot. Cab. t.* 695. *Willd.! Herb. fol.* 2. *P. plumosa, E. & Z.!* 1027, *ex pte.*; spike large, leaves lanceolate, glabrate, papillous-punctate. *P. pubescens, E. & Z.!* 1026. *Drege.* 6778.

VAR *γ.* **Thunbergiana**; spike oblong, lax; leaves lanceolate-linear, spreading, often secundate, floral ones sub-silky-pubescent. *P. plumosa, Thunb.! dissert de Phylica, p.* 7. *E. & Z.!* 1027, *ex pte.* *Drege!* 6777 *a.*

VAR. *δ.* **neglecta**; spike short, oblong, lax; leaves linear-lanceolate, secundate, as well as the branches sub-glabrous; floral leaves shorter.

HAB. Sandy and stony places, Cape flats, and in mountains near Capetown, in Hott. Holl., Zwartland, and distr. Worcester; var. *γ.* also in Nieuwekloof; var *δ.* near Hexriver, *Zey.!* 318. Jul.–Oct. (Herb. Thunb., Willd., Wendl., T.C.D., Sond.)

Shrub, 1–2 feet high, greyish brown, with erect or adscendent branches. Leaves on short petioles, 6–9 lines long, about 1 line or in some specimens 2 lines wide, younger hairy. Floral leaves horizontal, villous, 6–9 lines, in var. *δ.* about 3 lines long. Spike an inch long, or shorter. Calyx 2½ lines long; segments erect, at length somewhat spreading. Petals oblong, complicate-cucullate. Capsule the size of a pea, shining; area twice smaller than the top of the capsule.

16. P. velutina (Sond); stem erect; branches virgate, *downy;* leaves *sub-imbricate,* linear, acute, rigid, with revolute margins, silky-downy above, at length glabrate, tomentose beneath; floral leaves villous-pubescent, spreading, much longer than the flowers; spike oblong; bracteolæ pubescent, shorter than the cylindric-funnel-shaped *downy* calyx, hairs of the tube *not reversed,* segments ovate, acute, erect, twice shorter than the tube.

HAB. Mountains, Attaquaskloof, *Niven*. (No station given) *Drege!* 6776. (Herb. Sond.)

Shrub 1½–2 feet; branches brownish or greyish. Leaves shortly petiolate, erect, or a little spreading, 7–10 lines long, ¾ line wide. Spike about 1 inch long; floral leaves 6–8 lines long. Calyx 2 lines long, segments short. Petals as in the preceding. Differs from the preceding by the short, appressed pubescence, smaller leaves, and the calyx.

17. P. recurvifolia (E. & Z.! 1033); stem erect; branches virgate, *pubescent;* leaves crowded, spreading, sub-recurved, lanceolate-linear, rigid, mucronate, with revolute margins, *villous-pubescent,* at length glabrate, tomentose beneath; floral leaves villous-pubescent, spreading, *much longer* than the flowers; spike oblong; bracteolæ pubescent, *longer* than the *campanulate, woolly* calyx; segments of the calyx erect, ovate, acute.

HAB. Mountains, Puspasvalley, Swellendam, *E. & Z.!* Sept.–Oct. (Herb. Sond.)

Distinguished from *P. velutina* by the more woolly indumentum, recurved leaves, longer spike, and shorter, woolly flowers. Leaves 6–9 lines long, at base sub-appressed; floral ones silky-pubescent. Calyx about 1½ line long. Petals as in *P. plumosa.* Capsule sub-trilobed, the size of a large pea, pubescent, at length glabrous, brown, impressed at top with a small area, and often apiculate by a short style.

18. P. excelsa (Wendl.! collect. iii. p. 3, t. 74); branches virgate-pubescent, leaves spreading or erect, linear-lanceolate or linear with revolute margins, villous, at length glabrate; floral leaves *hirsute,* about *twice* as long as the flower; *spike cylindric, lax;* bracteolæ villous, shorter than the campanulate tomentose calyx; segments of the calyx short, erect. *P. spicata, Bot. Mag.* 2704? *Tylanthus excelsus, Presl.*

VAR. α. **laxa**; branches lax; leaves lanceolate, spreading, villous. *P. excelsa, Wendl. l. c. P. cylindrica, E. Mey.! a. P. imberbis, E. Mey.! a.*

VAR. β. **stricta**; branches erect; leaves linear, subacerose, imbricate, appressed, villous. *P. cylindrica, E. & Z.!* 1029, *non Wendl. Drege!* 6775.

VAR. γ. **papillosa**; branches and leaves sub-glabrous; leaves erect, linear acerose, papillose, slightly hairy. *P papillosa, Wendl.! l. c. p.* 5, *t.* 75.

VAR. δ. **brevifolia**; branches rigid; leaves shorter, erect, sub-imbricate, linear-lanceolate, the younger fulvid-hairy. *P. fulva, E. & Z.!* 1032.

HAB. Var. α. Dutoitskloof, 1–4000 f. *Drege!* Var. β. Somerset, Hott. Holl. and Winterhoeksberg, 2–3000. Var. γ. Hott. Holl. Var. δ. Tulbagh, near the cataract, *E. & Z.!* Oct.–Jan. (Herb. Wendl., T.C.D., Sond.)

Shrub, 2–4 feet high; branches brownish, twigs crowded. Leaves short, petiolate 6–8 lines, in var δ. 4 lines long. Spikes 1–1½ inch long, sometimes shorter. Floral leaves acuminate, 3 lines long, fulvid or yellow green. Calyx 2 lines long, limb acutish or sub-obtuse. Ovary tomentose or silky-pubescent. Capsule roundish, smooth, blackish, the size of a pea, crowned by the calyx.

19. P. Meyeri (Sond); branches pubescent; leaves approximate, spreading, *arcuate-incurved, linear, obtuse,* with revolute margins, convex and smooth above, tomentose beneath, the younger *villous-pubescent;* floral leaves *silky-plumose,* erect, *surrounding and surpassing the glabrous flowerhead;* bracteolæ hirsute, equalling the calyx; tube of the calyx smooth, shorter than the silky-villous segments; petals mitriform cucullate. *P. squarrosa, E. Mey.! in Hb. Drege! non Vent. Walpersia squarrosa, Presl.*

HAB. Zwartebergen near Vrolyk, 3-4000. *Drege!* Aug. (Herb. Sond., T.C.D.)

Shrub erect, 1-2 feet high, with sparing, erect, short branches. Leaves 3-4 lines long, ½ line wide. Petiole short, appressed. Head of flowers roundish, the size of a small walnut. Floral leaves or involucre 6-8 lines long, thin and villous, hairs spreading, silky. Calyx 2 lines long, greyish white. Known from *P. rigida* and *capitata* by the smaller leaves and heads, and by the petals.

20. **P. affinis** (Sond.); branches sub-glabrous; leaves approximate, spreading, arcuate-incurved, linear, acerose, obtuse, or acutish, with revolute margins, convex, glabrous, and *impressed punctate above*, tomentose beneath, the *younger rigid-ciliate;* exterior floral leaves surrounding the roundish flowerhead, *lanceolate, foliaceous, rigid-ciliate,* with revolute margins, the ciliæ *blackish;* intermediate *sub-foliaceous, ciliate-hirsute,* on longer petioles; interior somewhat longer, erect, *linear, not foliaceous,* without and at top plumose-ciliate; bracteolæ hirsute; about equalling the calyx; capsule globose, smooth; area pretty large.

HAB. Cape (no station given) *Drege!* 6783.

Distinguished from the simil. preceding by the rigidly-ciliate, yellowish upper leaves, smaller flower-heads, with different involucre. Flowers not perfect. Capsule the size of a pea.

21. **P. reflexa** (Lam.! ill. n. 2625); branched, crowded; twigs *yellowish-tomentose;* leaves spreading or reflexed, *ovate, sub-cordate, callous-obtuse,* flattish, smooth, punctate above, tomentose beneath, with slightly revolute margins, the younger fulvid-villous; heads of flowers hemispherical; involucre yellowish or fulvid; exterior scales foliaceous; interior somewhat longer, lanceolate, not foliaceous, plumose-hirsute; bracteolæ villous; calyx slightly conical, *fulvid-hirsute,* segments acute, erect, twice as short as the tube; petals cucullate. *Commel. prael. p. 62, f. 13. P. cordata, Herb. un. itin. n. 630. non L. Soulangia cordata, E. & Z. 1049.*

HAB. Rocky places on the summit of the Tablemountain, Ecklon. Nov. (Herb. Lam., Wendl., Lehm., Sond.)

A yellowish green shrub, 1-2 feet high. Lower leaves mostly sub-reflexed, about 5 lines long, 2½ lines wide; upper 4 lines long, 2 lines wide. Petiole 1-2 lines long, appressed; the leaves surrounding the flower-head or the exterior involucre not differing from the other leaves; the interior involucre gradually smaller, and destitute of lamina, surpassing the flower-head, 6-8 lines in diameter. Calyx short, pedicellate, 2 lines long, equalling the bracteoles. From *P. callosa L.*, with which it is confounded by authors, if differs by yellowish colour of heads and twigs, and large, feathery, and involucrated heads.

22. **P. tortuosa** (E. Meyer!); much branched; twigs fastigiate, hairy; leaves *sub-imbricate,* incurved-erect, *linear, acerose, callous-acute,* hairy-ciliate, at length glabrous, punctate-scabrous, with revolute margins; heads of flowers hemispherical; involucre with 3-4 rows of scales; the exterior foliaceous, as well as the intermediate ciliate; the interior longer, surpassing the flowers, villous-hirsute; bracteolæ small, villous; calyx tubular, *smooth,* 5-dentate, teeth acute, without hirsute, within *gibbous,* smooth; petals cucullate.

HAB. Rocky places, Zwartebergen near Klaarstrom, 2–3000 f. *Drege!* Nov. (Herb. Lehm., Sond., T.C.D.)

Stem and branches erect, greyish; twigs crowded, short, sometimes yellowish. Leaves green, 3–4 lines long, ⅓–½ line wide; petiole ½ line long. Head of flowers greenish, 4–6 lines in diameter. Exterior involucre leafy; intermediate scales longer petiolate, the interior linear, 4 lines long, somewhat longer than the villous heads. Calyx 2 lines long, slightly funnel-shaped, equalling the bracteolæ. Style short. Differs from *P. affinis* by the erect, shorter, scabrous leaves, much shorter flower heads, with yellowish not blackish cilia, &c.

** *Petals setaceous or wanting.* (Sp. 23–28.)

23. P. capitata (Thunb. ! prod. p. 45); stem erect; branches and twigs erect, *villous-pubescent*; leaves spreading, *lanceolate*, or linear-lanceolate, with revolute margins, villous or sub-glabrous above, whitish-tomentose beneath; floral ones a little spreading, very villous, heads of flowers *roundish;* bracteolæ villous; calyx *villous*, segments subulate, *glabrous above*, spreading, longer than the tube; petals minute, setaceous, or wanting.

VAR. α, **lanceolata**; leaves lanceolate, at length glabrous, shining above, heads of flowers large, greyish, white, or yellowish. *Burm. Afr. t.* 44, *f.* 3. *P. capitata, Thunb.! fl. cap. p.* 203. *Sieb. fl. cap. exs. n.* 67. *Fl. mixt. n.* 90. *E. & Z.!* 1025. *Wendl.! coll.* 2, *t.* 50. *Bot. Reg. t.* 711. *P. plumosa, Lodd. Bot. Cab. t.* 253. *P. pubescens, Ait. Kew.* 1. *p.* 268. *Willd. Herb. fol.* 1 *&* 3. *DC. l. c. p.* 36, *Zey.!* 2211.

VAR. β. **angustifolia**; leaves linear-lanceolate or sub-linear, spreading or curvate-erect, villous; heads of flowers large, mostly gold-coloured. *Drege.!* 6780.

VAR. γ. **brachycephala**; branches smooth, except the youngest twigs; leaves lanceolate, curvate-erect, glabrous and shining above, the younger sub-villous, the floral ones very villous; heads of flowers much smaller. *Drege.!* 6781.

HAB. Hills and Mountains near Capetown and Rondebosch, *Thunb.! E. & Z.!* Var. β. sides of Bergriver, Stellenbosch, *Drege, Niven, W.H.H.* Var. γ. Cape flats. Dec.–Aug. (Herb. Thunb., Holm., Wendl., T.C.D., Sond.)

Shrub, 2–6 feet high, branches alternate. Leaves on very short petioles, approximate, at base obtuse, the lower and uppermost shorter, about 1 inch long, in var. α. 1½–2 lines, in var. β. ½–1 line wide. Heads of flowers 1½–2 inches in diameter, in var γ. twice as small; leaves or scales of the involucre small, with spreading hairs. Bracteolæ small, equalling the calyx, or shorter. Calyx 4 lines long, segments very small. Petals often wanting, in some specimens very small, shorter than the inflexed filaments. Capsule roundish, smooth, the size of a large pea; area small. Seeds ovate, nut brown. Very similar to *P. plumosa*, differs by the larger heads of flowers, the calyx, and the petals.

24. P. rigida (E. & Z.!); branches, twigs, and petioles *tomentose;* leaves approximate, spreading, or slightly erect, *rigid*, lanceolate, from the *subcordate base tapering into a rigid mucro*, with revolute margins, glabrous and shining above, tomentose beneath; the younger villous; floral leaves *elongate, sub-filiform, villous, plumose*, surrounding and longer than the *globose* head of flowers; bracteolæ very villous, equalling the calyx; tube of the calyx at length glabrous; segments subulate, without villous, within glabrous, but *bearded in the middle;* petals wanting. *P. capitata, E. Mey.! in Hb. Drege.! non Thunb. Walpersia rigida, Presl.*

HAB. Rocky places, Ezelsbank, 3–5000 f. *Drege!* Cederberge, Clanwilliam, *E. & Z.* Nov.–March. (Herb. Sond.)

Shrub robust, 2 feet and higher, with rigid branches, tomentose, dense, fulvid or blackish. Leaves 8–10 lines long, 1–1½ line wide; petiole 1 line long. Head of flowers the size of a walnut; receptacle very villous; involucre feathery-villous, greyish white. Calyx 3 lines long. Style short.

25. P. cylindrica (Wendl.! collect. 1. p. 7); stem and branches erect; twigs pubescent; leaves *lanceolate* or *linear lanceolate*, villous, with revolute margins, greenish above, whitish tomentose beneath, patent, the upper erect; floral ones shorter, imbricate, woolly; *spike cylindrical, dense, white;* bracteolæ small, shorter than the tubular, very villous calyx; segments of the calyx lanceolate, spreading; petals setaceous. *Willd.! enum.* 1. *p.* 253. *R. & Sch. l. c.* v. 489. *Spreng. l. c. f.* 5. *DC. l. c. p.* 35. *P. cylindrica, Drege! c. Tylanthus hirtifolius, Presl.*

VAR. *β.* **glabrata**; leaves linear-lanceolate, acute, rigid, smooth, shining, with revolute margins; upper ones sub-villous.

HAB. Between Heerelogement and Olifantsriver, *Zey.!* 325. Between Langevallei and Bergvalley, *Zey.!* 320. June. Var. *β.* Piquetberg and Langevalley, and between Pikanierskloof and Markaskraal, *Drege!* 6788 *a.* Jan. (Herb. Wendl., T.C.D., Sond.)

Spontaneous specimens more slender than those cultivated in the Herrenhausen botanic gardens, and the flower-heads smaller. Shrub over 2 feet, with mostly crowded, brownish branches. Leaves short, petiolate, 6–8 lines long, 1 line wide. superior and floral ones shorter, villous, often penicillate at top. Spike terminal, woolly, 1–2 lines long. Calyx bi-bracteolate, whitish, woolly, 2 lines long, segments small, yellow green within. Petals very minute. Capsule smooth, the size of a pea, area small.

26. P. spicata (L. fil. suppl. p. 153); stem and branches erect, twigs greyish, pubescent; leaves *erect, oblong,* or *ovate-cordate, acuminate,* very acute, minutely downy or glabrous above, whitish tomentose beneath, with revolute margins; floral ones (bracteæ) shorter, woolly; spike ovate or ovate-cylindric, white, woolly; bracteolæ small, very villous, shorter than the woolly calyx, segments small, longer than the tube; petals setaceous; ovary glabrous. *Thunb. fl. cap. p.* 204. *Willd.! spec.* 1. *p.* 1111. *Lam. ill. t.* 127, *f.* 3. *Trichocephalus spicatus, Brogn. l. c. E. & Z.!* 997. *Walpersia capitata, Presl.*

HAB. Mount Paardeberg, *Thunberg!* Tulbagh near Waterfall, *E. & Z.! Niven.* Kruisrivier, *Zey.!* 319. Simonsbay, Paarl, Dutoitskloof, Piquetberg, *Gueinzius, Drege,* 6764, 6787. Fl. throughout the year. (Herb. Thunb., Wendl., T.C.D., Sond.)

1 feet high or more, leaves sagittate-cordate, the lower sometimes 1 inch long, 4–5 lines wide, the intermediate 7–9 lines long, 3 lines wide, the uppermost about 4 lines long, 2 lines wide, or shorter. Petiole about 1 line long. Spike 1 inch long, or shorter, rarely 2 inches long; floral leaves (bracteæ) about 4 lines long. Calyx 2½ lines long, segments spreading. Capsule globose, blackish, smooth, the size of a large pea, crowned by the calyx.

27. P. reclinata (Bernh.! in Krauss. Beyt. p. 44); branches and twigs hairy; leaves *spreading, reclinate, arcuate-curved, lanceolate,* callous, flattish, smooth and *shining above,* tomentose beneath, margins revolute, younger leaves slightly hairy; heads of flowers hemispherical; floral leaves in 2–3 rows, the exterior *glabrous,* the interior ones not foliaceous, lanceolate, feathery-hirsute, with villous margins, longer than the flower-head; bracteolæ small, villous, equalling the calyx; segments of the calyx *subulate;* petals setaceous. *Walpersia curvifolia, Presl.*

HAB. Mountain sides, Outeniqua, George, *Krauss.* Jan. *Drege!* 6790. Aug. (Herb. T.C.D., Sond.)

2 feet high or more, branches virgate. Leaves crowded, about 4 lines long, 1 line wide. Petiole erect, 1 line long. Head of flowers half an inch in diameter. Interior scales of the involucre 6 lines long, about 1 line wide, brownish, appressed-villous, at top whitish, hirsute. Receptacle woolly. Calyx 2½ lines long.

28. P. trachyphylla (Sond.); twigs pubescent; leaves *slightly spreading,* arcuate-curved, lanceolate, *callous-mucronate,* with revolute margins, *papillose-scabrous* above, tomentose beneath, the younger villous; heads of flowers hemispherical, exterior leaves of the involucre *villous,* the intermediate linear, the interior ones longer, plumoso-villous; bracteolæ small, villous, equalling the very villous calyx; segments of the calyx *lanceolate,* shorter than the tube, *bearded on the throat within;* petals setaceous. *Trichocephalus, trachyphyllus, E. & Z.!* 1001.

HAB. Amongst Rocks, Winterhoeksberge near Tulbagh, Worcester, *E. & Z.!* Nov. (Herb. Sond.)

Shrub, 2 feet high, branches umbellate. Leaves 4 lines long, about 1 line wide; petiole erect, ½ line long. Heads of flowers 6 lines in diameter, surrounded by the uppermost leaves. Interior scales of the flowerheads about 6 lines long, dense, whitish, villous at the top, longer than the flowerheads. Calyx 2½ lines long. Petals pilose.

β. *Ericoides.* Flowers in spherical or hemispherical heads; involucre leafy, flat, or hemispherical, shorter or not longer than the flowers. (Sp. 29–56.)

* Petals setaceous or wanting. (Sp. 29–31.)

29. P. retrorsa (E. Meyer!); twigs pubescent; leaves (minute) *reflexed, sessile, ovate, obtuse, cordate* at base, with revolute margins, glabrous and papillose above, tomentose beneath; heads of flowers woolly; scales of the involucre ovate or lanceolate, coloured; calyx turbinate, very villous, segments ovate, acute; petals setaceous or wanting. *Tylanthus retrorsus, Presl.*

HAB. Roodeberg & Ezelskop, 4–5000 f. *Drege!* Nov. (Herb. Sond, Lehm., T.C.D.)

Shrub, much branched; twigs umbellate, greyish or brownish. Leaves crowded, in about 4 ranks, 1 line long, or smaller Capitula about the size of a large pea. Scales of the involucre red, greenish or whitish, striate, about 1 line long. Bracteolæ subulate, villous, equalling the calyx. Calyx 1 line long, hairs of the segments reddish. I never found the petals, but according to W. Arnott, in Hook. Jour. Bot. vol. iii. p. 253, they are setaceous, as in *Trichocephalus.* Anthers 1-celled.

30. P. montana (Sond.); twigs umbellate, minutely pubescent; leaves *erecto-patent,* rigid, *linear-lanceolate,* callous, acute, convex, punctate-scabrous above, tomentose beneath, with revolute margins; heads of flowers roundish-ovate, surrounded by the uppermost leaves; calyx tubular, tomentose, equalling the subulate bracteolæ; segments of the calyx lanceolate-acuminate; petals setaceous, glabrous.

HAB. Interior regions, *Niven.* Leliefontein, at the foot of Mount Ezelskop, 4–5000 f. *Drege!* 6760. (Herb. Sond.)

Resembles in habit *P. stipularis.* Branches and twigs erect, greyish pubescent.

Leaves 3–4 lines long, ¾–1 line wide, often recurved at the top; the younger somewhat pubescent. Petiole ⅛ line long. Capitulum greyish-white, the size of a hazelnut. Bracteolæ tomentose. Calyx 2 lines long. Capsule tomentose, at length subglabrous, the size of a small pea, crowned by the calyx.

31. P. debilis (E. & Z.! 1016); twigs pubescent; leaves (minute) spreading, sub-reflexed, *cordate-lanceolate, apiculate,* sub-incurved, with revolute margins, convex, smooth, papillose above, tomentose beneath, the younger sub-villous; heads of flowers small; *involucre foliaceous;* calyx turbinate, villous; bracteolæ subulate, villous-barbate; petals setaceous, incurved, glabrous. *Tylanthus debilis, Presl.*

HAB. Mountains in Kannaland, near villa "Gideon Joubert," Swellendam. *E. & Z.!* Nov. (Herb. Sond., Lehm.)

Shrub ½–1 foot high, much branched from the base; branches alternate, twigs filiform, brownish-grey. Leaves approximate, 2 lines long, ¾–1 line wide, cordate-emarginate at the base, shining. Capitulum white, 10–15 flowers on the very villous receptacle. Calyx 1 line long, very villous; segments lanceolote. Ovary conical, soon glabrate.

** Petals roundish, concave, ciliate-barbate. (Sp. 32–34.)

32. P. brevifolia (E. & Z.! 1017); twigs *pubescent;* leaves (minute) crowded, spreading, sub-sessile, *cordate-lanceolate, hooked-inwards,* acute, smooth, shining, tomentose beneath; margins revolute; heads of flowers woolly, involucrated by the uppermost leaves; calyx *smooth;* segments subulate, spreading, very villous; petals roundish, fimbriate, ciliate.

HAB. Between rocks, Zwarteberg, near Caledon, *E. & Z.!* Aug. (Herb., Sond., T.C.D.)

1 foot high, greyish; branches and twigs umbellate. Leaves about 1–1½ line long, ⅓ line wide, sub-compressed, minutely punctate above. Capitulum white, the size of a large pea. Receptacle woolly. Calyx 1 line long; segments very small.

33. P. cuspidata (E. & Z.! 1031); twigs *villous-tomentose;* leaves sub-imbricate, erecto-patent, very short, petiolate, appressed at the base, *cordate, cuspidate,* or *lanceolate, callous-mucronate,* with revolute margins, *glabrous, shining above,* tomentose beneath; younger ones sub-villous; heads of flowers *globose,* spiked; inferior bracteæ foliaceous; *calyx woolly-tomentose,* segments lanceolate, subulate, bearded at the base within; petals concave, roundish, bearded ciliate. *P. imberbis? Thunb.! herbar. P. strigosa, Berg.! ex pte. Petalopogon cuspidatus, Reiss. in Decad. nov. stirp. mus. Vind.* 10, *n.* 92. *Trichocephalus Harveyi, Arn.! in Hook. Jour. Bot. vol. iii. p.* 253.

HAB. Near Capetown, *Grubb* in Herb. Bergius. Doornhoogde, near Bergvalley, Blauwberg, Zwellendam, *Thunb. Niven, E. & Z.! Zeyh.* 322. *Drege,* 6754, 6785, 6786. Sept. (Herb., Holm., Thunb., Sond., Wendl.)

Shrub, 1–2 feet high; branches dichotomous; twigs fastigiate, short. Petiole often scarcely conspicuous, sometimes ½–1 line long. Leaves 4 lines long, 1 line wide, rigid. Capitula white, globose, the size of a cherry; involucre formed by the uppermost leaves. Bracteæ usually shorter or as long as the flowers, in the lower part of the capitulum a little longer. Bracteolæ subulate, hirsute. Calyx tubular-sub-campanulate, very villous at base. Style short; stigmas lobed. Capsule blackish, at length glabrous, the size of a pea; area large.

34. .P Thunbergiana (E. Meyer!); twigs *thinly pubescent;* leaves crowded, sub-sessile, or short-petioled, spreading, sub-recurved, appressed at the base, cordate-lanceolate, callous-mucronate, with revolute margins, *pubescent*, older ones glabrate above, tomentose beneath; heads of flowers *ovate* or *ovate-oblong*, spiked; bracteæ foliaceous; calyx *appressed-silky*, segments lanceolate, within bearded at the base; petals concave, roundish, bearded-ciliate. *E. Mey. in herb. Drege, but not P. ericoides, Thunb. Tylanthus Thunbergianus, Presl.*

HAB. Hills between Paarl and Pont, and near Bergriver; sandy places at the foot of Simonsberg, *Drege.* Oct.–Dec. (Herb., Lehm., Sond., T.C.D.)

Shrub, about 1 foot high, greyish, with alternate branches. Leaves mostly imbricate, 2–3 lines long. ¾–1 line wide, in the twigs sub-distichous. Capitulum about the size of a hazel-nut, greyish-white; involucre formed by the uppermost leaves. Bracteæ cuspidate. Bracteolæ small, hirsute. Calyx 1 line long. Capsule blackish, at length glabrous, the size of a small pea.—Very similar to the preceding, but differs by the slender, shorter stem, with numerous branches, pubescent, patent recurved leaves, smaller, ovate or oblong capitula, and the calyx.

*** Petals roundish concave, mostly cucullate or mitriform, quite glabrous. (Sp. 35–56.)

35. P. callosa (Thunb.! prod. p. 45) twigs greyish-pubescent; leaves spreading or sub-reflexed, sub-petiolate; leaves *cordate, ovate* or *oblong, callous-acute*, hairy, sub-tomentose beneath, young ones pilose, adult ones smooth above; margins revolute; heads of flowers *sub-globose;* involucre leafy, as long or a little shorter than the head; bracteolæ villous short; calyx sub-turbinate, segments lanceolate, very villous outside, equalling the glabrous tube; petals *rotundate, cucullate;* capsule smooth. *Fl. cap. p.* 204. *Linn. f. suppl. p.* 153. *DC. l. c. p.* 36. *E. Mey. in Hb Drege. P. cordata L.! mant. p.* 342. *P. atrata, Bernh! in Krauss. Beit. p.* 45, *non Lichtenstein. Trichocephalus callosus, E. & Z.! n.* 998. *Tylanthas callosus Presl.*

HAB. Mountain sides near Groenekloof, Devilsmountain, *Thunb. E. & Z.! Krauss.* Paarl and Paarlberg, *Dr. Pappe, Drege, n.* 1917 *a.* 9534. Sept., Oct., and Mar. (Herb., Thunb., Holm., T.C.D., Lehm., Sond.)

Shrub, 1–3 feet high, greyish; leaves at length blackish. Branches rigid or virgate. Petiole ½ line long. Leaves 4–6 lines long, 2–3 lines wide, others 1½–1 line wide, deeply cordate, slightly convex, quite smooth or papillate above, tipped by a whitish callus; the upper ones smaller, or lanceolate. Capitula solitary or corymbose, aggregate, tomentose, the size of a small cherry. Involucre appressed. Bracteæ short. Calyx 2 lines long, with spreading segments, longer than the bracteolæ. Capsule black, shining. the size of a small pea, crowned by the calyx.—Drege n. 1917 *b.* is perhaps a different species, which has an hemispherical capitulum and more developed, lanceolote bracteæ; the flowers are not perfect.

36. P. atrata (Licht.! in R. & Sch. syst. veg. v. p. 490); twigs clothed with *dark villi;* leaves spreading, very short-petioled; *cordate-lanceolate, incurved,* callous, obtuse, *very smooth above,* tomentose beneath, with revolute margins; heads of flowers terminal, *hemispherical,* villous; leaves of the involucre sub-obtuse, equalling the capitulum or a little longer; calyx turbinate, glabrous; segments lanceolate, very villous outside; bracteolæ equalling the calyx, subulate, penicillate-hirsute; petals cucullate, *calyptriform;* capsule glabrous. *Trichocephalus ruber, E. & Z.!* 1005, *excl. syn. T. laevis, E. & Z.!* 1000. *P. cordata, var. lævis. Schlech. Linn.* 6, *p.* 195.

VAR. β. **litoralis**; twigs clothed with greyish villi; leaves pale green. *P. litoralis, Bernh.! l. c. Trichocephalus litoralis, E. & Z.!* 1002.

HAB. Hott. Holl., *Lichtenstein.* Near Palmietriver, Stellenbosch and Zwartebergskloof, Caledon, *E. & Z.!* Var. β. sea-coast near Cape Receif, Algoabay, *E. & Z.!* Zitz Kamma, *Krauss, Gueinzius.* Jun.–Sept. (Hb. r. Berol., Wendl., Sond., T.C.D.)

Shrub, 1–3 feet high. Leaves 4–5 lines long, 1–1½ line wide. Capitulum whitish, woolly. Bracteæ lanceolate, leafy. Bracteolæ very small, 3 lines long. Capsule small, shining. Distinguished from the preceding by the incurved, more obtuse. not punctate-papillose leaves, larger heads, larger and longer leaves of the involucre and the very villous flowers.

37. **P. alpina** (E. & Z.! 1018); branches and twigs *pubescent;* leaves small, sub-sessile, *erect, sub-imbricate, cordate-lanceolate,* shining and *scabrous* above, tomentose beneath, margins revolute; heads of flowers terminal, hemispherical (small); leaves of the involucre equalling the capitulum; calyx turbinate, *glabrous;* segments acute, very villous, shorter than the tube, but longer than the villous bracteolæ; petals *rotundate cucullate.*

HAB. Mountains near Brackfontein, Clanwilliam, *E. & Z.! Drege!* 6756. Aug. (Herb. Sond.)

½–1 foot high, branches often trichotomous, reddish. Leaves 2–3 lines long, 1 line wide, sometimes a little incurved. Capitulum the size of a pea. Leaves of the involucre cuspidate. Bracteæ equalling the flowers. Calyx about 1 line long.

38. **P. humilis** (Sond.); branches and twigs *filiform, pubescent;* leaves minute, spreading or reflexed, sub-sessile, cordate, *ovate, sub-obtuse,* convex, very *smooth,* shining above, tomentose beneath, with revolute margins; heads of about 8–12 lax, aggregated flowers, involucrated by the uppermost leaves; calyx *campanulate, woolly,* equalling the villous bracteolæ; petals cucullate; ovary sub-lanuginose.

HAB. Mountains near Grietjesgat, between Lowrypass and Palmietriver, Stellenbosch, 2–4000 feet, *Ecklon.* Jun. (Herb., Sond.)

Shrub, a span high, with ascending greyish branches. Leaves a line long, smooth and shining, but very minutely punctate above. Capitula roundish, the size of a small pea. Calyx about ¾ line long; segments acute. Capsule unripe, crowned with the calyx; area about as large as the capsule.—Very similar to *P. debilis,* distinguished by the obtuse, very smooth leaves, calyx, and petals.

39. **P. acmæphylla** (E. & Z.! 1019); twigs *villous-pubescent;* leaves small, *erecto-patent or sub-reflexed,* very short petiolate, *cordate-lanceolate,* callous-acute, *villous,* at length glabrate above, white-tomentose beneath, with revolute margins; heads of flowers hemispherical; leaves of the *villous* involucre ovate, acute, equalling the head; tube of the calyx glabrous; segments *lanceolate,* villous; bracteolæ villous, equalling the calyx; petals cucullate; capsule globose, shining. *P. parviflora, E. Mey. d.*

HAB. Stony places, mountain sides at Tulbagh, near waterfall, *E. & Z.!* Lowrypass, 1–2000 feet, and Strandfontein, near Olifant river, *Drege,* 6753. Nov. (Herb., Sond.)

1 foot high, with greyish or yellowish, often crowded branches. Lower leaves approximate, 4 lines long, 1 line wide, upper ones remote, 3–2 lines long. Capitulum woolly, the size of a large pea. Calyx about 1 line long. Capsule nut-brown, the size of a small pea, crowned with the calyx; area small.—Differs from *P. comosa* by the cordate, villous leaves and shining smaller capsule; from *P. alpina* by the villous, not scabrous (on the twigs), remote leaves, bracteolæ, and calyx.

40. P. gracilis (Sond.); twigs villous; leaves very short, petiolate, spreading, incurved-erect, *cordate*, linear-lanceolate, with revolute margins, villous-ciliate, convex and *glabrous* above, whitish tomentose beneath; heads of flowers terminal, solitary; involucre hemispherical, its scales petiolate foliaceous, *equalling* the capitulum; calyx turbinate, its tube smooth; segments lanceolate, very villous, longer than the villous bracteolæ; petals mitriform, *obtuse*. *Trichocephalus gracilis, E. & Z.!* 1006. *Tylanthus gracilis. Presl.*

HAB. Grassy places in mountains near Palmietriver, Stellenbosch, *E. & Z.!* Jun. (Herb. Sond., T.C.D.)

A slender shrub, 1–2 feet high, with umbellate, greyish branches. Leaves 4–6 lines long, 1 line wide, dark green, the upper ones smaller. Capitulum dense-flowered, white hirsute, 6 lines in diameter. Bracteolæ subulate. Calyx about 2 lines long; segments short.

41. P. virgata (Sond.); branches elongate; twigs villous; leaves very short, petiolate, erecto-patent, linear-lanceolate, acute, at the base *slightly cordate*, with revolute margins, villous-ciliate, *pilose*, at length glabrous above, tomentose beneath; heads of flowers small, *corymbose or sub-paniculate;* involucre hemispherical, its leaves ovate, acute, villous, *shorter than the capitulum;* calyx turbinate, its tube glabrous, segments lanceolate, very villous, a little longer than the villous bracteolæ; petals mitriform, *obtuse, acuminate. Trichocephalus virgatus, E. & Z.!* 1011. *Tylanthus virgatus, Presl.*

HAB. Moist places in mountains, Klynriviersberge, Caledon. *E. & Z.!* at Hemel en Aarde, *Zeyh.* 2217. Aug. (Herb., Sond., T.C.D.)

Shrub, very slender, about 2 feet high; twigs crowded, filiform, reddish. Lower leaves spreading, 4–5 lines long, 1 line wide, upper ones remote, erect, smaller. Corymbs 2–6–cephalous, or a great many peduncles disposed in a panicle. Capitula 8–12-flowered, woolly, the size of a pea. Flowers 1½ line long. Bracteolæ subulate, very villous at top.—Allied to *P. gracilis*, the leaves nearly the same, but not so evidently cordate, the heads much smaller and aggregated, and the petals acuminate.

42. P. diosmoides (Sond.); branches and twigs thinly *greyish-tomentose;* leaves petiolate, erecto-patent or sub-erect, *lanceolate, sub-incurved.* callous-apiculate, the younger sub-pilose, the adult ones glabrous, shining above, tomentose beneath, with revolute margins; heads of flowers terminal, corymbose; leaves of the hemispherical involucre *pluriseriate, lanceolate*, hirsute, equalling the capitulum; calyx cuneate, woolly, a little longer than the subulate feathery-hirsute bracteolæ; segments acute; petals cucullate.

HAB. Stony places between Great Houhoek and Potriver, *Zeyh.* Jul. (Herb., Sond.)

1 foot high, with greyish branches. Leaves 4–6 lines long, 1 line wide; petiole 1½ line long, appressed, tomentose. Calyx 2 lines long. Capitulum rarely solitary, mostly 3–1, corymbose, on greyish, 4–6 lines long peduncles, the size of a large pea or a small cherry, clothed with greyish-white hairs.—Easily known by the long petioles and greyish hirsute capitula.

43. P. propinqua (Sond.); twigs clothed with dark villi; leaves spreading, short-petioled, *oblong-lanceolate*, callous-acute, *at base obtuse*, glabrous, shining above, tomentose beneath, with sub-recurved margins; heads of flowers solitary, *hemispherical;* leaves of the involucre *ovate-*

oblong, acute, equalling the capitulum; calyx turbinate, tube glabrous, segments lanceolate, acuminate, very villous outside, longer than the villous bracteolæ; petals cucullate, obtuse. *Trichocephalus elongatus, E. & Z.!* 999. *excl. syn.*

HAB. Mountain sides near Zwellendam, *E. & Z.!* Oct. (Herb., Sond.)

1 foot, or higher. Branches and twigs rigid, angular. Leaves dense, 4–6 lines long, 1½ line wide, pale green, the upper ones sub-erect. Capitulum 6–8 lines in diameter. Involucre foliaceous, 2–3 seriate. Flowers 3 lines long, limb very white. Bracteolæ linear, very villous, 2 lines long.—Very similar to *P. atrata,* from which it differs by more oblong, not cordate, evidently petiolate leaves.

44. P. comosa (Sond.); branches *minutely pubescent;* leaves erecto-patent, short-petioled, *lanceolate, incurved,* callous-acute, obtuse at base, glabrous, *papillate-scabrous* above, tomentose beneath, with revolute margins, the *younger pubescent;* head of flowers *sub-globose;* leaves of the involucre ovate, acute, equalling the capitulum; calyx *downy,* segments lanceolate, villous; bracteolæ villous, a little shorter than the calyx; petals rotundate-cucullate; capsule *velvetty pubescent. Trichocephalus comosus, E. & Z.!* 1009. *Tylanthus comosus, Presl.*

HAB. Among rocks in mountains of the Tulbagh valley, near waterfall, *E. & Z.!* Dec. (Herb., Lehm., Sond.)

Shrub, erect, much branched, greyish; twigs short. Leaves approximate, scabrous, 4 lines long, about 1 line wide, the upper ones about 3 lines long. Petiole 1 line long. Capitulum the size of a small cherry Calyx 2 lines long; segments shorter than the tube. Capsule greyish, the size of a pea.

45. P. rubra (Willd.! in R. & Sch. l. c. p. 491); branches virgate, twigs *filiform, thinly pubescent,* the ultimate very short; leaves spreading, short-petioled, *lanceolate, acute,* at base obtuse, smooth above, tomentose beneath, with sub-revolute margins; heads of flowers minute, terminal, 6–10-flowered, *involucrated by the uppermost leaves;* calyx minute, *campanulate,* tomentose, segments acute; bracteolæ villous, shorter than the calyx; petals cucullate; capsule obovate, smooth.

VAR. β. **parvifolia**; leaves erecto-patent, smaller, linear-lanceolate. *P. empetroides, E. & Z.!* 1021, *ex pte.*

HAB. Hott. Holl. Lichtenstein. Var. β. Howhoeksberge, Stellenbosch, *E. & Z.!* (Herb., Willd., Holm., Hook., Sond.)

Shrub, over 2 feet high; branches red, in var. β. yellowish or greyish. Leaves 4 lines long, 1 line wide; in var. β. 2–3 lines long, ½ line wide. Petiole ½–1 line long. The numerous twigs, terminated by a head, represent a large panicle. Capitulum woolly, 1–1½ line in diameter. Flowers ½ line long, Capsule blackish, 2 lines long, cuneate, crowned by the persistent calyx; area small.—Var. β. is very similar to *P. parviflora,* but differs by the glabrous, not scabrous, leaves, with slightly revolute edges. *Phylica, Drege,* 6761, is also very similar, but distinguished by the globose capsule; capitula and flowers are wanting.

46. P. apiculata (Sond.); branches *nodose-scabrous;* twigs with *spreading hairs;* leaves spreading, sub-incurved, lanceolate, tipped with a *very acute, callous point,* at base obtuse, flattish and smooth above, whitish-tomentose beneath, with revolute, *ciliate* margins; heads of flowers solitary; involucre hemispherical, pubescent, its scales foliaceous cuspidate, the outer ones larger, *equalling the capitulum;* calyx turbinate, *smooth,* longer than the subulate, feathery-hirsute bracteolæ; segments hirsute; petals rotundate, cucullate.

HAB. Cape, *Gueinzius*. (Herb., Sond.)

Shrub, over 1 foot high, with ascending branches and mostly crowded, red twigs, clothed with longish hairs. Petiole 1 line long, appressed. Leaves 5–6 lines long, 1½ line wide. Capitula as large as those of *P. ericoides;* sometimes by the aggregate twigs corymbose, greyish-white. Calyx 1½ line long, segments short, acute, sometimes sub-lanceolate. Bracteolæ 1 line long. Capsule unripe, roundish, glabrous. Differs from *P. rubra* by the hairy twigs, larger leaves, and flower-heads; from *P. gracilis*, by larger, flat, not cordate, leaves and capitulum; from *P. atrata*, by the not cordate, ciliate leaves, and smaller heads.

47. P. bicolor (L. mant. 208); branches erect; twigs *hirsute;* leaves spreading or reflexed, upper ones sub-erect, short-petioled, *linear-lanceolate*, acutish, *hirsute*, tomentose beneath, with revolute margins, adult ones glabrous and punctate above; heads of flowers *sub-globose or ovate*, dense; involucre *shorter* than the head, as well as the linear-bracteolæ *fulvid or yellowish;* calyx longer than the bracteolæ, tube ovate-oblong with *reversed hairs*, segments ovate-lanceolate, very hirsute; petals complicately-cucullate. *R. & Sch. l. c. p.* 479. *DC. l. c. p.* 35. *P. strigosa, Berg.! cap. p.* 50. *ex pte. Thunb.! fl. cap. p.* 200. *Herb. Un. itin. n.* 631. *Sieb. fl. cap. exs. n.* 190. *E. & Z.!* 1030.

HAB. Rocky places on the Steenberg, near Muysenberg, and on the top of the Table-mountain, *Grubb., Thunb., E. & Z.! Pappe.* Paarlberg, *Drege.* Jun.–Jul. (Herb., Holm., Thunb., Wendl., T.C.D., Sond.)

Shrub, 2–3 feet high; branches often verticillate, sub-glabrous on the lower, but very villous on the upper part. Leaves 6–8 lines long, 1–1½ line wide, upper ones smaller. Petiole ½–1 line long, appressed. Capitulum as large as a cherry, very villous, many-flowered; outer leaves of the involucre greenish, the inner ones golden. Calyx about 3 lines long; segments thrice shorter than the tube. Anthers bilocular. Style short.

48. P. nigrita (Sond.); twigs *minutely pubescent;* leaves spreading, short-petioled, linear-lanceolate, acute, *smooth*, tomentose beneath, with revolute margins; heads of flowers solitary, *hemispherical;* involucre *flat*, its scales *roundish*, *acute*, midribbed; calyx turbinate, *glabrous*, longer than the subulate villous bracteolæ, segments lanceolate, villous; petals cucullate, roundish. *Trichocephalus atratus, E. & Z.!* 1004. *excl. syn.*

HAB. On mountain sides, Hauhoekberge, and in Knafflockskraal, Stellenbosch, in Mount Zwarteberg, Caledon, *E. & Z.!* Jul.–Aug. (Herb., Sond., T.C.D.)

1 foot high; branches erecto-patent; twigs erect, rigid, greyish. Leaves blackish when dry, 3 lines long, ½ line wide, rigid, mucronate, or sub-obtuse, convex above, sulcate beneath. Petiole ½ line long. Capitulum woolly, 4–5 lines in diameter. Involucre brownish. Calyx 1½ line long; segments spreading.—In habit it agrees nearly with *P. trachyphylla* and *P. gracilis;* in characters, with *P. ericoides*, from which it differs by the robust habit and larger heads.

49. P. strigulosa (Sond.); twigs minutely *appressed, pubescent;* leaves spreading or secundate, upper ones erect, very short-petioled, linear-lanceolate, acute, with a recurved point, *appressed, pubescent*, at length glabrate and punctate above, tomentose beneath, with revolute margins; heads of flowers solitary, *lax*, 5–8-*flowered;* involucre short, with appressed hairs; calyx tubular, clothed with *reversed hairs*, thrice longer than the bracteolæ, segments short, ovate, lanceolate, erect, at length somewhat spreading; petals cucullate. *P. eriophoros, Thunb.! fl. cap. p.* 200, *non Berg.*

HAB. Near Capetown, Thunb., Sieber; sandy flats near Krumrivier, *Zeyh.* 321. Jul. (Herb., Thunb., Hook., Sond.)

Shrub, erect, 1–1½ foot high, greyish; branches alternate; twigs erecto-patent, 1–3 inches long. Leaves 5–6 lines long, 1 line wide, upper ones smaller and more appressed. Flowers spicate-capitate. Bracteæ lanceolate, shorter than the calyx. Calyx 2 lines long, greyish, hairs rigid, retrorse, in the segments rarer. Anthers 2-celled.—Similar to *P. montana*, but differing by the hamulate leaves, the head and petals.

50. P. ericoides (L. spec. p. 283); branches fastigiate; twigs thinly pubescent; leaves short-petioled, spreading or erectish, *linear* or *linear-subulate*, bluntish or acute, smooth, downy beneath, margins revolute; heads of flowers terminal on the *sub-umbellate twigs*, small; involucre hemispherical, its scales ovate, foliaceous, cuspidate; calyx turbinate, smooth, longer than the very villous bracteolæ, segments ovate, acute, hirsute; petals concave, cucullate; capsule smooth. *Commel. hort. Amst.* 2. *t.* 1. *Berg. cap. p.* 49. *Thunb. fl. cap. p.* 199. *Bot. Mag. t.* 224. *Spreng. Berl. Mag. viii. p.* 103. *t.* 8. *f.* 1. *P. glabrata, Thunb.! l. c. P. acerosa, herb. Willd.! herb. n.* 4686. *fol.* 1 *&* 2. *P. microcephala, Willd.! herb. n.* 4695. *P. ericoides, E. Meyer in herb. Drege, e. P. imberbis, ericoides et eriophoros, E. & Z.!* 1010. 1014. 1015. *Tylanthus eriophorus, Presl.*

HAB. Sandy places near Capetown; at Vankampsbay, Caledon, Gnadenthal, and Cape Agulhas. Nov.–Aug. (Herb., Thunb., r. Berol., Wendl., T.C.D., Sond., &c.)

Shrub, erect, 1–3 feet high. Lower leaves about 4 lines, upper ones 3 lines long, ½–¾ lines wide, in cultivated specimens often wider, very smooth. Capitulum as large as a pea, at base pubescent, involucrated by the uppermost leaves. Flowers 1 line long. Capsule crowned by the calyx, as large as a very small pea; area very small.—*P. glabrata*, Thunb., Herb., is a specimen with a little larger leaves. The true *P. ericoides* has been confounded by Thunberg with *P. rubra*, var. *β. parvifolia.*

51. P. parviflora (L. Mant. 209); branches slender; twigs pubescent; leaves small, short-petioled, spreading or erect, *lanceolate, subulate*, acutish, *pilose, scabrous*, downy beneath, margins revolute; heads of flowers minute, terminal, solitary on *paniculated twigs;* involucre hemispherical, its scales rhomboid-ovate, acute, hirsute; calyx turbinate, smooth, longer than the hirsute bracteolæ, segments ovate, acute, hirsute; petals concave-cucullate; capsule smooth. *Berg. cap. p.* 48. *Thunb. l. c. Herb. Un. itin. n.* 634. *Sieb. flor. cap. exs. n.* 68 *and* 183. *E. & Z.!* 1020. *P. parviflora, E. Mey. in Herb. Drege, c. ex pte. Tylanthus parviflorus, Presl.*

HAB. Sandy places in the Cape Flats, common near Saldanhabay and Grooteport. Jan. and following months. (Herb., Sond., T.C.D., &c.)

Stem brownish or greyish, smooth, over 1 foot high. Branches aggregate; twigs filiform, the ultimate very short. Leaves 1–2 lines long. Capitula involucrated by the uppermost leaves, as large as a hempseed or a small pea. Flowers white, not 1 line long. Capsule small.—Distinguished from the preceding by the pilose scabrous leaves and small paniculated heads. *P. parviflora*, Willd.! Herb. n. 4678. fol. 1.= *P. rubra, Willd. Herb. n.* 4693; 2. = *Grubbia rosmarinifolia, Berg;* 3. = *Phylica*, doubtful species, similar to *P. alpina.*

52. P. selaginoides (Sond); branches crowded; twigs minutely pubescent; leaves spreading, incurved-erect, short-petioled, linear, with revolute margins, *terete, obtuse, with inflexed point, smooth*, downy beneath, the younger often villous; heads of flowers terminal, solitary; involucre hemispherical, exterior scales foliaceous, interior ones cuneate-ovate, sub-coloured, pubescent, shorter than the flowers; calyx turbinate, *villous*, on a *short, very villous stalk*, longer than the villous bracteolæ, segments acute; *petals cucullate.*

HAB. Swellendam, *Ecklon.* Feb. (Herb. Sond.)

A heath-like shrub, 1–2 feet high, with brown-greyish branches. Leaves 2 lines long; the upper ones approximate. Capitulum as large as a pea or somewhat larger. Calyx 1½ line long, with spreading segments. Without flowers it is very similar to *P. abietina.*

53. P. disticha (E. & Z.! 1022); branches virgate, 2–3-chotomous, *glabrous;* twigs *filiform*, minutely puberulous; leaves *erecto-patent* or *sub-appressed*, short-petioled, *lanceolate* or *linear-subulate*, with revolute margins, *sub-terete, obtuse or acutish*, very smooth, downy beneath, the younger minutely appressed-pubescent above; heads of flowers terminal (minute); involucre with ovate or obovate, acutish scales, puberulous; calyx turbinate, *glabrous*, twice longer than the hirsute bracteolæ, segments acute, hirsute; petals *mitriform-acuminate;* capsule turbinate, smooth. *P. empetroides, E. & Z.!* 1021. *ex pte. P. microcephala, E. & Z.!* 1023. *ex pte. non Willd.*

HAB. Stony places in mountains, Howhoeksberge, Stellenbosch, and Zwarteberge, and Babylons Toorenberg, Caledon, *E. & Z.! Zey.!* 2216. Jul. (Herb. Sond. T.C.D.)

A fine shrub, 1–2 feet high, with purplish branches and twigs. Leaves about 2 lines long, ¼ line wide. Capitula as large as in *P. parviflora*, 5–10 flowered. Calyx about 1 line long. Capsule shining, about as large as the capitulum.

54. P. brachycephala (Sond); branches elongate; twigs numerous, alternate, *puberulous;* leaves *spreading*, short-petioled, lanceolate or linear-subulate, with revolute margins, *sub-terete, obtuse*, smooth, downy beneath, the younger appressed, pilose above; heads of flowers (minute) solitary on scattered twigs, *only involucrated by the uppermost leaves;* calyx *campanulate*, longer than the bracteolæ, as well as the cuneate ovarium, *villous;* segments acute; petals cucullate, *obtuse. P. microcephala, E. & Z.!* 1023. *ex pte.*

HAB. Between rocks on mountain sides near villa "Gideon Joubert," Swellendam, *E. & Z.!* Nov. (Herb. Sond.)

Habit of the preceding, greyish. Twigs short. Leaves 1–2 lines long, very deciduous; the upper ones approximate, horizontal. Petioles pubescent. Capitulum like that of *P. parviflora*, woolly, but the involucre consists only of a few petiolate leaves. Flowers about 1 line long.

55. P. eriophoros (Berg.! pl. cap. p. 52); branches and twigs pubescent; leaves short-petioled, *linear, acute, sub-villous*, adult ones glabrous, white-tomentose beneath, flat or sub-convex, with revolute margins; heads of flowers terminal, solitary, *roundish;* involucre *shorter* than the head; calyx *pubescent*, longer than the linear, subulate, hirsute bracteolæ, segments acute, as long as the tube; petals cucullate; capsule *roundish*, appressed pubescent, at length glabrate.

VAR. α. **Bergiana**; leaves linear, acutish; calyx woolly, hirsute. *P. eriophoros, Berg. l. c. P. lanceolata, Thunb.! fl. cap. p.* 200. *P. rosmarinifolia, Lam.! ill. n.* 2614. *Lodd. Bott. Cab. t.* 849. *E. & Z.!* 1036. *P. secunda, Herb. un. itin. n.* 633. *ex pte.*

VAR. β. **imberbis**; leaves linear or linear subulate, bluntish; calyx appressed-pubescent. *P. imberbis, Berg.! l. c. p.* 51. *non Thunb. nec E. & Z.! P. bruniades, Lam. ill. n.* 2620.

VAR γ. **secunda**; leaves linear or linear subulate. secundate, acute, smooth or nearly so; calyx pubescent. *P. eriophoros, Berg. var.! P. secunda, Thunb.! fl. cap. p.* 201. *Herb. un. itin. n.* 633. *E. & Z.!* 1034. *P. tenuis, Spreng. P. ericoides, E. Mey. in Hb. Drege, a and b. Tylanthus secundus, Presl., Drege,* 6759, 6769.

HAB. Hills near Capetown and on the Cape flats, near Tulbagh, in Hott. Holl. Grasbergrivier and Waterwal. May.–Jun. (Herb. Holm., Thunb., Lam., Hook., T.C.D., Wendl., Sond.)

Shrub, about 1 foot high, greyish. Twigs short. Leaves approximate, 6–10 lines long, 1 line wide, slightly tapering at base, the lower smoothish, upper appressed-pubescent. Petiole ½–1 line long. Capitula many-flowered, sub-globose, woolly, about 3 lines in diameter. Bracteæ longer than the subulate bracteolæ. Calyx 1½–2 lines long; segments erect, at length spreading. Petals complicately cucullate. Anthers 2-celled. Style cylindrical. Capsule crowned with the calyx, as large as a small pea. Var. γ. is more slender and sub-glabrous, has smaller leaves, and often capitula ⅓ inch long, with longer bracteæ.

56. P. cephalantha (Sond.); twigs very minutely pubescent; leaves approximate, short-petioled, erecto-patent, linear, with revolute margins, *terete, obtuse, scabrous, hamate-recurved at top;* heads of flowers *spherical* (small), solitary, involucrated by the uppermost, small appressed leaves; calyx *campanulate, tomentose,* longer than the hirsute bracteolæ, segments acute; petals cucullate; capsule *ovate, tomentose. P. eriophoros, E. & Z.! ex pte. P. parviflora, E. Meyer in Herb. Drege. c. ex pte.*

HAB. Cape flats, *E. & Z.! Drege, Dr. Pappe.* May.–Sept. (Herb. T.C.D., Sond.)

Shrub, about 1 foot high. Branches and branchlets clothed with a yellowish or greyish down. Leaves 2 lines long. Capitulum as large as a pea, pubescent at base. Calyx about 1 line long, with yellowish down at top. Capsule crowned with the calyx, as large as a small pea; area small. Differs from *P. ericoides* by the obtuse, hamulate leaves and spherical capitulum; from *P. eriophoros* by the small, terete, obtuse, scabrous leaves, and much smaller capitulum.

b. Calyx cylindrical, segments about 4 times shorter than the tube. *Phylica sect. Eriophylica,* Meisn. *Calophylica,* Presl. (Sp. 57–58.)

57. P. gnidioides (E. & Z.! 1037); branches sub-corymbose; twigs very minutely pubescent; leaves sub-imbricate, *erecto-patent,* linear, *mucronate, compressed, sub-ancipital,* glabrous, downy beneath, younger ciliate; flowers capitate; leaves of the involucre acuminate, *sub-coloured,* ciliate, villous at base, equalling the calyx; bracteolæ ciliate; calyx white-tomentose, elongated, segments oblong-lanceolate, spreading; petals cucullate. *P. (eriophylica) gnidioides, Meisn. in Hook. Lond. Journ. Bot. vol.* 2, *p.* 58. *P. juniperifolia, E. & Z.!* 1038. *Calophylica gnidioides and juniperifolia, Presl. Bot. Bem. p.* 39. *Drege,* 7360 (*sub-Gnidia*). *Zey.!* 3746.

HAB. Hills near Vanstaadensriviersberge, Port Elizabeth, Bethelsdorp and Olifantshoek, Uitenh., Mount Bothasberg near Vishrivier, Albany, *E. & Z.!* Langekloof, George, *Dr. Krauss.* Jul. (Herb. Sond., Lehm.)

Shrub over 2 feet; branches brown-greyish; twigs short. Leaves short, petiolate, opposite, verticillate, mostly alternate, 3–5 lines long, ½ line wide, lower sub-remote, upper dense, straight or at point recurved. Capitulum as large as a cherry, solitary, or often several aggregated. Leaves of the involucre very similar to the other leaves, but thinner. Calyx 4 lines long, sub-silky-tomentose; tube 3 lines long, equalling the linear bracteolæ; limb reddish or purple within. Anthers 1-celled. Style filiform, as long as the calyx tube; stigma capitate, sub-lobed. Capsule roundish, sub-trigonous, velvetty-tomentose, the size of a pea. Very like some species of *Gnidia*.

58. P. abietina (E. & Z.! 1013); branches fastigiate; twigs umbellate, very minutely pubescent; leaves approximate, *spreading*, linear, *obtuse, apiculate with inflexed point, sub-terete*, with revolute margins, glabrous, downy beneath; flowers capitate; leaves of the involucre *greenish*, petiolate, inner petioles longer, hirsute; bracteolæ hirsute; calyx white-tomentose, elongated, segments *hirsute*, oblong-lanceolate, spreading; petals cucullate.

HAB. Stony places, sides of Mount Winterhoeksberg, near villa "Viljœn," Uit. *E. & Z.!* Feb. (Herb. Sond.)

Shrub erect, about 2 feet high, greyish. Leaves deciduous, 1½ line long, with rigid point, the uppermost imbricate. Capitulum as in the preceding, solitary, never several aggregate. Calyx 3 lines long. Anthers 1-celled. Style as long as the tube. Habit nearly that of *P. ericoides*, but leaves and heads very different.

Phylica spec. 6784, *Herb. Drege*, is similar to *P. affinis*; the specimens are insufficient. *P. nitida, Lam.! ill. n.* 2613, is not a Cape plant; in Herb. Wildenow and others are specimens of the same species from the Mauritius. *P. elongata, Willd.! P. squamosa, Willd.! P. globosa, Thunb.! P. trichotoma, Thunb.!* are *Bruniaceæ*. *Trichocephalus verticillatus*, E. & Z.! 1003 is a *Stilbacea*.

ORDER XLVI. **TEREBINTACEÆ**, Juss.

(By W. SONDER.)

(Terebintaceæ, Juss. DC. Prod. 2. p. 62. Anacardiaceæ and Burseraceæ, Endl. Gen. ccxlv. ccxlvi. Anacardiaceæ and Amyridaceæ, Lindl. Veg. Kingd. clxxiv. clxxi.)

Flowers regular, generally unisexual, rarely complete. *Calyx* free or connate with the lower half of the pistil, 3–5-parted, persistent, imbricate, sometimes enlarged in fruit. *Petals* 3–5, surrounding a fleshy disc, spreading imbricate, mostly deciduous. *Stamens* inserted with the petals, as many or twice as many, rarely more; filaments subulate; anthers introrse, 2-celled, opening lengthwise. *Ovary* mostly free, of *one* carpel or of two to five consolidated carpels, some abortive; *styles* or stigmas as many as the carpels; *ovule* solitary or in pairs, pendulous, *Fruit* drupaceous, rarely dry. *Seed* exalbuminous, with fleshy cotyledons and radicle next the hilum.

Trees or shrubs with resinous, often caustic and milky juices. Leaves alternate, rarely simple, mostly pinnate or trifoliate, with or without minute stipules. Flowers small or minute, in panicles, racemes, or spikes. An extensive tropical or sub-tropical Order, with outlying species in the temperate zones, chiefly of the northern hemisphere. The genus *Rhus* is almost cosmopolitan. Several have esculent fruit, of which the *Mango* is the most celebrated. The fruits of *Harpephyllum* and *Odina* may be eaten. Frankincense and myrrh, and other similar odoriferous resins are produced by trees of the Sub-order *Burseraceæ*.

TABLE OF THE SOUTH AFRICAN GENERA.

Sub-order 1. ANACARDIEÆ. *Ovary* of one unilocular, fertile carpel, with or without 3–4-abortive carpels; ovule solitary.

I. **Odina.**—*Petals* 4. *Stamens* 8. *Styles* 4.
II. **Rhus.**—*Petals* 5, ovate. *Stam.* 5. *Styles* 3. *Drupe* wingless, longer than the calyx.
III. **Smodingium.**—*Petals* 5, oblong. *Stam.* 5. *Styles* 3. *Samara* sub-orbicular, with a broad wing.
IV. **Botryceras.**—*Petals* 4–5, lanceolate. *Stam.* 4–5. *Style* 1; *stigma* trifid. *Drupe* oval, compressed, small, winged.
V. **Loxostylis.**—*Petals* 5, lanceolate. *Stam.* 5. *Styles* 1–4. *Drupe* oval, oblique, enclosed in an enlarged, leafy, and coloured calyx.

Sub-order 2. SPONDIEÆ. *Ovary* solid, 2–5-celled; styles 2–5; ovules solitary in each cell. *Fruit* plurilocular; 2–5-seeded.

VI. **Sclerocarya.**—Male-flowers spicate. *Stamens* 12–15.
VII. **Harpephyllum.**—Male-flowers panicled. *Stamens* 8–9.

Sub-order 3. BURSEREÆ. *Ovary* solid, 2–5-celled; ovules two in each cell. *Fruit* plurilocular, one or several seeded.

VIII. **Balsamodendron.**—*Sepals* and *petals* 4. *Stamens* 8. *Ovary* 2-celled.

APPENDIX.

IX. ? **Cathastrum.**—*Flowers* bisexual, pentamerous. *Stamens* 5. *Ovary* unilocular; ovules 6–8, in two rows, sutural; *stigma* peltate. *Leaves* simple, opposite. *Flowers* axillary.

I. ODINA, Roxburgh.

Polygamous. *Calyx* shortly 4-lobed, persistent, segments ovate or rounded. *Petals* 4, oblong, concave, spreading, æstivation imbricative. *Stamens* 8, inserted below the margin of the disc; *anthers* ovate. *Torus* discoid, fleshy, 8-crenated, the crenatures alternating with the stamens. Rudimentary pistillum (in the male) 4-partite; segments erect, compressed, clavate. *Ovary* (in the female) free, oblong, 1-celled; *ovule* solitary, pendulous from one side near the apex of the cell. *Styles* 4, from the top of the ovary, short, erect; *stigmas* simple. *Drupe* sub-baccate, with a hard, one-seeded putamen. *Seed* of the same shape as nut. *Embryo* slightly curved, inverted; cotyledons fleshy, flat. *Wight et Arnott, Prod. Flor. Pen. Ind. p.* 171.

Large trees. Leaves alternate, about the ends of the branches, unequally pinnated; leaflets 3–4 pair, opposite, almost sessile, oblong-ovate, entire, paler beneath. Racemes terminal, fascicled, filiform. Flowers small, fasciled.—Name unexplained, probably a native appellation in India.

1. **O. edulis** (Sond.); branchlets, petioles and leaves tomentose; leaves 2–3-jugate, leaflets sub-sessile, oblong-ovate or obovate-obtuse, or with a short obtuse acumen, entire, green and shining above, rufo-tomentose beneath, reticulate; *racemes* short, tomentose; drupe ovate, sub-globose.

HAB. On rocks at the north side of the Macalisberg, *Burke, Zeyh. No.* 349. Flow. May–Jun.; fruit, Dec. (Herb., Hook., Sond.)

Leaves ¾–1 foot long. Petiole 3–4 inches long, as well as the rachis furrowed above. Leaflets 7 or 5, inferior pair on very short petioles, intermediate pairs sessile, terminal leaflets on a half uncial stalk, 3–4 inches long, 2–2½ inches wide, obtuse, rarely emarginate, with revolute margins; the young leaflets densely stellate tomentose on both surfaces. Racemes 2–3 inches long, simple or with a few branches at the base. Flowers minute, about a line long, on very short pedicels, 3 or more fasciculate. Calyx 4-parted. Petals 4, oblong, longer than the 8 stamens. Drupe edible, 4–5 lines long, smooth, with 4 short styles on the top.—*Spondias microcarpa*, Rich. flor. Seneg. is a very similar plant, but the leaves are quite glabrous and the drupes different.

2. O.? discolor (Sond., Linn. xxiii. 1. p. 25); branchlets glabrous; leaves imparipinnate, 4–5-jugate; leaflets sessile, ovate-oblong, or oblong-obtuse, or with a short obtuse acumen, quite entire, glabrous and reticulate above, cano-tomentose beneath; male flowers in terminal *spikes;* rhachis pubescent; calyx puberulous at base; petals 4, patent; stamens 8.

HAB. Macalisberg, *Burke and Zeyh.* n. 1853, Sept. (Herb., Hook., Sond.)

Branchlets terete, yellowish, or somewhat purple. Leaves 1 foot. Petiole canaliculate, 2–3 uncial, a little broader at base. Leaflets at base obtuse or sub-cordate, 2–3 inches long, 1–1½ inch wide, the upper petiolulate. Male spikes 2–4 approximate, simple, about 3 inches long. Flowers sessile, minute. Segments of the calyx glabrous. Petals oblong, about 2 lines long. Rudimentary styles 4, very short; female flowers, and the fruits are wanting.—This plant has a great affinity to *Lannea velutina*, Rich.! in fl. Seneg. p. 155. t. xlii., and if the fruit is not different, it must be removed with *Lannea* from *Anacardieæ* to *Bursereæ*. I have an authentic specimen of *Lannea velutina* with half-ripe fruit; the drupe is 3-celled, with 2 or 1 seed in each cell. *Lannea* is incorrectly referred by Endlicher to *Odina*.

II. RHUS, L.

Flowers hermaphrodite, or by abortion diœcious. *Calyx* small, 5–6-partite, persistent. *Petals* 5–6, oblong or ovate, spreading. *Stamens* 5–6, all perfect. *Ovary* 1, sub-globose, 1-celled, abortive in the male flower. *Style* 1, short, with 3 stigmas, or 3 styles. *Drupe* nearly dry or sub-fleshy, 1-celled, containing a bony, 1-seeded (by abortion) nucleus. *Seeds* without albumen, sustained by a funicle rising from the bottom of the cell; cotyledons leafy or fleshy. *Endl. Gen.* 5905.

A large genus of shrubs or small trees, frequently resinous. Leaves alternate, usually (in the Cape species) trifoliolate, rarely simple. Flowers panicled or racemose, small, greenish or white. The name is derived from the Greek, *ρoos* or *ρous*, *flowing;* perhaps because of the resinous juices of several.

ANALYSIS OF THE SPECIES.

I. Leaves palmately trifoliate, leaflets rising from the top of the petiole.

A. Leaflets tomentose.

a. Leaflets paler, or white underneath:

α. quite entire, or sinuate-toothed.

leaflets narrow-linear, mucronate; panicle glabrous ... ... (1) **rosmarinifolia.**

leaflets linear-lanceolate, acuminate; panicle pubescent ... ... (2) **stenophylla.**

leaflets lanceolate, petiolulate; branchlets glabrous ... ... (3) **angustifolia.**

leaflets lanceolate, sessile; branchlets tomentose .. .. (4) **discolor.**

leaflets oval-cuneate; panicles pubescent, sparsi-flowered ... ... (5) **divaricata.**

leaflets obovate-cuneate ; panicles tomentose ; flowers glomerate ... ... (6) **obovata.**
leaflets elliptic or ovate acute ... ... (8) **tomentosa.**
leaflets rhomboid-suborbicular .. ... (7) **populifolia.**
β. pinnatifid, or deeply incised.
leaflets inciso-pinnatifid ; lobes obtuse ; drupe pubescent ... ... (9) **incisa.**
leaflets sharply cut, pinnatifid ; drupe smooth or muricate ... ... (10) **dissecta.**
b. Leaflets same colour on both sides.
branchlets, leaflets, and panicles thinly yellowish tomentose ... ... (11) **magalismontana**

B. Leaves and panicles villous or pubescent, rarely the leaflets sub-glabrous.
a. Leaflets 3 lobed, or 3–7 dentate, or sinuate.
leaflets cuneate-oblong, 3 lobed, or 3 dentate ... ... (16) **tridentata.**
leaflets obovate, 3–7 dentate, villous ... (12) **parvifolia.**
leaflets ovate, or obovate, sinuated, villous beneath ... ... (18) **sinuata.**
b. Leaflets entire.
leaflets obovate, hairy ; racemes shorter than the leaves ... ... (13) **villosa.**
leaflets obovate, pubescent, uncial ; panicles filiform, long ... ... (14) **refracta.**
leaflets obovate-oblong, pubescent, 2 uncial ; panicles ample, leafless, long ... (15) **pyroides.**
leaflets oblong, acute or elliptic-lanceolate, sub-sericeous ; panicles of the length of the leaves ... ... (17) **puberula.**

C. Leaves and panicles quite smooth, or the panicles slightly hairy.
a. petioles sub-terete, furrowed above.
α. leaflets dentate or serrate.
leaflets cuneate, with 3–7 large teeth ;
green on both sides ; branchlets puberulous ... ... (19) **cuneifolia.**
rufous beneath ; branchlets rufovillous (20) **crenata.**
paler beneath ; branchlets glabrous (21) **dentata.**
leaflets ovate-acuminate, 2–6 dentate ... (25) **laevigata,** var. *β.*
leaflets oval-obtuse, emarginate, crenate, serrate, quite glabrous ... ... (29) **natalensis.**
leaflets lanceolate, obtuse, mucronate, remotely denticulate ... ... (28) **Gueinzii.**
leaflets linear, acute, erosely-toothed ... (32) **erosa.**
β. leaflets quite entire.
* leaflets flat above.
leaflets obovate-cuneiform, mucronate, with revolute margins ; branchlets glabrous ... ... (22) **mucronata.**
leaflets oblong, cuneate at base, flat ; branchlets pubescent ... ... (23) **Burkeana.**
leaflets elliptic, cuspidate, margined ; branchlets glabrous .. .. (24) **Zeyheri.**
leaflets ovate, acuminate ... ... (25) **laevigata.**
leaflets lanceolate, acuminate ; panicles hirsute ... ... (27) **viminalis.**
leaflets lanceolate-linear ; panicles glabrous ... ... (26) **lancea.**
leaflets linear, very blunt, broadest towards the apex ... ... (31) **tridactyla.**

** leaflets concave above.
leaflets small linear, mucronate, convex below ... ... (33) **Dregeana.**
*** leaflets carinate above.
leaflets lanceolate-oblong or linear-oblong, mucronate, cuneate, whiteish margined ... ... (30) **Ecklonianа.**
b. petioles with small wings.
leaves sub-sessile ... ... (38) **scoparia.**
leaves on long or moderate petioles.
leaflets obcordate ... ... (34) **glauca.**
leaflets ovate, sub-acute, ribbed; drupe oblique ... ... (37) **Africana.**
leaflets obovate, flat, without prominent nerves ... ... (35) **lucida.**
leaflets cuneate-oblong, flat, ribbed on both sides ... ... (40) **excisa.**
leaflets obovate, with elevated veins and revolute margins ... ... (36) **scytophylla.**
leaflets obovate-oblong, white-margined, with revolute margins ... ... (41) **albomarginata.**
leaflets obovate-cuneate, undulate, denticulate or incised ... ... (39) **undulata.**
leaflets lanceolate, short, acuminate, with 1 or 2 teeth ... ... (44) **rigida.**
leaflets lanceolate, carinate, undulate, glabrous ... ... (43) **celastroides.**
leaflets linear-lanceolate, flat, ciliate ... (42) **ciliata.**

D. Young branches, leaves and panicles covered with red glands.
leaflets linear-cuneate ... ... (45) **horrida.**
leaflets cuneate-obovate ... ... (46) **longispina.**

II. Leaves simple, penni-nerved.
leaflets linear-oblong, quite glabrous, glaucous (53) **longifolia.**
leaflets obovate or elliptic-obovate, quite glabrous, green ... ... (49) **concolor.**
leaflets pulverulent, glaucous ... ... (47) **Thunbergii.**
leaflets whitish-tomentose beneath, glabrous, reticulate above ... ... (48) **dispar.**
leaflets silky-tomentose beneath.
alternate.
oblong or linear-oblong, with recurved margins ... ... (50) **mucronifolia.**
oblong-lanceolate or lanceolate undulate-crenate ... ... (52) **paniculosa.**
ternate.
oblong-lanceolate or lanceolate, flat (51) **salicina.**

I. Leaves palmately *trifoliate, leaflets rising* from the top of the petiole. (Sp. 1–46.)

A. *Leaflets tomentose.* (Sp. 1–11.)

1. R. rosmarinifolia (Vahl. Symb. 3. p. 50); branchlets *glabrous;* leaves sub-sessile or petiolate; leaflets *narrow-linear*, mucronate, with revolute margins, glabrous above, whitish-tomentose beneath; panicles axillary and terminal, *glabrous;* pedicels capillary. *Burm. Afr. t.* 91. *f.* 1. *Willd. spec.* 1. *p.* 1484. *Thunb.! herbar. fol. α. fl. cap. p.* 262 (*partim*). *DC. p.* 71. *E. & Z.! n.* 1088. *Herb. Un. itin. n.* 690.

HAB. Mountains round Capetown, in Stellenbosch, Uitenhage, and Caledon. *Thunb.! E. & Z.! Pappe.* Piquetberg, *Drege*, 6812. Mar.–Jul. (Herb. Thunb., Sond., T.C.D., etc.)

A shrub 2–4 feet high, the branches and branchlets virgate. Leaves nearly sessile or on a petiole 2–3 lines long. Leaflets straight or incurved, 1–2 inches long, ½–1 line wide, the terminal somewhat longer. Axillary panicles short, the terminal ones mostly longer; peduncles slender, patent, the uppermost pedicels about 1 line or as long as the flowers. Petals oblong, twice as long as the calyx. Drupe the size of a pea, pubescent or sub-glabrous.

2. R. stenophylla (E. & Z.! n. 1094); branchlets *puberulous;* leaves sub-sessile or petiolate; leaflets sessile, *linear-lanceolate, acuminate, tapering to the base,* mucronate, with revolute margins, entire, or with one or a few sharp teeth, glabrous above, whitish-tomentose beneath; panicles axillary and terminal *pubescent;* pedicels capillary *R. rosmarinifolia, Thunb.! fl. cap. p. 262 (partim) herbar. fol. β. and γ. Sieber, fl. cap. exs. n. 216. R. lavandulæfolia, Presl. Bot. Bem. p. 42.*

VAR. β. **brevifolia**; leaves elliptico-lanceolate; acute, tapering at the base, entire or paucidentate at the top.

HAB. Among shrubs on Tablemountain and Hott. Holl. *E. & Z.! Drege!* 6813. Var. β. in Vanstaadensriviersberge, *Zey.!* 2228. Oct. (Herb. Thunb., Lehm., Sond.)

Distinguished from the preceding, to which it is much allied (and perhaps a variety), by the more robust habit, the larger and generally longer leaves, and the pubescent panicle. Leaflets in var. α, 1–2½ inches long, 2–3 lines wide; in var. β. 8–9 lines long, 2 lines wide, at the top with 2–4 short but acute teeth. The flowers and fruit pubescent, as in the preceding. *Drege n.* 6811, is a radical shoot of this or *R. incisa.*

3. R. angustifolia (L. spec. 382); branchlets *glabrous;* leaves petiolate; leaflets *petiolulate, lanceolate, mucronate,* somewhat tapering at base, smooth, shining, reticulate above, densely tomentose beneath, on both sides penninerved, with quite *entire, flat,* or slightly recurved margins; panicles terminal, longer than the leaves, pubescent; drupe sub-globose, puberulous. *Pluk. Alm. t. 219. f. 6. Thunb.! fl. cap. p. 263. E. & Z.! n. 1092. Sieb. exs. n. 217. Drege! herb. n. 6210. Zey.! n. 2229. R. argentea, Will. dict. n. 11.*

HAB. Common among shrubs in the districts of Cape, Stellenbosch and Worcester. Oct.–Dec. (Herb. Thunb., Lehm., T.C.D., Sond.)

A shrub, with brown branches. Petiole ½ inch or longer, flat above, lateral petioles 2 lines, the terminal 3–4 lines long. Leaflets 2–2½ inches long, 4–6 lines wide, green, with impressed middle nerve above, whitish or yellowish tomentose below, somewhat red in the prominent middle nerve. Flowers white, ½ line long. Panicles 2–5 inches long, pyramidal. Pedicels capillary, bracteate at the base. Drupe the size of a pea. In a few specimens the ultimate branchlets and petioles are thinly pubescent. Very similar to *R. lancea,* Thunb., but distinguished by the tomentose leaflets and the drupe.

4. R. discolor (E. Mey.! in Herb. Drege.); branches and petioles *tomentose and hirsute;* leaves petiolate, leaflets *sessile, lanceolate,* mucronate, quite entire, *adpressed-pubescent,* or sub-glabrous above, *softly reddish-tomentose below,* flat or with revolute margins, the terminal tapering at base; panicles *hirsute,* the axillary shorter than the leaves, the terminal longer, many-flowered; drupe globose, glabrous, shining. *R. rufescens, E. & Z.! 1093. non Hamilton.*

HAB. Mountains Winterberge, and at Chumiberg, Caffr. *E. & Z.!* Vanstaadensberg, Katberg, between Buffelrivier and Key, *Drege!* Magalisberg, *Burke*, 328, *Zey*, 336. Dec.–Jan. (Herb. Hook., Lehm., Sond., T.C.D.)

Very near the preceding, but differs by the soft, reddish tomentum, the sessile leaflets, shorter racemes and drupe. Petioles 3–6 lines long. Leaves about 3 inches long, 6–8 lines wide, acute or obtuse, mucronate, the lateral ones unequal at base. Drupe yellowish brown, the size of a small pea.

5. R. divaricata (E. & Z.! n. 1106); branches spreading, terete, *smooth*, branchlets and petioles *minutely-pubescent;* leaves on *longish* petioles; leaflets sessile, *ovate*, obtuse or emarginate, mucronulate, the terminal narrowed at base, the laterals smaller, sub-oblique, puberulous above, albo- or reddish-tomentose beneath, quite entire, with the margin slightly recurved or paucidentate; panicles axillary, small, *sparsi-flowered*, pubescent, shorter than the leaves or sub-equal; drupe globose, smooth. *R. subferruginata, Presl.*

HAB. On sides of the mountains at Klipplattriver, Tambukiland, *E. & Z.!* Witbergen between Gorip and Caledon river, *Zey. Tereb., n.* 6; *Drege, n.* 6796. Uitvlugt between Limœnfontein and Buffelrivier, 3–4000 f. *Drege.* Nov.–Jan. (Herb. Lehm., Sond.)

A shrub, seemingly small, much branched, ultimate branchlets short. Petioles 6–10 lines long, furrowed. Terminal leaflets ¾–1 inch long, 5–6 lines wide; the lateral ones about half as long, regularly penninerved, puberulous, at last glabrous above. Panicle about an inch long, shorter or longer, the rhachis flexuose, lateral branches sometimes few, 3–5 flowered, pedicels 1 line long. Flowers about ¾ lines long. Petals twice as long as the ovate, acute calyx. Drupe tipped with the three styles. Known from the very similar *R. refracta*, by the tomentose, not villous or glabrous leaves.

6. R. obovata (Sond.); branchlets smooth or puberulous; leaves petiolate, leaflets on very short petiolules, *obovate-cuneate*, obtuse-mucronulate, shortly dentate or rarely entire, glabrous above, whitish-tomentose beneath; panicles terminal, elongated, *tomentose;* flowers *sessile-glomerate;* drupe pubescent. *R. sinuata, E. & Z.! n.* 1111. *not of Thunb.!*

HAB. In the forests of Adow and Olifantshoek, Uitenh., Gaurits and Camptoursriver, George; Hassigaybosch, Albany, *E. & Z.! Zey.!* 2240. Zuurebergen, *Drege*, 6794. Oct. (Herb. Lehm., T.C.D., Sond.)

A shrub, 10 feet high, with spreading branches. Petiole 3–6 lines long, sub-terete, puberulous. Leaflets sub-sessile, the terminal 1–1½ inch long, 6–10 lines wide, with obtuse or acute teeth. Panicle 2–4 inches long, the branches naked at base, the flowers sessile. The drupe is scarcely the size of a small pea.

7. R. populifolia (E. Mey. in Hb. Drege); branches smooth or thinly pubescent; leaves petiolate; leaflets on *very short* petiolules, *triangular* or *rhombeo-sub-orbiculate*, obtuse or acute, dentate, appressed, puberulous or glabrous above, whitish-tomentose beneath; panicles terminal, puberulous; drupe oblique, puberulous, tipped by the styles.

HAB. At the mouth of the Gariep, *Drege!* Sept. (Herb. E. Mey. T.C.D., Sond.)

A shrub, with the habit of the preceding, but differing by the leaves and fruits. Petiole about 6 lines long, scarcely canaliculate above. Leaflets 1½ inch long, 1–1¼ inch wide, glabrous above, with yellowish, impressed nerves, on the lower surface with elevated nerves, the terminal larger, obtuse at base, or tapering into a petiole 2 lines long. Fructiferous panicles not longer than the leaves. Drupe wider than long (3 lines wide, 2 lines long.)

8. R. tomentosa (L. spec. 382); branches puberulous or glabrous; leaves petiolate; leaflets *elliptic or ovate, acute, or acuminated at both* *ends*, entire or *coarsely serrated* from the middle to the apex, smoothish above, whitish-tomentose beneath; panicles terminal, downy, longer than the leaves; drupe depressed, globose, pubescent. *Comm. hort. Amst. I. t.* 92. *Thunb.! prod. p.* 52. *Flor. cap. p.* 266. *DC. l. c. p.* 73. *E. & Z.! n.* 1109. *Herb. Un. itin. n.* 686. *Sieb. exs. n.* 155. *R. Ecklonis Schrad.! hort. Gœtt. R. elliptica, E. Mey.! Zey.! n.* 2232, 2234. *Pappe. Sylv. cap. p.* 13.

VAR. *β.* **petiolaris**; leaflets elliptic-oblong, acute, or acuminate at both ends, with long petiolules. *R. ellipticum, Thunb.! l. c. p.* 263. *R. bicolor, Licht.! in Herb. Willd. R. & Sch. Syst.* 6. *p.* 661. *R. discolor, Schrad.! hort. Gœtting.*

HAB. Everywhere on the slopes of the mountains of the Cape, Stellenbosch, Worcester, and Uitenhage districts. Jul.–Sept. (Herb. Thunb., Sond., T.C.D., &c.)

A shrub or tree, height of trunk 2–3 feet, diameter 3–4 inches. Branches somewhat patent, branchlets brownish-purple. Leaflets green and costate above, ribbed below, quite entire or with 3–6 teeth, the lateral ones smaller, the terminal 1½–2 inches long, 1–1¼ inch wide, in var. *β.* 1½–2½ inches long, 6–8 lines wide. Common petiole ½–1 inch or longer, sub-terete; petiolules in var. *α.* very short or wanting, in var. *β.* 3–6 lines long. Panicles mostly terminal, elongated; flowers minute. Fruit the size of a small pea.

9. R. incisa (L. fil. suppl. 183); branches *puberulous;* leaves petiolate; leaflets sessile, obovate, wedge-shaped, obtuse, *incised pinnatifid, lobes obtuse*, pubescent above, whitish tomentose beneath; panicles terminal, tomentose, pendulous, *as long as the leaves;* flowers sub-sessile; drupe *pubescent. Thunb.! prod. p.* 52. *fl. cap. p.* 267. *DC. l. c. p.* 72. *E. & Z.! n.* 1112.

HAB. Sandy places, Paardeberg, Saldanhabay and St. Helenabay, *Thunb.!* Brackfontein, *E. & Z.!* Between Paarlberg and Paardeberg, *Drege*, 6793. Jul.–Sept. (Herb. Thunb., Lehm., T.C.D., Sond.)

Shrub, 1–2 feet high, with rigid branches and horizontal, very short branchlets. Petioles puberulous, somewhat flat above, 4–6 lines long. Leaflets obtuse, mucronulate or emarginate, with impressed nerves above, ribbed below, the terminal one very cuneate, 6–10 lines long, 4–6 lines wide. Panicle 1–1½ inche, the upper flowers spicate. Ripe fruit unknown.

10. R. dissecta (Thunb.! fl. cap. 267); branches *glabrous;* leaves petiolate; leaflets sessile, *sharply cut, pinnatifid*, with revolute margins, glabrous above, whitish-tomentose beneath; panicles terminal, smooth; peduncles divaricate, capillary; drupe oblique, *smooth*, or somewhat *muricate.*

VAR. *α.*; leaflets obovate or sub-orbiculate, cuneate, incised-dentate, teeth mucronate. *R. argentea, E. & Z.! n.* 1127. *R. dissecta, E. Mey.! e. in Hb. Drege.*

VAR. *β.*; leaflets pinnatifid, lobes lanceolate, acute. *R. dissecta, Thunb.! E. & Z.! n.* 1128. *E. Meyer, a. and f.*

HAB. Sides of the mountains, Brackfontein, *E. & Z.!* Heerelogement, *Z. n.* 332. Blaauewberg, Piquetberg, *Drege!* Var. *β.* in Zwartland, *E. & Z.!* Groenekloof, *Beil.* Langvallei and Bergvallei, *Drege.* Jun.–Nov. (Herb. Thunb., E. Mey., Lehm., T.C.D., Sond.)

A small shrub, branches brown; branchlets filiform. Petiole ½–1 inch long. Leaflets green, with whitish veins above; in var. *α.* 6–9 lines long, the lamina 3–6 lines wide, with 3–7 very sharp teeth; in var. *β.* the same size, but deeply pinnatifid, lobes ½–1 line wide. Panicle 1–1½ inch, about 12-flowered, the ultimate pedicels about 2 lines long. Flowers greenish-yellow, about 1 line long. Drupe somewhat compressed, brownish, shining, 3–4 lines wide, 2–3 lines long.

11. R. magalismontana (Sond.); *branchlets, leaves, and panicles yellowish-tomentose;* leaves petiolate; leaflets sessile, obovate-oblong or oblong, obtuse, mucronulate, wedge-shaped, reticulated and sub-glabrous above; panicles many-flowered, terminal, a little longer than the leaves.

HAB. Crocodileriver near Magalisberg, *Zey.! n.* 341. Dec. (Herb. Sond.)

The whole plant is covered with a pale, golden down. Branches terete, striate, branchlets angulate. Petiole 6–9 lines long furrowed above. Leaflets with a prominent middle nerve on both sides; terminal 1½–2 inches long, 6–9 lines wide, the lateral ones smaller. Panicle compound; branches somewhat patent, the ultimate pedicels very short. Flowers imperfect in our specimens.

B. Leaves and panicles villous or pubescent, rarely the leaflets sub-glabrous. (Sp. 12–18.)

12. R. parvifolia (Harv. in Hb., T.C.D.); villous, branchlets short; leaves petiolate; leaflets sessile, obovate or elliptic, wedge-shaped, 3–7 *dentate at the point*, teeth mucronulate; panicles axillary, and terminal, short, of the *length of the leaves* or a little longer; calyx villous. *R. mollis, E. Meyer, non H. B. Kth.*

HAB. Buffelrivier, 1–2000 f. *Drege!* Near Port Natal, 3–4000 f. *Dr. Sutherland.* Jan. (Herb. E. Mey., T.C.D., Sond.)

A small shrub. Flowering branches 1–2 inches, the sterile ones longer and more yellowish. Petioles in the sterile branches ½–1 inch long, with leaves of the same length, and 6–10 lines wide; petioles in the flowering branches 2–3 lines long, with greenish leaves 3–5 lines long, 3–4 lines wide. Panicles pauciflowered, about ½ inch long; pedicels capillary, ½–1 line long. Petals white, twice as long as the calyx.

13. R. villosa (L. fil. suppl. 183); leaves petiolate; leaflets sessile, *obovate, obtuse*, mucronulate, *quite entire*, with revolute margins, *hairy or villous* on both surfaces, as well as the petioles and branchlets; racemes axillary, much *shorter* than the leaves, the terminal paniculated, somewhat longer; drupe orbicular, compressed, glabrous. *Thunb.! prod. p.* 52. *fl. cap. p.* 265. *DC. l. c. p.* 70. *Herb. Un. itin. n.* 687, 688. *E. & Z.! n.* 1098. *R. atomaria, Jacq. hort. Schonbr. t.* 343. *E. & Z.!* 1099.

VAR. β. **glabrata**; leaves sub-glabrous or glabrous. *R. pubescens, Thunb. prod. p.* 52. *Fl. cap. p.* 264. *R. pubescens, δ. subglabra, E. & Z.! n.* 1100, *var δ.*

HAB. Cape flats, mountains near Capetown and Grootvadersbosch, *Thunb. E. & Z.!* Klipfontyn, *Zey,* 347. Somerset, *Mrs. F. W. Barber.* Paarlberg, *Drege!* Var. β. Cape Recief, near Port Elizabeth, *Zey.! n.* 2236. Krakakamma, *E. & Z.!* Aug.–Jan. (Herb. Thunb., Sond., T.C.D., etc.)

Shrub or tree. Petioles ½–1 inch, furrowed above. Leaflets prominent-veined on both sides, 1½ inch long, 1 inch wide, or 1 inch long, 6–7 lines wide. Panicles compound, the male usually longer; the ultimate pedicels ½–1 line long. Flowers minute. Drupe the size of a peppercorn.

14. R. refracta (E. & Z.! n. 1103); branches and branchlets *horizontal or reflexed*, sometimes sub-spinous, pubescent; leaves petiolate; leaflets sessile, obovate, obtuse or acute, emarginate, the terminal one with a cuneate, tapering base, with revolute margins, *pubescent* on both sides or sub-glabrous, prominently veined below; panicles axillary and terminal, *filiform, elongated, lax, villous,* the female ones shorter; drupe globose, glabrous. *R. villosa, c. E. Mey. in Hb. Drege.*

HAB. Zwartkopsrivier, *E. & Z.!* Boshmansriviershoogde, *Zey.!* 2237. Draakensteenberg, 1–2000, *Drege.* Port Natal, *Gueinzius, Sanderson.* March.–April. (Herb. Lehm., T.C.D., Sond.)

Branches filiform. Petioles furrowed, 4–6 lines long. Leaflets pubescent or villous, often glabrous above, and sparsely hairy and paler beneath, the terminal one ¾–1 inch long, 6 lines wide. Panicles slender, lateral ones 1–2 inches, terminal 3–4 inches long, pedicels very short. Flowers white, minute. Distinguished from *R. villosa* by the much smaller, emarginate, not reticulated leaves, and the long, slender panicle; from *R. obovata* by the smaller not tomentose leaves, smaller flowers and glabrous fruit; from *R. crenata,* by the four times longer petioles.

15. **R. pyroides** (Burch. Trav. I. p. 340); leaves petiolate; leaflets sessile or very short-petiolate, sub-equal, *obovate-oblong,* obtuse or sub-acute, quite entire, as well as the branchlets, covered with close-pressed pubescence; racemes axillary, shorter than the leaves, terminal ones longer, disposed in an *elongated, ample, leafless, pubescent panicle;* drupe sub-globose, glabrous. *DC. l. c. p.* 70. *R. villosa, fol. β. Thunb. herbar. R. pubescens, E. & Z.! n.* 1100, *non Thunb. Zey. n.* 344. *Drege!* 6800.

VAR. *β.* **glabrata**; branches and leaves sub-glabrous or glabrous, panicles pubescent. *R. pyroides, E. Meyer.*

HAB. Asbestos mountains, *Burchell;* in the districts of Caledon, Worcester, Uitenhage, Albany, in Caffraria, and near Port Natal, *E. & Z.! Drege, Col. Bolton, Sanderson, Williamson, Guienzius.* Dec.–Feb. (Herb. Thunb., T.C.D., Sond, &c.)

A shrub, 6–10 feet high; branches sometimes spinous. Petiole ½–1 inch, furrowed above. Leaflets sessile, or the terminal one shortly petiolate, ovate, ovate-oblong, or obovate-cuneate, obtuse, acute, or somewhat acuminate, prominent-veined on both surfaces, about 2 inches long, 1 inch wide. Terminal panicle sometimes 6 inches long or longer, branches elongate, patent, pedicels capillary, flowers very minute, white, drupe (unripe) the size of a peppercorn. In some specimens some of the leaflets are a litttle toothed = var. *β. subdentata, E. Meyer.* Distinguished from *R. villosa* by the elongated, ample panicle and minute flowers, but perhaps only a variety.

16. **R. tridentata** (Sond.); branches, petioles, and leaves covered with close-pressed pubescence; leaves petiolate; leaflets sessile, or the terminal one shortly petiolate, *oblong,* with revolute margins, the *terminal one cuneate at the dilated point, 3–lobed,* the medial lobe larger, acute, the lateral ones obtuse, mucronulate; panicles terminal, elongated, leafless, pubescent.

HAB. Port Natal, *Gueinzius.* (Herb. Sond.)

Very like the preceding, but differs by the cuneate, 3–lobed or 3–dentate, terminal leaflet. Petiole 6–9 lines long. Lateral leaflets 1½–2 inches long, 6–7 lines wide, terminal 2½–3 inches long, at the top 1 inch wide. Panicle as large as in the preceding; flowers very minute.

17. **R. puberula** (E. & Z.! 1104); branches, petioles, leaves, and panicles, *sub-sericeous, pubescent;* leaves petiolate; leaflets *oblong, acute,* or elliptic-*lanceolate,* mucronulate, narrowed at base, quite entire, with revolute margins, veined on both surfaces, the terminal one somewhat larger; panicles terminal, the length of the leaves, the axillary ones shorter; drupe globose, glabrous. *R. sericea, E. & Z.! n.* 1105. *R. Meyeriana, Presl. l. c.*

Var. β. **fastigiata**; branches very numerous, fastigiate; leaflets small, cuneate, oblong, mucronate or acute, elevated-veined on both surfaces; panicles numerous, few-flowered. *R. fastigiata, E. & Z.! n.* 1109. *R. angustifolia d. E. Mey.; and with glabrous leaves, R. excisa, Thunb.! herb. fol. α. R. humilis, E. & Z.! n.* 1108.

HAB. Mountains in Hassagaybosch, *Albany*, and Schiloh near Klipplaatrivier, *E. & Z.! Drege!* 6808. Grahamstown and Cradock, near Tarkarivier, *Zey.! Terebinth. n.* 2. *and No.* 2239. Orangerivier, *Burke, Zey.! n.* 343. Port Natal, *Miss Owen.* Var. β. Krakakamma and Hassagaybosch, *E. & Z.! Zey.! n.* 2238. Omsamwubo, *Drege.* Nov.–Dec. (Herb. T.C.D., Hook., Lehm., Sond.)

A shrub, 6–8 feet high; branches sometimes spinous, branchlets often climbing. Petioles furrowed above, ½–1 inch long. Leaves sessile or the terminal one shortly petiolate; leaflets puberulous, or sub-glabrous above, paler beneath, 1½–2 inches long, 6–8 lines wide; in var. β. 9–12 lines long, 2–4 lines wide, the lateral ones smaller and more oblong. Axillary panicles twice shorter, terminal a little longer than the leaves. Flowers minute, white. Drupe sub-compressed, the size of a peppercorn. Distinguished from the preceding by the smaller leaflets and shorter panicles.

18. R. sinuata (Thunb.! Prod. p. 52); branchlets pubescent or villous; leaves petiolate; leaflets sessile, obovate or ovate, obtuse, *glabrous, sinuated* above, *villous* beneath, the terminal ones large, tapering at base; racemes axillary, spicate, equalling leaves, terminal ones paniculated, small, pubescent. *Willd. spec. t.* 2. *p.* 1482. *Myrica trifoliata, L. syst. veg.* 14. *p.* 884?

HAB. Cape of Good Hope, *Thunb.* (Herb., Thunb.)

Habit of *R. refracta*. Branches horizontal. Petiole filiform, 4–6 lines long. Terminal leaflet about an inch long, 5–6 lines wide, sinuate-dentate, teeth obtuse, lateral leaflets twice or three times shorter. Axillary racemes an inch long, flowers on short pedicels, mostly 2 or 3-glomerate; terminal racemes branched, 2–3 inches long. Flowers white, ½ line long. Drupe lentiform, of the size of a peppercorn.—Of this plant, I only know the specimen in Herb. Thunberg. *R. Thunbergiana*, E. & Z.! n. 1110, is very similar, but the leaflets are deeply incised and acute; flowers and fruits are wanting.

C. Leaves and panicles quite smooth; or the panicles somewhat hairy. (Sp. 19-44.)

a. Petioles terete, furrowed above, but not winged. (Sp. 19–33.)

19. R. cuneifolia (Thunb.! Prod. p. 52); branches *puberulous;* leaves sessile; leaflets sessile, wedge-shaped, with 3–7 teeth at the apex, smooth, with prominent veins, and green on both surfaces; panicles axillary and terminal, the female as long as the leaves, the male longer; drupe oblique, ovate, glabrous, beaked with the styles. *Fl. cap. p.* 267. *E. & Z.! n.* 1131. *DC. l. c. p.* 71.

HAB. Hottentottsholland, *Thunb.*, Grietjesgat and Steenbrassenrivier, Langehoogde and Boutjeskraal, *E. & Z.!* Sept. (Herb., Thunb, Lehm., Hook., Sond.)

An erect shrub, with terete, purplish, somewhat patent branches. Leaflets rigid, coriaceous, strong ribbed, cuneate; teeth mostly acute, about one inch long, ½ inch wide. Panicles small, 1–3 inches long. Flowers minute, of the length of the pedicels. Drupe shining, nearly 3 lines long.

20. R. crenata (Thunb.! fl. cap. p. 266); branchlets striate, *rufous-villous;* leaves on very short petioles; leaflets sessile, obovate, cuneate, with 3–5 blunt teeth at the apex, glabrous on both surfaces, but *rufous beneath;* axillary racemes rufo-puberulous, shorter than the leaves, the terminal ones longer, paniculated. *E. & Z.! n.* 1123.

HAB. Algoabay, Cape Recief, *E. & Z.!* Oct.–Jul. (Herb., Thunb., Lehm., T.C.D. Sond.)

A much branched shrub, 2 feet high, with terete, shortly villous or tomentose branches. Petiole mostly 1 line, rarely 2–3 lines long. Leaflets green, somewhat glaucous above, thinly veined beneath, crenately trifid, rarely with 5 or 7 obtuse teeth at the top, with revolute margins, 6–9 lines long, 4 lines wide. Axillary panicles ½ inch, the terminal ones, 1–1½ inch long. Flowers very minute, very shortly pedicellate, in the branchlets sub-crowded. Drupe brown-purplish, shining, size of a small pea.

21. R. dentata (Thunb.! fl. cap. p. 266); branches glabrous; leaves petiolate; leaflets sessile, obovate, coarsely inciso-dentate, cuneate at base and quite entire, glabrous on both sides; axillary racemes short, terminal ones longer, paniculated; drupe globose, glabrous. *R. micrantha, Thunb.! l. c. E. & Z.! n.* 1126. *R. grandidentata, DC. l. c. p.* 72. *R. dentata and cuneifolia, E. Meyer.*

VAR. β. **puberula**; branchlets, petioles, and panicles puberulous; leaves glabrous or ciliate. *R. crenata, Thunb.? Drege, Herb.*

HAB. In woods, Vanstadensriviersberg, Uitenh. Zwartehoogdens at Salatkraal, Albany. Chumiberg, Caffr. Schilo, Tambukiland, *E. & Z.!* Hassagaybosch, *Zey. n.* 2231. Witbergen, 4–5000; *inter* Windvogelberg and Zwartkey, *Drege.* Var. β. in Zuurebergen, *Drege.* Oct.–Jan. (Herb. Thunb., Lehm., T.C.D., Sond.)

An erect shrub, with scabrous branches and short, rigid, or elongated, often unilateral branchlets. Petioles 6–9 lines long, or shorter. Leaflets sessile or a little petiolulate, of the same colour on both sides, or paler beneath, with prominent veins, the margins pale, whitish, teeth 3–7, mucronate; terminal leaflet very cuneate, in some specimens 1–1½ inch long, 8–10 lines wide, in others 5–6 lines long, 3–4 wide (var. parvifolia, E. & Z.!); lateral leaflets smaller, ovate or obovate. Axillary racemes about ½ inch, the terminal ones 1–2 inch, more compound, with patent lower branches; pedicels capillary. Flowers small. Drupe shining, the size of a small pea.

22. R. mucronata (Thunb.! fl. cap. p. 264); branches punctate-scabrous; branchlets, petioles, and leaves *quite glabrous;* leaves petiolate; leaflets sessile, *obovate-cuneiform,* obtuse, mucronate, or sub-acute, subcoriaceous, reticulate, paler or livid beneath, *quite entire,* with *revolute margins;* axillary panicles shorter than the leaves, terminal panicles longer and pyramidal; drupe globose, glabrous.

VAR. α. **Burmanni**; leaflets obovate, obtuse, mucronate; panicles quite glabrous. *Burm. Afr. t.* 91. *f.* 2. *R. mucronata, Herb. Thunb. fol.* α. *R. Burmanni, DC. l. c. p.* 69. *Herb. Un. itin. n.* 683. *R. Burmanni et pendulina, E. & Z.! n.* 1101. 1102. *R. Eckloniana, tenuiflora, and pilipes, Presl. l. c. R. lucidum* α. *Ait. Hort. Kew, ed.* 2. *p.* 166. *ex DC.*

VAR. β. **Jacquini**; leaflets obovate, shortly acuminate, acute or mucronulate; panicles glabrous or a little hairy. *R. mucronata, Herb. Thunb. fol.* β. *R. elongata, Jacquin, Hort. Schoenb. t.* 345.

HAB. Sides of the mountains, Tafelberg and Leeuwenberg; Krakakamma, Cape Recief, *E. & Z.! Zey. n.* 2236; Koeberg, *Dr. Pappe;* Vanstadensrivier, *Drege,* 6798; Paarlberg, *Drege,* 6801; Magalisberg, *Burke, Zeyh. n.* 340. Nov.–Jan. (Herb, Thunb., Jacq., Lehm., Hook., T.C.D., Sond.)

Shrub erect; branches greyish or purple; branchlets short or virgate. Petiole ½–1 inch. Leaflets pale or livid beneath, the terminal one 1–1½ inch long, 6–9 lines wide, rarely 2 inches long, 1 inch wide; larger and more cuneate than the lateral ones. Terminal panicle 2–4 inches long, pyramidal, glabrous; pedicels capillary, ultimate ½–1 line long. Flowers minute, whitish. Drupe the size of a small pea.

23. R. Burkeana (Sond.); branches terete, striate, *thinly pubescent;* leaves petiolate; leaflets sessile, *oblong, mucronate, cuneate at base*, glabrous, quite entire, *flat*, reticulated on both sides, paler beneath; axillary panicles small, shorter than the leaves, the terminal one longer, glabrous.

HAB. Aapgesrivier, among rocks, *Burke, Zeyh. n.* 335. Oct. (Herb. Hook., Sond.)

Apparently a small shrub, with erect, purple branches. Petioles puberulous, terete, striate below, furrowed above, ½ inch long. Leaflets sub-coriaceous, pale, with prominent veins, 1½–2 inches long, 4–6 lines wide. Terminal panicle compound, 1½–2 inch, pyramidal. Flowers minute, glabrous.—Fruit I have not seen.

24. R. Zeyheri (Sond.); quite glabrous, sub-glaucous; branches terete; branchlets striate-angulate; leaves petiolate; leaflets sessile, broadly *obovate or elliptic-orbiculate, shortly cuspidate*, acute, cuneate at base, *margin* penni-nerved on both surfaces; racemes glabrous, the terminal one longer, paniculated; drupe globose, glabrous.

HAB. Among shrubs at Magalisberg, *Zeyh. n.* 345. Nov. (Herb., Sond.)

Branches sub-papillose; branchlets purple, quite smooth. Petiole terete, a little furrowed above, ½–1 inch long. Leaflets sub-coriaceous, about 1½ inch long, an inch wide, penni-nerved but not reticulate, the lateral ones elliptic, acute at both ends. Axillary racemes pedunculate, 1–2 inches long, 8–12-flowered, pedicels 2 lines long; the terminal one as long as, or a little longer than, the leaves. Flowers unknown. Fruit the size of a small pea.

25. R. laevigata (L. spec. p. 1672); glabrous; leaves long-petioled; leaflets sessile or shortly petiolate, *ovate, acuminate*, mucronate, entire, with sub-undulate margins, paler beneath; panicles axillary and terminal, *elongated, lax;* drupe globose, glabrous. *Thunb. prod. p.* 52. *Fl. cap. p.* 264. *E. & Z.! n.* 1096. *Pappe, Sylv. cap. p.* 12. *R. laevigata et acuminata, E. Meyer in Herb. Drege. R. crassinervia, Presl. l. c.*

VAR. β. **dentata** (E. Meyer); leaves with 2–6 acute teeth.

HAB. Grootvadersbosch, *Thunb., Mundt., Zeyher.* Zitzikamma, Krakakamma, and Vanstadesberg, *E. & Z.! Zeyh. n.* 2233. Grahamstown and Hassagaybosch, Enon., Zuureberge, *Drege.* Port Natal, *Krauss.* β. near Gekau, *Drege.* Sep.–Jan. (Herb. Thunb., T.C.D., Lehm., Sond.)

Shrub, 5 feet or more. Bark thin, rough. Wood hard, reddish, called Essenhout (Ash-wood) in the Western districts, and Bosganna by the natives in the Eastern. Branches purple, branchlets elongate. Petiole 2–3 uncial, furrowed above; petiolules of the lateral leaflets ½–1–2 lines, of the intermediate leaflets 3–4 lines long. Leaflets elevated-veined, shining above, acute at base, 2–3 inches long, 1–1½ inch wide, the margins mostly quite entire, in var. β. with some acute, mucronate teeth. Panicles compound, loose; pedicels capillary. Flowers minute, yellowish-white. Drupe shining, the size of a pea.

26. R. lancea (L. f. suppl. p. 184); *quite glabrous;* leaves on long petioles; leaflets sessile, *lanceolate-linear*, acuminate, mucronate, quite entire, paler and parallel-veined beneath; panicles axillary, *shorter* than the leaves; drupe lentiform, glabrous. *Thunb.! prod. p.* 52. *Fl. cap. p.* 263. *DC. l. c. p.* 70. *E. & Z.!* 1091. *R. viminalis, Jacq. hort. Schoenb. t.* 344. *E. & Z.!* 1089. *E. Meyer in Hb. Drege. R. fragrans, Lichtenst.! in R. et Sch. syst. veg. VI. p.* 661 (*flowering specimen*). *R. denudata, Licht.! l. c.* (*fructiferous*).

HAB. Grootvadersbosch, Hantum, Roggeveld, and Bokkeveld, *Thunb.;* Bosjesmansrivier, Uit., Vishrivier, near Albany; Katrivier, near Fort Beaufort, *E. & Z.!*

Somerset and in Zwartebergen, *Zeyh. No.* 334, *and Terebinth. No.* 1. Sneeuwbergen, *Drege.* Sep.-Dec. (Herb. Thunb., Vind., Lehm., T.C.D., Sond.)

A little tree with the habit of a Willow. Bark grey and smooth. Wood reddish-brown, hard and very tough. Branches virgate. Petioles terete, furrowed above, 1-2 inches long. Leaflets 4-5 inches long, 4-6 lines wide or smaller, sessile or the intermediate one shortly petiolulate. Panicles glabrous, 1-2 uncial, of the length of the leaves or shorter; pedicels capillary. Flowers very minute. Drupe rather dry, size of a pea.

27. R. viminalis (Vahl. Symb. iii. p. 50); branches glabrous; leaves on long petioles, leaflets sessile, *lanceolate-acuminate,* mucronate, quite entire, glabrous or ciliate, *sub-concolorous* and parallel-veined on both surfaces; panicles racemose, axillary and terminal, *hirsute, shorter* than the leaves; drupe glabrous. *Thunb.! fl. cap. p.* 263. *R. denudata, E. & Z.! n.* 1090 (*non Lichtenstein*). *R. elongata, E. & Z.! n.* 1097, *not of Jacquin.*

VAR. β. **pendulina**; leaflets lanceolate, acute, or acuminate; panicles loose. *R. pendulina, Jacq.! Willd. enum. p.* 324. *R. pallida, E. Meyer. R. pendulina, glabrior, Presl. l. c.*

HAB. 24-Rivers, Clanw. and in gardens, Capetown, *E. & Z.!* Var. β. Giftberg, and near the mouth of the Gariep, *Drege.* Nov. (Herb. Thunb., Vind., Lehm., Sond.)

Very much resembling the preceding, but differs by the lax, often pendulous branches, the concolorous, or on the lower surface scarcely paler, and shorter leaflets, and by the hirsute peduncles and calyces. Leaflets 2-2½ inches long, 5-7 lines wide, attenuated to the point, acute at base; in var. β. 2½-3½ inches long, 4-6 lines wide, obtuse, apiculate, or attenuated. Panicles 1-1½ uncial, the terminal one longer, pyramidal. Drupe unripe, sub-globose.

28. R. Gueinzii (Sond.); *quite glabrous;* leaves petiolate; leaflets sessile, *lanceolate-obtuse, mucronulate,* with *revolute, remotely denticulate margins,* paler beneath, parallel-veined on both sides; panicles axillary and terminal, *longer* than the leaves, with slender branches and capillary pedicels.

HAB. Port Natal, *Dr. Gueinzius.* (Herb. T.C.D., Sond.)

A shrub, with terete, greyish branches; flowering branchlets short. Petioles sub-terete, furrowed above, ½-1 inch long. Leaflets 2-2½ inches long, 4-6 lines wide, particularly in the upper part denticulate, a few narrowed at base, the lateral ones mostly half as long. Lateral panicles about 3 inches long, ultimate pedicels ½ line long. Flowers very minute.—Differs from *R. viminalis* and *lancea* by the obtuse, denticulate leaves and longer panicles.

29. R. Natalensis (Bernh.! in Krauss. Beitr. p. 46); quite glabrous; leaves petiolate; leaflets sessile, oval, obtuse, emarginate, crenate-serrate, cuneate at base, penninerved, paler beneath; racemes axillary, simple, or sub-paniculate, shorter than the leaves; drupe sub-globose, glabrous. *Cissus Natalensis, Bernh. in Sched. pl. Krauss.*

HAB. Woods near Port Natal, *Krauss, Gueinzius, Williamson.* Aug. (Herb. T.C.D., Sond.)

Branches terete, greyish, punctate, scabrous. Leaves 1½-2 inches long, 9-12 lines wide. Petioles terete, furrowed above, ¾-1 inch. Racemes ½-1½ uncial, simple or at the top paniculated; pedicels capillary, 1-1½ lines long. Flowers minute, white. Calyx and petals obtuse. Drupe yellowish-brown, of the size of a small pea.

30. R. Eckloniana (Sond.); quite glabrous; leaves petiolate; leaflets sessile, oblong or *lanceolate-oblong,* mucronate, cuneate at base, quite entire, *whitish-marginated, carinate above,* paler beneath; axillary racemes sub-paniculate, of the length of the leaves or shorter, terminal panicle somewhat longer; drupe globose, glabrous. *R. angustifolia? a. Thunb.! Herb. R. tridactyla, E. & Z.! n.* 1095, *non Burchell.*

HAB. Zwartkopsrivier, *Zeyh. n.* 2230. Hassagaybosch, *Zeyh. Terebinth. n.* 4. Bothasberg, at Great Vishriver, *Ecklon.* Howisonspoort, *H. Hutton.* April. (Herb. T.C.D., Sond.)

Branches purple. Petioles sub-terete, furrowed above, 6–8 lines long. Leaflets nearly complicate, oblong-lanceolate, 1½ inches long, 4 lines wide, or linear-oblong, cuneate, obtuse, sub-emarginate or acute, inch long, 3 lines wide, coriaceous, penninerved, paler or livid beneath. Racemes ½–2 inches long, small, panicled. Drupe shining, yellowish, size of a pea, on a pedicel a little shorter.

31. R. tridactyla (Burch. trav. i. p. 340); branches stiff, spreading, unarmed, leaflets smooth, quite entire, linear, *very blunt, broadest towards the apex. DC l. c. p.* 71.

HAB. Asbestos mountains, *Burchell.* (Unknown to me.)

Shrub, 4–5 feet. Flowers greenish-yellow.

32. R. erosa (Thunb.! fl. cap. p. 263); quite glabrous; leaves long-petioled; leaflets sessile, *linear* or lanceolate-linear, acute, *erosely toothed, flat, shining;* panicles axillary, shorter than the leaves; drupe globose, glabrous. *R. serræfolia, Burch. trav. ii. p.* 100. *DC. l. c. p.* 71.

HAB. Klipplaatrivier, Tambukiland, *E. & Z.!* Winterfeld, between Beaufort and Graafreynet, *Drege.* Zuureberg, distr. Cradock, *Zeyh. n.* 346. Nov.–Dec. (Herb. Thunb., Lehm., T.C.D., Holm., Sond.)

Shrub, 5–6 feet, much branched, branches sub-flexuose, angulate. Petioles sub-terete, furrowed above, an inch long or longer. Leaflets 3–3½ inches long, 1½–3 lines wide, with the middle nerve on both sides prominent, reticulate, undulate-dentate, teeth short, acute; the lateral leaflets incurved, erect, somewhat shorter than the terminal. Panicles about 2 inches long, di- or trichotomous, pedicels sub-flexuose, ultimate short. Drupe yellowish, shining, size of a peppercorn.

33. R. Dregeana (Sond.); quite glabrous; leaves petiolate; leaflets sessile, *small, linear,* mucronate, with incurved, entire margins, *concave above, convex beneath;* panicles terminal, pyramidal, of the length of the leaves or longer; pedicels *pendulous;* drupe globose, glabrous. *R. lancea, E. Meyer, non Thunb.*

HAB. Stormberge and Sneeuwberge, 4–5000 feet, *Drege.* Graafreynet, *Mrs. F. W. Barber.* Sep.–Dec. (Herb. Hook., T.C.D., Sond.)

Shrub, much branched, pale, with greyish branches and angulate branchlets. Pedicels sub-terete, furrowed above, 4–6 lines long. Leaflets uninerved, 2–3 inches long, 1 line wide, the lateral ones shorter, somewhat spreading. Panicles in some specimens inch long and wide, in others 2 inches long, pyramidal; branches horizontal, flexuose; pedicels capillary, 1 line long. Flowers small, white. Drupe yellowish, shining, the size of a small pea.

b. Petioles with narrow wings. (Sp. 34–44.)

34. R. glauca (Desf.! arb. 2. p. 326); glabrous; branches short, sub-angulate; leaves petiolate; leaflets sessile, *obcordate,* quite entire, flat, covered with glaucous powder or smooth; panicles terminal, short; drupe globose, glabrous. *Pers. syn.* 1. *p.* 326. *Thunb.! fl. cap. p.* 265. *R. glauca et Thunbergiana, R. et Sch. syst. veg. vi. p.* 657. *DC. l. c. p.* 69, *E. & Z.! n.* 1120, 1121. *Herb. Un. itin. n.* 685. *Sieb. fl. cap. exs. n.* 218. *R. lucida, E. Meyer et n.* 116 *in Herb. Drege. Zeyh. n.* 2241.

HAB. Cape flats and in mountains near Capetown; Kampsbay and Zwartkopsrivier. May.–Jun. (Herb. Thunb., Sond., T.C.D., &c.)

A shrub, with greyish branches and sub-flexuous branchlets. Petioles 3–4 lines long, small-winged. Leaflets thick, parallel-nerved, flat or with sub-revolute margins, glaucous, pulverulent, or resinous, shining, green or livid beneath, the terminal one very cuneate, 6–8 lines long, 5–6 lines wide, emarginate, with or without a mucro, sometimes obtusely 3-dentate; the lateral ones smaller. Panicles 1–1½ inch or longer. Flowers minute. Drupe reddish, shining, the size of a small or large pea.—Specimens from the Paris bot. gardens, sent by Desfontaines to Prof. Schrader, are not different from those in the herbarium of Thunberg.

35. R. lucida (Linn. spec. 382); glabrous, or branches pulverulent-puberulous; leaves shortly petiolate; leaflets sessile, obovate, quite entire, very blunt, somewhat emarginate, quite smooth, glossy, *without prominent nerves* on both sides or above; panicles axillary and terminal, *shorter* or a *little* longer than the leaves, glabrous; flowers hermaphrodite; drupe globose, glabrous. *R. lucida, β. Ait. hort. Kew. ed. 2. vol. 2. p. 166. Commel. hort. i. t. 93. Thunb.! fl. cap. p. 264. Jacq. hort. Schoenb. t. 347. DC. Prod. 2. p. 69. E. & Z.! n.* 1113. *Drege, n.* 6802, 6807. *Zeyh. n.* 2235. *Herb. Un. itin. n.* 684. *Pappe. sylv. cap. p.* 13.

VAR. β. **subdentata** (DC. l. c.); some of the leaflets a little toothed. *Drege, n.* 6791.

VAR. γ. **elliptica**; leaflets elliptic, obtuse, cuneate at base, livid and veined beneath.

HAB. On hills and mountains, amongst shrubs, in the Cape and Swellendam districts; Klynriviersberge, Caledon; Howison's Poort, Paarlberg, Bockfontein, and Onrustrivier. β. Rondebosch. γ. Downs near Onrustrivier, *Zeyh. n.* 2248. Aug.–Nov. (Herb. Thunb., T.C.D., &c.)

Height 4–6 feet. Wood hard and tough. Branches terete, greyish. Petioles 3–6 lines long, carinate, with small wings. Leaflets coriaceous, resinous, shining, commonly without nerves and veins, rarely on the under surface somewhat ribbed, obtuse, retuse, rarely with a blunt, short acumen, very unequal, in var. γ. 1–1½ inch long, inch wide, some 2–2½ inches long, inch wide, others no more than an inch long and 6–8 lines wide. Axillary panicles or racemes shorter or of the length of the leaves, the terminal ones longer. Flowers minute, whitish. Drupe the size of a pea.

36. R. scytophylla (E. & Z.! n. 1130); glabrous or branches puberulent; leaves short-petiolate; leaflets sessile, *obovate-cuneate*, emarginate, mucronulate, quite entire, with *revolute* margins, coriaceous, shining, *with raised veins on both sides;* panicles axillary, *longer* than the leaves, glabrous or puberulous; flowers *diœcious;* drupe *globose*, glabrous.

HAB. Mountains near villa Grietjesgat near Palmietrivier, *E. & Z.!* Swellendam, *Mundt.!* Hott. Holl. *Zey.! n.* 2247. Mar.–Jun. (Herb. Lehm., T.C.D., Sond.)

Very similar to the preceding, differs by the more coriaceous ribbed leaflets, longer panicles, and very minute, diœcious flowers. Petioles 3–6 lines long, with small wings. Leaflets 1 inch long, 6 lines wide, or 1½–2 inches long, 9–12 lines wide, very obtuse or truncate-retuse, with a very short mucro; the lateral ones erect, smaller, unequal at base. Panicles 2 inches or longer, upper branches 2–4 lines long, the lower ones longer; pedicels capillary. Fruit as in *R. lucida.*

37. R. Africana (Mill. dict. n. 11.); branchlets thinly pubescent; leaves short-petiolate; leaflets sessile, *ovate*, sub-acute or emarginate, shortly mucronate, cuneate at base, coriaceous, with *prominent veins*, downy or glabrous, shining, with revolute margins, quite entire or *with*

3–7 *short teeth;* panicles axillary, puberulous, shorter than the leaves; flowers diœcious; drupe *ovate-oblique,* glabrous, beaked by the persistent styles. *DC. l. c. p.* 73.

VAR. β. **macrophylla**; leaflets 2–2½ inches long, 1–1½ inch wide, very large, ovate or obovate, cuneate, acute or obtuse, mucronate, entire or sub-dentate, downy or glabrous. *R. mucronata, E. & Z.! n.* 1129.

HAB. Witsenberg, *Ecklon!* Var. β. Brackfontein and Tulbaghsberge, *E. & Z.!* Heerelogement, *Zeyher!* Between Honigvalei and Blauwberg, *Drege! n.* 6806. Dec. (Herb. Sond.)

A low, much-branched shrub. Petioles small-winged, 2–3 lines long. Leaflets ¾–1 inch long, 4–7 lines wide, more or less acute, in some specimens quite entire, in others in the upper parts toothed; lateral leaves shorter, sub-horizontal. Flowers minute. Drupe yellowish, the size of a pea.

38. R. scoparia (E. & Z.! n. 1122); *branches, petioles, and panicles downy;* leaves very short-stalked; leaflets sessile, cuneate-obovate, emarginate, with revolute margins, penninerved beneath, glabrous; panicles *axillary and terminal,* longer than the leaves; drupe glabrous.

HAB. Woods in Olifantshoek, near Bosjesmansrivier, *Uitenh. E. & Z.!* Nieuweveldsberge, *Drege! n.* 6803. Oct. (Herb. Lehm., Sond.)

Very similar to *R. lucida,* but differs by the short-stalked or sub-sessile leaves, the smaller (about 1 inch long, 6 lines wide) leaflets and longer pubescent panicles. Panicles many and minutely-flowered, twice or three times longer than the leaves. In the specimens of Drege the leaflets are more coriaceous and smaller (6–8 lines long, 4–5 lines wide) than in those of E. & Z.

39. R. undulata (Jacq.! hort. Schœnb. t. 346); *glabrous;* branchlets angulate; leaves longish-stalked; leaflets sessile, *obovate, tapering to the base,* acute, obtuse, or emarginate, mucronulate, with *undulate-denticulate margins,* or *incised,* 5–7 *dentate, glossy* above, parallel-veined on both surfaces; panicles axillary and terminal, equalling the leaves or longer; drupe globose, glabrous. *R. aglaophylla, E. & Z.! n.* 1117. *R. spathulata, E. & Z.! n.* 1119. *R. micrantha, E. & Z.! n.* 1124. *R. excisa, E. Mey.! E. & Z.! n.* 1125. *non Thunb. Sieber, flor. cap. exs. n.* 154. *R. nervosa, Poir? Enc. suppl. V. p.* 264.

HAB. Among shrubs at Constantia, Tulbagh, Brackfontein, Heerelogement, Zwartkopsrivier, Bothasberg near Vishrivier, *E. & Z.! Zey.! n.* 338. Klein Winterhoek, inter Gekau and Basche, Fort Beaufort, *Drege!* 6799. Jun.–Aug. fruct. Dec. (Herb. Vind., Lehm., Sond.)

Very similar to *R. lucida,* of which it has the habit, but the leaves are longer-stalked, the leaflets thinner, smaller and undulate, mostly dentate, and the panicles longer. From the preceding it is distinguished by the glabrous, resinous, shining, long-stalked leaves. Branchlets patent. Petiole ½–1 inch long, winged. Leaflets membranaceous, with pellucid veins, wedge-shaped, obtuse, or sub-acuminate, 1–1¾ inches long, 6–10 lines wide, the lateral ones smaller and more oval. Male panicle 3–4 inches long, lower branches 1½–2 uncial; pedicels capillary; female or hermaphrodite panicles shorter and less compound. Drupe shining, sub-compressed, the size of a small pea.

40. R. excisa (Thunb.! fl. cap. p. 264); glabrous; branchlets *angulate or two-edged;* leaves petiolate; leaflets sessile, *cuneate-oblong* or obovate-oblong, obtuse, acute, or emarginate, *flat, quite entire,* ribbed on both surfaces, paler beneath, sub-coriaceous; panicles terminal, somewhat longer than the leaves; drupe globose, glabrous.

VAR. α. **Thunbergiana**; leaflets oblong, cuneate, mostly obtuse, mucronulate. *R. excisa, Thunb.! herbar. fol. β. R. nervosa, E. & Z.! n.* 1115. *non Poir. Zey.! n.* 2242, *et Tereb. No.* 4.

VAR. β. **pallens**; leaflets obovate-oblong, acute, or obtuse, mucronulate, paler beneath. *R. pallens, E. & Z.! n.* 1114. *R. mucronata, E. Meyer.*

VAR. γ. **emarginata**; leaflets obovate-oblong, obtuse, emarginate, mucronulate. *R. plicæfolia, E. & Z.! n.* 1118. *Z. n.* 2243.

HAB. Var. α. in the districts Uitenhage and Albany, *E. & Z.!* Nieuwe Hantam, *Drege!* 6804. *b.* non *a.* and 6809, *ex pte.* Var. β. between Coega and Zondagsrivier, Zwartkopsrivier and Langevalei. Var. γ. Zwartkopsrivier, Bethelsdorp, and Sondagsrivier. Aprl.–Jul. (Herb. Thunb., Hook., T.C.D., Lehm., Sond.)

Shrub grey or brown, erect, with rigid branches. Petioles winged, particularly in the upper part. Leaflets of var. α. 8–10 lines long, 3–4 lines wide; in var β. 1½ inch long, 6 lines wide; in var. γ. 1–1½ inch long, 5–6 lines wide; lateral leaflets smaller, somewhat patent, oblong. Panicles terminal, rarely axillary, many-flowered, the male ones longer. Flowers minute. Drupe white, shining, somewhat compressed, the size of a small pea. Differs from the preceding by the 3 times smaller, not undulate or incised-toothed leaflets.

41. R. albomarginata (Sond.); quite glabrous; branchlets angulate-compressed; leaves petiolate; leaflets *obovate-oblong or oblong, plaited, mucronate,* sub-cuneate at base, veined on both sides, coriaceous, *white-margined;* panicles axillary, as long as the leaves.

HAB. Slaaykraal, *Burke.* Kowie River, *Zeyher.* Nov. (Herb. Hook., Sond.)

Very similar to the preceding; differs in the thicker, coriaceous leaflets, usually three times larger, and always with a smooth, white, revolute margin, and particularly by the smaller axillary, very rarely terminal panicles. Petioles winged, 1 inch long or shorter, 1½ line wide. Intermediate leaflets 2–3 inches long, about an inch wide, others 1–1½ uncial. Flowers minute. Fruit wanting.

42. R. ciliata (Lichtenstein in herb. Willd.!); branchlets puberulous, spinescent; leaves petiolate; leaflets sessile, *linear-lanceolate,* flat, quite entire, as well as the petioles, *ciliate,* puberulous beneath; panicles terminal, *puberulous;* drupe glabrous. *R. & Sch.! syst. veg. VI. p.* 661. *DC. l. c. p.* 71.

HAB. Grooterivierspoort, *Licht.!* Rhinosterkop, near Vaalriver, *Burke,* 275. *Zey.!* 337. Orangeriver, *Zey.!* 339. Nieuwe Hantam, *Drege!* 6804. *a.* Dec.–Feb. (Herb. Hook., T.C.D., Sond.)

Habit of *Celastrus linearis.* Branches patent, scabro-punctate, branches nearly horizontal. Leaflets 1 inch long, 1½–3 lines wide, shortly acute, uninerved with inconspicuous veins, shining, the lateral ones smaller; petiole 4–6 lines long, small-winged. Flowers minute, sometimes diœcious, racemose-panicled or the lower racemes simple; pedicels patent, 1 line long; the upper racemes disposed in a compound panicle. Fruit unripe, sub-globose.

43. R. celastroides (Sond.); *quite glabrous;* branchlets spinescent, as well as the stalked leaves, resinous, shining; leaflets *lanceolate, subcarinate, undulate,* with a short, recurved or *twisted acumen,* cuneate at base; panicles terminal, *glabrous.*

HAB. On rocks, Kammas, Betchuanaland, *Zey.!* 333. Mar. (Herb. Sond.)

Habit of the preceding, differing by the shining, undulate, shorter leaflets. Leaves sometimes fasciculate; some of them 1 or 2 foliolate, sessile, the others stalked, perfect, 3-foliolate. Intermediate leaflet about 6 lines long, 1½ line wide, with quite entire, undulate margins, uninerved, few-veined beneath; lateral leaflets similar but smaller. Petioles 4–6 lines long, carinate, subulate. Panicle uncial, with patent branches. Flowers very minute.

44. R. rigida (Mill. dict. n. 14); quite glabrous; branches sulcate; leaves petiolate; leaflets sessile, *cuneate-lanceolate*, or linear-lanceolate, *shortly acuminate, stiff*, entire, or with 1–2 short, acute teeth at the apex, coriaceous, uninerved, concoloured on both surfaces; racemes axillary or sub-terminal, simple or panicled, glabrous, twice or three times *shorter than the leaves;* drupe *ovate*, glabrous, beaked with the persistent styles. *DC. l. c. p.* 71. *R. rimosa, E. & Z.! n.* 1134. *R. triceps, E. Meyer.*

VAR. β. **florida**; racemes panicled, as long as the leaves or longer.

HAB. Mountains near Heerelogement, *E. & Z.!* Vierentwintigriviers, *Zeyher.* Olifantsrivier, *Drege!* 1967. Var. β. Giftberg, *Drege!* 6797. Nov.–Dec. (Herb. Lehm., Hook., Sond.)

An erect, rigid shrub, branches grayish, branchlets purple, resinous. Leaflets somewhat glaucous, acute, mostly recurved, apiculate, 1½–2 inches long, 3–4 lines wide, the lateral ones similar, but smaller. Petioles small-winged, 4–6 lines long. Racemes 6–8 lines long, few-flowered, rarely branched at base. Flowers minute. Drupe about 3 lines long. Panicles in var. β. 1½–2 inches long. Flowers as in var. α. Fruits are wanting.

D. Young branches, *leaves and panicles* covered with red glands. (Sp. 45–46.)

45. R. horrida (E. & Z.! n. 1135); the patent, spinescent branches and petioles, leaves, and flowers covered with minute, red glands; leaves petiolate, fasciculate, palmately 3–foliolate, petioles dilatate, leaflets sessile, *linear-cuneate, obtuse*, quite entire, thickish, *sub-glaucous;* racemes axillary, 3–6–flowered, shorter than the leaves; drupe oblique, *oval*, sub-compressed, glabrous. *R. platypoda, E. Meyer.*

HAB. Mountains; Kamiesberge and in Namaqualand, *E. & Z.! Drege!* Sandy places, Kamos and at Springbokkeel, Bushmanland, *Zey.! n.* 348. Flor. Nov.–Dec. Fruct. Mar. (Herb. Lehm., Sond.)

Shrub much branched, branches and branchlets terete, rigid, spinous. Petioles 2–4 lines long, broader at top; younger leaves red coloured, old ones sub-glabrous; leaflets 3–4 lines long, obtuse, rarely emarginate at top, 1 line broad. Peduncles racemose, shorter than the petiole, as well as the calyx, ferruginous. Petals white, ½ line long. Drupe transversely broader, 1½ line long, 2 lines broad, beaked with the very short styles

46. R. longispina (E. & Z.! n. 116); branches spinescent; spines horizontal, bearing leaves and flowers; leaves mostly fasciculate, palmately 3–foliolate, petiolate; petioles broadly-winged; younger leaflets as well as the panicles covered with minute, red glands, the older ones glabrous or punctate beneath, shining above, veined on both surfaces, *cuneate-obovate, obtuse* or sub-emarginate, entire, the lateral ones oblong, unequal at base; racemes panicled, solitary or fasciculate, as long as the leaves or shorter; drupe oblique, *globose*, mucronate, sub-compressed, glabrous. *R. pterota, Presl. l. c.*

HAB. Zwartkoprivier and near Ado, Fort Beaufort near Katrivier, *E. & Z.! Zey.! n.* 2244. Krakakamma, *Zey.! n.* 2245. Swellendam, *Mundt.!* July. (Herb. Lehm., T.C.D., Sond.)

A tall shrub with grey branches and red-glandular branchlets. Petioles the length of the leaves, shorter or longer, at top 1 line broad. Leaflets coriaceous, flat or with recurved margins, mostly 1–1½ inch long, 5–8 lines wide, some smaller, others

2–2½ inches long, 10–12 lines wide; the lateral ones smaller. Racemes 1–1½ uncial, ultimate pedicels ½–1 line long. Petals 3 times longer than the calyx; drupe the size of a small pea, broader than long. *R. longispina*, Presl, is *R. undulata*, Jacq. *aglaophylla, E. & Z.!* confounded by *E. & Z.* in several collections.

II. Leaves simple, penninerved. *Heeria*, Meisn. gen. *Anaphrenium*, E. Meyer. (Sp. 47–53.)

47. R. Thunbergii (Hook. Icon. t. 595); leaves simple, stalked, alternate, obovate-elliptic, or obovate-oblong, retuse, coriaceous, penninerved, *pulverulent-glaucous* beneath or on both sides; panicles terminal; pedicels, sepals, and petals outwardly pubescent; stamens 5–10; drupe very large, oblique, *depressed-globose. Rœmeria argentea, Thunb.! fl. cap. p.* 194. *excl. syn. Burm. E. & Z.! n.* 1087. *Pappe. Sylv. cap. p.* 12. *Sideroxylon argenteum, Thunb. prod. p.* 36. *Bumelia? argentea, R. & Sch. syst. veg. IV. p.* 499. *excl. syn. Burm. Anaphrenium argenteum, E. Meyer. R. argyrophylla, Presl, Bot. Bem. p.* 42.

HAB. Rocky situations in the districts of Stellenbosch, Worcester, and Clanwilliam. Feb. (Herb. Thunb., T.C.D., Sond., &c.)

Kliphout or Klipesse of the colonists. Tree 12–15 feet high; diameter 3 to 4 feet. Bark rough, craggy. Wood resinous, hard. Leaves 1½–2 inches or longer. Hermaphrodite flowers in panicles, the female ones in racemes. Flowers small, white. Fruit almost dry, rugulose, 1-celled, 1-seeded, ¾–1 inch broad.

48. R. dispar (Presl. l. c.); leaves simple, stalked, alternate, obovate-elliptic or ovate-oblong, obtuse or retuse, coriaceous, penninerved, *glabrous and reticulate* above, *whitish-tomentose* with reddish veins beneath; panicles terminal; branches *divaricate;* pedicels and calyces thinly puberulous; drupe sub-fleshy, *reniform*, compressed. *Anaphrenium dispar, E. Meyer in herb. Drege.*

HAB. Modderfontein, 1,500–2,000 f. Camiesberge, at Kasparskloof, Ellebogfontein and Geelbekskraal, 3–4000 f., *Drege!* On rocks at Lislap and Kammapus, *Zey.! n.* 329. May.–Aug. (Herb. Lehm., T.C.D., Sond.

Branches rugose, grayish, alternate or opposite. Leaves 1½–2 inches long. Very like the preceding, but differs by the reticulated leaves, glabrous above, the shorter panicle with divaricate branches, the larger pentamerous flowers; and essentially by the smaller, compressed, reniform, sub-fleshy fruit, about 5–6 lines broad, 3–4 lines long.

49. R. concolor (Presl. l. c.); leaves simple, stalked, alternate, obovate-elliptic or obovate-obtuse, emarginate, coriaceous, *quite glabrous*, reticulate above, *paler and veined beneath;* panicles terminal, abbreviate; branches *erecto-patent*, as well as the calyx puberulous; petals outwardly pubescent, 4 times longer than the obtuse, ciliolate calyx. *Anaphrenium concolor, E. Meyer.*

HAB. Between Natvoet and Gariep, 1000–1500. *Drege.* Sept. (Herb. Lehm., Sond., T.C.D.)

Branches cicatrized and rugose, as in the preceding. Leaves 1½–2 inches long, somewhat attenuate at base; petioles 2–3 lines long. Flowers 5-merous. Petals oblong, 3 lines long. Stamens shorter than the petals. Ovarium 1-locular, 1-ovuled. Styles deeply trifid, branches terete. Distinguished from the preceding by the leaves on both surfaces glabrous, the more compact (1–1½ uncial) panicle, with larger flowers.

50. R. mucronifolia (Sond.); leaves simple, stalked, *alternate, linear-oblong* or oblong-obtuse, mucronate, with *recurved margins, glabrous* *and shining* above, *silky-tomentose* and penninerved beneath; axillary panicles shorter than the leaves, terminal ones longer, compound, many-flowered, canescent; pedicels, calyx, and petals silky; drupe sub-fleshy, reniform. *Anaphrenium s. Heeria mucronifolia, Bernh.! in Krauss. Beytr. R. salicifolia, Presl. l. c.*

HAB. Vishriviersberg, near Katrivier, *Zeyher*. Keiskamma and Buffelrivier, *Drege!* 3582. Dec.–Jan. (Herb. Lehm., T.C.D., Sond.)

Branches rugose. Branchlets as well as the (2–3 lines long) petioles thinly pubescent. Leaves 2½–3 inches long, 8–12 lines wide, in the upper branches smaller (1–1½ inches long, 4–6 lines wide), obtuse on both ends; middle nerve in the upper surface impressed, the mucro 1 line long. Terminal panicle about 3-uncial. Petals oblong, thrice longer than the obtuse calyx. Drupe somewhat smaller than in the preceding, rugose, shining.

51. R. salicina (Sond.); leaves simple, stalked, *ternate, oblong-lanceolate* or lanceolate, acute or sub-obtuse, mucronate, *flat*, margined, *adpressed-pubescent above*, silky-tomentose beneath, penninerved on both sides, lateral nerves prominent; panicles terminal, canescent, *shorter* than the leaves, drupe sub-globose, sub-fleshy.

HAB. Magalisberg, *Burke and Zeyher, n.* 330 *a*. May. (Herb. Hook., Sond.)

Perhaps a variety of the preceding, but differs by the more lanceolate, flat leaves, not glabrous on the upper surface, and the more glossy tomentum. Panicles and fruits are not different. Flowers are wanting.

52. R. paniculosa (Sond.); leaves simple, stalked, *alternate, oblong-lanceolate* or lanceolate, acute or obtuse, mucronate, margined, *undulate-crenate*, glabrous above save on the pubescent nerves, silky-tomentose beneath, penninerved on both sides; panicles terminal, *much longer* than the leaves; pedicels, calyx, and petals outwardly silky-canescent; drupe sub-globose, sub-fleshy.

HAB. Crocadilerivier, near Magalisberg, *Burke*, 280. *Zey.!* 330, *b*. Nov. (Herb. Hook., T.C.D., Sond.)

A shrub, very similar to the preceding, but known by the undulate and crenulate in the upper surface not adpressed-pubescent leaflets, and the compound, long panicle. Leaflets about the same size as those of *R. salicina*, 2½–3 inches long, 8–10 lines wide; the upper ones smaller. Panicles longish, pedunculate, silky-tomentose, flowers sub-glomerate, about 2 lines long. Petals oblong, twice longer than the acute calyx and the stamens. Ripe fruit wanting.

53. R. longifolia; (Sond.); leaves simple, long-stalked, alternate, *linear-oblong*, somewhat obtuse, at the plaited, sub-recurved apex emarginate, with *incrassate-undulate margins, glabrous, sub-glaucous*, livid beneath, with prominent veins on both sides; panicles terminal, ample; pedicels, calyx, and petals outwardly-canescent; drupe oblique, reniform, sub-fleshy. *Anaphrenium longifolium, Bernh.! in Krauss. Beytr.* p. 46.

HAB. Woods near Port Natal, *Krauss, Guienzius, Plant*. Aug. (Herb. Sond., T.C.D.)

A tree, 25–30 feet high, the panicle excepted, glabrous. Petioles furrowed above, uncial. Larger leaves 4–5 inches long, 1–1¼ inch wide, acute at both ends. Panicle about 6 inches long, erect or pendulous; ultimate pedicels short, angulate. Calyx ovate, acute. Petals ovate-oblong, obtuse, 1 line long, three times longer than the stamens. Style thick, 3-partite. Drupe glabrous, 3 lines long, 4 lines broad.

DOUBTFUL SPECIES.

R. Wildingii (Dehnh. Rivist. nap. i. 3. p. 172). Without doubt not different from *R. viminalis, Vahl.* (see No. 27.)

R. alatum, *Thunb. Herb.* = **Hippobromus alatus**, *E. & Z.! p.* 241.
R. cirrhiflorum, *Thunb. Herb.* = **Cissus Thunbergii**, *Harv.* p. 250, *ex pte.* (leaves ternate, leaflets obovate.)
R. digitatum, *Thunb. Herb.* = **Cissus Thunbergii**, *E. & Z.! Harv. p.* 250, *ex pte.* (leaves quinate, leaflets oblong-cuneate.)
R. dimidiatum, *Thunb. Herb.* = **Cissus dimidiata**, *E. & Z.! p.* 251.
R. obliquum, *Thunb. Herb.* = **Pteroxylon utile**, *E. & Z.! p.* 243.
R. pauciflorum, *Thunb. Herb.* = **Hippobromus alatus**, *E. & Z.! p.* 241.
R. spicatum, *Thunb. Herb.* = **Schmidelia decipiens**, *Arn. p.* 239.
R. tridentatum, *Thunb. Herb.* = **Cissus tridentata**, *E. & Z.!* = **C. pauciflora**, *var. β. Harv. p.* 251.

III. SMODINGIUM, E. Meyer.

Flowers polygamous. *Calyx* 5-cleft, persistent. *Petals* 5, oblong, deciduous. *Stamens* 5, opposite to the calyx; filaments subulate; anthers ovate, 2-celled. *Styles* 3, persistent, short; stigmas obtuse. *Samara* sub-orbicular, oblique, emarginate, multi-vittate on both sides of the disc; vittæ parallel, sigmoid-flexuose. *Seeds* solitary, testa adherent to the pericarp. *Albumen* none. *Radicle* superior. *Sond.*

A glabrous shrub, with long-stalked, trifoliolate, coarsely serrate, lanceolate, penninerved leaves, and elongated, terminal panicles of minute flowers. Differs from *Rhus* by the winged, vittate pericarp, with adherent seeds; from *Botryceras* by the 3-foliolate leaves and the larger-winged, oblique fruit. Name from σμοδιγξ, an *indurated mark;* from the callous fruit.

1. S. argutum (E. Mey.! in Herb. Drege.)

HAB. Shady valleys between Omtata and Omsamwubo, *Drege!* Feb.–Mar. (Herb. Sond.)

Shrub erect; branches striate. Petioles 2–3 inches long, sub-terete. Leaves palmately 3-foliate; leaflets on very short stalks, 4–5 inches long, 1 inch wide, lanceolate, narrowed at both ends, with large, 2 lines long, mucronate teeth, but entire at base, green above, paler beneath. Panicle ample, like the inflorescence of *R. pyroides;* rhachis and branches thinly pubescent. Flowers about 1 line long. Petals three times longer than the calyx. Fruit 4 lines long, 3 lines wide, wings 1 line wide; pericarp oblique, with 4–6 sigmoid vittæ. Styles lateral, shorter than the emarginate wing.

IV. BOTRYCERAS, Willd.

Flowers diœcious. *Male: calyx* patent, 4–5-cleft nearly to the base; segments oblong, imbricate in æstivation. *Petals* 4–5, lanceolate, reflexo-patent. *Stamens* 4–5, opposite to the divisions of the calyx, placed round an expanded, fleshy disc; anthers versatile, 2-celled, cells gibbous, dehiscing laterally. *Ovary* none. *Female: calyx* and *corolla* persistent, as in the male, but petals oblong, obtuse. *Ovary* ovate-compressed, 1-celled; ovule solitary, pendulous. *Style* 1, thick, oblique, rising from the upper margin of the ovary. *Stigma* 3-lobed, bristly. *Fruit* with a membranous pericarp, compressed, winged at the margin, rugose, tipped with the permanent, sub-lateral style, with a long, one-seeded nucleus. *Seed* without albumen. *Cotyledons* flat-convex. *Ra-*

dicle superior. *Willd. Berl. Magaz. V.* 396. *Endl. gen. n.* 5907. *Laurophyllus, Thunb.! Prod. cap.* 153. *Harv. gen. p.* 64. *Daphnitis, Spreng. gen. n.* 559.

A tree or large shrub, glabrous and resinous. Leaves simple, stalked, elliptic-oblong, more or less acute, penninerved, glabrous, serrate, the margins reflexed. Flowers small, in terminal panicles; those of the male plant lax, with slender ramifications; of the female very dense, the branches compressed, multifid inwards, with persistent bracteæ. Name, from *βοτρυς, a cluster*, and *κερας, a horn;* the female inflorescence resembles a cluster of much-branched antlers.

1. **B. laurinum** (Willd.! l. c.); *Laurophyllus capensis, Thunb.! prod. p.* 31. *fl. cap. p.* 153.

HAB. Woods in mountainous places, Houtniquakloof, *Thunb.!* Duivelsbosch near Zwellendam, Vanstaadensberg, Uitenhage, *E. & Z.! Drege!* Oct.–Feb. (Herb. Lehm., Sond., T.C.D.)

Leaves evergreen, pale or livid below, 3–5-uncial. Petals trigonous, ½–1 inch long. Drupe about 2 lines long, compressed like an umbelliferous fruit.

V. LOXOSTYLIS, Spreng. fil.

Flowers diœcious. *Male: calyx* deeply 5-parted; segments lanceolate, acuminate, coloured. *Petals* 5, three times longer than the calyx, unguiculate, lanceolate-acuminate, imbricate in æstivation. *Stamens* 5, free, unequal, as well as the petals inserted between 5 glandular scales in the tube of the calyx. *Ovary* none. *Female: calyx* 3-parted; the fructiferous large, foliaceous; segments linear-oblong, sub-obtuse, pale green, veined. *Petals* minute. *Ovary* ovate, oblique, sub-compressed. *Styles* 3–4, unequal, sub-lateral, often 2–3 abortive. *Stigmas* capitate. *Drupe* dry, sub-coriaceous, rugose, green, sub-oval, obtuse, oblique, abortively 1-celled, 1-seeded, included in the persistent calyx and twice as short as it. *Seed* compressed-subreniform, umbilicate in the sinus; testa hard, coriaceous. *Albumen* none. *Cotyledons* accumbent, carnose. *Embryo* curved, radicle obtuse, ascendent. *Reichb. Icon. exot. t.* 205. *E. & Z.! Enum. p.* 152. *Endl. Gen. n.* 5908. *Anasyllis, E. Meyer.*

A tree, 15 feet high, quite glabrous. Leaves alternate, stalked, imparipinnate, 2–6-jugate; leaflets opposite, sub-sessile, lanceolate, entire, the rhachis winged. Flowers terminal, panicled.—Name, *λοξος, oblique*, and *στυλος, a style;* the carpels are oblique.

1. **L. alata** (Spreng. Herb. Zeyh. n. 3), *E. & Z.! n.* 1137. *Anasyllis angustifolia and latifolia, E. Meyer in Herb. Drege.*

HAB. Winterhoeksberg, Elandskloof, *E. & Z.!* Sandfontein, *Zeyh. n.* 351. Enon and Zuurebergen, Albany, *H. Hutton*, &c.; and near Omtendo, *Drege.* Feb.–Jul. (Herb. Lehm., T.C.D., Sond.)

Branches terete, striate. Petiole an inch long. Rachis with cuneate wings. Leaflets in some specimens 1–1½ inch long, 3 lines wide; in others 2 inches long, 5–6 lines wide, lanceolate, obtuse, mucronate or acute, with recurved margins, sub-coriaceous. Panicle ample, compound, ultimate pedicels capillary. Calyx in the male 1 line long. Petals narrow, whitish, recurved. Stamens shorter than the petals. Fructiferous calyx 6 lines long, 1½–2 lines wide. Drupe 3 lines long, sub-compressed.

VI. SCLEROCARYA, Hochst.

Flowers polygamous, diœcious; males spicate, hermaphrodite, sub-solitary. *Sepals* 4, coloured, sub-orbicular, at base imbricate. *Petals*

4, ovate-oblong, obtuse, reflexed, patent, three times longer than the calyx, imbricate in æstivation. *Male flowers* with 12–15 stamens, inserted round a fleshy, depressed, entire disc; filaments elongate; anthers introrse, oblong, bilocular. *Hermaphrodite flowers,* with some sterile and anantherous stamens amongst the fertile. *Ovary* subglobose, 2–3-locular. *Styles* 2–3, distant, short, thick. *Stigmas* peltate. *Drupe* sub-fleshy, with a hard, woody nucleus, 2–3-celled, cells 1-seeded (1–2 abortive). *Seed* pendulous, without albumen. *Cotyledons* thick, plano-convex. *Hochst. Flora, Bot. Zeitz. t.* 27. 2. *Besond. Beil. p.* 1. *Spondias spec. Richard.*

Trees or shrubs. Leaves petiolate, exstipulate, imparipinnate; leaflets obovate, petiolulate, 6–10-jugate, glabrous.—Name from *σκληρος, hard,* and *καρυα,* the *walnut.*

1. S. caffra (Sond. Linn. xxiii. 1. p. 26); glabrous; petiole trigonous; leaflets 5–13 long-petiolulate, ovate or elliptic, shortly cuspidate, mucronate, entire or paucidentate, pale below; flowers diœcious; male spikes terminal, abbreviate; sepals and petals 4; drupe sub-orbicular, 2–3-celled.

HAB. North side of the Macalisberg, *Burke, Zeyh. n.* 1857. Port Natal, *Gueinzius, Drege, n.* 4063. Fl. Aug.; fruit, Dec. (Herb. Hook., T.C.D., Sond.)

Leaves aggregate at the ends of the branches, alternate, ½–1 foot long; leaflets 1½–2 inches long, 12–15 lines wide, with a short, often oblique acumen, at base acute, penninerved, shining above, very pale beneath. Male spikes 2–4, shortly pedunculate, glabrous, 1–2 inches long. Flowers very short-stalked, bracteolate, the dried stalk red. Stamens equalling the petals. Drupe of the size of a small walnut, sub-orbicular. Perfect seeds are wanting.—Nearly allied to *S. Birrea,* Hochst. (*Spondias Birrea, Rich. fl. Seneg. t.* 41.)

VII. HARPEPHYLLUM, Bernhardi.

Flowers diœcious. *Male: calyx* 4–5-cleft, segments obtuse. *Petals* 4, rarely 5, longer than the calyx, imbricate in æstivation. *Stamens* 8 or 9, as well as the petals inserted below the margin of the crenate disc. *Filaments* shortly subulate; anthers ovate, 2-celled. Rudimentary pistillum 4-lobed. *Female-flowers* unknown. *Fruit* an obovate, smooth, sub-fleshy drupe, with a bony, 2-celled putamen. *Seeds* . . . *Bernh. in Krauss. Beytr. p.* 47.

A glabrous tree. Leaves aggregate at the top of the branches, stalked, alternate, imparipinnate; leaflets sessile, falcate-lanceolate, entire, unequal sided, the terminal equal sided. Male flowers in a terminal panicle.—Name from *άρπη, a sickle,* and *φυλλον, a leaf:* alluding to the falcate leaflets.

1. H. caffrum (Bernh.) *Spondias? falcata, Meisn.*

HAB. Woods at Howisonspoort, near Grahamstad, *Zeyh. n.* 2028; in distr. Uitenhage, *Ecklon, Brehm.* Caffraria, *Krauss. Drege,* 8265. Apr.–May.; fruct. mat. Feb.–Mar. (Herb. T.C.D., Lehm., Sond.)

A tree, 20–30 feet high, quite resembling *Eckebergia capensis.* Branchlets nodose after the leaves have fallen. Leaves 6–10 inches long. Petiole furrowed above, 1–2 inches long. Leaflets 5–7-jugate, patent, 2–2½ inches long, 6–9 lines wide, lanceolate-acuminate, acute at base, the upper margin convex-arcuate, the lower straight or concave-arcuate; middle nerve prominent, primary veins conspicuous on the upper surface. Male panicle terminal, half the length of the leaves. Flowers small, whitish or yellowish, on very short pedicels. Drupe about an inch long, 6 lines broad.—The tree is called by the inhabitants "Eschenhout;" the edible fruit, "Zuurebesges."

VIII. **BALSAMODENDRON**, Kunth.

Flowers polygamous. *Calyx* 4-toothed, persistent. *Petals* 4, linear-oblong, inserted under the margin of the torus, induplicately valvate in æstivation. *Stamens* 8, shorter than the corolla, glabrous, inserted under the margin of the torus. *Torus* cup-shaped, fleshy, deeply 8-crenated. *Ovary* sessile, 2-celled; ovules 2 in each cell, collateral, suspended from the middle of the axis. *Style* very short. *Stigma* obtuse, 4-lobed. *Drupe* globose or ovate, acute, marked with 4 sutures; sarcocarp splitting into 2 valves; nut thick and very hard, long, 2-celled (one of the cells by abortion often obliterated), at length divisible into two nuts. *Seeds* solitary in each perfect cell, suborbicular, concave-convex, without albumen; testa membranaceous. *Cotyledons* suborbicular, fleshy, corrugate and plicate; radicle superior, cylindrical, thick. *Kunth. Annal. sc. nat. ii.* 348. *Endl. Gen. n.* 5930. *Protium, sec.* 2. *Wight and Arn. Prod. p.* 176. *Heudelotia, A. Rich. Flor. Seneg.* 150. *Amyridis spec. L.*

Balsamiferous trees. Leaves unequally pinnated; leaflets 3–5, sessile, without dots. Peduncles terminal, on very short branchlets, 1-flowered, solitary or aggregated.—The name signifies *Balsam-tree: Myrrh* is the product of a North African species, *B. myrrha.*

1. **B. Capense** (Sond.); glabrous; ultimate branches short, patent, somewhat spinescent, with small, very short branchlets; leaves palmately 3-foliolate; leaflets obcordate or sub-orbicular, crenulate, glabrous, terminal cuneate, equal-sided, laterals smaller, oblique at the base; peduncles 1-flowered, shorter than the leaves; calyx deeply 4-toothed, lobes ovate, acute; drupe sub-globose, mucronate, somewhat compressed.

HAB. Between Natvoet and Gariep, 1000–1500 feet, *Drege, n.* 6809, *ex parte.* Sept (Herb. Sond.)

Branches erect, terete, striate, purplish; branchlets 1–3 inches long or shorter, bearing the leaves, and at their top the flowers. Leaves on a short (1–1½ line long) petiole. Leaflets somewhat coriaceous, with a middle nerve and a few lateral veins, the terminal 4–5 lines long, 2–3½ lines wide, cuneate or narrowed in a short stalk; the lateral ones half as long, patent, quite sessile. Peduncle 1–1½ line, calyx 1 line long. Petals and stamens unknown. Drupe half an inch long, mucronate with the very short style, fleshy, bivalved. Putamen by abortion 1-celled, 1-seeded, rarely 2-celled, one of the cells very small, containing a rudimentary seed. Seed attached to the middle of the dissepiment. Testa very thin, brownish. Cotyledons exactly as in *Elaphrium.*—Very similar to *B. gileadense,* Kunth. (*Nees v. Esenb. in Duesseld. offic. Pflanz. tab.* 611; *Amyris gileadensis, L. Vahl. Symb. I. t.* 11); and differs only by the more orbicular, crenulate, shorter-stalked leaves, and the deeply-toothed or 4-fid calyx.

APPENDIX.

IX.? **CATHASTRUM**, Turcz.

[This genus appears to me to range much better with *Celastrineæ,* of which it has completely the habit, and from which it differs merely in having the ovary reduced to a single carpel. The opposite leaves and numerous superposed ovules seem to exclude it from *Terebintaceæ.*—W.H.H.]

Flowers hermaphrodite. *Calyx* 5-cleft, segments rounded, obtuse.

Petals 5, inserted under the margin of the torus, reflexed, obovate or rotundate, ciliolate, imbricate in æstivation. *Stamens* 5, alternate with the petals and a little shorter, inserted in the margin of the 5-angulate torus; filaments filiform; anthers sub-cordate, 2-celled. *Ovary* free, sessile, 1-locular; ovules 6–8, in two rows, rarely uniseriate, superposed on a parietal placenta. *Style* very short. *Stigma* peltate. *Fruit* unknown. *Turcz. Bullet. Mosc.* 1858. *No.* 2. *p.* 448.

A shrub or tree, with the aspect of a *Celastrus* or *Cassine*, quite glabrous. Twigs and leaves opposite. Leaves on short petioles, oblong-lanceolate or oblong, somewhat acuminate at both ends, obtuse or slightly emarginate, rarely acute at point, coriaceous, with undulate, entire margins. Flowers very small, in axillary short and few-flowered racemes.—Name signifies a plant resembling *Catha*, a genus now merged in *Celastrus*.

1. C. Capense (Turczan. l. c.)

HAB. Forests in Krakakamma, between Port Elizabeth and Vanstadesberg, *Zeyh. Celastrin. n.* 2. *Dr. Alexander Prior.* Dec.–Jan. (Herb. T.C.D., Sond.)

Branches terete, smooth; branchlets compressed, angular. Petiole 1–2 lines long. Leaves 2–2½ inches long, 6–9 lines wide, veined, paler beneath. Racemes ½–¾ inch long. Pedicels at base bracteolate as well as the petals, about 1 line long.

ORDER XLVII. **CONNARACEÆ**, R. Br.

(By W. SONDER.)

(Connaraceæ, R. Br. Congo, 431 (1818). Endl. Gen. ccxlvii. Lindl. Veg. Kingd. No. clxxv.)

Flowers mostly perfect, regular. *Calyx* 5-cleft or parted, persistent. *Petals* 5, inserted in the bottom of the calyx, sessile or clawed, imbricate in æstivation. *Stamens* 10, inserted within the petals; filaments free or slightly connate at base; anthers introrse, 2-celled. *Ovary* of 5 separate carpels, either all fertile or several abortive and reduced to styles; the fertile unilocular, with 2 collateral, erect or ascending ovules; styles terminal, as many as the carpels. *Capsules* follicular, 5 or fewer, coriaceous, dehiscing by the ventral suture, rarely indehiscent. *Seed* generally solitary, erect, with or without a fleshy arillus; embryo exalbuminous, with fleshy cotyledons; or albuminous, with leafy cotyledons; radicle remote from the hilum.

Tropical or sub-tropical trees or shrubs, sometimes scandent, natives of both hemispheres. Leaves alternate, trifoliolate or imparipinnate, the leaflets coriaceous, veiny, dotless, very entire. Stipules none. Flowers racemose or panicled, small, terminal or axillary.—This small Order is intermediate between *Terebintaceæ* and *Leguminosæ;* from some depauperated leguminous genera it can scarcely be known by any definite character, except by the radicle being remote from the hilum. *Zebra-wood* is the most important product of the Order.

I. **CNESTIS**, Juss.

Calyx 5-cleft, valvate in the bud. *Petals* 5, inserted in the bottom of the calyx as well as the 10 stamens, free. *Carpels* 5 (1–4 abortive), distinct, 1-styled, somewhat stipitate, coriaceous, 2-valved, pod-like, opening on the back, usually clothed with stinging hairs both inside and outside, with 2 ova in each carpel. *Seeds* solitary, rising from the base of the carpel, erect, destitute of aril. *Albumen* fleshy. *Embryo* straight. *Cotyledons* leafy. *Radicle* superior. *Juss. Gen.* 374. *Lam. t.* 387. *DC. Prod. ii.* 86. *Endl. Gen. n.* 5950.

Usually scandent shrubs. Leaves alternate, imparipinnate, coriaceous, entire. Flowers in racemose panicles. Name from *κνηθω, to scratch:* the hairs of the capsules excite itching.

1. C. Natalensis (Planch. & Sond.); leaves with 7–9 pairs of oblong leaflets, which are unequal, oblique at the base, obtuse or sub-retuse at the apex, glabrous above, pale and thinly tomentose beneath, reticulate on both sides; panicles terminal, tomentose, much shorter than the leaves; carpels ovate-oblong, sub recurved, tomentose. *Omphalobium? discolor, Sond. Linn. XXIII.* 1. *p.* 24. *Xanthoxylum Natalense, Hochst- in Krauss. Beyt. p.* 41.

HAB. Forests between Umloasrivier and Natalbai, *Dr. Krauss, Gueinzius, Sanderson.* Sep.–Nov. (Herb. Hook. T.C.D., Sond.)

A shrub, 4–5 feet high. Branches brown-red. Leaves about 4 inches long. Leaflets on short petiolules, 12–15 lines long, 4–6 lines wide, whitish-grey on the lower surface, with prominent rufous middle nerve. Panicle terminal, rarely axillary, 1–1½ inch long, ferruginous, thinly tomentose. Pedicels very short. Flowers 1–1½ lines long. Calyx downy, segments lanceolate, longer than the linear-oblong petals. Styles glabrous, with blunt stigmas. Carpels thinly rufo-tomentose, about 10 lines long, 5 lines broad. Seed black, shining.

END OF VOL. I.

INDEX.

[SYNONYMS IN *italics*].

Printed and bound in England by William Clowes and Sons Ltd, London and Beccles

Printed in the United States
By Bookmasters

Printed in the United States
By Bookmasters